UCF $90.00

GEOTECHNICAL AND FOUNDATION ENGINEERING

GEOTECHNICAL AND FOUNDATION ENGINEERING
Design and Construction

Robert W. Day

McGraw-Hill
New York San Francisco Washington, D.C. Auckland Bogotá
Caracas Lisbon London Madrid Mexico City Milan
Montreal New Delhi San Juan Singapore
Sydney Tokyo Toronto

Library of Congress Cataloging-in-Publication Data

Day, Robert W.
   Geotechnical and foundation engineering : design and construction
Robert W. Day.
     p.   cm.
   ISBN 0-07-134138-2
   1. Engineering geology. 2. Foundations. I. Title.
TA705.D32   1999
624.1'51—dc21   99-20164
                                                                                               CIP

McGraw-Hill
A Division of The McGraw·Hill Companies

Copyright © 1999 by The McGraw-Hill Companies, Inc. All rights reserved. Printed in the United States of America. Except as permitted under the United States Copyright Act of 1976, no part of this publication may be reproduced or distributed in any form or by any means, or stored in a data base or retrieval system, without the prior written permission of the publisher.

1 2 3 4 5 6 7 8 9 0   DOC/DOC   9 0 4 3 2 1 0 9

ISBN 0-07-134138-2

The sponsoring editor of this book was Larry Hager. The editing supervisor was Peggy Lamb, and the production supervisor was Tina Cameron. This book was set in the HB1 design in Times Roman by Joanne Morbit, Michele Zito, and Paul Scozzari of McGraw-Hill's Professional Book Group composition unit, Hightstown, New Jersey, in cooperation with Spring Point Publishing Services.

Printed and bound by R. R. Donnelley & Sons Company.

 This book is printed on recycled, acid-free paper containing a minimum of 50% recycled, de-inked fiber.

McGraw-Hill books are available at special quantity discounts to use as premiums and sales promotions, or for use in corporate training programs. For more information, please write to the Director of Special Sales, McGraw-Hill, 11 West 19th Street, New York, NY 10011. Or contact your local bookstore.

Information contained in this work has been obtained by The McGraw-Hill Companies, Inc. ("McGraw-Hill") from sources believed to be reliable. However, neither McGraw-Hill nor its authors guarantee the accuracy or completeness of any information published herein, and neither McGraw-Hill nor its authors shall be responsible for any errors, omissions, or damages arising out of use of this information. This work is published with the understanding that McGraw-Hill and its authors are supplying information but are not attempting to render engineering or other professional services. If such services are required, the assistance of an appropriate professional should be sought.

Dedicated with love to my wife Deborah and my parents.

CONTENTS

Preface   xiii
Acknowledgments   xv

Chapter 1. Introduction   1.1

1.1. Geotechnical Engineer / *1.1*
1.2. Engineering Geologist / *1.6*
1.3. Typical Clients / *1.9*
1.4. Example / *1.10*
1.5. Outline of Chapters / *1.16*

Part 1   Development of Programs of Geotechnical Investigation

Chapter 2. The Assignment   2.3

2.1. Preliminary Information / *2.3*
2.2. Type of Project / *2.3*
2.3. Project Requirements / *2.11*

Chapter 3. Proposal and Planning the Work   3.1

3.1. Proposal / *3.1*
   3.1.1. Cost Estimate / *3.1*
   3.1.2. Engineer's Agreement with the Client / *3.4*
3.2. Planning the Work / *3.5*
   3.2.1. Technical Guidelines / *3.6*

Part 2   Geotechnical Field and Laboratory Studies

Chapter 4. Field Exploration   4.3

4.1. Introduction / *4.3*
4.2. Document Review / *4.4*
4.3. General Purpose of Subsurface Exploration / *4.12*

vii

4.4. Borings / 4.17
    4.4.1. Soil Sampling / 4.21
    4.4.2. Field Tests / 4.25
    4.4.3. Boring Layout / 4.35
    4.4.4. Depth of Subsurface Exploration / 4.38
4.5. Test Pits and Trenches / 4.39
4.6. Preparation of Logs / 4.43
4.7. Role of the Engineering Geologist / 4.43
    4.7.1. Geophysical Techniques / 4.47
4.8. Subsoil Profile / 4.53
Problems / 4.54

Chapter 5. Laboratory Testing   5.1

5.1. Introduction / 5.2
5.2. Common Laboratory Tests / 5.3
    5.2.1. Index Tests / 5.3
    5.2.2. Oedometer / 5.6
    5.2.3. Shear Strength Tests / 5.8
    5.2.4. Permeability / 5.15
    5.2.5 Laboratory Tests for Pavements and Deterioration / 5.18
5.3. Sample Disturbance / 5.20
5.4. Soil Classification / 5.21
    5.4.1. Particle Size and Description / 5.25
    5.4.2. Clay Mineralogy / 5.29
    5.4.3. Unified Soil Classification System / 5.34
    5.4.4. Classification Based on Plasticity / 5.38
5.5. Rock Classification / 5.43
Problems / 5.45

Part 3   Analysis of Geotechnical Data and Engineering Computations

Chapter 6. Basic Geotechnical and Foundation Principles   6.3

6.1. Introduction / 6.5
6.2. Phase Relationships / 6.5
6.3. Effective Stress / 6.10
6.4. Stress Distribution / 6.12
6.5. Shear Strength / 6.23
    6.5.1. Cohesionless Soil / 6.23
    6.5.2. Cohesive Soil / 6.27
    6.5.3. Total Stress and Effective Stress Analyses / 6.45
6.6. Permeability and Seepage / 6.46
Problems / 6.58

Chapter 7. Settlement of Structures   7.1

7.1. Introduction / 7.2
7.2. Allowable Settlement / 7.3
7.3. Collapsible Soil / 7.8
7.4. Settlement of Cohesive and Organic Soils / 7.16
    7.4.1. Immediate or Initial / 7.16

          7.4.2. Primary Consolidation / *7.19*
          7.4.3. Secondary Compression / *7.32*
   7.5. Settlement of Cohesionless Soil / *7.33*
   7.6. Other Common Causes of Settlement / *7.36*
          7.6.1. Limestone Cavities or Sinkholes / *7.37*
          7.6.2. Underground Mines and Tunnels / *7.37*
          7.6.3. Subsidence Due to Extraction of Oil or Groundwater / *7.38*
          7.6.4. Landfills and Decomposition of Organic Matter / *7.39*
   7.7. Selection of Foundation Type / *7.42*
          7.7.1. Shallow Foundations / *7.43*
          7.7.2. Deep Foundations / *7.46*
   Problems / *7.49*

Chapter 8. Bearing Capacity   8.1

   8.1. Introduction / *8.2*
   8.2. Bearing Capacity for Shallow Foundations / *8.7*
          8.2.1. Bearing Capacity for Cohesionless Soil / *8.9*
          8.2.2. Bearing Capacity for Cohesive Soil / *8.10*
          8.2.3. Other Bearing Capacity Considerations / *8.12*
   8.3. Bearing Capacity for Deep Foundations / *8.14*
          8.3.1. Bearing Capacity for Cohesionless Soil / *8.16*
          8.3.2. Bearing Capacity for Cohesive Soil / *8.20*
   8.4. Pavement Design / *8.23*
          8.4.1. Pavement Section / *8.23*
          8.4.2. California Method / *8.24*
   Problems / *8.28*

Chapter 9. Expansive Soil   9.1

   9.1. Introduction / *9.1*
          9.1.1. Expansive Soil Factors / *9.2*
          9.1.2. Laboratory Testing / *9.4*
   9.2. Swelling of Desiccated Clay / *9.7*
          9.2.1. Identification of Desiccated Clay / *9.7*
          9.2.2. Hydraulic Conductivity and Rate of Swell / *9.10*
   9.3. Types of Expansive Soil Movement / *9.12*
          9.3.1. Lateral Movement / *9.12*
          9.3.2. Vertical Movement / *9.13*
          9.3.3. Effect of Vegetation / *9.14*
   9.4. Foundation Design for Expansive Soil / *9.16*
          9.4.1. Conventional Slab-on-Grade Foundation / *9.16*
          9.4.2. Posttensioned Slab-on-Grade / *9.16*
          9.4.3. Pier and Grade Beam Support / *9.21*
          9.4.4. Other Treatment Alternatives / *9.23*
   9.5. Pavements and Flatwork / *9.23*
   9.6. Construction on Expansive Rock / *9.32*
   Problems / *9.34*

Chapter 10. Slope Stability   10.1

   10.1. Typical Types of Slope Movement / *10.1*
   10.2. Allowable Lateral Movement / *10.2*
   10.3. Rockfall / *10.4*

10.4. Surficial Slope Stability / *10.10*
10.5. Gross Slope Stability / *10.27*
10.6. Landslides / *10.35*
10.7. Debris Flow / *10.48*
10.8. Slope Softening and Creep / *10.53*
10.9. Dams / *10.59*
    10.9.1. Large Dams / *10.59*
    10.9.2. Small Dams / *10.62*
    10.9.3. Landslide Dams / *10.63*
Problems / *10.63*

Chapter 11. Earthquakes   11.1

11.1. Introduction / *11.1*
11.2. Earthquakes / *11.3*
    11.2.1. Fault and Ground Rupture Zone / *11.3*
    11.2.2. Liquefaction / *11.5*
    11.2.3. Slope Movement and Settlement / *11.12*
    11.2.4. Translation and Rotation / *11.12*
11.3. Estimating Earthquake Ground Movement / *11.12*
11.4. Selection of Foundation Type / *11.15*
Problems / *11.17*

Chapter 12. Erosion   12.1

12.1. Introduction / *12.1*
12.2. Principles of Erosion Control / *12.3*
12.3. Erosion-Prone Landforms / *12.13*
    12.3.1. Sea Cliffs / *12.16*
    12.3.2. Badlands / *12.17*
Problems / *12.18*

Chapter 13. Deterioration   13.1

13.1. Introduction / *13.1*
13.2. Sulfate Attack of Concrete / *13.1*
13.3. Pavement Deterioration / *13.7*
13.4. Frost / *13.8*
13.5. Tree Roots / *13.9*
13.6. Historic Structures / *13.10*

Chapter 14. Unusual Soil   14.1

14.1. Introduction / *14.1*
14.2. Examples of Unusual Soil / *14.1*
14.3. Case Study of Unusual Soil / *14.5*

Chapter 15. Retaining Walls   15.1

15.1. Introduction / *15.1*
15.2. Basic Retaining Wall Analyses / *15.3*

CONTENTS xi

15.2.1. Simple Retaining Wall without Wall Friction / *15.3*
15.2.2. Simple Retaining Wall with Wall Friction / *15.7*
15.3. Design and Construction of Retaining Walls / *15.13*
15.4. Restrained Retaining Walls / *15.19*
15.5. Mechanically Stabilized Earth Retaining Walls / *15.22*
15.6. Sheet Pile Walls / *15.35*
15.7. Temporary Retaining Walls / *15.43*
  15.7.1. Braced / *15.43*
  15.7.2. Steel I beam and Wood Lagging / *15.45*
  15.7.3. Utility Trench Shoring / *15.45*
15.8. Pier Walls / *15.49*
Problems / *15.59*

Chapter 16. Groundwater and Moisture Migration   16.1

16.1. Introduction / *16.1*
16.2. Groundwater / *16.2*
  16.2.1. Pavements / *16.3*
  16.2.2. Slopes / *16.4*
16.3. Moisture Migration through Floor Slabs / *16.9*
  16.3.1. Design and Construction Details / *16.12*
16.4. Moisture Migration through Basement Walls / *16.15*
  16.4.1. Types of Damage / *16.15*
  16.4.2. Design and Construction Details / *16.18*
16.5. Surface Drainage / *16.19*
16.6. Pipe Breaks and Clogs / *16.19*
16.7. Percolation Tests for Sewage Disposal System / *16.22*
Problems / *16.24*

Part 4 Performance or Engineering Evaluation of Construction

Chapter 17. Grading   17.3

17.1. Introduction / *17.3*
17.2. Grading Specifications / *17.4*
17.3. Compaction Fundamentals / *17.7*
  17.3.1. Laboratory Compaction Testing / *17.8*
  17.3.2. Field Compaction / *17.9*
  17.3.3. Relative Compaction / *17.13*
17.4. Types of Fill / *17.14*
17.5. Utility Trench and Backfill Compaction / *17.18*
17.6. Pumping / *17.22*
17.7. Adjacent Property Damage / *17.23*
Problems / *17.27*

Chapter 18. Construction Services   18.1

18.1. Introduction / *18.1*
18.2. Monitoring / *18.4*
18.3. Observational Method / *18.7*
18.4. Underpinning / *18.10*

Part 5 Preparation or Engineering Evaluation of Geotechnical Reports

Chapter 19. Reports   19.3

19.1. Introduction / *19.3*
19.2. Preparation of Reports and File Management / *19.4*
    19.2.1. Report Preparation / *19.4*
    19.2.2. Daily Field Reports / *19.5*
    19.2.3. File Management / *19.6*
    19.2.4. Examples of Poor File Management / *19.6*
19.3. Engineering Jargon, Superlatives, and Technical Words / *19.7*
    19.3.1. Engineering Jargon / *19.7*
    19.3.2. Superlatives / *19.8*
    19.3.3. Technical Words / *19.9*
19.4. Engineering Evaluation of Geotechnical Reports / *19.9*
19.5. Strategies to Avoid Civil Liability / *19.10*
    19.5.1. Assessing Risk / *19.11*
    19.5.2. Insurance / *19.12*
    19.5.3. Limitation of Liability Clauses / *19.12*

APPENDIXES

    Appendix A. Glossary / *A.1*
      Introduction / *A.1*
      Glossary 1   Engineering Geology and Subsurface Exploration Terminology / *A.3*
      Glossary 2   Laboratory Testing Terminology / *A.8*
      Glossary 3   Terminology for Engineering Analysis and Computations / *A.12*
      Glossary 4   Construction and Grading Terminology / *A.17*
Appendix B. Technical Guidelines for Soil and Geology Reports / *B.1*
Appendix C. Example of Grading Specifications / *C.1*
Appendix D. Percolation Test Procedure / *D.1*
Appendix E. Example of a Preliminary Geotechnical Report / *E.1*
Appendix F. Solution to Problems / *F.1*
Appendix G. References / *G.1*
Index / *I.1*

PREFACE

The goal of the book is to present the practical aspects of geotechnical and foundation engineering. While the major emphasis of college education is engineering analyses, this often represents only a portion of the knowledge needed to practice geotechnical engineering. One objective of this book is to discuss the engineering judgment that needs to be acquired through experience. An example is the application of sufficient redundancy in the design and construction of the project.

Because of the assumptions and uncertainties associated with geotechnical engineering, it is often described as an "art," rather than an exact science. Thus simple analyses are prominent in this book, with complex and theoretical evaluations kept to an essential minimum.

The idea for the format of the book was to follow in a logical manner the typical process of a design project, from the first step of preparing a proposal for engineering services to the final step of preparing the "as-built" report upon completion of the project. The book is divided into five separate parts. Part 1 (Chaps. 2 and 3) provides a discussion of the development of programs of geotechnical investigation, which includes obtaining information on the assignment, preparation of a proposal, and planning the work. Part 2 (Chaps. 4 and 5) deals with geotechnical field and laboratory studies, which consists of subsurface exploration and laboratory testing of soil, rock, or groundwater samples. Part 3 (Chaps. 6 to 16) presents the analysis of geotechnical data and engineering computations for conditions commonly encountered by the design engineer, such as settlement, expansive soil, and slope stability. Part 4 (Chaps. 17 and 18) provides a discussion of the performance or engineering evaluation of construction and Part 5 (Chap. 19) is a concluding chapter dealing with the preparation or engineering evaluation of geotechnical reports.

The book presents the practical aspects of geotechnical and foundation engineering. The topics should be of interest to design engineers, especially Chap. 19 which provides information on the preparation of reports.

Robert W. Day

ACKNOWLEDGMENTS

I am grateful for the contributions of many people who helped to make this book possible. Special thanks is due Professor Charles C. Ladd, at the Massachusetts Institute of Technology, who reviewed the text and offered many helpful suggestions. I would also like to thank Professor Ladd for the opportunity to have worked on his Orinoco Clay project. Numerous figures, especially those in the section on shear strength, are reproduced from my M.I.T. thesis (*Engineering Properties of the Orinoco Clay*). In addition, I would like to thank Dr. Ladd for the opportunity to have worked with him on the project that is described in Sec. 14.3.

Numerous practicing engineers reviewed various portions of the text and provided valuable assistance during its development. In particular, I am indebted to Robert Brown, Edred Marsh, Rick Walsh, and Scott Thoeny. Thanks also to Dennis Poland, Ralph Jeffery, and Todd Page for their help with the geologic aspects of the book, and Rick Dorrah and Eric Noether for drafting the figures for the book.

I would also like to thank Professor Timothy Stark, at the University of Illinois, who performed the ring shear tests, provided a discussion of the test procedures, and prepared the ring shear test plots. Thanks also to Kean Tan who performed the triaxial compression tests and prepared the shear strength data plots. I am also indebted to Gregory Axten, president of American Geotechnical, who provided valuable support during the review and preparation of the book. I am also grateful for the input of John Pizzi, at Hardesty & Hanover, LLP, for the many conversations we have had over the years concerning his unique experiences with the design and construction of bridge foundations.

Tables and figures taken from other sources are acknowledged where they occur in the text. Finally, I wish to thank Larry Hager, Peggy Lamb, Tina Cameron, George F. Watson, and others on the McGraw-Hill editorial staff, who made this book possible and refined my rough draft into this finished product.

CHAPTER 1
INTRODUCTION

1.1 GEOTECHNICAL ENGINEER

In a broad sense, a geotechnical engineer is an individual who performs an engineering evaluation of earth materials. This typically includes soil, rock, and groundwater and their interaction with earth retention systems, structural foundations, and other civil engineering works. Geotechnical engineering is a subdiscipline of civil engineering and requires a knowledge of basic engineering principles, such as statics, dynamics, fluid mechanics, and the behavior of engineering materials. An understanding of construction techniques and the performance of civil engineering works influenced by earth materials is also required.

Geotechnical engineering is often divided into two categories: soil mechanics and rock mechanics.

1. *Soil mechanics.* The majority of geotechnical engineering deals with soil mechanics and in practice, the term *soils engineer* is synonymous with *geotechnical engineer*. Soil has many different meanings depending on the field of study. For example, in agronomy (application of science to farming) soil is defined as a surface deposit that contains mineral matter that originated from the original weathering of rock and also contains organic matter that has accumulated through the decomposition of plants and animals. To an agronomist, soil is that material that has been sufficiently altered and supplied with nutrients that it can support the growth of plant roots. But to a geotechnical engineer, soil has a much broader meaning and can include not only agronomic material, but also broken-up fragments of rock, volcanic ash, alluvium, aeolian sand, glacial material, and any other residual or transported product of rock weathering. Difficulties naturally arise because there is not a distinct dividing line between rock and soil. For example, to a geologist a given material may be classified as a formational rock because it belongs to a definite geologic environment, but to a geotechnical engineer it may be sufficiently weathered or friable that it should be classified as a soil.

2. *Rock mechanics.* To the geotechnical engineer, rock is a relatively solid mass that has permanent and strong bonds between the minerals. Rocks can be classified as sedimentary, igneous, or metamorphic. There are significant differences in the behavior of soil versus rock, and there is not much overlap between soil mechanics and rock mechanics. As will be discussed in Sec. 1.2, an engineering geologist is often involved with projects involving rock mechanics.

Table 1.1 presents a list of common soil and rock conditions that require special consideration of the geotechnical engineer.

TABLE 1.1 Problem Conditions Requiring Special Consideration

Problem type (1)	Description (2)	Comments (3)
Soil	Organic soil, highly plastic soil	Low strength and high compressibility
	Sensitive clay	Potentially large strength loss upon large straining
	Micaceous soil	Potentially high compressibility
	Expansive clay, silt, or slag	Potentially large expansion upon wetting
	Liquefiable soil	Complete strength loss and high deformations caused by earthquakes
	Collapsible soil	Potentially large deformations upon wetting
	Pyritic soil	Potentially large expansion upon oxidation
Rock	Laminated rock	Low strength when loaded parallel to bedding
	Expansive shale	Potentially large expansion upon wetting; degrades readily upon exposure to air and water
	Pyritic shale	Expands upon exposure to air and water
	Soluble rock	Rock such as limestone, limerock, and gypsum that is soluble in flowing and standing water
	Cretaceous shale	Indicator of potentially corrosive groundwater
	Weak claystone	Low strength and readily degradable upon exposure to air and water
	Gneiss and schist	Highly distorted with irregular weathering profiles and steep discontinuities
	Subsidence	Typical in areas of underground mining or high groundwater extraction
	Sinkholes	Areas underlain by carbonate rock (karst topography)
Condition	Negative skin friction	Additional compressive load on deep foundations due to settlement of soil
	Expansion loading	Additional uplift load on foundation due to swelling of soil
	Corrosive environment	Acid mine drainage and degradation of soil and rock
	Frost and permafrost	Typical in northern climates
	Capillary water	Rise in water level which leads to strength loss for silts and fine sands

Source: Reproduced with permission from *Standard Specifications for Highway Bridges*, 16th edition, AASHTO, 1996.

Foundation Engineering. A foundation is defined as that part of the structure that supports the weight of the structure and transmits the load to underlying soil or rock. Some engineers consider foundation engineering to be a part of geotechnical engineering (e.g., Cernica, 1995a), while others consider it to be a separate field of study (e.g., Holtz and Kovacs, 1981).

In general, foundation engineering applies the knowledge of geology, soil mechanics, rock mechanics, and structural engineering to the design and construction of foundations for buildings and other structures. The most basic aspect of foundation engineering deals with the selection of the type of foundation, such as using a shallow or deep foundation system. Another important aspect of foundation engineering involves the development of

INTRODUCTION

design parameters, such as the bearing capacity of the foundation. Foundation engineering could also include the actual foundation design, such as determining the type and spacing of steel reinforcement in concrete footings.

Foundations are commonly divided into two categories: shallow and deep foundations. Table 1.2 presents a list of common types of foundations.

TABLE 1.2 Common Types of Foundations

Category (1)	Common types (2)	Comments (3)
Shallow foundations	Spread footings	Spread footings (also called pad footings) are often square in plan view, are of uniform reinforced concrete thickness, and are used to support a single column load located directly in the center of the footing.
	Strip footings	Strip footings (also called wall footings) are often used for load-bearing walls. They are usually long reinforced concrete members of uniform width and shallow depth.
	Combined footings	Reinforced-concrete combined footings are often rectangular or trapezoidal in plan view, and carry more than one column load.
	Conventional slab-on-grade	A continuous reinforced-concrete foundation consisting of bearing wall footings and a slab-on-grade. Concrete reinforcement often consists of steel rebar in the footings and wire mesh in the concrete slab.
	Posttensioned slab-on-grade	A continuous posttensionsed concrete foundation. The posttensioning effect is created by tensioning steel tendons or cables embedded within the concrete. Common posttensioned foundations are the ribbed foundation, California slab, and PTI foundation.
	Raised wood floor	Perimeter footings that support wood beams and a floor system. Interior support is provided by pad or strip footings. There is a crawl space below the wood floor.
	Mat foundation	A large and thick reinforced-concrete foundation, often of uniform thickness, that is continuous and supports the entire structure. A mat foundation is considered to be a shallow foundation if it is constructed at or near ground surface.

TABLE 1.2 Common Types of Foundations (Continued)

Category (1)	Common types (2)	Comments (3)
Deep foundations	Driven piles	Driven piles are slender members, made of wood, steel, or precast concrete, that are driven into place by pile-driving equipment.
	Other types of piles	There are many other types of piles, such as bored piles, cast-in-place piles, and composite piles.
	Piers	Similar to cast-in-place piles, piers are often of large diameter and contain reinforced concrete. Pier and grade beam support are often used for foundation support on expansive soil.
	Caissons	Large piers are sometimes referred to as caissons. A caisson can also be a watertight underground structure within which construction work is carried on.
	Mat or raft foundation	If a mat or raft foundation is constructed below ground surface or if the mat or raft foundation is supported by piles or piers, then it should be considered to be a deep foundation system.
	Floating foundation	A special foundation type where the weight of the structure is balanced by the removal of soil and construction of an underground basement.
	Basement-type foundation	A common foundation for houses and other buildings in frost-prone areas. The foundation consists of perimeter footings and basement walls that support a wood floor system. The basement floor is usually a concrete slab.

Note: The terms *shallow* and *deep* foundations in this table refer to the depth of the soil or rock support of the foundation.

Typical Geotechnical Engineering Activities. Table 1.3, adapted from the *California Plain Language Pamphlet of the Professional Engineers Act and Board Rules* (1995), presents a list of items that are considered to be qualifying experience for geotechnical engineers. As indicated in Table 1.3, there are five basic aspects to geotechnical engineering:

1. Development of programs of geotechnical investigation
2. Geotechnical field and laboratory studies
3. Analysis of geotechnical data and engineering computations

TABLE 1.3 Qualifying Experience for Geotechnical Engineers

Qualifying experience (1)	Typical items (2)
Development of programs of geotechnical investigation	Communication with other design consultants to determine their geotechnical input needs. Performance of literature searches and site history analysis related to surface and subsurface conditions. Formulation or engineering evaluation of field exploration and laboratory testing programs to accomplish the scope of the investigation. Preparation or engineering evaluation of proposals.
Geotechnical field and laboratory studies	Direction and/or modification of field exploration programs, as required, upon evaluation of the conditions being encountered. Classification and evaluation of subsurface conditions. Understanding the purposes for and being qualified to perform routine field and laboratory tests for soil strength, bearing capacity, expansion properties, consolidation, soil collapse potential, erosion potential, compaction characteristics, material acceptability for use in fill, pavement support qualities, freeze-thaw properties, grain size, permeability/percolation properties, groundwater conditions, and soil dynamic properties.
Analysis of geotechnical data and engineering computations	Analysis of field and laboratory data. Performance of computations using test results and available data regarding bearing capacity; foundation type, depth, and dimensions; allowable soil bearing pressures; potential settlement; slope stability; retaining systems; soil treatment; dewatering and drainage; floor support; pavement design; site preparation; fill construction; liquefaction potential; ground response to seismic forces; groundwater problems and seepage; and underpinning.
Performance or engineering evaluation of construction	Performance or supervision of geotechnical testing and observation of site grading. Analysis, design, and evaluation of instrumentation programs to evaluate or monitor various phenomena in the field, such as settlement, slope creep, pore water pressures, and groundwater variations.
Preparation or engineering evaluation of geotechnical reports	Preparation of plans, logs, and test results. Documentation testing and observation. Preparation of written reports which present findings, conclusions, and recommendations of the investigation. Preparation of specifications and guidelines.

Source: Adapted from the *California Plain Language Pamphlet of the Professional Engineers Act and Board Rules*, 1995.

4. Performance or engineering evaluation of construction
5. Preparation or engineering evaluation of geotechnical reports

Table 1.3 also lists the typical types of projects encountered by geotechnical engineers, such as:

- Determining the type of foundation for the structure, including the depth and dimensions.
- Calculating the potential settlement of the structure.
- Determining design parameters for the foundation, such as the bearing capacity and allowable soil bearing pressures.
- Determining the expansion potential of a site.
- Designing pavements, such as for roads, highways, and airport runways.
- Investigating the stability of slopes, such as natural slopes, dams, levees, and embankments.
- Investigating the possibility of ground movement due to seismic forces, which would also include the possibility of liquefaction.
- Performing studies and tests to determine the erosion potential of slopes or possible deterioration of pavements and foundations.
- Evaluating possible soil treatment to improve certain properties of the soil.
- Determining design parameters for earth retaining systems, such as retaining walls, bulkheads, and sheet pile walls.
- Providing recommendations for dewatering and drainage of underground structures.
- Investigating groundwater and seepage problems and developing mitigation measures.
- Site preparation, including compaction specifications and density testing during grading.
- Underpinning of foundations.

Geotechnical Engineering Terms. Like most professions, geotechnical engineering has its own terminology with special words and definitions. Appendix A is divided into four separate glossaries:

Glossary 1   Engineering Geology and Subsurface Exploration Terminology
Glossary 2   Laboratory Testing Terminology
Glossary 3   Terminology for Engineering Analysis and Computations
Glossary 4   Construction and Grading Terminology

1.2 ENGINEERING GEOLOGIST

An engineering geologist is an individual who applies geologic data, principles, and interpretation so that geologic factors affecting planning, design, construction, and maintenance of civil engineering works are properly recognized and utilized (*Geologist and Geophysicist Act,* 1986). Table 1.4 (adapted from *Fields of Expertise,* undated) presents a summary of the education, training, and practice of the engineering geologist versus the geotechnical engineer.

In some areas of the United States, there may be minimal involvement of engineering geologists except for projects involving such items as rock slopes or earthquake fault studies. In other areas of the country, such as California, the geotechnical investigations are usually performed jointly by the geotechnical engineer and engineering geologist. The majority of geotechnical reports include both engineering and geologic

INTRODUCTION   1.7

TABLE 1.4 Areas of Education, Training, and Practice for Engineering Geologists and Geotechnical Engineers

Category (1)	Engineering geologist (2)	Geotechnical engineer (3)
Education	Physical sciences Natural sciences Scientific methods Observation Interpretation	Physical sciences Mathematics Research techniques Design methods Economics
Training and practice	Research Examination Description Explanation Analysis Opinion	Research Measurement Testing Calculation Analysis Design

Source: Adapted from *Fields of Expertise* (undated).

aspects of the project, and the report is signed by both the geotechnical engineer and the engineering geologist. As an example, App. B presents the *Technical Guidelines for Soil and Geology Reports* (part of the California *Orange County Grading Manual*, 1993). Note in App. B that preliminary geotechnical reports usually include an opinion by the geotechnical engineer and engineering geologist on the engineering and geologic adequacy of the site for the proposed development. As indicated in App. B, upon completion of the grading of the project, a compaction report is required that includes:

> A statement that the soil engineering and engineering geologic aspects of the grading have been inspected and are in compliance with the applicable conditions of the grading permit and the soil engineer's and engineering geologist's recommendations.

Fields of Expertise. Table 1.5 (adapted from *Fields of Expertise*, undated) presents a summary of the fields of expertise for the engineering geologist and geotechnical engineer, with the last column indicating the areas of overlapping expertise. Note in Table 1.5 that the engineering geologist should have considerable involvement with projects dealing with rock mechanics, field explorations (such as subsurface exploration and surface mapping), groundwater studies, earthquake analysis, and engineering geophysics. Since natural soil deposits are formed by geologic processes, the input of an engineering geologist can be invaluable for nearly all types of geotechnical engineering projects.

Areas of Responsibility. Because the geotechnical engineer and engineering geologist work as a team on most projects, it is important to have an understanding of each individual's area of responsibility. The area of responsibility is based on education and training (Table 1.4) and areas of expertise (Table 1.5). According to *Fields of Expertise* (undated), the individual responsibilities are as follows:

1. Responsibilities of the engineering geologist:
   - Description of the geologic environment pertaining to the engineering project.
   - Description of earth materials, such as their distribution and general physical and chemical characteristics.

TABLE 1.5 Fields of Expertise

Topic (1)	Engineering geologist (2)	Geotechnical engineer (3)	Overlapping areas of expertise (4)
Project planning	Development of geologic parameters Geologic feasibility	Design Material analysis Economics	Planning investigations Urban planning Environmental factors
Mapping	Geologic mapping Aerial photography Air photo interpretation Landforms Subsurface configurations	Topographic survey Surveying	Soil mapping Site selections
Exploration	Geologic aspects (fault studies, etc.)	Engineering aspects	Conducting field exploration Planning, observation, etc. Selecting samples for testing Describing and explaining site conditions
Engineering geophysics	Soil and rock hardness Mechanical properties Depth determinations	Engineering applications	Minimal overlapping of expertise
Classification and physical properties	Rock description Soil description (Modified Wentworth system)	Soil testing Earth materials Soil classification (USCS)	Soil description
Earthquakes	Location of faults Evaluation of active and inactive faults Historic record of earthquakes	Response of soil and rock materials to seismic activity Seismic design of structures	Seismicity Seismic conditions Earthquake probability
Rock mechanics	Rock mechanics Description of rock Rock structure, performance, and configuration	Rock testing Stability analysis Stress distribution	*In situ* studies Regional or local studies
Slope stability	Interpretative Geologic analyses and geometrics Spatial relationship	Engineering aspects of slope stability analysis and testing	Stability analyses Grading in mountainous terrain
Surface waters	Geologic aspects during design	Design of drainage systems Coastal and river engineering Hydrology	Volume of runoff Stream description Silting and erosion potential Source of material and flow Sedimentary processes
Groundwater	Occurrence Structural controls Direction of movement	Mathematical treatment of well systems Development concepts	Hydrology
Drainage	Underflow studies Storage computation Soil characteristics	Regulation of supply Economic factors Lab permeability	Well design, specific yield Field permeability Transmissibility

Source: Adapted from *Fields of Expertise* (undated).

- Deduction of the history of pertinent events affecting the earth materials.
- Forecast of future events and conditions that may develop.
- Recommendation of materials for representative sampling and testing.
- Recommendation of ways of handling and treating various earth materials and processes.
- Recommendation or providing criteria for excavation (particularly angle of cut slopes) in materials where engineering testing is inappropriate or where geologic elements control stability.
- Inspection during construction to confirm conditions.

2. Responsibilities of the geotechnical engineer:

- Directing and coordinating the team efforts where engineering is a predominant factor.
- Controlling the project in terms of time and money requirements and degree of safety desired.
- Engineering testing and analysis.
- Reviewing and evaluating data, conclusions, and recommendations of the team members.
- Deciding on optimum procedures.
- Developing designs consistent with data and recommendations of team members.
- Inspection during construction to assure compliance.
- Making final judgments on economy and safety matters.

1.3 TYPICAL CLIENTS

Typical clients who hire geotechnical engineers and engineering geologists are:

- *Mass builders and professional developers.* Individuals or companies who are in the business of developing housing tracts and industrial parks need the services of geotechnical engineers. The mass builder or professional developer may hire the geotechnical engineer to perform a site investigation and prepare a preliminary soils report on the feasibility of the proposed project. The geotechnical engineer could also be hired to perform density testing and observations during grading of the project as well as provide foundation recommendations and design parameters. The mass builder and professional developer are usually knowledgeable about the exact services that will be required of the geotechnical engineer.
- *Individual property owners.* This category could include individuals, corporations, financial institutions, or other types of owners in the private sector. The geotechnical services would be similar as described above, but the project often deals with the construction of a single structure. These clients are not developers, but are instead having a structure built to meet their needs. These clients usually only have a limited idea of the geotechnical engineer's role in the project.
- *Contractors and design professionals.* The geotechnical engineer could be hired by a building contractor or design professional, such as the architect, to work on the project. Working for these clients can sometimes result in an awkward relationship, as for example when the grading contractor hires the geotechnical engineer to check fill compaction. Difficulties can naturally arise when the geotechnical engineer criticizes the grading contractor (i.e., the client) for failure to obtain adequate compaction. In such cases, it would be best to have a contract with the project owner, rather than the grading contractor.

- *Governmental agencies.* The geotechnical engineer could be hired by numerous city, county, state, or national agencies. For example, the engineer may be involved with the construction or maintenance of publicly owned facilities. The geotechnical engineer may also be hired or work for governmental agencies, such as the state transportation department, public works division, or city engineer's office.

1.4 EXAMPLE

The previous sections have described the various types of projects involving geotechnical engineers. The purpose of this section is to present an example of the geotechnical engineering aspects pertaining to the construction of a building. The project is an office building constructed in a commercial business park. A series of 10 photographs was taken during the construction of the building and all 10 of the photographs show a general view to the north.

The large site initially consisted of undeveloped land. The first step was the rough grading of the entire project to create level building pads, and then the streets and utilities were installed. The individual lots within the commercial business park were then sold. The photographs in this section show the construction of an office building on one of the lots.

Figure 1.1 shows the site, which consists of a level building pad that had been previously rough graded. This figure shows the excavation of the individual footings for the building. Note that the footings are not being hand-dug, but rather are being excavated by large excavation machinery. At this stage of the project, the type of foundation has been selected and the actual dimensions and reinforcement conditions of the foundation have been designed. The foundation being constructed in Fig. 1.1 can be classified as a shallow foundation consisting of individual spread, strip, and combined footings.

Figure 1.2 shows the next step in the construction of the foundation. The footings have been excavated to the desired dimensions, and steel reinforcement is being posi-

FIGURE 1.1 Level building pad and excavation of footings.

INTRODUCTION 1.11

FIGURE 1.2 Installation of steel reinforcement in footings.

tioned within the footing. The arrow in Fig. 1.2 points to a team of workers installing the steel reinforcement.

Figure 1.3 shows a footing containing steel reinforcement being filled with concrete. On the right side of Fig. 1.3, the arrow is pointing to a surveyor who is making sure of the exact locations and dimensions of the footings per the building plans and will determine the final level of concrete in the footings. There are essentially no forms being used for the construction of the footings, and the size of the footings will correspond to the size of the excavations.

FIGURE 1.3 Filling of footings with concrete.

Figure 1.4 shows several of the footings filled with concrete. Other footings are still in the process of being filled with concrete. Note in Fig. 1.4 that steel bolts extend up through the top of the footings. These bolts will be used to attach the building frame to the foundation. The small arrows in Fig. 1.4 point to spread footings while the larger arrows point to combined footings.

Figure 1.5 is a view of the site several days after the footings have been installed. There was a heavy rainstorm prior to the taking of the photograph. Note that the top of the footings are flooded. Instead of being a detriment, the rain and flooding of the top of the footings will help in the curing of the concrete. However, as shown in Fig. 1.5, the heavy rainfall did cause erosion of the on-site sandy soil. Note in Fig. 1.5 that the side of one of the footings is exposed.

Figures 1.6 through 1.8 show the erection of the steel framing for the building. Figure 1.6 is an overview photograph that shows the construction of the structural frame of the building, which consists of steel beams and columns to carry vertical loads and steel cross-braces to provide lateral stability. Figure 1.7 shows a close-up view of two spread footings. Note that the steel columns have been attached to the center of the footings where the steel bolts extend up through the top of the footings. In Fig. 1.8, the arrow points to a combined footing. This combined footing is supporting two steel columns, and there are cross-braces between the two steel columns. This area of the foundation was designed as a combined footing because it will not only be subjected to downward vertical loads due to the weight of the building, but must also carry lateral loads due to wind or seismic forces which will be transmitted to the footing by the cross-braces. This combined loading (vertical and horizontal) is the reason for the much larger size of the combined footing in Fig. 1.8 compared to the spread footings shown in Fig. 1.7.

Figures 1.9 and 1.10 show two views of the construction of perimeter strip footings. The arrow in Fig. 1.9 points to wood forms that have been constructed prior to placement of

INTRODUCTION 1.13

FIGURE 1.4 Several footings filled with concrete.

FIGURE 1.5 View of site after rainstorm.

FIGURE 1.6 Overview of construction of steel frame.

FIGURE 1.7 Spread footings for steel columns.

INTRODUCTION 1.15

FIGURE 1.8 Combined footing supporting two steel columns with cross-bracing.

FIGURE 1.9 Wood forms for perimeter strip footings.

the concrete. The forms are built so that during placement, the fluid concrete will be confined by the forms. The wood forms are constructed so that the final size of the concrete strip footings will conform to the required dimensions per the building plans. In Fig. 1.10, concrete has been placed within the forms and the concrete strip footings have gained sufficient strength so that the wood forms were able to be removed. For this building, the perimeter strip footings will not be supporting much load because the exterior walls will be attached to the steel building frame, with loads transmitted to the steel columns. Note in Fig. 1.10 that steel decking is being installed on the upper floors. Although not shown, the final step in the construction of the foundation was the placement of a floor slab to fill the area between the strip footings.

1.5 OUTLINE OF CHAPTERS

As indicated in Fig. 1.11, the idea for the format of the book was to follow in a logical manner the typical process of a design project, from the first step of preparing a proposal for engineering services to the final step of preparing the "as-built" report upon completion of the project. The book is divided into five separate parts. Part 1 (Chaps. 2 and 3) provides a discussion of the development of programs of geotechnical investigation, which includes obtaining information on the assignment, preparation of a proposal, and planning the work. Part 2 (Chaps. 4 and 5) deals with geotechnical field and laboratory studies, which consist of subsurface exploration and laboratory testing of soil, rock, or groundwater samples. Part 3 (Chaps. 6 to 16) presents the analysis of geotechnical data and engineering computations for conditions commonly encountered by the design engineer, such as settlement, expansive soil, and slope stability. Part 4 (Chaps. 17 and 18) provides a discussion of the performance or

FIGURE 1.10 View of completed perimeter strip footings.

INTRODUCTION

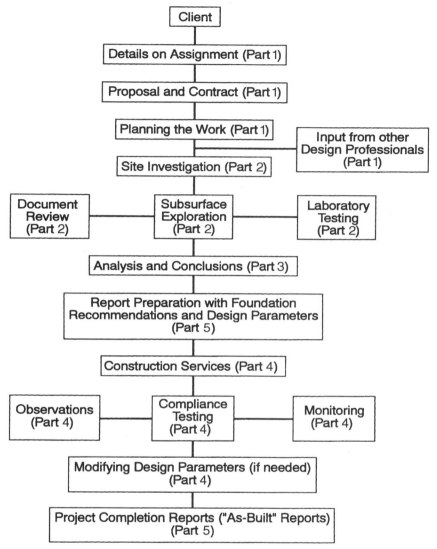

FIGURE 1.11 Outline of book.

engineering evaluation of construction, and Part 5 (Chap. 19) is a concluding chapter dealing with the preparation or engineering evaluation of geotechnical reports.

A list of symbols is provided at the beginning of some chapters. An attempt has been made to select those symbols most frequently listed in standard textbooks and used in practice. Dual units are used throughout the book, consisting of:

1. International System of Units (SI)
2. Inch-pound units (I-P units), which are also frequently referred to as the United States Customary System units (USCS)

In some cases, figures have been reproduced that use the old metric system (e.g., stress in kg/cm^2). These figures have not been revised to reflect SI units.

P · A · R · T · 1

DEVELOPMENT OF PROGRAMS OF GEOTECHNICAL INVESTIGATION

Tilt-up building construction, where the perimeter concrete walls are cast on the floor slab and then tilted up into position.

CHAPTER 2
THE ASSIGNMENT

2.1 PRELIMINARY INFORMATION

As indicated in Fig. 1.11, the first step in the project is to obtain details on the assignment and develop a proposal that is acceptable to the client. As with any project, basic information should be obtained before preparing the proposal, such as:

1. *Client information.* This includes the name, address, and telephone number of the client. It may also be appropriate to check the reputation of the client if it is not already known.
2. *Project location.* Basic information on the location of the project is required. The location of the project can be compared with known geologic hazards, such as active faults, landslides, or deposits of liquefaction-prone sand.
3. *Type of project.* The geotechnical engineer could be involved with all types of civil engineering construction projects, such as residential, commercial, or public works projects. It is important to obtain as much preliminary information about the project as possible. Such information could include the type of structure and use, size of the structure including the number of stories, type of construction and floor systems, preliminary foundation type (if known), and estimated structural loadings. Typical project types are further discussed in Sec. 2.2.
4. *Scope of work.* In order to prepare a proposal that is acceptable to the client, the geotechnical engineer must have specific details on the scope of work. For example, the scope of work could include a preliminary geotechnical investigation to determine the feasibility of the project, compaction testing and observations during grading of the site, and the preparation of foundation design recommendations.
5. *Conflict of interest.* The geotechnical engineer must investigate possible conflicts of interest. As an example of a conflict of interest, suppose you were hired by a contractor to help prepare a bid for a project. If a different contractor subsequently asked you to also prepare a bid on the same project, you could not accept the assignment because it would be a conflict of interest.

2.2 TYPE OF PROJECT

The geotechnical engineer may be hired for many different types of projects, such as:

Single-Family Dwellings and Condominiums.   As urban sprawl continues in the United States, the most common types of structures being built are single-family dwellings or condominiums and their associated roads and utilities. Especially for large housing tracts, the geotechnical engineer will usually have significant involvement with the project. As indicated in App. B, such projects are often divided into two basic categories: flatland and hillside. The guidelines in App. B indicate that a more rigorous geotechnical and geologic investigation is required for a hillside than a flatland site.

A common feature of single-family dwellings and low-rise condominiums is the use of lightweight construction, such as wood framing (Figs. 2.1 and 2.2) or even aluminum framing (Fig. 2.3). Usually footing widths and depths of single-family dwellings are governed by minimum building code requirements, rather than the loads applied to the foundation. As an example, Table 2.1 presents minimum footing widths and depths from the *Uniform Building Code* (1997). As indicated in Table 2.1, a foundation supporting two floors should have a minimum footing width of 15 in. (0.38 m) and a minimum footing depth of 18 in. (0.46 m).

The geotechnical engineer could also be involved with the design and construction of homeowner improvements. For example, Fig. 2.4 shows the construction of a pool in the rear yard of a single-family dwelling.

Commercial and Industrial Sites. The most common type of commercial project is office buildings, including skyscrapers, that are either being built specifically for the use of the client or will be rented out to various tenants. Common types of office buildings are steel-framed (Figs. 2.5 and 2.6), reinforced-concrete (Fig. 2.7), combined reinforced-concrete and steel (Fig. 2.8), and tilt-up concrete exterior panel buildings such as shown in Figs. 2.9 to 2.11. Industrial sites can contain a variety of projects such as factories and refineries. Commercial and industrial projects frequently have a variety of loading and performance criteria that require special geotechnical investigation and foundation design.

FIGURE 2.1 Wood framing for single-family dwelling.

FIGURE 2.2 Wood framing for low-rise condominium.

FIGURE 2.3 Aluminum framing for single-family dwelling.

TABLE 2.1 Minimum Requirements for Stud Bearing Walls

Number of floors supported by the foundation (1)	Width of footing, in. (2)	Depth of footing below undisturbed ground surface, in. (3)
1	12	12
2	15	18
3	18	24

Source: Information obtained from the *Uniform Building Code*, 1997.

FIGURE 2.4 Construction of rear yard pool.

Other Projects in the Private Sector. There are many other types of private-sector projects besides dwellings and commercial and industrial sites. Examples include the construction of small private dams, power plants, and energy transmission facilities, and transportation projects, such as privately owned roads.

Public Works Projects. This category of projects is very broad and includes all types of projects built with public money. Examples include levees and dams, harbors, airports, stadiums, and publicly owned buildings. This category also includes public transportation facilities, such as roads, highways, train beds, highway overpasses, bridges, and tunnels. Military projects are also included in this category, such as armories, waterway projects, military housing projects, and other military base facilities. While some geotechnical engineering firms may specialize in public works projects, the majority of geotechnical engineers are involved with both private and public works projects.

FIGURE 2.5 Steel-frame office building.

FIGURE 2.6 Steel-frame office building.

FIGURE 2.7 Reinforced-concrete office building.

FIGURE 2.8 Concrete-and-steel–frame office building.

FIGURE 2.9 Tilt-up concrete exterior panel building.

FIGURE 2.10 Tilt-up concrete exterior panel building with interior steel framing.

FIGURE 2.11 Tilt-up concrete exterior panel building with interior steel framing.

Essential Facilities. The last category is essential facilities, which can be defined as those structures or buildings that must be safe and usable for emergency purposes after an earthquake or other natural disaster in order to preserve the health and safety of the general public. Typical examples of essential facilities are as follows (*Uniform Building Code,* 1997):

- Hospitals and other medical facilities having surgery or emergency treatment areas
- Fire and police stations
- Municipal government disaster operations and communication centers deemed to be vital in emergencies

According to the *Standard Specifications for Highway Bridges* (AASHTO, 1996), other facilities that could be classified as essential are as follows:

- Military bases and supply depots and National Guard installations
- Facilities such as schools, arenas, etc., which could provide shelter or be converted to aid stations
- Major airports
- Defense industries and those that could easily or logically be converted to such
- Refineries, fuel storage, and distribution centers
- Major railroad terminals, railheads, docks, and truck terminals
- Major power plants, including nuclear power facilities and hydroelectric centers at major dams
- Other facilities that the state considers important for national defense or for emergencies resulting from natural disasters or other unforeseen circumstances.

According to AASHTO (1996), essential bridges are defined as those that must continue to function after an earthquake. Transportation routes to critical facilities such as hospitals, police, fire stations, and communication centers must continue to function and bridges required for this purpose should be classified as essential. In addition, a bridge that has the potential to impede traffic if it collapses onto an essential route should also be classified as essential.

These types of essential or critical facilities would require the most thorough geotechnical investigations and most rigorous standards of construction. These types of structures would also generally have the highest factors of safety and the most redundancy built into the project.

2.3 PROJECT REQUIREMENTS

For some projects, the project requirements will be quite specific and may even be in writing. For example, a public works project may require a geotechnical investigation consisting of a certain number, type, and depth of borings, and may also specify the types of laboratory tests to be performed. The more common situation is where the client is relying on the geotechnical engineer to prepare a proposal, perform an investigation, and provide design parameters that satisfy the needs of the project engineers and requirements of the local building officials or governing authority. The general requirements for foundation engineering projects are as follows (Tomlinson, 1986):

1. Knowledge of the general topography of the site as it affects foundation design and construction, e.g., surface configuration; adjacent property; the presence of watercourses, ponds, hedges, trees, rock outcrops, etc.; and the available access for construction vehicles and materials.
2. The location of buried utilities such as electric power and telephone cables, water mains, and sewers.
3. The general geology of the area with particular reference to the main geologic formations underlying the site and the possibility of subsidence from mineral extraction or other causes.
4. The previous history and use of the site including information on any defects or failures of existing or former buildings attributable to foundation conditions.
5. Any special features such as the possibility of earthquakes or climate factors such as flooding, seasonal swelling and shrinkage, permafrost, and soil erosion.
6. The availability and quality of local construction materials such as concrete aggregates, building and road stone, and water for construction purposes.
7. For maritime or river structures, information on tidal ranges and river levels, velocity of tidal and river currents, and other hydrographic and meteorological data.
8. A detailed record of the soil and rock strata and groundwater conditions within the zones affected by foundation bearing pressures and construction operations, or of any deeper strata affecting the site conditions in any way.
9. Results of laboratory tests on soil and rock samples appropriate to the particular foundation design or construction problems.
10. Results of chemical analyses on soil or groundwater to determine possible deleterious effects of foundation structures.

2.12 DEVELOPMENT OF PROGRAMS OF GEOTECHNICAL INVESTIGATION

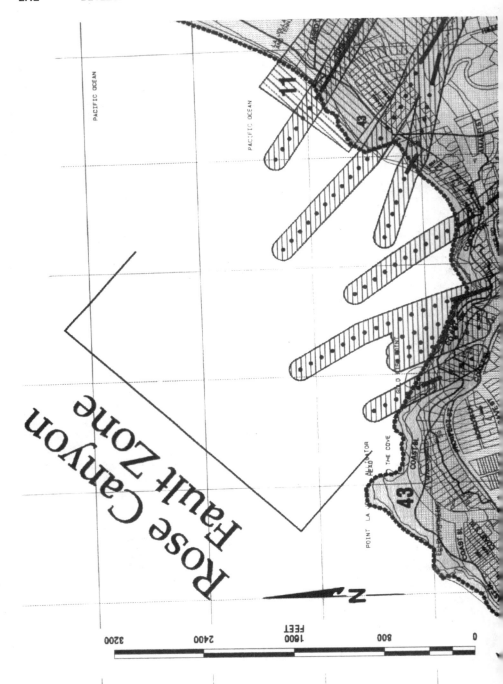

THE ASSIGNMENT

FIGURE 2.12 Portion of *Seismic Safety Study* (1995) developed by the City of San Diego.

As mentioned in Sec. 1.3, often the client lacks knowledge of the exact requirements of the geotechnical and foundation aspects of the project. For example, the client may have only a vague idea that the building needs a foundation, and therefore a geotechnical engineer must be hired. The owner assumes that you will prepare a proposal to satisfy all of the geotechnical requirements of the project.

Knowing the requirements of the local building department or governing authority is essential. For example, as indicated in "Technical Guidelines for Soil and Geology Reports" (App. B), the building department requires that specific items must be addressed by the geotechnical engineer and engineering geologist, such as settlement potential of the structure, grading recommendations, geologic aspects, and, for hillside projects, slope stability analyses. Even if these items will not impact the project, they nevertheless must be investigated and discussed in the geotechnical report.

There may be other important project requirements that the client is unaware of and is relying on the geotechnical engineer to furnish. For example, Fig. 2.12 presents a portion of the *Seismic Safety Study* (1995) which indicates the location of geologic hazards, faults, and deposits of liquefaction-prone soil. Projects to be located within these study areas would require investigations that address the specific hazards listed on the maps.

In summary, it is essential that the geotechnical engineer know the general requirements for the project (such as the 10 items listed above), as well as local building department or other regulatory requirements. If all required items are not investigated or addressed in the geotechnical report, then the building department or regulatory authority may refuse to issue a building permit. This will naturally result in an upset client because of the additional work that is required, delays in construction, and possible unanticipated design and construction expenses.

CHAPTER 3
PROPOSAL AND PLANNING THE WORK

3.1 PROPOSAL

A proposal is a document that lists the scope of services, estimated cost to perform the professional services, and the amount of time needed to complete the services. In some cases, the proposal may simply be a verbal cost estimate for services requested by the client or a one-page letter outlining the work to be performed on the project and the cost for the services. For other projects, the proposal can be quite extensive and may take a considerable amount of time and effort to prepare. An initial site visit may even be required to evaluate the scope and nature of the project.

There are several basic elements in preparing a proposal, such as having a thorough understanding of the scope of services, estimating the amount of time needed to complete the professional services, and finally preparing a cost estimate.

3.1.1 Cost Estimate

The appropriate type of compensation is an hourly fee or per diem rate for professional services. Table 3.1 presents an example of a schedule of fees that lists the hourly rate of various professionals and the costs for laboratory and field work. The easiest approach would be to have the client agree to reimburse for time and expenses, which is basically the number of hours worked on the project times the hourly or per diem rate listed on the schedule of fees plus expenses. Most clients, however, will not agree to such an open-ended contract and would rather be provided with a cost estimate or a dollar amount not to be exceeded.

For simple or repetitive projects, the cost estimate may be based on prior work. For example, a geotechnical engineer may have a set fee to perform a percolation test (needed for the design of the septic system) for a single-family residential property.

Especially for large or complicated projects, the development of a cost estimate can take considerable effort. Table 3.2 shows an example of the type of form that can be used to obtain the cost estimate. Usually the cost estimate is developed at the same time that the work is planned (Sec. 3.2). The cost estimate is based on multiplying the number of hours required for a specific task by the corresponding rate from the schedule of fees. Costs for laboratory tests, field work, and subcontract expenses are included in the cost estimate. While the cost estimate can be given verbally, it is often best to send the client a written document indicating the cost of the geotechnical work and the scope of services associated with the work.

TABLE 3.1 Example of a Schedule of Fees

Professional and staff hourly rates:
   Principal geotechnical engineer or principal engineering geologist . . . .$/h
   Chief geotechnical engineer or chief engineering geologist . . . .$/h
   Senior geotechnical engineer or senior engineering geologist . . . .$/h
   Project engineer or project geologist . . . .$/h
   Staff engineer or staff geologist . . . .$/h
   Associate engineer or associate geologist . . . .$/h
   Compaction testing technician . . . .$/h
   Drafting or CAD services . . . .$/h
   Office services. . . . .$/h

Subsurface exploration, compaction testing, and monitoring:
   Drill rig rental costs (24-in.-diameter bucket auger boring) . . . .$/h
   Drill rig rental costs (solid- or hollow-stem auger) . . . .$/h
   Drill rig rental costs (rotary coring) . . . .$/h
   Drill rig rental costs (air track) . . . .$/h
   Bulldozer for construction of drill rig access roads . . . .$/h
   Test pit excavation costs . . . .$/h
   Trench excavation costs (backhoe) . . . .$/h
   Mobile laboratory equipment for field compaction . . . .$/h
   Inclinometer pipe and materials . . . .$/ft
   Piezometer pipe and materials . . . .$/ft

Laboratory testing:
   Moisture content (ASTM D2216) . . . .$/test
   Wet density . . . .$/test
   Atterberg limits (ASTM D4318) liquid/plastic . . . .$/test
   Particle size analysis (ASTM D422) . . . .$/test
   Specific gravity—soils (ASTM D854) . . . .$/test
   Specific gravity—oversize particles (ASTM C127) . . . .$/test
   Sand equivalent (ASTM D2419) . . . .$/test
   Collapse test (ASTM D5333) . . . .$/test
   Swell test (ASTM D4546) . . . .$/test
   Expansion index (UBC Std. 18-2) . . . .$/test
   Modified Proctor compaction test (ASTM D1557) . . . .$/test
   R-value (ASTM D2844) . . . .$/test
   Unconfined compression (ASTM D2166) . . . .$/test
   Direct shear test (ASTM D3080) . . . .$/test
   Triaxial compression test (ASTM D4767) . . . .$/test
   Consolidation test (ASTM D2435) . . . .$/test
   Hydraulic conductivity (permeability, ASTM D2434 or D5084) . . . .$/test
   Special handling, storage, and/or disposal . . . . .Hourly rates
   Outside laboratory . . . . Cost+20%

It may also be appropriate to include a statement in the proposal to cover items not specifically listed in the scope of services. For example, the statement could be worded as follows:

> This proposal does not include costs for meetings, conferences, and/or items which have not been specifically noted in the scope of services. Such costs are difficult to estimate at present and have not been included. Work requested by responsible parties outside the scope will be billed as "extras" on a time-and-expense basis under purview of this proposal, unless another proposal is specifically requested.

TABLE 3.2 Example of a Cost Estimating Sheet

Category (1)	Description (2)	Hours (3)	Rate (4)	Cost (5)
Proposal	Planning and preparation of proposal			
Field exploration	In-house and agency research Review client's or other engineers' documents Subsurface exploration (drilling, test pits, trenches) Engineering geologist work (mapping, aerial photos, etc.) Preparation of logs and field paperwork Preparation of soil profile			
Laboratory testing	Soil classification tests (particle size, Atterberg limits) Moisture content and wet density determinations Settlement potential (consolidation, collapse tests) Expansion potential (expansion index, swell tests) Shear strength (direct shear, triaxial, etc.) Erosion and deterioration potential Compaction tests (Modified or Standard Proctor) Miscellaneous (specific gravity, sand equivalent, R-value)			
Analysis of data and engineering computations	Laboratory data reduction and analysis Engineering calculations (settlement, bearing capacity, expansive soil, slope stability, seismic analysis, etc.) Development of design parameters for foundations, retaining walls, effect of groundwater, etc. Computer analyses (slope stability, etc.) Design of geotechnical elements (foundations, etc.) Engineering geology analyses and recommendations			
Compaction testing and other construction services	Compaction testing (technician) Observations during grading by engineer and geologist Sampling and testing during grading operations Other construction services			
Report preparation	Report writing, editing, and review Preparation of laboratory and field data Drafting and graphics for report Word processing and report production Blueprinting and production of plans			
Subcontract expenses	Subcontract expenses for drill rig rental Subcontract expenses for test pits and trenches Expenses for monitoring equipment Other subcontract expenses			

Total estimate = $_____

Price quoted = $_____

Most geotechnical proposals for site development include subsurface exploration as part of the scope of services. The cost of performing subsurface exploration is often difficult to estimate because of the unknown subsurface conditions and access requirements for the drill rig. It may be appropriate to include wording in the proposal that indicates that, if unusual subsurface conditions are encountered, there could be increased excavation costs due to the unanticipated conditions, and the client would be immediately notified about the increased costs of the subsurface exploration.

3.1.2 Engineer's Agreement with the Client

In addition to sending the client a written cost estimate and scope of services, it is important to have the client sign an engineering agreement, which is basically a contract between the geotechnical engineer and the client. In some cases, the client will send the geotechnical engineer a contract to sign. A more common situation is that a packet containing both the written cost estimate and actual contract is sent to the client, with a request that one copy of the signed contract be returned to the geotechnical engineer.

The contract should contain such items as the client's name and address, location of the project, agreed-upon cost of services, invoicing and payment procedures, and protection of your work product. Some clients, such as governmental agencies, may refuse to sign the engineering agreement. In these cases, a letter of authorization for services may serve as the contract between the client and the geotechnical engineer.

It is always best to have an attorney prepare or review the contract. In a general sense, the most important items that should be included in the geotechnical engineer's agreement with the client are as follows:

1. *Contract title and introductory wording.* The contract should contain a title and introductory wording indicating that the document is the contract between the geotechnical engineer and the client.
2. *Project and client information.* The section of the contract where the project name and address and client's name and address are inserted.
3. *Type of services.* The section of the contract where a brief summary of the scope of services is inserted.
4. *Cost of services.* The section of the contract where the cost estimate or not-to-exceed dollar amount is inserted.
5. *Signature page.* A final section of the contract which states that both the geotechnical engineer and client have read the contract and agree to all the terms and conditions. Spaces should be provided for both the geotechnical engineer and the client to sign and date the contract.

Other items that could be included in the fine print of the contract are as follows:

Safety. A statement indicating that the geotechnical engineer will not be responsible for the general safety on the job or the work of other contractors and third parties.

Ownership of documents. A statement indicating ownership of documents could be included in the contract. For example, the contract could state that reports prepared for the client are the property of the client, but all other documents, such as excavation logs, field data, field notes, test data, calculations, estimates, plans, sections, and other original documents prepared by the geotechnical engineer shall remain the property of the geotechnical engineer. It may also be appropriate to indicate that reports prepared pursuant to the agreement may not be transferred or used by others without the expressed, written approval of the geotechnical engineer.

Disclaimer of warranties. An important consideration is to include a disclaimer in the contract. For example, the disclaimer could state that geotechnical engineering is not an exact science and no warranty or guaranty is expressed or implied in connection with the rendition of services under this agreement.

Termination and modification of agreement. The contract could indicate the procedure to terminate or modify the agreement. For example, it could state that in the event that either party desires to terminate the contract prior to completion of the project, written notification of such intention to terminate must be tendered to the other party. In terms of modifications, the contract could state that any subsequent change, alteration, addition, or modification must be mutually agreed upon, in writing, and signed by both parties. It would also be important to state that in the absence of written notification to terminate or modify the contract, the agreement remains in full force and effect.

Contract jurisdiction. It may be important to state the location where the contract was signed. In the event of legal action, the jurisdiction is often the location where the contract was signed. As an example, the contract could state that this agreement shall be deemed to have been entered into in the City of San Diego, California, and shall be governed by the laws of California. Other legal terminology could include a statement that if any provision of the agreement is determined to be in violation of any law or ordinance, it will not invalidate the whole of the agreement, but rather the provision shall be deemed stricken from the agreement.

Retainers/payments. Prompt payment for engineering services is always desirable. It may be appropriate to state that payments are due within 30 days upon the receipt of the invoice for engineering services. An interest charge for payments beyond the due date could also be listed.

Time limit for signing of contract. A time limit indicating how long the client has to sign the contract could be included in the agreement. For example, the contract could state that the client must sign it within 2 months from the date of submittal, and after that date, the contract is invalid.

Limitation of liability. The contract could include a limitation of liability clause. Geotechnical engineering is often described as a risky profession and these clauses are inserted in order to reduce the potential liability of the geotechnical engineer. Limitation of liability clauses will be further discussed in Chap. 19, Sec. 19.5.

3.2 PLANNING THE WORK

This section describes the steps necessary to complete the project with the ultimate goal of completing the project on time, within the budget, and to the satisfaction of the client. The elements of the work as described in this section are general in nature and they may not be inclusive for all types of geotechnical and foundation projects. Figure 1.11 shows the typical steps needed to complete the project. As indicated in Fig. 1.11, the work often culminates with the preparation of a final "as-built" report that summarizes the construction of the project.

The first step should be to plan the work for the project. For a minor project, the planning effort may be minimal. But for those large-scale projects, the plan can be quite extensive and could change as the design and construction progresses. As previously mentioned, in many cases the planning effort is performed at the same time the cost estimate is developed. The planning effort should include the following:

- Budget and scheduling considerations
- Selection of the interdisciplinary team (such as geotechnical engineer, engineering geologist, hydrogeologist) that will work on the project
- Preliminary subsurface exploration plan, such as the number, location, and depth of borings
- Document collection
- Laboratory testing requirements
- Types of engineering analyses and design parameters that will be required

3.2.1 Technical Guidelines

When planning the work, it may be useful to review technical guidelines such as listed in App. B. These guidelines present the items that must be addressed per the requirements of the local building department. Including all requirements at the planning stage is important to prevent delays and expenses at a later date.

PART · 2

GEOTECHNICAL FIELD AND LABORATORY STUDIES

Toe of a lava flow, Mojave Desert, California

CHAPTER 4
FIELD EXPLORATION

The following notation is introduced in this chapter:

C_b	Borehole diameter correction for the SPT N-value
C_r	Rod length correction for the SPT N-value
D	Diameter of the vane [Eq. (4.4)]
D	Smallest dimension of the actual footing [Eq. (4.6)]
D_1	Smallest dimension of the steel plate (plate load test)
D_e	Diameter of the sampler cutting tip
D_i	Inside diameter of the sampler
D_o	Outside diameter of the sampler
E_m	Hammer efficiency for SPT
H	Height of the vane
k_o	At-rest earth pressure
k_s	Subgrade modulus
N	SPT N-value
N_{60}	SPT N-value corrected for field testing procedures
q	Stress for plate load test
q_c	Cone resistance
S	Settlement of the actual footing
S_1	Depth of penetration of the steel plate (plate load test)
s_u	Undrained shear strength of clay
T_{max}	Maximum torque required to shear the clay for the field vane test
δ	Penetration of the plate (plate load test)
ϕ	Friction angle of soil

4.1 INTRODUCTION

As indicated in Fig. 1.11, Part 1 of the book deals with the assignment, proposal, and planning of the work. Part 2 of the book deals with the site investigation, which includes document review, subsurface exploration, and laboratory testing. The goal of the site investigation is to obtain a detailed understanding of the engineering and geologic properties of the soil and rock strata and groundwater conditions that could impact the proposed development.

Field exploration is often difficult and expensive, and, in many cases, it may be prudent to have an initial site meeting with the client and other design professionals, such as the project structural engineer, civil engineer, or architect. The purpose of the field meeting is to discuss the project and upcoming field exploration. The preliminary field meeting is often helpful in educating the client about the design and construction process, finalizing details on the proposed field exploration, and coordinating the work and time constraints of the various design professionals involved with the project.

4.2 DOCUMENT REVIEW

As listed in Table 4.1, the geotechnical engineer should review available documents during the design and construction of the project. The following is a brief summary of these types of documents:

Preliminary Design Information. The documents dealing with preliminary design and proposed construction of the project should be reviewed. For example, the structural engineer or architect may have design information, such as the building size, height, loads, and details on proposed construction materials and methods. Preliminary plans may even have been developed that show the proposed construction.

History of the Site. If the site had prior development, it is also important to obtain information on the history of the site. The site could contain old deposits of fill, abandoned septic systems and leach fields, buried storage tanks, seepage pits, cisterns, mining shafts,

TABLE 4.1 Typical Documents That May Need to Be Reviewed for the Project

Project phase (1)	Types of documents (2)
Design	Available design information, such as preliminary data on the type of project to be built at the site and typical foundation design loads
	If applicable, data on the history of the site, such as information on prior fill placement or construction at the site
	Data (if available) on the design and construction of adjacent property
	Local building code
	Special study data, such as Fig. 2.12, developed by the local building department or other governing agency
	Standard drawings issued by the local building department or other governing agency
	Standard specifications that may be applicable to the project, such as *Standard Specifications for Public Works Construction* or *Standard Specifications for Highway Bridges*
	Other reference material, such as seismic activity records, geologic and topographic maps, and aerial photographs
Construction	Reports and plans developed during the design phase
	Construction specifications
	Field change orders
	Information bulletins used during construction
	Project correspondence between different parties
	Building department reports or permits

tunnels, and other man-made surface and subsurface works that could impact the new proposed development. There may also be information concerning on-site utilities and underground pipelines, which may need to be capped or rerouted around the project.

Aerial Photographs and Geologic Maps. During the course of the work, it may be necessary for the engineering geologist to check reference materials, such as aerial photographs or geologic maps. Aerial photographs are taken from an aircraft flying at a prescribed altitude along preestablished lines. Interpretation of aerial photographs takes considerable judgment, and, for reasons of training and experience, it is usually the engineering geologist who interprets the aerial photographs. Viewing a pair of aerial photographs, with the aid of a stereoscope, provides a three-dimensional view of the land surface. This view may reveal important geologic information at the site, such as the presence of landslides, fault scarps, types of landforms (e.g., dunes, alluvial fans, and glacial deposits such as moraines and eskers), erosional features, general type and approximate thickness of vegetation, and drainage patterns. By comparing older aerial photographs with newer ones, the engineering geologist can also observe any man-made or natural changes that have occurred at the site.

Geologic maps can be especially useful to the geotechnical engineer and engineering geologist because they often indicate potential geologic hazards (e.g., faults and landslides) as well as the type of near surface soil or rock at the site. For example, Fig. 4.1 presents a portion of a geologic map and Fig. 4.2 shows cross sections through the area shown in Fig. 4.1 (from Kennedy, 1975). Note that the geologic map and cross sections indicate the location of several faults, the width of the faults, and often state whether the faults are active or inactive. For example, Fig. 4.2 shows the Rose Canyon fault zone, an active fault having a ground shear zone about 300 m (1000 ft) wide. The cross sections in Fig. 4.2 also show fault-related displacement of various rock layers. Symbols are used to identify various deposits, and Table 4.2 defines the geologic symbols shown in Figs. 4.1 and 4.2.

A major source for geologic maps in the United States is the United States Geological Survey (USGS). The USGS prepares many different geologic maps, books, and charts; a list of the USGS publications is provided in their *Index of Publications of the Geological Survey* (USGS, 1997). The USGS also provides an "Index to Geologic Mapping in the United States," which shows a map of each state and indicates the areas where a geologic map has been published.

Topographic Maps. Both old and recent topographic maps can provide valuable site information. Figure 4.3 presents a portion of the topographic map for the Encinitas Quadrangle, California (USGS, 1975). As shown in Fig. 4.3, the topographic map is to scale and indicates the locations of buildings, roads, freeways, train tracks, and other civil engineering works as well as natural features such as canyons, rivers, lagoons, sea cliffs, and beaches. The topographic map in Fig. 4.3 even shows the locations of sewage disposal ponds, water tanks, and by using different colors and shading, it indicates older versus newer development. But the main purpose of the topographic map is to indicate ground surface elevations or elevations of the sea floor, as shown in Fig. 4.3. This information can be used to determine the major topographic features at the site and to plan subsurface exploration, such as seeking available access to the site for drilling rigs.

Building Code and Other Specifications. A copy of the most recently adopted local building code should be reviewed. Usually only a few sections of the building code will be directly applicable to geotechnical projects. For example, the main applicable geotechnical sections in the *Uniform Building Code* (1997) are Chap. 18, "Foundations and Retaining Walls," and Chap. 33, "Excavation and Grading."

4.6 GEOTECHNICAL FIELD AND LABORATORY STUDIES

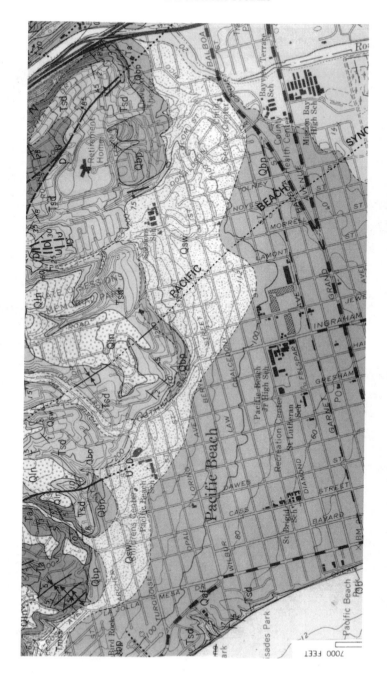

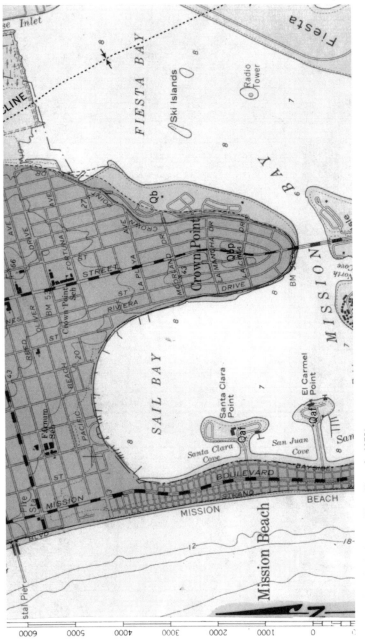

FIGURE 4.1 Geologic map. *(From Kennedy, 1975.)*

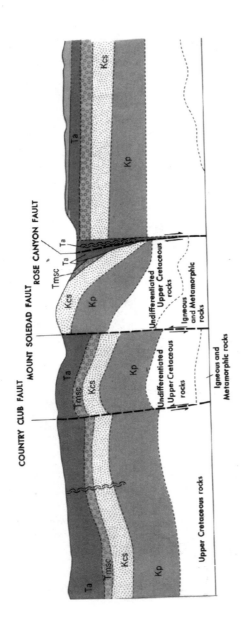
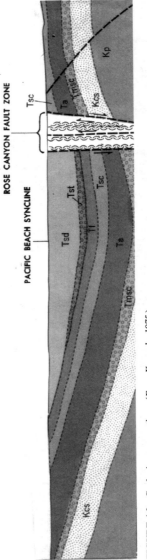

FIGURE 4.2 Geologic cross sections. *(From Kennedy, 1975.)*

TABLE 4.2 Symbols and Descriptions for Geologic Map and Cross Sections Shown in Fig. 4.1 and 4.2

Geologic symbol (1)	Type of material (2)	Description (3)
Q_{af}	Artificial fill	Artificial fill consists of compacted earth materials derived from many sources. Only large areas having artificial fill have been delineated on the geologic map.
Q_b	Beach sand	Sand deposited along the shoreline derived from many sources as a result of longshore drift and alluvial discharge from major stream courses.
Q_{al}	Alluvium	Soil deposited by flowing water, including sediments deposited in river beds, canyons, flood plains, lakes, fans at the foot of slopes, and estuaries.
Q_{sw}	Slope wash	Soil and/or rock material that has been transported down a slope by mass wasting assisted by runoff of water not confined to channels.
Q_{ls}	Landslide	Landslides are mass movement of soil or rock that involves shear displacement along one or several rupture surfaces, which are either visible or may be reasonably inferred.
Q_{bp}, Q_{lb}, Q_{ln}	Formational rock	Various sedimentary rock formations formed during the Pleistocene epoch (part of the Quaternary Period).
$T_a, T_f, T_{sc}, T_{sd}, T_{st}$	Formational rock	Various sedimentary rock formations formed during the Eocene epoch (part of the Tertiary Period).
K_{cs}, K_p	Formational rock	Various rock formations formed during the Cretaceous Period.

Note: For geologic symbols, Q represents soil or rock deposited during the Quaternary Period, T = Tertiary Period, and K = Cretaceous Period.

Depending on the type of project, there may be other specifications that are applicable for the project and need to be reviewed. Documents that may be needed for public works projects include the *Standard Specifications for Public Works Construction* (1997) and the *Standard Specifications for Highway Bridges* (AASHTO, 1996).

Documents at the Local Building Department. Other useful technical documents include geotechnical reports for adjacent properties, which can provide an idea of possible subsurface conditions. A copy of geotechnical reports on adjacent properties can often be obtained at the archives of public agencies, such as the local building department. Other valuable reference materials are standard drawings or standard specifications, which can also be obtained from the local building department. For example, Figs. 4.4 and 4.5 show two figures reproduced from the *City of San Diego Standard Drawings* (1986). Figure 4.4 provides specifications on the construction of storm drains, including the trench width, type of material to be used in the bedding zone (i.e., open graded aggregate having a maximum size of 1 in.), and the required compaction for the bedding and backfill material. Figure 4.5 provides specifications on the construction of gunite brow and terrace ditches, which are used to intercept ground surface runoff and then channel the flow to drainage facilities such as culverts, concrete-lined drainage ditches, or storm drain lines. Brow ditches are typically constructed at the top of fill or cut slopes, and terrace ditches are usually constructed on terraces built near the midheight of fill or cut slopes.

4.10 GEOTECHNICAL FIELD AND LABORATORY STUDIES

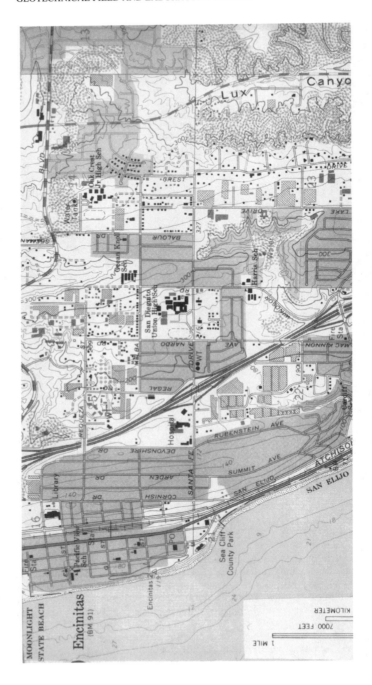

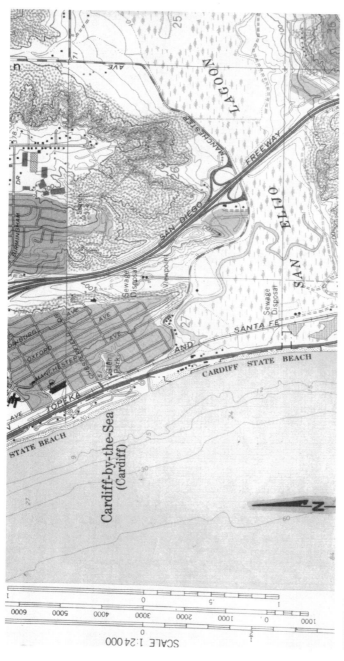

FIGURE 4.3 Topographic map. *(From USGS, 1975.)*

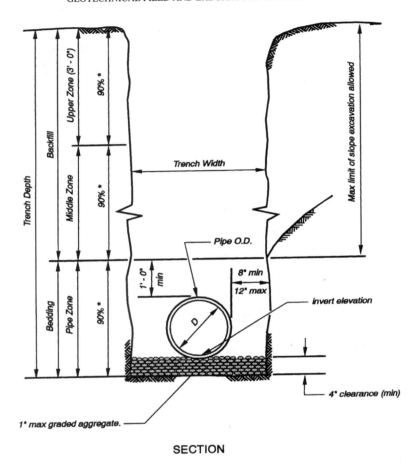

FIGURE 4.4 Pipe bedding and trench backfill for storm drains. (*From City of San Diego Standard Drawings*, 1986.)

Notes:
1- For trenching on improved streets see Standard Drawing G-24 or G-25 for resurfacing details.
2- (*) indicated minimum relative compaction.

4.3 GENERAL PURPOSE OF SUBSURFACE EXPLORATION

The general purpose of subsurface exploration is to determine the following (AASHTO, 1996):

Soil strata
- Depth, thickness, and variability
- Identification and classification
- Relevant engineering properties, such as shear strength, compressibility, stiffness, permeability, expansion or collapse potential, and frost susceptibility

FIELD EXPLORATION

Rock strata
- Depth to rock
- Identification and classification
- Quality, such as soundness, hardness, jointing and presence of joint filling, resistance to weathering (if exposed), and soluble nature of the rock

Groundwater elevation

Ground surface elevation

Local conditions requiring special consideration

There are different types of subsurface exploration, such as borings, test pits, and trenches. Table 4.3 (from Sowers and Royster, 1978; based on the work by ASTM; Lambe, 1951; Sanglerat, 1972; and Sowers and Sowers, 1970) summarizes the boring, core drilling, sampling, and other exploratory techniques that can be used by the geotechnical engineer.

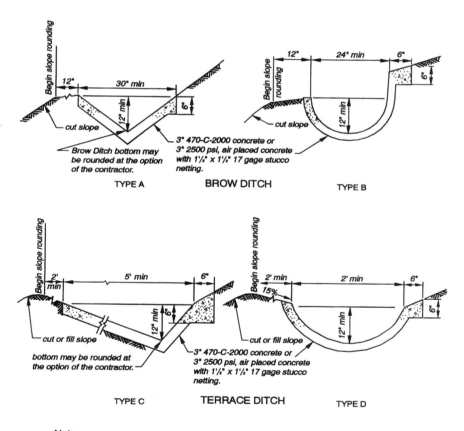

FIGURE 4.5 Drainage ditches. (*From City of San Diego Standard Drawings, 1986.*)

4.14   GEOTECHNICAL FIELD AND LABORATORY STUDIES

TABLE 4.3 Boring, Core Drilling, Sampling, and Other Exploratory Techniques

Method (1)	Procedure (2)	Type of sample (3)	Applications (4)	Limitations (5)
Auger boring, ASTM D 1452	Dry hole drilled with hand or power auger; samples preferably recovered from auger flutes	Auger cuttings, disturbed, ground up, partially dried from drill heat in hard materials	In soil and soft rock; to identify geologic units and water content above water table	Soil and rock stratification destroyed; sample mixed with water below the water table
Test boring, ASTM D 1586	Hole drilled with auger or rotary drill; at intervals samples taken 36-mm ID and 50-mm OD driven 0.45 m in three 150-mm increments by 64-kg hammer falling 0.76 m; hydrostatic balance of fluid maintained below water level	Intact but partially disturbed (number of hammer blows for second plus third increment of driving is standard penetration resistance or N)	To identify soil or soft rock; to determine water content; in classification tests and crude shear test of sample (N-value a crude index to density of cohesionless soil and undrained shear strength of cohesive soil)	Gaps between samples, 30 to 120 cm; sample too distorted for accurate shear and consolidation tests; sample limited by gravel; N-value subject to variations depending on free fall of hammer
Test boring of large samples	50- to 75-mm ID and 63- to 89-mm OD samplers driven by hammers up to 160 kg	Intact but partially disturbed (number of hammer blows for second plus third increment of driving is penetration resistance)	In gravelly soils	Sample limited by larger gravel
Test boring through hollow-stem auger	Hole advanced by hollow-stem auger; soil sampled below auger as in test boring above	Intact but partially disturbed (number of hammer blows for second plus third increment of driving is N-value)	In gravelly soils (not well adapted to harder soils or soft rock)	Sample limited by larger gravel; maintaining hydrostatic balance in hole below water table is difficult
Rotary coring of soil or soft rock	Outer tube with teeth rotated; soil protected and held stationary inner tube; cuttings flushed upward by drill fluid (examples: Denison, Pitcher, and Acker samplers)	Relatively undisturbed sample, 50 to 200 mm wide and 0.3 to 1.5 m long in liner tube	In firm to stiff cohesive soils and soft but coherent rock	Sample may twist in soft clays; sampling loose sand below water table is difficult; success in gravel seldom occurs
Rotary coring of swelling clay, soft rock	Similar to rotary coring of rock; swelling core retained by third inner plastic liner	Soil cylinder 28.5 to 53.2 mm wide and 600 to 1500 mm long encased in plastic tube	In soils and soft rocks that swell or disintegrate rapidly in air (protected by plastic tube)	Sample smaller; equipment more complex

FIELD EXPLORATION

Method	Procedure	Type of sample	Applications	Limitations
Rotary coring of rock, ASTM D 2113	Outer tube with diamond bit on lower end rotated to cut annular hole in rock; core protected by stationary inner tube; cuttings flushed upward by drill fluid	Rock cylinder 22 to 100 mm wide and as long as 6 m, depending on rock soundness	To obtain continuous core in sound rock (percent of core recovered depends on fractures, rock variability, equipment, and driller skill)	Core lost in fracture or variable rock; blockage prevents drilling in badly fractured rock; dip of bedding and joint evident but not strike
Rotary coring of rock, oriented core	Similar to rotary coring of rock above; continuous grooves scribed on rock core with compass direction	Rock cylinder, typically 54 mm wide and 1.5 m long with compass orientation	To determine strike of joints and bedding	Method may not be effective in fractured rock
Rotary coring of rock, wire line	Outer tube with diamond bit on lower end rotated to cut annular hole in rock; core protected by stationary inner tube; cuttings flushed upward by drill fluid; core and stationary inner tube retrieved from outer core barrel by lifting device or "overshot" suspended on thin cable (wire line) through special large-diameter drill rods and outer core barrel	Rock cylinder 36.5 to 85 mm wide and 1.5 to 4.6 m long	To recover core better in fractured rock, which has less tendency for caving during core removal; to obtain much faster cycle of core recovery and resumption of drilling in deep holes	Same as ASTM D 2113 but to lesser degree
Rotary coring of rock, integral sampling method	22-mm hole drilled for length of proposed core; steel rod grouted into hole; core drilled around grouted rod with 100- to 150-mm rock coring drill (same as for ASTM D 2113)	Continuous core reinforced by grouted steel rod	To obtain continuous core in badly fractured, soft, or weathered rock in which recovery is low by ASTM D 2113	Grout may not adhere in some badly weathered rock; fractures sometimes cause drift of diamond bit and cutting rod
Thin-wall tube, ASTM D 1587	75- to 1250-mm thin-wall tube forced into soil with static force (or driven in soft rock); retention of sample helped by drilling mud	Relatively undisturbed sample, length 10 to 20 diameters	In soft to firm clays, short (5-diameter) samples of stiff cohesive soil, soft rock, and, with aid of drilling mud, firm to dense sands	Cutting edge wrinkled by gravel; samples lost in loose sand or very soft clay below water table; more disturbance occurs if driven with hammer
Thin-wall tube, fixed piston	75- to 1250-mm thin-wall tube, which has internal piston controlled by rod and keeps loose cuttings from tube, remains stationary while outer thin wall tube forced ahead into soil; sample is held in tube by piston	Relatively undisturbed sample, length 10 to 20 diameters	To minimize disturbance of very soft clays (drilling mud aids in holding samples in loose sand below water table)	Method is slow and cumbersome

TABLE 4.3 Boring, Core Drilling, Sampling, and Other Exploratory Techniques (Continued)

Method (1)	Procedure (2)	Type of sample (3)	Applications (4)	Limitations (5)
Swedish foil	Samples surrounded by thin strips of stainless steel, stored above cutter, to prevent contact of soil with tube as it is forced into soil	Continuous samples 50 mm wide and as long as 12 m	In soft, sensitive clays	Samples sometimes damaged by coarse sand and fine gravel
Dynamic sounding	Enlarged disposable point on end of rod driven by weight falling fixed distance in increments of 100 to 300 mm	None	To identify significant differences in soil strength or density	Misleading in gravel or loose saturated fine cohesionless soils
Static penetration	Enlarged cone, 36 mm diameter and 60° angle forced into soil; force measured at regular intervals	None	To identify significant differences in soil strength or density; to identify soil by resistance of friction sleeve	Stopped by gravel or hard seams
Borehole camera	Inside of core hole viewed by circular photograph or scan	Visual representation	To examine stratification, fractures, and cavities in hole walls	Best above water table or when hole can be stabilized by clear water
Pits and trenches	Pit or trench excavated to exposure soils and rocks	Chunks cut from walls of trench; size not limited	To determine structure of complex formations; to obtain samples of thin critical seams such as failure surface	Moving excavation equipment to site, stabilizing excavation walls, and controlling groundwater may be difficult
Rotary or cable tool well drill	Toothed cutter rotated or chisel bit pounded and churned	Ground	To penetrate boulders, coarse gravel; to identify hardness from drilling rates	Identifying soils or rocks difficult
Percussion drilling (jack hammer or air track)	Impact drill used; cuttings removed by compressed air	Rock dust	To locate rock, soft seams, or cavities in sound rock	Drill becomes plugged by wet soil

Source: Adapted from Sowers and Royster, 1978. Reprinted with permission from *Landslides: Analysis and Control, Special Report 176.* Copyright 1978 by the National Academy of Sciences. Courtesy of the National Academy Press, Washington, D.C.

FIELD EXPLORATION 4.17

As mentioned above, the borings, test pits, or trenches are used to determine the thickness of soil and rock strata, estimate the depth to groundwater, obtain soil or rock specimens, and perform field tests such as standard penetration tests (SPT) or cone penetration tests (CPT). The Unified Soil Classification System (USCS) can be used to classify the soil exposed in the borings or test pits (Casagrande, 1948). The subsurface exploration and field sampling should be performed in accordance with standard procedures, such as those specified by the American Society for Testing and Materials (ASTM, 1970, 1971, and D 420-93, 1998) or other recognized sources (e.g., Hvorslev, 1949, and ASCE, 1972, 1976, and 1978). Appendix A (Glossary 1) presents a list of terms and definitions for engineering geology and subsurface exploration.

4.4 BORINGS

A boring is defined as a cylindrical hole drilled into the ground for the purposes of investigating subsurface conditions, performing field tests, and obtaining soil, rock, or groundwater specimens for testing. Borings can be excavated by hand (e.g., hand auger), although the usual procedure is to use mechanical equipment to excavate the borings.

There are many different types of equipment used to excavate borings. Typical types of borings are listed in Table 4.3 and include:

- *Auger boring.* A mechanical auger is a very fast method of excavating a boring. The hole is excavated by rotating the auger while at the same time applying a downward pressure on the auger to help obtain penetration of the soil or rock. There are basically two types of augers: flight augers and bucket augers (see Fig. 4.6). Common available diameters of flight augers are 5 cm to 1.2 m (2 in. to 4 ft) and of bucket augers are 0.3 to 2.4 m (1 to 8 ft). The auger is periodically removed from the hole, and the soil lodged in the groves of the flight auger or contained in the bucket of the bucket auger is removed. A casing is generally not used for auger borings and the hole may cave-in during the excavation of loose or soft soils or when the excavation is below the groundwater table. Augers are probably the most common type of equipment used to excavate borings.

- *Hollow-stem flight auger.* A hollow-stem flight auger has a circular hollow core which allows for sampling down the center of the auger. The hollow-stem auger acts like a casing and allows for sampling in loose or soft soils or when the excavation is below the groundwater table.

- *Wash-type borings.* Wash-type borings use circulating drilling fluid, which removes cuttings from the borehole. The cuttings are created by the chopping, twisting, and jetting action of the drill bit which breaks the soil or rock into small fragments. Casings are often used to prevent cave-in of the hole. Because drilling fluid is used during the excavation, it is difficult to classify the soil and obtain uncontaminated soil samples.

- *Rotary coring.* This type of boring equipment uses power rotation of the drilling bit as circulating fluid removes cuttings from the hole. Table 4.3 lists various types of rotary coring for soil and rock.

- *Percussion drilling.* This type of drilling equipment is often used to penetrate hard rock, for subsurface exploration or for the purpose of drilling wells. The drill bit works much like a jackhammer, rising and falling to break up and crush the rock material.

Figure 4.6 shows a bucket auger drilling rig which in southern California is routinely used to excavate a 0.76-m- (30-in.-) diameter boring that can then be downhole-logged by

FIGURE 4.6 A flight auger drill rig (top) and a bucket auger drill rig (bottom).

the geotechnical engineer or engineering geologist. Figure 4.7 shows a photograph of the top of the boring with the geologist descending into the hole in a steel cage. Note in Fig. 4.7 that a collar is placed around the top of the hole to prevent loose soil or rocks from being accidentally knocked down the hole. The process of downhole logging is a valuable technique because it allows the geotechnical engineer or engineering geologist to observe the subsurface materials as they exist in place. Usually the process of the excavation of the boring smears the side of the hole, and the surface must be chipped away to observe intact soil or rock. Going downhole is dangerous because of the possibility of a cave-in of the hole as well as "bad air" (presence of poisonous gases or lack of oxygen) and should be attempted only by an experienced geotechnical engineer or engineering geologist.

The downhole observation of soil and rock can lead to the discovery of important subsurface conditions. For example, Fig. 4.8 provides an example of the type of conditions observed downhole. Figure 4.8a shows a knife that has been placed in an open fracture in bedrock. The open fracture in the rock was caused by massive landslide movement. Figure 4.8b is a side view of the same condition.

It takes considerable experience to anticipate which type of drill rig and sampling equipment would be best suited to the site under investigation. In general, the most economical equipment for borings are truck-mounted rigs that can quickly and economically drill through hard or dense soil. It some cases, it is a trial-and-error process of using different drill rigs to overcome access problems or difficult subsurface conditions. For example, one deposit encountered by the author consisted of hard granite boulders surrounded by soft and highly plastic clay. The initial drill rig selected for the project was an auger drill rig, but the auger could not penetrate through the granite boulders. The next drill rig selected was an air track rig, which uses a percussion drill bit that easily penetrated through the granite boulders, but the soft clay plugged up the drill bit and it became stuck in the ground. Over

FIGURE 4.7 Downhole logging (arrow points to top of steel cage used for downhole logging).

4.20 GEOTECHNICAL FIELD AND LABORATORY STUDIES

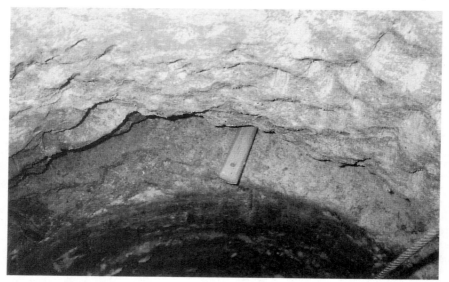

FIGURE 4.8a Knife placed in an open fracture in bedrock caused by landslide movement (photograph taken downhole in a large-diameter auger boring).

FIGURE 4.8b Side view of condition shown in Fig. 4.8a.

15 m (50 ft) of drill stem could not be removed from the ground and it had to be left in place, a very costly experience with difficult drilling conditions.

Some of my other memorable experiences with drilling are as follows:

Drilling accidents. Most experienced drillers handle their equipment safely, but accidents can happen to anyone. One day, as I observed a drill rig start to excavate the hole, the teeth of the auger bucket caught on a boulder. The torque of the auger bucket was transferred to the drill rig, and it flipped over. Fortunately, no one was injured.

Underground utilities. Before drilling, the local utility company, upon request, will locate their underground utilities by placing ground surface marks that delineate utility alignments. An incident involving a hidden gas line demonstrates that not even utility locators are perfect. On a particularly memorable day, I drove a Shelby tube sampler into a 10-cm- (4-in.-) diameter pressurized gas line. The noise of escaping gas was enough to warn of the danger. Fortunately, an experienced driller knew what to do: turn off the drill rig and call 911.

Downhole logging. As previously mentioned, a common form of subsurface exploration in southern California is to drill a large-diameter boring, usually 0.76 m (30 in.) in diameter. Then the geotechnical engineer or engineering geologist descends into the earth to get a close-up view of soil conditions. On this particular day, several individuals went down the hole and noticed a small trickle of water in the hole about 6 m (20 ft) down. The sudden and total collapse of the hole riveted the attention of the workers, especially the geologist who had moments before been down at the bottom of the hole.

Because subsurface exploration has a potential for serious or even fatal injury, it is especially important that young engineers and geologists be trained to evaluate the safety of engineering operations in the field. This must be done before they supervise field operations.

4.4.1 Soil Sampling

There are many different types of samplers used to retrieve soil and rock specimens from the borings. Common examples are indicated in Table 4.3. Figure 4.9 shows three types of samplers: the California sampler, Shelby tube sampler, and standard penetration test (SPT) sampler.

The most common type of soil sampler used in the United States is the Shelby tube, which is a thin-walled sampling tube. It can be manufactured to different diameters and lengths, with a typical diameter varying from 5 to 7.6 cm (2 to 3 in.) and a length of 0.6 to 0.9 m (2 to 3 ft). The Shelby tube should be manufactured to meet exact specifications, such as those stated by ASTM D 1587-94 (1998). The Shelby tube shown in Fig. 4.9 has an inside diameter of 6.35 cm (2.5 in.).

Many localities have developed samplers that have proven successful with local soil conditions. For example, in southern California, a common type of sampler is the California sampler, which is a split-spoon-type sampler that contains removable internal rings, 2.54 cm (1 in.) in height. Figure 4.9 shows the California sampler in an open condition, with the individual rings exposed. The California sampler has a 7.6-cm (3.0-in.) outside diameter and a 6.35-cm (2.50-in.) inside diameter. This sturdy sampler, which is considered to be a thick-walled sampler, has proven successful in sampling hard and desiccated soil and soft sedimentary rock common in southern California.

There are three types of soil samples that can be recovered from borings:

1. *Altered soil.* During the boring operations, soil can be altered by mixing or contamination. For example, if the boring is not cleaned out prior to sampling, a soil sample taken

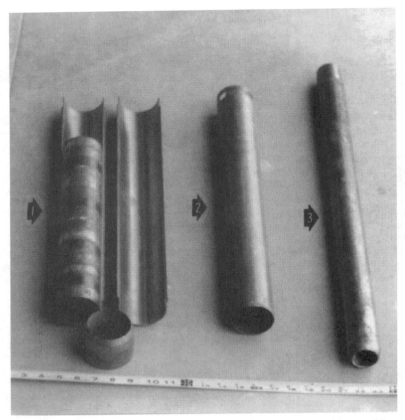

FIGURE 4.9 Soil samplers (no. 1 is the California sampler in an open condition, no. 2 is a Shelby tube, and no. 3 is the standard penetration test sampler).

from the bottom of the borehole may actually consist of cuttings from the side of the borehole. These borehole cuttings, which have fallen to the bottom of the borehole, will not represent *in situ* conditions at the depth sampled. In other cases, the soil sample may become contaminated with drilling fluid, which is used for wash-type borings. These types of soil samples that have been mixed or contaminated by the drilling process should not be used for laboratory tests because they will lead to incorrect conclusions regarding subsurface conditions. Soil that has a change in moisture content due to the drilling fluid or from heat generated during the drilling operations should also be classified as altered soil.

Soil that has been densified by overpushing or overdriving the soil sampler should also be considered as altered because the process of overpushing or overdriving could squeeze water from the soil. Figure 4.10 shows a photograph of the rear end of a Shelby tube sampler. The soil in the sampler has been densified by being overpushed as indicated by the smooth surface of the soil and the mark in the center of the soil (due to the sampler connectors).

2. *Disturbed samples.* Disturbed soil is defined as soil that has been remolded during the sampling process. For example, soil obtained from driven samplers, such as the standard penetration test split spoon sampler, or chunks of intact soil brought to the surface in an auger bucket (i.e., bulk samples) are considered disturbed soil. Disturbed soil can be used for numerous types of laboratory tests. Chapter 5 will further discuss the types of laboratory tests that can be performed on disturbed soil samples.

3. *Undisturbed sample.* It should be recognized that no soil sample can be taken from the ground and be in a perfectly undisturbed state. But this terminology has been applied to those soil samples taken by certain sampling methods. Undisturbed samples are often defined as those samples obtained by slowly pushing thin-walled tubes, having sharp cutting ends and tip relief, into the soil. Two parameters, the *inside clearance ratio* and the *area ratio,* are often used to evaluate the disturbance potential of different samplers. They are defined as follows:

$$\text{Inside clearance ratio (\%)} = 100 \frac{D_i - D_e}{D_e} \quad (4.1)$$

$$\text{Area ratio (\%)} = 100 \frac{D_o^2 - D_i^2}{D_i^2} \quad (4.2)$$

where D_e = diameter at the sampler cutting tip, D_i = inside diameter of the sampling tube, and D_o = outside diameter of the sampling tube, as shown in Fig. 4.11.

In general, a sampling tube for undisturbed soil specimens should have an inside clearance ratio of about 1 percent and an area ratio of about 10 percent or less. Having an inside clearance ratio of about 1 percent provides for tip relief of the soil and reduces the friction between the soil and inside of the sampling tube during the sampling process. A thin film of oil can be applied at the cutting edge to also reduce the friction between the soil and metal tube during sampling operations. The purpose of having a low area ratio and a sharp

FIGURE 4.10 Densified soil due to overpushing a Shelby tube.

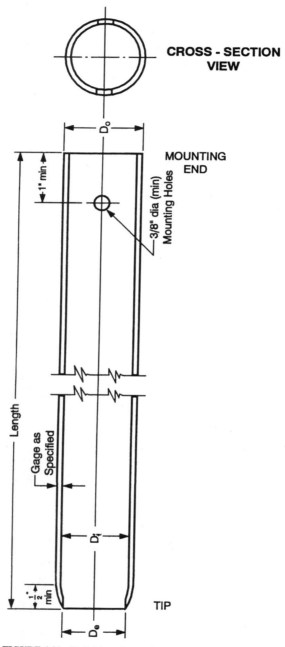

FIGURE 4.11 Definition of terms for sampling tube.

cutting end is to slice into the soil with as little disruption and displacement of the soil as possible. Shelby tubes are manufactured to meet these specifications and are considered to be undisturbed soil samplers. As a comparison, the California sampler has an area ratio of 44 percent and is considered to be a thick-walled sampler.

It should be mentioned that using a thin-wall tube, such as a Shelby tube, will not guarantee an undisturbed soil specimen. Many other factors can cause soil disturbance, such as:

- Pieces of hard gravel or shell fragments in the soil, which can cause voids to develop along the sides of the sampling tube during the sampling process
- Soil adjustment caused by stress relief in making a borehole
- Disruption of the soil structure due to hammering or pushing the sampling tube into the soil stratum
- Expansion of gas during retrieval of the sampling tube
- Jarring or banging the sampling tube during transportation to the laboratory
- Roughly removing the soil from the sampling tube
- Crudely cutting the soil specimen to a specific size for a laboratory test

The actions listed above cause a decrease in effective stress (Chap. 6), a reduction in the interparticle bonds, and a rearrangement of the soil particles. An "undisturbed" soil specimen will have little rearrangement of the soil particles and perhaps no disturbance except that caused by stress relief where there is a change from the *in situ* k_0 (at rest) condition to an isotropic "perfect sample" stress condition (Ladd and Lambe, 1963). A disturbed soil specimen will have a disrupted soil structure with perhaps a total rearrangement of soil particles. When measuring the shear strength or deformation characteristics of the soil, the results of laboratory tests run on undisturbed specimens obviously better represent *in situ* properties than laboratory tests run on disturbed specimens. Disturbance will be further discussed in Chap. 5.

Soil samples recovered from the borehole should be kept within the sampling tube or sampling rings. The soil sampling tube should be tightly sealed with end caps or the sampling rings thoroughly sealed in containers to prevent a loss of moisture during transportation to the laboratory. The soil samples should be marked with the file or project number, date of sampling, name of engineer or geologist who performed the sampling, and boring number and depth (e.g., B-1 @ 20–21 ft). It is standard practice to estimate the depth of drilling from the amount of drill rods or stem that are in the ground at any given time. Figure 4.12 shows a driller marking the drilling cable in order to estimate the depth of the boring excavation.

4.4.2 Field Tests

There are several different types of tests that can be performed at the time of drilling. Common types of field tests are discussed in the following sections:

Standard Penetration Test (SPT). The standard penetration test can be used for all types of soil, but in general, the SPT should only be used for sand deposits (Coduto, 1994). The SPT can be especially of value for clean sand deposits where the sand falls or flows out from the sampler when retrieved from the ground. Without a soil sample, other types of tests, such as the standard penetration test, must be used to assess the engineering properties of the sand. Often in drilling a borehole, if subsurface conditions indicate

FIGURE 4.12 Driller marking the drill cable in order to estimate the depth of the boring excavation.

a sand stratum and sampling tubes come up empty, the sampling gear can be quickly changed to perform standard penetration tests.

The standard penetration test consists of driving a thick-walled sampler into the sand deposit. Per ASTM D 1586-92 (1998), the SPT sampler must have an inside barrel diameter $D_i = 3.81$ cm (1.5 in.) and an outside diameter $D_o = 5.08$ cm (2 in.). The SPT sampler is shown in Figure 4.9. The SPT sampler is driven into the sand by using a 63.5-kg (140-lb) hammer falling a distance of 0.76 m (30 in.). The SPT sampler is driven a total of 45 cm (18 in.), with the number of blows recorded for each 15-cm (6-in.) interval. The "measured SPT N-value" (blows per foot) is defined as the penetration resistance of the sand, which equals the sum of the number of blows required to drive the SPT sampler over the depth interval of 15 cm to 45 cm (6 in. to 18 in.). The reason the number of blows required to drive the SPT sampler for the first 15 cm (6 in.) is not included in the N-value is because the drilling process often disturbs the soil at the bottom of the borehole and the readings at 15 cm to 45 cm (6 in. to 18 in.) are believed to be more representative of the *in situ* penetration resistance of the sand.

The measured SPT N-value can be influenced by the type of soil, such as the amount of fines and gravel-size particles in the soil. Saturated sands that contain appreciable fine soil particles, such as silty or clayey sands, could give abnormally high N-values if they have a tendency to dilate or abnormally low N-values if they have a tendency to contract during the undrained shear conditions associated with driving the SPT sampler. Gravel-size particles increase the driving resistance (hence increased N-value) by becoming stuck in the SPT sampler tip or barrel.

Table 4.4 presents a correlation between the measured SPT N-value (blows per foot) and the density condition of a clean sand deposit. Note that this correlation is very approximate and the boundaries between different density conditions are not as distinct as implied by the table. As indicated in Table 4.4, if it only takes four blows or less to drive the SPT sampler, then the sand should be considered to be very loose and could be subjected to significant settlement due to the weight of a structure or due to earthquake shaking. On the other hand, if it takes over 50 blows to drive the SPT sampler, then the sand is considered to be in a very dense condition and would be able to support high bearing loads and would be resistant to settlement from earthquake shaking.

A factor that could influence the measured SPT N-value is groundwater. It is important to maintain a level of water in the borehole at or above the *in situ* groundwater level. This is to prevent groundwater from rushing into the bottom of the borehole which could loosen the sand and result in low measured N-values.

Besides soil and groundwater conditions described above, there are many different testing factors that can influence the accuracy of the SPT readings. For example, the measured SPT N-value could be influenced by the hammer efficiency, rate at which the blows are applied, borehole diameter, and rod lengths. The following equation is used to compensate for these testing factors (Skempton, 1986):

$$N_{60} = 1.67 E_m C_b C_r N \qquad (4.3)$$

where N_{60} = SPT N-value corrected for field testing procedures
E_m = hammer efficiency (for U.S. equipment, E_m equals 0.6 for a safety hammer and equals 0.45 for a doughnut hammer)
C_b = borehole diameter correction (C_b = 1.0 for boreholes of 65- to 115-mm diameter, 1.05 for 150-mm diameter, and 1.15 for 200-mm diameter hole)
C_r = rod length correction (C_r = 0.75 for up to 4 m of drill rods, 0.85 for 4 to 6 m of drill rods, 0.95 for 6 to 10 m of drill rods, and 1.00 for drill rods in excess of 10 m)
N = measured SPT N-value

Figure 4.13 (adapted from DeMello, 1971) presents an empirical correlation between the SPT N_{60} value, vertical effective stress (Sec. 6.3), and friction angle (a measure of the shear strength) of clean quartz sand.

TABLE 4.4 Correlation between SPT N-Value and Density of Clean Sand

N-value, blows per foot (1)	Sand density (2)	Relative density, % (3)
0 to 4	Very Loose Condition	0– 15
4 to 10	Loose Condition	15– 35
10 to 30	Medium Condition	35– 65
30 to 50	Dense Condition	65– 85
Over 50	Very Dense Condition	85–100

Source: Data from Terzaghi and Peck, 1967, and Lambe and Whitman, 1969.
Note: See Sec. 6.2 for definition of relative density.

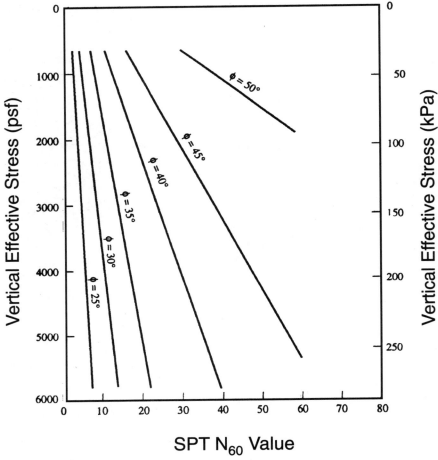

FIGURE 4.13 Empirical correlation between SPT N_{60} value, vertical effective stress, and friction angle for clean quartz sand deposits. (*Adapted from DeMello, 1971; reproduced from Coduto, 1994.*)

Even with the limitations and all of the corrections that must be applied to the measured SPT *N*-value, the standard penetration test is probably the most widely used field test in the United States. This is because it is relatively easy to use, the test is economical compared to other types of field testing, and the SPT equipment can be quickly adapted and included as part of almost any type of drilling rig.

Cone Penetration Test (CPT). The idea for the cone penetration test (CPT) is similar to the standard penetration test except that instead of driving a thick-walled sampler into the soil, a steel cone is pushed into the soil. There are many different types of cone penetration devices, such as the mechanical cone, mechanical-friction cone, electric cone, and piezocone (see App. A for descriptions). The simplest type of cone is shown in Fig. 4.14 (from ASTM D 3441-94, 1998). The cone is first pushed into the soil to the desired depth (initial position) and then a force is applied to the inner rods which moves the cone downward into the extended position. The force required to move the cone into the extended position (Fig. 4.14) divided by the horizontally projected area of the cone is defined as the cone resistance q_c. By continually repeating the two-step process shown in Fig. 4.14, the cone resistance data is obtained at increments of depth. A continuous record of the cone resistance versus depth can be obtained by using the electric cone, where the cone is pushed into the soil at a rate of 10 to 20 mm/s (2 to 4 ft/min).

Figure 4.15 (adapted from Robertson and Campanella, 1983) presents an empirical correlation between the cone resistance q_c, vertical effective stress (Sec. 6.3), and friction angle of clean quartz sand. Note that Fig. 4.15 is similar in appearance to Fig. 4.13, which should be the case because both the SPT and CPT involve basically the same process of

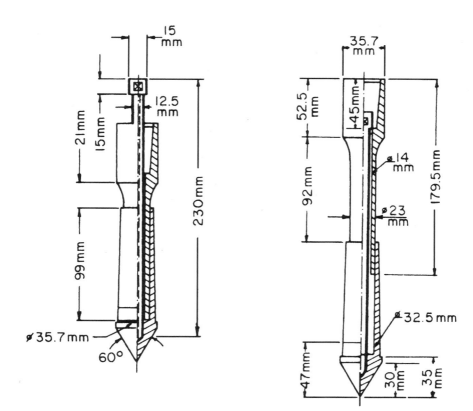

INITIAL POSITION   **EXTENDED POSITION**

FIGURE 4.14 Example of mechanical cone penetrometer tip (Dutch mantle cone). [*Reprinted with permission from the American Society for Testing and Materials (1998).*]

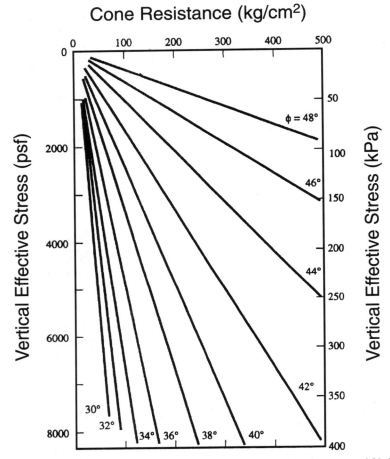

FIGURE 4.15 Empirical correlation between cone resistance, vertical effective stress, and friction angle for clean quartz sand deposits. Note: 1 kg/cm² approximately equals 1 tsf. (*Adapted from Robertson and Campanella, 1983; reproduced from Coduto, 1994.*)

forcing an object into the soil and then measuring the resistance of the soil to penetration by the object.

A considerable amount of work has been performed in correlating cone resistance q_c with subsurface conditions. Figure 4.16 (from Schmertmann, 1977) presents four examples, where the cone resistance q_c has been plotted versus depth below ground surface. The shape of the cone resistance q_c plots versus depth can be used to identify sands, clays, cavities, or rock. Some cones can be adapted to measure the frictional resistance f_s along a side sleeve or the development and dissipation of excess pore water pressure during penetration of the cone, and this data can also be useful in identifying different types of soil, such as sand or clay. The cone can even be equipped with a video camera (Raschke and Hryciw, 1997) to enable the type of soil to be viewed during the cone penetration test.

A major advantage of the cone penetration test is that by using the electric cone, a continuous subsurface record of the cone resistance q_c, such as shown in Fig. 4.16, can be obtained. This is in contrast to the standard penetration test, which obtains data at intervals in the soil deposit. Disadvantages of the cone penetration test are that soil samples can not be recovered and special equipment is required to produce a steady and slow penetration of the cone. Unlike the SPT, the ability to obtain a steady and slow penetration of the cone is not included as part of conventional drilling rigs. Because of these factors, in the United States, the CPT is used less frequently than the SPT.

Vane Shear Test (VST). The SPT and CPT are used to correlate the resistance of driving a sampler (N-value) or pushing a cone (q_c) with the engineering properties (such as density condition) of the soil. In contrast, the vane test is a different *in situ* field test because it directly measures a specific soil property, the undrained shear strength s_u of clay. Shear strength will be further discussed in Sec. 6.5.

The vane test consists of inserting a four-bladed vane, such as shown in Fig. 4.17, into the borehole and then pushing the vane into the clay deposit located at the bottom of the borehole. Once inserted into the clay, the maximum torque T_{max} required to rotate the vane and shear the clay is measured. The undrained shear strength s_u of the clay can then be calculated by using the following equation, which assumes uniform end shear for a rectangular vane:

$$s_u = \frac{T_{max}}{\pi(0.5\, D^2 H + 0.167 D^3)} \tag{4.4}$$

where, as shown in Fig. 4.17, T_{max} = maximum torque required to rotate the rod which shears the clay, H = height of the vane, and D = diameter of the vane. Standard field specifications, such as ASTM D 2573-94 (1998), have been developed to provide uniformity of the field testing procedure. The vane can provide an undrained shear strength s_u that is too high if the vane is rotated too rapidly. The vane test also gives unreliable results for clay strata that contain sand layers or lenses, varved clay, or clay that contains gravel or gravel-size shell fragments.

There are other types of vane testing equipment, such as the tapered vane (see ASTM D 2573-94, 1998). There are even miniature vane devices, which can be used in the field or laboratory. The miniature vane device is not inserted down a borehole, but rather the test is performed on clay specimens brought to the surface from undisturbed samplers. An example of a miniature vane is the torvane device, which is a handheld vane that is manually inserted into the clay surface and then rotated to induce a shear failure of the clay. On top of the torvane there is a calibrated scale that directly indicates the undrained shear strength s_u of the clay. The torvane device has a quick failure rate which could overestimate the undrained shear strength s_u. Because the miniature vane only tests a very small portion of the clay, the strength could be overestimated for fissured clay, varved clay, or clay containing slickensides. Also, ASTM D 4648-94 (1998) indicates that the miniature vane provides unreliable readings for clays having an undrained shear strength s_u in excess of 100 kPa (1.0 tsf) because the actual failure surface deviates from the assumed cylindrical failure surface, resulting in an overestimation of the undrained shear strength s_u.

Other Field Tests Performed in Boreholes. There are many other types of field tests that can be performed in boreholes. Examples include the pressuremeter test (PMT), screw plate compressometer (SPC), and Iowa borehole shear test (BST) (Holtz and

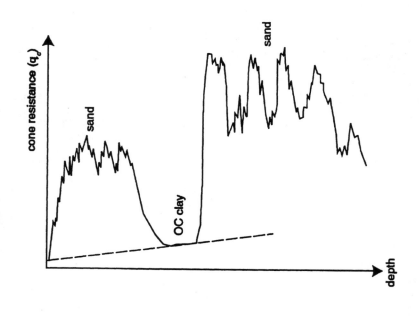

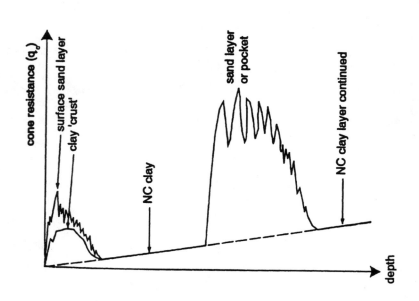

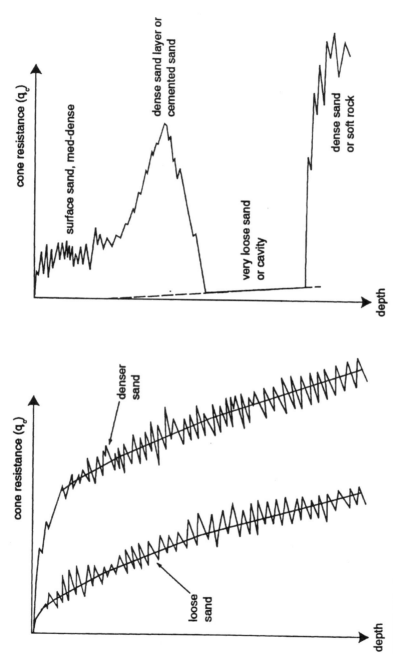

FIGURE 4.16 Simplified examples of CPT cone resistance q_c versus depth, showing possible interpretations of soil types and conditions. *(From Schmertmann, 1977.)*

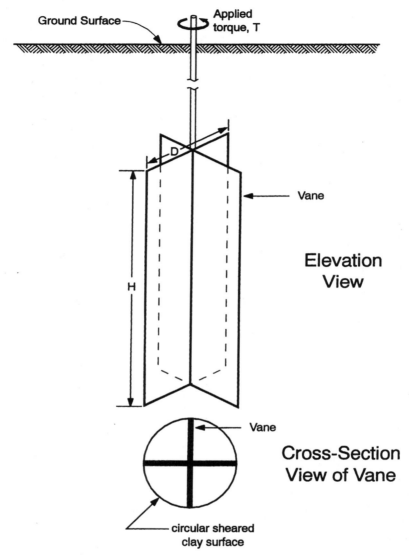

FIGURE 4.17 Diagram illustrating the field vane test.

Kovacs, 1981; Mitchell, 1978). These types of tests are used much less frequently than the SPT, CPT, and VST.

The pressuremeter test (PMT) is a field test that involves the expansion of a cylindrical probe within an uncased borehole. The screw plate compressometer (SPC) is a field test where a plate is screwed down to the desired depth, and then as pressure is applied, the settlement of the plate is measured. The Iowa borehole shear test (BST) is a field test where the device is lowered into an uncased borehole and then expanded against the sidewalls. The force required to pull the device toward ground surface is measured and, much as with a direct shear device, the shear strength properties of the *in situ* soil can then be determined.

4.4.3 Boring Layout

The required number and spacing of borings for a particular project must be based on judgment and experience. Obviously the more borings that are performed, the more knowledge that will be obtained about the subsurface conditions. This can result in a more economical foundation design and less risk of meeting unforeseen or difficult conditions during construction.

In general, boring layouts should not be random. Instead, if an approximate idea of the location of the proposed structure is known, then the borings should be concentrated in that area. For example, borings could be drilled at the four corners of a proposed building, with an additional (and deepest) boring located at the center of the proposed building. If the building location is unknown, then the borings should be located in lines, such as across the valley floor, in order to develop soil and geologic cross sections. Table 4.5 (from NAVFAC DM -7.1, 1982) provides guidelines on the typical boring layout for various types of projects.

If geologic features outside the building footprint could affect the structure, then they should also be investigated with borings. For example, if there is an adjacent landslide or fault zone that could impact the site, then it will also need to be investigated with subsurface exploration.

Some of the factors that influence the decisions on the number and spacing of borings are the following:

- *Relative costs of the investigation.* The cost of additional borings must be weighed against the value of additional subsurface information.
- *Type of project.* A more detailed and extensive subsurface investigation is required for an essential facility (Sec. 2.2) compared to a single-family dwelling.
- *Topography (flatland versus hillside).* As indicated in App. B, a hillside project usually requires more subsurface investigation than a flatland project because of the slope stability requirements.
- *Nature of soil deposits (uniform versus erratic).* Fewer borings may be needed when the soil deposits are uniform, compared to erratic deposits.
- *Geologic hazards.* The more known or potential geologic hazards at the site, the greater the need for subsurface exploration.
- *Access.* In many cases, the site may be inaccessible and access roads will have to be constructed, as shown in Figs. 4.18 and 4.19. In Fig. 4.18, the construction of the access road has resulted in considerable destruction of the site vegetation, while in Fig. 4.19 the access road was relatively easy to construct because the site is an open-pit mine. Creating access roads throughout the site can be expensive and disruptive and may influence decisions on the number and spacing of borings.
- *Governmental or local building department requirements.* For some projects, there may be specifications on the required number and spacing of borings. For example, the *Standard Specifications for Highway Bridges* (AASHTO, 1996) states:

> A minimum of one soil boring shall be made for each substructure unit. [*Author's note:* A substructure unit is defined as a pier, abutment, retaining wall, foundation, or similar item.] For substructure units over 30 m (100 ft) in width, a minimum of two borings shall be required.

TABLE 4.5 Guidelines for Boring Layout

Areas of investigation (1)	Boring layout (2)
New site of wide extent	Space preliminary borings 60 to 150 m (200 to 500 ft) apart so that area between any four borings includes approximately 10% of total area. In detailed exploration, add borings to establish geological sections at the most useful orientations.
Development of site on soft compressible soil	Space borings 30 to 60 m (100 to 200 ft) at possible building locations. Add intermediate borings when building site is determined.
Large structure with separate closely spaced footings	Space borings approximately 15 m (50 ft) in both directions, including borings at possible exterior foundation walls, at machinery or elevator pits, and to establish geologic sections at the most useful orientations.
Low-load warehouse building of large area	Minimum of four borings at corners plus intermediate borings at interior foundations sufficient to define subsoil profile.
Isolated rigid foundation	For foundation 230 to 930 m^2 (2500 to 10,000 ft^2) in area, minimum of three borings around perimeter. Add interior borings, depending on initial results.
Isolated rigid foundation	For foundation less than 230 m^2 (2500 ft^2) in area, minimum of two borings at opposite corners. Add more for erratic conditions.
Major waterfront structures, such as dry docks	If definite site is established, space borings generally not farther than 15 m (50 ft), adding intermediate borings at critical locations, such as deep pump well, gate seat, tunnel, or culverts.
Long bulkhead or wharf wall	Preliminary borings on line of wall at 60-m (200-ft) spacing. Add intermediate borings to decrease spacing to 15 m (50 ft). Place certain intermediate borings inboard and outboard of wall line to determine materials in scour zone at toe and in active wedge behind wall.
Cut stability, deep cuts, and high embankments	Provide three to five borings on line in the critical direction to provide geological section for analysis. Number of geologic sections depends on extent of stability problem. For an active slide, place at least one boring upslope of sliding area.
Dams and water-retention structures	Space preliminary borings approximately 60 m (200 ft) over foundation area. Decrease spacing on centerline to 30 m (100 ft) by intermediate borings. Include borings at location of cutoff, critical spots in abutment, spillway, and outlet works.

Source: From NAVFAC DM-7.1, 1982.

Often a preliminary subsurface plan is developed to perform a limited number of exploratory borings. The purpose is just to obtain a rough idea of the soil, rock, and groundwater conditions at the site. Then once the preliminary subsurface data is analyzed, additional borings as part of a detailed exploration are performed. The detailed subsurface exploration can be used to better define the soil profile, explore geologic hazards, and obtain further data on the critical subsurface conditions that will likely have the most impact on the design and construction of the project.

FIGURE 4.18 Construction of an access road for drilling equipment.

FIGURE 4.19 Access road constructed for subsurface investigation at an open-pit mine (arrow points to bucket auger drill rig).

4.4.4 Depth of Subsurface Exploration

Like the boring layout, the depth of subsurface exploration for a particular project must be based on judgment and experience. Borings should always be extended through unsuitable foundation bearing material, such as uncompacted fill, peat, soft clays and organic soil, and loose sands, and into dense soil or hard rock of adequate bearing capacity. For localized structures, such as commercial or industrial buildings, it is common practice to carry explorations to a depth beneath the loaded area of 1.5 to 2.0 times the least dimension of the building (Lowe and Zaccheo, 1975). Table 4.6 presents additional guidelines for different types of geotechnical and foundation projects according to NAVFAC DM-7.1 (1982).

Another commonly used rule of thumb is that for isolated square footings, the depth of subsurface exploration should be 2 times the width of the footing. For isolated strip footings, the depth of subsurface exploration should be 4 times the width of the footing. These recommendations are based on the knowledge that the pressure of surface loads dissipates with depth. Thus, at a certain depth, the effect of the surface load is very low. For example,

TABLE 4.6 Guidelines for Boring Depths

Areas of investigation (1)	Boring depth (2)
Large structure with separate closely space footings	Extend to depth where increase in vertical stress for combined foundations is less than 10% of effective overburden stress. Generally all boring should extend to no less than 9 m (30 ft) below lowest part of foundation unless rock is encountered at shallower depth.
Isolated rigid foundations	Extend to depth where vertical stress decreases to 10% of bearing pressure. Generally all borings should extend no less than 9 m (30 ft) below lowest part of foundation unless rock is encountered at shallower depth.
Long bulkhead or wharf wall	Extend to depth below dredge line between 0.75 and 1.5 times unbalanced height of wall. Where stratification indicates possible deep stability problem, selected borings should reach top of hard stratum.
Slope stability	Extend to an elevation below active or potential failure surface and into hard stratum, or to a depth for which failure is unlikely because of geometry of cross section.
Deep cuts	Extend to depth between 0.75 and 1 times base width of narrow cuts. Where cut is above groundwater in stable materials, depth of 1.2 to 2.4 m (4 to 8 ft) below base may suffice. Where base is below groundwater, determine extent of previous strata below base.
High embankments	Extend to depth between 0.5 to 1.25 times horizontal length of side slope in relatively homogeneous foundation. Where soft strata are encountered, borings should reach hard materials.
Dams and water retention structures	Extend to depth of 0.5 base width of earth dams or 1 to 1.5 times height of small concrete dams in relatively homogeneous foundations. Borings may terminate after penetration of 3 to 6 m (10 to 20 ft) in hard and impervious stratum if continuity of this stratum is known from reconnaissance.

Source: From NAVFAC DM-7.1, 1982.

a common guideline is to perform subsurface exploration to a depth where the increase in vertical pressure from the foundation is less than 10 percent of the applied pressure from the foundation. There could be problems with this approach because, as will be discussed in Chap. 7, there could be settlement of the structure that is independent of its weight or depth of influence. Settlement due to secondary influences, such as collapsible soil, is often unrelated to the weight of the structure. Especially when geologic conditions are not well established, it is always desirable to extend at least one boring into bedrock to guard against the possibility of a deeply buried soil stratum having poor support characteristics.

For some projects, there may be specifications on the required depth of borings. For example, the *Standard Specifications for Highway Bridges* (AASHTO, 1996) states:

> When substructure units will be supported on deep foundations, the depth of subsurface exploration shall extend a minimum of 6 m (20 ft) below the anticipated pile or shaft tip elevation. Where pile or shaft groups will be used, the subsurface exploration shall extend at least two times the maximum pile group dimension below the anticipated tip elevation, unless the foundation will be end bearing on or in rock. For piles bearing on rock, a minimum of 3 m (10 ft) of rock core shall be obtained at each exploration location to insure the exploration has not been terminated on a boulder.

At the completion of each boring, it should immediately be backfilled with on-site soil and compacted by using the drill rig equipment. In certain cases, the holes may need to be filled with a cement slurry or grout. For example, if the borehole is to be converted to an inclinometer (slope monitoring device), then it should be filled with a weak cement slurry. Likewise, if the hole is to be converted to a piezometer (pore water pressure monitoring device), then special backfill materials, such as a bentonite seal, will be required. It may also be necessary to seal the hole with grout or bentonite if there is the possibility of water movement from one stratum to another. For example, holes may need to be filled with cement or grout if they are excavated at the proposed locations of dams, levees, or reservoirs.

4.5 TEST PITS AND TRENCHES

In addition to borings, other methods for performing subsurface exploration include test pits and trenches. Test pits are often square in plan view with a typical dimension of 1.2 m by 1.2 m (4 ft by 4 ft). Trenches are long and narrow excavations usually made by a backhoe or bulldozer. Table 4.7 (from NAVFAC DM-7.1, 1982) presents the uses, capabilities, and limitations of test pits and trenches.

Like the downhole logging of large-diameter bucket auger borings, test pits and trenches provide for a visual observation of subsurface conditions. They can also be used to obtain undisturbed block samples of soil. The process consists of carving a block of soil from the side or bottom of the test pit or trench. Soil samples can also be obtained from the test pits or trenches by manually driving Shelby tubes, drive cylinders (ASTM D 2937-94, 1998), or other types of sampling tubes into the ground.

Backhoe trenches are an economical means of performing subsurface exploration. The backhoe can quickly excavate the trench which can then be used to observe and test the *in situ* soil. In many subsurface explorations, backhoe trenches are used to evaluate near-surface and geologic conditions (i.e., up to 15 ft deep), with borings being used to investigate deeper subsurface conditions. As indicated in Table 4.7, backhoe trenches are especially useful when performing fault studies. For example, Figs. 4.20 and 4.21 show two views of the excavation of a trench that is being used to investigate the possibility of an on-site active fault. Figure

TABLE 4.7 Use, Capabilities, and Limitations of Test Pits and Trenches

Exploration method (1)	General use (2)	Capabilities (3)	Limitations (4)
Hand-excavated test pits	Bulk sampling, *in situ* testing, visual inspection.	Provides data in inaccessible areas, less mechanical disturbance of surrounding ground.	Expensive, time-consuming, limited to depths above groundwater level.
Backhoe-excavated test pits and trenches	Bulk sampling, *in situ* testing, visual inspection, excavation rates, depth of bedrock and groundwater.	Fast, economical, generally less than 4.6 m (15 ft) deep, can be up to 9 m (30 ft) deep.	Equipment access, generally limited to depths above groundwater level, limited undisturbed sampling.
Dozer cuts	Bedrock characteristics, depth of bedrock and groundwater level, rippability, increase depth capability of backhoe, level area for other exploration equipment.	Relatively low cost, exposures for geologic mapping.	Exploration limited to depth above the groundwater table.
Trenches for fault investigations	Evaluation of presence and activity of faulting and sometimes landslide features.	Definitive location of faulting, subsurface observation up to 9 m (30 ft) deep.	Costly, time-consuming, requires shoring, only useful where datable materials are present, depth limited to zone above the groundwater level.

Source: From NAVFAC DM-7.1, 1982.

4.21 is a close-up view of the conditions in the trench and shows the fractured and disrupted nature of the rock. Note in Fig. 4.21 that metal shoring has been installed to prevent the trench from caving in.

California Bearing Ratio and Plate Load Tests. The California bearing ratio (CBR) test and the plate load test are two types of field tests that can be performed at ground surface, or on a level surface excavated in a test pit, trench, or bulldozer cut. The California bearing ratio test consists of pushing a 2.0-in.-diameter piston into the soil. The CBR is calculated as follows (see ASTM D 4429-93, 1998):

1. Record the load (pounds) versus depth of penetration (inches) of the piston as it is being pushed into the soil.
2. Calculate the stress (psi) on the piston by dividing the load by the area of the piston.
3. Plot the stress on the piston (psi) versus the depth of penetration (inches).
4. The bearing value is normally the stress (psi) corresponding to a depth of penetration of 0.10 in.

FIELD EXPLORATION

FIGURE 4.20 Backhoe trench for a fault study.

FIGURE 4.21 Close-up view of trench excavation.

5. This bearing value is converted to a ratio by dividing it by 1000 psi, which represents the penetration resistance of compacted crushed rock.
6. This bearing ratio, which is a fraction, is multiplied by 100 and reported as the California bearing ratio. Typical values of CBR range from less than 5 for soft clays up to 80 for dense sandy gravel.

When interpreting the CBR data, the geotechnical engineer must consider a possible increase in field moisture content of the soil which could soften fine-grained soils and lead to a lower CBR than originally measured. There is the possibility that the soil might be disturbed and loosened during the construction operations, also resulting in a lower CBR than originally measured.

The plate load test is similar to the California bearing ratio test, except that instead of a piston, a steel plate is pushed into the soil (ASTM D 1196-97, 1998). The usual procedure is to push a square or round steel plate into the soil and record the load versus depth of penetration. The load is converted to a stress by dividing the load by the area of the plate. The stress versus depth of penetration data is then used to calculate the subgrade modulus k_s, which is also known as the modulus of subgrade reaction. The procedure (according to NAVFAC DM-7.1, 1982) is to plot the stress on the plate versus the penetration of the plate. Using this plot, the yield point at which the penetration rapidly increases is estimated. Then the stress q and depth of penetration of the plate, δ, corresponding to half the yield point are determined from the plot, and the subgrade modulus is defined as

$$k_s = \frac{q}{\delta} \tag{4.5}$$

Table 4.8 (from U.S. Army Waterways Experiment Station, 1960) presents a summary of different soil types and various soil properties pertinent to roads and airfields. The last two columns (11 and 12) of Table 4.8 present field CBR and subgrade modulus k_s in pounds per cubic inch. Note in Table 4.8 that for given soil types, the subgrade modulus k_s in pounds per cubic inch is about 10 times the field CBR. Both the CBR and subgrade modulus k_s are often used for the design of pavements for roads, highways, and airfields.

The plate load test can also be used to directly estimate the settlement potential of a footing. For settlement of sands caused by an applied surface loading, an empirical equation that relates the depth of penetration of the steel plate, S_1, to the settlement of the actual footing, S, is as follows (Terzaghi and Peck, 1967):

$$S = \frac{4S_1}{(1 + D_1/D)^2} \tag{4.6}$$

where D_1 = smallest dimension of the steel plate and D = smallest dimension of the actual footing. In order to use Eq. (4.6), the stress exerted by the steel plate corresponding to S_1 must be the same bearing stress exerted by the actual footing.

The calculated settlement from Eq. (4.6) can significantly underestimate the actual settlement at the site in cases where there is settlement due to secondary influences (such as collapsible soil) or in cases where there is a deep looser or softer layer that is not affected by the small plate, but is loaded by the much larger footing.

4.6 PREPARATION OF LOGS

A log is defined as a written record prepared during the subsurface excavation of borings, test pits, or trenches that documents the observed conditions. Although logs are often prepared by technicians or even the driller, the most appropriate individuals to log the subsurface conditions are geotechnical engineers or engineering geologists who have considerable experience and judgment acquired by many years of field practice. It is especially important that the subsurface conditions likely to have the most impact on the proposed project be adequately described. Table 4.9 lists other items that should be included on the excavation log.

Figure 4.22 (from Day and Poland, 1996) presents a boring log prepared by an engineering geologist. Figure 4.22 is referred to as a *downhole log* because the engineering geologist prepared the written record as he was slowly lowered into a 0.6-m (24-in.) auger bucket drill hole. Note in Fig. 4.22 that the engineering geologist has discovered the presence of an upper zone of fill underlain by "landslide debris." The engineering geologist also discovered a basal "shear zone" at a depth of 25.1 ft (7.7 m) and has measured the strike (N70W, North 70° West) and dip (15NE, 15° Northeast) of the shear surface. This is the most important subsurface information, the fact that the site contains the remnants of an ancient landslide that has been obscured by old fill.

The boring log in Fig. 4.22 also lists the results of tests performed on undisturbed soil and rock samples obtained from Shelby tubes. These tests include the dry unit weight and moisture content of the soil or rock. These index properties are determined in the laboratory and are further discussed in Chap. 5. The low dry density and high moisture content of the soil and rock are due to the diatomaceous nature of the material.

An important part of the preparation of logs is to determine the geologic or man-made process that created the soil deposit. Table 4.10 presents a list of common soil deposits encountered during subsurface exploration. Usually the engineering geologist is most qualified to determine the type of soil deposit. As indicated in Table 4.10, the different soil deposits can have unique geotechnical and foundation implications and it is always of value to determine the geologic or man-made process that created the soil deposit.

4.7 ROLE OF THE ENGINEERING GEOLGIST

In California, the majority of subsurface exploration for civil engineering projects is performed by engineering geologists. This is probably due in part to the large number of California geologic hazards, such as active faults, landslides, debris flow, slope creep, rockfall, unstable soil and rock deposits, and even volcanic activity. The engineering geologist is also usually more qualified to classify rock types and perform geophysical studies. The consequences of a missed geologic hazard can be quite high, as described below.

Missed Geologic Features. During the design and construction of the project, it is essential that all the geologic features that could affect the site are discovered and analyzed. The purpose of this section is to describe the Laguna Niguel landslide, located in Laguna Niguel, California, which was not discovered during the design and construction of the project.

The Laguna Niguel landslide occurred during the El Niño winter of 1997–1998. The landslide was triggered by the heavy California rainfall due to the El Niño weather pattern. Rainfall can infiltrate into the ground where it can trigger landslides by lubricating slide

TABLE 4.8 Characteristics of Compacted Subgrade for Roads and Airfields

Major Divisions (1)	Subdivisions (2)	USCS symbol (3)	Name (4)	Value as subgrade (no frost action) (5)	Potential frost action (6)
Coarse-grained soils	Gravel and gravelly soils	GW	Well-graded gravels or gravel-sand mixtures, little or no fines	Excellent	None to very slight
		GP	Poorly graded gravels or gravelly sands, little or no fines	Good to excellent	None to very slight
		GM	Silty gravels, gravel-sand-silt mixtures	Good to excellent	Slight to medium
		GC	Clayey gravels, gravel-sand-clay mixtures	Good	Slight to medium
	Sand and sandy soils	SW	Well-graded sands or gravelly sands, little or no fines	Good	None to very slight
		SP	Poorly graded sands or gravelly sands, little or no fines	Fair to good	None to very slight
		SM	Silty sands, sand-silt mixtures	Fair to good	Slight to high
		SC	Clayey sands, sand-clay mixtures	Poor to fair	Slight to high
Fine-grained soils	Silts and clays with liquid limit less than 50	ML	Inorganic silts, rock flour, silts of low plasticity	Poor to fair	Medium to very high
		CL	Inorganic clays of low plasticity, gravelly clays, sandy clays, etc.	Poor to fair	Medium to high
		OL	Organic silts and organic clays of low plasticity	Poor	Medium to high
	Silts and clays with liquid limit greater than 50	MH	Inorganic silts, micaceous silts, silts of high plasticity	Poor	Medium to very high
		CH	Inorganic clays of high plasticity, fat clays, silty clays, etc.	Poor to fair	Medium
		OH	Organic silts and organic clays of high plasticity	Poor to very poor	Medium
Peat	Highly organic	PT	Peat and other highly organic soils	Not suitable	Slight

Source: U.S. Army Waterways Experiment Station, 1960.

TABLE 4.8 (Continued)

Compressibility (7)	Drainage properties (8)	Compaction equipment (9)	Typical dry densities (10) pcf	Typical dry densities (10) Mg/m³	CBR (11)	Sub. mod*, pci (12)
Almost none	Excellent	Crawler-type tractor, rubber-tired roller, steel-wheeled roller	125–140	2.00–2.24	40–80	300–500
Almost none	Excellent	Crawler-type tractor, rubber-tired roller, steel-wheeled roller	110–140	1.76–2.24	30–60	300–500
Very slight to slight	Fair to very poor	Rubber-tired roller, sheepsfoot roller	115–145	1.84–2.32	20–60	200–500
Slight	Poor to very poor	Rubber-tired roller, sheepsfoot roller	130–145	2.08–2.32	20–40	200–500
Almost none	Excellent	Crawler-type tractor, rubber-tired roller	110–130	1.76–2.08	20–40	200–400
Almost none	Excellent	Crawler-type tractor, rubber-tired roller	105–135	1.68–2.16	10–40	150–400
Very slight to medium	Fair to poor	Rubber-tired roller, sheepsfoot roller	100–135	1.60–2.16	10–40	100–400
Slight to medium	Poor to very poor	Rubber-tired roller, sheepsfoot roller	100–135	1.60–2.16	5–20	100–300
Slight to medium	Fair to poor	Rubber-tired roller, sheepsfoot roller	90–130	1.44–2.08	15 or less	100–200
Medium	Practically impervious	Rubber-tired roller, sheepsfoot roller	90–130	1.44–2.08	15 or less	50–150
Medium to high	Poor	Rubber-tired roller, sheepsfoot roller	90–105	1.44–1.68	5 or less	50–100
High	Fair to poor	Sheepsfoot roller, rubber-tired roller	80–105	1.28–1.68	10 or less	50–100
High	Practically impervious	Sheepsfoot roller, rubber-tired roller	90–115	1.44–1.84	15 or less	50–150
High	Practically impervious	Sheepsfoot roller, rubber-tired roller	80–110	1.28–1.76	5 or less	25–100
Very high	Fair to poor	Compaction not practical	—	—	—	—

*Subgrade Modulus.

TABLE 4.9 Types of Information That Should Be Recorded on Exploratory Logs

Item (1)	Description (2)
Excavation number	Each boring, test pit, or trench excavated at the site should be assigned an excavation number. For example, in Fig. 4.22, the boring has been labeled LB-1, for large-diameter boring number 1.
Project information	Project information should include the project name, file number, client, and site address. The individual preparing the log should also be noted.
Type of equipment	Include on the log the type of excavation (i.e., hand-dug pit, backhoe trench, etc.) and the total depth and size of the excavation. For borings, indicate type of drilling equipment, use of drilling fluid, and kelly bar weights. Also indicate if casing was used.
Site-specific information	The exploratory log should list the surface elevation, dates of excavation, and ground surface conditions.
Type of field tests	For borings, list all field tests, such as SPT, CPT, or vane test. Also indicate if the boring was converted to a monitoring device, such as a piezometer.
Type of sampler	Indicate type of sampler and depth of each soil or rock sample recovered from the excavation. For driven samplers, indicate type and weight of hammer and number of blows per foot to drive the sampler. Indicate sample recovery and RQD for rock strata.
Soil and rock descriptions	Classify the soil and rock exposed in the excavation (see Chap. 5). Also indicate moisture and density condition of the soil and rock.
Excavation problems	List excavation problems, such as instability (sloughing, groundwater-induced caving, squeezing of the hole, etc.), hard drilling, or boring termination due to refusal.
Groundwater	Indicate depth to groundwater or seepage zones. At the end of the subsurface exploration, indicate the depth of free-standing water in the excavation.
Geologic features and hazards	Identify geologic features and hazards. Geologic features include type of deposit (see Table 4.10), formation name, fracture condition of rock, etc. Geologic hazards include landslides, active fault shear zones, liquefaction-prone sand, bedding shear surfaces or slickensides, and underground voids or caverns.
Unusual conditions	Any unusual subsurface condition should be noted. Examples include artesian groundwater, boulders or other obstructions, and loss of drilling fluid which could indicate an underground void or cavity.

Note: Additional information may be required for subsurface explorations for mining or agricultural purposes, for the investigation of hazardous waste, or for other special types of subsurface exploration.

planes and by raising the groundwater table, which increases driving forces from seepage pressures and decreases resisting forces from the buoyancy effect.

The mass movement of the landslide occurred on March 19, 1998 (see Fig. 4.23a for a site plan). This landslide is very large and deep and caused extensive damage. Figure 4.23b shows a cross section through the landslide and the original predevelopment topography (dashed line), the final as-graded topography, and the failure condition. As shown in Fig. 4.23b, the mass movement of the landslide caused a dropping down at the top (head) and a bulging upward of the ground at the base (toe) of the landslide.

FIELD EXPLORATION

Figures 4.24 to 4.34 show pictures of the damage caused by the Laguna Niguel landslide. The figures show the following:

- *Figures 4.24 to 4.27.* These photographs show damage to condominiums at the toe of the landslide. The area shown in Fig. 4.24 was originally relatively level, but the toe of the landslide has uplifted both the road and the condominiums. The arrow in Fig. 4.24 indicates the original level of the road and condominiums. Note also in Fig. 4.24 that the upward thrust caused by the toe of the landslide has literally ripped the building in half. At other locations, the toe of the landslide uplifted the road by about 6 m (20 ft) as shown in Figs. 4.25 and 4.26. Many buildings located at the toe of the landslide were completely crushed by the force of the landslide movement (Fig. 4.27).
- *Figures 4.28 to 4.33.* These photographs show damage at the head of the landslide. The main scarp, which is shown in Fig. 4.28, is about 12 m (40 ft) in height. The vertical distance between the two arrows in Fig. 4.28 represents the down-dropping of the top of the landslide. Figure 4.29 shows a corner of a house that is suspended in midair because the landslide has dropped down and away. Figure 4.30 shows a different house where the rear is also suspended in midair because of the landslide movement. A common feature of all the houses located at the top of the landslide was the vertical and lateral movement caused by the landslide as it dropped down and away. For example, Fig. 4.31 shows one house that has dropped down relative to the driveway, Fig. 4.32 shows a second house where a gap opened up in the driveway as the house was pulled downslope, and Fig. 4.33 shows a third house (visible in the background) that was pulled downslope, leaving behind a groove in the ground, caused by the house footings as they dragged across the ground surface.
- *Figure 4.34.* This photograph shows the graben area that opened up at the head of the landslide.

As this example illustrates, landslides can be very destructive. It is especially important that the geotechnical engineer and engineering geologist identify such hazards at the site prior to design and construction of the project.

4.7.1 Geophysical Techniques

Geophysical techniques can be employed by the engineering geologist to obtain data on the subsurface conditions. Use of geophysical techniques involves considerable experience and judgment in the interpretation of results. Table 4.11 (adapted from NAVFAC DM-7.1, 1982) lists common types of geophysical techniques.

Probably the most commonly used geophysical technique is the seismic refraction method. This method is based on the fact that seismic waves travel at different velocities through different types of materials. For example, seismic waves will travel much faster in dense, solid rock than in soft clay. The test method commonly consists of placing a series of geophones in a line on the ground surface. Then a shock wave is produced usually by placing a metal plate on the ground surface and in line with the geophones and then striking the metal plate with a sledge hammer. This seismic energy is detected by the geophones, and, by analyzing the recorded data, the velocity of the seismic wave as it passes through the ground and the depth to bedrock can often be determined.

Very dense and hard rock will have a high seismic wave velocity, while soft and fractured rock will have a much lower seismic wave velocity. Based on this principle, the seismic wave velocity can be used to determine whether the underlying rock can be excavated by commonly available equipment, or is so dense and hard that it must be blasted apart. For

TEST EXCAVATION LOG No. LB-1

F.N. 60210.02
Sheet: 1 of 2
Start: 7 OCT 94
End: 7 OCT 94

Project Name: _____
Location: _____
Estimated Surface Elevation: 1050' ± **Total Depth:** 35.0' **Rig Type:** 24" Limited Access
By: DMP

Field Description

Surface Conditions: Grass landscaping approximately 8 feet from the northeast corner of the house.

Subsurface Conditions: **FORMATION: Classification, color, moisture, tightness, etc.**

FILL:
From 0.0'-3.5', Silty CLAY, dark gray/brown, moist, slightly stiff. Semi-abundant to abundant roots to 1/2" in diameter, slightly porous.

—— 3.50 ——

From 3.5'-6.5', becomes dark brown, moist, slightly stiff to stiff. Semi-abundant Siltstone gravel near basal contact with Landslide Debris (Qls). At 6.5', undulating near horizontal contact with Qls.

—— 6.50 ——

Depth-Feet	Sample Type	Blow Counts Per Half Foot	Dry Unit Weight pcf	Moisture Content %	Relative Compaction	Saturation %	USCS Symbol	Graphic Log
0								
	BULK							
	INTACT		107	18		87		
5	INTACT		73	28		59		

4.48

	76 / 34	77
9.00		
12.00	76 / 36	81
13.40		
16.00	81 / 37	94
18.60		

LANDSLIDE DEBRIS (Qls):
From 6.5'-9.0', SILTSTONE, light gray/brown, dry to damp, loose to very loose. Very fractured to angular gravel to cobble (3") sized pieces. Occasional infilled animal burrow, near-vertical infilling of topsoil (6" to 8" in width) within very fractured zone indicates ripping of rock during grading of pad. At 9.0', approximately 1" thick fine sand lense, light gray, oriented N25E 50E.

From 9.0'-12.0', increasing moisture, fractured rock generally cobble (to 4") sized.

From 12.0'-13.4', brecciated zone with randomly oriented sub-angular cobbles (6" to 18") of laminated Siltstone, light gray/brown, within clayey matrix.

From 13.4'-16.0', Fine Sandy SILTSTONE, gray/chocolate brown, moist to very moist, hard. Fractured.

From 16.0'-18.6', Fine Sandy Clayey SILTSTONE, gray/chocolate brown, moist to very moist, hard. Fractured, occasional rootlet to 1/16" in diameter, fractured surfaces sub-polished and stained orange, slight clay infilling of fractures which are near-vertical to sub-horizontal. At 18.6', SHEAR ZONE, poorly developed, approximately 1/2" to 1" thick brecciated clay, stained orange/brown, moist, slightly soft to slightly stiff, oriented N60W 32N.

From 18.6'-24.6', Clayey SILTSTONE, reddish/gray/brown, moist, hard to very hard. At 24.5', Fine Sandstone lense approximately 1/2" thick, light brown, dry to moist, soft to slightly hard. Oriented N70W 15N

NOTES: **Kelly Weights 0-12 feet: 850 lbs; 12-20.5 feet: 650 lbs; 20.5+ feet: 450 lbs.**

FIGURE 4.22 Boring log. *(From Day and Poland, 1996; reproduced with permission of ASCE.)*

TEST EXCAVATION LOG No. LB-1

F.N. 60210.02
Sheet: 2 of 2
By: DMP

Project Name: _____

Field Description

Surface Conditions: Grass landscaping approximately 8 feet from the northeast corner of the house.

Subsurface Conditions: FORMATION: Classification, color, moisture, tightness, etc.

Depth-Feet	Sample Type	Blow Counts Per Half Foot	Dry Unit Weight pcf	Moisture Content %	Relative Compaction	Saturation %	USCS Symbol	Graphic Log	Field Description
	BULK								
	INTACT		78	41		97			24.60
25									From 24.6'-25.1', Clayey SILTSTONE, reddish/gray/brown, moist, hard to very hard. Slightly fractured. 25.10
									25.50 From 25.1'-25.5', SHEAR ZONE, Silty CLAY, gray/brown, moist, slightly soft to slightly stiff, diatomaceous, brecciated. Basal shear zone approximately 1/2" thick oriented N70W 15NE.
			83	33		88			**BEDROCK-UNNAMED SHALE (Tush):** From 25.5'-28.5', Clayey SILTSTONE, reddish/gray/brown, moist, hard to very hard. Slightly fractured. At 28.5' contact with bedrock.
									28.50 From 28.5'-32.0', Clayey SILTSTONE, reddish/gray/brown, moist, very hard. Occasional sub-horizontal 1/2" gypsum vein, occasional dark gray/black staining to 4" in diameter.

4.50

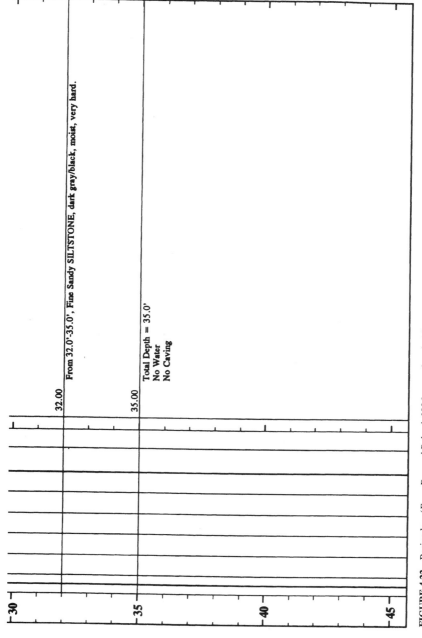

FIGURE 4.22 Boring log. *(From Day and Poland, 1996; reproduced with permission of ASCE.)(Continued)*

TABLE 4.10 Common Man-Made and Geologic Soil Deposits

Main category (1)	Common types of soil deposits (2)	Possible engineering problems (3)
Structural fill	Dense or hard fill. Often the individual fill lifts can be identified.	Upper surface of structural fill may have become loose or weathered.
Uncompacted fill	Random soil deposit that can contain chunks of different types and sizes of rock fragments.	Susceptible to compression and collapse.
Debris fill	Contains pieces of debris, such as concrete, brick, and wood fragments.	Susceptible to compression and collapse.
Municipal dump	Contains debris and waste products such as household garbage or yard trimmings.	Significant compression and gas from organic decomposition.
Residual soil deposit	Soil deposits formed by in-place weathering of rock.	Engineering properties are highly variable.
Organic deposit	Examples include peat and muck which form in bogs, marshes, and swamps.	Very compressible and unsuitable for foundation support.
Alluvial deposit	Soil transported and deposited by flowing water, such as streams and rivers.	All types of grain sizes, loose sandy deposits susceptible to liquefaction.
Aeolian deposit	Soil transported and deposited by wind. Examples include loess and dune sands.	Can have unstable soil structure that may be susceptible to collapse.
Glacial deposit	Soil transported and deposited by glaciers or their melt water. Examples include till.	Erratic till deposits and soft clay deposited by glacial melt water.
Lacustrine deposit	Soil deposited in lakes or other inland bodies of water.	Unusual soil deposits can form, such as varved silts or varved clays.
Marine deposit	Soil deposited in the ocean, often from rivers that empty into the ocean.	Granular shore deposits but offshore areas can contain soft clay deposits.
Colluvial deposit	Soil transported and deposited by gravity, such as talus, hill-wash, or landslide deposits	Can be geologically unstable deposit.
Pyroclastic deposit	Material ejected from volcanoes. Examples include ash, lapilli, and bombs.	Weathering can result in plastic clay. Ash can be susceptible to erosion.

Note: The first four soil deposits are man-made; all others are due to geologic processes.

example, Figs. 4.35 to 4.38 present data from the *Caterpillar Performance Handbook* (1997). The figures show the seismic velocity versus rock type and indicate at what seismic velocities a Caterpillar D8R, D9R, D10R, and D11R tractor/ripper can rip or not rip apart the rock. This information can be very important to the client because of the much higher costs and risks associated with blasting rock as compared to using a conventional piece of machinery to rip apart and excavate the rock.

4.8 SUBSOIL PROFILE

The final section of this chapter presents examples of subsoil profiles. The results of the subsurface exploration are often summarized on a subsoil profile. Usually the engineering geologist is the person most qualified to develop the subsoil profile on the basis of experience and judgment in extrapolating conditions between the borings, test pits, and trenches.

Figures 4.39 to 4.42 show four examples of subsoil profiles. As shown in the figures, the results of field and laboratory tests are often included on the subsoil profiles. The development of a subsoil profile is often a required element for geotechnical and foundation engineering analyses. For example, subsoil profiles are used to determine the foundation type (shallow versus deep foundation), to calculate the amount of settlement or heave of the structure, to evaluate the effect of groundwater on the project and

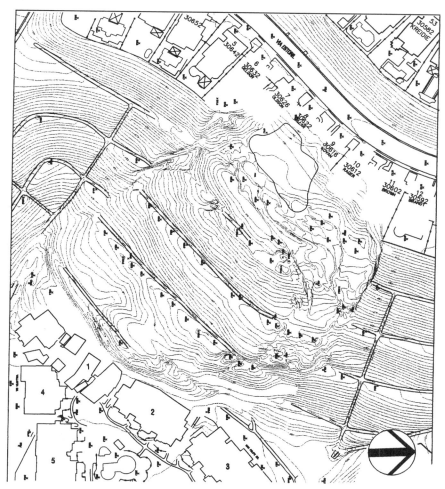

FIGURE 4.23a Laguna Niguel site plan. (*Note*: Approximate Scale: 1″ = 150′.)

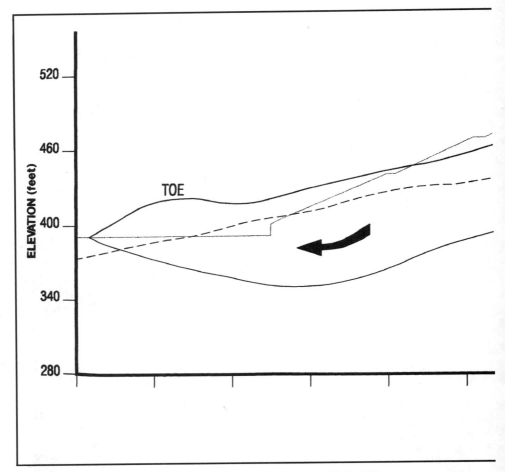

FIGURE 4.23*b* Laguna Niguel landslide: cross section through the center of the landslide.

develop recommendations for dewatering of underground structures, to perform slope stability analyses for projects having sloping topography, and to prepare site development recommendations.

PROBLEMS

1. A sampling tube has an outside diameter D_o of 3.00 in., a tip diameter D_e of 2.84 in., and a wall thickness of 0.065 in. Calculate the clearance ratio and area ratio, and indicate if the sampling tube meets the criteria for undisturbed soil sampling. Answers: Clearance ratio=1.06%, area ratio=9.26%, and yes, it meets the criteria for undisturbed soil sampling.

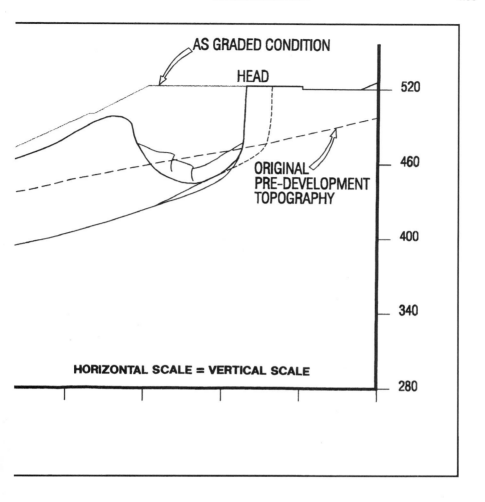

2. A standard penetration test (SPT) was performed on a near-surface deposit of clean sand where the number of blows to drive the sampler 18 in. was 5 for the first 6 in., 8 for the second 6 in., and 9 for the third 6 in. Calculate the measured SPT N-value (blows per foot) and indicate the *in situ* density condition of the sand by referring to Table 4.4. Answer: Measured SPT N-value = 17, indicating a medium-condition of the sand.

3. A field vane shear test was performed on a clay, where the vane had a length H of 4.0 in. and a diameter D of 2.0 in. The maximum torque T_{max} required to shear the soil was 8.5 ft-lb. Calculate the undrained shear strength s_u of the soil. Answer: 500 psf.

4. A plate load test was performed on a sand deposit. A 12-in. D_1 steel square plate was used for the plate load test. At a stress of 2000 psf, the penetration of the steel plate was 0.27 in. (S_1). If the estimated size of the actual square footing will be 5.0 ft (D) and will also subject the underlying sand to a bearing stress of 2000 psf, what is the estimated settlement? Answer: $3/4$ in.

FIGURE 4.24 Laguna Niguel landslide: view of the toe of the landslide (note that the arrow indicates the original level of this area, but the landslide uplifted both a portion of the road and the condominium).

5. A construction site in New England requires excavation of rock. The geologist has determined that the rock is granite and, from geophysical methods (i.e., refraction), the seismic velocity of the *in situ* granite is 12,000 to 15,000 ft/s. A Caterpillar D10R tractor/ripper is available. Can the granite be ripped apart? Answer: No, blasting will be required.

FIGURE 4.25 Laguna Niguel landslide: view of the toe of the landslide where the road has been uplifted 20 ft.

FIGURE 4.26 Laguna Niguel landslide: another view of the toe of the landslide where the road has been uplifted 20 ft.

FIGURE 4.27 Laguna Niguel landslide: View of damage at the toe of the landslide. The landslide has crushed the buildings located in this area.

FIGURE 4.28 Laguna Niguel landslide: head of the landslide. (*Note:* The vertical distance between the two arrows is the distance that the landslide dropped downward.)

FIGURE 4.29 Laguna Niguel landslide. Head of the landslide. (*Note:* The smaller arrow points to the corner of a house which is suspended in midair and the larger arrow points to the main scarp.)

FIGURE 4.30 Laguna Niguel landslide: head of the landslide showing another house with a portion suspended in midair. (*Note:* The arrow points to a column and footing that are suspended.)

FIGURE 4.31 Laguna Niguel landslide: head of the landslide where a house has dropped downward. (*Note:* The arrow points to the area where the house foundation has punched through the driveway on its way down.)

FIGURE 4.32 Laguna Niguel landslide: head of the landslide showing lateral movement of the entire house. (*Note:* The distance between the arrows indicates the amount of lateral movement.)

FIGURE 4.33 Laguna Niguel landslide: head of the landslide showing a house that has been pulled downslope.

FIGURE 4.34 Laguna Niguel landslide: graben area created as the landslide moved downslope.

TABLE 4.11 Geophysical Techniques

Name of method (1)	Procedure or principle utilized (2)	Applicability and limitations (3)
Seismic methods 1. Refraction	Based on time required for seismic waves to travel from source of energy to points on ground surface, as measured by geophones spaced at intervals on a line at the surface. Refraction of seismic waves at the interface between different strata gives a pattern of arrival times at the geophones versus distance to the source of seismic waves. Seismic velocity can be obtained from a single geophone and recorder with the impact of a sledge hammer on a steel plate as a source of seismic waves.	Utilized for preliminary site investigation to determine rippability, faulting, and depth to rock or other lower stratum substantially different in wave velocity than the overlying material. Generally limited to depths up to 30 m (100 ft) of a single stratum. Used only where wave velocity in successive layers becomes greater with depth.
2. High-resolution reflection	Geophones record travel time for the arrival of seismic waves reflected from the interface of adjoining strata.	Suitable for determining depths to deep rock strata. Generally applies to depths of a few thousand feet. Without special signal enhancement techniques, reflected impulses are weak and easily obscured by the direct surface and shallow refraction impulses. Method is useful for locating groundwater.
3. Vibration	The travel time of transverse or shear waves generated by a mechanical vibrator consisting of a pair of eccentrically weighted disks is recorded by seismic detectors placed at specific distances from the vibrator.	Velocity of wave travel and natural period of vibration gives some indication of soil type. Travel time plotted as a function of distance indicates depths or thickness of surface strata. Useful in determining dynamic modulus of subgrade reaction and obtaining information on the natural period of vibration for the design of foundations of vibrating structures.
4. Uphole, downhole, and crosshole surveys	(a) Uphole or downhole: Geophones on surface, energy source in borehole at various locations starting from hole bottom. Procedure can be revised with energy source on surface, detectors moved up or down the hole. (b) Downhole: Energy source at the surface (e.g., wooden plank struck by hammer), geophone probe in borehole. (c) Crosshole: Energy source in central hole, detectors in surrounding holes.	Obtain dynamic soil properties at very small strains, rock mass quality, and cavity detection. Unreliable for irregular strata or soft strata with large gravel content. Also unreliable for velocities decreasing with depth. Crosshole measurements best suited for *in situ* modulus determination.

Electrical methods		
1. Resistivity	Based on the difference in electrical conductivity or resistivity of strata. Resistivity is correlated to material type.	Used to determine horizontal extent and depths up to 30 m (100 ft) of subsurface strata. Principal applications are for investigating foundations of dams and other large structures, particularly in exploring granular river channel deposits or bedrock surfaces. Also used for locating fresh/salt water boundaries.
2. Drop in potential	Based on the determination of the drop in electrical potential.	Similar to resistivity methods but gives sharper indication of vertical or steeply inclined boundaries and more accurate depth determinations. More susceptible than resistivity method to surface interference and minor irregularities in surface soils.
3. E-logs	Based on differences in resistivity and conductivity measured in borings as the probe is lowered or raised.	Useful in correlating units between borings, and has been used to correlate materials having similar seismic velocities. Generally not suited to civil engineering exploration but valuable in geologic investigations.
Magnetic measurements	Highly sensitive proton magnetometer is used to measure the Earth's magnetic field at closely spaced stations along a traverse.	Difficult to interpret in quantitative terms but indicates the outline of faults, bedrock, buried utilities, or metallic trash in fills.
Gravity measurements	Based on differences in density of subsurface materials which affects the gravitational field at the various points being investigated.	Useful in tracing boundaries of steeply inclined subsurface irregularities such as faults, intrusions, or domes. Methods not suitable for shallow depth determination but useful in regional studies. Some application in locating limestone caverns.

Source: Adapted from NAVFAC DM-7.1, 1982.
Note: Also see AGI Data Sheets 59.1 to 60.2 (American Geological Institute, 1982) for a summary of the applications of geophysical methods.

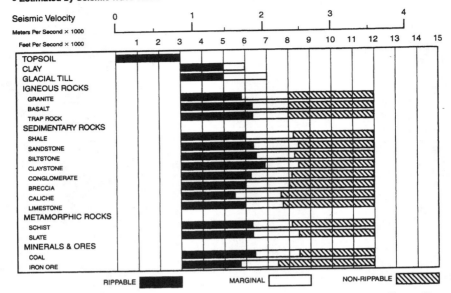

FIGURE 4.35 Rippability of rock versus seismic velocity for a Caterpillar D8R tractor/ripper. (*From Caterpillar Performance Handbook, 1977.*)

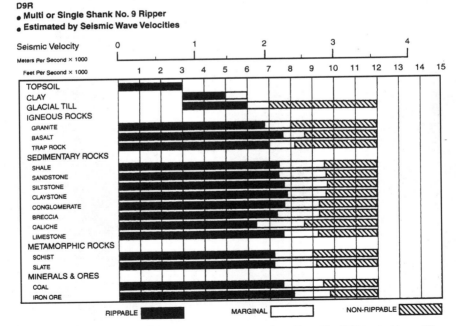

FIGURE 4.36 Rippability of rock versus seismic velocity for a Caterpillar D9R tractor/ripper. (*From Caterpillar Performance Handbook, 1997.*)

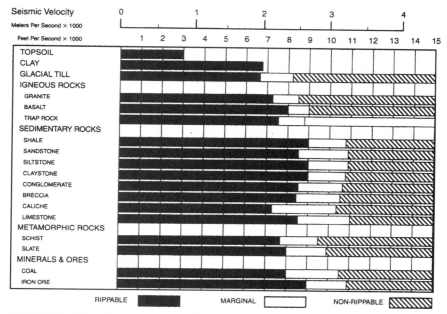

FIGURE 4.37 Rippability of rock versus seismic velocity for a Caterpillar D10R tractor/ripper. (*From Caterpillar Performance Handbook, 1997.*)

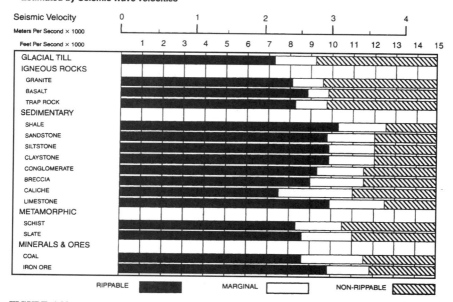

FIGURE 4.38 Rippability of rock versus seismic velocity for a Caterpillar D11R tractor/ripper. (*From Caterpillar Performance Handbook, 1997.*)

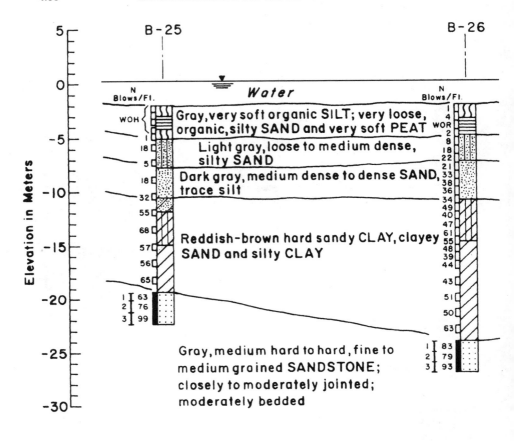

FIGURE 4.39 Subsoil profile. (*From Lowe and Zaccheo, 1975; copyright Van Nostrand Reinhold.*)

FIELD EXPLORATION

4.67

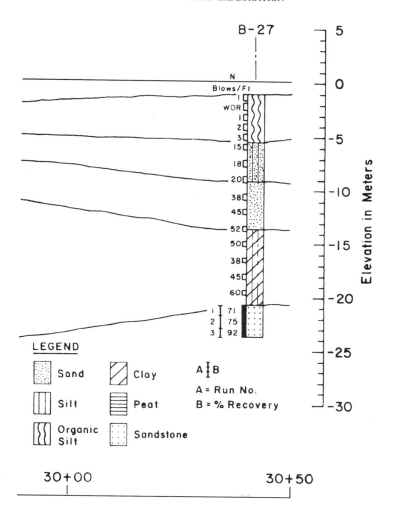

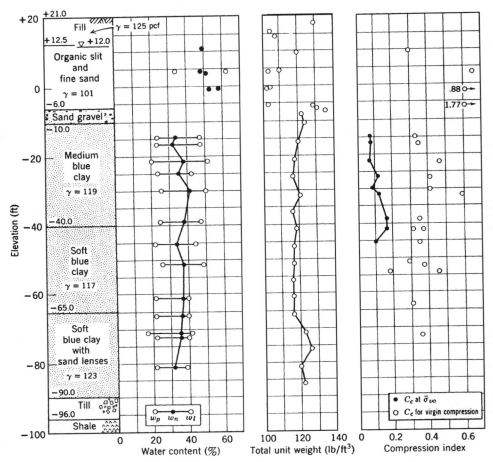

FIGURE 4.40 Subsoil profile, Cambridge, Mass. (*From Lambe and Whitman, 1969; reprinted with permission of John Wiley & Sons.*)

FIELD EXPLORATION **4.69**

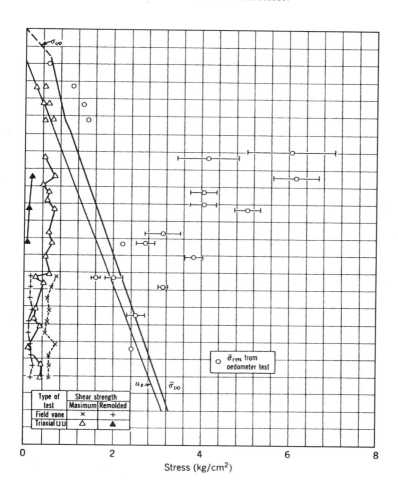

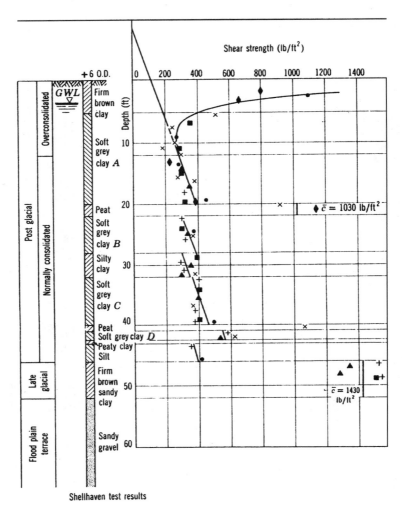

FIGURE 4.41 Subsoil profile, Thames estuary clay, England. (*From Skempton and Henkel, 1953; reprinted from Lambe and Whitman, 1969; reprinted with permission of John Wiley & Sons.*)

FIELD EXPLORATION

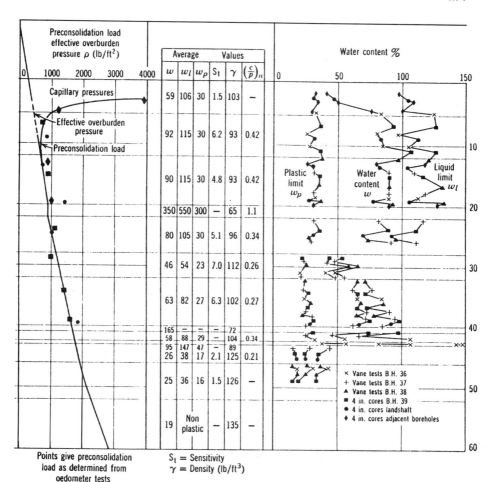

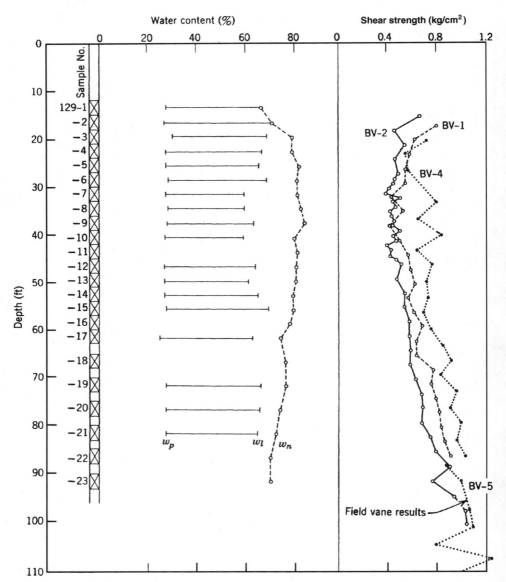

FIGURE 4.42 Subsoil profile, Canadian clay. (*From Lambe and Whitman, 1969; reprinted with permission of John Wiley & Sons.*)

FIELD EXPLORATION **4.73**

γ lb/ft³	% clay	S.C. gr/lt.	S_t
100.7	77	0.2	12
98.2	82	0.4	21
96.1	86		
96.2	78		48
96.7	80	0.6	37
95.6	82	0.8	54
95.5			150
94.9	80	0.5	127
94.4	85	0.9	100
96.1	83	0.8	128
95.6			74
95.3	88		
95.8	85	1.2	72
96.2	86		
95.7		1.5	76
			118
97.0	80	1.9	53
96.9			53
97.1		2.2	76
98.1	83	2.0	37
98.5	73	3.4	34
98.8			
98.9		2.5	

Stress history (kg/cm²)

Effective overburden stress

Max
Min Prob

Preconsolidation stress

CHAPTER 5
LABORATORY TESTING

The following notation is introduced in this chapter:

SYMBOL	DEFINITION
a	Area of standpipe (falling-head permeameter)
A	Cross-sectional area (constant- and falling-head permeameters)
A	Activity [Eq. (5.16)]
A_f	Cross-sectional area at failure (unconfined compression test)
A_0	Initial cross-sectional area (unconfined compression test)
c	Cohesion based on a total stress analysis
c'	Cohesion based on an effective stress analysis
C_c	Coefficient of curvature
C_u	Coefficient of uniformity
D_{60}, D_{30}, D_{10}	Particle sizes corresponding to percent passing for particle size analysis
Δh	Change in total head
h_0, h_f	Head measurements from a falling-head permeameter test
i	Hydraulic gradient
k	Hydraulic conductivity (also known as coefficient of permeability)
L	Length of soil specimen (constant- and falling-head permeameters)
LI	Liquidity index
LL	Liquid limit
L_0	Initial height of the clay specimen (unconfined compression test)
ΔL	Change in height of clay specimen at failure (unconfined compression test)
PI	Plasticity index
PL	Plastic limit
P_f	Vertical force causing failure (unconfined compression test)
q_u	Unconfined compressive strength
Q	Total discharge volume in a given time (constant-head permeameter)
SL	Shrinkage limit
S_t	Sensitivity of clay
t	Time
t_f	Time to failure (direct shear test)

t_{50}	Time for a silt or clay to achieve 50% consolidation
v	Velocity of flow (Darcy's law)
w	Water content (also known as moisture content)
ε_f	Axial strain at failure (unconfined compression test)
ϕ	Friction angle based on a total stress analysis
ϕ'	Friction angle based on an effective stress analysis
ϕ'_r	Drained residual friction angle

5.1 INTRODUCTION

As indicated in Fig. 1.11, in addition to document review and subsurface exploration, an important part of the site investigation is laboratory testing. The laboratory testing usually begins once the subsurface exploration is complete. The first step in the laboratory testing is to log in all of the materials (soil, rock, or groundwater) recovered from the subsurface exploration. Then the geotechnical engineer and engineering geologist prepare a laboratory testing program, which basically consists of assigning specific laboratory tests for the soil specimens. The actual laboratory testing of the soil specimens is often performed by experienced technicians, who are under the supervision of the geotechnical engineer. Because the soil samples can dry out or there could be changes in the soil structure with time, it is important to perform the laboratory tests as soon as possible.

Usually at the time of the laboratory testing, the geotechnical engineer and engineering geologist will have located the critical soil layers or subsurface conditions that will have the most impact on the design and construction of the project. The laboratory testing program should be oriented toward the testing of those critical soil layers or subsurface conditions. For many geotechnical projects, it is also important to determine the amount of ground surface movement due to construction of the project. In these cases, laboratory testing should model future expected conditions so that the amount of movement or stability of the ground can be analyzed. During the planning stage, specific types of laboratory tests may have been selected, but, depending on the results of the subsurface exploration, additional tests or a modification of the planned testing program may be required.

Laboratory tests should be performed in accordance with standard procedures, such as those recommended by the American Society for Testing and Materials (ASTM) or those procedures listed in standard textbooks or specification manuals (e.g., Lambe, 1951; Bishop and Henkel, 1962; Department of the Army, 1970; *Standard Specifications,* 1997).

For laboratory tests, Tomlinson (1986) states:

> It is important to keep in mind that natural soil deposits are variable in composition and state of consolidation; therefore it is necessary to use considerable judgment based on common sense and practical experience in assessing test results and knowing where reliance can be placed on the data and when they should be discarded. It is dangerous to put blind faith in laboratory tests, especially when they are few in number. The test data should be studied in conjunction with the borehole records and the site observations, and any estimations of bearing pressures or other engineering design data obtained from them should be checked as far as possible with known conditions and past experience.
>
> Laboratory testing should be as simple as possible. Tests using elaborate equipment are time-consuming and therefore costly, and are liable to serious error unless carefully and conscientiously carried out by highly experienced technicians. Such methods may be quite unjustified if the samples are few in number, or if the cost is high in relation to the cost of the project. Elaborate and costly tests are justified only if the increased accuracy of the data will give worthwhile savings in design or will eliminate the risk of a costly failure.

LABORATORY TESTING 5.3

5.2 COMMON LABORATORY TESTS

Table 5.1 presents a list of common soil laboratory tests used in geotechnical engineering. As indicted in Table 5.1, laboratory tests are often used to determine the index properties, shear strength, compressibility, and hydraulic conductivity of the soil.

The following is a brief summary of these commonly used laboratory tests. Appendix A (Glossary 2) presents a list of laboratory terms and definitions.

5.2.1 Index Tests

Index tests are the most basic types of laboratory tests performed on soil samples. Index tests include the following:

TABLE 5.1 Common Soil Laboratory Tests Used in Geotechnical Engineering

Type of condition (1)	Soil properties (2)	Specification (3)
Index tests	Classification Particle size Atterberg limits Water (or moisture) content Wet density Specific gravity Sand equivalent (SE)	ASTM D 2487-93 ASTM D 422-90 ASTM D 4318-95 ASTM D 2216-92 Block samples or sampling tubes ASTM D 854-92 ASTM D 2419-95*
Settlement (Chap. 7)	Consolidation Collapse Organic content Fill compaction: Standard Proctor Fill compaction: Modified Proctor	ASTM D 2435-96 ASTM D 5333-96† ASTM D 2974-95 ASTM D 698-91 ASTM D 1557-91
Expansive soil (Chap. 9)	Swell Expansion index test	ASTM D 4546-96 ASTM D 4829-95 or UBC 18-2
Shear strength for slope movement (Chap. 10)	Unconfined compressive strength Unconsolidated undrained Consolidated undrained Direct shear Ring shear Miniature vane (i.e., torvane)	ASTM D 2166-91 ASTM D 2850-95 ASTM D 4767-95 ASTM D 3080-90 Stark and Eid, 1994 ASTM D 4648-94
Erosion (Chap. 12)	Dispersive clay Erosion potential	ASTM D 4647-93 Day, 1990b
Pavements and deterioration (Chaps. 8, 13)	Pavements: CBR Pavements: R-value Sulfate	ASTM D 1883-94 ASTM D 2844-94 Chemical analysis
Permeability (Chap. 16)	Constant head Falling head	ASTM D 2434-94 ASTM D 5084-97†

*This specification is in the ASTM Standards Volume 04.03.
†These specifications are in the ASTM Standards Volume 04.09. All other ASTM standards are in Volume 04.08.

- Water content (also known as moisture content)
- Wet density determinations (also known as total density)
- Specific gravity tests
- Particle size distributions and Atterberg limits (used to classify the soil)
- Laboratory tests specifically labeled as index tests, such as the expansion index test, which is used to evaluate the potential expansiveness of a soil (Chap. 9).

Water Content Tests. The water content (also known as moisture content) test is probably the most common and simplest type of laboratory test. This test can be performed on disturbed or undisturbed soil specimens. The water content test consists of determining the mass of the wet soil specimen and then drying the soil in an oven overnight (12 to 16 hours) at a temperature of 110°C (ASTM D 2216-92, 1998). The water content w of a soil is defined as the mass of water in the soil divided by the dry mass of the soil, expressed as a percentage. Water content values are often reported to the nearest 0.1 or 1 percent of measured value. Table 5.2 provides guidelines on the amount of soil that should be used for a water content test based on the largest particle dimension.

Values of water content w can vary from essentially 0 up to 1200 percent. A water content of 0 percent indicates a dry soil. An example of a dry soil would be near-surface rubble, gravel, or clean sand located in a hot and dry climate, such as Death Valley, California. Soil having the highest water content is organic soil, such as fibrous peat, which has been reported to have a water content as high as 1200 percent (NAFVAC DM-7.1, 1982).

The water content data is often plotted with depth on the soil profile. Note in Fig. 4.40 (Chap. 4) that the water content versus depth has been plotted (darkened circles) and the water content varies from about 30 to 40 percent for the clay. Water content data is also plotted (open circles) in Fig. 4.42, and for this Canadian clay, the water content varies from about 60 to 80 percent.

Total Density and Total Unit Weight. The total density (also known as the wet density) should be obtained only from undisturbed soil specimens, such as those extruded from Shelby tubes or on undisturbed block samples obtained from test pits and trenches. When extruding the soil from sampler tubes, it is important to push the soil out the back of the sampler. Soil should not be pushed out the front of the tube (i.e., the cutting end) because this causes a reversal in direction of soil movement as well as the possible compression of the soil because D_e is less than D_i; see Fig. 4.11. In order to determine the total

TABLE 5.2 Minimum Soil Sample Mass for Water Content Determination

Largest particle dimension (1)	Sieve size corresponding to particle dimension (2)	Recommended minimum mass of wet soil specimen for water content to be reported to the nearest 0.1% (3)	Recommended minimum mass of wet soil specimen for water content to be reported to the nearest 1% (4)
2 mm or less	No. 10	20 grams	
4.7 mm	No. 4	100 grams	20 grams
9.5 mm	⅜ in.	500 grams	50 grams
19.0 mm	¾ in.	2500 grams	250 grams

Notes: Table based on ASTM D 2216-92 (1998). For soil containing particles larger than ¾ in., first sieve the sample on the ¾-in. sieve, record the mass of oversize particles (plus ¾-in. sieve material), and then determine the water content of the soil matrix (minus ¾-in. sieve soil).

density of a soil specimen, both the mass and corresponding volume of the soil specimen must be known. This can be accomplished by extruding the soil from the sampling tube directly into metal confining rings of known volume, such as shown in Fig. 5.1.

The total unit weight is defined as the wet soil weight per unit volume. In this and all subsequent unit weight definitions, the use of the term weight means force. In the International System of Units (SI), unit weight has units of kN/m^3 (kilonewtons per cubic meter), while in the United States Customary System, the unit weight has units of pcf, but this implies pounds of force (lbf) per cubic foot.

As with water content, it is common to plot the data versus depth on the soil profile. For example, in Fig. 4.40 (Chap. 4), the total unit weight (pcf) has been plotted (open circles) versus depth, and the total unit weight is about 120 pcf (19 kN/m^3) for the clay.

Specific Gravity. Specific gravity is a dimensionless parameter that relates the density of the soil particles to the density of water. For soil, the specific gravity is obtained by measuring the dry mass of the soil and then using a pycnometer to obtain the volume of the soil (see ASTM D 854-92, 1998). Table 5.3 presents typical values and ranges of specific gravity versus different types of soil minerals. Because quartz is the most abundant type of soil mineral, the specific gravity for inorganic soil is often assumed to be 2.65. For clays, the specific gravity is often assumed to be 2.70 because common clay particles, such as montmorillonite and illite, have slightly higher specific gravity values.

Oversize particles are often defined as gravel and cobbles that are retained on the 3/4-in. sieve (Day, 1989). For these large particles, there may be internal rock fractures, voids, or moisture trapped within the gravel and cobbles. The specific gravity test procedures and calculations can include these features and the test result is referred to as the *bulk specific gravity* (ASTM C 127-93, 1998).

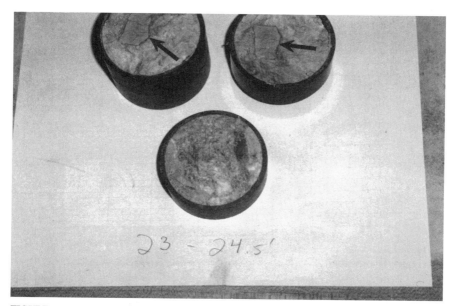

FIGURE 5.1 Weathered soil extruded from a Shelby tube directly into 2.5-in. diameter confining rings (arrows point to fragments of weathered rock).

TABLE 5.3 Formula and Specific Gravity of Common Soil Minerals

Type of mineral (1)	Formula (2)	Specific gravity (3)	Comments (4)
Quartz	SiO_2	2.65	Silicate, most common type of soil mineral
K feldspar	$KAlSi_3O_8$	2.54–2.57	Feldspars are also silicates and are the second
Na feldspar	$NaAlSi_3O_8$	2.62–2.76	most common type of soil mineral.
Calcite	$CaCO_3$	2.71	Basic constituent of carbonate rocks
Dolomite	$CaMg(CO_3)_2$	2.85	Basic constituent of carbonate rocks
Muscovite	Varies	2.76–3.0	Silicate sheet-type mineral (mica group)
Biotite	Complex	2.8–3.2	Silicate sheet-type mineral (mica group)
Hematite	Fe_2O_3	5.2–5.3	Frequent cause of reddish-brown color in soil
Gypsum	$CaSO_4 \cdot 2H_2O$	2.35	Can lead to sulfate attack of concrete
Serpentine	$Mg_3Si_2O_5(OH)_4$	2.5–2.6	Silicate sheet or fibrous type mineral
Kaolinite	$Al_2Si_2O_5(OH)_4$	2.61–2.66	Silicate clay mineral, low activity
Illite	Complex	2.60–2.86	Silicate clay mineral, intermediate activity
Montmorillonite	Complex	2.74–2.78	Silicate clay mineral, highest activity

Notes: Silicates are very common and account for about 80% of the minerals at the Earth's surface. Data accumulated from the following sources: Lambe and Whitman (1969) and Mottana et al. (1978).

Particle Size Distribution and Atterberg Limits. These types of index tests are used to classify a soil. They can also be used to correlate a soil type with basic engineering properties. Soil classification will be discussed in Sec. 5.4

5.2.2 Oedometer

The oedometer (also known as a *consolidometer*) is the primary laboratory equipment used to study the settlement or expansion behavior of soil. The oedometer test should only be performed on undisturbed soil specimens, or in the case of studies of fill behavior, on specimens compacted to anticipated field and moisture conditions.

Figure 5.2 shows a diagram of the two basic types of oedometer. The equipment can vary, but in general, an oedometer consists of the following:

1. A metal ring that laterally confines the soil specimen
2. Porous plates placed on the top and bottom of the soil specimen
3. A loading device that applies a vertical stress to the soil specimen
4. A dial gauge to measure the vertical deformation of the soil specimen as it is loaded or unloaded
5. A surrounding container to allow the soil specimen to be submerged in distilled water

For the fixed-ring oedometer, the soil specimen will be compressed from the top downward as the load is applied, while for the floating-ring oedometer, the soil specimen is compressed inward from the top and bottom. An advantage of the floating ring oedometer is that there is less friction between the confining ring and the soil. Disadvantages of the floating ring are that it is often more difficult to set up and soil may squeeze or fall out of the junction of the bottom porous plate and ring. Because of these disadvantages, the fixed-ring oedometer is the most popular testing setup.

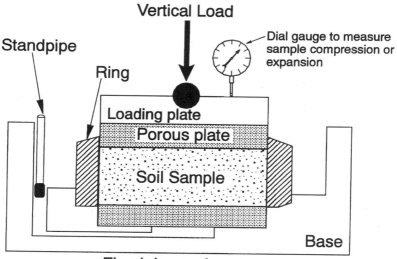

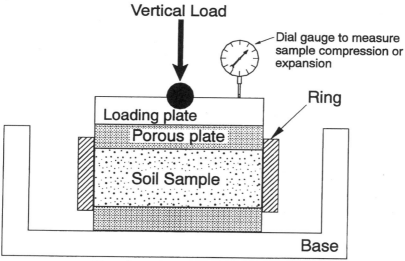

Note: Although not shown above, the dial gauge should be positioned at the center of the loading plate.

FIGURE 5.2 Fixed- and floating-ring oedometer apparatus.

The oedometer test is popular because of its simplicity, and it can be used to model and predict the behavior of the *in situ* soil. For example, a soil specimen can be placed in the oedometer and then subjected to an increase in pressure equivalent to the weight of the proposed structure. By analyzing the settlement versus load data, the geotechnical engineer can calculate the amount of expected settlement due to the weight of the proposed structure. As discussed in Chaps. 7 and 9, the oedometer can also be used to predict the consolidation behavior of soft saturated clay, settlement potential of collapsible soil, and the amount of heave of expansive soils.

5.2.3 Shear Strength Tests

The shear strength of a soil is a basic geotechnical engineering parameter and is required for the analysis of foundations, earthwork, and slope stability problems. For example, Fig. 4.23*b* (Chap. 4) shows a cross section of a massive landslide, which is basically a shear-type failure of the slope. The shear strength of the soil can be determined in the field (e.g., by a vane shear test) or in the laboratory.

There are many types of shear strength tests that can be performed in the laboratory. The objective is to obtain the shear strength of the soil. Laboratory tests are generally divided into two categories:

Shear strength tests based on total stress. The purpose of these laboratory tests is to obtain the undrained shear strength s_u of the soil or the failure envelope in terms of total stresses (total cohesion c and total friction angle ϕ). These types of shear strength tests are often referred to as *undrained* shear strength tests.

Shear strength tests based on effective stress. The purpose of these laboratory tests is to obtain the effective shear strength of the soil based on the failure envelope in terms of effective stress (effective cohesion c' and effective friction angle ϕ'). These types of shear strength tests are often referred to as *drained* shear strength tests.

The concept of total and effective stress as well as the failure envelope is discussed in Chap. 6. The types of common laboratory tests used to obtain the shear strength of the soil are as follows:

Vane Tests. As discussed in Sec. 4.4.2, the vane shear test (VST) can be performed in the field to obtain the undrained shear strength s_u of clay by using Eq. (4.4). The miniature vane or torvane device could also be used in the laboratory to obtain the undrained shear strength s_u of clay.

Unconfined Compression Test. The unconfined compression test is a very simple type of test that consists of applying a vertical compressive pressure to a cylinder of laterally unconfined clay. The vertical compressive pressure that causes shear failure of the clay is termed the unconfined compressive strength q_u. Some clays are so plastic that instead of shearing, they deform and bulge, and in these cases the unconfined compressive strength q_u is defined as the vertical compressive pressure at 15 percent axial strain. The unconfined compressive test is a shear strength test based on total stress, and it can be used to determine the undrained shear strength s_u of the clay, defined as:

$$s_u = \frac{q_u}{2} \tag{5.1}$$

The unconfined compressive test should not be performed on dry and crumbly soils, fissured or varved clays, silts, peat, and all types of granular material. The ratio of height to diameter of the cylinder of clay should be between 2 and 2.5. The clay specimen should also be tested in standard laboratory equipment that can accurately record the force applied to the top of the clay specimen as well as record the vertical deformation of the specimen as it is loaded. Typical loading rates of the clay specimen are 0.5 to 2 percent of the initial height L_0 per minute (ASTM D 2166-91, 1998). Because the area of the clay specimen changes as it is loaded, corrections should be made to the unconfined compressive strength, as follows:

$$\varepsilon_f = \frac{\Delta L}{L_0} \tag{5.2}$$

where ε_f = axial strain at failure, ΔL = change in height of the clay specimen at failure, and L_0 = original height of the test specimen. With this data, the cross-sectional area of the specimen at failure (A_f) can be calculated:

$$A_f = \frac{A_0}{1 - \varepsilon_f} \tag{5.3}$$

where A_0 = initial cross-sectional area of the cylinder of clay. The unconfined compressive strength q_u of the clay can then be calculated, as follows:

$$q_u = \frac{P_f}{A_f} \tag{5.4}$$

where P_f is the recorded vertical force that caused a shear failure of the cylinder of clay.

To determine the *in situ* undrained shear strength of the clay deposit, undisturbed soil specimens are required. The unconfined compression test could also be performed on compacted fill specimens to determine the shear strength of fill. The unconfined compression test could even be performed on completely remolded soil specimens in order to determine the sensitivity S_t of the clay, defined as:

$$S_t = \frac{q_u \text{ (for undisturbed specimen)}}{q_u \text{ (for remolded specimen)}} \tag{5.5}$$

The sensitivity of clay is important because it indicates the amount of shear strength loss if the clay is remolded. This could be very important for all types of geotechnical and foundation engineering analyses. For example, during construction, a highly sensitive clay could be disturbed and lose its shear strength, causing the construction equipment to become bogged down or sink into the sensitive clay, resulting in significantly higher costs. In the United States, the classification of sensitivity versus condition is as follows (Holtz and Kovacs, 1981):

- S_t less than 4 indicates a "low" sensitivity
- S_t from 4 to 8 indicates a "medium" sensitivity
- S_t from 8 to 16 indicates a "high" sensitivity
- S_t greater than 16 indicates a "quick" clay

When remolded, some quick clays are so sensitive that they actually turn into a liquid. Much like "quick assets" (a term meaning "liquid assets"), "quick" clays imply a "liquid" condition.

Note in the last column of Fig. 4.42 (Chap. 4) that sensitivity values have been listed, with S_t values up to 150. This Canadian clay is definitely a quick clay. A quick clay indicates an unstable soil structure, where the bonds between clay particles are destroyed upon remolding. Such unstable soil can be formed by the leaching of soft glacial clays deposited in salt water and subsequently uplifted, or by the creation of unstable soil due to weathering of volcanic ash (Terzaghi and Peck, 1967). A remolded quick clay may actually regain some of its lost shear strength, because the soil particles begin to re-form bonds; this increase in shear strength of a remolded soil is known as *thixotropy*.

The unconfined compression test can also be used to determine the consistency (i.e., degree of firmness) of the clay. The values are as follows:

- "Very soft" clay, s_u less than 12 kPa (250 psf)
- "Soft" clay, s_u between 12 to 25 kPa (250 to 500 psf)
- "Medium" clay, s_u between 25 to 50 kPa (500 to 1000 psf)
- "Stiff" clay, s_u between 50 to 100 kPa (1000 to 2000 psf)
- "Very stiff" clay, s_u between 100 to 200 kPa (2000 to 4000 psf)
- "Hard" clay, s_u greater than 200 kPa (4000 psf)

Figure 5.3 shows a photograph of the failed condition of a clay specimen from an unconfined compression test. The unconfined clay specimen was subjected to an increase in vertical compressive pressure, with a variation in test procedure in that the specimen was submerged in distilled water and loaded very slowly. At failure, portions of the soil specimen became detached from the main body. The arrows in Fig. 5.3 point to the shear surfaces that developed during failure of the unconfined clay specimen.

Unconsolidated Undrained Triaxial Compression Test. The unconsolidated undrained triaxial compression test is similar to the unconfined compression test, except that clay specimens are subjected to a confining fluid pressure in a triaxial chamber (ASTM D 2850-95, 1998). Figure 5.4 shows a diagram of the triaxial apparatus. The cylindrical clay specimen is placed in the center of the chamber, sealed with a rubber membrane, and subjected to a confining fluid pressure. A vertical pressure is slowly applied to the top of the specimen by slowly loading the piston. No drainage of the clay specimen is allowed during the test. The clay specimen is sheared in compression without drainage at a constant rate of axial deformation (strain controlled).

If the clay specimens are saturated, then the confining pressure will not change the wet density of the specimen. Therefore, if several identical clay specimens are tested, they will all have approximately the same undrained shear strength (failure envelope will be a horizontal line). Since this test does not provide any additional information for saturated clay specimens, it is used much less frequently than the unconfined compression test.

Consolidated Undrained Triaxial Compression Test. The consolidated undrained triaxial compression test is similar to the unconsolidated undrained compression test, except that the clay specimen is first consolidated prior to shearing (ASTM D 4767-95, 1998). This test can also be used to measure the excess pore water pressures that develop within the clay as it is sheared. The shear strength based on total stress or based on effective stress (by measuring the pore water pressure during shear) can be obtained from this laboratory test. The test is quite complicated and should be performed only by experienced personnel. This test has been used to study the shear behavior of all types of soil. Further details of this test are presented in Sec. 6.5.

Direct Shear Test. The direct shear test, first used by Coulomb in 1776, is the oldest type of shear testing equipment. The direct shear test is also the most common laboratory equip-

LABORATORY TESTING

FIGURE 5.3 Shear failure of a saturated specimen of clay due to an increase in vertical pressure (arrows point to failure surfaces).

ment used to obtain the drained shear strength (shear strength based on effective stress) of a soil. Figure 5.5 shows the direct shear testing device. The testing procedure is as follows (ASTM D 3080-90, 1998):

1. The soil specimen (typical diameter = 63.5 mm and height = 25.4 mm) is placed on top of a porous plate in the center of the direct shear apparatus.
2. A porous plate is placed on the top of the specimen and a dial gauge is set up to measure vertical deformation.
3. A loading device is used to apply a vertical force to the soil specimen. This vertical force, also known as the normal force, is converted to a stress by dividing the vertical force by the cross-sectional area of the soil specimen.
4. A surrounding container (not shown in Fig. 5.5) allows the soil specimen to be submerged in distilled water.
5. The soil specimen is allowed to equilibrate or fully consolidate under the applied vertical stress.

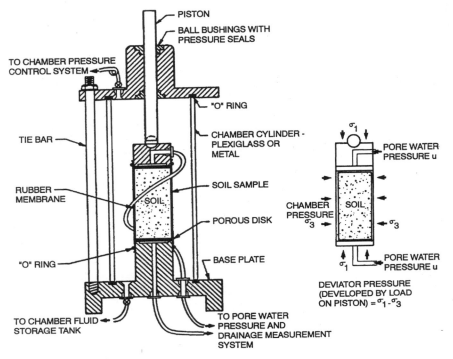

FIGURE 5.4 Triaxial apparatus.

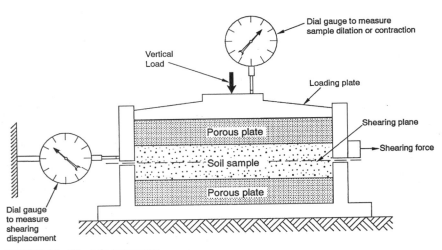

FIGURE 5.5 Direct shear apparatus.

6. A dial gauge is set up to measure horizontal deformation and then an increasing force is applied to the upper half of the direct shear apparatus. The shear force can be applied by using deadweights placed on a hanger (stress-controlled test) or by a motor acting through gears (strain-controlled test).

7. The soil specimen is slowly sheared in half (a single shear plane).

8. The maximum shear force required to shear the sample is recorded and the shear strength is defined as this shear force divided by the cross-sectional area of the soil specimen.

9. The shear strength versus the vertical stress can be plotted and a failure envelope is obtained.

It is important to shear the soil specimen slowly enough that excess pore water pressures do not develop within the soil specimen. For clean sands, the total elapsed time to failure should be about 10 minutes, while for silty sands the total elapsed time to failure should be about 60 minutes. For clean sands and silty sands, the setup and actual testing operation is relatively quick and simple, which is the main reason for the test's popularity. If sands contain gravel-size particles, then the gravel can artificially increase the shear strength of the sand because of the usual small height (25.4 mm) of the specimen. Other limitations of the device are that the shear stresses and displacements are nonuniformly distributed, and the failure plane may not be perfectly horizontal within the specimen (Dounias and Potts, 1993; Day, 1995a). Even with these limitations, the direct shear apparatus generally provides reasonably accurate values of the effective friction angle (ϕ') for cohesionless soil.

For silts and clays, the drained shear strength can also be determined, but the rate of shearing must be much slower. According to ASTM D 3080-90 (1998), the total elapsed time to failure (t_f) should be:

$$t_f = 50 t_{50} \tag{5.6}$$

where t_{50} = time for the silt or clay to achieve 50 percent consolidation under the vertical stress. This time t_{50} is obtained from a time versus vertical deformation plot (discussed in Sec. 7.4.2, Chap. 7). For silt or clay, Eq. (5.6) can require that the test be performed for a considerable amount of time. For example, for a clay that reaches 50 percent consolidation in 20 minutes, the total elapsed time to failure must be at least 1000 minutes (16.7 hours). This may require the installation of special equipment (special gears to slow down the strain rate) to allow the direct shear device to shear very slowly. Probably the two most common problems with the direct shear testing of a silt or clay are: (1) the soil is not saturated prior to shearing, and (2) the soil is sheared too quickly. Both of these conditions can result in the shear strength parameters being overestimated.

Drained Residual Shear Strength. The previous shear strength tests have described methods to obtain the undrained shear strength s_u or the shear strength parameters (c or c' and ϕ or ϕ') that define the failure envelope. These tests normally determine the shear strength of the soil, which is defined as the shear stress at failure (i.e., the peak shear strength).

For some projects, it may be important to obtain the residual shear strength of fine-grained soil, which is defined as the remaining (or residual) shear strength after a considerable amount of shear deformation has occurred. For example, a clay specimen could be placed in the direct shear box and then sheared back and forth several times to develop a well-defined shear failure surface. Once the shear surface is developed, the drained residual shear strength would be obtained by performing a final, slow shear of the specimen. The drained residual shear strength can be applicable to many types of soil or rock conditions where a considerable amount of shear deformation has already occurred. For example, the stability analysis of ancient landslides, slopes in overconsolidated fissured clays, and slopes in fissured shales will often be based on the drained residual shear strength of the failure surface (Bjerrum, 1967a; Skempton and Hutchinson, 1969; Skempton, 1985; Hawkins and Privett, 1985; Ehlig, 1992).

Skempton (1964) stated that the residual shear strength (friction angle) is independent of the original shear strength, water content, and liquidity index (Sec. 5.4.2), and depends only on the size, shape, and mineralogical composition of the constituent particles. Besides the direct shear equipment, the drained residual shear strength can be determined by using a modified Bromhead ring shear apparatus (Stark and Eid, 1994). Back calculations of landslide shear strength indicate that the residual shear strength from the ring shear test are reasonably representative of the slip surface (Watry and Ehlig, 1995). The ring shear specimen is annular with an inside diameter of 7 cm (2.8 in.) and an outside diameter of 10 cm (4 in.). Drainage is provided by annular bronze porous plates secured to the bottom of the specimen container and the loading platen.

Remolded specimens are often used for the ring shear testing. A typical testing procedure is as follows:

1. The remolded specimen is typically obtained by air drying the soil (such as slide plane clay seam material), crushing it with a mortar and pestle, ball milling the crushed material, and processing it through the U.S. Standard Sieve No. 200.
2. Distilled water is added to the processed soil until a water content approximately equal to the liquid limit is obtained. The specimen is then allowed to rehydrate for 7 days in a moist room. A spatula is used to place the remolded soil paste into the annular specimen container.
3. To measure the drained residual shear strength, the ring shear specimen is often consolidated at a high vertical pressure (such as 700 kPa) and then the specimen is unloaded to a much lower vertical pressure (such as 50 kPa).
4. The specimen is then presheared by slowly rotating the ring shear base for one complete revolution using the hand wheel.
5. After preshearing, the specimen is sheared at a slow drained displacement rate (such as 0.02 mm/min). This slow shear displacement rate has been successfully used to test soils that are very plastic.
6. After a drained residual strength condition is obtained at the low vertical pressure, shearing is stopped and the normal stress is increased to a higher pressure (such as 100 kPa). After consolidation at this higher pressure, the specimen is sheared again until a drained residual condition is obtained.
7. This procedure is also repeated for other effective normal stresses. The slow shear displacement rate (0.02 mm/min) should be used for all stages of the multistage test.

An example of the type of data obtained from the ring shear test is presented in Figs. 5.6 and 5.7. This data was obtained from the ring shear testing of clay obtained from the slide plane of an actual landslide. The index properties for the slide plane material are listed in Fig. 5.6. The landslide developed in the Friars formation which consists primarily of montmorillonite clay particles (Kennedy, 1975).

Figure 5.6 presents the drained residual failure envelope, and Fig. 5.7 shows the stress-displacement plots for the ring shear test on the slide plane sample. It can be seen in Fig. 5.6 that the failure envelope is nonlinear, which is a common occurrence for residual soil (Maksimovic, 1989a). If a linear failure envelope is assumed to pass through the origin and the shear stress at an effective normal stress of 100 kPa (2090 psf), the residual friction angle (ϕ'_r) is 8.2°. If a linear failure envelope is assumed to pass through the origin and the shear stress at an effective normal stress of 700 kPa (14,600 psf), the residual friction angle

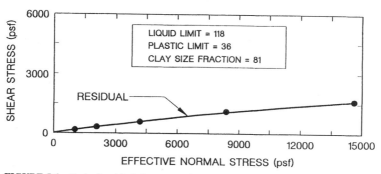

FIGURE 5.6 Drained residual shear strength envelope from ring shear test on slide plane material.

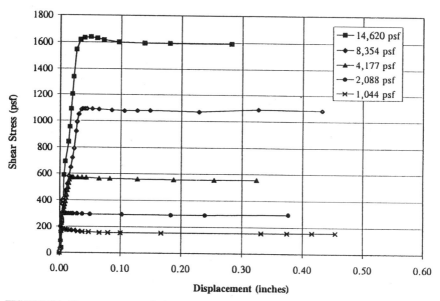

FIGURE 5.7 Shear stress versus displacement from ring shear test on slide plane material.

(ϕ'_r) is 6.2°. As will be discussed in Sec. 6.5, these drained residual friction angles are very low and are probably close to the lowest possible drained residual friction angles of soil.

5.2.4 Permeability

Permeability is defined as the ability of water to flow through a saturated soil. A high permeability indicates that water flows rapidly through the void spaces, and vice versa.

A measure of the soils permeability is the hydraulic conductivity, also known as the coefficient of permeability k. The hydraulic conductivity can be measured in the laboratory by using the constant-head (ASTM D 2434-94, 1998) or falling-head (ASTM D 5084-97, 1998) permeameter.

Constant-Head Permeameter. Figure 5.8 shows a constant-head permeameter apparatus. A saturated soil specimen is placed in the permeameter and then a head of water (Δh) is maintained. The hydraulic conductivity is based on Darcy's law, which states that the velocity v of flow in soil is proportional to the hydraulic gradient i, or:

$$v = ki \tag{5.7}$$

where k = hydraulic conductivity (also known as coefficient of permeability), which has units of cm/sec, and i = hydraulic gradient which is defined as the change in total head (Δh) divided by the length of the soil specimen (L), or $i = \Delta h/L$. From Eq. (5.7), the coefficient of permeability (k) from the constant-head permeameter can be calculated as follows:

$$k = \frac{QL}{\Delta h A t} \tag{5.8}$$

where Q = total discharged volume in milliliters in a given time t, A = area of the soil specimen, and L = length of the soil specimen. The total head loss Δh for the constant-head permeameter is defined in Fig. 5.8.

The constant-head permeameter is often used for sandy soil that has a high permeability. It is important that the porous plate which supports the soil specimen (Fig. 5.8) has a very high permeability. As an alternative, a reinforced permeable screen can be used in place of the porous plate. Another important consideration is that the soil specimen diameter should be at least 10 times larger than the size of the largest soil particle.

Falling-Head Permeameter. Figure 5.9 (from Department of the Army, 1970) shows a laboratory falling-head permeameter. This equipment is often used to determine the hydraulic conductivity of a saturated silt or clay specimen. Filter paper is often placed over the porous plate to prevent the migration of soil fines through the porous plate. Also, a frequent cause of inaccurate results is the inability to obtain a seal between the soil specimen and the side of the permeameter. Because of these factors (migration of fines and inadequate sealing), a greater degree of skill is required to perform the falling-head permeability test.

The objective of the falling-head permeability test is to allow the water level in a small diameter tube to fall from an initial position h_0 to a final position h_f. The amount of time it takes for the water level to fall from h_0 to h_f is recorded. From Darcy's law, the equation to determine the hydraulic conductivity k for a falling-head test is as follows:

$$k = 2.3 \frac{aL}{At} \log_{10} \frac{h_0}{h_f} \tag{5.9}$$

where a = area of the standpipe, L = length of the soil specimen, A = area of the soil specimen, and t = time for the water level in the standpipe to fall from h_0 to h_f.

Permeability Data. Figure 5.10 (adapted from Casagrande) presents a plot of the hydraulic conductivity (coefficient of permeability) versus drainage properties, type of soil, and type of permeameter apparatus best suited for the measurement of k. The bold lines in Fig. 5.10 indicate major divisions in the hydraulic conductivity. A hydraulic conductivity of about 1 cm/s is the approximate boundary between laminar and turbulent flow. A hydraulic conductivity of about 1×10^{-4} cm/s is the approximate dividing line between

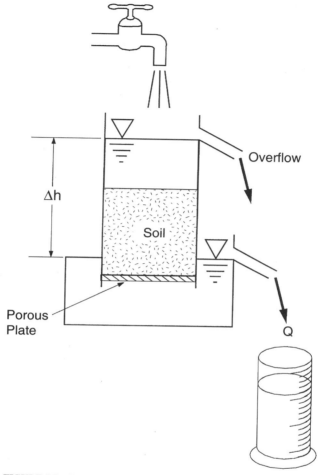

FIGURE 5.8 Constant-head permeameter.

good drainage and poorly drained soils. In general, those soils that contain more fines, such as clays, will have the lowest hydraulic conductivity. This is because clay particles provide very small drainage paths, with a resultant large resistance to fluid flow, even though a clay will often have significantly more void space than a sand. According to Terzaghi and Peck (1967), the classification of soil according to hydraulic conductivity k is as follows:

- High degree of permeability, k is over 0.1 cm/s
- Medium degree of permeability, k is between 0.1 and 0.001 cm/s
- Low permeability, k is between 0.001 and 1×10^{-5} cm/s
- Very low permeability, k is between 1×10^{-5} and 1×10^{-7} cm/s
- Practically impermeable, k is less than 1×10^{-7} cm/s

5.18 GEOTECHNICAL FIELD AND LABORATORY STUDIES

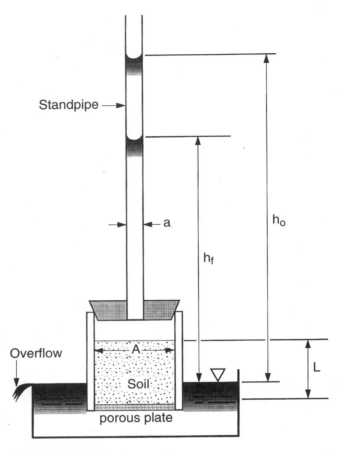

FIGURE 5.9 Falling-head permeameter.

In terms of engineering practice, the hydraulic conductivity is often reported to one or at most two significant figures. This is because in many cases the laboratory permeability will not represent *in situ* conditions. For example, Tomlinson (1986) states:

> There is a difference between the horizontal and vertical permeability of natural soil deposits due to the effects of stratification with alternating beds of finer or coarser grained soils. Thus the results of laboratory tests on a few samples from a vertical borehole are of rather doubtful value in assessing the representative permeability of the soil for calculating the quantity of water to be pumped from a foundation excavation.

5.2.5 Laboratory Tests for Pavements and Deterioration

Pavements. Two common types of laboratory tests for pavements are the laboratory California bearing ratio (CBR) and the *R*-value. As mentioned in Sec. 4.5, the California

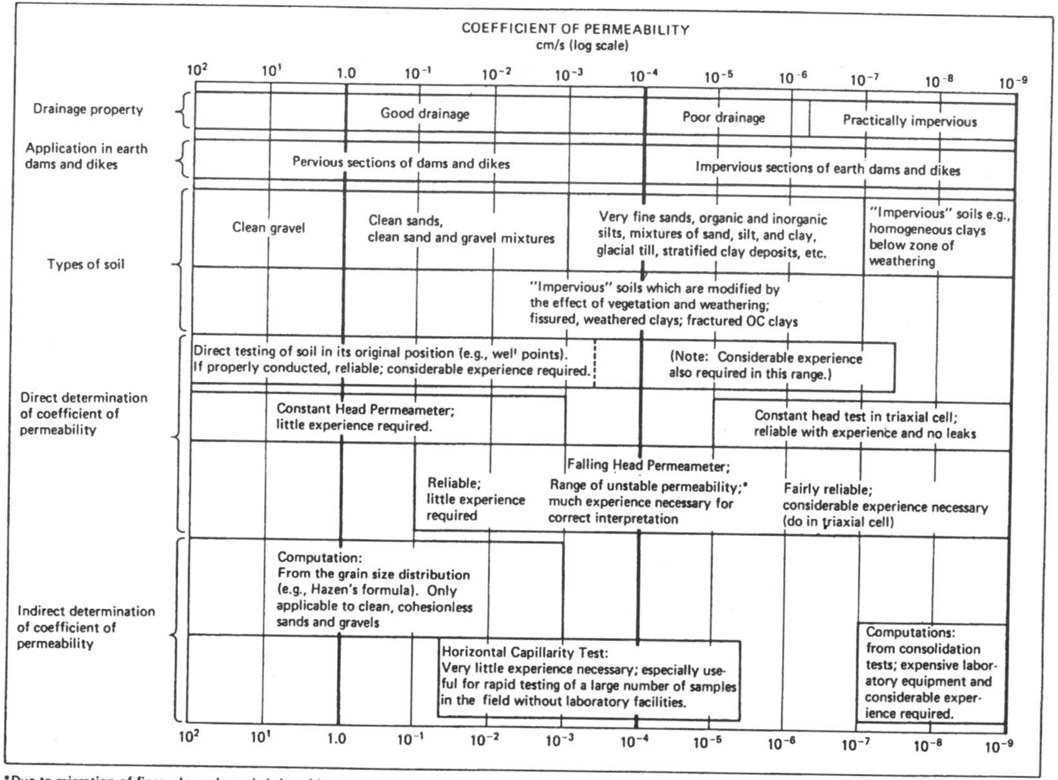

*Due to migration of fines, channels, and air in voids.

FIGURE 5.10 Coefficient of permeability versus drainage property, soil type, and method of determination. [*Developed by Casagrande, with minor additions by Holtz and Kovacs (1981); reproduced from Holtz and Kovacs (1981).*]

bearing ratio can be determined in the field. It can also be performed in the laboratory on compacted soil specimens (ASTM D 1883-94, 1998). The laboratory CBR could be used to assess the condition of base material. It can also be used for the design of roads and airfields where compacted fill is to be used as the subgrade. The testing procedure and calculations for the laboratory CBR are essentially the same as for the field CBR discussed in Sec. 4.5.

In California, the most common type of laboratory test used for flexible pavement design is the R-value (ASTM 2844-94, 1998). The R-value will be discussed in Sec. 8.4, "Pavement Design."

Deterioration. All man-made and natural materials are susceptible to deterioration. This topic is so broad that it is not possible to cover every type of geotechnical or foundation element susceptible to deterioration. Only the most common types of deterioration will be covered.

The most common types of deterioration of concrete are those due to frost action (freeze-thaw cycles) and sulfate attack. There could also be deterioration of concrete or the steel reinforcement due to chloride attack and acid attack. According to AASHTO (1996), laboratory testing of soil and groundwater samples for sulfates, chloride, and pH should be sufficient to assess deterioration potential. When chemical wastes are suspected, a more thorough chemical analysis of soil and groundwater samples would be required.

Timber piles can be especially susceptible to deterioration. They can be attacked by a variety of organisms. For example, the immersed portions of the wood piles in marine or river environments are liable to severe attack by marine organisms (marine borers, etc.). Timber piles above the groundwater table can also experience decay due to the growth of fungi, and they can be attacked by termites, ants, and beetles. To reduce the deterioration due to decay, fungi growth, and insect attack, timber piles should be treated with a preserving chemical. This process consists of placing the timber piles in a pressurized tank filled with creosote or some other preserving chemical. The pressure treatment forces the creosote into the wood pores and creates a thick coating on the pile perimeter. Creosote timber piles normally last as long as the design life of the structure. An exception is the case where they are subjected to prolonged high temperatures (such as supporting blast furnaces), because studies show that they lose strength with time in such an environment (Coduto, 1994).

A further discussion of deterioration will be presented in Chap. 13.

5.3 SAMPLE DISTURBANCE

As discussed in Sec. 4.4.1, there are basically three types of soil samples that can be recovered from the subsurface exploration: (1) altered soil, (2) disturbed soil, and (3) undisturbed soil. Altered soil should not be used for laboratory tests. Disturbed soil can be used for numerous types of laboratory tests, such as tests using compacted soil (laboratory CBR etc.) or for index tests such as water content, classification tests, and specific gravity. In order to evaluate the engineering properties (such as shear strength, compressibility, etc.) of *in situ* soil strata, undisturbed soil specimens must be used in the laboratory. Undisturbed soil samples can be obtained from thin-walled Shelby tubes (that meet inside clearance ratio and area ratio criteria). However, simply using a Shelby tube does not guarantee an undisturbed specimen.

Sample disturbance can occur during sampling, transporting the soil to the laboratory, and during the preparation of the soil specimens in the laboratory. As previously mentioned, the sampling, transporting, and trimming actions cause a decrease in effective stress

(Chap. 6), a reduction in the interparticle bonds, and a rearrangement of the soil particles. An "undisturbed" soil specimen will have little rearrangement of the soil particles and perhaps no disturbance except that caused by stress relief where there is a change from the *in situ* k_0 (at-rest) condition to an isotropic "perfect sample" stress condition (Ladd and Lambe, 1963). A disturbed soil specimen will have a disrupted soil structure with perhaps a total rearrangement of soil particles.

Although rarely used in practice, one method of assessing the quality of soil samples is to obtain an x-ray radiograph of the soil contained in the sampling tube (ASTM D 4452-95, 1998). A radiograph is a photographic record produced by the passage of x-rays through an object and onto photographic film. Denser objects absorb the x-rays and can appear as dark areas on the radiograph. Worm holes, coral fragments, cracks, gravel inclusions, and sand or silt seams can easily be identified by radiography (Allen et al., 1978).

Figures 5.11 to 5.13 (from Day, 1980, and Ladd et al., 1980) present three radiographs taken of Orinoco clay contained within Shelby tubes. These three radiographs illustrate the common types of soil disturbance:

- *Voids.* The arrow in Fig. 5.11 points to a soil void. Such voids are often caused by the sampling and transporting process. The open voids can be caused by many different factors, such as gravel or shells which impact with the cutting end of the sampling tube and/or scrape along the inside of the sampling tube and create voids. The voids and highly disturbed clay shown in Fig. 5.11 are possibly due to cuttings inadvertently left at the bottom of the borehole. Some of the disturbance could also be caused by tube friction during sampling as the clay near the tube wall becomes remolded while it travels up the tube.

- *Soil cracks.* Figures 5.12 and 5.13 show numerous cracks in the clay. For example, the arrows labeled 1 point to some of the soil cracks in Figs. 5.12 and 5.13. The soil cracks probably developed during the sampling process. A contributing factor in the development of the soil cracks may have been gas coming out of solution which fractured the clay.

- *Turning of edges.* Turning or bending of edges of various thin layers shows as curved-down edges on the sides of the specimen. This effect is caused by the friction between the soil and sampler. There could be other distortion-type effects at the soil/sampling tube interface. For example, the arrow labeled 2 in Fig. 5.12 shows cracking and distortion of the soil at the tube interface.

- *Gas-related voids.* The circular voids (labeled 3) shown in Fig. 5.12 were caused by gas coming out of solution during the sampling process when the confining pressures were essentially reduced to zero.

In contrast to soil disturbance, the arrow labeled 4 in Fig. 5.13 indicates an undisturbed section of the soil sample. Note in Fig. 5.13 that the individual fine layering of the soil sample can even be observed.

5.4 SOIL CLASSIFICATION

The purpose of a soil classification system is to provide the geotechnical engineer with a way to predict the behavior of the soil for engineering projects. In the United States, the most widely used soil classification system is the Unified Soil Classification System (U.S. Army Engineer Waterways Experiment Station, 1960; Howard, 1977). The Unified Soil Classification System is abbreviated USCS (not to be confused with the United States Customary System of units, which has the same abbreviation). The Unified Soil

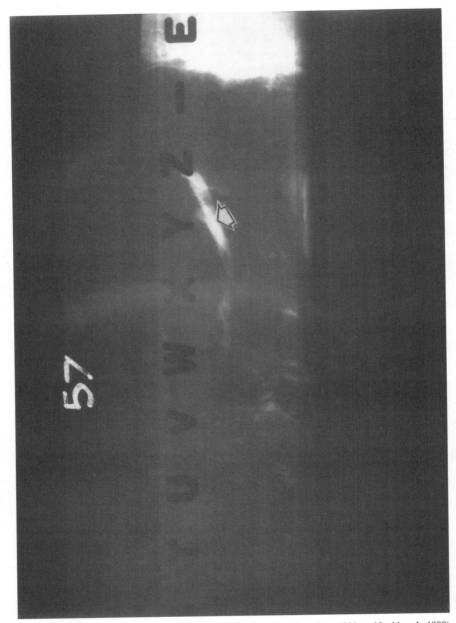

FIGURE 5.11 Radiograph of Orinoco clay within a Shelby tube. (From Day, 1980, and Ladd et al., 1980)

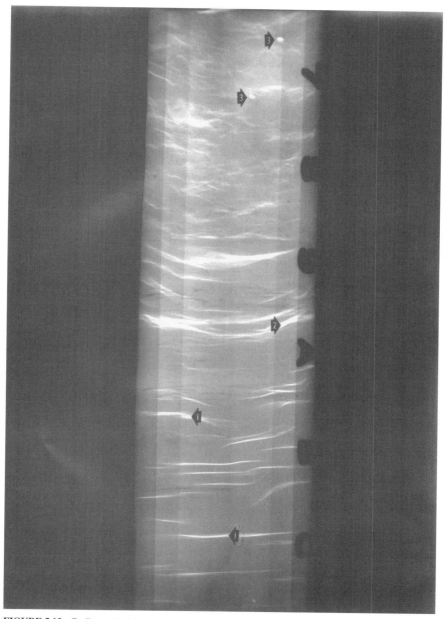

FIGURE 5.12 Radiograph of Orinoco clay within a Shelby tube. (From Day, 1980, and Ladd et al., 1980)

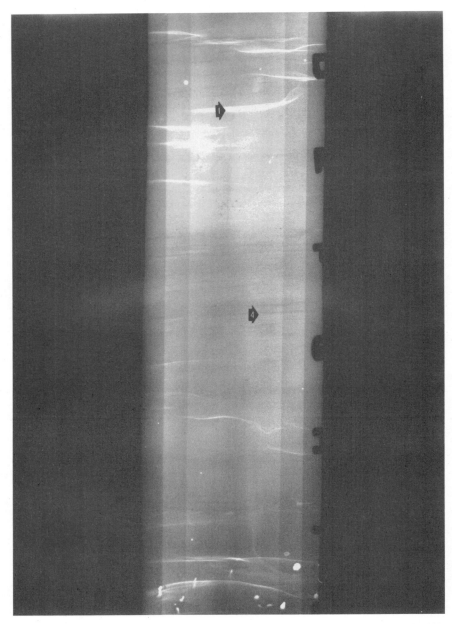

FIGURE 5.13 Radiograph of Orinoco clay within a Shelby tube. (From Day, 1980, and Ladd et al., 1980)

Classification System was initially developed by Casagrande (1948) and then later modified by Casagrande in 1952.

5.4.1 Particle Size and Description

A basic element of the Unified Soil Classification System is the determination of the amount and distribution of the particle sizes of the soil. The distribution of particle sizes larger than 0.075 mm (No. 200 sieve) is determined by sieving, while the distribution of particle sizes smaller than 0.075 mm is determined by a sedimentation process (hydrometer). For the Unified Soil Classification System, the rock fragments or soil particles versus size are defined as follows (from largest to smallest particle sizes):

- *Boulders.* Rocks that have an average diameter greater than 300 mm (12 in.).
- *Cobbles.* Rocks that are smaller than 300 mm (12 in.) and are retained on the 75-mm (3-in.) U.S. standard sieve.
- *Gravel-size particles.* Rock fragments or soil particles that will pass a 75-mm (3-in.) sieve and be retained on a No. 4 (4.75-mm) U.S. standard sieve. Gravel-size particles are subdivided into coarse gravel sizes or fine gravel sizes.
- *Sand-size particles.* Soil particles that will pass a No. 4 (4.75-mm) sieve and be retained on a No. 200 (0.075-mm) U.S. standard sieve. Sand size particles are subdivided into coarse sand size, medium sand size, or fine sand size.
- *Silt-size particles.* Fine soil particles that pass the No. 200 (0.075-mm) U.S. standard sieve and are larger than 0.002 mm.
- *Clay-size particles.* Fine soil particles that are smaller than 0.002 mm.

It is very important to distinguish between the size of a soil particle and the classification of the soil. For example, a soil could have a certain fraction of particles that are of clay size. The same soil could also be classified as a clay. But the classification of a clay does not necessarily mean that the majority of the soil particles are of clay size (smaller than 0.002 mm). In fact, it is not unusual for a soil to be classified as a clay and have more silt-size particles than clay-size particles. Throughout the book, when reference is given to particle size, the terminology *clay-size particles* or *silt-size particles* will be used. When reference is given to a particular soil, then the terms such as *silt* or *clay* will be used.

Sieve Analysis. A sieve is a piece of laboratory equipment that consists of a pan with a screen (square woven wire mesh) at the bottom. U.S. standard sieves are used to separate particles of a soil sample into various sizes. A sieve analysis is performed on dry soil particles that are larger than the No. 200 U.S. standard sieve (i.e., sand-size, gravel-size, and cobble-size particles).

The first step in a sieve analysis is to oven-dry the soil and obtain its initial mass. Then the soil is washed on the No. 200 sieve to remove all the fines (silt and clay-size particles). The portion retained on the No. 200 sieve is oven-dried. A series of sieves, arranged in descending size of openings, are stacked, and the dry soil is poured into the top as shown in Fig. 5.14. The sieves are shaken (usually with a mechanical shaker) and the amount of dry soil retained on each sieve is recorded. From this data, the percent finer is calculated as the dry mass of soil passing through the sieve divided by the initial dry mass of the soil specimen, expressed as a percentage.

Figure 5.15 shows the results of a grain size analysis (sieve and hydrometer tests). The sieve portion of the analysis is shown on the top of the graph. Note the two labels, "U.S. Sieve Openings in Inches" and "U.S. Sieve Numbers." The U.S. sieve openings in inches

5.26 GEOTECHNICAL FIELD AND LABORATORY STUDIES

FIGURE 5.14 Laboratory sieves (view at the top of a stack of sieves, where dry soil has been poured into the top of the sieves).

refer to sieves that are manufactured with specific size openings. For example, the 3-in. U.S. standard size sieve has square openings that are 3 in. (75 mm) wide. The U.S. sieve number refers to the number of opening per inch. For example, a No. 4 sieve has 4 openings per inch that are 0.19 in. (4.75 mm) wide. A common mistake is that the No. 4 sieve has openings 0.25 in. wide (i.e., 1.0 in. divided by 4), but because of the wire mesh, the openings are actually less than 0.25 inch. Sizes of commonly used U.S. standard sieve numbers are as follows:

U.S. standard sieve number	Sieve opening, mm
4	4.75
10	2.00
20	0.85
40	0.425

60	0.25
100	0.15
140	0.106
200	0.075

Hydrometer Analysis. The particle distribution for fines (silt and clay size particles finer than the No. 200 sieve) are determined by a sedimentation process. A hydrometer (see Fig. 5.16) is used to obtain the necessary data during the sedimentation process (ASTM D 422-90, 1998). The test procedure consists of mixing a known mass (usually about 50 g) of soil with a dispersing agent (sodium hexametaphosphate). The dispersing agent prevents the clay-size particles from forming flocs during the hydrometer test. The soil specimen, dispersing agent,

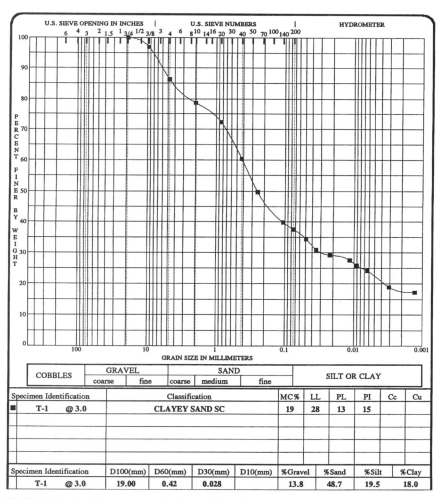

FIGURE 5.15 Grain size curve, particle size analysis, and soil classification.

and distilled water are thoroughly mixed and then transferred to a 1000-mL glass cylinder. Hydrometer readings versus time from the beginning of sedimentation are then recorded. The hydrometer is calibrated so that it records the actual mass of soil particles and dispersing agent that is in suspension.

The hydrometer test is based on Stokes law, which relates the diameter of a single sphere to the time required for the sphere to fall a certain distance in a liquid of known viscosity. The idea for the hydrometer analysis is that a larger, and hence heavier, soil particle will fall faster through distilled water than a smaller, and hence lighter, soil particle. The test procedure is approximate because many fine soil particles are not spheres, but are rather of a plate-like shape.

The top of Fig. 5.15 (labeled "Hydrometer") indicates that part of the grain size curve that was determined from the hydrometer analysis. If the sieve and the hydrometer tests are performed correctly, the portion of the grain size curve from the sieve analysis should flow smoothly into the portion of the curve from the hydrometer analysis, such as shown in Fig. 5.15. A large and abrupt jump in the grain size curve from the sieve to the hydrometer test indicates errors in the laboratory testing procedure.

Particle Size Analysis. The grain size curve (also known as the particle size distribution) shown in Fig. 5.15 was actually obtained from a computer program (*gINT*, 1991). The raw data from the sieve and hydrometer tests were the input to the computer program and the grain size curve and analysis of data was the output. Note in Fig. 5.15 that the particle sizes for cobbles, gravel, sand, and silt or clay are listed on the plot for easy reference.

At the bottom of Fig. 5.15, the computer program (*gINT*, 1991) has performed an analysis of the sieve and hydrometer tests, and based on dry weight, indicates the percent gravel-size particles (13.8 percent), sand-size particles (48.7 percent), silt-size particles (19.5 percent), and clay-size particles (18.0 percent). The computer program also determines the D_{100}, D_{60}, D_{30}, and D_{10} particle sizes. In Fig. 5.15, the D_{100} is the largest particle size recorded (19.00 mm), the D_{60} is the particle size corresponding to 60 percent finer by dry weight (0.42 mm), D_{30} is the particle size corresponding to 30 percent finer by dry weight (0.028 mm), and D_{10} is the particle size corresponding to 10 percent finer by dry weight. Because of the presence of over 10 percent clay size particles for the soil data shown in Fig. 5.15, D_{10} could not be obtained for this soil.

Using the particle size dimension data, the coefficient of uniformity C_u and coefficient of curvature C_c can be calculated as follows:

$$C_u = \frac{D_{60}}{D_{10}} \tag{5.10}$$

$$C_c = \frac{(D_{30})^2}{(D_{10})(D_{60})} \tag{5.11}$$

As will be discussed in Sec. 5.4.3, these two parameters are used in the Unified Soil Classification System to determine whether a soil is well graded (many different particle sizes) or poorly graded (many particles of about the same size).

Other Grain Size Scales. As indicated above, the Unified Soil Classification System has specific size dimensions for boulders, cobbles, gravel, etc. There are many other classification systems that use different particle size dimensions and terminology. For example, the Modified Wentworth scale is frequently used by geologists. It includes such terms as *pebbles* and *mud* in the classification system and uses different particle size dimensions to define sand, silt, and clay (Lane et al., 1947).

FIGURE 5.16 Laboratory hydrometer equipment. The 1000-mL cylinder on the left contains the hydrometer suspended in water; the 1000-mL cylinder on the right contains a dispersed soil specimen to be tested.

Soil scientists also use their own classification system, which subdivides sand-size particles into categories of very coarse, coarse, medium, fine, and very fine. Soil scientists also use different particle size dimension definitions (U.S. Department of Agriculture, 1975). The geotechnical engineer should be aware that terms or soil descriptions used by geologists or soil scientists may not match the engineer's classification because they are using a different grain size scale.

5.4.2 Clay Mineralogy

Clay Mineral Structure. The amount and type of clay minerals present in a soil have a significant effect on soil engineering properties such as plasticity, swelling, shrinkage, shear strength, consolidation, and permeability. This is due in large part to their very small flat or plate-like shape which enables them to attract water to their surfaces, also known as the *double-layer effect*. The double layer is a grossly simplified interpretation of the positively charged water layer and the negatively charged surface of the clay particle itself. Two reasons for the attraction of water to the clay particle (double layer) are:

1. The dipolar structure of the water molecule which causes it to be electrostatically attracted to the surface of the clay particle.
2. The clay particles attract cations, which contribute to the attraction of water by the hydration process. Ion exchange can occur in the double layer, where under certain conditions sodium, potassium, and calcium cations can be replaced by other cations present in the water. This property is known as *cation exchange capacity*.

In addition to the double layer, there is an absorbed water layer that consists of water molecules tightly held to the clay particle face by, for example, hydrogen bonding. The presence of the very small clay particles surrounded by water helps explain their impact on the engineering properties of soil. For example, clays that have been deposited in lakes or marine environments often have a very high water content and are very compressible with a low shear strength because of this attracted and bonded water. Another example is desiccated clays, which have been dried, but have a strong desire for water and will swell significantly upon wetting (Chap. 9).

Plasticity. The term *plasticity* is applied to silts and clays and indicates an ability to be rolled and molded without breaking apart. The Atterberg limits are defined as the water content corresponding to different behavior conditions of silts and clays. Although originally six limits were defined by Albert Atterberg (1911), in geotechnical engineering, the term Atterberg limits refers only to the liquid limit (LL), plastic limit (PL), and shrinkage limit (SL), defined as follows:

- *Liquid limit (LL)*. The water content corresponding to the behavior change between the liquid and plastic state of a silt or clay. The liquid limit is arbitrarily defined as the water content at which a pat of soil, cut by a groove of standard dimensions, will flow together for a distance of 12.7 mm (0.5 in.) under the impact of 25 blows in a standard liquid-limit device (ASTM D 4318-95, 1998). Figure 5.17 shows the liquid-limit device (containing soil) and the grooving tool.
- *Plastic limit (PL)*. The water content corresponding to the behavior change between the plastic and semisolid state of a silt or clay. The plastic limit is arbitrarily defined as the water content at which a silt or clay will just begin to crumble when rolled into a tread approximately 3.2 mm ($^1/_8$ in.) in diameter (ASTM D 4318-95, 1998).
- *Shrinkage limit (SL)*. The water content corresponding to the behavior change between the semisolid to solid state of a silt or clay. The shrinkage limit is also defined as the water content at which any further reduction in water content will not result in a decrease in volume of the soil mass (ASTM D 427-93 or D 4943-95, 1998). The shrinkage limit is rarely obtained in practice because of laboratory testing difficulties and limited use of the data.

In accordance with ASTM, the liquid limit, plastic limit, and shrinkage limit are performed on that portion of the soil that passes the No. 40 sieve (0.425 mm). For many soils, a significant part of the soil specimen (i.e., those soil particles larger than the No. 40 sieve) will be excluded during testing. This can cause problems when classifying the soil and will be further discussed in Sec. 5.4.4.

FIGURE 5.17 Liquid-limit device containing tested soil (the grooving tool is on the right side of the photograph).

A measure of a soils plasticity is the plasticity index (PI), defined as:

$$PI = LL - PL \tag{5.12}$$

where LL = liquid limit and PL = plastic limit. The PI is often expressed as a whole number. The plasticity index is important because it has been correlated with numerous soil engineering properties (Sec. 5.4.4). Note in Fig. 5.15 that the liquid and plastic limits were performed for this soil, and the data (LL, PL, and PI) has been included on the figure.

Another useful parameter is the liquidity index (LI), defined as:

$$LI = \frac{w - PL}{PI} \tag{5.13}$$

The liquidity index can be used to identify sensitive clays. For example, quick clays often have a water content w that is greater than the liquid limit, and thus the liquidity index is greater than 1.0. At the other extreme are clays that have liquidity index values that are zero or even negative. These liquidity index values indicate a soil that is desiccated and could have significant expansion potential (Chap. 9). According to ASTM, the Atterberg limits are performed on soil that is finer than the No. 40 sieve, but the water content can be performed on soil containing larger soil particles (Table 5.2) and thus the liquidity index should be calculated only for soil that has all its particles finer than the No. 40 sieve.

Plasticity Chart and Activity

Plasticity Chart. Using the Atterberg limits, Casagrande (1932, 1948) developed the plasticity chart. It is used in the Unified Soil Classification System to classify soils. As shown in Fig. 5.18, the plasticity chart is a plot of the liquid limit (LL) versus the plasticity index (PI). Also shown on Fig. 5.18 are the locations where common clay minerals plot on the plasticity chart (from Mitchell, 1976, and Holtz and Kovacs, 1981). Casagrande (1932) defined two basic dividing lines on the plasticity chart:

LL = 50 line: This line is used to divide silts and clays into high plasticity (LL > 50) and low to medium plasticity (LL < 50) categories.

A-line: The A-line is defined as:

$$PI = 0.73 (LL - 20) \tag{5.14}$$

where PI = plasticity index and LL = liquid limit. The A-line is used to separate clays, which plot above the A-line, from silts, which plot below the A-line.

An additional line has been added to the Casagrande plasticity chart, known as the U-line (see Fig. 5.18). The U-line (or upper-limit line) is defined as:

$$PI = 0.90 (LL - 8) \tag{5.15}$$

The U-line is valuable because it represents the uppermost boundary of test data found thus far for natural soils. The U-line is a good check on erroneous data, and any test results that plot above the U-line should be rechecked.

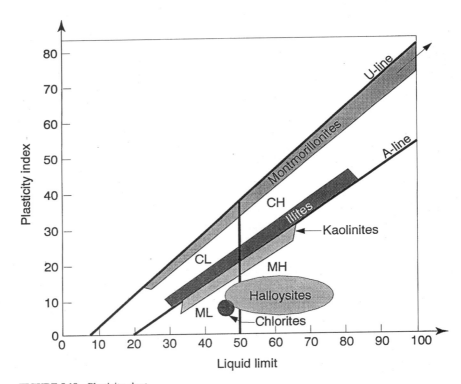

FIGURE 5.18 Plasticity chart.

There have been other minor changes proposed for Casagrande's original plasticity chart. For example, at very low PI values, the A-line and U-line are defined differently than shown in Fig. 5.18 (see ASTM D 2487-93, 1998).

Activity. In 1953, Skempton defined the activity A of a clay as:

$$A = \frac{\text{PI}}{\text{clay fraction}} \tag{5.16}$$

where the clay fraction = that part of the soil specimen finer than 0.002 mm, based on dry weight. Clays that are *inactive* are defined as those clays that have an activity less than 0.75, *normal activity* is defined as those clays having an activity between 0.75 and 1.25, and an *active* clay is defined as those clays having an activity greater than 1.25. Quartz has an activity of zero, while at the other extreme is sodium montmorillonite, which can have an activity from 4 to 7.

Because the PI is determined from Atterberg limits that are performed on soil that passes the No. 40 sieve (0.425 mm), a correction is required for soils that contain a large fraction of particles coarser than the No. 40 sieve. For example, suppose a clayey gravel contains 70 percent gravel particles (particles coarser than No. 40 sieve), 20 percent silt-size particles, and 10 percent clay-size particles (silt- and clay-size particles are finer than No. 40 sieve). If the PI = 40 for the soil particles finer than the No. 40 sieve, then the activity for the clayey gravel would be 1.2 (i.e., 40/33.3).

Types of Clay Particles. Clay minerals present in a soil can be identified by their x-ray diffraction patterns. This process is rather complicated, expensive, and involves special equipment that is not readily available to the geotechnical engineer. A more common approach is to use the location of clay particles as they plot on the plasticity chart (Fig. 5.18) to estimate the type of clay mineral in the soil. This approach is often inaccurate, because soil can contain more than one type of clay mineral.

The three most common clay minerals are listed below, with their respective activity (A) values (from Skempton, 1953, and Mitchell, 1976):

1. *Kaolinite* (A = 0.3 to 0.5). The kaolin minerals are a group of clay minerals consisting of hydrous aluminum silicates. A common kaolin mineral is kaolinite, having the general formula $Al_2Si_2O_5(OH)_4$. Kaolinite is usually formed by alteration of feldspars and other aluminum-bearing minerals. Kaolinite is usually a large clay mineral of low activity and often plots below the A-line (Fig. 5.18). Holtz and Kovacs (1981) state that kaolinite is a relatively inactive clay mineral, and, even though it is technically a clay, it behaves more like a silt material. Kaolinite has many industrial uses including the production of china, medicines, and cosmetics.

2. *Montmorillonite* (Na-montmorillonite, A = 4 to 7 and Ca-montmorillonite, A = 1.5). A group of clay minerals that are characterized by weakly bonded layers. Each layer consists of two silica sheets with an aluminum (gibbsite) sheet in the middle. Water and exchangeable cations (e.g., Na or Ca) can enter and separate the layers, creating a very small crystal that has a strong attraction for water. Montmorillonite has the highest activity and it can have the highest water content, greatest compressibility, and lowest shear strength of all the clay minerals. As shown in Fig. 5.18, montmorillonite plots just below the U-line.

Montmorillonite often forms as the result of the weathering of ferromagnesian minerals, calcic feldspars, and volcanic materials (Coduto, 1994). For example, sodium montmorillonite is often formed from the weathering of volcanic ash. Other environments that are likely to form montmorillonite are alkaline conditions with a supply of magnesium ions and a lack of leaching (Coduto, 1994). Such conditions are often present in semiarid regions.

3. *Illite* (A = 0.5 to 1.3). This clay mineral has a structure similar to montmorillonite, but the layers are more strongly bonded together. In terms of cation exchange capacity, in ability to absorb and retain water, and in physical characteristics such as plasticity index, illite is intermediate in activity between clays of the kaolin and montmorillonite groups. As shown in Fig. 5.18, illite often plots just above the A-line.

There are many other types of clay minerals. Even within a clay mineral category, there can be different crystal components because of isomorphous substitution. This is the process where ions of approximately the same size are substituted in the crystalline framework. Also shown on Fig. 5.18 are two other less common clay minerals, chlorite and halloysite, both of which have less activity than the three clay minerals previously described. Although not very common, halloysite is an interesting clay mineral because instead of the usual flat particle shape, it has a tubular shape that can affect engineering properties in unusual ways. It has been observed that classification and compaction tests made on air-dried halloysite samples give markedly different results than tests on samples at their natural water content (Holtz and Kovacs, 1981).

5.4.3 Unified Soil Classification System

The Unified Soil Classification System (USCS) separates soils into two main groups: coarse-grained soils and fine-grained soils. The basis of the USCS is that the engineering behavior of coarse-grained soils depends on their grain size distributions and the engineering behavior of fine-grained soil depends on their plasticity characteristics. Table 5.4 (adapted from ASTM D 2487-93, 1998) presents a summary of the Unified Soil Classification System (USCS). As indicated in Table 5.4, the two main groups of soil are defined as follows:

- *Coarse-grained soils.* Defined as having more than 50 percent (by dry mass) of soil particles retained on the No. 200 sieve.
- *Fine-grained soils.* Defined as having 50 percent or more (by dry mass) of soil particles passing the No. 200 sieve.

As indicated in Table 5.4, the coarse-grained soils are divided into gravels and sands. Both gravels and sands are further subdivided into four secondary groups as indicated in Table 5.4. The four secondary classifications are based on whether the soil is well graded, is poorly graded, contains silt-size particles, or contains clay-size particles.

As indicated in Table 5.4, the fine-grained soils are divided into soils of low or high plasticity. The three secondary classifications are based on liquid limit (LL) and plasticity characteristics (PI).

Note in column 3 of Table 5.4 that symbols (known as *group symbols*) are used to identify different soil types. The group symbols consist of two capital letters. The first letter indicates the following:

G   Gravel
S   Sand
M   Silt
C   Clay
O   Organic

The second letter indicates the following:

W   Well graded, which indicates that a coarse-grained soil has particles of all sizes.
P   Poorly graded, which indicates a coarse-grained soil has particles of the same size, or the soil is skip-graded or gap-graded.
M   A coarse-grained soil that has silt-size particles.
C   A coarse-grained soil that has clay-size particles.
L   A fine-grained soil of low plasticity.
H   A fine-grained soil of high plasticity.

An exception is peat, where the group symbol is PT. Also note in Table 5.4 that certain soils require the use of dual symbols.

In addition to the classification of a soil, other items should also be included in the field or laboratory description of a soil, such as:

1. *Soil color.* Usually the standard primary color (red, orange, yellow, etc.) of the soil is listed. Although not frequently used in geotechnical engineering, color charts have been developed. For example, the *Munsell Soil Color Charts* (1975) display 199 different standard color chips systematically arranged according to their Munsell notations, on cards carried in a loose-leaf notebook. The arrangement is by the three variables that combine to describe all colors and are known in the Munsell system as hue, value, and chroma.

Color can be very important in identifying different types of soil. For example, the Friars formation, which is a stiff-fissured clay and is a frequent cause of geotechnical problems such as landslides and expansive soil, can often be identified by its dark green color. Another example is the Sweetwater formation, which is also a stiff-fissured clay, and has a bright pink color due to the presence of montmorillonite.

2. *Soil structure.* In some cases the structure of the soil may be evident. Definitions vary, but in general, the soil structure refers to both the geometric arrangement of the soil particles and the interparticle forces which may act between them (Holtz and Kovacs, 1981). There are many different types of soil structure, such as cluster, dispersed, flocculated, honeycomb, single-grained, and skeleton (see App. A, Glossary 2, for definitions). In some cases, the soil structure may be visible under a magnifying glass, or in other cases the soil structure may be reasonably inferred from laboratory testing results.

3. *Soil texture.* The texture of a soil refers to the degree of fineness of the soil. For example, terms such as smooth, gritty, or sharp can be used to describe the texture of the soil when it is rubbed between the fingers.

4. *Soil porosity.* The soil classification should also include the *in situ* condition of the soil. For example, numerous small voids may be observed in the soil, and this is referred to as *pin-hole porosity*.

5. *Clay consistency.* For clays, the consistency (i.e., degree of firmness) should be listed. As indicated in Section 5.2.3, the consistency of a clay varies from "very soft" to "hard" according to the undrained shear strength of the clay. The undrained shear strength can be determined from the unconfined compression test or from field or laboratory vane tests.

If the shear strength of the soil has not been determined, then the consistency of the clay can be estimated in the field or laboratory from the following:

- *Very soft.* The clay is easily penetrated several centimeters by the thumb. The clay oozes out between the fingers when squeezed in the hand.

TABLE 5.4 Unified Soil Classification System (USCS)

Major divisions (1)	Subdivisions (2)	USCS symbol (3)	Typical names (4)	Laboratory classification criteria (5)	
Coarse-grained soils (More than 50% retained on No. 200 sieve)	Gravels (More than 50% of coarse fraction retained on No. 4 sieve)	GW	Well-graded gravels or gravel-sand mixtures, little or no fines	Less than 5% fines*	$C_u \geq 4$ and $1 \leq C_c \leq 3$
		GP	Poorly graded gravels or gravelly sands, little or no fines	Less than 5% fines*	Does not meet C_u and/or C_c criteria listed above
		GM	Silty gravels, gravel-sand-silt mixtures	More than 12% fines*	Minus No. 40 soil plots below the A-line
		GC	Clayey gravels, gravel-sand-clay mixtures	More than 12% fines*	Minus No. 40 soil plots on or above the A-line
	Sands (50% or more of coarse fraction passes No. 4 sieve)	SW	Well-graded sands or gravelly sands, little or no fines	Less than 5% fines*	$C_u \geq 6$ and $1 \leq C_c \leq 3$
		SP	Poorly graded sands or gravelly sands, little or no fines	Less than 5% fines*	Does not meet C_u and/or C_c criteria listed above
		SM	Silty sands, sand-silt mixtures	More than 12% fines*	Minus No. 40 soil plots below the A-line
		SC	Clayey sands, sand-clay mixtures	More than 12% fines*	Minus No. 40 soil plots on or above the A-line

Fine-grained soils (50% or more passes the No. 200 sieve)	Silts and clays (liquid limit less than 50)	ML	Inorganic silts, rock flour, silts of low plasticity	Inorganic soil	PI < 4 or plots below A-line
		CL	Inorganic clays of low plasticity, gravelly clays, sandy clays, etc.	Inorganic soil	PI > 7 and plots on or above A-line†
		OL	Organic silts and organic clays of low plasticity	Organic soil	LL (oven dried)/LL (not dried) < 0.75
	Silts and clays (liquid limit 50 or more)	MH	Inorganic silts, micaceous silts, silts of high plasticity	Inorganic soil	Plots below A-line
		CH	Inorganic highly plastic clays, fat clays, silty clays, etc.	Inorganic soil	Plots on or above A-line
		OH	Organic silts and organic clays of high plasticity	Organic soil	LL (oven dried)/LL (not dried) < 0.75
Peat	Highly organic	PT	Peat and other highly organic soils	Primarily organic matter, dark in color, and organic odor	

*Fines are those soil particles that pass the No. 200 sieve. For gravels and sands with between 5 and 12% fines, use of dual symbols is required (i.e., GW-GM, GW-GC, GP-GM, or GP-GC).

†If 4 ≤ PI ≤ 7 and PI plots above A-line, then dual symbols (i.e., CL-ML) are required.

- *Soft.* The clay is easily penetrated 2 to 3 cm (1 in.) by the thumb. The clay can be molded by slight finger pressure.
- *Medium.* The clay can be penetrated about 1 cm (0.4 in.) by the thumb with moderate effort. The clay can be molded by strong finger pressure.
- *Stiff.* The clay can be indented about 0.5 cm (0.2 in.) by the thumb with great effort.
- *Very stiff.* The clay can not be indented by the thumb, but can be readily indented with the thumbnail.
- *Hard.* With great difficulty, the clay can only be indented with the thumbnail.

The estimates of the consistency of clay as listed above should be used only for classification purposes and never as the basis for engineering design parameters.

6. *Sand density condition.* For sands, the density state of the soil varies from "very loose" to "very dense" (see Table 4.4). The determination of the density condition of sands will be further discussed in Sec. 6.2.

7. *Soil moisture condition.* The moisture condition of the soil should also be listed. Moisture conditions vary from a "dry" soil to a "saturated" soil. The moisture condition of a soil (i.e., the degree of saturation) will be further discussed in Sec. 6.2.

Especially in the arid climate of the southwestern United States, soil may be in a dry and powdery state. Often these soils are misclassified as silts, when in fact they are highly plastic clays. It is always important to add water to dry or powdery soils in order to assess their plasticity characteristics.

8. *Additional descriptive items.* The Unified Soil Classification System is applicable only for soil and rock particles passing the 75-mm (3-in.) sieve. Cobbles and boulders are larger than 75 mm (3 in.), and if applicable, the words *with cobbles* or *with boulders* should be added to the soil classification. Other descriptive terminology includes the presence of rock fragments, such as "crushed shale, claystone, sandstone, siltstone, or mudstone fragments," and unusual constituents such as "shells, slag, glass fragments, and construction debris."

An example of a complete soil classification and description for a soil (from Fig. 5.15), is as follows:

> Clayey Sand (SC). Based on dry mass, the soil contains 13.8% gravel size particles, 48.7% sand size particles, 19.5% silt size particles, and 18.0% clay size particles. The gravel size particles are predominately hard and angular rock fragments. The sand size particles are predominately composed of angular quartz grains. Atterberg limits performed on the soil passing the No. 40 sieve indicate a LL = 28 and a PI = 15. The *in situ* soil has a reddish-brown color, soft consistency, gritty texture, and is wet. The *in situ* soil is a residual soil and is part of an old vernal pool.

5.4.4 Classification Based on Plasticity

It is often believed that there is a distinct division in soil behavior between a coarse-grained soil and a fine-grained soil, where, as previously mentioned, the engineering behavior of a coarse-grained soil is based on grain size distribution while the engineering behavior of a fine-grained soil is based on plasticity characteristics. In most cases this will be true, such as the difference in engineering behavior between a clean sand and gravel versus a highly plastic clay. But there are also many situations where a coarse-grained soil actually has an engineering behavior identical to fine-grained soils, and vice versa.

For example, the grain size distribution shown in Fig. 5.15 is for a soil classified as a clayey sand (coarse-grained soil). This soil was obtained during the construction of a building foundation in San Diego, California. The soil was observed to have a soft consistency. Figures 5.19 to 5.22 show four views of the site, where construction equipment attempted to compact this soil. It was observed that the soil was simply too wet and plastic to be effectively compacted. In essence, although this soil was classified as a coarse-grained soil according to the Unified Soil Classification System, it actually behaved as a fine-grained soil with the engineering behavior governed by the plasticity characteristics. The *in situ* soil was a residual soil and was part of an old vernal pool. Figure 5.23 shows a photograph of a nearby existing vernal pool, which has standing water from the El Niño winter rains of 1997–1998. Such vernal pools develop on flat-topped mesas and will usually dry out during the summer season.

An example of an opposite situation is a fine quartz silt, also known as rock flour, created by the grinding actions of glaciers. According to the Unified Soil Classification System, this soil is a fine-grained soil, but it is not plastic and its engineering behavior is closer to a coarse-grained soil than a fine-grained soil.

Table 5.5 presents an alternate classification system known as the inorganic soil classification based on plasticity (ISBP) [Day (1994a)]. According to the ISBP, a nonplastic soil is defined as a soil where the minus No. 40 fraction can not be rolled at any water content, or the plastic limit is equal to or greater than the liquid limit. The primary ISBP divisions are plastic versus nonplastic soils. For nonplastic soils, the subdivisions are based on grain size distributions, while for plastic soils, the subdivisions are based on plasticity characteristics (LL and PI).

For the soil shown in Figs. 5.19 to 5.22, which has the particle size distribution shown in Fig. 5.15, the major difference is that, according to the USCS the soil is classified as a coarse-grained soil (engineering behavior based on grain size distribution), but according to the ISBP, the soil is classified as a soil of low plasticity (engineering behavior correctly based on its plasticity).

Another advantage of the ISBP classification system is that the soil, in terms of shear strength, compressibility, and expansion potential, is generally arranged from the best to

FIGURE 5.19 Construction site, view to the north.

FIGURE 5.20 Construction site, view to the south.

FIGURE 5.21 Construction site, view to the east.

FIGURE 5.22 Construction site, close-up view of plastic nature of soil.

FIGURE 5.23 Vernal pool.

TABLE 5.5 Inorganic Soil Classification Based on Plasticity (ISBP)

Major divisions (1)	Subdivisions (2)	ISBP symbol (3)	Typical names (4)	Laboratory classification criteria (5)
Nonplastic soils	Gravels (Greater fraction of total sample is retained on No. 4 sieve)	GW	Well-graded gravels, sandy gravels, silty-sandy gravels	$C_u \geq 4$ and $1 \leq C_c \leq 3$
		GP	Poorly graded gravels, gravel-sand-silt mixtures	Does not meet C_u and/or C_c criteria listed above. In addition, the % passing No. 200 sieve $< 15\%$
		GM	Poorly graded, nonplastic silty gravels, gravel-silt mixtures	Does not meet C_u and/or C_c criteria listed above. In addition, the % passing No. 200 sieve $\geq 15\%$
	Sands (Greater fraction of total sample is between No. 4 and No. 200 sieves)	SW	Well-graded sands and gravelly sands	$C_u \geq 6$ and $1 \leq C_c \leq 3$
		SP	Poorly graded sands or sand-gravel-silt mixtures	Does not meet C_u and/or C_c criteria listed above. In addition, the % passing No. 200 sieve $< 15\%$
		SM	Poorly graded, nonplastic silty sands, sand-silt mixtures	Does not meet C_u and/or C_c criteria listed above. In addition, the % passing No. 200 sieve $\geq 15\%$
	NP silt	MN	Nonplastic silts, rock flour. Gravelly silts and sandy nonplastic silts	Greater fraction of the total sample passes the No. 200 sieve. Silts are nonplastic
Plastic soils	(Minus No. 40 fraction plots below A-line)	GM*	Plastic silty gravels, gravel-silt mixtures	50% or more particles retained on the No. 200 sieve with the greater fraction of gravel size
	Plastic silts	SM*	Plastic silty sands, sand-silt mixtures	50% or more particles retained on the No. 200 sieve with the greater fraction of sand size
		ML MI MH	Plastic silts, sandy silts, and clayey silts	For silt of low plasticity (ML) PI ≤ 10 For silt of intermediate plasticity (MI) $10 < PI \leq 30$ For silt of high plasticity (MH) PI > 30
	Clays (Minus No. 40 fraction plots on or above A-line)	GC*	Clayey gravels, gravel-clay mixtures	50% or more particles retained on the No. 200 sieve with the greater fraction of gravel size
		SC*	Clayey sands, sand-clay mixtures	50% or more particles retained on the No. 200 sieve with the greater fraction of sand size
		CL CI CH	Clay, sandy clays, and silty clays	For clay of low plasticity (CL) PI ≤ 10 For clay of intermediate plasticity (CI) $10 < PI \leq 30$ For clay of high plasticity (CH) PI > 30

*Must state whether plasticity is high, intermediate, or low, according to PI calculated by using Eq. (5.17).

worst inorganic soil type. For example, a well-graded gravel (GW) is often considered the best soil type and is often used for roadway base material, while a clay of high plasticity (CH) is often the worst type of soil in terms of having the lowest shear strength and highest compressibility. The ISBP classification system also does not have dual symbols. A disadvantage of the ISBP classification system is that it does not include organic soils, but because of their unique engineering properties, they should probably be separately classified.

As mentioned in Sec. 5.4.2, the Atterberg limits are performed on soil that passes the No. 40 sieve. Coarse clayey sands (SC) and clayey gravel (GC) may contain a significant amount of particles retained on the No. 40 sieve. In these cases, the plasticity index (PI) used for classification in Table 5.5 should be based on the following equation:

$$\text{PI (for Table 5.5)} = (\text{PI}) \text{ (fraction of soil passing No. 40 sieve)} \qquad (5.17)$$

For plastic soils, the ISBP subdivisions are based on the PI value from Eq. (5.17). The reason for using the plasticity index (PI) as the subdivision for plastic soils is because to date the plasticity index seems to be a better indicator of soil behavior than the liquid limit. Table 5.6 presents some of the correlations between the plasticity index (PI) and certain engineering properties. For example, the potential for swelling or shrinkage of plastic soil is routinely based on the plasticity index, not the liquid limit.

5.5 ROCK CLASSIFICATION

The purpose of this last section of Chap. 5 is to provide a brief introduction to rock classification. There are three basic types of rocks: igneous, sedimentary, and metamorphic. Because of special education and training, usually the best person to classify rock is the engineering geologist. Table 5.7 presents a simplified rock classification and common rock types.

In addition to determining the type of rock, it is often important to determine the quality of the rock. One measure of the quality of rock is its hardness, which has been correlated with the unconfined compressive strength of rock specimens. Table 5.8 lists hardness of rock as a function of the unconfined compressive strength. Because the unconfined compressive strength is performed on small rock specimens, in most cases, it will not represent the actual condition of *in situ* rock. The reason is the presence of joints, fractures, fissures, and planes of weakness in the actual rock mass, which govern its engineering properties,

TABLE 5.6 Correlations between Soil Behavior and Plasticity Index

Correlation (1)	References (2)
Coefficient of earth pressure at rest versus PI	Massarsch, 1979
Effective friction angle for normally consolidated clay versus PI	Ladd et al., 1977; NAVFAC DM-7, 1971
Normalized undrained modulus versus PI	Numerous authors
Expansion potential versus PI	Holtz, 1959
Normalized undrained shear strength versus PI	Numerous authors
Precompression of glacial clays attributed to aging versus PI	Bjerrum, 1972
Anisotropic undrained strength ratio versus PI	Ladd et al., 1977

TABLE 5.7 Simplified Rock Classification

Common igneous rocks		
Major division (1)	Secondary divisions (2)	Rock types (3)
Extrusive	Volcanic explosion debris (fragmental)	Tuff (lithified ash) and volcanic breccia
	Lava flows and hot siliceous clouds	Obsidian (glass), pumice, and scoria
	Lava flows (fine-grained texture)	Basalt, andesite, and rhyolite
Intrusive	Dark minerals dominant	Gabbro
	Intermediate (25–50% dark minerals)	Diorite
	Light color (quartz and feldspar)	Granite

Common sedimentary rocks		
Major division (1)	Texture (grain size) or chemical composition (2)	Rock types (3)
Clastic rocks*	Grain sizes larger than 2 mm (pebbles, gravel, cobbles, and boulders)	Conglomerate (rounded cobbles) or breccia (angular rock fragments)
	Sand-size grains, 0.062 to 2 mm	Sandstone
	Silt-size grains, 0.004 to 0.062 mm	Siltstone
	Clay-size grains, less than 0.004 mm	Claystone and shale
Chemical and organic rocks	Carbonate minerals (e.g., calcite)	Limestone
	Halite minerals	Rock salt
	Sulfate minerals	Gypsum
	Iron-rich minerals	Hematite
	Siliceous minerals	Chert
	Organic products	Coal

Common metamorphic rocks		
Major division (1)	Structure (foliated or massive) (2)	Rock types (3)
Coarse crystalline	Foliated	Gneiss
	Massive	Metaquartzite
Medium crystalline	Foliated	Schist
	Massive	Marble, quartzite, serpentine, soapstone
Fine to microscopic	Foliated	Phyllite, slate
	Massive	Hornfels, anthracite coal

*Grain sizes correspond to the Modified Wentworth scale.

such as deformation characteristics, shear strength, and permeability. The unconfined compressive test also does not consider other rock quality factors, such as its resistance to weathering or behavior when submerged in water. For example, the author has observed complete disintegration (in only a few seconds) of the Friars formation claystone, which contains montmorillonite, when it is submerged in water.

Another measure of the quality of the rock is the RQD (rock quality designation), which is computed by summing the lengths of all pieces of the core (NX size) equal to or longer than 10 cm (4 in.) and dividing by the total length of the core run. The RQD is multiplied by 100 and expressed as a percentage. The mass rock quality can be defined as follows:

RQD, %	Rock quality
0–25	Very poor
25–50	Poor
50–75	Fair
75–90	Good
90–100	Excellent

In calculating the RQD, only the natural fractures should be counted, and any fresh fractures due to the sampling process should be ignored. RQD measurements can provide valuable data on the quality of the *in situ* rock mass, and can be used to locate zones of extensively fractured or weathered rock.

There are many excellent publications on the identification, sampling, and laboratory testing of rock. For example, the *Engineering Geology Field Manual* (U.S. Department of the Interior, 1987), provides guidelines on the performance of field work leading to the development of geologic concepts and reports. This manual also provides a discussion of geologic mapping, discontinuity rock surveys, sampling and testing methods of rock, and provides instructions for core logging.

PROBLEMS

The problems have been divided into basic categories as indicated below:

Laboratory Testing

1. An unconfined compression test is performed on an undisturbed specimen of clay. The clay specimen has an initial diameter of 2.50 in. and an initial height of 6.00 in. The vertical force at the shear failure is 24.8 lb and the dial gauge records 0.80 in. of axial deformation from the initial condition to the shear failure condition. Calculate the undrained shear strength s_u of the clay and indicate its consistency. *Answer:* $s_u = 315$ psf, and the clay is considered to have a soft consistency.

2. For the clay in Prob. 1, an unconfined compression test was performed on a remolded specimen and the undrained shear strength s_u was found to be 45 psf. Calculate the sensitivity value and classification of the clay. *Answer:* $S_t = 7$; therefore a clay of medium sensitivity.

3. In a constant-head permeameter test, the outflow $Q = 782$ mL in a measured time of 31 seconds. The sand specimen has a diameter of 6.35 cm and a length L of 2.54 cm.

TABLE 5.8 Hardness of Rock versus Unconfined Compressive Strength

Hardness (1)	q_u (2)	Rock description (3)
Very soft	10 to 250	Material disintegrates upon single blow of a geologic hammer
Soft	250 to 500	Rock can be scraped or peeled with a knife
Hard	500 to 1000	Rock can not be scraped or peeled and breaks with one hammer blow
Very hard	1000 to 2000	Can break rock with several solid blows of a geologic hammer
Extremely hard	>2000	Numerous solid blows from a geologic hammer to break rock

Notes: q_u = unconfined compressive strength of the rock, in kg/cm^2; 1 kg/cm^2 is approximately equal to 1 tsf. Data obtained from *Basic Soils Engineering* (Hough, 1969).

The total head loss Δh for the permeameter is 2.0 m. Calculate the hydraulic conductivity (also known as the coefficient of permeability). *Answer:* $k = 0.01$ cm/s.

4. In a falling-head permeameter test, the time required for the water in a standpipe to fall from $h_0 = 1.58$ m to $h_f = 1.35$ m is 11.0 hours. The clay specimen has a diameter of 6.35 cm and a length L of 2.54 cm. The diameter of the standpipe is 0.635 cm. Calculate the hydraulic conductivity. *Answer:* $k = 1 \times 10^{-7}$ cm/s.

5. A clay has a liquid limit = 60 and a plastic limit = 20. Determine the predominate clay mineral in the soil. *Answer:* Montmorillonite

Soil Classification

6. A soil has the following particle size gradation based on dry mass:

Gravel- and sand-size particles coarser than No. 40 sieve	= 65%
Sand-size particles finer than No. 40 sieve	= 10%
Silt-size particles	= 5%
Clay-size particles (finer than 0.002 mm)	= 20%
Total	= 100%

   In accordance with ASTM, the Atterberg limits were performed on the soil finer than the No. 40 sieve, and the results are LL = 93 and PL = 18. Calculate the activity A of this soil. *Answer:* $A = 1.3$

7. From a particle size analysis, the following values were obtained: $D_{60} = 15$ mm, $D_{50} = 12$ mm, $D_{30} = 2.5$ mm, and $D_{10} = 0.075$ mm. Calculate C_u and C_c. *Answer:* $C_u = 200$ and $C_c = 5.6$.

8. For the soil in Prob. 7, assume that 18 percent of the soil passes the No. 40 sieve and LL = 68 and PI = 34 for this soil fraction. What are the USCS and ISBP group symbols? *Answers:* GP-GM (USCS) and GM of low plasticity (ISBP).

9. A sand has $C_u = 7$ and $C_c = 1.5$. The sand contains 4 percent nonplastic fines (by dry mass). What are the USCS and ISBP group symbols? *Answer:* SW in both cases.

10. An inorganic soil has 100 percent passing (by dry mass) the No. 40 sieve. LL = 43 and PI = 16. What are the USCS and ISBP group symbols? *Answers:* ML (USCS) and MI (ISBP).

11. An inorganic clay has LL = 60 and PL = 20. What are the USCS and ISBP group symbols? *Answer:* CH in both cases.

12. A fine-grained soil has a black color and organic odor. LL = 65 on non-oven-dried soil and LL = 40 on oven-dried soil. What is the USCS group symbol? *Answer:* OH.
13. From its dry mass, a soil has 20 percent gravel-size particles, 40 percent sand-size particles, and 40 percent fines. On the basis of dry mass, 48 percent of the soil particles pass the No. 40 sieve, and LL = 85 and PL = 18 for soil passing the No. 40 sieve. What are the USCS and ISBP group symbols? *Answers:* SC (USCS) and SC of high plasticity (ISBP).

P · A · R · T · 3

ANALYSIS OF GEOTECHNICAL DATA AND ENGINEERING COMPUTATIONS

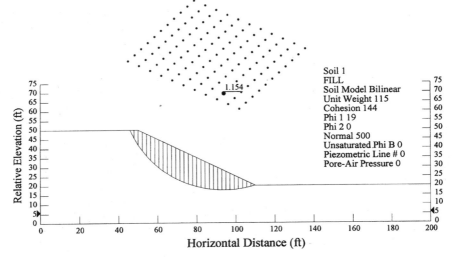

Slope stability analysis using the Geo-Slope (1991) computer program.

CHAPTER 6
BASIC GEOTECHNICAL AND FOUNDATION PRINCIPLES

The following notation is used in this chapter:

SYMBOL	DEFINITION
A	Cross-sectional area
B	Width of the footing
c	Cohesion based on a total stress analysis
c'	Cohesion based on an effective stress analysis
D_r	Relative density
D_{100}, D_0	Largest and smallest particle sizes
e	Void ratio of the soil
e_{max}	Void ratio corresponding to the loosest state of the soil
e_{min}	Void ratio corresponding to the densest state of the soil
G	Specific gravity
h	Distance above the groundwater table [Eq. (6.19)]
h	Thickness of fill layer [Eq. (6.23)]
h	Total head (Sec. 6.6)
h'	Equipotential drop, i.e., change in total head (Sec. 6.6)
Δh	Change in total head (Sec. 6.6)
h_c	Height of capillary rise
h_e	Elevation head
h_p	Pressure head
h_w	Height of water behind the sheet pile wall (Fig. 6.25)
i	Hydraulic gradient ($i = \Delta h/L$)
i_c	Critical hydraulic gradient
i_e	Exit hydraulic gradient
I	Influence value
j	Seepage force per unit volume of soil
k	Coefficient of permeability (also known as hydraulic conductivity)
k_0	Coefficient of lateral earth pressure at rest
L	Length of the footing [Eqs. (6.27) and (6.28)]
L	Length between the total head differences (Sec. 6.6)
m, n	Parameters used to determine σ_z below the corner of a loaded area
M	Mass of water plus mass of soil solids

6.4 ANALYSIS OF GEOTECHNICAL DATA AND ENGINEERING COMPUTATIONS

M_s	Dry mass of the soil solids
M_w	Mass of water in the soil
n	Porosity
n_d	Number of equipotential drops
n_f	Number of flow channels
N	Number of blocks [Eq. (6.31) and Fig. 6.9]
OCR	Overconsolidation ratio
P	Vertical load per unit length of footing [Eqs. (6.25) to (6.28)]
P	For Fig. 6.2b, P equals the vertical pressure exerted by the footing
Q	Surface point or line load [Eqs. (6.29) and (6.30)]
Q	Quantity of water that exits or enters the soil (Sec. 6.6)
r	Horizontal distance from the point load
s_u	Undrained shear strength of the soil
S	Degree of saturation of the soil
t	Time
u	Pore water pressure in the soil
u_s	Hydrostatic pore water pressure in the soil
v	Superficial velocity, which equals v in Darcy's Equation ($v = ki$)
v_s	Seepage velocity
V	Total volume of the soil
V_g	Volume of gas (air) in the soil
V_s	Volume of soil solids
V_v	Volume of voids in the soil, which equals $V_g + V_w$
V_w	Volume of water in the soil
w	Water content (also known as moisture content)
x	Width of the loaded area (Fig. 6.4)
y	Length of the loaded area (Fig. 6.4)
z	Depth below ground surface
z_w	Depth below groundwater table
ϕ	Friction angle based on a total stress analysis
ϕ'	Friction angle based on an effective stress analysis
ϕ'_u	Effective friction angle at the ultimate shear strength state
γ_b	Buoyant unit weight of saturated soil below the groundwater table
γ_d	Dry unit weight of soil
γ_t	Total unit weight of soil
γ_w	Unit weight of water
ρ_s	Density of the soil solids
ρ_t	Total density of the soil
ρ_w	Density of water
σ'	Effective stress
σ	Total stress
σ_0, q_0	Uniform pressure applied by the footing
σ'_v, σ'_{v0}	Vertical effective stress
σ_v	Vertical total stress
σ'_h	Horizontal effective stress
σ_h	Horizontal total stress

σ'_n Effective normal stress
σ_n Total normal stress
σ'_p Preconsolidation pressure (largest σ'_v ever experienced by the soil)
$\sigma_z, \Delta\sigma_v$ Increase in vertical stress in the soil due to the applied load
τ_f Shear strength of the soil

6.1 INTRODUCTION

Part 2 of the book describes the field and laboratory studies needed for a detailed understanding of soil, rock, and groundwater conditions that exist at the site. Part 3 of the book (Chaps. 6 through 16) deals with the engineering analysis of this data. Figure 1.11 (Chap. 1) shows the engineering analysis in relation to the other steps needed to complete the geotechnical aspects of the project.

Specific items that are included in Part 3 of the book are as follows:

- Basic Geotechnical and Foundation Principles (Chap. 6)
- Settlement of Structures (Chap. 7)
- Bearing Capacity (Chap. 8)
- Expansive Soil (Chap. 9)
- Slope Stability (Chap. 10)
- Earthquakes (Chap. 11)
- Erosion (Chap. 12)
- Deterioration (Chap. 13)
- Unusual Soil (Chap. 14)
- Retaining Walls (Chap. 15)
- Groundwater and Moisture Migration (Chap. 16)

The purpose of the engineering analysis is often to develop site development and foundation design parameters required for the project. It is important to recognize that without adequate and meaningful data from the subsurface exploration (Chap. 4) and laboratory testing (Chap. 5), the engineering analysis presented in the next 11 chapters will be of doubtful value and may even lead to erroneous conclusions.

The engineering calculations can vary from simple phase relationships to sophisticated finite element analyses that model the geotechnical elements (e.g., Duncan, 1996; Poh et al., 1997). In performing the analysis and in the development of foundation design parameters, the geotechnical engineer may need to rely on the expertise of other specialists. For example, as mentioned in Sec. 4.7, geologic analyses are often essential for projects having landslide or rockfall potential or seismic activity (Norris and Webb, 1990).

6.2 PHASE RELATIONSHIPS

Phase relationships are the basic soil relationships used in geotechnical engineering. Figure 6.1 shows an element of soil that can be divided into three basic parts:

1. *Solids.*   The mineral soil particles
2. *Liquids.*   Usually water that is contained in the void spaces between the solid mineral particles

3. *Gas.* Air, for example, that is also contained in the void spaces between the solid mineral particles

As indicated on the right side of Fig. 6.1, the three basic parts of soil can be rearranged into their relative proportions based on volume and mass.

Phase Relationships Directly from Laboratory Testing. Certain phase relationships can be determined directly from laboratory testing. Three of these laboratory tests (index properties) are as follows:

1. *Water content w.* The water content (also known as moisture content) test is described in Sec. 5.2.1. The water content w is defined as follows (ASTM D 2216-92, 1998):

$$w(\%) = \frac{100 M_w}{M_s} \tag{6.1}$$

where M_w = mass of the water in the soil and M_s = dry mass of the solids. As previously mentioned, a near-surface gravel or sand in a desert environment can have a water content of essentially zero, while a peat can have a water content as high as 1200 percent.

2. *Total density ρ_t and total unit weight γ_t.* The total density, also known as wet density, must be obtained from undisturbed block samples or soil samples extruded from thin-walled samplers (Sec. 4.4.1). The wet density is defined as:

$$\rho_t = \frac{M}{V} \tag{6.2}$$

where M = total mass of the soil (which is the sum of the mass of water M_w and mass of solids M_s), and V = total volume of the soil sample as shown in Fig. 6.1. Since most laboratories use balances that record in grams, and the gram is a unit of mass in the

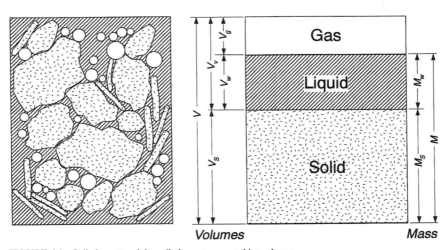

FIGURE 6.1 Soil element and the soil element separated into phases.

International System of Units (SI), the correct terminology for Eq. (6.2) is *density* (mass per unit volume).

When dealing with engineering calculations, it is usually easier to deal with total unit weight γ_t. In order to convert total density to total unit weight in the International System of Units (SI), the density is multiplied by g (where g = acceleration of gravity = 9.81 m/s²) to obtain the unit weight, which has units of kN/m³. For example, in the International System of Units (SI), the density of water ρ_w = 1.0 g/cm³ or 1.0 Mg/m³, while the unit weight of water γ_w = 9.807 kN/m³.

In the United States Customary System, density and unit weight have exactly the same value. Thus the density of water and the unit weight of water are both 62.4 pcf. However, for the density of water ρ_w, the units should be thought of as pounds-mass (lbm) per cubic foot, while for unit weight γ_w, the units are pounds-force (lbf) per cubic foot. In the United States Customary System, it is common to assume that 1 lbm is equal to 1 lbf.

3. *Specific gravity* G. As mentioned in Sec. 5.2.1, the specific gravity of soil particles can be measured by using a pycnometer (ASTM D 854-92, 1998). The specific gravity G is the density of solids ρ_s divided by the density of water ρ_w = 1.0 g/cm³, or:

$$G = \frac{\rho_s}{\rho_w} \tag{6.3}$$

The density of solids ρ_s is defined as the mass of solids M_s divided by the volume of solids V_s. As previously mentioned, inorganic soil often has a specific gravity of 2.65 while clays often have a specific gravity of 2.70.

Indirect Phase Relationships. This section describes those phase relationships that can not be determined in a laboratory, but instead must be calculated. For example, in Fig. 6.1, the volume of voids V_v is defined as:

$$V_v = V_g + V_w \tag{6.4}$$

where V_g = volume of gas (air) and V_w = volume of water. Other phase relationships that can not be determined in a laboratory, but instead must be calculated, are as follows:

1. *Void ratio* e *and porosity* n. The void ratio e is defined as follows:

$$e = \frac{V_v}{V_s} \tag{6.5}$$

and porosity n is defined as:

$$n = \frac{V_v}{V} \tag{6.6}$$

The void ratio e and porosity n are related as follows:

$$e = \frac{n}{1-n} \quad \text{and} \quad n = \frac{e}{1+e} \tag{6.7}$$

6.8 ANALYSIS OF GEOTECHNICAL DATA AND ENGINEERING COMPUTATIONS

The void ratio and porosity indicate the relative amount of void space in a soil. The lower the void ratio and porosity, the denser the soil (and vice versa). The natural soil having the lowest void ratio is probably till. For example, a typical value of dry density for till is 2.34 Mg/m³ (146 pcf), which corresponds to a void ratio of 0.14 (NAVFAC DM-7.1, 1982). A typical till consists of a well-graded soil ranging in particle sizes from clay to gravel and boulders. The high density and low void ratio are due to the extremely high stress exerted by glaciers (Winterkorn and Fang, 1975). For compacted soil, the soil type with typically the lowest void ratio is a well-graded decomposed granite (DG). A typical value of maximum dry density (Modified Proctor) for a well-graded DG is 2.20 Mg/m³ (137 pcf), which corresponds to a void ratio of 0.21. In general, the factors needed for a very low void ratio for compacted and naturally deposited soil are (Aberg, 1996; Day, 1997a):

- A well-graded grain-size distribution
- A high ratio of D_{100}/D_0 (ratio of the largest and smallest grain sizes)
- Clay particles (having low activity) to fill in the smallest void spaces
- A process, such as compaction or the weight of glaciers, to compress the soil particles into dense arrangements

At the other extreme are clays, such as sodium montmorillonite, which at low confining pressures can have a void ratio of more than 25. Highly organic soil, such as peat, can have even higher void ratios.

2. *Relative density.* The relative density is a measure of the density state of a non-plastic soil. The relative density can be used only for soil that is nonplastic, such as sands and gravels (see Table 5.5). The relative density (D_r in percent) is defined as:

$$D_r (\%) = 100 \frac{e_{max} - e}{e_{max} - e_{min}} \tag{6.8}$$

where e_{max} = loosest possible state of the soil, usually obtained by pouring the soil into a mold of known volume (ASTM D 4254-96, 1998); e_{min} = densest possible state of the soil, usually obtained by vibrating the soil particles into a dense state (ASTM D 4253-96, 1998); and e = the natural void ratio of the soil. The density state of the natural soil can be described as follows:

Condition	D_r, %
Very loose	0–15
Loose	15–35
Medium	35–65
Dense	65–85
Very dense	85–100

The relative density D_r should not be confused with the relative compaction RC which will be discussed in Chap. 17.

3. *Dry unit weight* γ_d. By using the water content w of the soil and the total unit weight γ_t, the dry unit weight γ_d can be calculated, as follows:

$$\gamma_d = \frac{\gamma_t}{1 + w} \tag{6.9}$$

In Eq. (6.9), the water content w must be expressed as a decimal (not as a percentage).

4. *Buoyant (or submerged) unit weight* γ_b. The buoyant unit weight γ_b can be defined as follows:

$$\gamma_b = \gamma_t - \gamma_w \qquad (6.10)$$

The buoyant unit weight can be used to determine the vertical effective stress which will be discussed in the next section. Note that the total unit weight γ_t used in Eq. (6.10) must be for the special case where the void spaces are completely filled with water (such as a soil specimen obtained from below the groundwater table).

5. *Degree of saturation* S. The degree of saturation S is defined as:

$$S\,(\%) = \frac{100\, V_w}{V_v} \qquad (6.11)$$

The degree of saturation indicates the degree to which the soil voids are filled with water. A totally dry soil will have a degree of saturation of 0 percent, while a saturated soil, such as a soil below the groundwater table, will have a degree of saturation of 100 percent. Typical ranges of degree of saturation versus soil condition are as follows (Terzaghi and Peck, 1967):

Condition	S, %
Dry	0
Humid	1–25
Damp	26–50
Moist	51–75
Wet	76–99
Saturated	100

If a soil is obtained below the groundwater table or after submergence in the laboratory, the degree of saturation is often assumed to be 100 percent and then phase relationships (such as the void ratio) are back-calculated. However, for soil below the groundwater table, a better approach is to use the degree of saturation as a final check on the accuracy of the laboratory test data (i.e., γ_t, w, and G).

Useful Relationships. A frequently used method of solving phase relationships is to first fill in the phase diagram shown in Fig. 6.1. Once the different masses and volumes are known, then the various phase relationships can be determined. Another approach is to use equations that relate different parameters. Some commonly used equations are as follows:

$$\gamma_t = \frac{G\,\gamma_w(1 + w)}{1 + e} \qquad (6.12)$$

$$\gamma_d = \frac{G\gamma_w}{1 + e} \qquad (6.13)$$

$$\gamma_b = \frac{\gamma_w(G - 1)}{1 + e} \qquad (6.14)$$

$$Gw = Se \qquad (6.15)$$

Summary. As previously mentioned in Sec. 5.4, the soil classification provides a written description of the soil. In essence, the phase relationships provide a mathematical description of the soil which is used in engineering analyses. For the soil shown in Figs. 5.19 to 5.22, the written and mathematical descriptions of the soil are as follows:

Written description (soil classification). Clayey sand (SC) of low plasticity. On the basis of dry mass, the soil contains 13.8 percent gravel-size particles, 48.7 percent sand-size particles, 19.5 percent silt-size particles, and 18.0 percent clay-size particles. The gravel-size particles are predominately hard and angular rock fragments. The sand-size particles are predominately composed of angular quartz grains. Atterberg limits performed on the soil passing the No. 40 sieve indicate a LL = 28 and a PI = 15. The *in situ* soil has a reddish-brown color, soft consistency, gritty texture, and is wet. The *in situ* soil is a residual soil and is part of an old vernal pool.

Mathematical description (phase relationships). Water content w = 19 percent, total unit weight γ_t = 19 kN/m^3 (120 pcf), dry unit weight γ_d = 16 kN/m^3 (100 pcf), specific gravity G = 2.65, void ratio e = 0.63, porosity n = 0.39 (or 39 percent), and degree of saturation S = 80 percent.

6.3 EFFECTIVE STRESS

Definition. With the soil characteristics expressed in terms of a written description (soil classification) and mathematical description (phase relationships), the next step in the analysis is often to determine the stresses acting on the soil. This is important because most geotechnical projects deal with a change in stress of the soil. For example, the construction of a building applies an additional stress on the soil supporting the foundation, which results in settlement of the building.

Stress is defined as the load divided by the area over which it acts. In geotechnical engineering, a compressive stress is considered positive and tensile stress is negative. Stress and pressure are often used interchangeably in geotechnical engineering. In using the International System of Units (SI), the units for stress are kPa. In the United States Customary System, the units for stress are psf (pounds-force per square foot). Stress expressed in units of kg/cm^2 has been used in the past and is still in use (e.g., see Figs. 4.15, 4.40, and 4.42). One kg/cm^2 is approximately equal to one ton per square foot (tsf).

An important concept in geotechnical engineering is effective stress. The effective stress σ' is defined as follows:

$$\sigma' = \sigma - u \tag{6.16}$$

where σ = total stress and u = pore water pressure. Many engineering analyses use the vertical effective stress, also known as the effective overburden stress, which is designated σ'_v or σ'_{v0}.

Total Stress. For the condition of a uniform soil and a level ground surface (geostatic condition), the total vertical stress σ_v at a depth z below the ground surface is:

$$\sigma_v = \gamma_t z \tag{6.17}$$

where γ_t = total unit weight of the soil. For soil deposits having layers with different total unit weights, the total vertical stress is the sum of the vertical stress for each individual soil layer. For some projects, total pressure cells (Sec. 18.2) can be installed to measure the total stress in the soil or the total stress of the soil acting on an earth structure, such as a retaining wall.

Pore Water Pressure u and Calculation of Vertical Effective Stress σ'_v. For the condition of a hydrostatic groundwater table (i.e., no groundwater flow or excess pore water pressures), the static pore water pressure (u or u_s) is:

$$u = \gamma_w z_w \qquad (6.18)$$

where γ_w = unit weight of water and z_w = depth below the groundwater table. From the total unit weight of the soil [Eq. (6.17)] and the pore water pressure [Eq. (6.18)], the vertical effective stress σ'_v can be calculated. An alternative method is to use the buoyant unit weight γ_b to calculate the vertical effective stress. For example, suppose that a groundwater table corresponds with the ground surface. In this case, the vertical effective stress σ'_v is simply the buoyant unit weight γ_b times the depth below the ground surface. More often, the groundwater table is below the ground surface, in which case the vertical total stress of the soil layer above the groundwater table must be added to the buoyant unit weight calculations. The vertical effective stress σ'_v is often plotted versus depth and included with the subsoil profile. For example, in Fig. 4.40, the vertical effective stress σ'_{v0} and the static pore water pressure u_s are plotted versus depth. Likewise in Figs. 4.41 and 4.42, the vertical effective stress (also known as effective overburden stress) has been plotted versus depth.

For cases where there is flowing groundwater or excess pore water pressure due to the consolidation of clay, the pore water will not be hydrostatic. Engineering analyses, such as seepage analyses (Sec. 6.6) or those based on the theory of consolidation (Sec. 7.4) can be used to predict the pore water pressure. For some projects, piezometers (Sec. 18.2) can be installed to measure the pore water pressure u in the ground.

Capillarity. Soil above the groundwater table can be subjected to negative pore water pressure. This is known as *capillarity*, also as *capillary action*, which is the rise of water through soil due to the fluid property known as surface tension. Because of capillarity, the pore water pressures are less than atmospheric values produced by the surface tension of pore water acting on the meniscus formed in the void spaces between the soil particles. The height of capillary rise h_c is related to the pore size of the soil, as follows (Hansbo, 1975):

Soil	h_c
Open graded gravel	0
Coarse sand	0.03–0.15 m (0.1–0.5 ft)
Medium sand	0.12–1.1 m (0.4–3.6 ft)
Fine sand	0.3–3.5 m (1.0–12 ft)
Silt	1.5–12 m (5–40 ft)
Clay	≥10 m (≥ 33 ft)

As the above data indicates, there will be no capillary rise for open graded gravel because of the large void spaces between the individual gravel-size particles. But for clay, which has very small void spaces, the capillary rise can be in excess of 10 m (33 ft). If the soil is saturated above the groundwater table, then the pore water pressure u is negative and can be calculated as the distance above the groundwater table h times the unit weight of water γ_w, or:

$$u = -\gamma_w h \qquad (6.19)$$

Capillarity is important in the understanding of soil behavior. Because of the negative value of pore water pressure due to capillarity, it essentially holds together the soil particles. For large-size soil particles, such as gravel and coarse sand-size particles, the effect of capillarity is negligible and the soil particles simply fall apart (hence they are cohesionless). Medium to fine sands do have a low capillarity, which is enough to build a sand castle at the beach, but this small capillarity is lost when the sand becomes submerged in water or the sand completely dries.

Silt- and clay-size particles are so small that they are strongly influenced by capillarity. These fine soil–size particles can be strongly held together by capillarity, which gives the soil the ability to be remolded and rolled without falling apart (hence they are cohesive). Capillarity is the mechanism that gives a silt or clay its plasticity, the ability to be remolded and rolled without falling apart. When the remolded silt or clay is submerged in water, the capillary tension is slowly eliminated and the soil particles will often disperse. As will be discussed in Sec. 6.5, the undrained shear strength of silts and clays is strongly influenced by capillarity.

Coefficient of Earth Pressure at Rest (k_0). The preceding sections discuss the vertical effective stress σ'_v for soil deposits. For many geotechnical projects, it may be important to determine the *in situ* horizontal stress σ_h. Similar to measurements of the vertical stress, the horizontal total stress can be measured by using total pressure cells, and by knowing the pore water pressure, the horizontal effective stress σ'_h can be calculated by using Eq. (6.16). The horizontal effective stress σ'_h can also be calculated as follows:

$$\sigma'_h = k_0 \sigma'_v \tag{6.20}$$

where k_0 = coefficient of lateral earth pressure at rest. The value of this coefficient depends on many factors, such as the soil type, density condition (loose versus dense), geological depositional environment (i.e., alluvial, glacial, etc.), and the stress history of the site (Massarsch et al., 1975; Massarsch, 1979). The value of k_0 in natural soils can be as low as 0.4 for soils formed by sedimentation and never preloaded, up to 3.0 or greater for some heavily preloaded soil deposits (Holtz and Kovacs, 1981). For soil deposits that have not been significantly preloaded, a value of $k_0 = 0.5$ is often assumed in practice, or the following equation is used (Jaky, 1944, 1948; Brooker and Ireland, 1965):

$$k_0 = 1 - \sin \phi' \tag{6.21}$$

where ϕ' = effective friction angle of the soil.

As an approximation, the value of k_0 for preloaded soil can be determined from the following equation (adapted from Alphan, 1967; Schmertmann, 1975; Ladd et al., 1977):

$$k_0 = 0.5(\text{OCR})^{0.5} \tag{6.22}$$

where OCR = the overconsolidation ratio, defined as the largest vertical effective stress ever experienced by the soil deposit σ'_p divided by the existing vertical effective stress σ'_v. The overconsolidation ratio and the preconsolidation pressure σ'_p will be further discussed in Sec. 7.4.

6.4 STRESS DISTRIBUTION

Introduction. The previous section described methods that are used to determine the existing stresses within the soil mass. This section describes commonly used methods to determine the increase in stress in the soil deposit due to applied loads. This is naturally important in settlement analyses because the settlement of the structure is due directly to its weight which causes an increase in stress in the underlying soil. In most cases, it is the increase in vertical stress that is of most importance in settlement analyses. The symbol σ_z is often used to denote an increase in vertical stress in the soil, although $\Delta\sigma_v$ (change in total vertical stress) is also used.

In dealing with stress distribution, a distinction must be made between one-dimensional and two- or three-dimensional loading. A one-dimensional loading applies a stress increase at depth that is 100 percent of the applied surface stress. An example of a one-dimensional loading would be the placement of a fill layer of uniform thickness and large areal extent at ground surface. Beneath the center of the uniform fill, the *in situ* soil is subjected to an increase in vertical stress that equals the following:

$$\sigma_z = \Delta\sigma_v = h\gamma_t \qquad (6.23)$$

where h = thickness of the fill layer and γ_t = total unit weight of the fill. In this case of one-dimensional loading, the soil would be compressed only in the vertical direction (i.e., strain only in the vertical direction).

Another example of one-dimensional loading is the uniform lowering of a groundwater table. If the total unit weight of the soil does not change as the groundwater table is lowered, then the one-dimensional increase in vertical stress for the *in situ* soil located below the groundwater table would equal the following:

$$\sigma_z = \Delta\sigma_v = h\gamma_w \qquad (6.24)$$

where h = vertical distance that the groundwater table is uniformly lowered and γ_w = unit weight of water.

Surface loadings can cause both vertical and horizontal strains, and this is referred to as *two-* or *three-dimensional loading*. Common examples of two-dimensional loading are loads from strip footings or long embankments (i.e., plane strain conditions). Examples of three-dimensional loading are loads from square and rectangular footings (spread footings) and round storage tanks. The following two sections describe methods that can be used to determine the change in vertical stress for two-dimensional (strip footings and long embankments) and three-dimensional (spread footings and round storage tanks) loading conditions. In these cases, the load usually dissipates rapidly with depth.

2:1 Approximation. A simple method to determine the increase in vertical stress with depth is the 2:1 approximation (also known as the *2:1 method*). Figure 6.2a illustrates the basic principle of the 2:1 approximation. This method assumes that the stress dissipates with depth in the form of a trapezoid that has 2:1 (vertical:horizontal) inclined sides as shown in Fig. 6.2a. The purpose of this method is to approximate the actual "pressure bulb" stress increase (see Fig. 6.2b) beneath a footing.

If there is a strip footing of width B that has a vertical load P per unit length of footing, then as indicated in Fig. 6.2a, the stress applied by the footing (σ_0) would be:

$$\sigma_0 = \frac{P}{B} \qquad (6.25)$$

where B = width of the strip footing. As indicated in Fig. 6.2a, at a depth z below the footing, the vertical stress increase σ_z due to the strip footing load would be:

$$\sigma_z = \Delta\sigma_v = \frac{P}{B+z} \qquad (6.26)$$

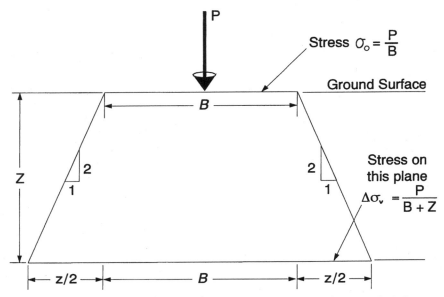

FIGURE 6.2a 2:1 approximation for the calculation of the increase in vertical stress at depth due to an applied load P.

If the footing is a rectangular spread footing having a length = L and a width = B, then the stress applied by the rectangular footing (σ_0) would be:

$$\sigma_0 = \frac{P}{BL} \qquad (6.27)$$

where P = entire load of the rectangular spread footing. According to the 2:1 approximation, the vertical stress increase (σ_z) at a depth = z below the rectangular spread footing would be:

$$\sigma_z = \Delta\sigma_v = \frac{P}{(B + z)(L + z)} \qquad (6.28)$$

A major advantage of the 2:1 approximation is its simplicity, and for this reason, it is probably used more often than any other type of stress distribution method. The main disadvantage with the 2:1 approximation is that the stress increase under the center of the loaded area equals the stress increase under the corner or side of the loaded area. The usual situation is that the soil underlying the center of the loaded area is subjected to a higher vertical stress increase than the soil underneath a corner or edge of the loaded area. Thus the 2:1 approximation is often only used to estimate the average settlement of the loaded area. Different methods, such as stress distribution based on the theory of elasticity, can be used to calculate the change in vertical stress between the center and corner of the loaded area.

Equations and Charts Based on the Theory of Elasticity. Equations and charts have been developed to determine the change in stress due to applied loads based on the theory of elasticity. The solutions assume an elastic and homogeneous soil that is continuous and in static equilibrium. The elastic solutions also use a specific type of applied load, such as a point load, uniform load, or linearly increasing load (triangular distribution). For loads where the length of the footing is greater than 5 times the width, such as for strip footings, the stress distribution is considered to be plane strain. This means that the horizontal strain of the elastic soil occurs only in the direction perpendicular to the long axis of the footing.

Although equations and charts based on the theory of elasticity are often used to determine the change in soil stress, soil is not an elastic material. For example, if a heavy foundation load is applied to a soil deposit, there will be vertical deformation of the soil in response to this load. If this heavy load is removed, the soil will rebound but not return to its original height because soil is not elastic. However, it has been stated that as long as the factor of safety against shear failure exceeds about 3, then stresses imposed by the

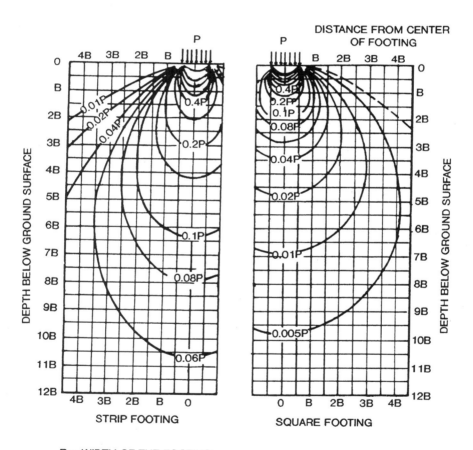

B = WIDTH OF THE FOOTING
P = UNIFORM STRESS APPLIED BY THE FOOTING

FIGURE 6.2b Pressure bulb beneath strip footing and square footing. The curves indicate the value of $\Delta\sigma_v$ beneath the footings due to the uniform pressure equal to P. (*Adapted from NAVFAC DM-7.1, 1982.*)

foundation load are roughly equal to the values computed from elastic theory (NAVFAC DM-7.2, 1982).

In 1885, Boussinesq published equations based on the theory of elasticity. For a surface point load Q applied at the ground surface such as shown in Fig. 6.3, the vertical stress increase at any depth z and distance r from the point load can be calculated by using the following Boussinesq (1885) equation:

$$\sigma_z = \Delta\sigma_v = \frac{3Qz^3}{2\pi(r^2 + z^2)^{5/2}} \tag{6.29}$$

If there is a uniform line load Q (force per unit length), the vertical stress increase at a depth z and distance r from the line load would be:

$$\sigma_z = \Delta\sigma_v = \frac{2Qz^3}{\pi(r^2 + z^2)^2} \tag{6.30}$$

In comparing the equation for a point load [Eq. (6.29)] with the equation for a line load [Eq. (6.30)] for the same load Q, the line load equation will usually give a higher value of σ_z than the point load equation.

In 1935, Newmark performed an integration of Eq. (6.30) and derived an equation to determine the vertical stress increase σ_z under the corner of a loaded area. Convenient charts have been developed, based on the Newmark (1935) equation. For example, the chart shown in Fig. 6.4 is easy to use and consists of first calculating m and n. The value m is defined as the width of the loaded area (x) divided by the depth z to where the vertical stress increase σ_z is to be calculated. The value n is defined as the length of the loaded area

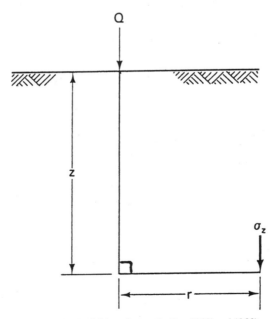

FIGURE 6.3 Definition of terms for Eqs. (6.29) and (6.30).

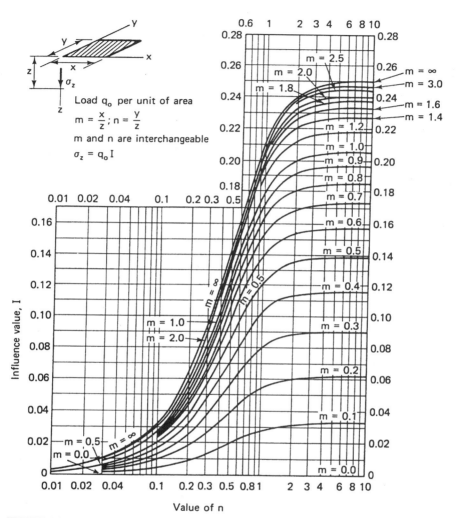

FIGURE 6.4 Chart for calculating the increase in vertical stress beneath the corner of a uniformly loaded rectangular area. (*From NAVFAC DM-7.1, 1982; reproduced from Holtz and Kovacs, 1981.*)

(*y*) divided by the depth *z*. The user enters the chart with the value of *n* and, upon intersecting the desired *m* curve, obtains the influence value *I* from the vertical axis. As indicated in Fig. 6.4, vertical stress increase σ_z is then calculated as the loaded area pressure q_0 times the influence value *I*.

Figure 6.4 can also be used to determine the vertical stress increase σ_z below the center of a rectangular loaded area. In this case, the rectangular loaded area would be divided into four parts and then Fig. 6.4 would be used to find the stress increase below the corner of one of the parts. By multiplying this stress by 4 (i.e., 4 parts), the vertical stress increase σ_z below the center of the total loaded area is obtained. This type of analysis is possible because of the principle of superposition for elastic materials. To find the vertical stress increase σ_z outside the loaded area, additional rectangular areas can be added and subtracted as needed in order to model the loading condition.

Figure 6.5 presents the Westergaard (1938) analysis for a soft elastic material reinforced by numerous strong horizontal sheets. This chart was based on an elastic material that contains numerous thin and perfectly rigid layers that allow for only vertical strain but not horizontal strain. This chart may represent a better model for the increase in vertical stress σ_z for layered soils, such as a soft clay deposit that contains numerous horizontal layers of sand.

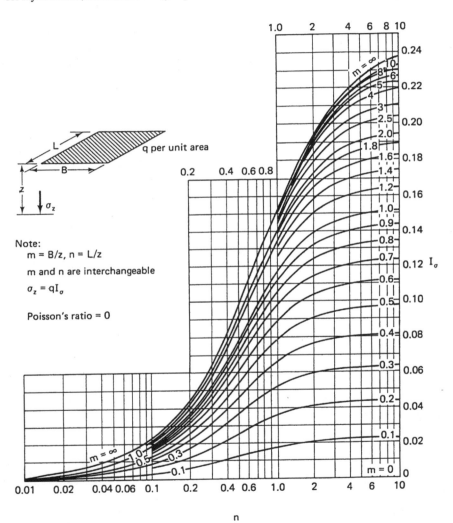

FIGURE 6.5 Chart based on Westergaard theory for calculating the increase in vertical stress beneath the corner of a uniformly loaded rectangular area. (*From Duncan and Buchignani, 1976; reproduced from Holtz and Kovacs, 1981.*)

Figures 6.6 to 6.8 present additional charts for different types of loading conditions based on the integration of the original Boussinesq equations. Note that Fig. 6.8 can be used to determine the vertical stress increase σ_z beneath the center of a long embankment by splitting the embankment down the middle and then multiplying the final result by 2 (i.e., two parts of the embankment).

Figure 6.9 presents a Newmark (1942) chart which can be used to determine the vertical stress increase σ_z beneath a uniformly loaded area of any shape. There are numerous influence charts, each having a different influence value. Note that the chart in Fig. 6.9 has an influence value I of 0.005. The first step is to draw the loaded area onto the chart, using a scale where AB equals the depth z. The center of the chart must correspond to the point where the increase in vertical stress σ_z is desired. The increase in vertical stress σ_z is then calculated as:

$$\sigma_z = q_0 IN \qquad (6.31)$$

where q_0 = applied stress from the irregular area, I = influence value (0.005 for Fig. 6.9) and N = number of blocks within the irregularly shaped area plotted on Fig. 6.9. In obtaining the value of N, portions of blocks are also counted. Note that the entire procedure must be repeated if the increase in vertical stress σ_z is needed at a different depth.

As an example of using Fig. 6.9, suppose a round storage tank will impose a surface load of 1000 psf and the tank has a diameter of 100 ft. It is desired to determine the increase in vertical stress σ_z at a depth z of 100 ft directly beneath the center of the round storage tank. In this simple case, the diameter of the tank equals the depth z, and thus a circle having a diameter of AB is drawn on Fig. 6.9 with the center of the circle corresponding to the center of Fig. 6.9. The number of blocks (N) within the circle drawn on Fig. 6.9 is 60, and thus

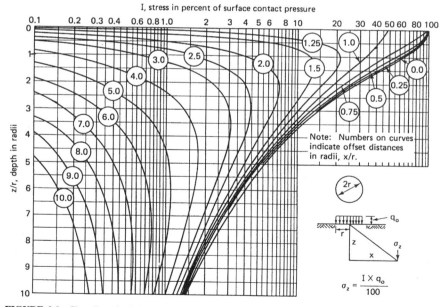

FIGURE 6.6 Chart for calculating the increase in vertical stress beneath a uniformly loaded circular area. (*From NAVFAC DM-7.1, 1982, and Foster and Ahlvin, 1954; reproduced from Holtz and Kovacs, 1981.*)

6.20 ANALYSIS OF GEOTECHNICAL DATA AND ENGINEERING COMPUTATIONS

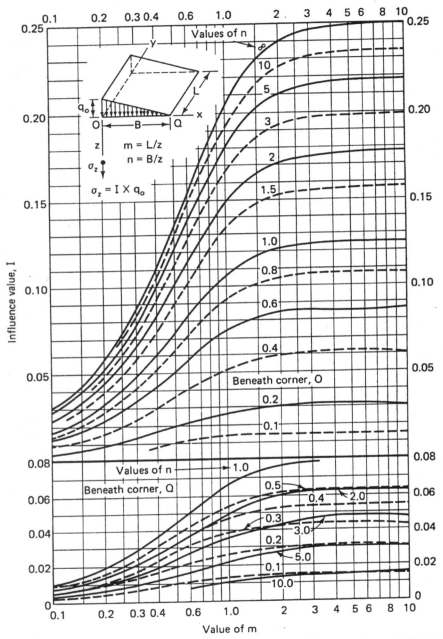

FIGURE 6.7 Chart for calculating the increase in vertical stress beneath the corner of a rectangular area that has a triangular load. (*From NAVFAC DM-7.1, 1982; reproduced from Holtz and Kovacs, 1981.*)

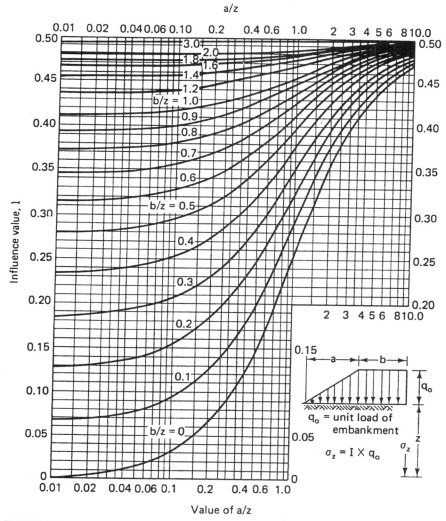

FIGURE 6.8 Chart for calculating the increase in vertical stress beneath the center of a very long embankment. (*From NAVFAC DM-7.1, 1982, and Osterberg, 1957; reproduced from Holtz and Kovacs, 1981.*)

the increase in vertical stress σ_z at a depth of 100 ft below the center of the tank is 300 psf (i.e., 1000 psf × 0.005 × 60 = 300 psf).

Figure 6.6 can be used as a check on these calculations. For Fig. 6.6, $z/r = 2$ and the offset distance = 0. Using these values, the influence value I from Fig. 6.6 is 30, and by using the equation listed in Fig. 6.6, the increase in vertical stress σ_z is once again 300 psf (i.e., 30 × 1000 psf ÷ 100 = 300 psf).

In addition to the equations and charts described above, the theory of elasticity has been applied to many other types of loading conditions (e.g., see Poulos and Davis, 1974).

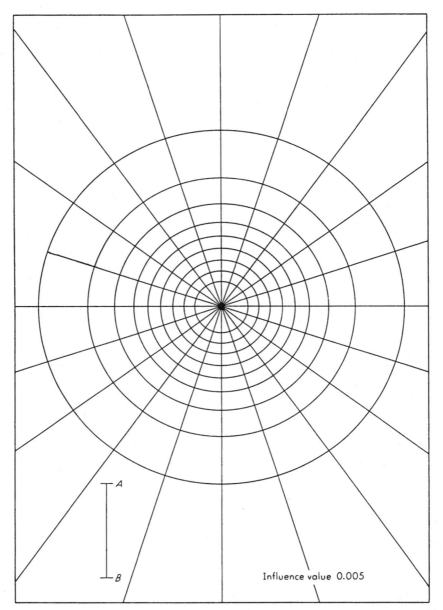

FIGURE 6.9 Newmark chart for calculating the increase in vertical stress beneath a uniformly loaded area of any shape. (*From Newmark, 1942; reproduced from Bowles, 1982.*)

6.5 SHEAR STRENGTH

An understanding of shear strength is essential in geotechnical engineering. This is because most geotechnical failures involve a shear-type failure of the soil. This is because of the nature of soil, which is composed of individual soil particles that slide (i.e., shear past each other) when the soil is loaded. An example of the importance of shear strength is shown in Figs. 4.24 to 4.34, where a landslide has slid down the hillside. As shown in Fig. 4.23b, this massive shear failure of the ground developed because the shear stresses exceeded the shear strength of the soil.

Section 5.2.3 presents a discussion of shear strength tests performed in the laboratory. As discussed in Sec. 5.2.3, there are many types of shear strength tests that can be performed in the laboratory. Laboratory tests are generally divided into two categories:

1. *Shear strength tests based on total stress.* The purpose of these laboratory tests is to obtain the undrained shear strength s_u of the soil or the failure envelope in terms of total stresses (total cohesion c and total friction angle ϕ). These types of shear strength tests are often referred to as *undrained* shear strength tests. Examples include the laboratory vane test, unconfined compression test, unconsolidated undrained triaxial test, and consolidated undrained triaxial test (see Sec. 5.2.3). These types of laboratory tests are performed exclusively on plastic (cohesive) soils.

2. *Shear strength tests based on effective stress.* The purpose of these laboratory tests is to obtain the effective shear strength of the soil from the failure envelope in terms of effective stress (effective cohesion c' and effective friction angle ϕ'). These types of shear strength tests are often referred to as *drained* shear strength tests. Examples include the direct shear test, consolidated drained triaxial test, consolidated undrained triaxial test with pore water pressure measurements, and the drained residual shear strength (see Sec. 5.2.3). These types of laboratory tests can be performed on plastic (cohesive) soil and nonplastic (cohesionless) soil.

The mechanisms that control the shear strength of soil are complex, but in simple terms the shear strength of soils can be divided into two broad categories: cohesionless (nonplastic) soils and cohesive (plastic) soils.

6.5.1 Cohesionless Soil

These types of soil are nonplastic. As indicated in Table 5.5, nonplastic soils include gravels, sands, and nonplastic silt such as rock flour. A cohesionless soil develops its shear strength as a result of the frictional and interlocking resistance between the individual soil particles. Cohesionless soils can be held together only by confining pressures and will fall apart when the confining pressure is released.

During shear of cohesionless soils, those soils in a dense state will tend to dilate (increase in volume), while those soils in a loose state tend to contract (decrease in volume). As indicated in Fig. 5.10, cohesionless soils have a high permeability, and for the shear strength testing of saturated cohesionless soils, water usually flows quickly into the soil when it dilates or out of the soil when it contracts. Thus the drained shear strength is of most importance for cohesionless soils. An important exception is the liquefaction of saturated, loose cohesionless soils, which is discussed in Chap. 11.

As previously mentioned, the shear strength of cohesionless soils can be measured in the direct shear apparatus (Fig. 5.5). As mentioned in Sec. 6.3, there can be a small capillary tension in cohesionless soils and thus the soil specimen is saturated prior to shearing (ASTM D

6.24 ANALYSIS OF GEOTECHNICAL DATA AND ENGINEERING COMPUTATIONS

3080-90, 1998). Because of these test specifications, which require the direct shear testing of soil in a saturated and drained state, the shear strength of the soil is expressed in terms of the effective friction angle ϕ'. Cohesionless soils can also be tested in a dry state and the shear strength of the soil is then expressed in terms of the friction angle ϕ. In a comparison of the effective friction angle ϕ' from drained direct shear tests on saturated cohesionless soil and the friction angle ϕ from direct shear tests on the same soil in a dry state, it has been determined that ϕ' is only 1 to 2° lower than ϕ (Terzaghi and Peck, 1967; Holtz and Kovacs, 1981). This slight difference is usually ignored and the friction angle ϕ and effective friction angle ϕ' are typically considered to mean the same thing for cohesionless (nonplastic) soils.

Figure 6.10 presents the results of drained direct shear tests performed on undisturbed samples of silty sand. Undisturbed samples were obtained from two borings (B-1 and B-2) excavated in the silty sand deposit. A total of six direct shear tests were performed. For each direct shear test, the effective normal pressure (i.e., the vertical load divided by the specimen area; see Fig. 5.5) was based on the *in situ* vertical overburden pressure. The undisturbed soil specimen location and the effective normal pressure σ'_n during the direct shear tests were as follows:

- Boring B-1 at a depth of 3.5 ft. Effective normal pressure = 425 psf.
- Boring B-1 at a depth of 6 ft. Effective normal pressure = 725 psf.
- Boring B-2 at a depth of 2.5 ft. Effective normal pressure = 300 psf.
- Boring B-2 at a depth of 4.5 ft. Effective normal pressure = 550 psf.
- Boring B-2 at a depth of 6.5 ft. Effective normal pressure = 775 psf.
- Boring B-2 at a depth of 10 ft. Effective normal pressure = 1250 psf.

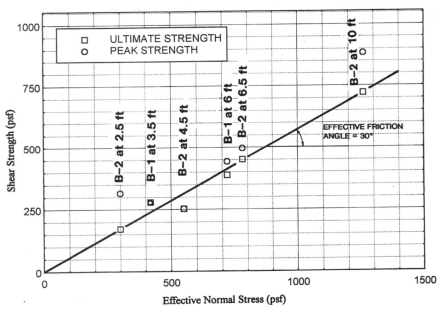

FIGURE 6.10 Effective shear strength versus effective normal pressure for drained direct shear tests on silty sand specimens.

During the shearing of the silty sand specimens in the direct shear apparatus, the shear stress (which equals the shearing force divided by the specimen area, see Fig. 5.5) versus lateral deformation was recorded and is plotted for two of the tests in Fig. 6.11. Note in Fig. 6.11 that the drained direct shear test performed on the silty sand at B-2 at 10 ft has a distinct peak in the curve, while the drained direct shear test on the silty sand at B-1 at 6 ft does not exhibit this distinctive peaking of the curve. This is a result of the dry density of the soil specimens, where the specimen at 10 ft is in a denser state (γ_d = 104 pcf, 16.4 kN/m³) than the silty sand at a depth of 6 ft (γ_d = 93.4 pcf, 14.7 kN/m³). Both the peak (highest point) and ultimate (final value) from Fig. 6.11 have been plotted in Fig. 6.10 and are designated as the *ultimate shear strength* (squares) and the *peak shear strength* (circles). In those cases where the peak and ultimate values coincide, only one point is shown.

There is considerable scatter of the data in Fig. 6.10, but this is not unusual for soil which usually has variable engineering properties. The straight line drawn in Fig. 6.10 is the *shear strength envelope,* also known as the *failure envelope,* for the silty sand deposit. Considerable experience is needed in determining this envelope. For example, taking a conservative approach, the envelope was drawn by ignoring the two high peak shear strength points (B-2 at 2.5 ft and B-2 at 10 ft) and then a best-fit line was drawn through the rest of the data points. In addition, the angle of inclination of the failure envelope, which is known as the effective friction angle ϕ', is 30°, which is a typical value for silty sands. Thus the failure envelope drawn in Fig. 6.10 was based on engineering judgment and experience.

Note in Fig. 6.10 that the failure envelope (straight line) is through the origin of the plot. This means that the effective cohesion c' is zero. If the line had intersected the vertical axis,

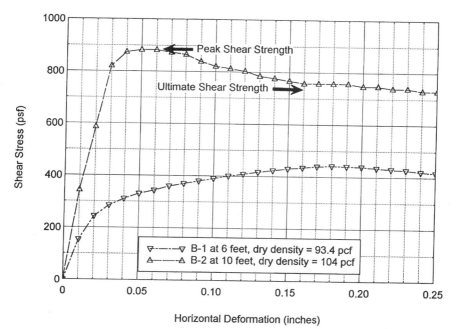

FIGURE 6.11 Shear stress versus horizontal deformation from the drained direct shear tests on silty sand specimens from depths of 6 ft and 10 ft.

then that value would be defined as the effective cohesion value c'. Therefore the shear strength of the soil can be defined by the Mohr-Coulomb failure law:

$$\tau_f = c' + \sigma'_n \tan \phi' \quad (6.32)$$

where τ_f = shear strength of the soil (straight line, as shown in Fig. 6.10), c' = effective cohesion, which is the intercept of the straight line on the y axis, σ'_n = effective normal stress during the direct shear testing, and ϕ' = effective friction angle, which is the angle of inclination of the straight line. When used for the analysis of field situations, the value of σ'_n in Eq. (6.32) would equal the effective normal stress on the shear surface under investigation.

As previously mentioned and as illustrated in Fig. 6.10, the effective cohesion c' is zero. This is the typical case for the shear strength of cohesionless soils. Since they are cohesionless, they should not have any cohesion and hence $c' = 0$. An exception is the testing of cohesionless soil at high normal pressures, where the shear strength envelope may actually be curved because of particle crushing (Holtz and Gibbs, 1956a). In this case, a straight line approximation at high normal stresses may indicate a cohesion intercept, but this value should be regarded as an extrapolated value that is not representative of the shear strength of cohesionless soils at low values of σ'_n.

Table 6.1 (data from Hough, 1969) presents values of effective friction angles for different types of cohesionless (nonplastic) soils. An exception to the values presented in Table 6.1 are cohesionless soils that contain appreciable mica flakes. A micaceous sand will often have a high void ratio and hence little interlocking and a lower friction angle (Horn and Deere, 1962).

For many projects, the effective friction angle of a sand deposit is determined from indirect means, such as the standard penetration test (SPT) and the cone penetration test (CPT; see Sec. 4.4.2). As indicated in Figs. 4.13 and 4.15, the effective friction angle ϕ' for clean quartz sand can be estimated from the results of standard penetration tests or cone penetration tests for various values of effective overburden pressure σ'_v.

In summary, for cohesionless soils, $c' = 0$ and the effective friction angle ϕ' depends on:

- *Soil type (Table 6.1).* Sand and gravel mixtures have a higher effective friction angle than nonplastic silts.

TABLE 6.1 Typical Effective Friction Angle ϕ' for Different Cohesionless Soils

Soil type (1)	Effective friction angle ϕ' at peak strength		Effective friction angle ϕ'_u at ultimate strength* (4)
	Medium (2)	Dense (3)	
Silt (nonplastic)	28 to 32°	30 to 34°	26 to 30°
Uniform fine to medium sand	30 to 34°	32 to 36°	26 to 30°
Well-graded sand	34 to 40°	38 to 46°	30 to 34°
Sand and gravel mixtures	36 to 42°	40 to 48°	32 to 36°

Source: Data obtained from Hough, 1969.
*The effective friction angle ϕ'_u at the ultimate shear strength state could be considered to be the same as the friction angle ϕ' for the same soil in a loose state.

- *Soil density.* For a given cohesionless soil, the denser the soil, the higher the effective friction angle. This is a result of the interlocking of soil particles, where at a denser state the soil particles are interlocked to a higher degree and hence the effective friction angle is greater than in a loose state. It has been observed that, in the ultimate shear strength state, the shear strength and density of a loose and dense sand tend to approach each other.

- *Grain size distribution.* A well-graded cohesionless soil will usually have a higher friction angle than a uniform soil. With more soil particles to fill in the small spaces between soil particles, there is more interlocking and frictional resistance developed for a well-graded soil than for a uniform cohesionless soil.

- *Mineral type, angularity, and particle size.* Soil particles composed of quartz tend to have a higher friction angle than soil particles composed of weak carbonate. Angular soil particles tend to have rougher surfaces and better interlocking ability. Larger-size particles, such as gravel-size particles, typically have higher friction angles than sand.

- *Deposit variability.* Because of variations in soil types, gradations, particle arrangements, and dry density values, the effective friction angle is rarely uniform with depth. It takes considerable judgment and experience in selecting an effective friction angle on the basis of an analysis of data such as shown in Fig. 6.10.

- *Indirect methods.* For many projects, the effective friction angle of the sand is determined from indirect means, such as the standard penetration test (SPT; Fig. 4.13) and the cone penetration test (CPT; Fig. 4.15).

6.5.2 Cohesive Soil

The shear strength of cohesive (plastic) soil is much more complicated than the shear strength of cohesionless soils. Also, in general the shear strength of cohesive (plastic) soils tends to be lower than the shear strength of granular soils. As a result, more shear-induced failures occur in cohesive soils, such as clays, than in granular (nonplastic) soils.

As indicated in Table 5.5, cohesive soils have "fines," which are silt- and clay-size particles that give the soil plasticity, or ability to be molded and rolled. The shear strength of cohesive (plastic) soil can be divided into three broad groups: undrained shear strength, drained shear strength, and drained residual shear strength.

Undrained Shear Strength. As the name implies, the undrained shear strength s_u refers to a shear condition where water does not enter or leave the cohesive soil during the shearing process. In essence, the water content of the soil must remain constant during the shearing process. There are many projects where the undrained shear strength is used in the design analysis. In general, these field situations must involve loading or unloading of the plastic soil at a rate that is much faster than the shear-induced pore water pressures can dissipate. During rapid loading of saturated plastic soils, the shear-induced pore water pressures can dissipate only by the flow of water into (negative shear-induced pore water pressures) or out of (positive shear-induced pore water pressures) the soil. But as indicated in Fig. 5.10, cohesive (plastic) soil has a low permeability, and if the load is applied quickly enough, there will not be enough time for water to enter or leave the plastic soil.

Section 5.2.3 describes various laboratory tests that can be performed to determine the undrained shear strength s_u of plastic soil. For example, the undrained shear strength could be determined from the unconfined compression test. The undrained shear strength could also be determined by using the miniature lab vane or the torvane device. The undrained shear strength could even be determined in the field, by using the vane shear test (VST, see Fig. 4.17).

Figures 6.12 and 6.13 (from Day, 1980, and Ladd et al., 1980) present examples of the undrained shear strength s_u versus depth for Borings E1 and F1 excavated in an offshore deposit of Orinoco clay (created by sediments from the Orinoco River, Venezuela). This data was needed for the design of offshore oil platforms to be constructed off the coast of Venezuela. For this project, the undrained shear strength s_u of the clay deposit was needed for the design of the oil platforms because of the critical storm wave loading condition, which is a quick loading condition compared to the drainage ability of 140 ft of clay.

The Orinoco clay can be generally classified as a clay of high plasticity (CH). This offshore clay deposit can be considered to be a relatively uniform soil deposit. The plot on the left side of Figs. 6.12 and 6.13 shows the water content w of the Orinoco clay versus depth. At Boring E1, the water content decreases from about 80 percent to 50–55 percent at a depth of 140 ft. At Boring F1, the water content decreases from about 90 percent to 60–70 percent at a depth of 130 ft.

The plot on the right side of Figs. 6.12 and 6.13 shows the undrained shear strength s_u versus depth. The undrained shear strength was obtained from the torvane device, laboratory vane, and unconfined compression test (UUC). Note in Figs. 6.12 and 6.13 that there is a distinct discontinuity in the undrained shear strength s_u at a depth of 60 ft for Boring E1 and 40 ft for Boring F1. This discontinuity is a result of using different sampling procedures. Above a depth of 60 ft at Boring E1 and 40 ft at Boring F1, samplers were hammered into the clay deposit, causing sample disturbance and a lower shear strength value for the upper zone of clay. For the deeper zone of clay, a WIP sampling procedure was utilized, which produced less sample disturbance and hence a higher undrained shear strength.

There is a considerable amount of scatter in the undrained shear strength s_u shown in Figs. 6.12 and 6.13. This is not unusual for soil deposits, and nonuniform deposits often

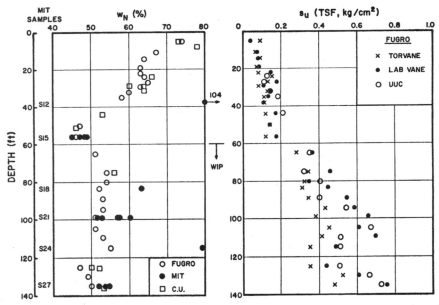

FIGURE 6.12 Water content and undrained shear strength s_u versus depth for Orinoco clay at Boring E1. *(From Day, 1980, and Ladd et al., 1980.)*

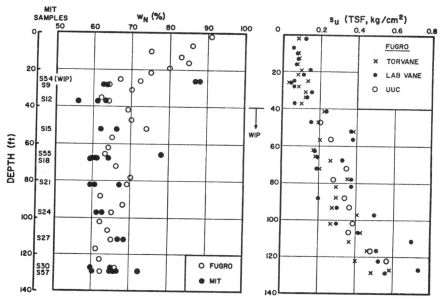

FIGURE 6.13 Water content and undrained shear strength s_u versus depth for Orinoco clay at Boring F1. (*From Day, 1980, and Ladd et al., 1980.*)

have much larger scatter in data than that shown in Figs. 6.12 and 6.13. It takes considerable experience and judgment to decide what value of undrained shear strength s_u to use on the basis of the data shown in Figs. 6.12 and 6.13. One approach is to use an average line through the data points, such as:

$$s_u = 0.0043\, z \qquad (6.33)$$

where s_u = undrained shear strength in tons per square foot (or kg/cm²) and z = depth in feet.

For the Orinoco clay deposit, the average total unit weight γ_t of the Orinoco clay is 105 pcf (16.5 kN/m³) at Boring E1 and 101 pcf (15.9 kN/m³) at Boring F1. This offshore clay is submerged, in a saturated state, and has hydrostatic pore water pressures u. The vertical effective stress σ'_v would be:

$$\sigma'_v\,(\text{psf}) = \gamma_b\, z = (\gamma_t - \gamma_w)\, z = (105 - 62.4)\, z = 43\, z$$

or in tsf,

$$\sigma'_v = 0.022\, z \qquad \text{for Boring E1} \qquad (6.34a)$$

$$\sigma'_v\,(\text{psf}) = \gamma_b\, z = (\gamma_t - \gamma_w)\, z = (101 - 62.4)\, z = 39\, z$$

or in tsf,

$$\sigma'_v = 0.020\, z \qquad \text{for Boring F1} \qquad (6.34b)$$

Substituting Eqs. (6.34) into Eq. (6.33), gives

$$\frac{s_u}{\sigma'_v} = 0.20 \quad \text{for Boring E1} \qquad (6.35a)$$

$$\frac{s_u}{\sigma'_v} = 0.22 \quad \text{for Boring F1} \qquad (6.35b)$$

The parameter s_u/σ'_v is known as the *normalized undrained shear strength*. It is a valuable parameter for the analysis of the undrained shear strength of saturated clay deposits. For example, note in Fig. 4.41 that the values of s_u/σ'_v (labeled *c/p* in Fig. 4.41) have been listed for the various clay deposits. As a check on the undrained shear strength tests, the value of s_u/σ'_v can be compared with published results, such as Jamiolkowski et al., 1985:

$$\frac{s_u}{\sigma'_v} = (0.23 \pm 0.04)\, \text{OCR}^{0.8} \qquad (6.36)$$

where OCR = the overconsolidation ratio, defined as the largest vertical effective stress ever experienced by the soil deposit (σ'_p) divided by the existing vertical effective stress σ'_v. The overconsolidation ratio OCR and the preconsolidation pressure σ'_p are further discussed in Sec. 7.4. The Orinoco clay deposit is essentially normally consolidated (OCR = 1), and thus Eq. (6.36) indicates that s_u/σ'_v should be approximately equal to 0.23 ± 0.04, which is very close to the values indicated in Eq. (6.35). Thus assuming an average undrained shear strength line through the data points in Figs. 6.12 and 6.13 seems to be a reasonable approach on the basis of a comparison of s_u/σ'_v with published results.

Undrained Shear Strength in Terms of Total Stress Parameters

Unconsolidated Undrained Triaxial Compression Test (ASTM D 2850-95). An alternative approach to determining the undrained shear strength s_u of the plastic soil is to obtain the undrained shear strength in terms of the total shear strength parameters c and ϕ. The usual process is to perform triaxial compression tests on specimens of the plastic soil. The triaxial test is discussed in Sec. 5.2.3. For the unconsolidated undrained triaxial compression test (ASTM D 2850-95, 1998), the cylindrical specimen of cohesive soil is placed in the center of the triaxial apparatus, sealed in a rubber membrane, and then subjected to a confining pressure without allowing the soil specimen to have a change in water content at any time during the triaxial testing. The soil is sheared in the triaxial apparatus (see Fig. 5.4) by increasing the vertical stress on the soil specimen.

The results of a triaxial test can be graphically represented by the Mohr circle, which is illustrated in Fig. 6.14 (adapted from Lambe and Whitman, 1969). In geotechnical engineering, compressive stresses are considered to be positive. In the triaxial compression test, the vertical stress at failure is the major principal stress σ_1 and is equal to the cell pressure plus the vertical load from the piston divided by the area of the specimen [using an area correction, Eq. (5.3)]. The horizontal stress is the minor principal stress σ_3 and is equal to the cell pressure. By knowing the major (σ_1) and minor (σ_3) principal stresses at failure, a Mohr circle can be drawn such as shown in Fig. 6.14. The major principal stress minus the minor principal stress (i.e., $\sigma_1 - \sigma_3$) is referred to as the *deviator* or *deviatoric stress* or *stress difference*. Note that the maximum shear stress τ_{max} is equal to one-half the deviator stress and equals the radius of the Mohr circle.

In geotechnical engineering, usually only the upper half of the Mohr circle is used in the analysis. Figure 6.15 shows the results of four unconsolidated undrained triaxial compression tests (ASTM D 2850-95, 1998) performed on a plastic soil (LL = 64, PI = 33) having a dry unit weight $\gamma_d = 11.0$ kN/m^3 (70 pcf), a water content $w = 49.3$ percent, and a

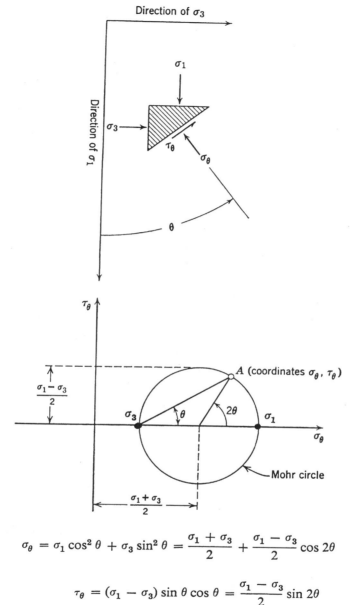

$$\sigma_\theta = \sigma_1 \cos^2 \theta + \sigma_3 \sin^2 \theta = \frac{\sigma_1 + \sigma_3}{2} + \frac{\sigma_1 - \sigma_3}{2} \cos 2\theta$$

$$\tau_\theta = (\sigma_1 - \sigma_3) \sin \theta \cos \theta = \frac{\sigma_1 - \sigma_3}{2} \sin 2\theta$$

FIGURE 6.14 Mohr circle. Upper diagram shows the definition of terms and the lower diagram shows the graphical construction of the Mohr circle. The equations for the calculation of normal stress σ_θ and shear stress τ_θ on a given plane are also indicated in this figure. (*Adapted from Lambe and Whitman, 1969.*)

degree of saturation $S = 95$ percent. Plots of the deviator stress versus axial strain for these four tests are also shown in Fig. 6.15. Note that this stress-strain plot is similar to the shear stress-deformation plot shown in Fig. 6.11. The stress-strain plots shown in Fig. 6.15 do not have a peak point and are indicative of normally consolidated (OCR = 1) clays.

These four triaxial tests are labeled 1 through 4 in Fig. 6.15, and the data is summarized below:

- *Test No. 1.* Confining pressure = 0 (hence this is identical to an unconfined compression test). The vertical stress at failure $(\sigma_1) = 2.87$ psi (19.8 kPa). Thus the Mohr circle drawn in Fig. 6.15 has $\sigma_1 = 2.87$ psi, $\sigma_3 = 0$, and a radius of 1.44 psi.
- *Test No. 2.* Confining pressure = 1.70 psi. The vertical stress at failure $(\sigma_1) = 5.87$ psi (40.5 kPa). Thus the Mohr circle drawn in Fig. 6.15 has $\sigma_1 = 5.87$ psi, $\sigma_3 = 1.70$ psi, and a radius of 2.1 psi.
- *Test No. 3.* Confining pressure = 3.50 psi. The vertical stress at failure $(\sigma_1) = 7.72$ psi (53.2 kPa). Thus the Mohr circle drawn in Fig. 6.15 has $\sigma_1 = 7.72$ psi, $\sigma_3 = 3.50$ psi, and a radius of 2.1 psi.
- *Test No. 4.* Confining pressure = 6.90 psi. The vertical stress at failure $(\sigma_1) = 10.93$ psi (75.4 kPa). Thus the Mohr circle drawn in Fig. 6.15 has $\sigma_1 = 10.93$ psi, $\sigma_3 = 6.90$ psi, and a radius of 2.0 psi.

For Test No. 1, the undrained shear strength s_u, which is the peak of the Mohr circle, is equal to 1.44 psi (9.9 kPa). Test No. 2 has a higher shear strength of 2.1 psi (14.5 kPa). This higher shear strength for Test No. 2 can be attributed to the slight compression of the clay when a confining pressure of 1.70 psi (11.7 kPa) was applied to the soil specimen. Initially, the soil specimen was not saturated ($S = 95$ percent), but the confining pressure compressed the clay and created a saturated condition for Tests No. 2 through 4.

Note that Tests No. 2 through 4 have essentially the same shear strength (peak point on the Mohr circle). The reason is that the increased cell pressure for Tests No. 3 and 4 did not densify the soil specimen, but rather only increased the pore water pressure of the clay specimens. In essence, the effective stress did not change for Tests No. 2 through 4 and thus the shear strength could not increase. This is an important concept in geotechnical engineering, that the shear strength of saturated soil can not increase unless the effective stress of the soil first increases.

In Fig. 6.16, the failure envelope has been drawn for Tests No. 2 through 4. Using Eq. (6.32) (Mohr-Coulomb failure law), the total stress parameters are $\phi = 0$ and $c = 2.1$ psi (14.5 kPa). These types of shear strength results have been termed the "$\phi = 0$ concept." It is important to recognize that this concept does not say that the clay has zero frictional resistance (i.e., $\phi = 0$). Instead, this concept indicates that there is no shear strength increase in the clay for the condition of a rapidly applied load. The clay must first consolidate (flow of water from the clay) before it can increase its shear strength.

An example of the use of the shear strength data shown in Fig. 6.16 would be the quick construction of a slope composed of this clay. Because the construction is quick, there would be no drainage (i.e., consolidation) of the clay during construction of the slope. To model this end-of-construction condition of the slope, it would be appropriate to perform a slope stability analysis using the undrained shear strength parameters from Fig. 6.16, i.e., $\phi = 0$ and $c = 2.1$ psi (14.5 kPa).

Consolidated Undrained Triaxial Compression Test (ASTM D 4767-95, 1998). In some cases, it may be appropriate to obtain the total stress shear strength parameters c and ϕ from consolidated undrained triaxial compression tests. For example, a structure (such as an oil tank or grain elevator) could be constructed and then sufficient time might elapse so that the saturated plastic soil consolidates under this load. If the oil tank or grain elevator is then quickly filled, the saturated plastic soil would be subjected to an undrained loading.

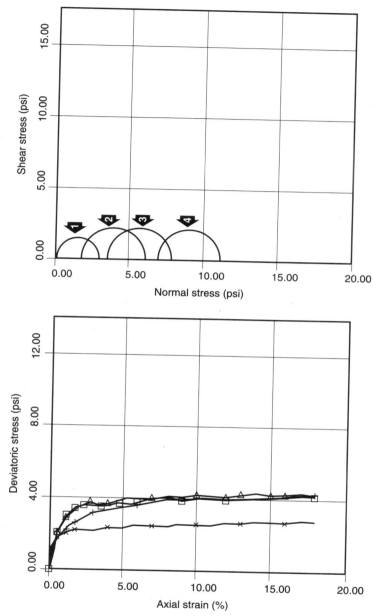

FIGURE 6.15 Upper plot is the Mohr circles and lower plot is the stress-strain curves for four unconsolidated undrained triaxial compression tests on a plastic clay.

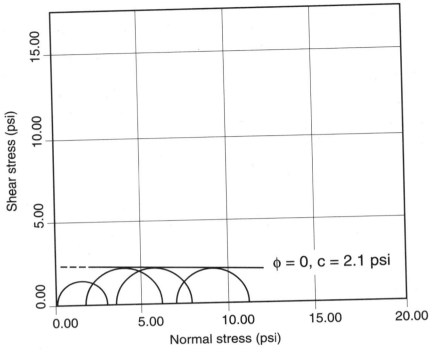

FIGURE 6.16 Upper plot from Fig. 6.15, with the failure envelope plotted for tests No. 2 through No. 4.

This condition can be modeled by performing consolidated undrained triaxial compression tests (ASTM D 4767-95, 1998) in order to determine the total stress parameters c and ϕ. The procedure consists of placing a cylindrical specimen of plastic (cohesive) soil in the center of the triaxial apparatus (Fig. 5.4), sealing the specimen in a rubber membrane, applying a confining pressure, and then allowing enough time for the soil specimen to consolidate. At the completion of consolidation, an axial load is applied to the soil specimen without allowing a change in water content (i.e., undrained loading). At failure, the minor principal stress σ_3 = the confining pressure and the major principal stress σ_1 = the vertical pressure at failure. By knowing σ_3 and σ_1, we can draw the Mohr circle at failure.

Figure 6.17 presents data from three triaxial tests performed in accordance with ASTM D 4767-95 (1998) test specifications (Day and Marsh, 1995). The three specimens used for the triaxial tests were composed of silty clay, having an LL = 54, PI = 28, with a grain size distribution (based on dry weight) of 15 percent fine sand–size particles, 33 percent silt-size particles, and 52 percent clay-size particles finer than 0.002 mm. On the basis of these index properties, the silty clay can be classified as a clay of intermediate plasticity (CI). In order to create the three triaxial specimens, the silty clay was compacted into a cylindrical mold to a dry unit weight of 13.4 kN/m³ (85 pcf, relative compaction = 80 percent) at a water content of 20 percent. These three triaxial tests are labeled 1 through 3 in Fig. 6.17, and the data is summarized below:

- *Test No. 1.* The first silty clay specimen was saturated and allowed to consolidate at an effective confining pressure of 3.5 psi (24 kPa). The total vertical stress at failure (σ_1) = 9.4 psi (65 kPa). Thus the Mohr circle drawn in Fig. 6.17 has σ_1 = 9.4 psi, σ_3 = 3.5 psi, and a radius of 3.0 psi.

- *Test No. 2.* The second silty clay specimen was saturated and allowed to consolidate at an effective confining pressure of 13.9 psi (96 kPa). The total vertical stress at failure (σ_1) = 26.6 psi (183 kPa). Thus the Mohr circle drawn in Fig. 6.17 has $\sigma_1 = 26.6$ psi, $\sigma_3 = 13.9$ psi, and a radius of 6.4 psi.
- *Test No. 3.* The third silty clay specimen was saturated and allowed to consolidate at an effective confining pressure of 27.8 psi (192 kPa). The total vertical stress at failure (σ_1) = 52.2 psi (360 kPa). Thus the Mohr circle drawn in Fig. 6.17 has $\sigma_1 = 52.2$ psi, $\sigma_3 = 27.8$ psi, and a radius of 12.2 psi.

A failure envelope (straight line) has been drawn tangent to the three Mohr circles and the cohesion (y-axis intercept) and friction angle (angle of inclination of the straight line) are $c = 1.2$ psi (8.3 kPa) and $\phi = 16°$. Note in Fig. 6.17 that the failure envelope does not pass through the peak point on the Mohr circle, but is rather tangent to the three Mohr circles. The shear strength τ_f of the silty clay from the consolidated undrained triaxial compression tests can be expressed as follows (Mohr-Coulomb failure law):

$$\tau_f = c + \sigma_n \tan \phi \qquad (6.37)$$

where σ_n = normal stress on the failure (or slip) surface. Note that Eq. (6.37) is identical to Eq. (6.32), except that the former is expressed in terms of total stress and the latter is expressed in terms of effective stress.

Drained Shear Strength. As previously mentioned, in a comparison of the effective friction angle ϕ' from drained direct shear tests on saturated cohesionless soil and the friction

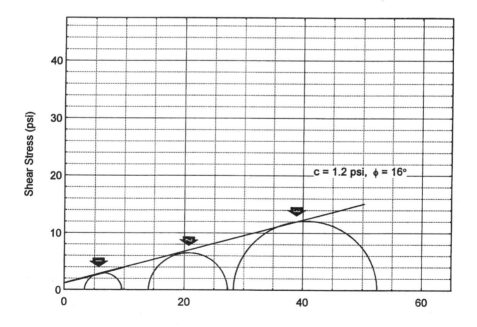

FIGURE 6.17 Consolidated undrained triaxial compression tests performed on a silty clay.

angle ϕ from direct shear tests on the same soil in a dry state, it has been determined that ϕ' is only 1 to 2° lower than ϕ (Terzaghi and Peck, 1967; Holtz and Kovacs, 1981). This slight difference is usually ignored and the friction angle ϕ and effective friction angle ϕ' are typically considered to mean the same thing for cohesionless (nonplastic) soils. Thus in an engineering analysis using cohesionless soil, the shear strength could be expressed as either ϕ (friction angle) or ϕ' (effective friction angle) and they will essentially have the same meaning and value.

For plastic (cohesive) soils, the friction angle ϕ and effective friction angle ϕ' not only have different values, but also have significantly different meaning. Friction angle ϕ and cohesion c determined from undrained shear strength tests should be used only in total stress analyses, while effective friction angle ϕ' and effective cohesion c' should be used only in effective stress analyses. These two types of engineering analyses will be further discussed in Sec. 6.5.3.

The effective friction angle ϕ' and effective cohesion c' of a plastic (cohesive) soil could be obtained by performing drained shear strength tests. But as indicated in Fig. 5.10, plastic soils usually have a very low permeability, and it often takes a considerable amount of time to perform drained shear strength tests.

An alternative procedure which is routinely used in practice is to perform triaxial compression tests with pore pressure measurements. As previously mentioned, consolidated undrained triaxial compression tests can be used to obtain the total stress shear strength parameters (ASTM D 4767-95, 1998). If the pore water pressures are measured during the shearing of the saturated cohesive soil, then the effective stress shear strength parameters c' and ϕ' could also be determined (also ASTM D 4767-95, 1998). This type of triaxial test (ASTM D 4767-95, 1998) is very practical because it can be used to determine the undrained shear strength in terms of the total stress parameters c and ϕ as well as the shear strength in terms of the effective stress parameters c' and ϕ' when the pore water pressures are measured during the shearing of the saturated cohesive soil specimen. For example, Fig. 6.18 presents the Mohr circles at failure in terms of effective stress for the three triaxial tests on compacted silty clay (i.e., same triaxial tests used to produce Fig. 6.17). During the undrained shearing of the three saturated silty clay specimens, the pore water pressures u were measured. In order to plot the Mohr circles at failure in terms of effective stress, the following equations [which are identical to Eq. (6.16)] were used:

$$\sigma'_1 = \sigma_1 - u \qquad (6.38a)$$

$$\sigma'_3 = \sigma_3 - u \qquad (6.38b)$$

where σ'_1 and σ'_3 are the major and minor principal effective stresses at failure. Failure is often defined as the largest vertical load (maximum σ_1) that the cohesive specimen can sustain. Since σ_3 is constant during the triaxial compression test, the maximum σ_1 will occur at exactly the same time as the *maximum deviator stress,* also known as *maximum stress difference* or *maximum value* of $\sigma_1 - \sigma_3$. In dealing with effective stresses, failure could also be defined as the maximum deviator stress (maximum value of $\sigma'_1 - \sigma'_3$) or as the maximum obliquity (maximum value of σ'_1/σ'_3).

The three triaxial tests on the compacted silty clay specimens are labeled 1 through 3 in Fig. 6.18, and the data is summarized below:

- *Test No. 1.* As previously mentioned, the first silty clay specimen was saturated and allowed to consolidate at an effective confining pressure of 3.5 psi (24 kPa). The measured change in pore water pressure u from the start of the undrained loading to failure equals 1.7 psi (12 kPa). From Eq. (6.38), the Mohr circle for Test No. 1 in Fig. 6.18 has a vertical effective stress at failure $(\sigma'_1) = 7.7$ psi (53 kPa) and a horizontal effective stress at failure $(\sigma'_3) = 1.8$ psi (12 kPa).

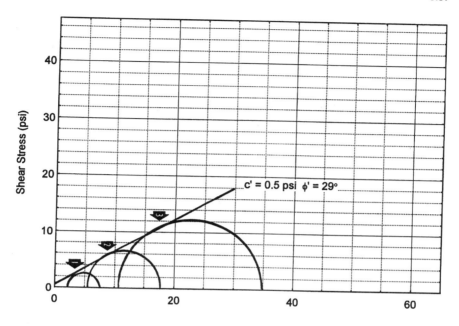

FIGURE 6.18 Mohr circles in terms of effective stress (same triaxial tests used to produce Fig. 6.17).

- *Test No. 2.* As previously mentioned, the second silty clay specimen was saturated and allowed to consolidate at an effective confining pressure of 13.9 psi (96 kPa). The measured change in pore water pressure u from the start of the undrained loading to failure equals 8.5 psi (59 kPa). From Eq. (6.38), the Mohr circle for Test No. 2 in Fig. 6.18 has a vertical effective stress at failure (σ'_1) = 18.1 psi (125 kPa) and a horizontal effective stress at failure (σ'_3) = 5.4 psi (37 kPa).

- *Test No. 3.* As previously mentioned, the third silty clay specimen was saturated and allowed to consolidate at an effective confining pressure of 27.8 psi (192 kPa). The measured change in pore water pressure u from the start of the undrained loading to failure equals 16.8 psi (116 kPa). From Eq. (6.38), the Mohr circle for Test No. 3 in Fig. 6.18 has a vertical effective stress at failure (σ'_1) = 35.4 psi (244 kPa) and a horizontal effective stress at failure (σ'_3) = 11.0 psi (76 kPa).

In order to determine the effective shear strength envelope, a straight line was drawn tangent to the Mohr circles in Fig. 6.18. The effective shear strength envelope can be defined by c' = 0.5 psi (3.4 kPa) and ϕ' = 29°.

By using the triaxial test with pore water pressure measurements (ASTM D 4767-95, 1998), both the shear strength parameters in terms of total stress (Fig. 6.17) and the shear strength parameters in terms of effective stress (Fig. 6.18) were obtained for the compacted silty clay. For this plastic soil, the shear strength parameters are as follows:

- Total stress analysis (Fig. 6.17):

$$c = 1.2 \text{ psi (8.3 kPa)} \quad \text{and} \quad \phi = 16°$$

- Effective stress analysis (Fig. 6.18):

$$c' = 0.5 \text{ psi } (3.4 \text{ kPa}) \quad \text{and} \quad \phi' = 29°$$

In Figs. 6.17 and 6.18, note that the sizes of the Mohr circles for each test number are identical (same radius). Note also, in comparing Fig. 6.18 to 6.17, that there has been a shift to the left of the Mohr circles. The amount of shifting of the Mohr circles is equal to the pore water pressure developed during the triaxial shearing of the three specimens. This can be deduced from Eq. (6.38), which indicates that both σ'_1 and σ'_3 (i.e., the x-axis coordinates of the Mohr circle) are adjusted by the pore water pressure u.

Holtz and Kovacs (1981) discuss the typical values of the effective friction angle ϕ' for plastic soils. They indicate that for undisturbed natural clays, the values of ϕ' range from around 20° for normally consolidated, highly plastic clays up to 30° or more for other types of plastic (cohesive) soil. The value of ϕ' for compacted clay is typically in the range of 25 to 30° and occasionally as high as 35°. In terms of effective cohesion for plastic soil, Holtz and Kovacs (1981) indicate that the value of c' for normally consolidated noncemented clays is very small and can be assumed to be zero for practical work.

If the plastic soil is overconsolidated (e.g., by compaction), then there may be an effective cohesion intercept c', such as shown in Fig. 6.18. Often the higher the overconsolidation ratio (i.e., the higher the preloaded condition of the plastic soil), the higher the effective cohesion intercept. It should be mentioned that the failure envelope at low effective stresses is often curved and passes through the origin, and the effective cohesion c' is usually an extrapolated value that does not represent the actual effective shear strength at low effective stresses (Maksimovic, 1989b). This will be further discussed in Sec. 10.4.

Pore Water Pressure Parameters A and B. The Skempton (1954) pore water pressure parameters relate the change in total stresses for an undrained loading. For an undrained triaxial test, the general expression is as follows:

$$\Delta u = B[\Delta\sigma_3 + A(\Delta\sigma_1 - \Delta\sigma_3)] \tag{6.39}$$

where Δu = change in pore water pressure generated during undrained shear (kPa or psf)
  $\Delta\sigma_1$ = change in total major principal stress during undrained shear (kPa or psf)
  $\Delta\sigma_3$ = change in total minor principal stress during undrained shear (kPa or psf)
  A, B = Skempton pore water pressure coefficients (dimensionless)

In performing a consolidated undrained triaxial compression test on a saturated plastic soil (ASTM D 4767-95, 1998), a high back pressure is used to be sure of saturation of the triaxial specimen. After the cohesive soil has consolidated and prior to triaxial shearing, the B-value should be calculated. The process consists of preventing drainage of the plastic soil and then applying a confining pressure $\Delta\sigma_c$ to the specimen by increasing the cell pressure. In this case, $\Delta\sigma_c = \Delta\sigma_1 = \Delta\sigma_3$ and Eq. (6.39) reduces to:

$$B = \frac{\Delta u}{\Delta\sigma_3} \tag{6.40}$$

By determining the increase in confining pressure ($\Delta\sigma_c = \Delta\sigma_3$) and measuring the increase in pore water pressure Δu during application of the confining pressure, the B-value can be calculated from Eq. (6.40). Table 6.2 lists values of measured B-values for various types of clay. Note in Table 6.2 that if the clay is saturated ($S = 100$ percent), the B-value is essentially equal to 1.0. This data shows that a stress applied to a saturated cohesive soil will be carried by an increase in pore water pressure and not the soil particle skeleton (i.e.,

$\Delta u = \Delta\sigma_c$). The measurements of the B-value on saturated clays are further confirmation that the shear strength of a saturated soil can not increase unless the effective stress of the soil first increases. The increase in pore water pressure due to applied loads will be further discussed in Sec. 7.4 (consolidation).

Note in Table 6.2 that if the soil is not saturated, the B-value is less than 1.0. The lower the degree of saturation, the lower the B-value. Thus the lower the degree of saturation, the greater the portion of the load that will be immediately carried by the soil particle skeleton. In the case of a completely dry soil ($S = 0$ percent), the B-value is zero and all the load will be immediately carried by the soil particle skeleton.

While the B-value is calculated before the triaxial shearing part of the test is performed, the A-value is calculated during the actual shearing of the cohesive specimen. For the conventional triaxial compression test in which the confining pressure is held constant throughout shearing, $\Delta\sigma_3 = 0$, and if the degree of saturation is 100 percent ($B = 1.0$), then Eq. (6.39) reduces to the following:

$$A = \frac{\Delta u}{\Delta\sigma_1} \tag{6.41}$$

Based on triaxial compression tests and using Eq. (6.41), Skempton and Bjerrum (1957) provided typical A-values at failure for various plastic soils, as follows:

- Very sensitive soft clay: A-value at failure is greater than 1
- Normally consolidated clay: A-value at failure is 0.5 to 1
- Overconsolidated clay: A-value at failure is 0.25 to 0.5
- Heavily overconsolidated clay: A-value at failure is 0 to 0.25

The reason for high A-values for very soft sensitive clay and normally consolidated clay is the contraction of the soil structure during shear, and because it is an undrained loading, positive pore water pressures will develop. The reason for zero or even negative A-values for highly overconsolidated soil is the dilation of the soil structure during shear, and because it is an undrained loading, negative pore water pressures will develop. The A-value can indicate a normally consolidated soil versus a heavily overconsolidated soil.

For soil compacted at optimum moisture content, the A-value at failure from a consolidated undrained triaxial compression test (ASTM D 4767-95, 1998) can indicate the type

TABLE 6.2 B-Values for Various Clays

Type of soil (1)	Degree of saturation, % (2)	B-value (3)
London clay (OC)	100	0.9981
London clay (NC)	100	0.9998
Vicksburg clay	100	0.9990
Kawasaki clay	100	0.999
Boulder clay	93	0.69
Boulder clay	87	0.33
Boulder clay	76	0.10

Notes: OC = overconsolidated clay, NC = normally consolidated clay.
Sources: Skempton (1954, 1961) and Lambe and Whitman (1969).

of soil behavior. For example, Fig. 6.19 shows a plot of the A-value at failure versus the percentage of swell or collapse measured in an oedometer apparatus (Day and Marsh, 1995). Table 6.3, which indicates the type of soil behavior in terms of swell or collapse versus the triaxial A-value, was developed from Fig. 6.19. As indicated in Fig. 6.19, soil that collapsed in the oedometer apparatus had a high A-value from the triaxial test, which is indicative of contractive behavior, and soil that tended to swell conversely had a low-to-negative A-value from the triaxial test, which is indicative of dilative behavior.

Stress Paths. A stress path is defined as a series of stress points that are connected together to form a line or curve. Fig. 6.20a and b illustrates the stress path for a triaxial compression test. The state of stress for the triaxial test starts at point A, where the cohesive specimen is subjected to isotropic compression (i.e., $\sigma_c = \sigma_1 = \sigma_3$). Then as the specimen is axially loaded, σ_1 increases while σ_3 remains constant. This results in a series of ever larger Mohr circles such as shown in Fig. 6.20a. Mohr circle E represents stress conditions at failure.

Instead of drawing a series of Mohr circles to represent the change in stress during the triaxial shear test, a stress path can be drawn as shown in Fig. 6.20b. The peak point of each Mohr circle (i.e., points A, B, C, D, and E) is used to draw the stress path on a p-q plot. The peak point of each Mohr circle is defined as:

In terms of total stresses:

$$p = \frac{\sigma_1 + \sigma_3}{2} \quad \text{and} \quad q = \frac{\sigma_1 - \sigma_3}{2} \qquad (6.42a)$$

In terms of effective stresses:

$$p' = \frac{\sigma'_1 + \sigma'_3}{2} \quad \text{and} \quad q = \frac{\sigma'_1 - \sigma'_3}{2} \qquad (6.42b)$$

Note that p represents the center of the Mohr circle and q is the radius of the Mohr circle. For a given stress state defined by σ_1, σ_3, and u, the value of q is the same for Eqs. (6.42a) and (6.42b). As previously discussed for Figs. 6.17 and 6.18, the Mohr circles for a given total and effective stress state are only shifted and the radius (i.e., value of q) of the Mohr circle does not change.

TABLE 6.3 Soil Behavior versus A-value

Soil type (1)	Triaxial A-value at failure*			
	A-value less than 0 (2)	A-value = 0 to 0.4 (3)	A-value = 0.4 to 0.6 (4)	A-value greater than 0.6 (5)
Compacted granular soil	No collapse	No appreciable collapse	Transition: possible collapse	Significant collapse
Compacted clay	Very high swell	Significant swell	Transition: some swell or collapse	Significant collapse

Notes: This table is valid for soil compacted at optimum moisture content. A-values at failure determined from consolidated undrained triaxial compression tests (ASTM D 4767-95, 1998).
*Shear failure defined as maximum obliquity (σ'_1/σ'_3).

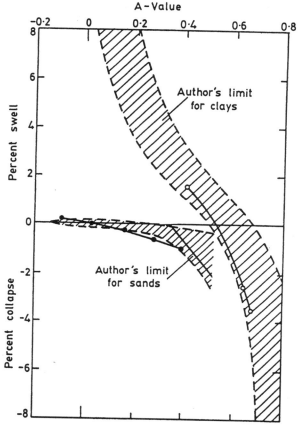

FIGURE 6.19 Percentage swell or collapse versus triaxial A-value at failure.

The shear strength parameters from the conventional plot (Figs. 6.17 and 6.18) are related to the shear strength parameters from the p-q plot by the following relationships:
In terms of total stresses:

$$\sin \phi = \tan \alpha \quad \text{and} \quad c \cos \phi = a \qquad (6.43a)$$

In terms of effective stresses:

$$\sin \phi' = \tan \alpha' \quad \text{and} \quad c' \cos \phi' = a' \qquad (6.43b)$$

Total stress paths, such as shown in Fig. 6.20b, or effective stress paths can be drawn for the consolidated undrained triaxial compression test (ASTM D 4767-95, 1998). In drawing effective stress paths, the effective principal stresses [Eq. (6.38)] are first determined during the actual shearing process of the specimen. Then Eq. (6.42b) is used to calculate the p-q points that are connected together to form the effective stress paths.

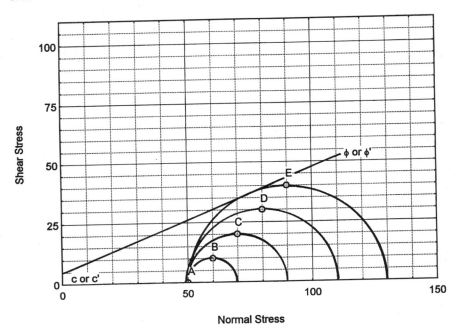

FIGURE 6.20a Series of Mohr circles representing the state of stress during a triaxial compression test (Mohr circle E represents the Mohr circle at failure). Figure 6.20b shows the stress path for these Mohr circles.

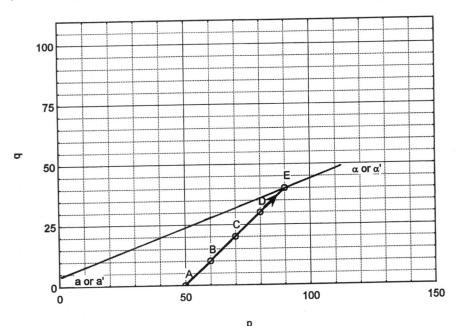

FIGURE 6.20b Peak points of the Mohr circles from Fig. 6.20a connected together to form a stress path.

Figure 6.21 shows the effective stress paths for the three tests performed on the compacted silty clay (i.e., same triaxial tests used to produce Figs. 6.17 and 6.18). Based on a straight line drawn through the peak points in the *p-q* plot, the parameters are $a' = 0.4$ psi and $\alpha' = 26°$. Using the relationships given in Eq. (6.43*b*), the shear strength parameters in terms of effective stress are identical to the parameters from Fig. 6.18 ($c' = 0.5$ psi and $\phi' = 29°$).

Note in Fig. 6.21 how the effective stress path for Test No. 1 bends to the right. This is indicative of overconsolidated soil behavior (precompression to a higher stress) which is a result of the initial compaction of the specimen. For Tests No. 2 and No. 3, the effective stress paths bend to the left, which is indicative of normally consolidated soil behavior. For Tests No. 2 and No. 3, the initial precompression stress due to compaction has been negated by the higher confining pressures associated with these two tests. This is one of the advantages of stress paths: they can visually illustrate different types of soil behavior.

Drained Residual Shear Strength. The beginning of this section on the shear strength of cohesive soil (Sec. 6.5.2) states that the shear strength can be divided into three broad groups:

1. *Undrained shear strength.* Examples are the undrained shear strength measured directly from field or laboratory tests (s_u) and the undrained shear strength in terms of the total stress parameters c and ϕ (Fig. 6.17).

2. *Drained shear strength.* The drained shear strength can be obtained from drained shear strength tests, although a more common approach is to perform consolidated undrained triaxial compression tests with pore water pressure measurements (ASTM D 4767-95, 1998) which are used to obtain the shear strength in terms of the effective stress parameters c' and ϕ' (Fig. 6.18).

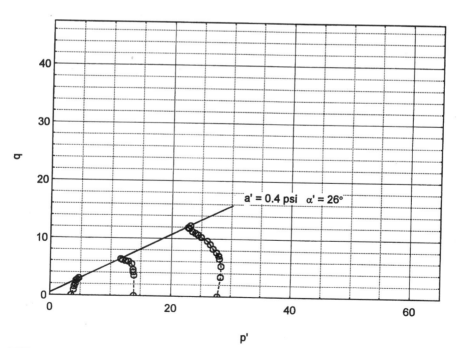

FIGURE 6.21 Effective stress paths (same triaxial tests used to produce Figs. 6.17 and 6.18).

3. *Drained residual shear strength.* The drained residual shear strength is discussed in Sec. 5.2.3. The residual shear strength is defined as the remaining (or residual) shear strength of plastic soil after a considerable amount of shear deformation has occurred. As previously mentioned in Sec. 5.2.3, the drained residual shear strength ϕ'_r can be very low. For example, the data presented in Fig. 5.6 indicates a ϕ'_r of 6° at high effective normal stresses. As shown in Fig. 5.6, the failure envelope in terms of the drained residual shear strength is often nonlinear.

Factors that Affect Shear Strength. There can be many factors that affect the shear strength of plastic soils, such as:

1. *Sample disturbance.* As discussed in Sec. 4.4.1, a soil sample can not be taken from the ground and be in a perfectly undisturbed state. Disturbance causes a decrease in effective stress, a reduction in the interparticle bonds, and a rearrangement of the soil particles. Sample disturbance can cause the greatest reduction in shear strength than any other factor. As an example of the effects of sample disturbance, Fig. 6.22 shows the undrained shear strength s_u of "undisturbed" and remolded Orinoco clay and indicates about a 75 percent reduction in the undrained shear strength (Ladd et al., 1980; Day, 1980).

For clays having a high sensitivity, such as quick clays, disturbance can severely affect the shear strength. In severe instances, quick clays can be disturbed to such an extent that they even become liquid (essentially no shear strength).

2. *Strain rate.* The faster a soil specimen is sheared, i.e., subjected to a fast strain rate, the higher the value of the undrained shear strength s_u. For torvane, lab vane, and UUC tests, the strain rate is very fast, with failure occurring in only a few minutes or less. In Fig. 6.22, torvane tests were performed on the Orinoco clay with the tests having different times to failure. Note in this figure that for both the undisturbed specimens and the remolded specimens that a slower strain rate results in a lower undrained shear strength. This is because a slower strain rate allows the soil particles more time to slide, deform, and creep around each other, resulting in a lower undrained shear strength.

3. *Anisotropy.* Soils, especially clays, have a natural strength variation where s_u depends on the orientation of the failure plane, thus s_u along a horizontal failure plane will not equal s_u along a vertical failure plane. For example, note the very fine layering of the Orinoco clay that is visible from the radiograph shown in Fig. 5.13. Lab vane and torvane tests have simultaneous horizontal and vertical failure planes with an undrained shear strength that rarely equals s_u from a UUC test which has an oblique failure plane.

These three factors listed above can affect soil in different ways. For example, it has been stated that for the vane shear test, the values of the undrained shear strength from field vane tests are likely to be higher than can be mobilized in practice (Bjerrum, 1973). This has been attributed to a combination of anisotropy of the soil and the fast rate of shearing involved with the vane shear test. Because of these factors, Bjerrum (1973) has proposed for field vane tests performed on saturated normally consolidated clays that the undrained shear strength s_u be reduced on the basis of the plasticity index of the clay. No reduction is required for clay having a plasticity index of 20, but for a PI in excess of 20, Bjerrum (1973) recommends that the undrained shear strength s_u should be progressively decreased (40 percent reduction at a PI = 100).

Like the vane shear test, the unconfined compression test can provide values of undrained shear strength that are too high because the saturated plastic soil is sheared too quickly. In many cases, the unconfined compression test has provided reasonable values of undrained shear strength because of compensating factors: too high an undrained shear strength (due to the fast strain rate) that is compensated by a reduction in undrained shear strength (due to disturbance) (Ladd, 1971).

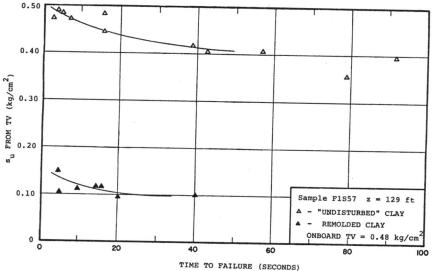

FIGURE 6.22 Effects of sample disturbance and strain rate on the undrained shear strength of Orinoco clay. (*From Day, 1980.*)

Because of all of these factors (sample disturbance, strain rate, and anisotropy), considerable experience and judgment are needed for the selection of shear strength parameters for cohesive soil.

6.5.3 Total Stress and Effective Stress Analyses

An understanding of total stress analysis and effective stress analysis is essential in geotechnical engineering, since these two methods are used in all types of earth projects, in activities such as slope stability evaluation, earth pressure calculations, and foundation design. Total stress analysis and effective stress analysis are defined as follows:

1. *Total stress analysis.* Total stress analysis uses the undrained shear strength of the soil. Total stress analysis is typically used only for cohesive (i.e., plastic) soil. It is often used for the evaluation of foundations and embankments to be supported by cohesive soil. The actual analysis is performed for rapid loading or unloading conditions often encountered during the construction phase or just at the end of construction. This analysis is applicable to field situations where there is a change in shear stress which occurs quickly enough that soft cohesive soil does not have time to consolidate, or in the case of heavily overconsolidated cohesive soils, the negative pore water pressures do not have time to dissipate. For this reason, the total stress analysis is often termed a *short-term analysis*. The total stress analysis uses the total unit weight γ_t of the soil, and the location of the groundwater table is not considered in the analysis.

Total stress analysis is also applicable when there is a sudden change in loading condition of a cohesive soil. Examples include wind or earthquake loadings that exert compression or tension forces on piles supported by clay strata, or the rapid drawdown of a reservoir which induces shear stresses on the clay core of the dam. The total stress analysis uses the undrained shear strength s_u of the plastic soil obtained from vane tests or unconfined compression tests. An alternative approach is to use the total stress parameters c and ϕ from unconsolidated undrained triaxial compression tests (ASTM D 2850-95, 1998) or the consolidated undrained triaxial test (ASTM D 4767-95, 1998).

An advantage of the total stress analysis is that the undrained shear strength is obtained from tests (such as the vane test or unconfined compression test) that are easy to perform. A major disadvantage of this approach is that the accuracy of the undrained shear strength is always in doubt because it depends on the shear-induced pore water pressures (which are not measured), which in turn depend on the many details (e.g., sample disturbance, strain rate effects, and anisotropy) of the test procedures (Lambe and Whitman, 1969).

2. Effective stress analysis. Effective stress analysis uses the drained shear strength parameters c' and ϕ'. Except for earthquake loading (i.e., liquefaction), essentially all analyses of cohesionless soils are made by using effective stress analysis. For cohesionless soil, $c' = 0$ and the effective friction angle ϕ' is often obtained from drained direct shear tests, drained triaxial tests, or from empirical correlations, such as the standard penetration test (Fig. 4.13) or the cone penetration test (Fig. 4.15).

For cohesive soils, the effective shear strength could be obtained from drained shear strength tests, although a more common procedure is to perform consolidated undrained triaxial compression tests with pore water pressure measurements performed on saturated cohesive soil in order to determine the effective shear strength parameters c' and ϕ' (ASTM D 4767-95, 1998).

Effective stress shear strength parameters are used for all long-term analysis where conditions are relatively constant. Examples include the long-term stability of slopes, embankments, earth supporting structures, and bearing capacity of foundations for cohesionless and cohesive soil. Effective stress analysis can also be used for any situation where the pore water pressures induced by loading can be estimated or measured. The effective stress analysis can use either one of the following approaches: (1) total unit weight of the soil (γ_t) and the boundary pore water pressures or (2) buoyant unit weight of the soil (γ_b) and the seepage forces.

An advantage of effective stress analysis is that it more fundamentally models the shear strength of the soil, because shear strength is directly related to effective stress. A major disadvantage of the effective stress analysis is that the pore water pressures must be included in the analysis. The accuracy of the pore water pressure is often in doubt because of the many factors which affect the magnitude of pore water pressure changes, such as the determination of changes in pore water pressure resulting from changes in external loads (Lambe and Whitman, 1969). For effective stress analysis, assumptions are frequently required concerning the pore water pressures that will be used in the analysis.

Throughout the book, total stress analysis and effective stress analysis will be discussed as they relate to the different geotechnical engineering projects.

6.6 PERMEABILITY AND SEEPAGE

This last section of Chap. 6 presents the basic principles of permeability and groundwater seepage through soil. In the previous sections, the discussion of groundwater includes the following:

- *Subsurface exploration (Sec. 4.3):* One of the main purposes of subsurface exploration is to determine the location of the groundwater table.

- *Laboratory testing (Sec. 5.2.4):* The rate of flow through soil is dependent on its permeability. The constant-head permeameter (Fig. 5.8) or the falling-head permeameter (Fig. 5.9) can be used to determine the coefficient of permeability (also known as the *hydraulic conductivity*). Equations to calculate the coefficient of permeability in the laboratory are based on Darcy's Law ($v = ki$).

- *Pore water pressure (Sec. 6.3):* The pore water pressure u for a level groundwater table with no flow and hydrostatic conditions is calculated by using Eq. (6.18) for soil below the groundwater table and Eq. (6.19) for soil above the groundwater table that is saturated because of capillary rise.

- *Shear strength (Sec. 6.5):* The shear strength of saturated soil is directly related to the water that fills the soil pores. Normally consolidated plastic soils have an increase in pore water pressure as the soil structure contracts, while heavily overconsolidated plastic soils have a decrease in pore water pressure as the soil structure dilates during shear.

The main items that will be discussed in this section are permeability, seepage velocity, seepage force, and two-dimensional flow nets. In the following chapters, these principles will be used for all types of engineering analyses, such as consolidation, slope stability, and the design of earth dams. Chapter 16 will specifically discuss groundwater and moisture migration.

Permeability. As mentioned in Sec. 5.2.4, cohesive soils such as clays have very low values of the coefficient of permeability k, while cohesionless soils such as clean sands and gravels can have a coefficient of permeability k that is a billion times larger than that of cohesive soils (see Fig. 5.10). The reason lies in the size of the drainage paths through the soil. Open graded gravel has very large, interconnected void spaces. At the other extreme are clay-size particles, which produce minuscule void spaces with a large resistance to flow and hence a very low permeability.

In many cases, the determination of the coefficient of permeability k from laboratory tests on small soil specimens may not be representative of the overall field condition. A common method of determining the field coefficient of permeability is through the measurements of the change in water levels in open standpipes. Figure 6.23a presents various standpipe conditions, and Fig. 6.23b presents the equations used to calculate the field permeability k.

Although the type of soil is the most important parameter that governs the coefficient of permeability k of saturated soil, there are many other factors that affect k, such as:

Void ratio. For a given soil, the higher the void ratio, the higher the coefficient of permeability. For example, Fig. 6.24 presents data from various soils and shows how the coefficient of permeability k increases as the void ratio increases.

Particle size distribution. A well-graded soil will tend to have a lower coefficient of permeability k than a uniform soil. This is because a well-graded soil will have more particles to fill in the void spaces and make the flow paths smaller.

Soil structure. Some types of soil structure, such as the honeycomb structure or cardhouse structure, have larger interconnected void spaces with a corresponding higher permeability.

Layering of soil. Natural soils are often stratified and have layers or lenses of permeable soil, resulting in a much higher horizontal permeability than vertical permeability. This is important because often it is only the vertical permeability that is measured in the falling-head and constant-head permeameters (see Figs. 5.8 and 5.9).

Soil imperfections or discontinuities. Natural soil has numerous imperfections, such as root holes, animal burrows, joints, fissures, seams, and soil cracks, that significantly

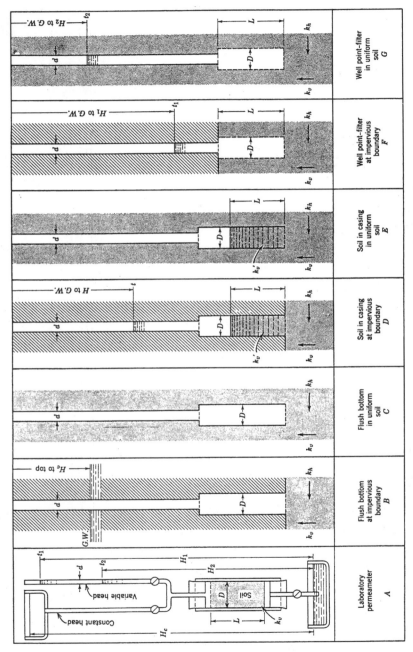

FIGURE 6.23a Methods for determining permeability. (*From Lambe and Whitman, 1969; reproduced with permission from John Wiley & Sons.*)

increase the permeability of the soil mass. For compacted clay liners, there may be incomplete bonding between fill lifts, which allows water to infiltrate through the fill mass.

Seepage Velocity. From Bernoulli's energy equation, the total head h is equal to the sum of the velocity head, pressure head, and elevation head. Head has units of length. For seepage problems in soil, the velocity head is usually small enough to be neglected, and thus for laminar flow in soil, the total head h is equal to the sum of the pressure head h_p and the elevation head h_e.

In order for there to be flow of water through soil, there must be a change in total head Δh. For example, given a level groundwater table with hydrostatic pore water pressures, the total head will be identical at all points below the groundwater table. Since there is no change in total head ($\Delta h = 0$), there can be no flow of water.

Darcy's law ($v = ki$) is used for the analysis of laminar flow in soil. It relates the velocity v to the change in total head ($i = \Delta h/L$, where L = length between the total head differences). Since the coefficient of permeability k is measured in a permeameter where the quantity of water is measured flowing out of (constant-head permeameter) or into (falling-head permeameter) the soil, the velocity v is actually a "superficial velocity" and not the actual velocity of flow in the soil voids. The "seepage velocity" v_s is related to the superficial velocity v, by the following equation:

$$v_s = \frac{v}{n} = \frac{ki}{n} = \frac{k \Delta h}{nL} \qquad (6.44)$$

where v_s = seepage velocity, i.e., velocity of flow of water in the soil (cm/s or in./min)
v = superficial velocity, i.e., velocity of flow of water into or out of the soil (cm/s or in./min)
k = the coefficient of permeability, also known as the hydraulic conductivity (cm/s or in./min)
i = hydraulic gradient ($i = \Delta h/L$), a dimensionless parameter
n = porosity of the soil, Eq. (6.6), expressed as a decimal (dimensionless parameter)
Δh = change in total head between two points in the soil mass (m or ft)
L = length between the two points in the soil mass (m or ft)

Between any two points in a saturated soil mass below a sloping groundwater table or below a level groundwater table that does not have hydrostatic pore water pressure (such as an artesian condition), Eq. (6.44) can be used to determine the velocity of flow v_s in the soil.

Seepage Forces. When water flows through a soil, it exerts a drag force on the individual soil particles. This force has been termed the *seepage force*. For example, an artesian condition can cause an upward flow of water to the ground surface. If this upward flow of water exerts enough of a seepage force upon the individual soil particles, they can actually lose contact with each other. This fluid condition of a cohesionless soil has been termed *quicksand*. In order to create this fluid condition, the vertical effective stress σ'_v in the soil must equal zero.

There are two basic approaches to evaluating a seepage condition, and each approach gives the same answer. The first approach is to use the total unit weight of the soil (γ_t) and the boundary pore water pressures. The second approach is to use the buoyant unit weight of the soil (γ_b) and the seepage force j, which is equal to the unit weight of water γ_w times the hydraulic gradient i, or $j = \gamma_w i$. The seepage force j has units of force per unit volume of soil (kN/m³ or pcf).

6.50 ANALYSIS OF GEOTECHNICAL DATA AND ENGINEERING COMPUTATIONS

Case	Constant Head	Variable Head
A	$k_v = \dfrac{4 \cdot q \cdot L}{\pi \cdot D^2 \cdot H_c}$	$k_v = \dfrac{d^2 \cdot L}{D^2 \cdot (t_2 - t_1)} \ln \dfrac{H_1}{H_2}$ $k_v = \dfrac{L}{t_2 - t_1} \ln \dfrac{H_1}{H_2}$ for $d = D$
B	$k_m = \dfrac{q}{2 \cdot D \cdot H_c}$	$k_m = \dfrac{\pi \cdot d^2}{8 \cdot D \cdot (t_2 - t_1)} \ln \dfrac{H_1}{H_2}$ $k_m = \dfrac{\pi \cdot D}{8 \cdot (t_2 - t_1)} \ln \dfrac{H_1}{H_2}$ for $d = D$
C	$k_m = \dfrac{q}{2.75 \cdot D \cdot H_c}$	$k_m = \dfrac{\pi \cdot d^2}{11 \cdot D \cdot (t_2 - t_1)} \ln \dfrac{H_1}{H_2}$ $k_m = \dfrac{\pi \cdot D}{11 \cdot (t_2 - t_1)} \ln \dfrac{H_1}{H_2}$ for $d = D$
D	$k_v' = \dfrac{4 \cdot q \left(\dfrac{\pi}{8} \cdot \dfrac{k_v'}{k_v} \cdot \dfrac{D}{m} + L \right)}{\pi \cdot D^2 \cdot H_c}$	$k_v' = \dfrac{d^2 \cdot \left(\dfrac{\pi}{8} \cdot \dfrac{k_v'}{k_v} \cdot \dfrac{D}{m} + L \right)}{D^2 \cdot (t_2 - t_1)} \ln \dfrac{H_1}{H_2}$ $k_v = \dfrac{\dfrac{\pi}{8} \cdot \dfrac{D}{m} + L}{t_2 - t_1} \ln \dfrac{H_1}{H_2}$ for $\begin{cases} k_v' = k_v \\ d = D \end{cases}$
E	$k_v' = \dfrac{4 \cdot q \cdot \left(\dfrac{\pi}{11} \cdot \dfrac{k_v'}{k_v} \cdot \dfrac{D}{m} + L \right)}{\pi \cdot D^2 \cdot H_c}$	$k_v' = \dfrac{d^2 \cdot \left(\dfrac{\pi}{11} \cdot \dfrac{k_v'}{k_v} \cdot \dfrac{D}{m} + L \right)}{D^2 \cdot (t_2 - t_1)} \ln \dfrac{H_1}{H_2}$ $k_v = \dfrac{\dfrac{\pi}{11} \cdot \dfrac{D}{m} + L}{t_2 - t_1} \ln \dfrac{H_1}{H_2}$ for $\begin{cases} k_v' = k_v \\ d = D \end{cases}$
F	$k_h = \dfrac{q \cdot \ln \left[\dfrac{2mL}{D} + \sqrt{1 + \left(\dfrac{2mL}{D} \right)^2} \right]}{2 \cdot \pi \cdot L \cdot H_c}$	$k_h = \dfrac{d^2 \cdot \ln \left[\dfrac{2mL}{D} + \sqrt{1 + \left(\dfrac{2mL}{D} \right)^2} \right]}{8 \cdot L \cdot (t_2 - t_1)} \ln \dfrac{H_1}{H_2}$ $k_h = \dfrac{d^2 \cdot \ln \left(\dfrac{4mL}{D} \right)}{8 \cdot L \cdot (t_2 - t_1)} \ln \dfrac{H_1}{H_2}$ for $\dfrac{2mL}{D} > 4$
G	$k_h = \dfrac{q \cdot \ln \left[\dfrac{mL}{D} + \sqrt{1 + \left(\dfrac{mL}{D} \right)^2} \right]}{2 \cdot \pi \cdot L \cdot H_c}$	$k_h = \dfrac{d^2 \cdot \ln \left[\dfrac{mL}{D} + \sqrt{1 + \left(\dfrac{mL}{D} \right)^2} \right]}{8 \cdot L \cdot (t_2 - t_1)} \ln \dfrac{H_1}{H_2}$ $k_h = \dfrac{d^2 \cdot \ln \left(\dfrac{2mL}{D} \right)}{8 \cdot L \cdot (t_2 - t_1)} \ln \dfrac{H_1}{H_2}$ for $\dfrac{mL}{D} > 4$

ASSUMPTIONS

Soil at intake, infinite depth, and directional isotropy (k_v and k_n constant). No disturbance, segregation, swelling, or consolidation of soil. No sedimentation or leakage. No air or gas in soil, well point, or pipe. Hydraulic losses in pipes, well point, or filter negligible.

FIGURE 6.23b Formulas for determination of permeability. (*From Hvorslev, 1951; reproduced from Lambe and Whitman, 1969, with permission from John Wiley & Sons.*)

Basic Time Lag	Notation
$k_v = \dfrac{d^2 \cdot L}{D^2 \cdot T}$ $k_v = \dfrac{L}{T}$ for $d = D$	D = Diam, intake, sample (cm) d = Diameter, standpipe (cm) L = Length, intake, sample (cm)
$k_m = \dfrac{\pi d^2}{8 \cdot D \cdot T}$ $k_m = \dfrac{\pi \cdot D}{8 \cdot T}$ for $d = D$	H_c = Constant piez. head (cm) H_1 = Piez. head for $t = t_1$ (cm) H_2 = Piez. head for $t = t_2$ (cm)
$k_m = \dfrac{\pi \cdot d^2}{11 \cdot D \cdot T}$ $k_m = \dfrac{\pi \cdot D}{11 \cdot T}$ for $d = D$	q = Flow of water (cm³/sec) t = Time (sec) T = Basic time lag (sec) k_v' = Vert. perm. casing (cm/sec)
$k_v' = \dfrac{d^2 \cdot \left(\dfrac{\pi}{8} \cdot \dfrac{k_v'}{k_v} \cdot \dfrac{D}{m} \right) + L}{D^2 \cdot T}$ $k_v = \dfrac{\dfrac{\pi}{8} \cdot \dfrac{D}{m} + L}{T}$ for $\begin{cases} k_v' = k_v \\ d = D \end{cases}$	k_v = Vert. perm. ground (cm/sec) k_h = Horz. perm. ground (cm/sec) k_m = Mean coeff. perm. (cm/sec) m = Transformation ratio $k_m = \sqrt{k_h \cdot k_v}$ $m = \sqrt{k_h/k_v}$ $\ln = \log_e = 2.3 \log_{10}$
$k_v' = \dfrac{d^2 \cdot \left(\dfrac{\pi}{11} \cdot \dfrac{k_v'}{k_v} \cdot \dfrac{D}{m} + L \right)}{D^2 \cdot T}$ $k_v' = \dfrac{\dfrac{\pi}{11} \cdot \dfrac{D}{m} + L}{T}$ for $\begin{cases} k_v' = k_v \\ d = D \end{cases}$	
$k_h = \dfrac{d^2 \cdot \ln \left[\dfrac{2mL}{D} + \sqrt{1 + \left(\dfrac{2mL}{D} \right)^2} \right]}{8 \cdot L \cdot T}$ $k_h = \dfrac{d^2 \cdot \ln \left(\dfrac{4mL}{D} \right)}{8 \cdot L \cdot T}$ for $\dfrac{2mL}{D} > 4$	Determination basic time lag T (graph: Piezometric head H on log scale vs. Time t on linear scale; H_0 at top, $0.37 H_0$ at $t = T$, $0.10 H_0$ at $t = 2.3 \cdot T$)
$k_h = \dfrac{d^2 \cdot \ln \left[\dfrac{mL}{D} + \sqrt{1 + \left(\dfrac{mL}{D} \right)^2} \right]}{8 \cdot L \cdot T}$ $k_h = \dfrac{d^2 \cdot \ln \left(\dfrac{2mL}{D} \right)}{8 \cdot L \cdot T}$ for $\dfrac{mL}{D} > 4$	

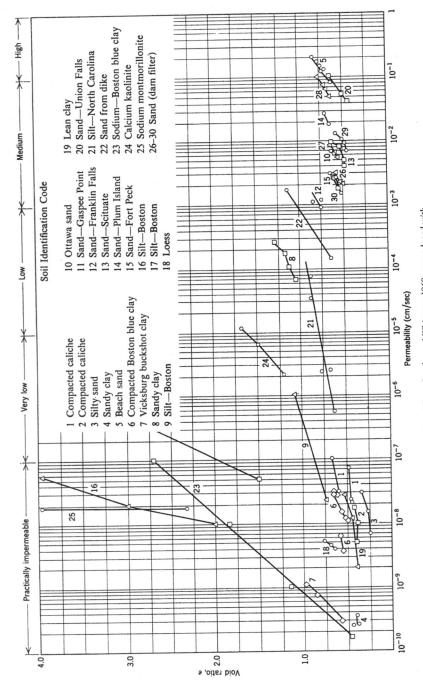

FIGURE 6.24 Coefficient of permeability k versus void ratio e. (From Lambe and Whitman, 1969; reproduced with permission from John Wiley & Sons.)

Because it often easier to measure or predict the pore water pressure, the first approach, based on the total unit weight of the soil and the boundary pore water pressures, is used more often in engineering analyses. The most common method used to predict the pore water pressures in the ground due to flowing groundwater employs a flow net. An example of an engineering analysis that may need to include the seepage effects of groundwater is the slope stability of an earth dam (Chap. 10).

Two-Dimensional Flow Nets. As shown in Figs. 5.8 and 5.9, the constant-head and falling-head permeameters have a one-dimensional flow condition (i.e., water flows vertically through the soil). For one-dimensional flow conditions, the quantity of water Q that exits the soil can be obtained by multiplying the superficial velocity v times the cross-sectional area A and time t, or:

$$Q = vAt = kiAt \tag{6.45}$$

For two-dimensional flow conditions, where, for example, the groundwater is flowing horizontally and vertically, the situation is more complex. The most common method of estimating the quantity of water Q and pore water conditions in the soil for a two-dimensional case is to construct a flow net. A *flow net* is defined as a graphical representation used to study the flow of water through soil. A flow net is composed of two types of lines: (1) flow lines, which are lines that indicate the path of flow of water through the soil, and (2) equipotential lines, which are lines that connect points of equal total head—i.e., all along an equipotential line, the numerical value of the total head h is constant.

Two-Dimensional Flow Net: Foundation Pit. Figure 6.25 presents a cross section depicting the seepage of water underneath a sheet pile wall and into a foundation pit. On one side of the sheet pile wall, the level of water is at a height h_w. On the other side of the sheet pile wall (i.e., the foundation pit), the water has been pumped out and the groundwater table corresponds to the ground surface. In this case, the change in total head Δh between the outside water level and the inside foundation pit water level is equal to h_w (i.e., $\Delta h = h_w$). Because there is a difference in total head, there will be the flow of water from the outside to the inside of the foundation pit. Note in Fig. 6.25 that the sheet pile wall is embedded a depth of 12 m into the soil. The site consists of 25 m of pervious sand overlying a relatively impervious clay stratum.

Assuming isotropic and homogeneous soil, the flow net has been drawn as shown in Fig. 6.25. The flow lines are those lines that have arrows which indicate the direction of flow. The equipotential lines are those lines that are generally perpendicular to the flow lines. Note in Fig. 6.25 that the flow lines and equipotential lines intersect at right angles and tend to form squares. By using the flow net, the quantity of seepage Q entering the foundation pit, per unit length of sheet pile wall, can be estimated by using Eq. (6.45), or:

$$Q = kiAt = k\,\Delta ht\,(A/L) = k\,\Delta ht(n_f/n_d) \tag{6.46}$$

where Q = quantity of water that enters the foundation pit per unit length of sheet pile wall (m³ for a 1-m-length of wall or ft³ for a 1-ft-length of wall)
  k = coefficient of permeability of the pervious stratum (m/day or ft/day)
  Δh = change in total head (m or ft)
  t = time (days)
  A/L = cross-sectional area of flow divided by the length of flow which is equal to n_f/n_d, i.e., the number of flow channels n_f divided by the number of equipotential drops n_d

Example. Using the flow net shown in Fig. 6.25, assume the coefficient of permeability $k = 0.1$ m/day (0.33 ft/day), the height of the water level outside the foundation pit

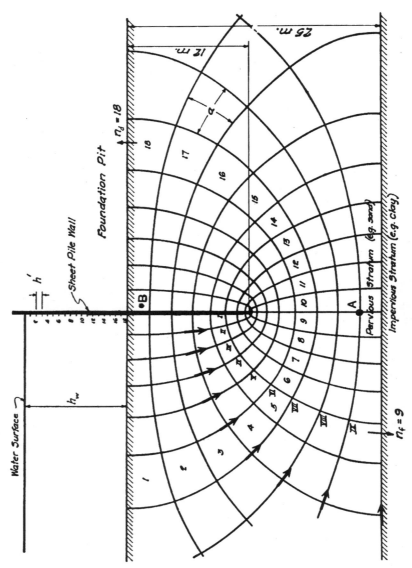

FIGURE 6.25 Flow net depicting the flow of water underneath a sheet pile wall and into a foundation pit. (Adapted from Casagrande, 1940.)

$(h_w) = 10$ m (33 ft), and the length of the sheet pile wall is 100 m (328 ft). Calculate the quantity of water that enters the foundation pit per week ($t = 7$ days).

Solution. The solution is obtained by using Eq. (6.46) with the following values:

$k = 0.1$ m/day (0.33 ft/day)

$\Delta h = h_w = 10$ m (33 ft)

$t = 7$ days

n_f = number of flow channels from the flow net = 9 (see Fig. 6.25)

n_d = number of equipotential drops from the flow net = 18 (see Fig. 6.25)

For these values in Eq. (6.46), the quantity of water Q that enters the foundation pit is 3.5 m³ per linear meter of wall length (38 ft³ per linear foot of wall length). For a 100-m-long sheet pile wall, the total amount of water that enters the foundation pit per week is 350 m³ (12,500 ft³). This is a considerable amount of water, and extensive pumping would be required to keep the foundation pit dry.

Using the Flow Net to Calculate Pore Water Pressure. In addition to determining the quantity of water Q that enters the foundation pit, the flow net can be used to determine the pore water pressures u in the ground. The equipotential lines are used to determine the pore water pressure in the soil. In Fig. 6.25, the value of h' is shown. This value represents the drop in total head between each successive equipotential line. Because there are 18 equipotential drops for the flow net shown in Fig. 6.25, $h' = \Delta h/18 = h_w/18$. Thus each successive equipotential line represents an additional drop in total head = h'. By establishing a datum to obtain the elevation head h_e, and knowing the value of the total head h along each equipotential line, we can calculate the pressure head ($h_p = h - h_e$) and hence pore water pressure ($u = h_p \gamma_w$).

Example. Assume the same conditions as the previous example. Calculate the pore water pressure u at point A in Fig. 6.25, if point A is located 2 m (6.6 ft) above the impervious stratum.

Solution. For convenience, let the datum be at the top of the impervious stratum. Then the total head h corresponding to the ground surface on the left side of the sheet pile wall is equal to the elevation head ($h_e = 25$ m) plus the pressure head ($h_p = h_w = 10$ m) or a total head h of 35 m (115 ft). In Fig. 6.25, the value of $h' = \Delta h/n_d = h_w/18 = 0.556$ m (1.82 ft). Since there are 9 equipotential drops for the flow of water to point A (Fig. 6.25), the reduction in total head is 9 times 0.556 m, or 5.0 m (16.5 ft).

The total head along the equipotential line that contains point A can now be calculated. The total head h at point A would equal 35 m (115 ft) minus the reduction in total head due to the 9 equipotential drops (5.0 m), or 30 m (98.5 ft). Since the elevation head at point A = 2 m (6.6 ft), and because the total head h is equal to the sum of the pressure head h_p and the elevation head h_e, the pressure head is equal to 28 m (i.e., $h_p = h - h_e = 30$ m − 2 m = 28 m). Converting head to pore water pressure ($u = h_p \gamma_w$), the pore water pressure u at point A = 275 kPa (5730 psf).

Check. Because point A is located at exactly the midpoint in the flow net, a quick check can be performed on the above calculations. At the midpoint, the drop in head of water would equal $1/2\ h_w$, or 5 m. Installing a piezometer at point A, the water level would rise 5 m above ground surface, or a total of 28 m (23 m plus 5 m).

Using the Flow Net to Determine the Exit Gradient. In a sense, the reduction in total head (i.e., loss of energy) as the water flows underneath the sheet pile wall is absorbed by the frictional drag resistance of the soil particles. If this frictional drag resistance of the soil particles becomes too great, piping may occur. Piping is the physical transport of soil particles by the groundwater seepage. Piping can result in the opening of underground voids which could lead to the sudden failure of the sheet pile wall.

The exit hydraulic gradient i_e can be used to assess the piping potential of soil into the foundation pit. The most likely location for a piping failure is at point B (Fig. 6.25), because this is where the flow of groundwater is directly upward and the distance between equipotential lines is least. If the distance between the two equipotential lines at point B is 2 m, then the exit hydraulic gradient at point B (i_e) would equal $h'/L = 0.556$ m/2 m $= 0.28$.

It has been determined that for a quicksand condition (also known as a *blowout* or *boiling* condition), the critical hydraulic gradient i_c is equal to the buoyant unit weight γ_b divided by the unit weight of water γ_w, or $i_c = \gamma_b/\gamma_w$. The buoyant unit weight is often about equal to the unit weight of water, and thus the critical hydraulic gradient i_c is about equal to 1. Certainly if the soil at point B in Fig. 6.25 should turn to quicksand, then there would be significant piping (loss of soil particles) at this location. Thus it is often useful to compare the exit hydraulic gradient i_e to the critical hydraulic gradient which will cause quicksand i_c, or:

$$F = \frac{i_c}{i_e} \tag{6.47}$$

where F = the factor of safety for the development of a quicksand condition. For the above example, i_e (point B) = 0.28, and for $i_c = 1.0$, the factor of safety is 3.6. Because of the catastrophic effects of a piping failure, usually a high factor of safety is required for the exit hydraulic gradient (according to Cedergren, 1989, the factor of safety should be at least 2 to 2.5).

Two-Dimensional Flow Net for an Earth Dam. Figure 6.26 shows an earth dam with the seepage of water through the dam and into a longitudinal drainage filter. In Fig. 6.26, the flow lines are those lines that have arrows indicating the direction of flow. The equipotential lines are those lines that are generally perpendicular to the flow lines. The vertical difference between the water level on the upstream side of the earth dam and the level of water in the drainage filter (i.e., dashed line) is the change in total head Δh. Note in Fig. 6.26 that a series of horizontal lines has been drawn adjacent the line representing the groundwater table in the earth dam. The distances between these horizontal lines are all equal and represent equal drops in total head for successive equipotential lines.

The quantity of flow Q collected by the drainage filter can be determined by using Eq. (6.46). Another use for the flow net shown in Fig. 6.26 would be to determine the stability of earth dam slopes. Based on the flow net, the pore water pressures u could be determined, and then the effective shear strength could be used in conjunction with a slope stability analysis to determine the stability of the earth dam (Chap. 10).

Construction of a Flow Net. The usual procedure in the preparation of a flow net is trial-and-error sketching. A flow net is first sketched in pencil with a selected number of flow channels, and then the equipotential lines are inserted. If the flow net does not have flow lines and equipotential lines intersecting at right angles, or if the flow net is not composed of approximate squares, then the lines are adjusted. Computer programs have been developed to aid in the construction of a flow net. For example, the SEEP/W (Geo-Slope, 1992) computer program can be used to determine the total head h distributions and the flow vectors (i.e., flow lines) for both simple and highly complex seepage problems.

The flow nets presented in this section are based on the assumption of isotropic and homogeneous soil. Flow nets can also be drawn for soils having different coefficients of permeability or for anisotropic soil where the coefficient of permeability is higher in the horizontal direction than the vertical direction (Casagrande, 1940).

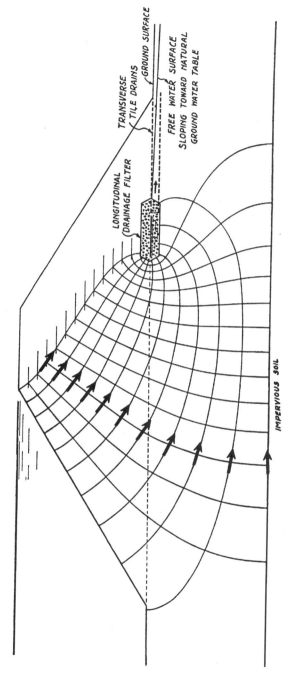

FIGURE 6.26 Flow net depicting the flow of water through an earth dam. (*Adapted from Casagrande, 1940.*)

PROBLEMS

The problems have been divided into basic categories as indicated below:

Phase Relationships

1. A soil specimen has the following measured properties:

    Total unit weight = 19.0 kN/m³ (121 pcf)
    Water content = 18.0%
    Specific gravity = 2.65

    Calculate the parameters shown in Fig. 6.1 assuming that the total volume (V) = 1m³ and 1 ft³. *Answers:* For $V = 1$ m³: $V_s = 0.62$ m³, $V_w = 0.30$ m³, $V_g = 0.08$ m³, $M_s = 1.64$ Mg, $M_w = 0.3$ Mg. For $V = 1$ ft³: $V_s = 0.62$ ft³, $V_w = 0.30$ ft³, $V_g = 0.08$ ft³, $M_s = 103$ lb, $M_w = 18.0$ lb.

2. For the conditions in Prob. 1, calculate the dry unit weight γ_d, void ratio e, porosity n, and degree of saturation S. *Answers:* $\gamma_d = 16.1$ kN/m³ (103 pcf), $e = 0.61$, $n = 38$ percent, $S = 78\%$.

3. Assume the soil specimen in Prob. 2 is a sand having a maximum void ratio = 0.85 and a minimum void ratio = 0.30. Calculate the relative density and indicate the density state of the sand. *Answer:* 44%, medium state.

4. A clay specimen has been obtained from below the groundwater table, and the total unit weight γ_t of the soil specimen is 19.5 kN/m³ (124 pcf). Calculate the buoyant unit weight of the clay. *Answer:* $\gamma_b = 9.7$ kN/m³ (61.6 pcf).

Total Stress, Effective Stress, and Pore Water Pressure

5. Assume the clay from Prob. 4 is uniform with depth and that the groundwater table coincides with the ground surface. For a level ground surface and hydrostatic pore water pressures in the ground, calculate the total vertical stress σ_v, pore water pressure u, and vertical effective stress σ'_v at a depth of 6 m (20 ft). *Answers:* $\sigma_v = 117$ kPa (2480 psf), $u = 59$ kPa (1250 psf), and $\sigma'_v = 58$ kPa (1230 psf).

6. Assume the clay from Prob. 4 is uniform with depth, there is a level ground surface, and the groundwater table is at a depth of 1.5 m (5 ft) below the ground surface. Above the groundwater table, assume the clay is saturated. For hydrostatic pore water pressures in the ground, calculate the total vertical stress σ_v, pore water pressure u, and vertical effective stress σ'_v at a depth of 6 m (20 ft). *Answers:* $\sigma_v = 117$ kPa (2480 psf), $u = 44$ kPa (940 psf), and $\sigma'_v = 73$ kPa (1540 psf).

7. Assume the clay from Prob. 4 is uniform with depth and the site is a lake where the water is 3 m (10 ft) deep. For hydrostatic pore water pressures in the ground and a level lake bottom, calculate the total vertical stress σ_v, pore water pressure u, and vertical effective stress σ'_v at a depth of 6 m (20 ft) below the lake bottom. *Answers:* $\sigma_v = 146$ kPa (3100 psf), $u = 88$ kPa (1870 psf), and $\sigma'_v = 58$ kPa (1230 psf). Note that the effective stresses for Probs. 5 and 7 are identical.

8. Assume the clay from Prob. 4 is uniform with depth, there is a level ground surface, and the groundwater table is at a depth of 3 m (10 ft) below the ground surface. Above the groundwater table, assume the clay is saturated. For the condition of capillary rise in the clay, calculate the total vertical stress σ_v, pore water pressure u, and vertical effective stress σ'_v at a depth of 1.5 m (5 ft) below the ground surface. *Answers:* $\sigma_v = 29$ kPa (620 psf), $u = -14.7$ kPa (-310 psf), and $\sigma'_v = 44$ kPa (930 psf).

BASIC GEOTECHNICAL AND FOUNDATION PRINCIPLES

9. Assume a piezometer is installed in a clay deposit and it records a pressure head of 3 m (10 ft) of water at a depth of 6 m (20 ft) below ground surface. If the total vertical stress (σ_v) = 117 kPa (2480 psf) at this depth, calculate the vertical effective stress σ'_v. *Answer:* σ'_v = 88 kPa (1860 psf).

10. For the soil profile shown in Fig. 4.40, calculate the total vertical stress σ_v, pore water pressure u, and vertical effective stress σ'_v at elevation −90 ft. *Answers:* σ_v = 623 kPa (13,000 psf or 6.3 kg/cm²), u = 307 kPa (6400 psf or 3.1 kg/cm²), and σ'_v = 316 kPa (6600 psf or 3.2 kg/cm²).

Stress Distribution

11. A uniform thickness fill of large areal extent will be placed at the ground surface. The thickness of the fill layer will be 3 m (10 ft). The total unit weight of the fill is 18.7 kN/m³ (119 pcf). If the groundwater table is below the ground surface, calculate the increase in vertical stress $\Delta\sigma_v$ beneath the center of the constructed fill mass. *Answer:* $\Delta\sigma_v$ = 56 kPa (1190 psf).

12. Assume the same conditions as Prob. 11 except that the fill mass has a width of 6 m (20 ft) and a length of 10 m (33 ft). If the groundwater table is below the ground surface, calculate the increase in vertical stress $\Delta\sigma_v$ beneath the center of the constructed fill mass at a depth of 12 m (39 ft) below original ground surface using the 2:1 approximation. *Answer:* $\Delta\sigma_v$ = 8.5 kPa (180 psf).

13. Solve Prob. 12 by using Eq. (6.29) and assuming the fill is a concentrated load (Q). *Answer:* $\Delta\sigma_v$ = 11 kPa (250 psf).

14. Solve Prob. 12 by using the Newmark chart (Fig. 6.4). *Answer:* $\Delta\sigma_v$ = 9.9 kPa (210 psf).

15. Solve Prob. 12 by using the Westergaard chart (Fig. 6.5). *Answer:* $\Delta\sigma_v$ = 6.1 kPa (130 psf).

16. Solve Prob. 12 by using Fig. 6.9. *Answer:* Number of blocks = 34, $\Delta\sigma_v$ = 9.5 kPa (200 psf).

17. For the proposed fill mass in Prob. 12, subsurface exploration will be performed at the center of the fill mass. If it is proposed to excavate the boring to a depth where the change in vertical stress $\Delta\sigma_v$ is 10 percent of the applied stress, determine the depth of subsurface exploration below the original ground surface. *Answer:* depth = 16.6 m (54 ft) based on the 2:1 approximation.

Shear Strength

18. A clean sand deposit has a level ground surface, a total unit weight γ_t above the groundwater table of 18.9 kN/m³ (120 pcf), and a submerged unit weight γ_b of 9.84 kN/m³ (62.6 pcf). The groundwater table is located 1.5 m (5 ft) below ground surface. Standard penetration tests were performed in a 10-cm- (4-in.-) diameter borehole. At a depth of 3 m (10 ft) below ground surface, a standard penetration test (SPT) was performed using a donut hammer with a blow count of 3 blows for the first 15 cm (6 in.), 4 blows for the second 15 cm (6 in.), and 5 blows for the third 15 cm (6 in.) of driving penetration. Assuming hydrostatic pore water pressures, determine the vertical effective stress σ'_v at a depth of 3 m (10 ft), the corrected N-value [i.e., N_{60}, Eq. (4.3)], and the friction angle ϕ of the sand. *Answers:* σ'_v = 43 kPa (910 psf), N_{60} = 5, ϕ = 30° (Fig. 4.13).

19. For Prob. 18, assume a cone penetration test (CPT) was performed at a depth of 3 m (10 ft) and the cone resistance (q_c) = 40 kg/cm² (3900 kPa). Determine the friction angle ϕ of the sand. *Answer:* ϕ = 40° (Fig. 4.15).

20. A drained direct shear test was performed on a nonplastic cohesionless soil. The specimen diameter = 6.35 cm (2.5 in.). At a vertical load of 150 N (34 lb), the peak shear force = 94 N (21 lb). For a second specimen tested at a vertical load of 300 N (68 lb), the peak shear force = 188 N (42 lb). Determine the effective friction angle ϕ' for the cohesionless soil. *Answer:* $\phi' = 32°$.

21. A consolidated undrained triaxial compression test was performed on a saturated cohesive soil specimen. The soil specimen was first consolidated at an effective confining pressure = 100 kPa (i.e., $\sigma'_1 = \sigma'_3$). At the end of consolidation, the soil specimen had an area = 9.68 cm² (1.50 in²) and a height = 11.7 cm (4.60 in.). The specimen was then subjected to an axial load, and at failure, the axial deformation = 1.48 cm (0.583 in.), the axial load = 48.4 N (10.9 lb), and the change in pore water pressure (Δu) = 45.6 kPa (950 psf). Calculate the area of the specimen at failure, the major principal effective stress at failure (σ'_1), the minor principal effective stress at failure (σ'_3), q and p' (in terms of effective stresses), the effective friction angle ϕ' assuming $c' = 0$, and the A-value at failure. *Answers:* area at failure = 11.1 cm² (1.72 in.²) from Eq. (5.3), σ'_1 at failure = 98 kPa (2050 psf), σ'_3 at failure = 54.4 kPa (1140 psf), $p' = 76.2$ kPa (1590 psf) and $q = 21.8$ kPa (455 psf), $\phi' = 17°$, and A-value at failure = 1.05.

22. On the basis of the results from Prob. 21, what type of inorganic soil would most likely have this shear strength? *Answer:* Normally consolidated clay of high plasticity (CH).

23. Assume the data shown in Fig. 6.20a was obtained from a consolidated undrained triaxial compression test performed on a saturated cohesive soil. The specimen was first consolidated at an effective confining pressure of 50 kPa (i.e., $\sigma'_1 = \sigma'_3$). After consolidation, the specimen height = 10.67 cm and the specimen diameter = 3.68 cm. The B-value was checked, and the applied cell pressure ($\Delta \sigma_c$) = 100 kPa and the measured change in pore water pressure (Δu) = 99.8 kPa.

During the undrained triaxial shearing of the specimen, the following data was recorded:

Point	Axial deformation, cm	$\Delta \sigma_1$, kPa	Δu, kPa
A	0	0	0
B	0.13	20	1.5
C	0.30	40	3.0
D	0.53	60	4.3
E	0.95	80	6.7

(*Note:* Point E represents the failure condition.)

Calculate the undrained modulus E_u for the test data. *Answer:* $E_u = 1600$ kPa.

24. For Prob. 23, calculate the total stress friction angle ϕ, assuming $c = 5$ kPa. *Answer:* $\phi = 23°$.

25. For Prob. 23, plot the effective stress path and determine α', assuming that $a' = 2$ kPa. *Answer:* $\alpha' = 24.5°$.

26. Using the data from Prob. 25, calculate the effective stress friction angle ϕ' and effective stress cohesion c'. *Answers:* $\phi' = 27°$, $c' = 2.2$ kPa.

27. For Prob. 23, calculate the B-value and the A-value at failure. *Answers:* B-value = 0.998, A-value at failure = 0.08.

Permeability and Seepage

28. In a uniform soil deposit having a level ground surface, an open standpipe with a constant interior diameter of 5.0 cm is installed to a depth of 10 m. The standpipe has a flush bottom (Case C, Fig. 6.23a). The groundwater table is located 4.0 m below the ground surface. The standpipe is filled with water and the water level is maintained at ground surface by adding 1.0 L every 33 seconds. Calculate the mean coefficient of permeability k for the soil deposit. *Answer:* $k = 0.006$ cm/s.

29. Assume the same conditions as Prob. 28, except that after the standpipe is filled with water to ground surface, the water level is allowed to drop without adding any water to the standpipe. If the water level in the standpipe drops from ground surface to a depth of 0.9 m below ground surface in 60 seconds, calculate the mean coefficient of permeability k of the soil deposit. *Answer:* $k = 0.006$ cm/s.

30. For the flow net shown in Fig. 6.25, what would happen to the quantity of water Q per unit time that enters the foundation pit if h_w were doubled. *Answer:* A doubling of h_w (i.e., Δh) doubles Q, from Eq. (6.46).

31. At the middle of the flow net square labeled 15 in Fig. 6.25, calculate the pore water pressure (u). Assume that $h_w = 10$ m and the middle of the flow net square labeled 15 in Fig. 6.25 is 13 m above the impervious stratum. *Answer:* $u = 137$ kPa.

32. Assume that the sand stratum shown in Fig. 6.25 has a total unit weight (γ_t) = 19.8 kN/m³. Using the data from Prob. 31, calculate the vertical effective stress σ'_v at the middle of the flow net square labeled 15 in Fig. 6.25. *Answer:* $\sigma'_v = 101$ kPa.

33. Using the data from Probs. 31 and 32, calculate the exit gradient i_e and the factor of safety for piping failure (F) for the flow net square labeled 18 in Fig. 6.25. To estimate the length L, scale from the distances shown on the right side of Fig. 6.25. *Answers:* $i_e = 0.14$, factor of safety = 7.

34. Using the data from Prob. 33, determine the seepage velocity v_s and direction of seepage at the flow net square labeled 18 in Fig. 6.25. Assume the coefficient of permeability $k = 0.1$ m/day. *Answers:* $v_s = 0.037$ m/day in an approximately upward direction.

35. For the flow net shown in Fig. 6.25, where does the seepage velocity v_s have the highest value? *Answer:* Since the equipotential drops are all equal, the highest seepage velocity occurs where the length of the flow net squares are the smallest, or at the sheet pile tip.

36. In Fig. 6.26, assume the height of water behind the earth dam (h_w) = 20 m, the soil composing the earth dam has a coefficient of permeability (k) = 1×10^{-6} cm/s, and the earth dam is 200 m long. Determine the quantity of water Q that will be collected by the drainage system each day. Assume water from the lowest flow channel also enters the drainage system. *Answer:* $Q = 2.5$ m³ of water per day.

37. Using the data from Prob. 36, determine the number of equipotential drops and the head drop h' for each equipotential drop. *Answer:* number of equipotential head drops = 14 and $h' = 1.43$ m. (*Note:* For the uppermost flow channel, one of the equipotential drops occurs in the drainage filter.)

38. On the basis of the flow net in Fig. 6.26, at what location will the seepage velocity v_s be the highest? *Answer:* Since the equipotential drops are all equal, the highest seepage velocity occurs where the length of the flow net squares are the smallest, or in the soil that is located in front of the longitudinal drainage filter.

39. On the basis of the flow net in Fig. 6.26, at what location will there most likely be piping of soil from the earth dam? *Answer:* The soil located in front of the longitudinal

drainage filter has the highest seepage velocity v_s, and this is the most likely location for piping of soil into the drainage filter.

40. Assuming the same conditions as Prob. 36, calculate the pore water pressure u at a point located at the centerline of the dam and 20 m below the dam top. *Answer:* $u = 110$ kPa.
41. For Prob. 40, assume $\gamma_t = 20.0$ kN/m³ for the earth dam soil. Determine the vertical effective stress σ'_v at the same point referred to in Prob. 40. *Answer:* $\sigma'_v = 290$ kPa.
42. Assume the effective shear strength parameters are $\phi' = 28°$ and $c' = 2$ kPa for the soil composing the earth dam. Calculate the shear strength τ on a horizontal plane at the point referred to in Prob. 41. *Answer:* $\tau = 156$ kPa.
43. A slope stability analysis is to be performed for the earth dam shown in Fig. 6.26. Assume the shear strength calculated in Prob. 42 represents an average shear strength along the critical slip surface. If the average existing shear stress in the soil along the same slip surface = 83 kPa, calculate the factor of safety F for slope stability of the earth dam. *Answer:* $F = 1.88$.

CHAPTER 7
SETTLEMENT OF STRUCTURES

The following notation is used in this chapter:

SYMBOL   DEFINITION

a_v	Coefficient of compressibility
B	Width of the foundation
c_v	Coefficient of consolidation
C_c	Compression index
C_r	Recompression index
$C_{c\varepsilon}$	Modified compression index
$C_{r\varepsilon}$	Modified recompression index
C_α	Secondary compression ratio
$\%C$	Percent collapse
e	Void ratio
e_0	Natural or initial void ratio
Δe	Change in void ratio
Δe_c	Change in void ratio upon wetting
E_s	Drained modulus of the soil
E_u	Undrained modulus of the soil
H	One-half the thickness of the clay layer (double drainage, Fig. 7.14)
H_{dr}	Height of the drainage path
H_0	Initial thickness of soil specimen (Fig. 7.6b)
H_0	Initial thickness of the *in situ* soil layer
H/D	Height of soil specimen (H) divided by the diameter of the loaded area (D)
ΔH_c	Change in height upon wetting
I_p	Dimensionless parameter derived from the theory of elasticity
k	Coefficient of permeability, also known as the hydraulic conductivity
L	Horizontal distance for calculation of maximum angular distortion
M_0	Initial dry mass of soil (solubility test)
M_f	Final dry mass of soil (solubility test)

7.1

N	Vertical pressure for consolidation test (Fig. 7.15)
N	SPT N-value
OCR	Overconsolidation ratio
q	Uniform pressure applied by the foundation
s_c	Primary consolidation settlement
s_i	Initial settlement, also known as immediate settlement
s_s	Settlement due to secondary compression
t	Time
T	Time factor
u_e	Excess pore water pressure
u_0	Initial excess pore water pressure
U_{avg}	Average degree of consolidation settlement
U_z	Consolidation ratio
z	Depth below the top of the clay layer (Fig. 7.14)
Z	Vertical axis in Fig. 7.14 ($Z = z/H$)
Δ	Maximum differential settlement of the foundation
δ	Vertical displacement for calculation of maximum angular distortion
δ/L	Maximum angular distortion of the foundation
ε_v	Vertical strain
$\Delta\varepsilon_v$	Change in vertical strain
μ	Poisson's ratio
ρ_{max}	Maximum total settlement of the foundation
σ'_v, σ'_{v0}	Vertical effective stress
σ'_{vm}, σ'_p	Maximum past pressure, also known as preconsolidation pressure
$\Delta\sigma_v$	Change in vertical stress

7.1 INTRODUCTION

Two Types of Settlement. Settlement can be defined as the permanent downward displacement of the foundation. There are two basic types of settlement:

1. *Settlement due directly to the weight of the structure.* The weight of a building may cause compression of an underlying sand deposit (Sec. 7.5) or consolidation of an underlying clay layer (Sec. 7.4).

Often the settlement analysis is based on the actual dead load of the structure. The dead load is defined as the structural weight due to beams, columns, floors, roofs, and other fixed members. The dead load does not include nonstructural items.

Live loads are defined as the weight of nonstructural members, such as furniture, occupants, inventory, and snow. Live loads can also result in settlement of the structure. For example, if the proposed structure is a library, then the actual weight of the books (a live load) should be included in the settlement analyses. Likewise, for a proposed warehouse, it may be appropriate to include the actual weight of anticipated stored items in the settlement analyses. In other projects where the live loads represent a significant part of the loading, such as large electrical transmission towers that will be subjected to wind loads, the live load (wind) may also be included in the settlement analysis.

As mentioned, the load used for settlement analyses should include the actual dead weight of the structure, and in some cases, live loads. Considerable experience and judgment are required to determine the load that is to be used in the settlement analyses.

2. Settlement due to secondary influences. The second basic type of settlement of a building is caused by secondary influence, which may develop at a time long after the completion of the structure. This type of settlement is not directly caused by the weight of the structure. For example, the foundation may settle as water infiltrates the ground and causes unstable soils to collapse (i.e., collapsible soil, Sec. 7.3). The foundation may also settle because of yielding of adjacent excavations or the collapse of limestone cavities or underground mines and tunnels (Sec. 7.6). Other causes of settlement that would be included in this category are natural disasters, such as settlement caused by earthquakes (Chap. 11) or undermining of the foundation from floods.

Subsidence. Subsidence is usually defined as a sinking down of a large area of the ground surface. Subsidence could be caused by the extraction of oil or groundwater which leads to a compression of the underlying porous soil or rock structure. Since subsidence is a result of a secondary influence (extraction of oil or groundwater), its effect on the structure would be included in the second basic type of settlement described above.

Expansive Soil. A special case is the downward displacement of a foundation because of the drying of underlying wet clays, which will be discussed in Chap. 9. Often this downward displacement of the foundation caused by the desiccation of clays is referred to as "settlement." But upon the introduction of moisture—during the rainy season, for example—the desiccated clay will swell and the downward displacement will be reversed. The foundation could even heave more than it initially settled. When dealing with expansive clays, it is best to consider the downward displacement of the foundation as part of the cyclic heave and shrinkage of expansive soil and not as permanent settlement.

Structural Design. The structural design of foundations, such as determining the thickness of the concrete foundation and type and spacing of steel reinforcement, is not covered in this book. There are many excellent references, such as *Foundation Analysis and Design* (Bowles, 1982) and *Foundation Design, Principles and Practices* (Coduto, 1994), that present the methods and procedures for the analysis and design of foundations.

Foundation Design Parameters. As indicated in the *Technical Guidelines for Geotechnical Reports* (App. B, Sec. D, subitem *c*), a soils report must discuss the "consolidation or settlement potential" of the structure. Determining the settlement of the structure is one of the primary obligations of the geotechnical engineer. In general, three parameters are required: the total settlement, differential settlement, and rate of settlement.

One approach for the design of the foundation is to first obtain the allowable settlement (total and differential) of the structure (Sec. 7.2). Then, from the results of the field exploration and laboratory testing, the total and differential settlement of the structure can be calculated (Secs. 7.3 to 7.6). As discussed in the final section (7.7) of the chapter, the foundation type may have to be changed or mitigation measures adopted if the calculated settlement of the structure exceeds the allowable settlement.

7.2 ALLOWABLE SETTLEMENT

The *allowable settlement* is defined as the acceptable amount of settlement, and it usually includes a factor of safety. In many cases the allowable settlement of the project will be provided by the structural engineer. As previously mentioned, at the start of a project the structural engineer should be consulted about the anticipated loading conditions, allowable settlement of the structure, and preliminary thoughts on the type of foundation. The ideal situation would be to obtain this information directly from the structural engineer.

If this information is unavailable, then the geotechnical engineer will have to estimate the allowable settlement of the structure in order to determine an appropriate type of foundation. There is a considerable amount of data available on the allowable settlement of structures (e.g., Leonards, 1962; ASCE, 1964; Feld, 1965; Peck et al., 1974; Bromhead, 1984; Wahls, 1994). For example, it has been stated that the allowable differential and total settlement should depend on the flexibility and complexity of the structure including the construction materials and type of connections (*Foundation Engineering Handbook*, Winterkorn and Fang, 1975).

In terms of the allowable settlement, Coduto (1994) states that it depends on many factors, including the following:

- *Type of construction.* For example, wood-frame buildings with wood siding would be much more tolerant than unreinforced brick buildings.
- *Use of the structure.* Even small cracks in a house might be considered unacceptable, whereas much larger cracks in an industrial building might not even be noticed.
- *Presence of sensitive finishes.* Tile or other sensitive finishes are much less tolerant of movements.
- *Rigidity of the structure.* If a footing beneath part of a very rigid structure settles more than the others, the structure will transfer some of the load away from the footing. However, footings beneath flexible structures must settle much more before any significant load transfer occurs. Therefore, a rigid structure will have less differential settlement than a flexible one.

Coduto (1994) also states that the allowable settlement for most structures, especially buildings, will be governed by aesthetic and serviceability requirements, not structural requirements. Unsightly cracks, jamming doors and windows, and other similar problems will develop long before the integrity of the structure is in danger. Because the determination of the allowable settlement is so complex, engineers often rely on empirical correlations between observed behavior of structures and the settlement that results in damage. The following sections provide a brief discussion of additional studies on the settlement of structures:

Skempton and MacDonald (1956). A major reference for the allowable settlement of structures is the paper by Skempton and MacDonald (1956) titled "The Allowable Settlement of Buildings". As shown in Fig. 7.1, Skempton and MacDonald defined the maximum angular distortion δ/L and the maximum differential settlement Δ for a building with no tilt. The angular distortion δ/L is defined as the differential settlement between two points divided by the distance between them less the tilt, where tilt equals rotation of the entire building. As shown in Fig. 7.1, the maximum angular distortion δ/L does not necessarily occur at the location of maximum differential settlement Δ.

Skempton and MacDonald studied 98 buildings, where 58 had suffered no damage and 40 had been damaged in varying degrees as a consequence of settlement. From a study of these 98 buildings, Skempton and MacDonald in part concluded the following:

- The cracking of the brick panels in frame buildings or load-bearing brick walls is likely to occur if the angular distortion of the foundation exceeds 1/300. Structural damage to columns and beams is likely to occur if the angular distortion of the foundation exceeds 1/150.

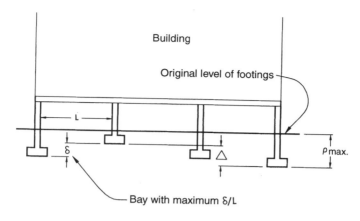

FIGURE 7.1 Diagram illustrating the definitions of maximum angular distortion and maximum differential settlement.

- By plotting the maximum angular distortion δ/L versus the maximum differential settlement Δ such as shown in Fig. 7.2, a correlation was obtained that is defined as $\Delta = 350\ \delta/L$. (*Note:* Δ is in inches.) On the basis of this relationship and an angular distortion (δ/L) of 1/300, cracking of brick panels in frame buildings or load-bearing brick walls is likely to occur if the maximum differential settlement Δ exceeds 32 mm (1¼ in.).

- The angular distortion criteria of 1/150 and 1/300 were derived from an observational study of buildings of load-bearing-wall construction, and steel and reinforced-concrete-frame buildings with conventional brick panel walls but without diagonal bracing. The criteria are intended as no more than a guide for day-to-day work in designing typical foundations for such buildings. In certain cases they may be overruled by visual or other considerations.

Grant et al. (1974). The paper by Grant et al. (1974) updated the Skempton and MacDonald data pool and also evaluated the rate of settlement with respect to the amount of damage incurred. Grant et al. (1974) in part concluded the following:

- A building foundation that experiences a maximum value of deflection slope δ/L greater than 1/300 will probably suffer some damage. However, damage does not necessarily occur at the point where the local deflection slope exceeds 1/300.

- For any type of foundation on sand or fill, new data tend to support Skempton and MacDonald's suggested correlation of $\Delta = 350\ \delta/L$ (see Fig. 7.2).

- Consideration of the rate of settlement is important only for the extreme situations of either very slow or very rapid settlement. On the basis of the limited data available, the values of maximum δ/L corresponding to building damage appear to be essentially the same for cases involving slow and fast settlements.

Slab-on-Grade Foundations. Data concerning the behavior of lightly reinforced, conventional slab-on-grade foundations has also been included in Fig. 7.2. This data indicates that cracking of gypsum wallboard panels is likely to occur if the angular distortion of the slab-on-grade foundation exceeds 1/300 (Day, 1990a). The ratio of 1/300 appears to be useful for both woodframe gypsum wallboard panels and the brick panels studied by

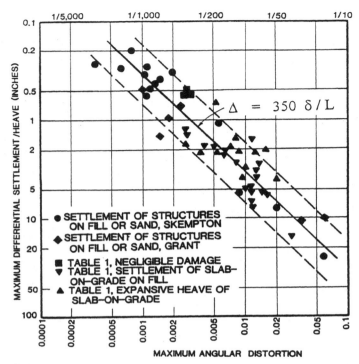

FIGURE 7.2 Maximum differential settlement versus maximum angular distortion. (*Initial data from Skempton and MacDonald, 1956; Table 1 in Day, 1990a.*)

Skempton and MacDonald (1956). The data plotted in Fig. 7.2 would indicate that the relationship $\Delta = 350\ \delta/L$ can also be used for buildings supported by lightly reinforced slab-on-grade foundations. Using $\delta/L = 1/300$ as the boundary where cracking of panels in woodframe residences supported by concrete slab-on-grade is likely to occur, and substituting this value into the relationship $\Delta = 350\ \delta/L$ (Fig. 7.2), the calculated differential slab displacement is 32 mm (1 1/4 in.). For buildings on lightly reinforced slabs-on-grade, cracking of gypsum wallboard panels is likely to occur when the maximum slab differential exceeds 32 mm (1 1/4 in.).

Terzaghi (1938). Concerning allowable settlement, Terzaghi (1938) stated:

> Differential settlement must be considered inevitable for every foundation, unless the foundation is supported by solid rock. The effect of the differential settlement on the building depends to a large extent on the type of construction.

Terzaghi summarized his studies on several buildings in Europe where he found that walls 18 m (60 ft) and 23 m (75 ft) long with differential settlements over 2.5 cm (1 in.) were all cracked, but four buildings with walls 12 m (40 ft) to 30 m (100 ft) long were undamaged when the differential settlement was 2 cm (3/4 in.) or less. This is probably the basis for the general design guide that building foundations should be designed so that the differential settlement is 2 cm (3/4 in.) or less.

Sowers (1962).   Another example of allowable settlements for buildings is Table 7.1 (from Sowers, 1962). In this table, the allowable foundation displacement has been divided into three categories: total settlement, tilting, and differential movement. Table 7.1 indicates that those structures that are more flexible (such as simple steel frame buildings) or have more rigid foundations (such as mat foundations) can sustain larger values of total settlement and differential movement.

Bjerrum (1963).   Figure 7.3 presents data from Bjerrum (1963). Like the studies previously mentioned, this figure indicates that cracking in panel walls is to be expected at an angular distortion δ/L of 1/300 and that structural damage of buildings is to be expected at δ/L of 1/150. This figure also provides other limiting values of angular distortion, such as those for buildings containing sensitive machinery or overhead cranes.

Settlement versus Cracking Damage.   Table 7.2 summarizes the severity of cracking damage versus approximate crack widths, typical values of maximum differential movement Δ, and maximum angular distortion δ/L of the foundation (Burland et al., 1977; Boone, 1996; Day, 1998). The relationship between differential settlement Δ and angular distortion δ/L is based on the equation $\Delta = 350\ \delta/L$ (from Fig. 7.2).

In assessing the severity of damage for an existing structure, the damage category (Table 7.2) should be based on multiple factors, including crack widths, differential settlement, and the angular distortion of the foundation. Relying on only one parameter, such as crack width, can be inaccurate in cases where cracking has been hidden or patched, or in cases where other factors (such as concrete shrinkage) contribute to crack widths.

TABLE 7.1   Allowable Settlement

Type of movement (1)	Limiting factor (2)	Maximum settlement (3)
Total settlement	Drainage	15–30 cm (6–12 in.)
	Access	30–60 cm (12–24 in.)
	Probability of nonuniform settlement:	
	Masonry walled structure	2.5–5 cm (1–2 in.)
	Framed structures	5–10 cm (2–4 in.)
	Smokestacks, silos, mats	8–30 cm (3–12 in.)
Tilting	Stability against overturning	Depends on H and W
	Tilting of smokestacks, towers	$0.004L$
	Rolling of trucks, etc.	$0.01L$
	Stacking of goods	$0.01L$
	Machine operation—cotton loom	$0.003L$
	Machine operation—turbogenerator	$0.0002L$
	Crane rails	$0.003L$
	Drainage of floors	0.01–$0.02L$
Differential movement	High continuous brick walls	0.0005–$0.001L$
	One-story brick mill building, wall cracking	0.001–$0.002L$
	Plaster cracking (gypsum)	$0.001L$
	Reinforced-concrete-building frame	0.0025–$0.004L$
	Reinforced-concrete-building curtain walls	$0.003L$
	Steel frame, continuous	$0.002L$
	Simple steel frame	$0.005L$

Source: Sowers (1962).
Notes. L = distance between adjacent columns that settle different amounts, or between any two points that settle differently. Higher values are for regular settlements and more tolerant structures. Lower values are for irregular settlement and critical structures. H = height and W = width of structure.

7.8 ANALYSIS OF GEOTECHNICAL DATA AND ENGINEERING COMPUTATIONS

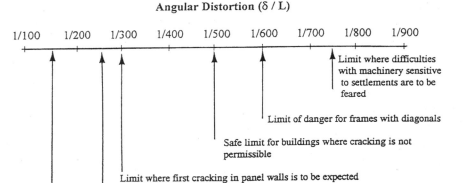

FIGURE 7.3 Damage criteria. (*After Bjerrum, 1963.*)

Component of Lateral Movement. Foundations subjected to settlement can be damaged by a combination of both vertical and horizontal movements. For example, a common cause of foundation damage is fill settlement. Figure 7.4 shows an illustration of the settlement of fill in a canyon environment. Over the sidewalls of the canyon, there tends to be a pulling or stretching of the ground surface (tensional features), with compression effects near the canyon centerline. This type of damage is due to two-dimensional settlement, where the fill compresses in both the vertical and horizontal directions (Lawton et al., 1991; Day, 1991a).

Another common situation where both vertical and horizontal foundation displacement occurs is at cut-fill transitions. A cut-fill transition occurs when a building pad has some rock removed (the cut portion), with a level building pad being created by filling in (with soil) the remaining portion. If the cut side of the building pad contains nonexpansive rock that is dense and unweathered, then very little settlement would be expected for that part of the building on cut. But the fill portion could settle under its own weight and cause damage. For example, a slab crack will typically open at the location of the cut-fill transition as illustrated in Fig. 7.5. The building is damaged by both the vertical foundation movement (settlement) and the horizontal movement, which manifests itself as a slab crack and drag effect on the structure (Fig. 7.5). A typical measure to prevent this type of damage is shown in Standard Detail No. 6 (App. C), where the cut portion of the building pad in undercut and replaced with fill to eliminate the abrupt change in bearing resistance at a cut-fill transition.

In the cases described above, the lateral movement is a secondary result of the primary vertical movement due to settlement of the foundation. Table 7.2 can therefore be used as a guide to correlate damage category with Δ and δ/L. In cases where lateral movement is the predominant or critical mode of foundation displacement, Table 7.2 may underestimate the severity of cracking damage for values of Δ and δ/L (Day, 1998; Boone, 1998).

7.3 COLLAPSIBLE SOIL

The next four sections (Secs. 7.3 to 7.6) of this chapter describe typical causes of settlement of structures and provides basic engineering procedures that can be used to calculate the settlement of the structure.

TABLE 7.2 Severity of Cracking Damage

Damage category (1)	Description of typical damage (2)	Approx. crack width (3)	Δ (4)	δ/L (5)
Negligible	Hairline cracks	< 0.1 mm	< 3 cm (1.2 in.)	< 1/300
Very slight	Very slight damage includes fine cracks that can be easily treated during normal decoration, perhaps an isolated slight fracture in building, and cracks in external brickwork visible on close inspection	1 mm	3–4 cm (1.2–1.5 in.)	1/300 to 1/240
Slight	Slight damage includes cracks that can be easily filled and redecoration would probably be required; several slight fractures may appear showing on the inside of the building; cracks that are visible externally and some repointing may be required; doors and windows may stick	3 mm	4–5 cm (1.5–2.0 in.)	1/240 to 1/175
Moderate	Moderate damage includes cracks that require some opening up and can be patched by a mason; recurrent cracks that can be masked by suitable linings; repointing of external brickwork and possibly a small amount of brickwork replacement may be required; doors and windows stick; service pipes may fracture; weathertightness is often impaired	5 to 15 mm or a number of cracks > 3 mm	5–8 cm (2.0–3.0 in.)	1/175 to 1/120
Severe	Severe damage includes large cracks requiring extensive repair work involving breaking out and replacing sections of walls (especially over doors and windows); distorted windows and door frames; noticeably sloping floors; leaning or bulging walls; some loss of bearing in beams; disrupted service pipes	15 to 25 mm but also depends on number of cracks	8–13 cm (3.0–5.0 in.)	1/120 to 1/70
Very severe	Very severe damage often requires a major repair job involving partial or complete rebuilding; beams lose bearing; walls lean and require shoring; windows are broken with distortion; there is danger of structural instability	Usually > 5 mm but also depends on number of cracks	> 13 cm (> 5 in.)	> 1/70

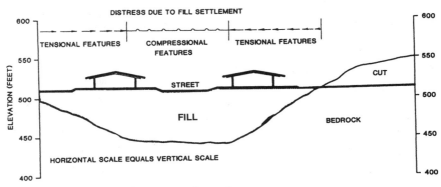

FIGURE 7.4 Fill settlement in a canyon environment.

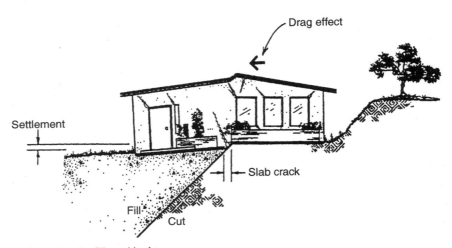

FIGURE 7.5 Cut-fill transition lot.

In the southwestern United States, probably the most common cause of settlement is collapsible soil. For example, Johnpeer (1986) states that ground subsidence from collapsing soils is a common occurrence in New Mexico. The category of collapsible soil would include loose debris, uncontrolled fill, deep fill, and natural soil, such as alluvium or colluvium.

In general, there has been an increase in damage because of collapsible soil, probably because of the lack of available land in many urban areas. This causes development of marginal land, which may contain deposits of dumped fill or deposits of natural collapsible soil. Also, substantial grading can be required to develop level building pads, which results in more areas having deep fill.

Collapsible soil can be broadly classified as soil that is susceptible to a large and sudden reduction in volume upon wetting. Collapsible soil usually has a low dry density and low moisture content. Such soil can withstand a large applied vertical stress with a small compression, but then experience much larger settlements after wetting, with no increase

SETTLEMENT OF STRUCTURES   7.11

in vertical pressure (Jennings and Knight, 1957). Accordingly, collapsible soil falls within the second basic category of settlement (see Sec. 7.1), which is settlement of the structure due to secondary influences.

Fill. Deep fill has been defined as fill that has a thickness greater than 6 m (20 ft) (Greenfield and Shen, 1992). Uncontrolled fills include fills that were not documented with compaction testing as they were placed. These include dumped fills, fills dumped under water, hydraulically placed fills, and fills that may have been compacted but have no documentation of testing or the amount of effort that was used to perform the compaction (Greenfield and Shen, 1992). These conditions may exist in rural areas where inspections are lax or for structures built many years ago when the standards for fill compaction were less rigorous.

For collapsible fill, compression will occur as the overburden pressure increases. The increase in overburden pressure could result from the placement of overlying fill or the construction of a building on top of the fill. The compression due to this increase in overburden pressure involves a decrease in void ratio of the fill due to expulsion of air. The compression usually occurs at constant moisture content. After completion of the fill mass, water may infiltrate the soil from irrigation, rainfall, or leaky water pipes. The mechanism that usually causes the collapse of the loose soil structure is a decrease in negative pore water pressure (capillary tension) as the fill becomes wet.

For a fill specimen submerged in distilled water, the main variables that govern the amount of one-dimensional collapse are the soil type, compacted moisture content, compacted dry density, and the vertical pressure (Dudley, 1970; Lawton et al., 1989, 1991, 1992; Tadepalli and Fredlund, 1991; Day, 1994b). In general, the one-dimensional collapse of fill will increase as the dry density decreases, the moisture content decreases, or the vertical pressure increases. For a constant dry density and moisture content, the one-dimensional collapse will decrease as the clay fraction increases once the optimum clay content (usually a low percentage) is exceeded (Rollins et al., 1994).

Alluvium or Colluvium. For natural deposits of collapsible soil in the arid climate of the southwest, a common mechanism involved in rapid volume reduction entails breaking of bonds at coarse particle contacts by weakening of fine-grained materials brought there by surface tension in evaporating water. In other cases, the alluvium or colluvium may have an unstable soil structure which collapses as the wetting front passes through the soil.

Laboratory Testing. If the results of field exploration indicate the possible presence of collapsible soil at the site, then soil specimens should be obtained and tested in the laboratory. One-dimensional collapse is usually measured in the oedometer (ASTM D 5333-96, "Standard Test Method for Measurement of Collapse Potential of Soils," 1998). After the soil specimen is placed in the oedometer, the vertical pressure is increased until it approximately equals the anticipated overburden pressure after completion of the structure. At this vertical pressure, distilled water is added to the oedometer to measure the amount of collapse of the soil specimen. Percent collapse (%C) is defined as the change in height of the specimen due to inundation divided by the initial height of the specimen (ASTM D 5333-96, 1998).

For projects where it is anticipated that there will be fill placement, the fill specimens can be prepared by compacting them to the anticipated field as-compacted density and moisture condition. The fill specimens would then be subjected to the anticipated overburden pressure. The procedure for measuring the one-dimensional collapse of the fill specimen would then be same as outlined above (ASTM D 5333-96, 1998).

Figure 7.6a presents the results of a one-dimensional collapse test performed on a fill specimen. The fill specimen contains 60 percent sand-size particles, 30 percent silt-size particles, and 10 percent clay-size particles and is classified as a silty sand (SM). To model field conditions, the silty sand was compacted at a dry unit weight γ_d of 14.5 kN/m³ (92.4 pcf) and water content w of 14.8 percent.

The silty sand specimen, having an initial height of 25.4 mm (1.0 in.), was subjected to a vertical stress of 144 kPa (3000 psf) and then inundated with distilled water. Figure 7.6a shows the amount of vertical deformation (collapse) as a function of time after inundation. Figure 7.6b presents the equations that are used to determine the percent collapse (%C). For the collapse test on the silty sand (fill type number 1, Fig. 7.6a), the percent collapse (%C) = (2.62 mm)/(25.4 mm) = 10.3%.

The collapse potential (CP) of a soil can be determined by applying a vertical stress of 200 kPa (2 tsf) to the soil specimen, submerging it in distilled water, and then determining the percent collapse, which is designated the collapse potential (CP). This collapse potential can be considered as an index test to compare the susceptibility of collapse for different soils. In Fig. 7.6b, the collapse potential (CP) versus severity of the problem has been listed in a tabular form.

Settlement Analyses. For collapsible soil, the settlement analysis is usually as follows:

1. The amount of collapse for the different soil layers underlying the site are obtained from the laboratory oedometer tests on undisturbed samples (ASTM D 5333-96, 1998). The vertical stress used in the laboratory testing should equal the overburden pressure plus the weight of the structure (using stress distribution, Sec. 6.4).
2. To obtain the settlement of each collapsible soil layer, the percent collapse is multiplied by the thickness of the soil layer.
3. The total settlement is the sum of the collapse value from the different soil layers.

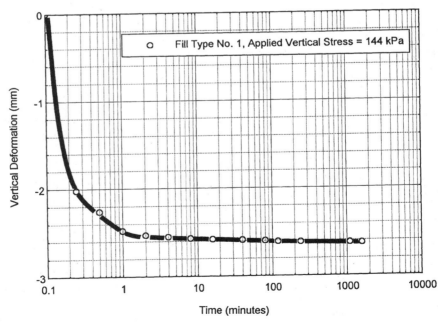

FIGURE 7.6a Typical collapse potential test results.

Percent collapse (%C) defined as:

$$\%C = \frac{100\ \Delta e_c}{1 + e_o} \quad \text{or} \quad \%C = \frac{100\ \Delta H_c}{H_o}$$

Δe_c = Change in void ratio upon wetting, e_o = Initial void ratio
ΔH_c = Change in height upon wetting, H_o = Initial height

Collapse Potential Values (Based on ASTM D 5333-96, 1998)*

Collapse Potential (CP) (1)	Severity of Problem (2)
0	No problem
0.1 to 2%	Slight
2.1 to 6%	Moderate
6.1 to 10%	Moderately severe
>10%	Severe

*Note: For collapse potential, the vertical pressure upon inundation must equal 200 kPa.

FIGURE 7.6b Equations to determine percent collapse and collapse potential (CP) versus severity of problem.

For example, suppose the data shown in the Fig. 7.6a represents a collapse test on an undisturbed soil specimen taken from the middle of a 3 m (10 ft) uniform layer of collapsible soil. Then the estimated settlement for this layer of collapsible soil would be 0.103 (10.3 percent) times 3 m (10 ft), or a settlement of 0.31 m (1.0 ft).

The above analysis could be performed for each boring excavated within the proposed footprint of the building. The settlement could then be calculated for each boring and the largest value would be the maximum settlement ρ_{max}. The maximum differential settlement Δ is more difficult to determine because it depends to a large extent on how water infiltrates the collapsible soil. For example, for the wetting of the collapsible soil only under a corner or portion of a building, the maximum differential settlement Δ could approach in value the maximum total settlement ρ_{max}. Usually, because of the unknown wetting conditions that will prevail in the field, the maximum differential settlement Δ is assumed to be from 50 to 75 percent of the maximum total settlement ρ_{max}. The amount of time for this settlement to occur depends on the availability of water and the rate of infiltration of the water. Usually the collapse process in the field is a slow process and may take many years to complete.

Design and Construction for Collapsible Soil. There are many different methods to deal with collapsible soil. If there is a shallow deposit of natural collapsible soil, then the deposit can be removed and recompacted during the grading of the site. In some cases, the soil can be densified (by compaction grouting, for example) to reduce the collapse potential of the soil. Another method to deal with collapsible soil is to flood the building footprint or force water into the collapsible soil stratum by using wells. As the wetting front moves through the ground, the collapsible soil will densify and reach an equilibrium state. Flooding or forcing water into collapsible soil should not be performed if there are adjacent buildings because of the possibility of damaging these structures. Also, after the completion of the flooding process, subsurface exploration and laboratory testing should be performed to evaluate the effectiveness of the process.

There are also foundation options that can be used for sites containing collapsible soil. A deep foundation system, one that derives support from strata below the collapsible soil, could be constructed. Also, posttensioned foundations or mat slabs can be designed and installed to resist the larger anticipated settlement from the collapsible soil.

As previously mentioned, the triggering mechanism for the collapse of fill or natural soil is the introduction of moisture. Common reasons for the infiltration of moisture include water from irrigation, broken or leaky water lines, and an altering of surface drainage which allows rainwater to pond near the foundation. Another source of moisture infiltration is from leaking pools. For example, Fig. 7.7 shows a photograph of a severely damaged pool shell which cracked because of the collapse of an underlying uncompacted debris fill. For sites with collapsible soil, it is good practice to emphasize the importance of positive drainage (no ponding of water at the site) and an immediate repair of any leaking utilities.

Collapse and Settlement Caused by Soluble Soil Particles. Especially in arid parts of the world, the soil may contain soluble soil particles, such as halite (salt). Deposits of halite can form in salt playas, sabkhas (coastal salt marshes), and salinas (Bell, 1983). Besides halite, the soil may contain other minerals that are soluble, such as magnesium or calcium carbonate (caliche) and gypsum (gypsiferous soil).

FIGURE 7.7 Damage to a pool shell due to collapse of underlying uncompacted fill (arrows point to cracks in the pool shell).

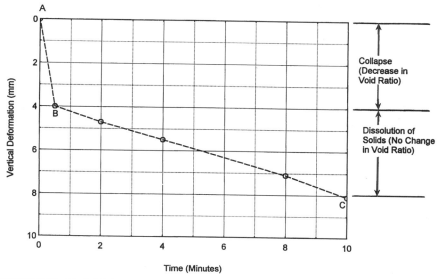

FIGURE 7.8 Vertical deformation versus time for laterally confined salt submerged in distilled water.

These soluble soil particles are dense and hard enough to carry the overburden pressure. But after the site is developed, there can be infiltration of water into the ground from irrigation or leaky water pipes. As this water penetrates the soil containing soluble minerals, two types of settlement can occur: (1) the collapse of the soil structure due to weakening of salt cemented bonds at particle contacts and (2) the water can dissolve away the soluble minerals (i.e., a loss of solids), resulting in ground surface settlement.

The oedometer apparatus can be used to study these two types of settlement. In one experiment (Day, 1996c), dry salt (106 g), which has a specific gravity of 2.2, was poured into the oedometer apparatus. The laterally confined salt, which had an initial diameter of 64 mm and a height of 25 mm, was subjected to a vertical pressure of 100 kPa. Point A in Fig. 7.8 represents the condition of the salt when subjected to the vertical pressure of 100 kPa. The void ratio of the salt at point A was 0.65. Distilled water was then added to the oedometer apparatus, and about 4 mm (16 percent) vertical collapse occurred as indicated by point B in Fig. 7.8. Assuming negligible salt dissolution for the first few seconds of submergence, the void ratio at point B in Fig. 7.8 is 0.38. Dial versus time readings were taken and the vertical deformation versus time was plotted in Fig. 7.8. After 10 minutes of inundation, the vertical pressure was removed, the oedometer was disassembled, and the salt was oven-dried. It was determined that 21 grams of salt had dissolved during the 10 minutes of submergence. The void ratio at the end of the test (point C, Fig. 7.8) was 0.39. As indicated in Fig. 7.8, it can be concluded that although the sample compressed from points B to C, the void ratio did not change as the salt dissolved. This is because the decrease in volume for the sample was compensated by a proportional loss of solids (due to dissolution), and thus the void ratio remained constant from points B to C. For denser conditions, the void ratio may increase as the soluble soil particles dissolve.

A simple method to determine the presence of soluble minerals is to perform a permeameter test. A specimen of the soil, having a dry mass M_0 of about 100 g, is placed in the permeameter apparatus. Filter paper should be used to prevent the loss of fines during the test. Distilled water is slowly flushed through the soil specimen. Usually about 2 L of distilled water is flushed through the soil specimen. After flushing, the soil is dried and the percent soluble soil particles (% soluble) is determined as the initial (M_0) minus final (M_f) dry mass divided by the initial dry mass of soil (M_0), expressed as a percentage, i.e., % soluble = 100 $(M_0-M_f)/M_0$.

As a check on the amount of soluble minerals, the water flushed through the soil specimen can be collected, placed in a sedimentation cylinder (1000 mL), and a hydrometer can be used to determine the amount (grams) of dissolved minerals. As an alternative, the water flushed through the soil specimen can be boiled and the residue collected and weighed.

A common recommendation is that soil which contains above 6 percent soluble soil particles cannot be used as structural fill (Converse Consultants Southwest, Inc., 1990). Soil having between 2 to 6 percent soluble minerals should be blended with nonsoluble soil in accordance with the following ratios (according to Converse Consultants Southwest, Inc., 1990):

Percent soluble soil particles	Blending proportions (nonsoluble soil:soluble soil)
2 to 4	1:1
4 to 6	2:1
Above 6	Can not be used as structural fill

7.4 SETTLEMENT OF COHESIVE AND ORGANIC SOILS

Cohesive and organic soil can be susceptible to a large amount of settlement from structural loads. It is usually the direct weight of the structure that causes settlement of the cohesive or organic soil (i.e., first basic category of settlement; see Sec. 7.1). The settlement of saturated clay or organic soil can have three different components: immediate (or initial), consolidation, and secondary compression.

7.4.1 Immediate or Initial

In most situations, surface loadings causes both vertical and horizontal strains, and this is referred to as *two- or three-dimensional loading*. Immediate settlement is a result of undrained shear deformations, or in some cases contained plastic flow, caused by the two-or three-dimensional loading (Ladd et al., 1977; Foott and Ladd, 1981). Common examples of three-dimensional loading are from square footings and round storage tanks.

Figure 7.9 shows the amount of vertical strain from laboratory tests performed on highly plastic (LL = 93, PI = 71) saturated soil specimens (Day, 1995b). Three conditions were modeled in the laboratory: (1) a one-dimensional loading (oedometer), (2) a three-dimensional loading where the ratio of the height of the soil specimen to the width of the loaded area (H/D ratio) equals 0.10, and (3) a three-dimensional loading where H/D = 0.30. The arrow in Fig. 7.9 indicates the maximum past pressure (i.e., the largest vertical effective stress the soil specimens were ever subjected to). Note in Fig. 7.9 the substantial increase in vertical strain when the vertical stress exceeds the maximum past pressure. For the three-dimensional loadings (H/D = 0.10 and 0.30), the plastic soil squeezed out from underneath the load (i.e., plastic flow occurred) and there was more vertical strain compared to the one-dimensional loading. In Fig. 7.9, the amount of immediate settlement is approximately equal to the difference between the one-dimensional loading curve and the three-dimensional loading curves. As Fig. 7.9 shows, the amount of immediate settlement can even exceed the amount of one-dimensional consolidation. As in these laboratory tests, a field condition that often leads to a substantial amount of immediate settlement is a thick near- or at-surface clay layer that is normally consolidated (OCR = 1) or is subjected to a quick loading ($\Delta\sigma_v$) that results in a vertical stress exceeding the maximum past pressure (σ'_p), as illustrated in Fig. 7.10.

There are many different methods available to determine the amount of immediate settlement for two- or three-dimensional loadings. Some of the more commonly used methods are described below:

Method Based on the Theory of Elasticity. One approach is to use the undrained modulus E_u (also known as the *stress-strain modulus* or *modulus of elasticity of the soil*) from triaxial compression tests in order to estimate the immediate settlement based on elastic theory. A commonly used equation based on the theory of elasticity is as follows:

$$s_i = qBI_p \frac{1-\mu^2}{E_u} \quad (7.1)$$

where s_i = immediate settlement of the loaded area (m or ft)
q = uniform pressure applied by the foundation (kPa or psf)

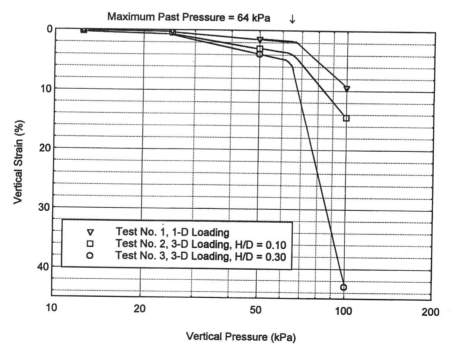

FIGURE 7.9 Vertical strain versus vertical pressure for one-dimensional and three-dimensional loading of a highly plastic saturated soil.

7.18 ANALYSIS OF GEOTECHNICAL DATA AND ENGINEERING COMPUTATIONS

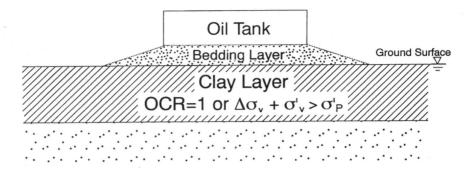

Note: Thickness of Bedding Layer Has Been Exaggerated For Viewing Purposes

FIGURE 7.10 Example of a condition causing significant immediate settlement.

B = width of the foundation (m or ft)
I_p = dimensionless parameter derived from the theory of elasticity to account for the thickness of the compressible layer, shape of the foundation, and flexibility of the foundation.*
μ = Poisson's ratio, which is often assumed to be 0.5 for saturated plastic soil subjected to undrained loading (dimensionless parameter)
E_u = Undrained modulus of the clay (kPa or psf)†

This approach based on the theory of elasticity will often provide approximate results provided the entire soil deposit underlying the foundation is overconsolidated and the load does not cause the vertical effective stress to exceed the maximum past pressure. However, using the theory of elasticity could significantly underestimate the amount of immediate settlement for situations where there is plastic flow of the soil (e.g., the example shown in Fig. 7.10).

Plate Load Tests. A second approach is to perform field plate load tests (see Sec. 4.5) to measure the amount of immediate settlement due to an applied load. The plate load test could significantly underestimate the immediate settlement if the test is performed on a near-surface sandy layer or surface crust of clay that is overconsolidated. Low values of immediate settlement would be recorded because the pressure bulb of the plate load test is very small. But when a large structure is built, the pressure bulb is much larger and could result in significant plastic flow if there is a normally consolidated clay layer underlying the stiff surface layer.

Stress-Path Method. A third approach is to determine the amount of immediate settlement from actual laboratory tests that model the field loading conditions. An undisturbed

---

*At the center of a circular flexible loaded area on an elastic half-space of infinite depth, I_p = 1.0. For a square flexible loaded area on an elastic half-space of infinite depth, I_p = 1.12 beneath the center of the loaded area and I_p = 0.56 beneath the corner of the loaded area. These theoretical influence values have been applied to the situation of an underlying very thick compressible layer relative to the size of the foundation. Numerous other influence values I_p for various conditions are presented in NAVFAC DM-7.1, 1982 (pages 7.1-212 and 7.1-213).

†E_u is often obtained from an undrained triaxial compression test performed on an undisturbed specimen where the stress-strain curve is plotted (i.e., lower plot of Fig. 6.15). The initial tangent modulus to the stress-strain curve is often assumed to be the value of E_u.

soil specimen could be set up in the triaxial apparatus and then the specimen could be subjected to vertical and horizontal stresses that are equivalent to the anticipated loading condition. The undrained vertical deformation (i.e., immediate settlement) due to the applied loading could then be measured. By measuring the amount of vertical deformation from a series of specimens taken from various depths below the proposed structure, the total amount of immediate settlement could be calculated. This approach has been termed the *stress-path method* (Lambe, 1967).

7.4.2 Primary Consolidation

Definition of Primary Consolidation. The increase in vertical pressure due to the weight of the structure constructed on top of saturated soft clays and organic soil will initially be carried by the pore water in the soil. This increase in pore water pressure is known as excess pore water pressure u_e. The excess pore water pressure will decrease with time, as water slowly flows out of the cohesive soil. This flow of water from cohesive soil (which has a low permeability) as the excess pore water pressures slowly dissipate is known as *primary consolidation,* or simply *consolidation.* As the water slowly flows from the cohesive soil, the structure settles as the load is transferred to the soil particle skeleton, thereby increasing the effective stress of the soil. Consolidation is a time-dependent process that may take many years to complete.

The typical one-dimensional case of settlement involves strain in only the vertical direction. Common examples of one-dimensional loading include the lowering of the groundwater table or a uniform fill surcharge applied over a very large area. In the case of a one-dimensional loading, both the strain and flow of water from the cohesive soil as it consolidates will be in the vertical direction only. Consolidation can also occur for two- or three-dimensional loadings, in which case there would be both horizontal and vertical flow of water from the cohesive soil as it consolidates.

Geologists also use the term *consolidation,* but it has a totally different meaning. In geology, consolidation is defined as the processes, such as cementation and crystallization, that transform a soil into a rock. The term *consolidation* can also be used to describe the change of lava or magma into firm rock (Stokes and Varnes, 1955).

On the basis of their stress history, saturated cohesive soils are considered to be either underconsolidated, normally consolidated, or overconsolidated. The overconsolidation ratio (OCR) is used to describe the stress history of cohesive soil, and it is defined as:

$$\text{OCR} = \frac{\sigma'_{vm}}{\sigma'_{v0}} \tag{7.2}$$

where OCR = overconsolidation ratio (dimensionless parameter)
σ'_{vm} or σ'_p = maximum past pressure (σ_{vm}), also known as the preconsolidation pressure (σ'_p), equal to the highest previous vertical effective stress that the cohesive soil was subjected to and consolidated under (kPa or psf)
σ'_{v0} or σ'_v = existing vertical effective stress (kPa or psf)

In terms of the stress history of a cohesive soil, there are three possible conditions, as follows:

1. *Underconsolidated* (OCR < 1). A saturated cohesive soil is considered underconsolidated if the soil is not fully consolidated under the existing overburden pressure and excess pore water pressures u_e exist within the soil. Underconsolidation occurs in areas where a cohesive soil is being deposited very rapidly and not enough time has elapsed for the soil to consolidate under its own weight.

2. *Normally consolidated* (OCR = 1). A saturated cohesive soil is considered normally consolidated if it has never been subjected to an effective stress greater than the existing overburden pressure and if the deposit is completely consolidated under the existing overburden pressure.

3. *Overconsolidated or preconsolidated* (OCR > 1). A saturated cohesive soil is considered overconsolidated if it has been subjected in the past to a vertical effective stress greater than the existing vertical effective stress. An example of a situation that creates an overconsolidated soil is where a thick overburden layer of soil has been removed by erosion over time. As indicated in Table 7.3, there are other mechanisms that can cause a cohesive soil to become overconsolidated.

For structures constructed on top of saturated cohesive soil, determining the overconsolidation ratio of the soil is very important in the settlement analysis. For example, if the cohesive soil is underconsolidated, then considerable settlement due to continued consolidation by the soil's own weight as well as the applied structural load would be expected. On the other hand, if the cohesive soil is highly overconsolidated, then a load can often be applied to the cohesive soil without significant settlement.

The oedometer apparatus (Fig. 5.2) is used to determine the consolidation properties of saturated cohesive soil (ASTM D 2435-96, 1998). Testing of soil in the oedometer apparatus has been discussed in Section 5.2.2. The typical testing procedure consists of placing an undisturbed specimen (usual diameter = 2.5 in., height = 1.0 in.) within the apparatus, applying a vertical seating pressure to the laterally confined specimen, and then submerging the specimen in distilled water. The specimen is then subjected to a incremental increase in vertical pressure, with each pressure remaining on the specimen for a period of 24 hours. Because of the small thickness of the soil specimen, a time period of 24 hours is

TABLE 7.3 Mechanisms That Can Create an Overconsolidated Soil

Main mechanism (1)	Item creating overconsolidated soil (2)	Remarks or references (3)
Change in total stress	Removal of overburden Past structures Weight of past glaciers	Soil erosion Human-induced factors Melting of glaciers
Change in pore water pressure	Change in groundwater table elevation Artesian pressures Deep pumping of groundwater Desiccation due to drying Desiccation due to plant life	Sea-level changes (Kenney, 1964) Common in glaciated areas Common in many cities May have occurred during deposition
Change in soil structure	Secondary compression (aging) Changes in environment, such as pH, temperature, and salt concentration Chemical alterations due to weathering, precipitation of cementing agents, and ion exchange	Lambe, 1958a, 1958b; Bjerrum, 1967b; Leonards and Altschaeffl, 1964 Bjerrum, 1967b; Cox, 1968

Source: Ladd (1973).

ordinarily sufficient to allow the cohesive soil to consolidate and come to equilibrium with little or no further deformation and with the excess pore water pressures u_e within the specimen approximately equal to zero. Dial readings of the vertical deformation versus time are recorded for each vertical pressure applied to the specimen.

Figure 7.11 presents results of consolidation tests performed on two specimens of Orinoco clay at a depth of 40 m (128 ft) (Ladd et al., 1980; Day, 1980). The upper plot is known as the *consolidation curve* for the two Orinoco clay soil specimens. The horizontal axis is the effective consolidation stress (i.e., the applied vertical pressure) which is designated σ'_{vc}, with the subscripts vc referring to vertical consolidation pressure and the prime mark indicating that it is an effective stress. Note that the horizontal axis is a logarithmic scale. In Fig. 7.11, the vertical axis is percent vertical strain ε_v. The vertical axis could also be in terms of the void ratio e.

The two consolidation tests shown in Fig. 7.11 start out at zero vertical strain. Then as the vertical pressure is increased in increments, the soil consolidates and the vertical strain increases. The usual procedure is to apply a vertical pressure that is double the previously applied pressure (load increment ratio = 1.0). However, as shown in Fig. 7.11, in order to better define the consolidation curve, the loading increments can be adjusted to better define the breaking point of the consolidation curve. For the two consolidation tests performed on Orinoco clay (Fig. 7.11), the specimens were loaded up to a vertical pressure of 12 kg/cm² (24,600 psf; 1180 kPa) and then unloaded.

In Fig. 7.11, the arrows indicate the numerical values of the vertical effective stress σ'_{v0} and the maximum past pressure σ'_{vm} for both consolidation tests. The vertical effective stress σ'_{v0} was determined by using Eq. (6.16). The most commonly used procedure for determining the maximum past pressure σ'_{vm} is to use the Casagrande construction technique (1936). Figure 7.12 illustrates this procedure, which is performed as follows:

1. Locate the point of minimum radius of the consolidation curve (point A, Fig. 7.12).
2. Draw a line tangent to the consolidation curve at point A.
3. Draw a horizontal line from point A.
4. Bisect the angle made by steps 2 and 3.
5. Extend the straight line portion of the virgin consolidation curve up to where it meets the bisect line. The point of intersection (point B) of these two lines is the maximum past pressure (also known as preconsolidation pressure σ'_p).

This construction procedure will give the same results if the vertical axis is in terms of vertical strain ε_v or void ratio e. Since the Casagrande technique is only an approximate method of determining the maximum past pressure, it is often useful to obtain a range in values. Note in Fig. 7.12 that the range in possible values has been determined as varying from point D to point E.

On the basis of the Casagrande construction technique, the maximum past pressure σ'_{vm} was determined for the two consolidation tests on Orinoco clay (Fig. 7.11). Note that these two tests have significantly different values of maximum past pressure (2.75 versus 1.35 kg/cm²). The reason for the difference was sample disturbance. Test No. 12 (solid circles) was disturbed during sampling and only had a vane shear strength (TV) of 0.30 kg/cm² (610 psf), as compared to the undisturbed specimen (Test No. 18) which had a vane shear strength of 0.51 kg/cm² (1040 psf). Thus sample disturbance can significantly lower the value of the maximum past pressure of saturated cohesive soil.

In Fig. 7.11, the vertical effective stress σ'_{v0} has a numerical value that is close to the maximum past pressure σ'_{vm} for the undisturbed soil specimen (open circles). This means that the Orinoco clay is essentially normally consolidated. In Fig. 7.13, all of the consolidation test data for the Orinoco clay is summarized on one sheet. The vertical axis is depth

7.22 ANALYSIS OF GEOTECHNICAL DATA AND ENGINEERING COMPUTATIONS

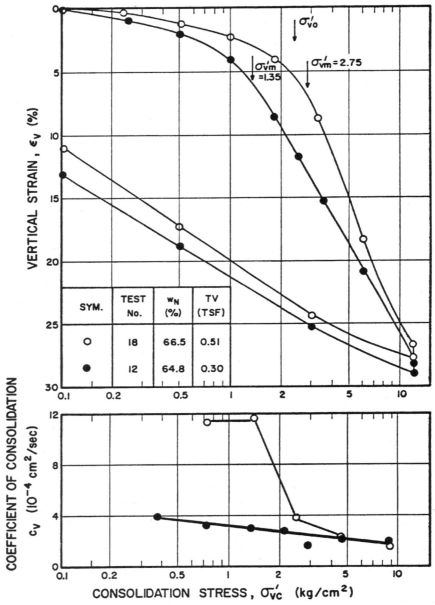

FIGURE 7.11 Consolidation test data for Orinoco clay at a depth of 40 m. (*From Ladd et al., 1980; Day, 1980.*)

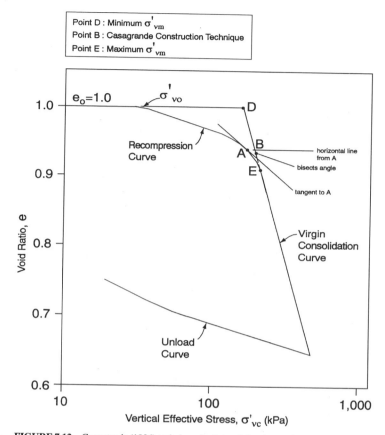

FIGURE 7.12 Casagrande (1936) technique for determining the maximum past pressure.

(feet) and the horizontal axis is effective stress in kg/cm². In Fig. 7.13, the vertical effective stress σ'_{v0} versus depth for the Orinoco clay at Borings E1 and F1 has been plotted. The maximum past pressure σ'_{vm} for each consolidation test has also been plotted with possible ranges in values based on the procedure shown in Fig. 7.12. The open symbols represent consolidation tests performed on undisturbed specimens, while the solid symbols represent consolidation tests performed on disturbed specimens. In general, the maximum past pressure data points σ'_{vm} are close to the vertical effective stress σ'_{v0} and for this offshore deposit of clay, it could be concluded that the clay is essentially normally consolidated (OCR = 1) to very slightly overconsolidated.

This information on the stress history of saturated clay deposits is very important and it is often added to the subsoil profile. For example, note in Figs. 4.40 to 4.42 that the maximum past pressures (i.e., preconsolidation pressures) obtained from the results of oedometer tests on the cohesive soils have been included on these subsoil profiles.

Calculating the Amount of Primary Consolidation Settlement. In addition to using the consolidation curve to determine the maximum past pressure σ'_{vm} of a cohesive soil,

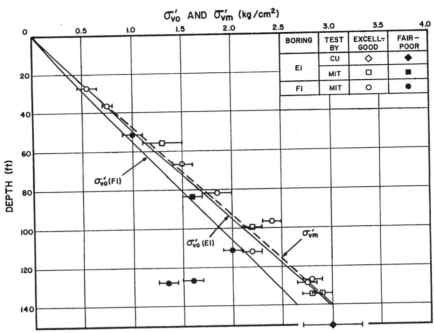

FIGURE 7.13 Stress history of Orinoco clay. (*From Ladd et al., 1980; Day, 1980.*)

the consolidation curve obtained from the oedometer test can also be used to estimate the primary consolidation settlement s_c due to a loading of the cohesive soil, such as from the construction of a building. Note in Fig. 7.12 that the consolidation curve can often be approximated as two straight line segments. The recompression index C_r represents the reloading of the saturated cohesive soil, and the compression index C_c represents the loading of the saturated cohesive soil beyond the maximum past pressure σ'_{vm}. The compression index is often referred to as the slope of the virgin consolidation curve because the *in situ* saturated cohesive soil has never experienced this loading condition. The reason for the relatively steep virgin consolidation curve, as compared to the recompression curve, is because there is a tendency for the soil structure to break down and contract once the pressure exceeds the maximum past pressure σ'_{vm}.

The values of the recompression index C_r and the compression index C_c are simply the slope of the lines shown in Fig. 7.12 and can be calculated as a change in void ratio Δe divided by the corresponding change in the effective pressures ($\log \sigma'_{vc2} - \log \sigma'_{vc1}$), or $C_r = \Delta e/\log (\sigma'_{vc2}/\sigma'_{vc1})$ for the recompression curve and $C_c = \Delta e/\log (\sigma'_{vc2}/\sigma'_{vc1})$ for the virgin consolidation curve. An easier method to obtain C_r and C_c is to determine Δe over one log cycle; for example, if $\sigma'_{vc2} = 100$ and $\sigma'_{vc1} = 10$, then $\log (\sigma'_{vc2}/\sigma'_{vc1}) = \log (100/10) = 1$. By using one log cycle, the values of $C_r = \Delta e$ for the recompression curve and $C_c = \Delta e$ for the virgin consolidation curve.

Using the calculated values or C_r and C_c, the primary consolidation settlement s_c due to an increase in load $\Delta \sigma_v$ can be determined from the following equations:

For Underconsolidated Soil (OCR < 1):

$$s_c = C_c \frac{H_0}{1 + e_0} \log \frac{\sigma'_{v0} + \Delta\sigma'_v + \Delta\sigma_v}{\sigma'_{v0}} \quad (7.3)$$

For Normally Consolidated Soil (OCR = 1):

$$s_c = C_c \frac{H_0}{1 + e_0} \log \frac{\sigma'_{v0} + \Delta\sigma_v}{\sigma'_{v0}} \quad (7.4)$$

For Overconsolidated Soil (OCR > 1):
Case I: $\sigma'_{v0} + \Delta\sigma_v \leq \sigma'_{vm}$

$$s_c = C_r \frac{H_0}{1 + e_0} \log \frac{\sigma'_{v0} + \Delta\sigma_v}{\sigma'_{v0}} \quad (7.5)$$

Case II: $\sigma'_{v0} + \Delta\sigma_v > \sigma'_{vm}$

$$s_c = C_r \frac{H_0}{1 + e_0} \log \frac{\sigma'_{vm}}{\sigma'_{v0}} + C_c \frac{H_0}{1 + e_0} \log \frac{\sigma'_{v0} + \Delta\sigma_v}{\sigma'_{vm}} \quad (7.6)$$

where s_c = settlement due to primary consolidation caused by an increase in load (m or ft)
C_c = compression index, obtained from the virgin consolidation curve (dimensionless)
C_r = recompression index, obtained from the recompression portion of the consolidation curve (dimensionless)
H_0 = initial thickness of the *in situ* saturated cohesive soil layer (m or ft)
e_0 = initial void ratio of the *in situ* saturated cohesive soil layer (dimensionless)
σ'_{v0} = initial vertical effective stress of the *in situ* soil (kPa or psf)
$\Delta\sigma'_v$ = for an underconsolidated soil, this represents the increase in vertical effective stress that will occur as the cohesive soil consolidates under its own weight (kPa or psf)
$\Delta\sigma_v$ = increase in load, typically due to the construction of a building or the construction of a fill layer at ground surface (kPa or psf)*
σ'_{vm} = maximum past pressure (kPa or psf), also known as the preconsolidation pressure σ'_p. It is obtained from the consolidation curve by using the Casagrande (1936) construction technique (see Fig. 7.12).

For overconsolidated soil, there are two possible cases that can be used to calculate the amount of settlement. The first case occurs when the existing vertical effective stress σ'_{v0} plus the increase in vertical stress $\Delta\sigma_v$ due to the proposed building weight does not exceed the maximum past pressure σ'_{vm}. For this first case, there will only be recompression of the cohesive soil.

For the second case, the sum of the existing vertical effective stress σ'_{v0} plus the increase in vertical stress $\Delta\sigma_v$ due to the proposed building weight exceeds the maximum

---

*The value of $\Delta\sigma_v$ (also known as σ_z) can be obtained from stress distribution theory as discussed in Sec. 6.4. Note that a drop in the groundwater table or a reduction in pore water pressure can also result in an increase in load on the cohesive soil.

7.26 ANALYSIS OF GEOTECHNICAL DATA AND ENGINEERING COMPUTATIONS

past pressure σ'_{vm}. For the second case, there will be virgin consolidation of the cohesive soil. Given the same cohesive soil and identical field conditions, the settlement due to the second case will be significantly more than the first case.

If the consolidation data is plotted on a vertical strain ε_v versus consolidation stress σ'_{vc} graph such as shown in Fig. 7.11, then the recompression curve and virgin consolidation curve can also be approximated as straight lines. When using a vertical strain ε_v versus consolidation stress σ'_{vc} plot, the slope of the recompression curve is designated the *modified recompression index* $C_{r\varepsilon}$ and the slope of the virgin consolidation curve is designated the *modified compression index* $C_{c\varepsilon}$. These indices are related as follows:

$$C_{r\varepsilon} = \frac{C_r}{1 + e_0} \tag{7.7}$$

$$C_{c\varepsilon} = \frac{C_c}{1 + e_0} \tag{7.8}$$

The values from Eqs. (7.7) and (7.8) can be substituted into Eqs. (7.3) to (7.6).

The value of H_0 in Eqs. (7.3) to (7.6) represents the initial thickness of the cohesive soil layer. Because σ'_{v0} and $\Delta\sigma_v$ can both change with depth, the cohesive soil layer may need to be broken into several horizontal layers in order to obtain an accurate value of the primary consolidation settlement s_c.

Example. A site has a level ground surface and a level groundwater table located 5 m below the ground surface. Below the groundwater table, the pore water pressures are hydrostatic. Subsurface exploration has discovered that the site is underlain with sand, except for a uniform and continuous clay layer that is located at a depth of 10 to 12 m below ground surface. A consolidation test performed on an undisturbed specimen obtained from the center of the clay layer indicates that it has a maximum past pressure $(\sigma'_{vm}) = 150$ kPa and the compression index $(C_c) = 0.83$. The average void ratio e_0 of the clay layer is 1.10 and the buoyant unit weight γ_b of the clay layer = 7.9 kN/m³. The total unit weight γ_t of the sand above the groundwater table is 18.9 kN/m³ and the total unit weight γ_t of the sand below the groundwater table is 19.7 kN/m³. A building having a width = 20 m and a length = 30 m is proposed for the site and will be constructed at ground surface. The structural engineer has determined that the building weight can be approximated as a uniform pressure applied at ground surface equivalent to 50 kPa. Calculate the primary consolidation settlement s_c of the clay layer due to the building weight.

Solution. The first step is to determine the vertical effective stress (σ'_{v0}) at the center of the clay layer, or

$$\sigma'_{v0} = (5 \text{ m})(18.9 \text{ kN/m}^3) + (5 \text{ m})(19.7 - 9.81 \text{ kN/m}^3) + (1 \text{ m})(7.9 \text{ kN/m}^3) = 152 \text{ kPa}$$

Since σ'_{v0} is approximately equal to σ'_{vm}, the clay layer is essentially normally consolidated (OCR = 1). Using the 2:1 approximation [Eq. 6.28)] with $B = 20$ m, $L = 30$ m, $z = 11$ m, and $\sigma_0 = 50$ kPa, then $\Delta\sigma_v = 24$ kPa.

Using Eq. (7.4) with the following values:

- $\sigma'_{v0} = 152$ kPa
- $\Delta\sigma_v = 24$ kPa
- $H_0 = 2$ m
- $C_c = 0.83$
- $e_0 = 1.10$

we find the calculated primary consolidation settlement $(s_c) = 0.050$ m (2 in.).

SETTLEMENT OF STRUCTURES

Rate of One-Dimensional Consolidation. As previously mentioned, when a soft clay is loaded, there is an increase in the pore water pressure (known as excess pore water pressure u_e). As water flows out of the clay, this excess pore water pressure slowly dissipates. In order to determine how fast this consolidation process will take, Terzaghi (1925) developed the one-dimensional consolidation equation to describe the time-dependent settlement behavior of clays. The Terzaghi equation is a form of the diffusion equation from mathematical physics. There are other phenomena that can be described by the diffusion equation, such as the heat flow through solids. In order to derive Terzaghi's diffusion equation, the following assumptions are made (Terzaghi, 1925; Taylor, 1948; Holtz and Kovacs, 1981):

1. The clay is homogeneous and the degree of saturation S is 100 percent (saturated soil).
2. Drainage is provided at the top and/or bottom of the compressible layer.
3. Darcy's law ($v = ki$) is valid.
4. The soil grains and pore water are incompressible.
5. Both the compression and flow of water are one-dimensional.
6. The load increment results in small strains so that the coefficient of permeability k and the coefficient of compressibility a_v remain constant.
7. There is no secondary compression.

Terzaghi's diffusion equation is derived by considering the continuity of flow out of a soil element, and is:

$$c_v \frac{\partial^2 u_e}{\partial z^2} = \frac{\partial u_e}{\partial t} \tag{7.9}$$

where u_e = excess pore water pressure, z = vertical height to the nearest drainage boundary, t = time, and c_v = coefficient of consolidation, defined as:

$$c_v = \frac{k(1 + e_0)}{\gamma_w a_v} \tag{7.10}$$

where k = coefficient of permeability (hydraulic conductivity)
e_0 = initial void ratio
γ_w = unit weight of water
a_v = coefficient of compressibility

It is commonly defined as:

$$a_v = \frac{-\Delta e}{\Delta \sigma'_v} \tag{7.11}$$

where Δe is the change in void ratio corresponding to the change in vertical effective stress $\Delta \sigma'_v$.

The mathematical solution to Eq. (7.9) can be expressed in graphical form, as shown in Fig. 7.14. The vertical axis is defined as Z, which equals:

$$Z = \frac{z}{H} \tag{7.12}$$

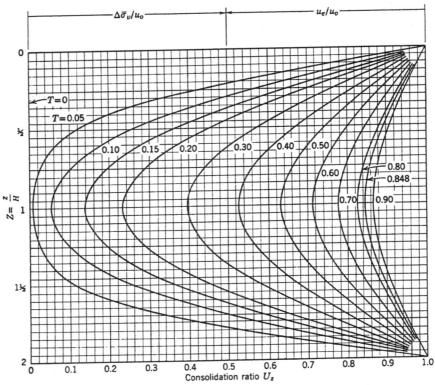

FIGURE 7.14 Consolidation ratio as a function of depth and time factor for uniform initial excess pore water pressure. (*From Lambe and Whitman, 1969; reproduced with permission of John Wiley & Sons.*)

where z = vertical depth below the top of the clay layer and H = one-half the thickness of the clay layer (double drainage). The horizontal axis is the consolidation ratio U_z, defined as:

$$U_z = 1 - \frac{u_e}{u_0} \qquad (7.13)$$

where u_e = excess pore water pressures in the cohesive soil and u_0 = initial excess pore water pressure which is equal to the applied load on the saturated cohesive soil (i.e., $u_0 = \Delta\sigma_v$).

In Fig. 7.14, the value of U_z can be determined for different time factors T, defined as:

$$T = \frac{c_v t}{H_{dr}^2} \qquad (7.14)$$

where T = time factor from Table 7.4 (dimensionless)
c_v = coefficient of consolidation from laboratory testing (cm²/s or ft²/day)

TABLE 7.4 Average Degree of Consolidation U_{avg} versus Time Factor T

Average degree of consolidation U_{avg}, % (1)	Time factor T (2)
0	0
5	0.002
10	0.008
15	0.017
20	0.031
25	0.049
30	0.071
35	0.092
40	0.126
45	0.159
50	0.197
55	0.238
60	0.286
65	0.340
70	0.403
75	0.477
80	0.567
85	0.683
90	0.848
95	1.13
100	∞

Assumptions: Terzaghi theory of consolidation, linear initial excess pore water pressures, and instantaneous loading.
Note: The above values are based on the following approximate equations by Casagrande (unpublished notes) and Taylor (1948):

For $U_{avg} < 60\%$, the time factor $T = \frac{1}{4}\pi(U_{avg}/100)^2$

For $U_{avg} \geq 60\%$, the time factor $T = 1.781 - 0.933 \log(100 - U_{avg})$

H_{dr} = height of the drainage path (cm or ft)*
t = time since the application of the load (the load is assumed to be applied instantaneously)

The purpose of the Terzaghi theory of consolidation is to estimate the settlement versus time relationship after loading of the cohesive soil. The amount of settlement, otherwise known as the *average degree of consolidation settlement* U_{avg}, is related to the time factor T as follows:

$$U_{avg} = 100 - \frac{100 \text{ (area within a } T \text{ curve)}}{\text{(total area of Fig. 7.14)}} \quad (7.15)$$

---

*If water can drain out through the top and bottom of the cohesive soil layer (double drainage), then $H_{dr} = \frac{1}{2}H$, where H = thickness of the cohesive soil layer. An example of double drainage would be if there are sand layers located on top of and below the cohesive soil layer. If water can drain only out of the top or bottom of the cohesive soil layer (single drainage), then $H_{dr} = H$. An example of single drainage would be if the clay layer is underlain by dense shale that is essentially impervious.

From Fig. 7.14 and Eq. (7.15), the relationship between the average degree of consolidation settlement U_{avg} and the time factor T has been calculated and is summarized in Table 7.4. When the load is applied, the average degree of consolidation U_{avg} is zero (no settlement) and the time factor T is also zero. When one-half of the consolidation settlement has occurred (i.e., $U_{avg} = 50$ percent), the time factor $(T) = 0.197$. The Terzaghi theory of consolidation predicts that the time (t) for complete consolidation ($U_{avg} = 100$ percent, i.e., total settlement) is infinity, but as a practical matter, a time factor T equal to 1.0 is often assumed for $U_{avg} = 100$ percent.

The steps in using the Terzaghi theory of consolidation are as follows:

1. **Coefficient of consolidation.** The first step is to determine the coefficient of consolidation c_v of the cohesive soil. The usual procedure is to calculate c_v from the results of laboratory consolidation tests. During each incremental loading of the saturated cohesive soil in the oedometer apparatus, dial readings of vertical deformation can be recorded and plotted versus time (log scale), as shown in Fig. 7.15. This vertical deformation versus time data was recorded from a laboratory consolidation test performed on a highly plastic soil (LL = 93, PI = 71, e_0 = 3.05). The vertical deformation versus time data was recorded when the vertical load on the soil (σ'_{vc}) = 50 kPa. The laboratory data shown in Fig. 7.15 can be used to determine the coefficient of consolidation. The procedure is as follows:

    a. *Determine end of primary consolidation.* The end of primary consolidation (location of arrow) is estimated as the intersection of the two straight line segments. The value of d_{100} = the end of primary consolidation (d_{100} = 2.0 mm).

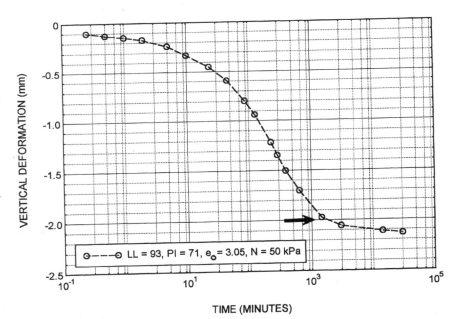

FIGURE 7.15 Data recorded from a consolidation test performed on saturated cohesive soil. At a vertical pressure N of 50 kPa, the vertical deformation as a function of time after loading was recorded and plotted above. The arrow indicates the end of primary consolidation.

b. Determine d_0. As an approximation, d_0 can be assumed to be the dial reading at $t = 0.1$ minutes (i.e., the curve is extended back to $t = 0.1$ minutes), or $d_0 = 0.1$ mm.

c. Determine d_{50} and t_{50}. The value of d_{50} is equal to the d_0 plus d_{100} divided by 2 [i.e., $d_{50} = 1/2 \, (d_0 + d_{100}) = 1/2 \, (0.1 + 2.0) = 1.05$]. Using the curve shown in Fig. 7.15, at a vertical deformation of 1.05 mm, the corresponding time $(t_{50}) = 150$ min. The value of t_{50} (150 min) is the length of time it took for the saturated cohesive soil specimen to experience 50 percent of its primary consolidation (i.e., $U_{avg} = 50$ percent) when subjected to a vertical pressure of 50 kPa.

d. Calculate the coefficient of consolidation c_v. Equation (7.14) is used to calculate the coefficient of consolidation c_v. According to Table 7.4, the time factor T for an average degree of consolidation U_{avg} of 50 percent is 0.197. The height H of the specimen at $d_{50} = 9.0$ mm, and since the specimen has double drainage in the laboratory oedometer apparatus, $H_{dr} = 4.5$ mm. Inserting $t_{50} = 150$ minutes, $H_{dr} = 4.5$ mm, and $T = 0.197$ into Eq. (7.14), we find the coefficient of consolidation $(c_v) = 0.027$ mm^2/min (4.4×10^{-6} cm^2/s). This very low coefficient of consolidation is a result of the clay particles (montmorillonite) in the soil (i.e., PI = 71).

The coefficient of consolidation c_v for laboratory specimens under different sample heights, load increment ratios, and load durations has been determined for numerous cohesive soils (Taylor, 1948; Leonards and Ramiah, 1959). Ladd (1973) states that it can be assumed that variations in sample height, load increment ratio, and load duration will generally lead to insignificant difference in c_v, provided that the resulting dial-versus-time reading has the characteristic shape shown in Fig. 7.15 (i.e., a Type 1 curve; Leonards and Altschaeffl, 1964) when plotted on a log-time scale. This means that there must be appreciable primary consolidation during the loading increment. A factor that can significantly affect the coefficient of consolidation is sample disturbance. For example, in Fig. 7.11 the coefficient of consolidation is shown for both the disturbed Orinoco clay specimen (solid circles) and the undisturbed Orinoco clay specimen (open circles). Especially for that part of the consolidation test that involves recompression of the cohesive soil, sample disturbance can significantly reduce the values of the coefficient of consolidation.

2. Determine the drainage height H_{dr} of the in situ cohesive soil. After the coefficient of consolidation c_v has been determined from laboratory testing of undisturbed soil specimens, the next step is to determine the drainage height of the *in situ* cohesive soil layer. As previously mentioned, if water can drain from the cohesive soil layer at both the top and bottom of the clay layer (double drainage), then $H_{dr} = 1/2 H$, where H = thickness of the cohesive soil layer. If water can only drain from the top or bottom of the cohesive soil layer (single drainage), then $H_{dr} = H$.

3. Determine the time factor T. On the basis of an average degree of consolidation U_{avg}, the time factor T can be obtained from Table 7.4.

4. Use Eq. (7.14) to determine the time t. Once c_v, T, and H_{dr} are known, the time t corresponding to a certain amount of settlement U_{avg} can be calculated from Eq. (7.14).

In summary, the steps listed above basically consist of determining the coefficient of consolidation c_v from laboratory consolidation tests performed on undisturbed soil specimens, and then this data is used to predict the time-settlement behavior of the *in situ* cohesive soil layer.

Example. Use the same conditions as the previous example. Assuming quick construction of the building and using the Terzaghi theory of consolidation, predict how long after construction it will take for 50 and 90 percent of the primary consolidation settlement to occur. From laboratory oedometer testing of an undisturbed clay specimen, where dial readings versus time were recorded at a vertical pressure of 200 kPa, the coefficient of consolidation c_v was calculated to be 1×10^{-4} cm^2/s.

Solution. Since there is sand on both on the top and bottom of the 2-m-thick clay layer, the clay layer has double drainage and $H_{dr} = 1$ m. For $U_{avg} = 50$ percent, the time factor (T) = 0.197 and for $U_{avg} = 90$ percent, the time factor (T) = 0.848 (Table 7.4). Using Eq. (7.14) with a value of $c_v = 1 \times 10^{-4}$ cm²/s (0.32 m²/year), we find it will take 0.62 years for 50 percent of the primary consolidation to occur (i.e., 0.62 years for 1 in. of settlement) and it will take 2.7 years for 90 percent of the primary consolidation to occur (i.e., 2.7 years for 1.8 in. of settlement).

Terzaghi's consolidation equation is one of the most widely taught and applied theories in geotechnical engineering. Some of the limitations of this theory are as follows (Duncan, 1993):

- c_v is commonly assumed to be constant, which is often not the case in the field.
- The stress-strain behavior of the soil skeleton is assumed to be linear and elastic.
- The strains are assumed to be uniform.
- There are often permeable sand lenses within the cohesive soil that can significantly decrease the length of time for primary consolidation as predicted by assuming a constant c_v.
- The strain often decreases with depth because the stress increase caused by surface loads decreases with depth, or the clay compressibility decreases with depth, or both.
- A final factor is that the theory is based on the vertical flow of water from the cohesive soil, but many cases involve two- or three-dimensional loading (such as the example problem), which would allow the clay to drain partially from its sides.

Because of all these factors, the Terzaghi theory of consolidation often overpredicts the time required to reach a certain average degree of consolidation. The theory should be used only as an approximation of the time-settlement behavior for loading of saturated cohesive soil. Field measurements (Sec. 18.2) are often essential in comparing the actual time-settlement behavior with the predicted behavior.

7.4.3 Secondary Compression

The final component of settlement is secondary compression, which is that part of the settlement that occurs after essentially all of the excess pore water pressures have dissipated (i.e., settlement that occurs at constant effective stress). The usual assumption is that secondary compression does not start until after primary consolidation is complete.

In Fig. 7.15, the vertical deformation that occurs above the arrow is primary consolidation, while the vertical deformation that occurs below the arrow is secondary compression. As previously mentioned, this end of primary consolidation (arrow in Fig. 7.15) was obtained as the intersection of the two straight lines of the deformation versus time curve (log time plot). The amount of secondary compression is often neglected because, as shown in Fig. 7.15, it is rather small compared to the primary consolidation settlement. However, secondary compression can constitute a major part of the total settlement for peat or other highly organic soil (Holtz and Kovacs, 1981).

Ladd (1973) describes secondary compression as a process where the particle contacts are still rather unstable at the end of primary consolidation and the particles will continue to move until they find a stable arrangement. This would explain why the rate of secondary compression often increases with compressibility. The more compressible the soil, the greater the tendency for a larger number of particles to be unstable at the end of primary consolidation (Ladd, 1973). Because there are no excess pore water pressures during secondary compression, it is often described as *drained creep*.

The amount of settlement s_s due to secondary compression can be calculated as follows:

$$s_s = C_\alpha H_0 \Delta \log t \qquad (7.16)$$

where s_s = settlement due to secondary compression (occurs after the end of primary consolidation) (m or ft)
C_α = secondary compression ratio (dimensionless parameter) which is defined as the slope of the secondary compression curve*
H_0 = initial thickness of the *in situ* cohesive soil layer (m or ft)
$\Delta \log t$ = change in log of time from the end of primary consolidation to the end of the design life of the structure†

As indicated in Table 7.3, secondary compression (aging) can create a slight overconsolidation of a cohesive soil, even though the effective stress has not changed. While the site history may indicate a normally consolidated soil, it is possible that a slight load can be applied to the cohesive soil without triggering virgin consolidation because of this secondary compression aging effect.

The final calculation for estimating the maximum settlement (ρ_{max}) of the *in situ* cohesive soil would be to add together the three components of settlement, or:

$$\rho_{max} = s_i + s_c + s_s \qquad (7.17)$$

where ρ_{max} = maximum settlement over the life of the structure
s_i = immediate settlement (Sec. 7.4.1)
s_c = primary consolidation settlement (Sec. 7.4.2)
s_s = secondary compression settlement [Eq. (7.16)]

7.5 SETTLEMENT OF COHESIONLESS SOIL

The previous section discusses the settlement of cohesive soil. This section deals with the settlement of cohesionless (nonplastic) soil caused by the structural load. A major difference between saturated cohesive soil and cohesionless soil is that the settlement of cohesionless soil is not time-dependent. Because of the generally high permeability of cohesionless soil, the settlement of a saturated cohesionless soil usually occurs as the load is applied during the construction of the building. Cohesionless soils typically do not have long-term settlement (i.e., such as secondary compression). Exceptions would be if the cohesionless soil is susceptible to collapse due to moisture infiltration (Sec. 7.3) or if it is subjected to seismic loading (Chap. 11).

For cohesionless soil, there are many different methods that can be used to determine the settlement due to structural loading. Some of the more commonly used methods are described below:

Plate Load Tests. The plate load test can be used to directly estimate the settlement potential of the footing (Sec. 4.5). For settlement of sands caused by an applied surface

---
*For example, the secondary compression curve shown in Fig. 7.15 (portion of curve below the arrow) can be approximated as a straight line. If the vertical axis in Fig. 7.15 is converted to strain (ε_v), then C_α equals the change in strain ($\Delta\varepsilon_v$) divided by the change in time (log scale), or $C_\alpha = \Delta\varepsilon_v/\Delta \log t$. Using one log cycle of time (i.e., $\Delta \log t = 1$), then the value of secondary compression $C_\alpha = \Delta\varepsilon_v$.

†For example, if primary consolidation is complete at a time = 10 years, and the design life of the structure is 100 years, then $\Delta \log t = \log 100 - \log 10 = 1$.

loading, Eq. (4.6) is an empirical equation from Terzaghi and Peck (1967) that relates the depth of penetration of the steel plate (S_1) to the settlement of the actual footing (S, or ρ_{max} for that footing having the largest settlement). As previously mentioned in Sec. 4.5, if Eq. (4.6) is to be used, the stress exerted by the steel plate corresponding to S_1 must be the same bearing stress exerted by the actual footing.

The calculated settlement from Eq. (4.6) can significantly underestimate the actual settlement at the site in cases where there is settlement due to secondary influences (such as collapsible soil) or in cases where there is a deep looser or softer layer that is not affected by the small plate, but is loaded by the much larger footing.

Laboratory Oedometer Tests. As discussed in the previous section, the main method of determining the settlement of cohesive soil is to test undisturbed soil specimens in the oedometer apparatus. If undisturbed soil specimens of the cohesionless soil can be obtained, such as from test pits or Shelby tube samplers, then the compression curve of cohesionless soil could be obtained by using the oedometer apparatus. Usually the compression curve for cohesionless soil does not have linear recompression and virgin compression indices such as shown in Fig. 7.12. As an alternative to using Eqs. (7.3) to (7.6), the following equation can be used to estimate the settlement ρ_{max} of the cohesionless soil:

For data plotted on a void ratio (e) versus log σ'_{vc} plot:

$$\rho_{max} = \frac{\Delta e \, H_0}{1 + e_0} \tag{7.18a}$$

For data plotted on a vertical strain (ε_v) versus log σ'_{vc} plot:

$$\rho_{max} = \Delta \varepsilon_v \, H_0 \tag{7.18b}$$

where ρ_{max} = maximum settlement of the building (m or ft)
Δe = change in void ratio (dimensionless) from the void ratio e versus log σ'_{vc} plot*
H_0 = initial height of the *in situ* cohesionless soil layer (m or ft)†
e_0 = initial void ratio of the cohesionless soil layer (dimensionless)
$\Delta \varepsilon_v$ = change in vertical strain (dimensionless) from the vertical strain (ε_v) versus log σ'_{vc} plot‡

Empirical Correlations. In many cases, undisturbed specimens of the cohesionless soil can not be obtained. Thus a common approach is to use empirical correlations in order to determine the settlement of the cohesionless soil. For example, Fig. 7.16 shows a chart by Terzaghi and Peck (1967) that presents an empirical correlation between the measured N-value (obtained from the standard penetration test; see Sec. 4.4.2) and the allowable soil pressure (tsf) that will produce a settlement of the footing of 1 in. (2.5 cm).

As an example of the use of Fig. 7.16, suppose a site contains a sand deposit and the proposed structure can be subjected to a maximum settlement (ρ_{max}) of 1.0 in. (2.5 cm). If the

---

*Enter the plot at the vertical effective stress (σ'_{v0}) and upon intersecting the compression curve, determine the initial void ratio e_i. Also enter the plot at the final vertical effective stress (i.e., $\sigma'_{v0} + \Delta\sigma_v$), and on intersecting the compression curve, determine the final void ratio e_f. And then $\Delta e = e_i - e_f$

†Often several undisturbed soil specimens are obtained and tested in the laboratory, and H_0 represents the thickness of the *in situ* layer that is represented by an individual oedometer test. For the analysis of several soil layers, Eq. (7.18) is used for each layer and then the maximum settlement is the sum of the settlements calculated for each soil layer.

‡Enter the plot at the vertical effective stress σ'_{v0} and on intersecting the compression curve, determine the initial vertical strain ε_i. Also enter the plot at the final vertical effective stress (i.e., $\sigma'_{v0} + \Delta\sigma_v$) and on intersecting the compression curve, determine the final vertical strain ε_f. And then $\Delta \varepsilon = \varepsilon_f - \varepsilon_i$

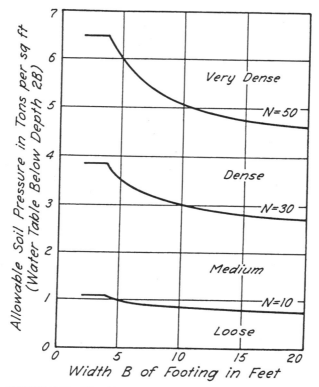

FIGURE 7.16 Allowable soil bearing pressures for footings on sand based on the standard penetration test. *(From Terzaghi and Peck, 1967; reprinted with permission of John Wiley & Sons.)*

measured N-value from the standard penetration test is 10 and the width of the proposed footings = 5 ft (1.5 m), then the allowable soil pressure = 1 tsf (100 kPa).

For measured N-values other than those for which the curves are drawn in Fig. 7.16, the allowable soil pressure can be obtained by linear interpolation between curves. According to Terzaghi and Peck (1967), if all of the footings are proportioned in accordance with the allowable soil pressure corresponding to Fig. 7.16, then the maximum settlement (ρ_{max}) of the foundation should not exceed 1 in. (2.5 cm) and the maximum differential settlement (Δ) should not exceed ¾ in. (2 cm). Figure 7.16 was developed for the groundwater table located at a depth equal to or greater than a depth of $2B$ below the bottom of the footing. For a high groundwater table close to the bottom of the shallow foundation, the values obtained from Fig. 7.16 should be reduced by 50 percent.

Elastic Method. As in determining the immediate settlement for cohesive soil (Sec. 7.4.1), the theory of elasticity can be used to estimate the settlement of cohesionless soil. The settlement would be determined from Eq. (7.1), using the drained modulus E_s of the cohesionless soil. The drained modulus is often obtained from a drained triaxial compression test performed on an undisturbed soil specimen where the stress-strain curve is plotted. The initial tangent modulus to the stress-strain curve is often assumed to be the value of E_s.

The drained modulus E_s of the cohesionless soil can also be obtained from empirical correlations with the standard penetration test (SPT) or the cone penetration test (CPT). Approximate correlations are as follows (NAVFAC DM-7.1, 1982):

Cohesionless soil type	E_s/N
Nonplastic silty sands or silt-sand mixtures	4
Clean, fine, to medium sands	7
Coarse sands and sand with a little gravel	10
Sandy gravel and gravel	12

The units for E_s/N are tons per square foot (1 tsf ≅ 100 kPa). For example, if a clean medium sand has a measured N-value = 10, then E_s = 70 tsf (7 MPa). For the cone penetration test, an approximate correlation (NAVFAC DM-7.1, 1982) is $E_s = 2 q_c$, where q_c = cone resistance in kg/cm² (1 kg/cm² ≅ 1 tsf ≅ 100 kPa).

In many cases, the soil modulus E_s will vary with depth. Schmertmann's method (Schmertmann, 1970; Schmertmann et al., 1978) is ideally suited to determining the settlement of footings when the soil modulus (E_s) varies with depth.

For complex foundation problems, computer programs have been developed to estimate the settlement based on the finite-element analysis. For example, the SIGMA/W (Geo-Slope, 1993) computer program can determine the settlement, assuming the soil is an elastic or an inelastic material.

Other Considerations. There are many other factors that may have to be considered in the settlement analysis. For example, an important assumption in all of these methods is that the soil remains undisturbed during the construction process. But often during the excavation of the foundation, the bottom of the excavation becomes disturbed, creating a loose soil zone. This disturbance can occur during the actual excavation or when workers are in the excavation installing the steel reinforcement. Also, debris such as loose soil may be inadvertently knocked into the footing excavation after completion of the excavation, but before the concrete is placed. Even a thin disturbed and loose zone can lead to settlement that is greatly in excess of calculated values.

It is important that undisturbed soil (i.e., natural ground) or adequately compacted soil be present at the bottom of the footing excavations and that the footings be cleaned of all loose debris prior to placement of concrete. Any loose rocks at the bottom of the footing excavation should also be removed and the holes filled with concrete (during placement of concrete for the footings). As will be discussed in Chap. 18, the geotechnical engineer often observes and prepares a report on the bearing conditions exposed in the footing excavations.

7.6 OTHER COMMON CAUSES OF SETTLEMENT

There are many other causes of settlement of structures. Some of the more common causes are discussed in this section. The types of settlement discussed in this section are within the second basic category of settlement (see Sec. 7.1), settlement of the structure primarily due to secondary influences. An exception is the construction of a structure atop a landfill, where the weight of the structure could directly cause compression of the underlying loose material.

7.6.1 Limestone Cavities or Sinkholes

Settlement related to limestone cavities or sinkholes will usually be limited to areas having karst topography. Karst topography is a type of landform developed in a region of easily soluble limestone bedrock. It is characterized by vast numbers of depressions of all sizes, sometimes by great outcrops of limestone, sinks and other solution passages, an almost total lack of surface streams, and larger springs in the deeper valleys (Stokes and Varnes, 1955).

Identification techniques and foundations constructed on karst topography are discussed by Sowers (1997). Methods to investigate the presence of sinkholes are geophysical techniques and the cone penetration device (*Earth Manual*, 1985; Foshee and Bixler, 1994). A low cone penetration resistance could indicate the presence of raveling, which is a slow process where granular soil particles migrate into the underlying porous limestone. An advanced state of raveling will result in subsidence of the ground, commonly referred to as *sinkhole activity*.

7.6.2 Underground Mines and Tunnels

According to Gray (1988), damage to residential structures in the United States caused by the collapse of underground mines is estimated to be between $25 million and $35 million each year, with another $3 million to $4 million in damage to roads, utilities, and services. There are approximately 2 million hectares of abandoned or inactive coal mines, with 200,000 of these hectares in populated urban areas (Dyni and Burnett, 1993).

It has been stated that ground subsidence associated with longwall mining can be predicted fairly well with respect to magnitude, time, and areal position (Lin et al., 1995). Once the amount of ground subsidence has been estimated, there are measures that can be taken to mitigate the effects of mine-related subsidence (National Coal Board, 1975; Kratzsch, 1983; Peng, 1986, 1992). For example, in a study of different foundations subjected to mining-induced subsidence, it was concluded that posttensioning of the foundation was most effective, because it prevented the footings from cracking (Lin et al., 1995).

As in the discussion presented in Sec. 7.2, the collapse of underground mines and tunnels can produce tension- and compression-type features within the buildings. Figure 7.17 (from Marino et al., 1988) shows that the location of the compression zone will be in the center of the subsided area. The tension zone is located along the perimeter of the subsided area. These tension and compression zones are similar to fill settlement in a canyon environment (Fig. 7.4).

Besides the collapse of underground mines and tunnels, there can be settlement of buildings constructed on spoil extracted from the mines. Mine operators often dispose of other debris, such as trees, scrap metal, and tires, within the mine spoil. In many cases, the mine spoil is dumped (no compaction) and can be susceptible to large amounts of settlement. For example, Cheeks (1996) describes an interesting case of a motel unknowingly built on spoil that had been used to fill in a strip-mining operation. The motel building experienced about 1 m (3 ft) of settlement within the monitoring period (5 years). The settlement and damage for this building actually started during construction and the motel owners could never place the building into service. A lesson from this case study is the importance of subsurface exploration. In many cases, the borings may encounter refusal on boulders generated from the mining operation and the geotechnical engineer may believe that the depth to solid rock is shallow. It is important when dealing with spoil generated from mining operations that the thickness of the spoil and the compression or collapse behavior of the material be adequately investigated. This may require use of rock coring or geophysical techniques to accurately define the limits and depth of the spoil pile.

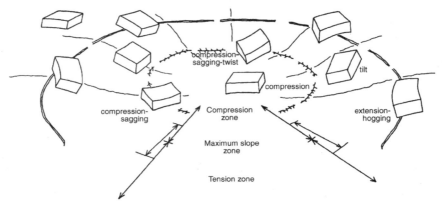

FIGURE 7.17 Location of tension and compression zones due to collapse of underground mines. (*From Marino et al., 1988; reprinted with permission from the American Society of Civil Engineers.*)

7.6.3 Subsidence Due to Extraction of Oil or Groundwater

Large-scale pumping of water or oil from the ground can cause settlement of the ground surface over a large area. The pumping can cause a lowering of the groundwater table which then increases the overburden pressure on underlying sediments. This can cause consolidation of soft clay deposits. In other cases, the removal of water or oil can lead to compression of the soil or porous rock structure, which then results in ground subsidence.

Lambe and Whitman (1969) describe two famous cases of ground surface subsidence due to oil or groundwater extraction. The first is oil pumping from Long Beach, California, which affected a 65-km^2 (25-mi^2) area and caused 8 m (25 ft) of ground surface subsidence. Because of this ground surface subsidence, the Long Beach Naval Shipyard had to construct sea walls to keep the ocean from flooding the facilities. A second famous example is ground surface subsidence caused by pumping of water for domestic and industrial use in Mexico City. Rutledge (1944) shows that the underlying Mexico City clay, which contains a porous structure of microfossils and diatoms, has a very high void ratio (up to $e = 14$) and is very compressible. Ground surface subsidence in Mexico City has been reported to be 9 m (30 ft) since the beginning of the twentieth century.

Besides ground surface subsidence, the extraction of groundwater or oil can also cause the opening of ground fissures. For example, Fig. 7.18 shows the ground surface subsidence in Las Vegas Valley between 1963 and 1987 due primarily to groundwater extraction. It has been stated (Purkey et al., 1994) that the subsidence has been focused on preexisting geologic faults, which serve as points of weakness for ground movement. Figure 7.19 shows one of these fissures that ran beneath the foundation of a home.

The theory of consolidation (Sec. 7.4) can often be used to estimate the ground surface subsidence caused by the pumping of shallow groundwater. In this case, the lowering of the groundwater table will increase the overburden pressure acting on the clay layer. For example, if pumping causes the groundwater table to be permanently lowered, then the increase in overburden stress $\Delta\sigma_v$ will be equal to the difference between the final (σ'_{vf}) and initial (σ'_{v0}) vertical effective stress in the clay layer (i.e., $\Delta\sigma_v = \sigma'_{vf} - \sigma'_{v0}$). In other cases, pumping of shallow groundwater may not cause a significant lowering of the groundwater table, but instead the pumping can cause a decrease in the pore water pressure in a permeable layer located below the clay layer. In this case, the reduction in pore water pressure will also lead to consolidation of the clay, and once again the increase in overburden stress

($\Delta\sigma_v$) will be equal to the difference between the final (σ'_{vf}) and initial (σ'_{v0}) vertical effective stress in the clay layer.

7.6.4 Landfills and Decomposition of Organic Matter

As previously mentioned, one common cause of settlement is consolidation and secondary compression of soft and saturated organic soils. This type of settlement is a result of the drainage of water from the void spaces because of an increase in overburden pressure. This is a different mechanism from settlement caused by the decomposition of organic matter.

Decomposition of Organic Matter. Organic matter consists of a mixture of plant and animal products in various stages of decomposition, of substances formed biologically and/or chemically from the breakdown of products, and of microorganisms and small animals and their decaying remains. For this very complex system, organic matter is generally divided

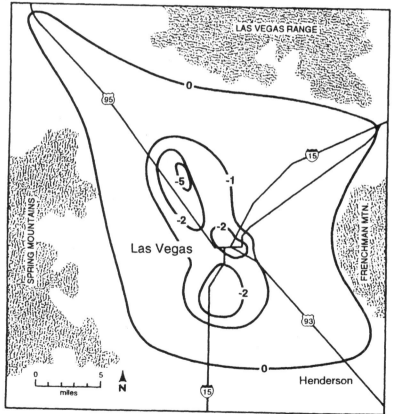

FIGURE 7.18 Land surface subsidence (in feet) in Las Vegas valley between 1963 and 1987. (*From Purkey et al., 1994; reprinted with permission of the Nevada Bureau of Mines and Geology.*)

FIGURE 7.19 Fissure, widened by erosion, running beneath the foundation of an abandoned home near Simmons Street in North Las Vegas. (*From Purkey et al., 1994; reprinted with permission of the Nevada Bureau of Mines and Geology.*)

into two groups: nonhumic and humic substances (Schnitzer and Khan, 1972). Nonhumic substances include numerous compounds such as carbohydrates, proteins, and fats that are easily attacked by microorganisms in the soil and normally have a short existence. Most organic matter in soils consists of humic substances, defined as amorphous, hydrophilic, acidic, and polydisperse substances (Schnitzer and Khan, 1972). An example of humic substances is the brown or black part of organic matter, which is so well decomposed that the original source cannot be identified. The important characteristics exhibited by humic substances are a resistance to microbic deterioration and the ability to form stable compounds (Kononova, 1966).

The geotechnical engineer can identify organic matter by its brown or black color, pungent odor, spongy feel, and in some cases its fibrous texture. The change in character of organic matter due to decomposition has been studied by Al-Khafaji and Andersland

(1981). They present visual evidence of the changes in pulp fibers due to microbic activity, using a scanning electron microscope (SEM). Decomposition of pulp fibers includes a reduction in length, diameter, and the development of rougher surface features. The ignition test is commonly used to determine the percent organics (ASTM D 2974-95, 1998). In this test the humic and nonhumic substances are destroyed using high ignition temperatures. The main source of error in the ignition test is the loss of surface hydration water from the clay minerals. Franklin et al. (1973) indicate that large errors can be produced if certain minerals, such as montmorillonite, are present in quantity. The ignition test also does not distinguish between the humic and nonhumic fraction of organics.

The problem with decomposing organics is the development of voids and corresponding settlement. The rate of settlement will depend on how fast the nonhumic substances decompose, and the compression characteristics of the organics. For large-mass graded projects it is difficult to keep all organic matter out of the fill. The detrimental effects of organic matter are generally recognized by most engineers. Common organic matter inadvertently placed in fill includes branches, shrubs, leaves, grass, and construction debris such as pieces of wood and paper. Figure 7.20 shows a photograph of decomposing organics placed in a structural fill.

Landfills. Landfills can settle due to compression of the loose waste products and decomposition of organic matter that was placed within the landfill. Landfills are a special case of settlement because they can have both basic types of settlement: settlement due directly to the weight of the structure which causes an initial compression of the loose waste products and settlement due to secondary effects, such as when the organic matter in the landfill slowly decomposes.

The study of landfills provides data on the rate of decomposition of large volumes of organic matter. An excavation at the Mallard North Landfill in Hanover Park, Illinois,

FIGURE 7.20 Decomposing organic matter placed in a structural fill.

unearthed a 15-year-old steak (dated by a legible newspaper buried nearby) that "had bone, fat, meat, everything" (Rathje and Psihoyos, 1991). The lack of decomposition of these nonhumic substances can be attributed to the fact that a typical landfill admits no light and little air or moisture, so the organic matter in the trash decomposes slowly. The nonhumic substances will decompose eventually, but because little air (and hence oxygen) circulates around the organics, the slower-working anaerobic microorganisms thrive.

Rathje and Psihoyos (1991) have found that 20 to 50 percent of food and yard waste biodegrades in the first 15 years. There can be exceptions such as the Fresh Kills landfill, opened by New York City in 1948, on Staten Island's tidal marshland. Below a certain level in the Fresh Kills landfill, there is no food debris or yard wastes, and practically no paper. The reason is, apparently, that water from the tidal wetlands has seeped into the landfill, causing the anaerobic microorganisms to flourish. The study by Rathje and Psihoyos (1991) indicates that the type of environment is very important in terms of how fast the nonhumic substances decompose.

There are many old abandoned landfills throughout the United States. In some cases it may be a large municipal landfill that has been abandoned, but a more frequent case is that the site contains old debris and other waste products that have been dumped at a site. Settlement of structures constructed atop municipal landfill could be due to compression of the underlying loose waste products or the decomposition of any organic matter remaining in the landfill. During decomposition of the organic matter, there will also be the generation of methane (a flammable gas), which must be safely vented to the atmosphere.

If possible, the ideal situation would be to remove the debris from the site and dispose of it off-site. In some cases, the debris can be incorporated into the fill. For example, if deep fill is to be placed at the site, then the concrete, brick, or glass debris can be mixed in with the fill and compacted as deep structural fill. It is important that nesting of concrete debris not be allowed and that no degradable material (such as organic waste products) or compressible material (such as plastic containers) be mixed in with the fill.

7.7 SELECTION OF FOUNDATION TYPE

This final section of Chap. 7 deals with the selection of the type of foundation. The selection of a particular type of foundation is often based on a number of factors, such as:

1. *Adequate depth.* It must have an adequate depth to prevent frost damage. For such foundations as bridge piers, the depth of the foundation must be sufficient to prevent undermining by scour.

2. *Bearing capacity failure.* The foundation must be safe against a bearing capacity failure (Chap. 8).

3. *Settlement.* The foundation must not settle to such an extent that it damages the structure.

4. *Quality.* The foundation must be of adequate quality so that it is not subjected to deterioration, such as from sulfate attack (Chap. 13).

5. *Adequate strength.* The foundation must be designed with sufficient strength that it does not fracture or break apart under the applied superstructure loads. The foundation must also be properly constructed in conformance with the design specifications. Figure 7.21 shows an example of an improperly constructed concrete slab-on-grade. Note in Fig. 7.21 that the slab reinforcement (welded wire mesh) was actually placed below the slab, rather than positioned in the concrete.

6. *Adverse soil changes.* The foundation must be able to resist long-term adverse soil changes. An example is expansive soil (Chap. 9), which could expand or shrink, causing movement of the foundation and damage to the structure.

7. *Seismic forces.* The foundation must be able to support the structure during an earthquake without excessive settlement or lateral movement (Chap. 11).

This chapter focuses on item 3, the impact of settlement on the structure. Later chapters deal with the other requirements that the foundation must meet. On the basis of an analysis of all of the factors listed above, a specific type of foundation (i.e., shallow versus deep; see Table 1.2) would be recommended by the geotechnical engineer. The following sections discuss various types of foundations commonly used for settlement-prone areas.

7.7.1 Shallow Foundations

A shallow foundation is often selected when the structural load will not cause excessive settlement of the underlying soil layers. In general, shallow foundations are more economical to construct than deep foundations. Common types of shallow foundations are listed in Table 1.2 and described below:

Spread Footings, Combined Footings, and Strip Footings. These types of shallow foundations are probably the most common types of building foundations. Section 1.4 has described the construction and provided details (Figs. 1.1 to 1.10) on the construction of these foundation elements. Figure 7.22 shows the construction of spread and combined footings for a building located in Las Vegas, Nevada.

FIGURE 7.21 Concrete slab-on-grade where the steel reinforcement (welded wire mesh) was actually placed below the concrete, rather than positioned within the concrete.

7.44 ANALYSIS OF GEOTECHNICAL DATA AND ENGINEERING COMPUTATIONS

FIGURE 7.22 Construction of spread and combined footings for a building in Las Vegas, Nevada.

Mat Foundation. Depending on economic considerations, mat foundations are often constructed for the following reasons (NAVFAC DM-7.2, 1982):

1. *Large individual footings.* A mat foundation is often constructed when the sum of individual footing areas exceeds about one-half the total foundation area.
2. *Cavities or compressible lenses.* A mat foundation can be used when the subsurface exploration indicates that there will be unequal settlement caused by small cavities or compressible lenses below the foundation. A mat foundation would tend to span over the small cavities or weak lenses and create a more uniform settlement condition.
3. *Shallow settlements.* A mat foundation can be recommended when shallow settlements predominate and the mat foundation would minimize differential settlements.
4. *Unequal distribution of loads.* For some structures, there can be a large difference in building loads acting on different areas of the foundation. Conventional spread footings could be subjected to excessive differential settlement, but a mat foundation would tend to distribute the unequal building loads and reduce the differential settlements.
5. *Hydrostatic uplift.* When the foundation will be subjected to hydrostatic uplift due to a high groundwater table, a mat foundation could be used to resist the uplift forces.

Posttensioned Slabs-on-Grade. Posttensioned slabs-on-grade are common in southern California and other parts of the United States. They are an economical foundation when there is no ground freezing or the depth of frost penetration is low. Posttensioned slabs-on-grade are most commonly used when it is necessary to resist expansive soil forces or when the projected differential settlement exceeds the tolerable value for a conventional (lightly reinforced) slab-on-grade. For example, posttensioned slabs-on-grade are frequently recommended if the projected differential settlement is expected to exceed 2 cm (0.75 in.).

Installation and field inspection procedures for posttensioned slab-on-grade have been prepared by the Post-Tensioning Institute (1996). Posttensioned slab-on-grade consists of concrete with embedded steel tendons that are encased in thick plastic sheaths. The plastic sheath prevents the tendon from coming in contact with the concrete and permits the tendon to slide within the hardened concrete during the tensioning operations. Usually tendons have a dead end (anchoring plate) in the perimeter (edge) beam and a stressing end at the opposite perimeter beam to enable the tendons to be stressed from one end. However, the Post-Tensioning Institute (1996) does recommend that the tendons in excess of 30 m (100 ft) be stressed from both ends. The Post-Tensioning Institute (1996) also provides typical anchorage details for the tendons.

Because posttensioned slab-on-grade performs better (i.e., less shrinkage = related concrete cracking) than conventional slab-on-grade, it is more popular even for situations where low levels of settlement are expected. For example, Figs. 7.23 and 7.24 show two views of the construction of a posttensioned slab-on-grade that is being constructed on a pad cut into sandstone.

Shallow Foundation Alternatives. If the expected settlement for a proposed shallow foundation is too large, then other options for foundation support or soil stabilization must be evaluated. Some commonly used alternatives are as follows:

1. *Grading.* Grading operations can be used to remove the compressible soil layer and replace it with structural fill. Usually the grading option is economical only if the compressible soil layer is near the ground surface and the groundwater table is below the compressible soil layer or the groundwater table can be economically lowered.

FIGURE 7.23 Construction of a posttensioned slab-on-grade for a single-family residence. (Note that the tensioning cables have been elevated on concrete chairs and the wood forms have been constructed.)

FIGURE 7.24 Construction of a posttensioned slab-on-grade for a single-family residence. (Note that the concrete has been placed, except for the garage area slab.)

2. *Surcharge.* If the site contains an underlying compressible cohesive soil layer, the site can be surcharged with a fill layer placed at the ground surface. Vertical drains (such as wick drains or sand drains) can be installed in the compressible soil layer to reduce the drainage paths and speed up the consolidation process. Once the compressible cohesive soil layer has had sufficient consolidation, the fill surcharge layer is removed and the building is constructed.
3. *Densification of soil.* There are many different methods that can be used to densify loose or soft soil. For example, vibroflotation and dynamic compaction are often effective at increasing the density of loose sand deposits. Another option is compaction grouting, which consists of introducing a mass of very thick consistency grout into the soil, which both displaces and compacts the loose soil.
4. *Floating foundation.* A floating foundation is a special type of deep foundation where the weight of the structure is balanced by the removal of soil and construction of an underground basement.

7.7.2 Deep Foundations

Pile Foundations. Probably the most common type of deep foundation is the pile foundation. Piles can consist of wood (timber), steel H sections, precast concrete, cast-in-place concrete, pressure-injected concrete, concrete-filled steel pipe piles, and composite-type piles. Piles are either driven into place or installed in predrilled holes. Piles that are driven into place are generally considered to be low-displacement or high-displacement, depending on the amount of soil that must be pushed out of the way as the pile is driven. Examples of low-displacement piles are steel H sections and open-ended steel pipe piles that do not form a soil plug at the end. Examples of high-displacement piles are solid-section piles, such as round timber piles or square precast concrete piles, and steel pipe piles with a closed end.

Various types of piles are as follows:

- *Batter pile.* A pile driven in at an angle inclined to the vertical to provide high resistance to lateral loads.

- *End-bearing pile.* A pile whose support capacity is derived principally from the resistance of the foundation material on which the pile tip rests. End-bearing piles are often used when a soft upper layer is underlain by a dense or hard stratum. If the upper soft layer should settle, the pile could be subjected to down-drag forces, and the pile must be designed to resist these soil-induced forces.

- *Friction pile.* A pile whose support capacity is derived principally from the resistance of the soil friction and/or adhesion mobilized along the side of the pile. Friction piles are often used in soft clays where the end-bearing resistance is small because of punching shear at the pile tip. A pile that resists upward loads (i.e., tension forces) would also be considered a friction pile.

- *Combined end-bearing and friction pile.* A pile that derives its support capacity from combined end-bearing resistance developed at the pile tip and frictional and/or adhesion resistance on the pile perimeter.

Section 8.3 will present the engineering analyses that are used to determine the load-carrying capacity of these different types of piles.

Piles are usually driven in specific arrangements and are used to support reinforced-concrete pile caps or a mat foundation. For example, the building load from a steel column may be supported by a concrete pile cap that is in turn supported by four piles located near the corners of the concrete pile cap.

Figure 7.25 shows an example of floor slab settlement due to consolidation and secondary compression of peat at the Meadowlands, a marshy area in New Jersey west of New York City. Piles have been used to support bearing walls, but as shown in Fig. 7.25, there is often cracking and deformation of the floor slabs around the piles (Whitlock and Moosa, 1996). The reason for the settlement and damage to the floor slab shown in Fig. 7.25 is that the floor slab was not designed to span unsupported between the piles. The construction of a structural floor slab that can transfer loads to the piles is an important design feature for sites having settling compressible soil, such as the peat layer shown in Fig. 7.25.

Piers. A pier is defined as a deep foundation system, similar to a cast-in-place pile, that consists of a column-like reinforced-concrete member. Piers are often of large enough diameter to enable downhole inspection. Piers are also commonly referred to as *drilled shafts, bored piles,* or *drilled caissons.* Piers will be discussed in Secs. 8.3 and 15.8.

Soil Report Recommendations. Upon completion of the settlement analyses, the foundation recommendations would normally be included in a soils report. An example of typical wording concerning foundation recommendations and settlement analyses at a proposed building site is as follows (see App. E):

> It is recommended that the buildings be supported by foundations anchored in bedrock. It is anticipated that piers will be needed for the Administrative Building. For foundations bearing on intact bedrock, the settlement (ρ_{max}) for the anticipated maximum column load of 550 kips is estimated to be about 0.5 inches (13 mm). As indicated by the calculations presented in

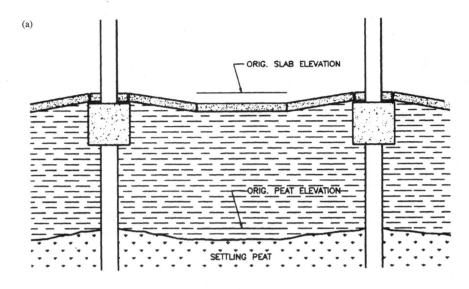

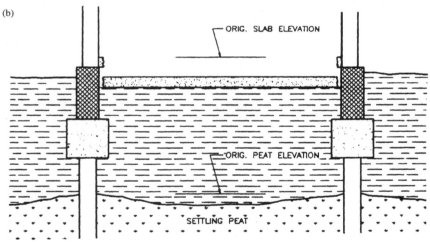

FIGURE 7.25 Slab displacement mechanisms caused by settling peat. (*From Whitlock and Moosa, 1996; reprinted with permission of the American Society of Civil Engineers.*)

the appendix, the maximum differential settlement (Δ) between columns is expected to be 0.25 inch (6.4 mm) or less. Settlement should consist of deformation of the bedrock and should occur during construction.

PROBLEMS

The problems have been divided into basic categories as indicated below:

Loadings and Allowable Settlement

1. A proposed project will consist of a industrial building (30 m wide and 42 m long) that will be used by a furniture moving company to store household items. The structural engineer indicates that preliminary plans call for exterior tilt-up walls, interior isolated columns on spread footings, and a floor slab. Perimeter tilt-up wall dead loads are 60 kN per linear meter. Interior columns will be spaced 6 m on center in both directions and each column will support 900 kN of dead load. The interior floor slab dead load = 6 kPa. The structural engineer also indicates that it is likely that the industrial building could be full of household items for a considerable length of time and this anticipated live load = 30 kPa.

Assume that subsurface exploration has discovered the presence of a soft clay layer and for the purposes of the settlement analysis, the weight of the industrial building can be assumed to be a uniform stress σ_0 applied at ground surface. Determine the uniform stress σ_0 that should be used in the settlement analysis. *Answer:* σ_0 = 60 kPa (1250 psf).

2. A building will be constructed with a continuous steel frame. If the length L between adjacent columns will be 6 m (20 ft), determine the maximum allowable differential settlement Δ. *Answer:* From Table 7.1, Δ = 1.2 cm (0.5 in.).

3. A building will be constructed to house sensitive machinery. Determine the maximum allowable angular distortion δ/L and maximum differential settlement Δ if the supporting columns are spaced 6 m (20 ft) apart. *Answers:* From Fig. 7.3, δ/L = 1/750 and thus Δ = 0.8 cm (0.3 in.).

4. It is proposed to construct a house with a conventional slab-on-grade foundation. It is decided that the house should have a factor of safety of 2.5 against gypsum wallboard cracking. What is the maximum allowable differential settlement? *Answer:* Δ = 0.5 in.

5. It is proposed to construct a flexible oil storage tank. If the tank has a diameter of 30 m and the maximum allowable angular distortion (δ/L) = 1/200, determine the maximum allowable differential settlement Δ. *Answer:* Δ = 7.5 cm (3 in.).

Collapsible Soil

6. An undisturbed soil specimen, having an initial height = 25.4 mm, is placed in an oedometer apparatus. During loading to a vertical pressure of 200 kPa, the soil specimen compresses 0.3 mm. When submerged in distilled water, the soil structure collapses and an additional 2.8 mm of vertical deformation occurs. Determine the collapse potential (CP) and collapse classification. *Answers:* CP = 11% ("severe" collapse potential).

7. A canyon contains collapsible alluvium. One side of the proposed building will be constructed over the canyon centerline, where the thickness of collapsible alluvium = 5 ft. Along the opposite side of the building, the thickness of the alluvium = 1 ft. If undisturbed specimens of the alluvium experience 5 percent collapse under the overburden pressure and

7.50 ANALYSIS OF GEOTECHNICAL DATA AND ENGINEERING COMPUTATIONS

proposed building loads, determine the maximum differential settlement Δ of the building after saturation of the alluvium. *Answer:* $\Delta = 2.4$ in.

8. It is proposed to construct a building on top of an existing fill mass. Undisturbed specimens of the fill were obtained and tested in the laboratory oedometer apparatus in order to determine the percent collapse. The data is summarized below:

Depth of fill layer below ground surface, ft	Laboratory oedometer testing (vertical pressure, psf)	Percent collapse
0–4	250	0
4–8	750	0.10
8–12	1250	0.25
12–16	1750	0.63
16–20	2250	1.14

Assume the fill weight = 125 pcf and there is 20 ft of fill underneath the building. Neglecting the weight of the building, determine the maximum settlement (ρ_{max}) assuming the entire fill mass becomes saturated. *Answer:* $\rho_{max} = 1.0$ in.

Immediate Settlement

9. A building site consists of a thick and saturated, heavily overconsolidated, cohesive soil layer. Assume that when the cohesive soil is subjected to the building load, it will still be in an overconsolidated state. On the basis of triaxial compression tests, the average undrained modulus E_u of the cohesive soil = 20,000 kPa. If the applied surface stress (q) = 30 kPa and if the square building has a width (B) = 20 m, calculate the immediate settlement s_i beneath the center and corner of the building. *Answer:* $s_i = 2.5$ cm (center) and $s_i = 1.3$ cm (corner).

10. A site consists of a thick, unsaturated cohesive fill layer. Assume that an oil tank, having a diameter of 10 m, will be constructed on top of the fill. On the basis of triaxial compression tests on the unsaturated cohesive fill, the average undrained modulus E_u of the cohesive fill = 40,000 kPa. When the tank is filled, the applied surface stress q = 50 kPa. Calculate the immediate settlement s_i beneath the center of the oil tank assuming Poisson's ratio = 0.4. *Answer:* $s_i = 1.0$ cm.

11. An oil tank is constructed at a site that contains a thick deposit of soft saturated clay. Figure 7.26 shows the time versus projected settlement of the oil tank, determined by the design engineer. The actual time versus settlement behavior of the tank (once it is filled with oil) is also plotted in Fig. 7.26. There is a substantial difference between the predicted and actual settlement behavior of the oil tank. What is the most likely cause of the substantial difference between the predicted and measured settlement behavior of the oil tank? *Answer:* Immediate settlement s_i due to undrained creep of the soft saturated clay was not included in the original settlement analysis by the design engineer.

Primary Consolidation Settlement

12. Figure 7.27 shows the consolidation curves for two tests performed on saturated cohesive soil specimens. Using the Casagrande construction technique, determine the maximum past pressure σ'_{vm} for both tests. *Answers:* $\sigma'_{vm} = 20$ kPa (solid line) and $\sigma'_{vm} = 30$ kPa (dashed line).

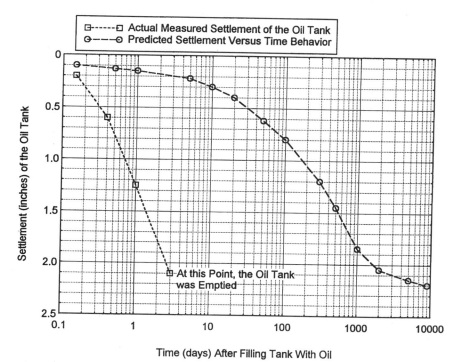

FIGURE 7.26 Settlement data for Prob. 11.

13. For the two consolidation curves shown in Fig. 7.27, calculate the compression index C_c. *Answers:* $C_c = 2.8$ (solid line), $C_c = 1.8$ (dashed line).

14. For the Orinoco clay data shown in Fig. 7.11, calculate the modified compression index $C_{c\varepsilon}$ for the undisturbed and disturbed soil specimens. *Answers:* $C_{c\varepsilon} = 0.36$ for the undisturbed Orinoco clay specimen and $C_{c\varepsilon} = 0.24$ for the disturbed Orinoco clay specimen.

15. Using the one-dimensional compression curve shown in Fig. 7.9, calculate the modified recompression index $C_{r\varepsilon}$ and the modified compression index $C_{c\varepsilon}$. *Answers:* $C_{r\varepsilon} = 0.04$, $C_{c\varepsilon} = 0.45$.

16. Use the data from the examples in Sec. 7.4.2 (i.e., the site underlain with a 2-m-thick clay layer) and assume the building is constructed very quickly. Calculate the primary consolidation settlement s_c by dividing the 2-m-thick clay layer into two layers, each having a thickness of 1 m, and then summing the consolidation settlement calculated for each layer. *Answer:* 4.9 cm.

17. Use the data from the examples in Sec. 7.4.2. Assume that undisturbed soil specimens from the clay layer indicate that the average maximum past pressure $(\sigma'_{vm}) = 125$ kPa, which is less than the equilibrium vertical effective stress (152 kPa) and hence the clay layer is underconsolidated (OCR < 1). At the time of construction, assume that the average vertical effective stress σ'_{v0} in the clay layer = 125 kPa. Calculate the primary consolidation settlement of the building due to consolidation of the clay layer from its own weight and consolidation due to the weight of the building. *Answer:* $s_c = 12$ cm.

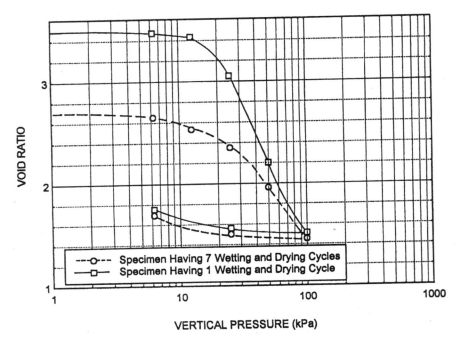

FIGURE 7.27 Consolidation curves for Prob. 12.

18. Use the data from the examples in Sec. 7.4.2. Assume that the undisturbed soil specimen from the center of the clay layer indicates a maximum past pressure σ'_{vm} of 170 kPa and that the recompression index $(C_r) = 0.10$. Determine the primary consolidation settlement s_c of the clay layer due to the building weight. *Answer:* $s_c = 1.7$ cm.

19. Use the data from the examples in Sec. 7.4.2. Assume that enough time has elapsed that primary consolidation due to the building weight is complete. At this time, an adjacent property owner initiates pumping of the groundwater table and the groundwater table is permanently lowered 4 m (i.e., the level of the groundwater table is now 9 m below ground surface). Determine the primary consolidation settlement s_c due to the permanent lowering of the groundwater table. *Answer:* $s_c = 6.4$ cm.

20. Use the data from the examples in Sec. 7.4.2. Assume that enough time has elapsed that primary consolidation due to the building weight is complete. At this time, an adjacent property owner initiates pumping of the sand layer below the clay layer and the pore water pressure is permanently lowered by an amount of 40 kPa. Assume that at the top of the clay layer, the pore water pressures are still hydrostatic. Using two layers (each 1 m thick), determine the primary consolidation settlement s_c due to the permanent reduction in pore water pressure of the sand layer located below the clay layer. *Answer:* 3.6 cm.

Rate of One-Dimensional Consolidation

21. A consolidation test is performed on a highly plastic, saturated cohesive soil specimen by using the oedometer apparatus (drainage provided on the top and bottom of the soil specimen). At a vertical stress of 50 kPa, the vertical deformation of the soil specimen as a function of time after the start of loading is recorded and the data is plotted on a semilog plot as shown in Fig. 7.28. Assume that at zero vertical deformation in Fig. 7.28, the specimen height = 10.0 mm. Calculate the coefficient of consolidation c_v. *Answer:* $c_v = 1.3 \times 10^{-5}$ cm^2/s.

22. Use the data from the examples in Sec. 7.4.2. Predict how long after construction it will take for 50 percent and 90 percent of the primary consolidation settlement to occur if the clay layer can only drain from the top (i.e., single drainage). *Answers:* For $U_{avg} = 50\%$, time = 2.5 years, and for $U_{avg} = 90\%$, time = 10.6 years (or 4 times the time calculated for double drainage).

23. Use the data from the examples in Sec. 7.4.2. At a time of 180 days after the completion of the building, it is determined that the center of the building has settled 2.5 cm. Assuming all of this 2.5 cm of settlement is due to one-dimensional primary consolidation, determine the field coefficient of consolidation c_v. *Answer:* $c_v = 0.4$ m^2/year (1.3×10^{-4} cm^2/s).

24. Use the data from the examples in Sec. 7.4.2. At a time of 180 days after construction of the building, a piezometer located in the middle of the clay layer and situated beneath the center of the building records a pressure head h_p of 7.88 m. On the basis of this pressure head h_p measurement, determine the field coefficient of consolidation c_v. *Answer:* $c_v = 0.4$ m^2/year (1.3×10^{-4} cm^2/s).

Secondary Compression

25. Using the data from Prob. 21, calculate the secondary compression ratio C_α. *Answer:* $C_\alpha = 0.01$.

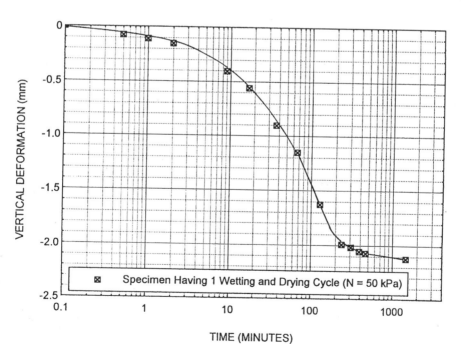

FIGURE 7.28 Time versus vertical deformation plot for Prob. 21.

7.54 ANALYSIS OF GEOTECHNICAL DATA AND ENGINEERING COMPUTATIONS

26. Use the data from the examples in Sec. 7.4.2. Assume the design life of the structure is 50 years and $C_\alpha = 0.01$. Calculate the secondary compression settlement s_s over the design life of the structure, using a time factor T equal to 1.0 to determine the time corresponding to the end of primary consolidation. What is the maximum total settlement ρ_{max} at the end of the design life of the structure if initial settlement s_i is neglected? *Answers:* $s_s = 2.4$ cm, $\rho_{max} = 7.4$ cm (3 in.).

Settlement of Cohesionless Soil

27. Use the data from Prob. 9, but instead of a clay, assume the site consists of a uniform clean coarse sand having an N-value = 20 and Poisson's ratio $(\mu) = 0.3$. Calculate the maximum total settlement ρ_{max} and the maximum differential settlement Δ. *Answers:* $\rho_{max} = 3$ cm (center) and $\Delta = 1.5$ cm.

28. Use the data from Prob. 10, but instead of a clayey fill, assume the site consists of compacted nonplastic silty sand that has an average N-value = 40 and Poisson's ratio $(\mu) = 0.3$. Calculate the maximum total settlement ρ_{max}. *Answer:* $\rho_{max} = 2.8$ cm.

29. Assume the same conditions as Prob. 28, but the cone penetration test (CPT) is performed where $q_c = 50$ kg/cm². Calculate the maximum total settlement ρ_{max}. *Answer:* $\rho_{max} = 4.6$ cm.

30. A square footing (6 ft by 6 ft) will be subjected to a vertical load of 230 kips (230 kips includes the weight of the footing). For the underlying clean fine to medium sand deposit, assume $N = 30$ and Poisson's ratio $(\mu) = 0.3$. Calculate the maximum total settlement ρ_{max}. *Answer:* $\rho_{max} = 1.2$ cm.

31. Assume the same conditions as Prob. 30. From Fig. 7.16, calculate the maximum total settlement ρ_{max}. *Answer:* From Fig. 7.16, for $B = 6$ ft, $q = 3.4$ tsf for 1-in. settlement. Since actual $q = 3.2$ tsf, ρ_{max} will be slightly less than 1 in. (*Note:* for this problem, the theory of elasticity and Fig. 7.16 provide similar answers.)

32. Assume the same conditions as Probs. 30 and 31, except that the groundwater table rises to near the bottom of the footing. From Fig. 7.16, calculate the size of the square footing so that the maximum total settlement ρ_{max} is approximately 1 in. *Answer:* By trial and error, the size of the footing should be 8.5 ft by 8.5 ft.

33. Subsurface exploration indicates that a level site has 20 m of sand overlying dense rock. From ground surface to a depth of 4 m, the sand is loose with a total unit weight $(\gamma_t) = 19.3$ kN/m³. From a depth of 4 to 12 m below ground surface, the sand has a medium density with a total unit weight $(\gamma_t) = 20.6$ kN/m³. From a depth of 12 to 20 m, the sand is dense with a total unit weight $(\gamma_t) = 21.4$ kN/m³. The groundwater table is located 5 m below ground surface. It is proposed to construct a large building at the site which will be supported by a mat foundation (30 m by 40 m). The bottom of the mat foundation will be located 4 m below ground surface, and the structural engineer has indicated that the building load can be approximated as a uniform pressure of 200 kPa that the bottom of the mat applies to the sand at a depth of 4 m. Undisturbed soil specimens of the sand were obtained at various depths, and the results of one-dimensional compression tests performed on saturated sand specimens are presented in Fig. 7.29. Using the data from Fig. 7.29, calculate the net stress applied by the foundation at a depth of 4 m (σ_0), the increase in stress $\Delta\sigma_v$ at the center of the sand layers using the 2:1 approximation, and the total settlement ρ_{max} of the mat foundation. *Answers:* $\sigma_0 = 123$ kPa, $\Delta\sigma_v = 98.5$ kPa at the center of the medium density sand layer and 67.5 kPa at the center of the dense sand layer, and $\rho_{max} = 12$ cm.

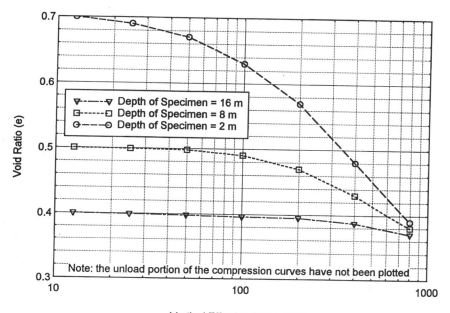

FIGURE 7.29 Compression curves for Prob. 33.

Selection of Foundation Type

34. Assume that the subsoil conditions at a site are as shown in Fig. 4.39. It is proposed to build a state highway bridge (essential facility) at this location. It is anticipated that during major flood conditions, there may be scour that could remove sediments down to an elevation of -10 m. What type of bridge foundation should be installed for the highway bridge? *Answer:* A deep foundation system consisting of piles or piers embedded in the sandstone.

35. Assume that the subsoil conditions at a site are as shown in Fig. 4.40. It is proposed to construct a two-story research facility that will also contain an underground basement to be used as the laboratory. On the basis of these requirements, what type of foundation should be installed for the research facility? *Answer:* Assuming the weight of soil excavated for the basement is approximately equal to the weight of the two-story structure, a floating foundation would be desirable.

36. Assume that the subsoil conditions at a site are as shown in Fig. 4.41. It is proposed to construct single-family detached housing at this site. What type of foundation should be used? *Answer:* Because the upper 10 ft of the site consists of overconsolidated clay, it would be desirable to use a shallow foundation system, on the assumption that building loads are light.

37. Assume the subsoil conditions at a site are as shown in Fig. 4.42. The structural engineer is recommending that driven, high-displacement piles be installed to a depth of 50 ft in the clay deposit. Are these type of piles appropriate? *Answer:* No. Because of the very high sensitivity of the clay, high-displacement piles will remold the clay and result in a loss of shear strength. The preferred option is to install low-displacement piles or use predrilled, cast-in-place concrete piles.

38. Assume the subsoil conditions at a site are as shown in Fig. 4.40. Suppose a large building, 400 ft long and 200 ft wide, has been constructed at ground surface (elevation +21 ft) without performing adequate subsurface exploration. The weight of the building can be represented as a uniform pressure = 1000 psf applied at ground surface. Consider just the clay from elevation −50 to −90 ft and, for the analysis, divide this soil into four 10-ft-thick layers. Determine the primary consolidation settlement s_c of the building. *Answer:* The clay from elevation −50 to −90 ft is essentially normally consolidated, and for $C_c = 0.35$, $e_0 = 0.9$, and the 2:1 approximation, $s_c = 4$ in.

CHAPTER 8
BEARING CAPACITY

The following notation is used in this chapter:

SYMBOL	DEFINITION
B	Width of the footing (Sec. 8.2)
B	Width of the pile (Sec. 8.3)
c	Cohesion based on a total stress analysis
c'	Cohesion based on an effective stress analysis
c_A	Adhesion between the cohesive soil and the pile or pier perimeter
D	Depth of the pile group
D_f	Depth below ground surface to the bottom of the footing
e	Eccentricity of the vertical load Q
F	Factor of safety
G_f	Gravel equivalent factor
h'	Depth of the groundwater table below ground surface [Eq. (8.4)]
k	Ratio of σ'_h divided by σ'_v
L	Length of the footing
N_c, N_γ, N_q	Dimensionless bearing capacity factors (Table 8.2)
q_{all}	Allowable bearing pressure
q_{ult}	Ultimate bearing capacity
q'	Largest bearing pressure exerted by an eccentrically loaded footing
q''	Lowest bearing pressure exerted by an eccentrically loaded footing
Q	Applied load to the footing
Q_{all}	Maximum allowable footing, pile, or pier load
Q_p	Ultimate pile or pier tip resistance force
Q_s	Ultimate skin friction resistance force for the pile or pier
Q_{ult}	Load causing a bearing capacity failure (Sec. 8.2)
Q_{ult}	Ultimate load capacity of a pile or pier (Sec. 8.3)
r, R	Radius of the pile or pier
R	R-value (Sec. 8.4.2)
s_u	Undrained shear strength of the soil
T	Flexible pavement thickness
T_{ult}	Ultimate uplift load of the pile or pier
TI	Traffic index
z	Depth below the pile group (Sec. 8.3.1)
z	Embedment depth of the pile (Sec. 8.3.2)

ϕ	Friction angle based on a total stress analysis
ϕ'	Friction angle based on an effective stress analysis
ϕ_w	Friction angle between the cohesionless soil and the pile or pier perimeter
γ_a	Adjusted unit weight for a groundwater table
γ_b	Buoyant unit weight of saturated soil below the groundwater table
γ_t	Total unit weight of the soil
σ'_h	Horizontal effective stress
σ'_n	Effective normal stress on the shear surface
σ'_v	Vertical effective stress
τ	Shear strength of the soil

8.1 INTRODUCTION

A bearing capacity failure is defined as a foundation failure that occurs when the shear stresses in the soil exceed the shear strength of the soil. Bearing capacity failures of foundations can be grouped into three categories, as follows (Vesic, 1963, 1975):

1. *General shear.* As shown in Fig. 8.1, a general shear failure involves total rupture of the underlying soil. There is a continuous shear failure of the soil (solid lines) from below the footing to the ground surface. When the load is plotted versus settlement of the footing, there is a distinct load at which the foundation fails (solid circle), and this is designated Q_{ult}. The value of Q_{ult} divided by the width B and length L of the footing is considered to be the ultimate bearing capacity q_{ult} of the footing. The ultimate bearing capacity has been defined as the bearing stress that causes a sudden catastrophic failure of the foundation (Lambe and Whitman, 1969).

Note in Fig. 8.1 that a general shear failure ruptures and pushes up the soil on both sides of the footing. For actual failures in the field, the soil is often pushed up on only one side of the footing, with subsequent tilting of the structure. A general shear failure occurs for soils that are in a dense or hard state.

2. *Punching shear.* As shown in Fig. 8.2, a punching shear failure does not develop the distinct shear surfaces associated with a general shear failure. For punching shear, the soil outside the loaded area remains relatively uninvolved and there is minimal movement of soil on both sides of the footing.

The process of deformation of the footing involves compression of soil directly below the footing as well as the vertical shearing of soil around the footing perimeter. As shown in Fig. 8.2, the load settlement curve does not have a dramatic break and for punching shear, the bearing capacity is often defined as the first major nonlinearity in the load-settlement curve (open circle). A punching shear failure occurs for soils that are in a loose or soft state.

3. *Local shear failure.* As shown in Fig. 8.3, local shear failure involves rupture of the soil only immediately below the footing. There is soil bulging on both sides of the footing, but the bulging is not as significant as in general shear. Local shear failure can be considered as a transitional phase between general shear and punching shear. Because of the transitional nature of local shear failure, the bearing capacity could be defined as the first major nonlinearity in the load-settlement curve (open circle) or at the point where the settlement rapidly increases (solid circle). A local shear failure occurs for soils that have a medium density or are in a firm state.

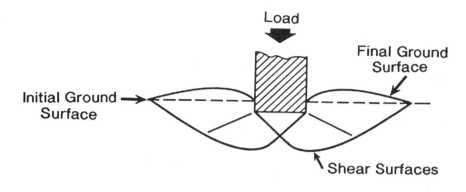
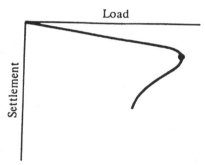

FIGURE 8.1 General shear foundation failure. (*After Vesic, 1963.*)

Compared to the number of structures damaged by settlement, there are far fewer structures that have bearing capacity failures. This is because of the following factors:

- *Settlement governs.* The foundation design is based on several requirements (see Sec. 7.7), and two of the main considerations are: (1) settlement due to the building loads must not exceed tolerable values (Chap. 7) and (2) there must be an adequate factor of safety against a bearing capacity failure (discussed in this chapter). In most cases, settlement governs, and the foundation bearing pressures recommended by the geotechnical engineer are based on limiting the amount of settlement.
- *Extensive studies.* Extensive studies of bearing capacity failures have led to the development of bearing capacity equations that are routinely used in practice to determine the ultimate bearing capacity of the foundation.
- *Factor of safety.* In order to determine the allowable bearing pressure q_{all}, the ultimate bearing capacity q_{ult} is divided by a factor of safety. The normal factor of safety used for bearing capacity analyses is 3. This is a high factor of safety compared to other factors of safety, such as only 1.5 for slope stability analyses (Chap. 10).
- *Minimum footing sizes.* Building codes often require minimum footing sizes and embedment depths. For example, Table 2.1 presents the minimum footing requirements for stud bearing walls according to the *Uniform Building Code* (1997).

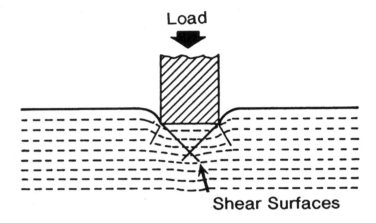

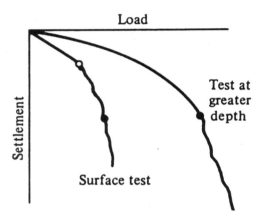

FIGURE 8.2 Punching shear foundation failure. (*After Vesic, 1963.*)

- *Allowable bearing pressures.* In addition, building codes often have maximum allowable bearing pressures for different soil and rock conditions. Table 8.1 presents maximum allowable bearing pressures based on the *Uniform Building Code* (Table 18-I-A, 1997). Especially in the case of dense or stiff soils, these allowable bearing pressures often have adequate factors of safety.
- *Footing dimensions.* Usually the structural engineer will determine the size of the footings by dividing the maximum footing load (dead load plus live load) by the allowable bearing pressure. Typically the structural engineer uses values of dead and live loads that also contain factors of safety. For example, the live load may be from the local building code which specifies minimum live load requirements for specific building uses. Such building code values often contain a factor of safety, which is in addition to the factor of safety of 3 that was used to determine the allowable bearing pressure.

Because the bearing capacity failure involves a shear failure of the underlying soil (Figs. 8.1 to 8.3), the analysis will naturally include the shear strength of the soil (Sec. 6.5). As indicated in Figs. 8.1 to 8.3, the depth of the bearing capacity failure is rather shallow. It is often assumed that the soil involved in the bearing capacity failure can extend to a depth equal to B (footing width) below the bottom of the footing. Thus for bearing capacity analysis, this zone of soil should be evaluated for its shear strength properties.

The documented cases of bearing capacity failures indicate that usually the following three factors (separately or in combination) are the cause of the failure: (1) there was an overestimation of the shear strength of the underlying soil, (2) the actual structural load at the time of the bearing capacity failure was greater than that assumed during the design phase, or (3) the site was subjected to alteration—such as the construction of an adjacent excavation—which resulted in a reduction in support and a bearing capacity failure.

A famous case of a bearing capacity failure is the Transcona grain elevator, located at Transcona, near Winnipeg, Canada. Figure 8.4 shows the October 1913 failure of the grain elevator. At the time of failure, the grain elevator was essentially fully loaded. The foundation had been constructed on clay which was described as a stiff clay. Note in Fig. 8.4 that the soil has been pushed up on only one side of the foundation, with subsequent tilting of the structure.

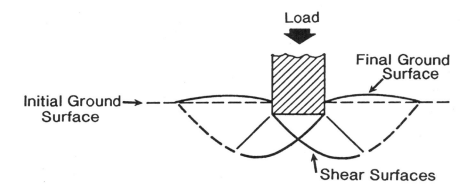

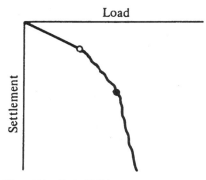

FIGURE 8.3 Local shear foundation failure. (*After Vesic, 1963.*)

8.6 ANALYSIS OF GEOTECHNICAL DATA AND ENGINEERING COMPUTATIONS

TABLE 8.1 Allowable Bearing Pressures

Material type (1)	Allowable bearing pressure* (2)	Maximum allowable bearing pressure† (3)
Massive crystalline bedrock	4000 psf (200 kPa)	12,000 psf (600 kPa)
Sedimentary and foliated rock	2000 psf (100 kPa)	6000 psf (300 kPa)
Gravel and sandy gravel (GW, GP)	2000 psf (100 kPa)	6000 psf (300 kPa)
Nonplastic soil: sands, silts, and NP silt (GM, SW, SP, SM, MN)‡	1500 psf (75 kPa)	4500 psf (220 kPa)
Plastic soil: silts and clays (ML, MI, MH, SC, CL, CI, CH)‡	1000 psf (50 kPa)	3000 psf (150 kPa)§

*Minimum footing width and embedment depth equals 1 ft (0.3 m).
†An increase of 20% of the allowable bearing pressure is allowed for each additional foot (0.3 m) of width or depth up to the maximum allowable bearing pressures listed in column 3. An exception is plastic soil.
‡Group symbols from the ISBP.
§No increase in the allowable bearing pressure is allowed for an increase in width of the footing.
Source: Data from the *Uniform Building Code* (1997).

FIGURE 8.4 Transcona grain elevator bearing capacity failure.

8.2 BEARING CAPACITY FOR SHALLOW FOUNDATIONS

As indicated in Table 1.2, common types of shallow foundation include spread footings for isolated columns (Fig. 1.7), combined footings for supporting the load from more than one structural unit (Fig. 1.8), strip footings for walls (Fig. 1.10), and mats or raft foundations constructed at or near ground surface. Shallow footings often have an embedment that is less than the footing width.

Bearing Capacity Equation. The most commonly used bearing capacity equation is that equation developed by Terzaghi (1943). For a uniform vertical loading of a strip footing, Terzaghi (1943) assumed a general shear failure (Fig. 8.1) in order to develop the following bearing capacity equation:

$$q_{ult} = \frac{Q_{ult}}{BL} = c\,N_c + \tfrac{1}{2}\,\gamma_t\,B\,N_\gamma + \gamma_t\,D_f\,N_q \tag{8.1}$$

where q_{ult} = ultimate bearing capacity for a strip footing (kPa or psf)
Q_{ult} = vertical load causing a general shear failure of the underlying soil (Fig. 8.1)
B = width of the strip footing (m or ft)
L = length of the strip footing (m or ft)
γ_t = total unit weight of the soil (kN/m³ or pcf)
D_f = vertical distance from the ground surface to the bottom of the strip footing (m or ft)
c = cohesion of the soil underlying the strip footing (kPa or psf)
N_c, N_γ, N_q = dimensionless bearing capacity factors

As indicated in Eq. (8.1), there are three terms that are added together to obtain the ultimate bearing capacity of the strip footing. These terms represent the following:

$c\,N_c$ The first term accounts for the cohesive shear strength of the soil located below the strip footing. If the soil below the footing is cohesionless (i.e., $c = 0$), then this term is zero.

$\tfrac{1}{2}\,\gamma_t\,B\,N_\gamma$ The second term accounts for the frictional shear strength of the soil located below the strip footing. The friction angle ϕ is not included in this term, but is accounted for by the bearing capacity factor N_γ. Note that γ_t represents the total unit weight of the soil located below the footing.

$\gamma_t\,D_f\,N_q$ This third term accounts for the soil located above the bottom of the footing. The value of γ_t times D_f represents a surcharge pressure that helps to increase the bearing capacity of the footing. If the footing was constructed at ground surface (i.e., $D_f = 0$), then this term would equal zero. This third term indicates that the deeper the footing, the greater the ultimate bearing capacity of the footing. In this term, γ_t represents the total unit weight of the soil located above the bottom of the footing. The total unit weight above and below the footing bottom may be different, in which case different values are used in the second and third terms of Eq. (8.1).

In order to calculate the allowable bearing pressure q_{all}, which is used to determine the size of the footings, the following equation is used:

$$q_{all} = \frac{q_{ult}}{F} \tag{8.2}$$

8.8 ANALYSIS OF GEOTECHNICAL DATA AND ENGINEERING COMPUTATIONS

where q_{all} = allowable bearing pressure (kPa or psf), q_{ult} = ultimate bearing capacity from Eq. (8.1), and F = factor of safety. For bearing capacity analysis, the commonly used factor of safety is equal to 3. Building codes often list allowable bearing pressures versus soil or rock types; Table 8.1, for example, presents the allowable bearing pressures q_{all} from the *Uniform Building Code* (1997).

Bearing Capacity Factors. Table 8.2 presents bearing capacity factors N_c, N_γ, and N_q (based on Terzaghi, 1943, and Peck, Hansen, and Thornburn, 1974). There are many other charts, graphs, and figures that present bearing capacity factors developed by other engineers and researchers based on varying assumptions (Vesic, 1975; Myslivec and Kysela, 1978). Some of these bearing capacity factors are even listed to an accuracy of five significant figures (e.g., Table 4.4.7.1A, Bearing Capacity Factors, AASHTO, 1996). These bearing capacity factors imply an accuracy which simply does not exist because the bearing capacity equation is only an approximation of the actual bearing failure. In Table 8.2, up to two significant figures are provided for the dimensionless bearing capacity factors.

As indicated in Table 8.2, the bearing capacity factors are directly related to the friction angle ϕ of the soil. A dense cohesionless soil would tend to have a high friction angle and high bearing capacity factors, resulting in a large ultimate bearing capacity. On the other hand, a loose cohesionless soil would tend to have a lower friction angle and lower ultimate bearing capacity. Thus a major disadvantage of building code values (such as those in Table 8.1) is that they consider only the material type, and not the density condition of the soil which influences the friction angle ϕ and bearing capacity factors. Building code values

TABLE 8.2 Bearing Capacity Factors (N_c, N_γ, and N_q) That Automatically Incorporate Allowances for Local Shear and Punching Shear Failure

Friction angle (ϕ or ϕ'), degrees (1)	N_c (2)	N_γ (3)	N_q (4)
0	5	0	1
10	8	1	3
20	15	4	5
30	30	8	10
31	33	10	12
32	35	15	18
33	39	20	21
34	42	28	28
35	46	35	34
36	51	48	45
37	56	65	55
38	61	80	67
39	68	100	80
40	75	120	93
41	84	140	110

Note: At high friction angles, bearing capacity factors increase rapidly, and these values should be used with caution.
Source: Based on Terzaghi, 1943, and Peck, Hansen, and Thornburn, 1974.

tend to underestimate the allowable bearing pressure for dense cohesionless soil, but may overestimate the allowable bearing pressure for loose cohesionless soil.

Note in Table 8.2 that the bearing capacity factors rapidly increase at high friction angles ϕ. These bearing capacity factors should be used with caution, because natural soils are not homogeneous and the natural variability of such soil will result in weaker layers that will be exploited during a bearing capacity failure.

Terzaghi (1943) originally developed the bearing capacity equation [Eq. (8.1)] for a general shear bearing capacity failure. This type of bearing capacity failure is shown in Fig. 8.1 and will develop for dense or stiff soil. For loose or soft soil, there will be a punching shear failure as shown in Fig. 8.2. The bearing capacity factors presented in Table 8.2 have been empirically adjusted and automatically incorporate allowances for local shear and punching shear.

Spread and Combined Footings. Equation (8.1) was developed by Terzaghi (1943) for strip footings. For other types of footings and loading conditions, corrections need to be applied to the bearing capacity equation. Many different types of corrections have been proposed (e.g., Meyerhof, 1951, 1953, 1965). One commonly used form of the bearing capacity equation for spread (square footings) and combined footings (rectangular footings) subjected to uniform vertical loading, is as follows (NAVFAC DM-7.2, 1982):

$$q_{ult} = \frac{Q_{ult}}{BL} = cN_c[1 + 0.3(B/L)] + 0.4\gamma_t BN_\gamma + \gamma_t D_f N_q \quad (8.3)$$

Equation (8.3) is similar to Eq. (8.1) and the terms have the same definitions. An important consideration is that, for the strip footing, the shear strength is actually based on a plane strain condition (soil is confined along the long axis of the footing). It has been stated that the friction angle ϕ is about 10 percent higher in the plane strain condition than the friction angle ϕ measured in the triaxial apparatus (Meyerhof, 1961; Perloff and Baron, 1976). Ladd et al. (1977) indicate that the friction angle ϕ in plane strain is larger than ϕ in triaxial shear by 4° to 9° for dense sands. A difference in friction angle of 4° to 9° has a significant impact on the bearing capacity factors (see Table 8.2). In practice, plane strain shear strength tests are not performed and thus there is an added factor of safety for the strip footing as compared to the analysis for spread or combined footings.

8.2.1 Bearing Capacity for Cohesionless Soil

As discussed in Sec. 6.5.1, cohesionless soil is nonplastic and includes gravels, sands, and nonplastic silt such as rock flour (see Table 5.5). A cohesionless soil develops its shear strength as a result of the frictional and interlocking resistance between the soil particles. For cohesionless soil, $c = 0$ and the first term in Eqs. (8.1) and (8.3) is zero.

For cohesionless soil, the location of the groundwater table can affect the ultimate bearing capacity. As mentioned in Sec. 6.5.1, the saturation of sand usually does not have much effect on the friction angle (i.e., $\phi' \cong \phi$). But a groundwater table creates a buoyant condition in the soil which results in less resistance of the soil and hence a lower ultimate bearing capacity. As previously mentioned, the depth of the bearing capacity failure is rather shallow (Figs. 8.1 to 8.3), and it is often assumed that the soil involved in the bearing capacity failure can extend to a depth equal to B (footing width) below the bottom of the footing. Thus for a groundwater table located in this zone, an adjustment to the ultimate bearing capacity is performed by adjusting the unit weight of the second term of Eqs. (8.1) and (8.3) by using the following equation (Myslivec and Kysela, 1978):

8.10 ANALYSIS OF GEOTECHNICAL DATA AND ENGINEERING COMPUTATIONS

$$\gamma_a = \gamma_b + \frac{h' - D_f}{B}(\gamma_t - \gamma_b) \tag{8.4}$$

where γ_a = adjusted unit weight that is used in place of γ_t for the second term in Eqs. (8.1) and (8.3) (kN/m³ or pcf)
γ_b = buoyant unit weight of the soil (kN/m³ or pcf)
γ_t = total unit weight of the soil (kN/m³ or pcf)
h' = depth of the groundwater table below ground surface (m or ft)
D_f = depth below ground surface to the bottom of the footing (m or ft)
B = width of the footing (m or ft)

Equation (8.4) is valid for $0 \le (h' - D_f) \le B$.

Note that no correction to the unit weight of the soil is required if the groundwater table is at a depth h' that is equal to or greater than $D_f + B$. The following example illustrates the use of the bearing capacity equation.

Example. A strip footing will be constructed on a nonplastic silty sand deposit which has the shear strength properties shown in Fig. 6.10 (i.e., $c' = 0$ and $\phi' = 30°$) and a total unit weight of 19 kN/m³ (120 pcf). Assume the strip footing will be 1.2 m (4 ft) wide and embedded 0.9 m (3 ft) below the ground surface. Using a factor of safety of 3 and assuming the groundwater table is well below the bottom of the footing, determine the allowable bearing pressure q_{all} for the nonplastic silty sand.

Solution

$$c' = 0$$
$$\gamma_t = 19 \text{ kN/m}^3 \text{ (120 pcf)}$$
$$B = 1.2 \text{ m (4 ft)}$$
$$N_\gamma = 8 \text{ (Table 8.2 for } \phi' = 30°)$$
$$N_q = 10 \text{ (Table 8.2 for } \phi' = 30°)$$
$$D_f = 0.9 \text{ m (3 ft)}$$

Substituting these values into Eq. (8.1), we find the ultimate bearing capacity q_{ult} = 260 kPa (5500 psf). Using a factor of safety = 3, the allowable bearing pressure q_{all} = 90 kPa (1800 psf). This 1.2-m- (4-ft-) wide strip footing could carry a maximum vertical load (Q_{all}) of 110 kN per linear meter of wall (7.2 kips per linear foot of wall).

8.2.2 Bearing Capacity for Cohesive Soil

The bearing capacity of cohesive (plastic) soil, such as silts and clays, is more complicated than the bearing capacity of cohesionless (nonplastic) soil. In the southwestern United States, surface deposits of clay are common and can have a hard and rocklike appearance when they become dried out during the summer. Instead of being susceptible to bearing capacity failure, desiccated clays can cause heave (upward movement) of lightly loaded foundations when the clays get wet during the rainy season. Expansion of clay will be discussed in Chap. 9.

Plastic saturated soils (silts and clays) usually have a lower shear strength than nonplastic cohesionless soil, and are more susceptible to bearing capacity failure, as in the example of the Transcona grain elevator discussed in Sec. 8.1. For saturated plastic soils, the bearing capacity usually has to be calculated for two different conditions: (1) total stress analysis (short-term condition) which uses the undrained shear strength of the plastic soil and (2) effective stress analysis (long-term condition) which uses the drained shear strength parameters (c' and ϕ') of the plastic soil.

Total Stress Analysis. This type of analysis uses the undrained shear strength of the plastic soil (Sec. 6.5.2). The undrained shear strength s_u could be determined from field tests, such as the vane shear test (VST), or in the laboratory from unconfined compression tests. If the undrained shear strength is approximately constant with depth, then $s_u = c$ and $\phi = 0$ for Eqs. (8.1) and (8.3).

If unconsolidated undrained triaxial compression tests (ASTM D 2850-95, 1998) are performed, then an envelope similar to Fig. 6.16 should be obtained for the saturated plastic soil and the $\phi = 0$ concept should be utilized (Sec. 6.5.2). For example, as shown in Fig. 6.16, the undrained shear strength parameters are $c = 2.1$ psi (14.5 kPa) and $\phi = 0$.

Note in Table 8.2 that for $\phi = 0$, the bearing capacity factors are $N_c = 5$, $N_\gamma = 0$, and $N_q = 1$. For both of these cases—undrained shear strength s_u and shear strength from unconsolidated undrained triaxial compression tests on saturated cohesive soil ($\phi = 0$ concept)—the bearing capacity Eqs. (8.1) and (8.3) reduce to the following:

For strip footings:

$$q_{ult} = 5c + \gamma_t D_f = 5S_u + \gamma_t D_f \tag{8.5a}$$

For spread footings:

$$q_{ult} = 5c\,[1 + 0.3\,(B/L)] + \gamma_t D_f = 5s_u\,[1 + 0.3\,(B/L)] + \gamma_t D_f \tag{8.5b}$$

Because of the use of total stress parameters, the groundwater table does not affect Eq. (8.5). An example of the use of the bearing capacity equation for cohesive soil is presented below:

Example. Assume the same conditions as the prior example, except that the strip footing will be constructed in a clay deposit having the shear strength parameters shown in Fig. 6.16.

Solution. For the strip footing, use Eq. (8.5a) where $c = 2.1$ psi (300 psf, 14.5 kPa). Using $\gamma_t = 19$ kN/m³ (120 pcf) and $D_f = 0.9$ m (3 ft) from the prior example and substituting these values into Eq. (8.5a), the ultimate bearing capacity $q_{ult} = 90$ kPa (1800 psf). Using a factor of safety of 3, the allowable bearing pressure $q_{all} = 30$ kPa (600 psf).

As is the case for the above two examples, the ultimate bearing capacity of plastic soil is often much less than the ultimate bearing capacity of cohesionless soil. This is the reason that local building codes (such as Table 8.1) allow higher allowable bearing pressures for cohesionless soil (such as sand) than plastic soil (such as clay). Also, because the ultimate bearing capacity does not increase with footing width for saturated plastic soils, there is often no increase allowed for an increase in footing width (see footnote marked § in Table 8.1).

In some cases, it may be appropriate to use total stress parameters c and ϕ in order to calculate the ultimate bearing capacity. For example, suppose a structure (such as an oil tank or grain elevator) is constructed and then sufficient time elapses that the saturated plastic soil consolidates under this load. If the oil tank or grain elevator is then quickly filled, the saturated plastic soil would be subjected to an undrained loading. This condition can be modeled by performing consolidated undrained triaxial compression tests (ASTM D 4767-95, 1998) in order to determine the total stress parameters c and ϕ. On the basis of the ϕ value, the bearing capacity factors would be obtained from Table 8.2 and then the ultimate bearing capacity would be calculated from Eqs. (8.1) or (8.3).

Effective Stress Analysis. This type of analysis uses the drained shear strength c' and ϕ' of the plastic soil. The drained shear strength could be obtained from triaxial compression tests with pore water pressure measurements performed on saturated specimens of the plastic soil. This analysis is called a *long-term analysis* because the shear induced pore water pressures (positive or negative) from the loading have dissipated and the hydrostatic pore water conditions now prevail in the field. Because an effective stress analysis is being performed, the location of the groundwater table must be considered in the analysis [use Eq. (8.4)].

The first step to perform the bearing capacity analysis would be to obtain the bearing capacity factors (N_c, N_γ, and N_q) from Table 8.2 using the value of ϕ'. An adjustment to the total unit weight may be required, depending on the location of the groundwater table [Eq. (8.4)]. Then Eq. (8.1) or (8.3) would be utilized (with c' substituted for c) to obtain the ultimate bearing capacity, with a factor of safety of 3 applied in order to calculate the allowable bearing pressure by Eq. (8.2).

Governing Case. Usually the total stress analysis will provide a lower allowable bearing capacity for soft or very soft saturated plastic soils. This is because the foundation load will consolidate the plastic soil, leading to an increase in the shear strength as time passes. For the long-term case (effective stress analysis), the shear strength of the plastic soil is higher, with a resulting higher bearing capacity.

Usually the effective stress analysis will provide a lower allowable bearing capacity for very stiff or hard saturated plastic soils. This is because such plastic soils are usually heavily overconsolidated and they tend to dilate (increase in volume) during undrained shear deformation. A portion of the undrained shear strength is a result of the development of negative pore water pressures during shear deformation. As these negative pore water pressures dissipate with time, the shear strength of the heavily overconsolidated plastic soil decreases. For the long-term case (effective stress analysis), the shear strength will be lower resulting in a lower bearing capacity.

Firm to stiff saturated plastic soils are intermediate conditions. The overconsolidation ratio (OCR) and the tendency of the saturated plastic soil to consolidate (gain shear strength) will determine whether the short-term condition (total stress parameters) or the long-term condition (effective stress parameters) provide the lower bearing capacity.

8.2.3 Other Bearing Capacity Considerations

There are many other possible considerations in the evaluation of bearing capacity. Some common items are as follows:

Lateral Loads. In addition to the vertical load acting of the footing, it may also be subjected to a lateral load. A common procedure is to treat lateral loads separately and resist the lateral loads by using the soil pressure acting on the sides of the footing (passive pressure) and by using the frictional resistance along the bottom of the footing. Foundations subjected to lateral loads will be further discussed in Sec. 15.2.

Moments and Eccentric Loads. It is always desirable to design and construct shallow footings so that the vertical load is applied at the center of gravity of the footing (see Fig. 1.7). For combined footings that carry more than one vertical load (see Fig. 1.8), the combined footing should be designed and constructed so that the vertical loads are symmetric. There may be design situations where the footing is subjected to a moment, such as where there is a fixed end connection between the building frame and the footing. This moment can be represented by a load P that is offset a certain distance (known as the eccentricity) from the center of gravity of the footing. For other projects, there may be property line constraints and the load must be offset a certain distance (eccentricity) from the center of gravity of the footing.

There are many different methods to evaluate eccentrically loaded footings. Because an eccentrically loaded footing will create a higher bearing pressure under one side compared to the opposite side, the best approach is to evaluate the actual pressure distribution beneath the footing. The usual procedure is to assume a rigid footing (hence linear pressure distribution) and use the section modulus ($1/6\, B^2$) in order to calculate the largest and lowest bearing pressure. For a footing having a width B, the largest (q') and lowest (q'') bearing pressures are as follows:

$$q' = \frac{Q(B + 6e)}{B^2} \qquad (8.6a)$$

$$q'' = \frac{Q(B - 6e)}{B^2} \qquad (8.6b)$$

where q' = largest bearing pressure underneath the footing, which is located along the same side of the footing as the eccentricity (kPa or psf)
q'' = lowest bearing pressure underneath the footing, which is located at the opposite side of the footing (kPa or psf)
Q = load applied to the footing (kN per linear meter of footing length or pounds per linear foot of footing length)
e = eccentricity of the load Q, i.e., the lateral distance from Q to the center of gravity of the footing (m or ft)
B = width of the footing (m or ft)

A usual requirement is that the load Q must be located within the middle third of the footing and the above equations are valid only for this condition. The value of q' must not exceed the allowable bearing pressure q_{all}. Eccentrically loaded footings will be further discussed in Chap. 15.

Footings at the Top of Slopes. Although methods have been developed to determine the allowable bearing capacity of foundations at the top of slopes (e.g., NAVFAC DM-7.2, 1982, page 7.2-135), these methods should be used with caution when dealing with plastic (cohesive) soil. This is because the outer face of slopes composed of plastic soils may creep downslope, leading to a loss in support to the footing constructed at the top of such slopes. Structures constructed at the top of clayey slopes will be further discussed in Sec. 10.8.

Inclined Base of Footing. Charts have been developed to determine the bearing capacity factors for footings having inclined bottoms. However, it has been stated that inclined bases should never be constructed for footings (AASHTO, 1996). Sometimes a sloping contact of underlying hard material will be encountered during excavation of the footing. Instead of using an inclined footing base along the sloping contact, the hard material should be excavated in order to construct a level footing that is entirely founded within the hard material.

Earthquake Loading. The effect of earthquakes on foundations will be discussed in Chap. 11. When performing foundation design for earthquake loadings, it is common to use a larger allowable bearing pressure in the analysis. For example, a common recommendation by the geotechnical engineer is that the allowable bearing pressure q_{all} can be increased by a factor of $1/3$ for earthquake analyses.

Soil Report Recommendations. Upon completion of the bearing capacity analyses, the recommendations would normally be included in a soils report. An example of typical wording concerning bearing capacity analyses at a proposed building site is as follows (see App. E):

> For intact Mission Valley formation (siltstone and sandstone) bedrock, the allowable bearing pressure for spread footings is 8000 pounds per square foot (400 kPa) provided that the footing is at least 5 ft (1.5 m) wide with a minimum of 2 ft (0.6 m) embedment in firm, intact bedrock. For continuous wall footings, the allowable bearing pressure is 4000 pounds per square foot (200 kPa) provided the footing is at least 2 ft (0.6 m) wide with a minimum of 2 ft (0.6 m) embedment in firm, intact bedrock. It is recommended that the structures be entirely supported by bedrock. It is anticipated that piers will be needed for the Administrative Building. Belled piers can be designed for an allowable end bearing pressure of 12,000 pounds per square foot (600 kPa) provided that the piers have a diameter of at least 2 ft (0.6 m), length of at least 10 ft (3 m), with a minimum embedment of 3 ft (0.9 m) in firm, intact bedrock. It is recommended that the geotechnical engineer observe pier installation to confirm embedment requirements.
>
> In designing to resist lateral loads, passive resistance of 1200 psf per foot of depth (200 kPa per meter of depth) to a maximum value of 6000 psf per foot of depth (900 kPa per meter of depth) and a coefficient of friction equal to 0.35 may be utilized for embedment within firm bedrock.
>
> For the analysis of earthquake loading, the above values of allowable bearing pressure and passive resistance may be increased by a factor of $\frac{1}{3}$.

8.3 BEARING CAPACITY FOR DEEP FOUNDATIONS

Deep foundations are used when the upper soil strata are too soft, weak, or compressible to support the foundation loads. Deep foundations are also used when there is a possibility of the undermining of the foundation. For example, bridge piers are often founded on deep foundations to prevent a loss of support due to flood conditions which could cause river bottom scour.

Types of Deep Foundations. The most common types of deep foundations are piles and piers that support individual footings or mat foundations (Table 1.2). Piles are defined as relatively long, slender, column-like members often made of steel, concrete, or wood that are either driven into place or cast in place in predrilled holes. Common types of piles are as follows:

- *Batter pile.* A pile driven in at an angle inclined to the vertical to provide high resistance to lateral loads.
- *End-bearing pile.* A pile whose support capacity is derived principally from the resistance of the foundation material on which the pile tip rests. End-bearing piles are often used when a soft upper layer is underlain by a dense or hard stratum. If the upper soft layer should settle, the pile could be subjected to down-drag forces, and the pile must be designed to resist these soil-induced forces.
- *Friction pile.* A pile whose support capacity is derived principally from the resistance of the soil friction and/or adhesion mobilized along the side of the pile. Friction piles are often used in soft clays where the end-bearing resistance is small because of punching shear at the pile tip.
- *Combined end-bearing and friction pile.* A pile that derives its support capacity from combined end-bearing resistance developed at the pile tip and frictional and/or adhesion resistance on the pile perimeter.

A pier is defined as a deep foundation system, similar to a cast-in-place pile, that consists of a column-like reinforced-concrete member. Piers are often of large enough diameter to enable downhole inspection. Piers are also commonly referred to as *drilled shafts, bored piles,* or *drilled caissons.*

There are many other methods available for forming deep foundation elements. Examples include earth stabilization columns, such as (NAVFAC DM-7.2, 1982):

- *Mixed-in-place piles.* A mixed-in-place soil-cement or soil-lime pile.
- *Vibroreplacement stone columns.* Vibroflotation or other method is used to make a cylindrical, vertical hole which is filled with compacted gravel or crushed rock.
- *Grouted stone columns.* This is similar to the above but includes filling voids with bentonite-cement or water-sand-bentonite cement mixtures.
- *Concrete vibro columns.* Similar to stone columns, but concrete is used instead of gravel.

Design Criteria. There are several different items that are used in the design and construction of piles. These items include the following:

1. *Engineering analysis.* From the results of subsurface exploration and laboratory testing, the bearing capacity of the deep foundation can be calculated in a way similar to that of shallow foundations in the previous section. The next two sections (Secs. 8.3.1 and 8.3.2) describe the engineering analyses for deep foundations in cohesionless and cohesive soil.

2. *Field load tests.* Prior to the construction of the foundation, a pile or pier could be load-tested in the field to determine its carrying capacity. Because of the uncertainties in the design of piles based on engineering analyses, pile load tests are common. The pile load test can often result in a more economical foundation than one based solely on engineering analyses. Pile load tests will be further discussed in Chap. 18.

3. *Application of pile-driving resistance.* In the past, the pile capacity was estimated from the driving resistance during the installation of the pile. Pile driving equations, such as the *Engineering News formula* (Wellington, 1888), were developed that related the pile capacity to the energy of the pile-driving hammer and the average net penetration of the pile per blow of the pile hammer. But studies have shown that there is no satisfactory relationship between the pile capacity from pile-driving equations and the pile capacity measured from load tests. On the basis of these studies, it has been concluded that use of pile-driving equations is no longer justified (Terzaghi and Peck, 1967).

Often the pile-driving resistance (i.e., blows per foot) is recorded as the pile is driven into place. When the anticipated bearing layer is encountered, the driving resistance (blows per foot) should substantially increase.

4. *Specifications and experience.* Other factors that should be considered in the deep foundation design include governing building code or agency requirements and local experience. Local experience is very important in the design and construction of pile foundations. For example, for the construction of a bridge pier, it was determined that 24-in.- (0.6-m-) diameter concrete-filled steel pipe piles would be required for the design loading conditions. Once driven into place and load-tested, the 24-in.- (0.6-m-) diameter pipe piles were unable to develop the necessary load capacity. On the basis of local experience advice, 12-in.- (0.3-m-) diameter steel pipe piles were installed, and, when field-tested, they developed a higher load capacity even though they have less surface area and less end-bearing area than the larger 24-in.- (0.6-m-) diameter pipe piles. The reason for the higher load capacity of the smaller driven piles could be that they caused less remolding and disturbance of the cohesive soil than the larger piles. This project demonstrated the importance of local experience in the selection of pile types and sizes.

8.3.1 Bearing Capacity for Cohesionless Soil

End-Bearing Pile or Pier. For an end-bearing pile or pier, the bearing capacity equations can be used to determine the ultimate bearing capacity q_{ult}. As mentioned in Sec. 8.2, Eq. (8.3) is often used for square footings. Assuming a pile that has a square cross section (i.e., $B = L$), and since $c = 0$ (cohesionless soil), Eq. (8.3) reduces to:

$$q_{ult} = \frac{Q_p}{B^2} = 0.4\,\gamma_t\,B\,N_\gamma + \gamma_t\,D_f\,N_q \tag{8.7}$$

In Eq. (8.7), the value of B (width of pile) is much less than the embedment depth D_f of the pile. Therefore, the first term on the right side of Eq. (8.7) can be neglected.

The value of $\gamma_t\,D_f$ in the second term in Eq. (8.7) is equivalent to the total vertical stress σ_v at the pile tip. For cohesionless soil (effective stress analysis), the groundwater table must be included in the analysis and the vertical effective stress σ'_v can be substituted for σ_v, and Eq. (8.7) reduces to:

For end-bearing piles having a square cross section:

$$q_{ult} = \frac{Q_p}{B^2} = \sigma'_v\,N_q \tag{8.8a}$$

For end-bearing piles or piers having a circular cross section:

$$q_{ult} = \frac{Q_p}{\pi\,r^2} = \sigma'_v\,N_q \tag{8.8b}$$

where q_{ult} = the ultimate bearing capacity of the end-bearing pile or pier (kPa or psf)
Q_p = point resistance force (kN or pounds)
B = width of the piles having a square cross section (m or ft)
r = radius of the piles or piers having a round cross section (m or ft)
σ'_v = vertical effective stress at the pile tip (kPa or psf)
N_q = dimensionless bearing capacity factor

For drilled piers or piles placed in predrilled holes, the value of N_q can be obtained from Table 8.2 according to the friction angle ϕ of the cohesionless soil located at the pile tip. However, for driven piles, the values of N_q listed in Table 8.2 are generally too conservative. Figure 8.5 presents a chart prepared by Vesic (1967) that provides the bearing capacity factor N_q from several different sources. Note in Fig. 8.5 that at $\phi = 30°$, N_q varies from about 30 to 150, while at $\phi = 40°$, N_q varies from about 100 to 1000. This is a tremendous variation in N_q values and is related to the different approaches used by the various researchers, where in some cases the basis of the relationship shown in Fig. 8.5 is theoretical, while in other cases the relationship is based on analysis of field data such as pile load tests.

There is a general belief that the bearing capacity factor N_q is higher for driven piles than for shallow foundations. One reason for a higher N_q value is the effect of driving the pile, which displaces and densifies the cohesionless soil at the bottom of the pile. The densification could be due to both the physical process of displacing the soil and the driving vibrations. These actions would tend to increase the friction angle of the cohesionless soil in the vicinity of the driven pile. Large-diameter piles would tend to displace and densify more soil than smaller diameter piles.

Example. Assume a pile having a diameter of 0.3 m (1 ft) and a length of 6 m (20 ft) will be driven into a nonplastic silty sand deposit which has the shear strength properties

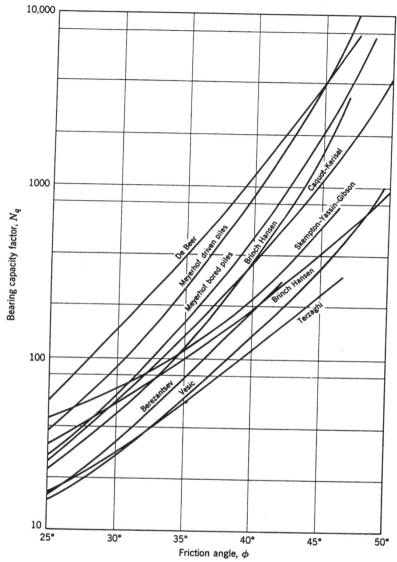

FIGURE 8.5 Bearing capacity factor N_q as recommended by various researchers for deep foundations. (*From Vesic, 1967; reproduced from Lambe and Whitman, 1969.*)

shown in Fig. 6.10 (i.e., $c' = 0$ and $\phi' = 30°$). Assume that the location of the groundwater table is located 3 m (10 ft) below the ground surface and the total unit weight above the groundwater table is 19 kN/m³ (120 pcf) and the buoyant unit weight γ_b below the groundwater table is 9.9 kN/m³ (63 pcf). Using the Terzaghi correlation shown in Fig. 8.5, calculate the allowable end-bearing capacity using a factor of safety of 3.

Solution. The vertical effective stress σ'_v at the pile tip equals:

$$\sigma'_v = (3\ m)(19\ kN/m^3) + (3\ m)(9.9\ kN/m^3) = 87\ kPa\ or\ 1800\ psf$$

From Fig. 8.5, for the Terzaghi relationship, $N_q = 30$. From Eq. (8.8b), the ultimate end-bearing capacity $(q_{ult}) = 2600$ kPa (54,000 psf). Multiplying q_{ult} by the area of the pile tip (πr^2), the ultimate pile tip resistance $(Q_p) = 180$ kN (42 kips). Using a factor of safety of 3, the allowable pile tip resistance $= 60$ kN (14 kips).

Friction Pile. As the name implies, a friction pile develops its load-carrying capacity due to the frictional resistance between the cohesionless soil and the pile perimeter. Piles subjected to vertical uplift forces would be designed as friction piles because there would be no end-bearing resistance as the pile is pulled from the ground.

For a linear increase in frictional resistance with confining pressure, the average ultimate frictional capacity q_{ult} can be calculated as follows:

For piles having a square cross section:

$$q_{ult} = \frac{Q_s}{4BL} = \sigma'_h \tan \phi_w = \sigma'_v k \tan \phi_w \qquad (8.9a)$$

For piles or piers having a circular cross section:

$$q_{ult} = \frac{Q_s}{2\pi rL} = \sigma'_h \tan \phi_w = \sigma'_v k \tan \phi_w \qquad (8.9b)$$

where q_{ult} = average ultimate frictional capacity for the pile or pier (kPa or psf)
Q_s = ultimate skin friction resistance force (kN or pounds)
B = width of the piles having a square cross section (m or ft)
r = radius of the piles or piers having a round cross section (m or ft)
L = length of the pile or pier (m or ft)
σ'_h = average horizontal effective stress over the length of the pile or pier (kPa or psf)
σ'_v = average vertical effective stress over the length of the pile or pier (kPa or psf)
k = dimensionless parameter equal to σ'_h divided by σ'_v [i.e., similar to Eq. (6.20)]*
ϕ_w = friction angle between the cohesionless soil and the perimeter of the pile or pier (degrees); commonly used friction angles are $\phi_w = \frac{3}{4} \phi$ for wood and concrete piles and $\phi_w = 20°$ for steel piles

Note in Eq. (8.9) that the term $4BL$ (for square piles) and $2\pi rL$ (for circular piles) is the perimeter surface area of the pile or pier. In Eq. (8.9), the term $\sigma'_h \tan \phi_w$ equals the shear strength τ between the pile or pier surface and the cohesionless soil. This term is identical to Eq. (6.32) (with $c' = 0$); i.e., $\tau = \sigma'_n \tan \phi'$. Thus the frictional resistance force Q_s in Eq. (8.9) is equal to the perimeter surface area times the shear strength of the soil at the pile or pier surface.

Example. Assume the same conditions as the previous example and that the pile is made of concrete. For driven displacement piles and using $k = 1$, calculate the allowable frictional capacity of the pile.

---

*Equation (6.21) can be used to estimate the value of k for loose sand deposits. Because of the densification of the cohesionless soil associated with driven displacement piles, values of k between 1 and 2 are often assumed.

Solution. The easiest solution consists of dividing the pile into two sections. The first section is located above the groundwater table ($z = 0$ to 3 m) and the second section is that part of the pile below the groundwater table ($z = 3$ to 6 m). The average vertical stress will be at the midpoint of these two sections, or:

σ_v (at $z = 1.5$ m) $= (1.5\text{ m})(19\text{ kN/m}^3) = 29$ kPa or 600 psf

σ'_v (at $z = 4.5$ m) $= (3\text{ m})(19\text{ kN/m}^3) + (1.5\text{ m})(9.9\text{ kN/m}^3) = 72$ kPa or 1500 psf

For a concrete pile:

$$\phi_w = {}^3\!/_4 \phi = {}^3\!/_4 (30°) = 22.5°$$

Substitute values into Eq. (8.9b): For $z = 0$ to 3 m, $L = 3$ m and

$Q_s = (2\pi rL)(\sigma'_v k \tan \phi_w) = (2\pi)(0.15\text{ m})(3\text{ m})(29\text{ kPa})(1)(\tan 22.5°) = 34$ kN or 7.6 kips

For $z = 3$ m to 6 m, $L = 3$ m and

$Q_s = (2\pi rL)(\sigma'_v k \tan \phi_w) = (2\pi)(0.15\text{ m})(3\text{ m})(72\text{ kPa})(1)(\tan 22.5°) = 84$ kN or 19 kips

Adding together both values of Q_s gives the total frictional resistance force = 34 kN + 84 kN = 118 kN or 26.6 kips. Applying a factor of safety of 3, we find the allowable frictional capacity of the pile is approximately equal to 40 kN (9 kips).

Combined End-Bearing and Friction Pile. Piles and piers subjected to vertical compressive loads and embedded in a deposit of cohesionless soil are usually treated in the design analysis as combined end-bearing and friction piles or piers. This is because the pile or pier can develop substantial load-carrying capacity from both end-bearing and frictional resistance. To calculate the ultimate pile or pier capacity for a condition of combined end-bearing and friction, the value of Q_p from Eq. (8.8) is added to the value of Q_s from Eq. (8.9).

In the previous example of a 6-m- (20-ft-) long pile embedded in a silty sand deposit, the allowable tip resistance (60 kN) and the allowable frictional capacity (40 kN) are added together to obtain the allowable combined end-bearing and frictional resistance capacity of the pile (100 kN).

Pile Groups. The previous discussion has dealt with the load capacity of a single pile in cohesionless soil. Usually pile groups are used to support the foundation elements, such as a group of piles supporting a spread footing (pile cap) or a mat slab. In loose sand and gravel deposits, the load-carrying capacity of each pile in the group may be greater than a single pile because of the densification effect due to driving the piles. Because of this densification effect, the load capacity of the group is often taken as the load capacity of a single pile times the number of piles in the group. An exception would be a situation where a weak layer underlies the cohesionless soil. In this case, group action of the piles could cause them to punch through the cohesionless soil and into the weaker layer or cause excessive settlement of the weak layer located below the pile tips.

In order to determine the settlement of the strata underlying the pile group, the 2:1 approximation (see Sec. 6.4) can be used to determine the increase in vertical stress $\Delta\sigma_v$ for those soil layers located below the pile tip. If the piles in the group are principally end-bearing, then the 2:1 approximation starts at the tip of the piles [L = length of the pile group, B = width of the pile group, and z = depth below the tip of the piles; see Eq. (6.28)]. If the pile group develops its load-carrying capacity principally through side friction, then the 2:1 approximation starts at a depth of ${}^2\!/_3 D$, where D = depth of the pile group.

8.20 ANALYSIS OF GEOTECHNICAL DATA AND ENGINEERING COMPUTATIONS

8.3.2 Bearing Capacity for Cohesive Soil

The analysis of the load-carrying capacity of piles and piers in cohesive soil is more complex than the analysis for cohesionless soil. Some of the factors that may need to be considered in the analysis are as follows (AASHTO, 1996):

- A lower load-carrying capacity of a pile in a pile group compared to that of a single pile.
- The settlement of the underlying cohesive soil due to the load of the pile group.
- The effects of driving piles on adjacent structures or slopes. The ground will often heave around piles driven into soft and saturated cohesive soil.
- The increase in load on the pile due to negative skin friction (i.e., down-drag loads) from consolidating soil.
- The effects of uplift loads from expansive and swelling clays.
- The reduction in shear strength of the cohesive soil due to construction effects such as the disturbance of sensitive clays or development of excess pore water pressures during the driving of the pile. There is often an increase in load-carrying capacity of a pile after it has been driven into a soft and saturated clay deposit. This increase in load-carrying capacity with time is known as *freeze* or *setup* and is caused primarily by the dissipation of excess pore water pressures.
- The influence of fluctuations in the elevation of the groundwater table on the load-carrying capacity when analyzed in terms of effective stresses.

Total Stress Analysis. The ultimate load capacity of a single pile or pier in cohesive soil is often determined by performing a total stress analysis. This is because the critical load on the pile, such as from wind or earthquake loads, is a short-term loading condition and thus the undrained shear strength of the cohesive soil will govern. The total stress analysis for a single pile or pier in cohesive soil typically is based on the undrained shear strength s_u of the cohesive soil or the value of cohesion c determined from unconsolidated undrained triaxial compression tests (i.e., $\phi = 0$ analysis; see Fig. 6.16).

The ultimate load capacity of the pile or pier in cohesive soil would equal the sum of the ultimate end-bearing and ultimate side adhesion components. In order to determine the ultimate end-bearing capacity, Eq. (8.3) can be utilized with $B = L$ and $\phi = 0$, in which case $N_c = 5$, $N_\gamma = 0$, and $N_q = 1$ (Table 8.2). The term $\gamma_t D_f N_q = \gamma_t D_f$ (for $\phi = 0$) is the weight of overburden, which is often assumed to be balanced by the pile weight and thus this term is not included in the analysis. Note in Eq. (8.3) that the term $c\,N_c\,(1 + 0.3\,B/L) = c\,(5)\,(1.3) = 6.5\,c$ (or $N_c = 6.5$). However, N_c is commonly assumed to be equal to 9 for deep foundations (Mabsout et al., 1995). Thus the ultimate load capacity Q_{ult} of a single pile or pier in cohesive soil equals:

$$Q_{ult} = \text{end bearing} + \text{side adhesion} = c\,N_c(\text{area of tip}) + c_A(\text{surface area})$$

or

$$Q_{ult} = c\,9(\pi R^2) + c_A(2\pi R z) = 9\pi c R^2 + 2\pi c_A R z \tag{8.10}$$

where Q_{ult} = ultimate load capacity of the pile or pier (kN or kips)
c = cohesion of the cohesive soil at the pile tip (kPa or psf)*
R = radius of the uniform pile or pier (m or ft)†

---

*Because it is a total stress analysis, the undrained shear strength ($s_u = c$) is used, or the undrained shear strength is obtained from unconsolidated undrained triaxial compression tests (i.e., $\phi = 0$ analysis, c = peak point of Mohr circles; see Fig. 6.16).

†If the pier bottom is belled or a tapered pile is used, then R at the tip would be different from the radius of the shaft.

z = embedment depth of the pile (m or ft)
c_A = adhesion between the cohesive soil and pile or pier perimeter (kPa or psf)‡

If the pile or pier is subjected to an uplift force, then the ultimate capacity T_{ult} for the pile or pier in tension is equal to:

$$T_{ult} = 2\pi c_A R z \tag{8.11}$$

where c_A is the adhesion between the cohesive soil and the pile or pier perimeter (Fig. 8.6). In order to determine the allowable capacity of a pile or pier in cohesive soil, the values calculated from Eqs. (8.10) and (8.11) would be divided by a factor of safety. A commonly used factor of safety is 3.

Effective Stress Analysis. For long-term loading conditions of piles or piers, an effective stress analysis could be performed. In this case, the effective cohesion c' and effective friction angle ϕ' would be used in the analysis for end bearing. The location of the groundwater

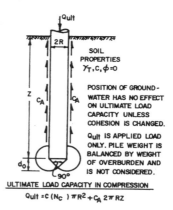

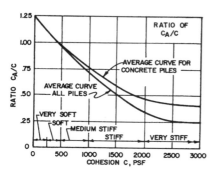

FIGURE 8.6 Ultimate capacity for a single pile or pier in cohesive soil. (*Reproduced from NAVFAC DM-7.2, 1982.*)

---

‡Figure 8.6 can be used to determine the value of the adhesion c_A for different types of piles and cohesive soil conditions.

8.22 ANALYSIS OF GEOTECHNICAL DATA AND ENGINEERING COMPUTATIONS

table would also have to be considered in the analysis. Along the pile perimeter, the ultimate resistance could be based on the effective shear strength between the pile or pier perimeter and the cohesive soil.

Pile Groups. The bearing capacity of pile groups in cohesive soils is normally less than the sum of individual piles in the group, and this reduction in group capacity must be considered in the analysis. The *group efficiency* is defined as the ratio of the ultimate load capacity of each pile in the group to the ultimate load capacity of a single isolated pile. If the spacing between piles in the group is greater than about 7 times the pile diameter, then the group efficiency is equal to one (i.e., no reduction in pile capacity for group action). The group efficiency decreases as the piles become closer together in the pile group. For example, a 9 × 9 pile group with a pile spacing equal to 1.5 times the pile diameter has a group efficiency of only 0.3. Figure 8.7 can be used to determine the ultimate load capacity of a pile group in cohesive soil.

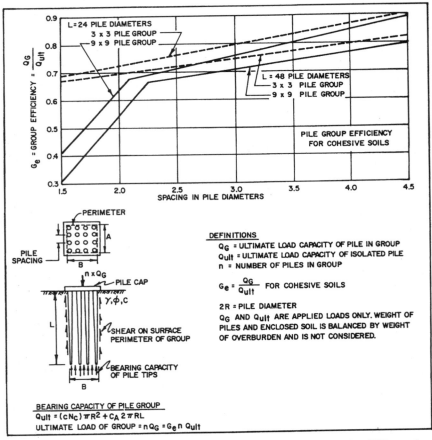

FIGURE 8.7 Ultimate capacity of a pile group in cohesive soil. (*Developed by Whitaker, 1957; reproduced from NAVFAC DM-7.2, 1982.*)

As with pile groups in cohesionless soil, the settlement of the strata underlying the pile group can be evaluated by using the 2:1 approximation (see Sec. 6.4) to calculate the increase in vertical stress $\Delta\sigma_v$ for those soil layers located below the pile tip. If the piles in the group develop their load-carrying capacity principally by end bearing in cohesive soil, then the 2:1 approximation starts at the tip of the piles [L = length of the pile group, B = width of the pile group, and z = depth below the tip of the piles; see Eq. (6.28)]. If the pile group develops its load-carrying capacity principally through cohesive soil adhesion along the pile perimeter, then the 2:1 approximation starts at a depth of $2/3\ D$, where D = depth of the pile group.

8.4 PAVEMENT DESIGN

Besides shallow and deep foundations, other structures can be susceptible to bearing capacity failures. For example, unpaved roads and roads with an inadequate pavement section or weak subgrade can also be susceptible to bearing capacity failures caused by heavy wheel loads. The heavy wheel loads can cause a general bearing capacity failure or a punching-type shear failure. These bearing capacity failures are commonly known as *rutting*, and they develop when the unpaved road or weak pavement section is unable to support the heavy wheel load.

Because the thickness of the pavement design is governed by the shear strength of the soil supporting the road, it is usually the geotechnical engineer who tests the soil and determines the pavement design thickness. The transportation engineer often provides design data to the geotechnical engineer, such as the estimated traffic loading, required width of pavement, and design life of the pavement.

8.4.1 Pavement Section

Pavements are usually classified as either *rigid* or *flexible,* depending on how the surface loads are distributed. A rigid pavement consists of Portland cement concrete slabs, which tend to distribute the loads over a fairly wide area. A flexible pavement is defined as a pavement having a sufficiently low bending resistance, yet having the required stability to support the traffic loads, e.g., macadam, crushed stone, gravel, and asphalt (California Division of Highways, 1973).

The most common type of flexible pavement consists of the following:

1. *Asphalt concrete.* The uppermost layer (surface course) is typically "asphalt concrete" that distributes the vehicle load in a cone-shape area under the wheel and acts as the wearing surface. The ingredients in asphalt concrete are asphalt (the cementing agent), coarse and fine aggregates, mineral filler (i.e., fines such as limestone dust), and air. Asphalt concrete is usually hot-mixed in an asphalt plant and then hot-laid and compacted by using smooth-wheeled rollers. Other common names for asphalt concrete are *black-top, hot mix,* or simply *asphalt* (Atkins, 1983).

2. *Base.* Although not always a requirement, in many cases there is a base material that supports the asphalt concrete. The base typically consists of aggregates that are well-graded, hard, and resistant to degradation from traffic loads. The base material is compacted into a dense layer that has a high frictional resistance and good load distribution qualities. The base can be mixed with up to 6 percent Portland cement to give it more strength and this is termed a *cement-treated base* (CTB).

8.24 ANALYSIS OF GEOTECHNICAL DATA AND ENGINEERING COMPUTATIONS

Figure 8.8 shows an example of base material that has been placed and compacted. Note in Fig. 8.8 that the concrete curbs are in place, the top of the base has been smoothed, and that the final step will be the placement of the asphalt concrete. Figure 8.9 shows a second example of the placement of base for a runway at the Reno/Tahoe International Airport.

3. *Subbase*. In some cases, a subbase will be used to support the base and asphalt concrete layers. The subbase often consists of a lesser quality aggregate that is lower-priced than base material.

4. *Subgrade*. The subgrade supports the pavement section (i.e., the overlying subbase, base, and asphalt concrete). The subgrade could be native soil or rock, a compacted fill, or soil that has been strengthened by the addition of lime or other cementing agents. Instead of strengthening the subgrade, a geotextile could be placed on top of the subgrade in order to improve its load-carrying capacity. Table 4.8 lists soil type versus value as subgrade, potential frost action, drainage properties, CBR values, and subgrade modulus.

8.4.2 California Method

The design of flexible pavement in California is usually based on the following equation (California Division of Highways, 1973):

$$T = \frac{0.0032 \, \text{TI}(100 - R)}{G_f} \qquad (8.12)$$

where T = flexible pavement thickness for a 10-year design life (ft)
       TI = traffic index (dimensionless parameter)

FIGURE 8.8 Compacted base material for a parking lot.

FIGURE 8.9 Compacted base material for the Reno/Tahoe International Airport.

R = R-value (dimensionless parameter)
G_f = gravel equivalent factor (dimensionless parameter)

This flexible pavement design equation was developed from empirical relationships and the performance of pavements in actual service. The terms in Eq. (8.12) are defined as follows:

1. *Traffic index (TI).* The effect of traffic on the roadway over the design life of the pavement is expressed by the traffic index (TI). The higher the equivalent wheel loads (EWL) during the life of the pavement, the higher the traffic index. Traffic index values for different types of facilities are summarized in Table 8.3, and the design traffic index will often be provided to the geotechnical engineer by the transportation engineer. If the traffic index is 8 or more, the design should be based on the same criteria used for state highways.

2. *R-value.* The R-value is the resistance of the subgrade or base and refers to the ability of the soil to resist lateral deformation when acted upon by a vertical load. In essence, the R-value is a relative measure of the shear strength of the soil. The R-value is determined by using standardized test procedures and equipment, such as specified by ASTM D 2844-94, "Standard Test Method for Resistance R-Value and Expansion Pressure of Compacted Soils" (1998). The R-value is determined on specimens compacted to anticipated field conditions. Prior to testing, the specimens are soaked in water to achieve saturation. Soaking the soil prior to testing is supposed to produce an R-value that represents the worst possible state that the soil might attain in the field. A crushed rock base can have an R-value of 75 to 85, while a clay subgrade will usually have an R-value of less than 10.

3. *Gravel equivalent factor G_f.* The gravel equivalent G_f represents the strength factor of the materials in the pavement structural section. The gravel equivalent is an empirical factor developed through research and field experience and it relates the relative strength of a unit thickness of a particular material in terms of an equivalent thickness of gravel. An

8.26 ANALYSIS OF GEOTECHNICAL DATA AND ENGINEERING COMPUTATIONS

TABLE 8.3 Traffic Index (TI) for Various Types of Facilities

Type of facility (1)	Traffic index (2)
Minor residential streets and cul-de-sacs	4
Average residential streets	4.5
Residential collectors and minor or secondary collectors	5
Major or primary collectors providing for traffic movement between minor collectors and major arterials	6
Farm-to-market roads providing for the movement of traffic through agricultural areas to major arterials	5–7
Commercial roads (arterials serving areas which are primarily commercial in nature)	7–9
Connector roads (highways and arterials connecting two areas of relatively high population density)	7–9
Major city streets and thoroughfares and county highways	7–9

Note: If the traffic index (TI) is 8 or more, the design should be based on the same criteria used for state highways.
Source: California Division of Highways, 1973.

aggregate base and subbase will have $G_f = 1.1$ and $G_f = 1.0$, respectively, while a class A cement-treated aggregate base has $G_f = 1.7$. Asphalt concrete (AC) has the following gravel equivalent factors (California Division of Highways, 1973):

TI	Asphalt concrete G_f
≤ 5	2.50
6	2.32
7	2.14

These gravel equivalent factors relate the relative strength of the asphalt concrete to the gravel base. Thus at a TI=5 or less, 1 in. of asphalt concrete ($G_f = 2.50$) is considered to be equivalent to 2.5 in. of subbase ($G_f = 1.0$).

In terms of the minimum design thickness of asphalt concrete and base, it has been stated (California Division of Highways, 1973) that in general, it is difficult to obtain good results over an untreated base with less than 6.4 cm (2.5 in.) of asphalt concrete surfacing. Also, it is very difficult to properly place and compact layers of aggregate base or subbase when these layers measure less than 10 cm (4 in.). Therefore the geotechnical engineer should not recommend thickness values less than these minimum values (i.e., 2.5 in. of asphalt concrete and 4 in. of base or subbase).

Example. A pavement design (10-year life) for a secondary collector road is required. The design TI = 5, R-value of the base = 65, and R-value of the subgrade = 20. Determine the thickness of the base and asphalt concrete.

Solution. The first step is to determine the required thickness of asphalt concrete using the R-value of the base (65). Note that for asphalt concrete with a TI = 5, the gravel equivalent factor (G_f) = 2.50.

BEARING CAPACITY

$$T = \frac{0.0032 \text{ TI}(100 - R)}{G_f} = \frac{0.0032(5)(100 - 65)}{2.50} = 0.22 \text{ ft} = 2.7 \text{ in.}$$

Usually ½-in. increments of asphalt concrete are specified, so use 3 in. of asphalt concrete. The second step is to determine the thickness of base material. Using the R-value of the subgrade (20) and rearranging Eq. (8.12) gives

$$TG_f = 0.0032 \text{ TI}(100 - R) = 0.0032(5)(100 - 20) = 1.28 \text{ ft}$$

The thickness of the asphalt concrete has already been selected (3 in.), therefore for asphalt concrete:

$$TG_f = \left(\frac{3}{12}\right)(2.50) = 0.62 \text{ ft}$$

Subtracting the asphalt concrete portion (0.62 ft) from the total required (1.28 ft) leaves a balance of 0.66 ft. Using this value to calculate the thickness of the base and recognizing that $G_f = 1.1$ for the base material, we obtain

$$TG_f = 0.66 \text{ ft} \quad \text{or} \quad T = \frac{0.66 \text{ ft}}{G_f} = \frac{0.66}{1.1} = 0.6 \text{ ft}$$

Thus the base must be 0.6 ft (7.2 in.) thick. Usually it is best to use 1-in. increments of base thickness, and therefore the recommended thickness would be 8 inches. The final recommended values are:

- Thickness of asphalt concrete: 3 in.
- Thickness of base material: 8 in.

The third step is to check the recommended values. This is accomplished by multiplying the thickness of each layer by the gravel equivalent factors, or:

Asphalt concrete: $TG_f = (3/12)(2.50) = 0.62$

Base: $TG_f = (8/12)(1.1) = \underline{0.73}$

Total $= 1.35$

Note that 1.35 is larger than 1.28, and thus the design checks.

Although Eq. (8.12) is an empirical equation, it does contain the major factors that influence pavement performance, such as the amount of traffic loads (TI) and the strength of the base and subgrade (R-values), and it includes a factor that relates the relative strengths of different types of materials in the pavement section (G_f). One limitation of Eq. (8.12) is that it does not have a factor to account for the expansiveness of the subgrade. This limitation will be further discussed in Sec. 9.5, "Pavements and Flatwork."

Besides Eq. (8.12), there are many other methods that are used for the design of pavements. Instead of using the R-value, some methods utilize the California bearing ratio (CBR) as a measure of the shear strength of the base and subgrade. Numerous charts have also been developed that relate the shear strength of the subgrade and the traffic loads to a recommended pavement thickness (e.g., Asphalt Institute, 1984). When designing pavements, the geotechnical engineer should always check with the local transportation authority for design requirements as well as the local building department or governing agency for possible specifications on the type of method that must be used for the design.

PROBLEMS

The problems have been divided into basic categories as indicated below:

Bearing Capacity of Shallow Foundations on Cohesionless Soil

1. Use the data from the example problem in Sec. 8.2.1 (i.e., the 1.2-m-wide strip footing). Assume the groundwater table is located at the bottom of the footing. Also assume that the total unit weight γ_t is the same above and below the groundwater table ($\gamma_t = 19$ kN/m^3). Calculate the ultimate bearing capacity q_{ult} and the allowable load Q_{all}. *Answers:* $q_{ult} = 215$ kPa and $Q_{all} = 86$ kN/m.

2. For Prob. 1, assume the groundwater table is located 1.5 m below the ground surface. Also assume that the total unit weight γ_t is the same above and below the groundwater table ($\gamma_t = 19$ kN/m^3). Calculate the ultimate bearing capacity q_{ult} and the allowable load Q_{all}. *Answers:* $q_{ult} = 240$ kPa and $Q_{all} = 95$ kN/m.

3. Use the data from the example in Sec. 8.2.1. Assume the soil located from ground surface to the bottom of the footing has a total unit weight (γ_t) = 15.7 kN/m^3 and, below the footing, the soil has a total unit weight (γ_t) = 19 kN/m^3. Calculate the ultimate bearing capacity q_{ult} and the allowable load Q_{all}. *Answers:* $q_{ult} = 230$ kPa and $Q_{all} = 93$ kN/m.

4. Use the data from the example in Sec. 8.2.1. Calculate the allowable load Q_{all} according to the *Uniform Building Code* (i.e., Table 8.1). *Answer:* $Q_{all} = 180$ kN/m (note that the UBC q_{all} is much greater than the q_{all} determined from the bearing capacity equation).

5. Use the data from the example in Sec. 8.2.1. Instead of a strip footing, assume a square footing (1.2 m by 1.2 m). Calculate the ultimate bearing capacity q_{ult} and the allowable load Q_{all}. *Answers:* $q_{ult} = 244$ kPa and $Q_{all} = 117$ kN.

6. For Prob. 5, calculate the allowable load Q_{all} according to the *Uniform Building Code* (i.e., Table 8.1). *Answer:* $Q_{all} = 216$ kN.

7. Use the data from Probs. 30 and 31 in Chap. 7 (i.e., the 6-ft by 6-ft footing). Assume the footing is constructed at ground surface ($D_f = 0$) and the total unit weight (γ_t) of the sand is 125 pcf. From Fig. 4.13, for an N-value = 30, $\phi' > 40°$ (use $\phi' = 40°$ in the analysis). Using the bearing capacity equation and a factor of safety = 3, determine the allowable load Q_{all} for this footing. What governs in the design of the footing (i.e., settlement or allowable bearing capacity)? *Answers:* $Q_{all} = 430$ kips. The allowable load based on a bearing capacity analysis (430 kips) is much greater than the load (230 kips) that will cause about 1-in. settlement. Therefore, settlement governs in the design of the footing.

8. For Prob. 7, calculate the allowable load Q_{all} according to the *Uniform Building Code* (i.e., Table 8.1) assuming $D_f = 1$ ft. *Answer:* $Q_{all} = 108$ kips. Note that for this dense sand, the UBC Q_{all} (108 kips) is much less than the Q_{all} that will cause about 1 in. of settlement (230 kips).

9. Use the data from the example in Sec. 8.2.1. Assume the strip footing must support a vertical load of 150 kN per linear meter of wall. Determine the minimum width of the strip footing based on a factor of safety = 3. *Answer:* $B = 1.56$ m.

Bearing Capacity for Shallow Foundations on Cohesive Soil

10. Use the data from the example problem in Sec. 8.2.1 and assume the subsoil consists of clay that has an undrained shear strength (s_u) = 20 kPa. For a total stress analysis,

BEARING CAPACITY   8.29

calculate the ultimate bearing capacity q_{ult} and the allowable load Q_{all}. *Answers:* $q_{ult} = 117$ kPa and $Q_{all} = 47$ kN/m.

11. Using the data from Prob. 10, assume the groundwater table is located at a depth of 0.9 m and that the clay has an overconsolidation ratio (OCR) = 1.5 that is approximately constant with depth. What governs in the design of the strip footing (i.e., long-term settlement or allowable bearing capacity based on a total stress analysis)? *Answers:* At a depth of 0.9 m, $\sigma'_{vm} = 26$ kPa and $q_{all} = 39$ kPa, therefore settlement governs. It is recommended that the bearing pressure not exceed 26 kPa to prevent virgin consolidation.

12. Use the data from the example in Sec. 8.2.1 and the shear strength parameters from Prob. 10. Assume the strip footing must support a vertical load of 50 kN per linear meter of wall. Determine the minimum width B of the strip footing based on a factor of safety of 3. *Answer:* $B = 1.28$ m.

13. Use the data from the example in Sec. 8.2.1, except that the site is underlain by heavily overconsolidated clay that has an undrained shear strength $(s_u) = 200$ kPa and a drained shear strength of $\phi' = 28°$ and $c' = 5$ kPa. Assume the groundwater table is located at a depth of 0.9 m. Performing both a total stress analysis and an effective stress analysis, determine the allowable load Q_{all}. *Answer:* The effective stress analysis governs and $Q_{all} = 131$ kN/m.

Eccentrically Loaded Footings

14. Use the data from the example in Sec. 8.2.1. Assume the vertical load exerted by the strip footing = 100 kN per linear meter of wall length and that this load is offset from the centerline of the strip footing by 0.15 m (i.e., $e = 0.15$ m). Determine the largest bearing pressure q' and the least bearing pressure q'' exerted by the eccentrically loaded footing. Is q' acceptable from an allowable bearing capacity standpoint? *Answers:* $q' = 146$ kPa and $q'' = 21$ kPa, and since $q' > q_{all}$, then q' is unacceptable.

15. Use the data from Prob. 14, but assume the footing is a square footing (1.2 m by 1.2 m) and the load of 100 kN exerted by the square footing is offset from the center of the footing by 0.15 m (i.e., $e = 0.15$ m). Determine the largest bearing pressure q' and the least bearing pressure q'' exerted by the eccentrically loaded footing. Is q' acceptable from an allowable bearing capacity standpoint? *Answers:* $q' = 122$ kPa and $q'' = 17$ kPa, and since $q' > q_{all}$, then q' is unacceptable.

Bearing Capacity for Deep Foundations (Cohesionless Soil)

16. Use the data from the example in Sec. 8.3.1 (i.e., the 0.3-m-diameter pile with a length of 6 m). A field load test on this pile indicates an ultimate load capacity Q_{ult} of 250 kN. Assume all of the load capacity is developed by end bearing. Also assume at the time of construction that 0.4-m-diameter piles are used instead of 0.3-m-diameter piles. What is the ultimate end-bearing capacity Q_p and the allowable pile capacity Q_{all} for the 0.4-m-diameter pile for a factor of safety = 3? *Answers:* $Q_p = 444$ kN and $Q_{all} = 148$ kN.

17. Use the data from Prob. 16, but instead of end bearing, assume all of the 250 kN is carried by skin friction. What is the ultimate frictional capacity Q_s and the allowable frictional capacity Q_{all} for the 0.4-m-diameter pile for a factor of safety = 3? *Answers:* $Q_s = 333$ kN and $Q_{all} = 111$ kN.

18. Use the data from Prob. 16, but assume that 60 percent of the 250-kN load is carried by end bearing and 40 percent of the 250-kN load is carried by skin friction. What is the ultimate load capacity Q_{ult} and the allowable load capacity Q_{all} for the 0.4-m-diameter pile using a factor of safety = 3? *Answers:* $Q_{ult} = 400$ kN and $Q_{all} = 133$ kN.

8.30 ANALYSIS OF GEOTECHNICAL DATA AND ENGINEERING COMPUTATIONS

19. Use the data from the example in Sec. 8.3.1. Determine the allowable pile tip resistance based on the *Uniform Building Code* (i.e., Table 8.1). *Answer:* Allowable pile tip resistance = 16 kN.

20. Assume a site has the subsoil profile shown in Fig. 4.39 and a bridge pier foundation consisting of piers will be installed at the location of B-25. Assume that each pier will be embedded 3 m into the sandstone. For the sediments above the sandstone, use an average buoyant unit weight (γ_b) = 9.2 kN/m³. For the sandstone, use an average buoyant unit weight (γ_b) = 11.7 kN/m³ and effective shear strength parameters c' = 50 kPa and ϕ' = 41°. For scour conditions (scour to elevation −10 m) and considering only end bearing, determine the allowable pier capacity Q_{all} if the piers are 1.0 m in diameter, for a factor of safety = 3. *Answer:* Q_{all} = 5000 kN [Eq. (8.3) in terms of effective stress; N_c and N_q from Table 8.2].

21. For Prob. 20, what would be the allowable pier capacity Q_{all} according to the *Uniform Building Code* (i.e., Table 8.1)? *Answer:* Q_{all} = 240 kN.

22. In Fig. 4.39, assume at Boring B-25 that the uppermost soil layer (silt, silty sand, and peat layer) extends from elevation −2 to −5 m. Use the data from Prob. 20, and assume that after construction of the bridge pier, a 0.5-m-thick sand layer (γ_b = 9.2 kN/m³) is deposited on the river bottom. Also assume that k = 0.5 and ϕ_w = 20° for the 0.5-m-thick sand sediment layer, and k = 0.4 and ϕ_w = 15° for the 3-m-thick silt-peat layer. Determine the down-drag load on each pier due to consolidation and compression of the silt-peat layer. *Answer:* Down-drag load = 19.3 kN.

Bearing Capacity of Deep Foundations (Cohesive Soil)

23. Assume the subsoil conditions at a site are as shown in Fig. 4.40. A 9 by 9 pile group, having a spacing of 3 pile diameters (Fig. 8.7), supports a pile cap at ground surface (elevation +20 feet). Assume each pile in the group has a diameter = 1.5 ft and the piles are 70 ft long. Neglecting skin friction resistance in the upper 30 ft, determine the load capacity of the pile group using a total stress analysis and a factor of safety=3. *Answers:* For an average s_u = 0.6 kg/cm² from elevation −10 to −50 ft, Q_s = 140 kips. For an average s_u = 0.4 kg/cm² at elevation −50 ft, Q_p = 13 kips. The allowable load capacity of the pile group = 2900 kips.

24. For Prob. 23, assume the pile cap supports a load of 2900 kips and this load is resisted by side friction and end bearing of the soil from elevation −10 to −50 ft. By dividing the clay from elevation −50 to −90 ft into four 10-ft-thick layers, calculate the primary consolidation settlement s_c due to the load of 2900 kips. Use the 2:1 approximation and assume it starts at an elevation of −37 ft (i.e., 2/3 D, where D = depth of that portion of the pile supported by soil adhesion). *Answer:* Since the clay from elevation −50 to −90 ft is essentially normally consolidated, and for C_c = 0.35, e_0 = 0.9, and the 2:1 approximation, s_c = 4.4 in.

25. Assume that the subsoil conditions at a site are as shown in Fig. 4.41. Assume that 1.5-ft-diameter piles are installed so that their tip is at a depth of 35 ft below ground surface. Using a total stress analysis, determine the ultimate skin adhesion capacity Q_s, neglecting the shear strength in the upper 5 ft, the ultimate end-bearing capacity Q_p, and the allowable load capacity Q_{all} using a factor of safety = 3. *Answers:* For an average undrained shear strength (s_u) = 350 psf, Q_s = 50 kips. For an average undrained shear strength (s_u) = 400 psf at a depth of 35 ft, Q_p = 6 kips. Q_{all} = 19 kips.

26. Use the data from Prob. 25. For a 9 by 9 pile group with a spacing of 1.5 pile diameters (Fig. 8.7), determine the allowable load capacity of the pile group using a factor of safety = 3. *Answer:* Allowable load capacity of pile group = 630 kips.

BEARING CAPACITY

27. Assume that the subsoil conditions at a site are as shown in Fig. 4.42. Predrilled cast-in-place concrete piles that have a diameter of 1 ft will be installed to a depth of 50 ft. Ignoring the soil adhesion in the upper 10 ft of the pile, calculate the allowable load capacity Q_{all} of the pile using a total stress analysis and a factor of safety = 3. *Answer:* For an average $s_u = 0.6$ kg/cm² (1200 psf) from 10 to 50 ft depth, $Q_s = 100$ kips and $Q_p = 8$ kips. $Q_{all} = 36$ kips.

Pavement Design

28. Solve the example in Sec. 8.4.2 for a traffic index (TI) = 6. *Answers:* Thickness of asphalt concrete = 3.5 in., thickness of aggregate base = 10 in.

29. Use the data from the example in Sec. 8.4.2. Instead of an aggregate base, assume a cement-treated base ($G_f = 1.7$) will be used. Determine the thickness of the asphalt concrete and CTB base. *Answers:* Thickness of asphalt concrete = 3 in., thickness of CTB base = 5 in.

30. Use the data from the example in Sec. 8.4.2. Assume that in addition to the asphalt concrete and aggregate base, a subbase having an R-value = 45 will be utilized. Assume that for the aggregate subbase that $G_f = 1.0$. Determine the thickness of asphalt concrete, aggregate base, and aggregate subbase. *Answers:* Thickness of asphalt concrete = 3 in., thickness of base = 4 in., and thickness of subbase = 4 in.

CHAPTER 9
EXPANSIVE SOIL

The following notation is used in this chapter:

SYMBOL	DEFINITION
c_A	Adhesion between pier and clay
c_s	Coefficient of swell
C_s	Swelling index (Fig. 9.12)
e_m	Edge moisture variation distance
EI	Expansion index
h_i	Initial height of specimen (expansion index test)
h_p	Height of specimen at the end of primary swell (expansion index test)
H_{dr}	Swell height
N	Vertical pressure for one-dimensional swell test
P	Pier weight plus dead load applied at the top of the pier
P_f	Final overburden pressure (Fig. 9.12)
P_0	Equivalent to P'_s (Fig. 9.12)
P'_s	Swell pressure (Fig. 9.12)
R	Radius of pier
t	Time
T	Time factor (from Table 7.4)
T_r	Resisting forces for pier
T_u	Uplift force for pier
y_m	Maximum anticipated vertical differential movement
Z_a	Depth of seasonal moisture change
Z_{na}	Portion of pier below the active zone

9.1 INTRODUCTION

Expansive soils are a worldwide problem, causing extensive damage to civil engineering structures. Jones and Holtz estimated in 1973 that the annual cost of damage in the United States due to expansive soil movement was $2.3 billion (Jones and Holtz, 1973). A more up-to-date figure is about $9 billion in damages annually to buildings, roads, airports, pipelines, and other facilities (Jones and Jones, 1987).

Section 7.4 discussed the consolidation of clay, which is basically the compression of soft clays that have a high water content. Expansive clays are different in that the near-surface clay often varies in density and moisture condition from the wet season to the dry season. For example, near- or at-surface clays often dry out during periods of drought but then expand during the rainy season or when they get wet by irrigation water or water from leaky pipes.

Although most states have expansive soil, Chen (1988) reported that certain areas of the United States, such as Colorado, Texas, Wyoming, and California, are more susceptible to damage from expansive soils than others. These areas have large surface deposits of clay and have climates characterized by alternating periods of rainfall and drought.

9.1.1 Expansive Soil Factors

There are many factors that govern the expansion behavior of soil. The primary factors are the availability of moisture and the amount and type of the clay size particles in the soil. For example, Seed et al. (1962) developed a classification chart based solely on the amount and type (activity) of clay-size particles (see Fig. 9.1). Other factors affecting the expansion behavior include the type of soil (natural or fill), the condition of the soil in terms of dry density and moisture content, the magnitude of the surcharge pressure, the amount of non-expansive material (gravel- or cobble-size particles), and the amount of aging (Ladd and Lambe, 1961; Kassiff and Baker, 1971; Chen, 1988; Day, 1991b, 1992a). In general, expansion potential increases as the dry density increases and the moisture content decreases. Also, the expansion potential increases as the surcharge pressure decreases. Some of these factors are individually discussed below:

1. *Amount of clay-size particles.* As shown in Fig. 9.1, the more clay-size particles of a particular type a soil has, the more swell there will be (all other factors being the same). As mentioned in Sec. 5.4.2, clay-size particles attract water to their particle faces (double-layer effect). Thus the more clay-size particles in a dry soil, the greater the need for water to be drawn into the soil and, hence, the higher the swell potential of the soil.

2. *Type of clay-size particles.* Also shown in Fig. 9.1, the type of clay-size particles significantly affects swell potential. Given the same dry weight, kaolinite clay particles (activity between 0.3 and 0.5) are much less expansive than sodium montmorillonite clay particles (activity between 4 and 7) (Holtz and Kovacs, 1981). This effect is also related to the attraction of water to the clay-size particle faces. Montmorillonite is a much smaller and more active clay mineral than kaolinite, and this results in much more attracted water per unit dry mass of clay particles. Once again, the need for more water results in a greater amount of water drawn into the dry soil and hence higher swell potential of the soil.

Using such factors as the clay-particle content, Holtz and Gibbs (1956b) developed a system to classify soils as having either a low, medium, high, or very high expansion potential. Table 9.1 lists typical soil properties versus the expansion potential (Holtz and Gibbs, 1956b; Holtz and Kovacs, 1981; Meehan and Karp, 1994; and *Uniform Building Code,* 1997).

3. *Surcharge pressure.* The bottom three rows of Table 9.1 list typical values of percent swell versus expansion potential. Note in Table 9.1 the importance of surcharge pressure on percent swell. At a surcharge pressure of 31 kPa (650 psf) the percent swell is much less than at a surcharge pressure of 2.8 kPa (60 psf). For example, for "highly" expansive soil, the percent swell for a surcharge pressure of 2.8 kPa (60 psf) is typically 10 to 15 percent, while at a surcharge pressure of 31 kPa (650 psf), the percent swell is 4 to 6 percent. The effect of surcharge is important because it is usually the lightly loaded structures, such as concrete flatwork, pavements, slab-on-grade foundations, and concrete canal liners, that are often impacted by expansive soil.

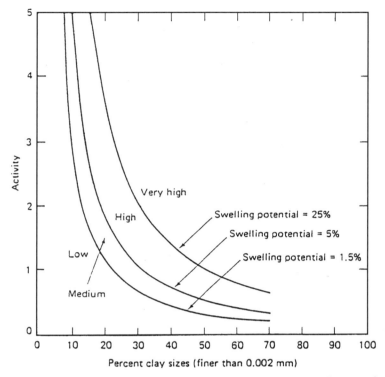

FIGURE 9.1 Classification chart for swelling potential. (*From Seed et al., 1962; reprinted with permission from the American Society of Civil Engineers*). *Note:* soil specimens were compacted using standard Proctor energy and tested with a normal stress of 6.9 kPa (1 psi).

TABLE 9.1 Typical Soil Properties versus Expansion Potential

Expansion potential (1)	Very low (2)	Low (3)	Medium (4)	High (5)	Very high (6)
Expansion index	0–20	21–50	51–90	91–130	130+
Clay content ($< 2\ \mu m$)	0–10%	10–15%	15–25%	25–35%	35–100%
Plasticity index	0–10	10–15	15–25	25–35	35+
% swell @ 2.8 kPa (60 psf)	0–3	3–5	5–10	10–15	15+
% swell @ 6.9 kPa (144 psf)	0–2	2–4	4–7	7–12	12+
% swell @ 31 kPa (650 psf)	0	0–1	1–4	4–6	6+

Note: % swell for specimens at moisture and density conditions according to HUD criteria.

9.1.2 Laboratory Testing

The geotechnical engineer can determine the presence of expansive soil by performing subsurface exploration and laboratory testing. Various types of laboratory tests for expansive soil are discussed below.

Expansion Index (EI) Test. A common laboratory test used to determine the expansion potential of the soil is the expansion index test. The test provisions are stated in the *Uniform Building Code* (1997), titled "Uniform Building Code Standard 18-2, Expansion Index Test," and in ASTM (1998), which has a nearly identical test specification (ASTM D 4829-95, 1998). The purpose of this laboratory test is to determine the expansion index, which is then used to classify the soil as having either a very low, low, medium, high, or very high expansion potential as shown in Table 9.1.

The expansion index test is performed on soil passing the No. 4 sieve. The usual testing procedure is as follows:

1. If particles larger than the No. 4 sieve are possibly expansive (such as claystone and shale), these particles are to be broken down so that they pass the No. 4 sieve.
2. The soil is then compacted into a standard mold having an internal diameter of 10.2 cm (4 in.) using two layers, with each layer receiving 15 blows from a 2.5-kg (5.5-lb) tamper having a 30.5-cm (12-in.) drop.
3. After compaction, the thickness of the compacted soil should be about 5.1 cm (2 in.). After removal from the compaction mold, the specimen is trimmed into a 10.2-cm (4-in.) diameter by a 2.5-cm- (1.0-in.-) thick oedometer ring.
4. The degree of saturation of the test specimen is then determined by utilizing either the known water content of the compacted material or that determined on trimmings.
5. If the degree of saturation is between 49 and 51 percent, the specimen is ready for testing, or else the test specimen preparation procedure is repeated after adjustment of initial water content.
6. The test specimen is mounted in the oedometer apparatus with porous plates on the top and bottom of the specimen.
7. The specimen is then subjected to a vertical stress of 6.89 kPa (144 psf).
8. A dial gauge (part of the oedometer apparatus) is used to measure the vertical movement, an initial dial reading is taken, and the sample is submerged in distilled water.
9. After the specimen has swelled, the expansion index (EI) can be calculated as follows (Day, 1993a):

$$\text{EI} = \frac{1000\,(h_p - h_i)}{h_i} = (10)(\%\text{ primary swell}) \qquad (9.1)$$

where h_p = height of the specimen at the end of primary swell and h_i = initial height of the specimen. On the basis of the expansion index, the expansion potential of the soil is determined as indicated in Table 9.1.

In order to obtain the specimen height at the end of primary swell (h_p), dial versus time readings can be recorded. The dial readings can be converted to percent swell and plotted versus time from the start of the test on a semi-log plot. Figure 9.2 presents an expansion index test performed on a clay of low plasticity (CL). The shape of the swelling versus time curve (Fig. 9.2) is similar to the consolidation of a saturated clay (Fig. 7.15), except that swelling has been plotted as positive values and consolidation as negative values. From this similarity, the end of primary swell can be determined as the intersection of the straight line portions of the swell curve. The arrow in Fig. 9.2 indicates a primary swell of 6.2 percent (or EI = 62), indicating a medium expansion potential (Table 9.1).

In some cases, the expansion index test can yield misleading results for soils classified as clayey gravels (GC). The reason is that clayey gravels will have a significant fraction retained on the No. 4 sieve, but the expansion index test uses only those particles passing the No. 4 sieve. There is no correction in the test specification to account for nonexpansive (i.e., hard rock) gravel particles retained on the No. 4 sieve. One approach used in practice is to reduce the expansion index (EI), depending on the percentage of particles by dry mass that pass the No. 4 sieve. This correction is as follows (Day, 1993a):

$$\text{EI (corrected)} = \frac{(\text{EI}) \, (\% \text{ passing No. 4 sieve})}{100} \quad (9.2)$$

Suppose a clayey gravel has 40 percent by dry mass passing the No. 4 sieve and for the particles passing the No. 4 sieve, the EI is 100. Then according to Eq. (9.2), the corrected EI would be 100 times 0.4, or 40 ("low" expansion potential). This procedure is not exact since it uses the dry mass, rather than the volume, of the oversize particles and does not consider the compaction energy (Day, 1992b).

As previously mentioned, there is a similarity of the shape of time versus deformation curves for consolidation and swell of clay in the oedometer apparatus. The rate of swell can

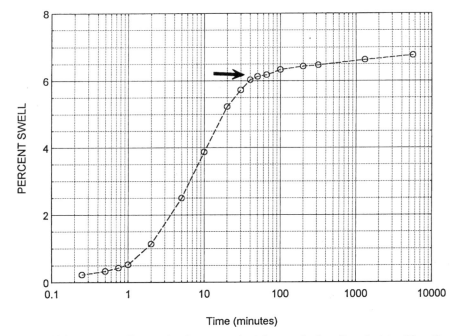

FIGURE 9.2 Percent swell versus time for an expansion index test of a clay of low plasticity. (*Note:* The arrow indicates the end of primary swell.)

be estimated from Terzaghi's diffusion equation where the coefficient of consolidation [c_v; see Eq. (7.14)] is replaced by the coefficient of swell c_s, or (Blight, 1965):

$$c_s = \frac{TH_{dr}^2}{t} \tag{9.3}$$

where T = time factor (Table 7.4), H_{dr} = swell height, and t = time. The coefficient of swell c_s can be determined from laboratory time versus swell data such as shown in Fig. 9.2. For example, for the data shown in Fig. 9.2, the time for 50 percent primary swell (t_{50}) is about 6.5 minutes. The specimen height corresponding to 50 percent primary swell is 2.62 cm (1.03 in.) and, since water can be drawn into the soil from the top and bottom, the coefficient of swell c_s from Eq. (9.3) is 9×10^{-4} cm^2/s.

Correlations with Other Index Tests. Other laboratory index tests, such as hydrometer analyses and Atterberg limits, can be used to classify the soil and estimate its expansiveness. Those soils having a high clay-size content and plasticity index, such as clays of high plasticity (CH), will usually be classified as having a high to very high expansion potential (see Table 9.1).

In Table 9.1, the clay content (percent clay-size particles) is usually a less reliable indicator of expansion potential than the plasticity index. This is because the type of clay mineral (i.e., kaolinite versus sodium montmorillonite) has such a large effect on expansion potential. A better correlation is the expansion potential versus plasticity index. When correlating the plasticity index (PI) and expansion potential in Table 9.1, the PI should be obtained from Eq. (5.17).

Swell Tests on Undisturbed Specimens. The most direct method of determining the amount of swelling is by performing a one-dimensional swell test by utilizing the oedometer apparatus (ASTM D 4546-96, 1998). The undisturbed soil specimen is placed in the oedometer and a vertical pressure (also referred to as *surcharge pressure*) is applied. Then the soil specimen is inundated with distilled water and the one-dimensional vertical swell is calculated as the increase in height of the soil specimen divided by the initial height, often expressed as a percentage. Such a test offers an easy and accurate method of determining the percent swell of the soil. After the soil has completed its swelling, the vertical pressure can be increased to determine the swelling pressure, which is defined as that pressure required to return the soil specimen to its original (initial) height (Chen, 1988).

As will be discussed in Sec. 9.4, the swell tests are often used in the design of foundations. Concerning the use of the swell test, ASTM D 4546-96 (1998) states:

> Estimates of the swell of soil determined by this test method [ASTM D 4546-96] are often of key importance in design of floor slabs on grade and evaluation of their performance. However, when using these estimates, it is recognized that swell parameters determined from these test methods for the purpose of estimating *in situ* heave of foundations and compacted soils may not be representative of many field conditions because:
>
> 1. Lateral swell and lateral confining pressure are not simulated.
> 2. Rates of swell indicated by swell tests are not always reliable indicators of field rates of heave due to fissures in the *in situ* soil mass and inadequate simulation of the actual availability of water to the soil. The actual availability of water to the foundation may be cyclic, intermittent, or depend on in-place situations, such as pervious soil-filled trenches and broken water and drain lines.
> 3. Secondary or long-term swell may be significant for some soils and should be added to primary swell.

4. Chemical content of the inundation water affects volume changes and swell pressure; that is, field water containing large concentrations of calcium ions will produce less swelling than field water containing large concentrations of sodium ions or even rain water.
5. Disturbance of naturally occurring soil samples greatly diminishes the meaningfulness of the results.

Another major limitation of the swell test is that the water content of the tested soil may not be the same water content at the time of construction of the foundation. For example, if a drought occurs and the clay dries out, the percent expansion will be underestimated. On the other hand, if a rainy season occurs just prior to construction and the clay absorbs moisture, the percent swell will be overestimated. But even with all of these limitations, swell tests on undisturbed soil specimens are generally considered to be the most reliable method of predicting the future potential heave of the foundation.

9.2 SWELLING OF DESICCATED CLAY

Desiccated clays are common in areas having near-surface clay deposits and periods of drought. Structures constructed on top of desiccated clay can be severely damaged by expansive soil heave (Jennings, 1953). For example, Chen (1988) states: "Very dry clays with natural moisture content below 15 percent usually indicate danger. Such clays will easily absorb moisture to as high as 35 percent with resultant damaging expansion to structures." There can also be desiccation and damage to final clay cover systems for landfills and site remediation projects, and for shallow clay landfill liners (Boardman and Daniel, 1996).

9.2.1 Identification of Desiccated Clay

The geotechnical engineer can often visually identify desiccated clay because of the numerous ground surface cracks, such as shown in Fig. 9.3. Desiccated clay deposits also generally have a distinct water content profile, where the water content increases with depth. For example, Fig. 9.4 shows the water content versus depth for two clay deposits located in Irbid, Jordan (Al-Homoud et al., 1997). Soil deposit A has a liquid limit of 35 and a plasticity index of 22, while soil deposit B has a liquid limit of 79 and a plasticity index of 27 (Al-Homoud et al., 1995).

Note in Fig. 9.4 that during the hot and dry summer, the water content of the soil is significantly lower than during the wet winter. During the summer, the lowest water contents are recorded near ground surface, and the water contents are below the shrinkage limit (SL). A near-surface water content below the shrinkage limit is indicative of severe desiccation of the clay.

Below a depth of about 3.2 m (10 ft) for soil deposit A and a depth of about 4.5 m (15 ft) for soil deposit B, the water content is relatively unchanged between the summer and winter monitoring period, and this depth is commonly known as the *depth of seasonal moisture change*. The depth of seasonal moisture change is also sometimes referred to as the *depth of the active zone*. The depth of seasonal moisture change would depend on many factors such as the temperature and humidity, length of the drying season, presence of vegetation that can extract soil moisture, depth of the water table, and the nature of the soil in terms of clay content, clay mineralogy, and density. As shown in Fig. 9.4, soil deposit B has a greater variation in water content from the dry summer to wet winter and a greater depth of seasonal moisture change. This is probably because soil deposit B has a higher clay content than soil deposit A.

9.8 ANALYSIS OF GEOTECHNICAL DATA AND ENGINEERING COMPUTATIONS

FIGURE 9.3 Two views of deposits of desiccated clay located in Death Valley, California. (For the upper photograph, note the hat in the center of the photograph, which provides a scale for the size of the desiccation cracks.)

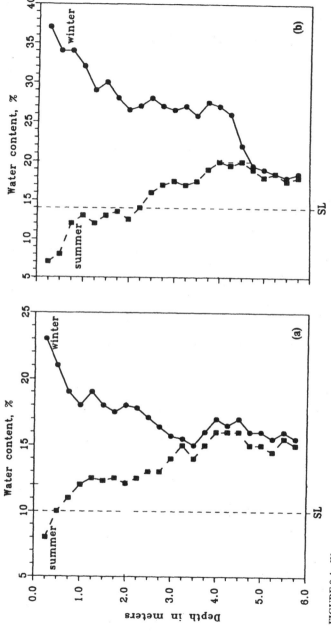

FIGURE 9.4 Water content versus depth for: (a) soil A; (b) soil B. (*From Al-Homoud et al., 1997; reprinted with permission from the American Society of Civil Engineers.*)

9.2.2 Hydraulic Conductivity and Rate of Swell

Figure 9.5 shows a specimen of desiccated natural clay obtained from Otay Mesa, California. The near-surface deposit of natural clay has caused extensive damage to structures, pavements, and flatwork constructed in this area. The clay particles in this soil are almost exclusively montmorillonite (Kennedy and Tan, 1977; Cleveland, 1960).

Figure 9.6 presents the results of a one-dimensional swell test (lower half of Fig. 9.6) and a falling head permeameter test (upper half of Fig. 9.6) performed on a specimen of desiccated Otay Mesa clay (Day, 1997b). At time zero, the desiccated clay specimen was inundated with distilled water. The data in Fig. 9.6 indicates three separate phases of swelling of the clay, as follows:

1. *Primary swell.* The first phase of swelling of the desiccated clay was primary swell. The primary swell occurs from time zero (start of wetting) to about 100 minutes. The end of primary swell (100 minutes) was estimated from Casagrande's log-time method, which can also be applied to the swelling of clays (Day, 1992b).

Figure 9.6 shows that during primary swell, there was a rapid decrease in the hydraulic conductivity (also known as permeability) of the clay. The rapid decrease in hydraulic conductivity was due to the closing of soil cracks as the clay swells. At the end of primary swell, the main soil cracks have probably closed and the hydraulic conductivity was about 7×10^{-7} cm/s.

2. *Secondary swell.* The second phase of swelling was secondary swell. The secondary swell occurs from a time of about 100 minutes to 20,000 minutes after wetting. Figure 9.6 shows that during secondary swell, the hydraulic conductivity continues to decrease as the clay continues to swell and the microcracks close up. The lowest hydraulic

FIGURE 9.5 Specimen of desiccated Otay Mesa clay.

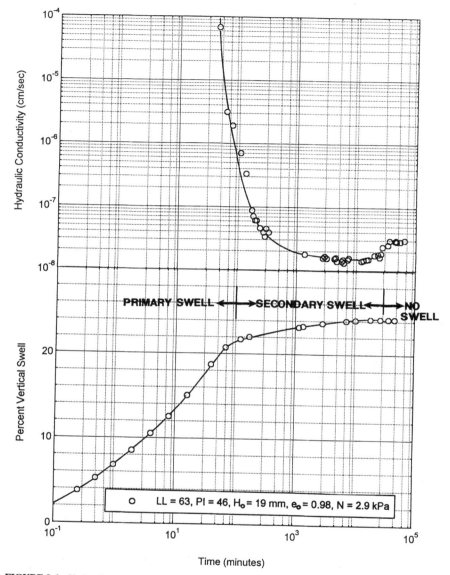

FIGURE 9.6 Hydraulic conductivity and percent swell versus time.

conductivity of about 1.5×10^{-8} cm/s occurs at a time of about 5000 minutes, when most of the microcracks have probably sealed up. From a time of 5000 to 20,000 minutes after wetting, there was a slight increase in hydraulic conductivity. This is probably a result of a combination of additional secondary swell which increases the void ratio and a reduction in entrapped air.

3. Steady state. The third phase started when the clay stopped swelling. This occurred at about 20,000 minutes after inundation with distilled water. No swell was recorded from a time of 20,000 minutes after wetting to the end of the test (50,000 minutes). As shown in Fig. 9.6, the hydraulic conductivity is constant once the clay has stopped swelling. From a time of 20,000 minutes after wetting to the end of the test (50,000 minutes), the hydraulic conductivity of the clay was constant at about 3×10^{-8} cm/s.

There appear to be three factors that govern the rate of swelling of desiccated clay:

- *Development of desiccation cracks.* The amount and distribution of desiccation cracks, such as shown in Figs. 9.3 and 9.5, are probably the greatest factors in the rate of swelling. Clays will shrink until the shrinkage limit (usually a low water content) is reached. Even as the moisture content decreases below the shrinkage limit, there is probably still the development of additional microcracks as the clay dries. The more cracks in the clay, the greater the pathways for water to penetrate the soil, and the quicker the rate of swelling.
- *Increased suction at a lower water content.* It is well known that as the water content decreases, the suction pressure of the clay increases (Fredlund and Rahardjo, 1993). At low water contents, the water is drawn into the clay by the suction pressures. The combination of both shrinkage cracks and high suction pressures allows water to be quickly sucked into the clay, resulting in a higher rate of swell.
- *Slaking.* Slaking is defined as the breaking of dried clay when submerged in water, due either to compression of entrapped air by inwardly migrating capillary water or to the progressive swelling and sloughing off of the outer layers (Stokes and Varnes, 1955). Slaking breaks apart the dried clay clods and allows water to quickly penetrate all portions of the desiccated clay. The process of slaking is quicker and more disruptive for clays having the most drying time and lowest initial water content.

In summary, a structure that is constructed during a hot and dry season when the near-surface water content of the desiccated clay is below the shrinkage limit, such as shown in Fig. 9.4, will have the greatest potential increase in water content and resulting heave of the foundation. The hydraulic conductivity (permeability) and rate of swell are important because they determine how fast the water will penetrate the soil.

9.3 TYPES OF EXPANSIVE SOIL MOVEMENT

9.3.1 Lateral Movement

Expansive soil movement can affect all types of civil engineering projects. For many retaining and basement walls, especially if the clay backfill is compacted below optimum moisture content, seepage of water into the clay backfill causes horizontal swelling pressures well in excess of at-rest values. Fourie (1989) measured the swell pressure of a compacted clay for zero lateral strain to be 420 kPa (8800 psf). Besides the swelling pressure induced by the expansive soil, there can also be groundwater or perched water pressure on the retaining or basement wall because of the poor drainage of clayey soils. Figure 9.7 shows the collapse of a retaining wall that has a clay backfill. Because of these detrimental effects of clay backfill, a common recommendation is to use only free-draining, granular material (clean sand or clean gravel) as retaining or basement wall backfill.

FIGURE 9.7 Collapse of a retaining wall that has a clay backfill.

9.3.2 Vertical Movement

If a structure having a large area, such as a pavement or foundation, is constructed on top of a desiccated clay, there are usually two main types of expansive soil movement. The first is the cyclic heave and shrinkage around the perimeter of the structure. The second is the long-term progressive swell beneath the center of the structure.

Cyclic Heave and Shrinkage. Cyclic heave and shrinkage commonly affects the perimeter of the foundation, by uplifting the edge of the structure or shrinking away from it.

Clays are characterized by their moisture sensitivity. They will expand when given access to water and shrink when they are dried out. A soil classified as having a very high expansion potential will swell or shrink much more than a soil classified as having a very low expansion potential (Table 9.1). For example, the perimeter of a pavement or slab-on-grade foundation will heave during the rainy season and then deform downward during the drought if the clay dries out. This causes cycles of up and down movement, causing cracking and damage to the structure. Field measurements of this up-and-down cyclic movement have been recorded by Johnson (1980).

The amount of cyclic heave and shrinkage depends on the change in moisture content of the clays below the perimeter of the structure. The moisture change in turn depends on the severity of the drought and rainy seasons, the influence of drainage and irrigation, and the presence of live tree roots, which can extract moisture and cause clays to shrink. The cyclic heave and shrinkage around the perimeter of a structure is generally described as a seasonal or short-term condition.

Progressive Swelling beneath the Center of the Structure (Center Lift). Two ways that moisture can accumulate underneath structures are by thermal osmosis and capillary action.

It has been stated that water at a higher temperature than its surroundings will migrate in the soil towards the cooler area to equalize the thermal energy of the two areas (Chen, 1988; Nelson and Miller, 1992). This process has been termed *thermal osmosis* (Sowers, 1979; Day, 1996a). Especially during the summer months, the temperature under the center of a structure tends to be much cooler than at the exterior ground surface.

Because of capillary action, moisture can move upward through soil, where it will evaporate at the ground surface. But when a structure is constructed, it acts as a ground surface barrier, reducing or preventing the evaporation of moisture. It is the effect of thermal osmosis and the evaporation barrier due to the structure that causes moisture to accumulate underneath the center of the structure. A moisture increase will result in swelling of expansive soils. The progressive heave of the center of the structure is generally described as a long-term condition, because the maximum value may not be reached until many years after construction. Figure 9.8 illustrates center lift beneath a house foundation and Fig. 9.9 shows the typical crack pattern in the concrete slab-on-grade due to expansive soil center lift.

9.3.3 Effect of Vegetation

One frequent cause of expansive soil damage is heave of the foundation (center lift) as shown in Fig. 9.8. Foundation movement can also be caused by the shrinkage of clay. For example, tree roots and rootlets can extract moisture from the ground, which can cause the near-surface clay to shrink and the foundation to deform downward.

There are cases (Cheney and Burford, 1975) where the opposite can also occur, where large trees have been removed and the clay has expanded as the soil moisture increases to its natural state. In the United States, Holtz (1984) states that large, broadleaf, deciduous trees located near the structure cause the greatest changes in moisture and the greatest resulting damage in both arid and humid areas. However, the most dramatic effects are during periods of drought, such as the severe drought in Britain from 1975 to 1976, when the amount of water used by the trees during transpiration greatly exceeded the amount of rainfall within the area containing tree roots. Biddle (1979, 1983) investigated 36 different trees, covering a range of tree species and clay types and concluded that poplars have much

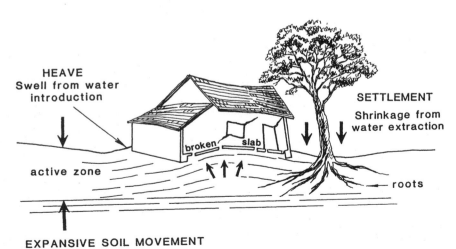

FIGURE 9.8 Illustration of center lift of foundation.

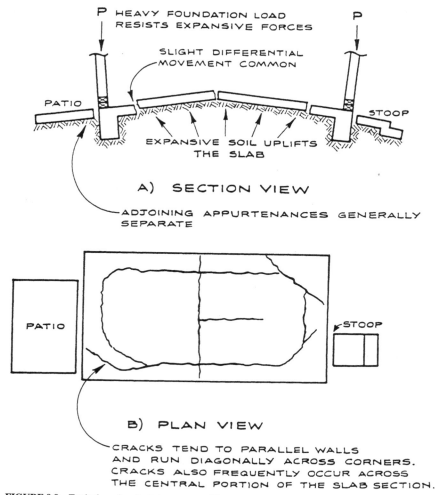

FIGURE 9.9 Typical crack pattern due to center lift.

greater effects than other species and that the amount of soil movement will depend on the clay shrinkage characteristics. Ravina (1984) states that it is the nonuniform moisture changes and soil heterogeneities that cause the uneven soil movements that damage shallow foundations, structures, and pavements.

Driscoll (1983) has suggested relationships between water content w of the soil and the liquid limit (LL), which indicate the level of desiccation, where

$$w = 0.5 \text{LL} \tag{9.4}$$

would indicate the onset of desiccation, and

$$w = 0.4 \text{LL} \tag{9.5}$$

would indicate when desiccation becomes significant. As mentioned in Sec. 9.2.1, a near-surface water content w below the shrinkage limit (SL) would be indicative of severe desiccation of the clay.

Tucker and Poor (1978) indicated that trees located at distances closer than their heights to structures caused significantly larger movements due to clay shrinkage than those trees located at greater distances. Hammer and Thompson (1966) stated that trees should not be planted a minimum of one-half their anticipated mature height from a shallow foundation and slow-growing, shallow-rooted varieties of trees were preferred. However, Cutler and Richardson (1989) indicated that total crown volume (hence leaf area) is generally more important than absolute height in relation to water demand. Cutler and Richardson (1989) used case histories to determine the frequency of damage as a function of tree-trunk distance from the structure for different species. Biddle (1983) recommended that for very high shrinkage clay, the perimeter footings should be at least 1.5 m (5 ft) deep and this would be sufficient to accommodate most tree-planting designs.

In summary, when dealing with expansive clays, it is important for the geotechnical engineer to consider the possibility of damage due to clay shrinkage caused by the extraction of moisture by tree roots. Shallow foundations, such as slab-on-grade or raised wood floor foundations having shallow perimeter footings are especially vulnerable to damage from clay shrinkage.

9.4 FOUNDATION DESIGN FOR EXPANSIVE SOIL

9.4.1 Conventional Slab-on-Grade Foundation

A conventional slab-on-grade consists of interior and exterior bearing wall footings and an interior slab. The concrete for the footings is usually placed at the same time as the concrete for the slab to create a monolithic foundation. For the conventional slab-on-grade foundation, the soils engineer rather than the structural engineer usually provides the construction details. The purpose of the design is to provide deepened perimeter footings that are below the primary zone of cyclic heave and shrinkage and to soak the subgrade soils (prior to construction) in order to reduce the long-term progressive swelling beneath the center of the slab.

Table 9.2 presents an example of expansive soil foundation specifications for conventional slab-on-grade on expansive soils in southern California (Day, 1994c). The design is empirical and depends on the expansion index (from the expansion index test). Table 9.2 indicates that both the depth of perimeter footings and the depth of presaturation should be increased as the soil becomes more expansive.

The purpose of the conventional slab-on-grade design, which is based on the use of deepened footings and presoaking, is to reduce the effects of expansive soil forces rather than make the foundation strong enough to resist such forces. Because the design is empirical, damage can develop because: (1) the perimeter footings are not deep enough to resist the cyclic heave and shrinkage and (2) the presoaking beneath the slab is not deep enough or it was not done properly (the soil is not saturated), resulting in long-term progressive swelling beneath the center of the slab.

9.4.2 Posttensioned Slab-on-Grade

A second type of foundation for expansive soils is the posttensioned slab-on-grade. There can be many different types of posttensioned designs. In Texas and Louisiana, the early

TABLE 9.2 Expansive Soil Foundation Recommendations

Expansive classification (1)	Depth of footing below adjacent grade (2)	Footing reinforcement (3)	Slab thickness and reinforcement conditions (4)	Presaturation below slabs (5)	Rock base below slabs (6)
None to low	0.5 m (18 in.) exterior 0.3 m (12 in.) interior	Exterior: 4 #4 bars, 2 top and 2 bottom Interior: 2 #4 bars, 1 top and 1 bottom	0.10 m (4 in.) nominal with #3 bars at 0.4 m (16 in.) on center, each way	to 0.3 m (12 in.)	Optional
Medium	0.6 m (24 in.) exterior 0.5 m (18 in.) interior	Exterior: 4 #5 bars, 2 top and 2 bottom Interior: 4 #4 bars, 2 top and 2 bottom	0.10 m (4 in.) net with #3 bars at 0.3 m (12 in.) on center, each way	to 0.5 m (18 in.)	0.10 m (4 in.)
High	0.8 m (30 in.) exterior 0.5 m (18 in.) interior	Exterior: 4 #5 bars, 2 top and 2 bottom Interior: 4 #5 bars, 2 top and 2 bottom	0.13 m (5 in.) net with #4 bars at 0.4 m (16 in.) on center, each way	to 0.6 m (24 in.)	0.15 m (6 in.)
Very high	0.9 m (36 in.) exterior 0.8 m (30 in.) interior	Exterior: 6 #5 bars, 3 top and 3 bottom Interior: 4 #5 bars, 2 top and 2 bottom	0.15 m (6 in.) nominal with #4 bars at 0.3 m (12 in.) on center, each way	to 0.8 m (30 in.)	0.20 m (8 in.)

posttensioned foundations consisted of a uniform-thickness slab with stiffening beams in both directions, which became known as the *ribbed foundation*. As previously mentioned, in California, a commonly used type of posttensioned slab consists of a uniform-thickness slab with an edge beam at the entire perimeter, but no or minimal interior stiffening beams or ribs. This type of posttensioned slab has been termed the *California slab* (Post-Tensioning Institute, 1996).

In "Design and Construction of Post-Tensioned Slabs-on-Ground," prepared by the Post-Tensioning Institute (1996), the design moments, shears, and differential deflections under the action of soil loading resulting from changes in water contents of expansive soils are predicted by equations developed from empirical data and a computer study of a plate on an elastic foundation. The *Uniform Building Code* (1997) presents nearly identical equations for the design of posttensioned slabs-on-grade. The idea for the design of a posttensioned slab-on-grade in accordance with the Post-Tensioning Institute is to construct a slab foundation that is strong enough and rigid enough to resist the expansive soil forces. To get the required rigidity to reduce foundation deflections, the stiffening beams (perimeter and interior footings) can be deepened. Although the differential movement to be expected for a given expansive soil is supplied by a soils engineer, the actual design of the foundation is usually by the structural engineer.

The posttensioned slab-on-grade should be designed for two conditions: (1) center lift (also called *center heave* or *doming*) and (2) edge lift (also called *edge heave* or *dishing*). These two conditions are illustrated in Fig. 9.10 (from Post-Tensioning Institute, 1996). Center lift is the long-term progressive swelling beneath the center of the slab, or because the soil around the perimeter of the slab dries and shrinks (causing the perimeter to deform downward), or a combination of both. Edge lift is the cyclic heave beneath the perimeter of the foundation. In order to complete the design, the soils engineer usually provides the maximum anticipated vertical differential soil movement y_m and the horizontal distance of moisture variation from the slab perimeter (e_m) for both the center-lift and edge-lift conditions (see Fig. 9.10). Values of moisture variation from the slab perimeter (e_m) are usually obtained from the Post-Tensioning Institute (1996), with typical values of e_m equal to 0.9–1.8 m (3–6 ft) for center lift and 0.6–1.8 m (2–6 ft) for edge lift.

There are many design considerations for posttensioned slabs-on-grade. Three important considerations are as follows:

1. The design of the foundation is based on static values of y_m, but the actual movement is cyclic. To mitigate leaks from utilities, flexible utility lines that enter the slab should be used. Also, the posttensioned slabs-on-grade must be rigid enough so that cracks do not develop in interior wallboard due to cyclic movement.

2. The values of e_m are difficult to determine because they are dependent on soil and structural interaction. The structural parameters that govern e_m include the magnitude and distribution of dead loads, the rigidity of the foundation, and the depth of the perimeter footings. The soil parameters include the amount and specific limits of the heave and shrinkage.

3. The geotechnical engineer might test the expansive clays during the rainy season, but the foundation might be built during a drought. In this case, the values of y_m for center lift could be considerably underestimated. The geotechnical engineer must test the clay under the conditions anticipated at the time of construction.

Calculating Foundation Heave. Figures 9.11 and 9.12 present two commonly used methods to calculate the total heave (also known as total swell) of a foundation on expansive soil. Both of these methods are based on the testing of undisturbed soil specimens in the oedometer apparatus. If the project will consist of compacted fill, then the soil specimens could be prepared by compacting the soil to anticipated field dry density and water content conditions.

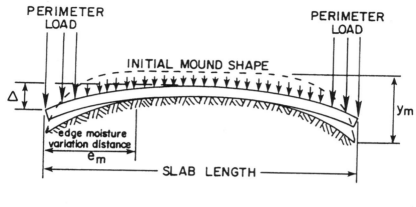

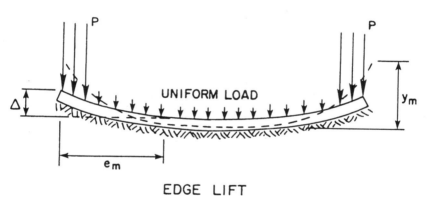

FIGURE 9.10 Depiction of center lift and edge lift for posttensioned slab-on-grade. (*Reproduced with permission from the Post-Tensioning Institute, 1996.*)

As discussed in Fig. 9.11, this method is based on testing undisturbed soil or compacted fill specimens by loading them in the oedometer apparatus (also known as the consolidometer) to a pressure that equals the ultimate value of the effective overburden pressure σ'_v plus the weight of the structure $\Delta\sigma_v$. The soil specimens are then submerged in distilled water and the percent swell is measured.

The approach shown in Fig. 9.12 is different in that the soil specimen is actually loaded in the oedometer apparatus much as in a consolidation test. At a low seating pressure, the soil specimen is inundated with distilled water. The vertical pressure is increased in order to prevent the soil specimen from swelling. At a certain pressure, known as the swelling pressure P'_s, the soil specimen will begin to compress. As shown in Fig. 9.12 (void ratio versus log pressure plot), the soil specimen is loaded to a high pressure and then the soil specimen is unloaded and allowed to swell. Often the swell curve will be linear on a void ratio versus log of the vertical pressure plot, and the swell index C_s can be calculated in the

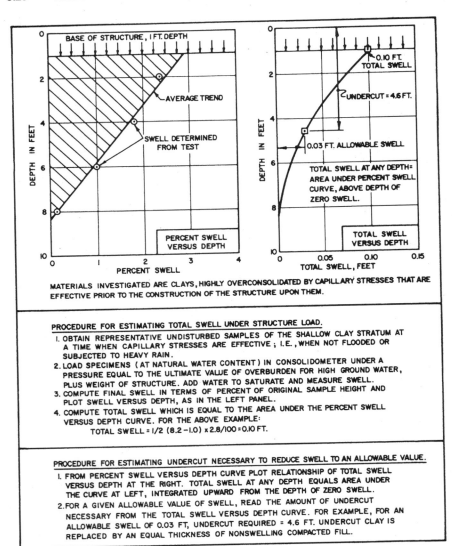

FIGURE 9.11 Computation of foundation heave for expansive clays. (*From NAVFAC DM-7.1, 1982.*)

same manner as the compression index C_c was calculated in Sec. 7.4.2. Note that the equation in Fig. 9.12 used to calculate the heave of the foundation is similar to Eq. (7.4), with h_i = the initial height of the *in situ* soil layer ($h_i = H_0$), the swelling index C_s used in place of the compression index C_c, the value of P_f = the final vertical effective stress in the soil after swelling has occurred, and P_0 = the swelling pressure (i.e., $P_0 = P'_s$). Note that in Fig. 9.12 the swelling soil was divided into three layers and the swell was calculated for each layer and then the total heave was calculated as the sum of the swell from the three layers.

9.4.3 Pier and Grade Beam Support

A third common foundation type for expansive soil is pier and grade beam support, as shown in Fig. 9.13 (from Chen, 1988). The basic principle is to construct the piers such that they are below the depth of seasonal moisture changes. The piers can be belled at the bottom to increase their uplift resistance. Grade beams and structural floor systems that are free of the ground are supported by the piers.

Chen (1988) provides several examples of proper construction details for pier and grade beam foundations. The design considerations for the pier and grade beam support are as follows (Woodward et al., 1972; Jubenville and Hepworth, 1981; Chen, 1988):

1. Sufficient pier length. A model test for pier uplift indicated that the expansive soil uplift on a pier is similar to the uplift capacity required for the extraction of a pile from the ground (Chen, 1988). From this test data, the uplift force T_u on the pier due to swelling of soil in the active zone can be estimated as follows:

$$T_u = c_A 2\pi R Z_a \tag{9.6}$$

where c_A = adhesion between the clay and pier, R = radius of the pier, and Z_a = depth of wetting, which usually corresponds to the depth of seasonal moisture change (the active zone). From the type of pier and the undrained shear strength of the clay, the value for c_A can be obtained from Fig. 8.6.

There are methods to reduce the uplift force T_u on the pier. For example, an air gap can be provided around the pier (Lambe and Whitman, 1969, p. 7). Another option is to enlarge the pier hole and install an easily deformable material between the pier and expansive clay.

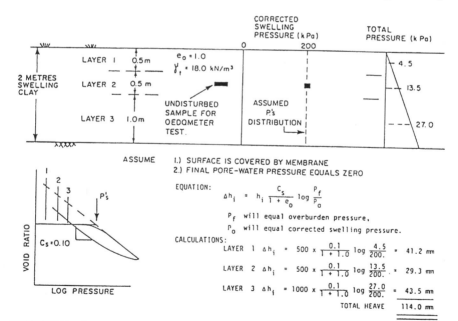

FIGURE 9.12 Computation of foundation heave for expansive clays. (*Method by Fredlund, 1983; reproduced from Chen, 1988.*)

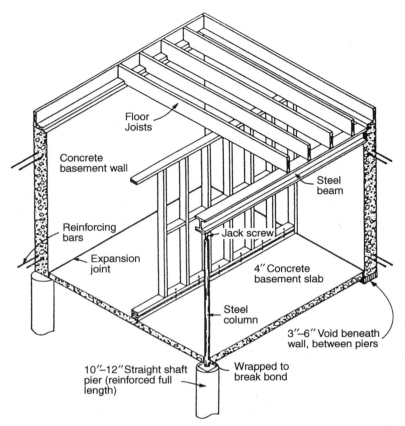

FIGURE 9.13 Typical detail of the grade beam and pier system. (*From Chen, 1988; copyright Elsevier Scientific Publishers, Netherlands.*)

The resisting force T_r for straight concrete piers can be calculated as follows:

$$T_r = P + c_A 2\pi R Z_{na} \tag{9.7}$$

where P = dead load of the pier, which includes both the weight of the pier and the dead load applied to the top of the pier, and Z_{na} = portion of the pier below the active zone. Note that the total length of the pier = $Z_a + Z_{na}$. If the ends of the piers are belled, then there would be additional resisting forces. By equating Eqs. (9.6) and (9.7), the value of Z_{na} can be obtained. By multiplying Z_{na} by an appropriate factor of safety (at least 1.5), the total depth of the pier can be determined.

As an alternative to the above calculations, it has been stated that the depth of piles or piers should be 1.5 times the depth where the swelling pressure is equal to the overburden pressure (David and Komornik, 1980).

2. *Pier diameter.* In terms of pier diameter, Chen (1988) states that to exert enough dead load pressure on the piers, it is necessary to use small-diameter piers in combination with long spans of the grade beams. Piers for expansive soil typically have a diameter of 0.3 m (1 ft).

3. *Pier reinforcement.* Because a large uplift force T_u can be developed on the pier as the clay swells in the active zone, steel reinforcement is required to prevent the piers from failing in tension. Reinforcement of the full length of the pier is essential to avoid tensile failures.

4. *Construction process.* It is important to use proper construction procedures when constructing the piers. For example, because the piers are often heavily loaded, the bottom of the piers must be cleaned of all loose debris or slough. After the pier hole is drilled and the concrete has been placed, excess concrete is often not removed from the top of the pier. This excess concrete has a mushroom shape, and because of its large area, the expansive soil can exert a substantial and unanticipated uplift force onto the top of the pier (Chen, 1988).

5. *Void space below grade beams.* A common procedure is to use a void-forming material (such as cardboard) to create a void space below the grade beam. It is important to create a complete and open void below the grade beam so that when the soil expands, it will not come in contact with the grade beam and cause damaging uplift forces.

9.4.4 Other Treatment Alternatives

Besides the construction of special foundations to resist expansive soils, there are other treatment alternatives as indicated in Table 9.3 (from Nelson and Miller, 1992). Each of these methods has advantages and disadvantages as indicated in Table 9.3.

Another treatment alternative is commonly known as *compaction control* (Gromko, 1974). This process is based on the observation that by compacting a clay at a water content that is wet of the optimum moisture content, the initial percent swell will be less than for the same soil compacted dry of optimum (Holtz and Gibbs, 1956b). Thus by compacting a clay wet of optimum, the swell potential of the soil will be reduced. However, if the clay should dry out prior to placement of the structure, then the beneficial effect of compacting the clay wet of optimum will be destroyed, causing the soil to become significantly more expansive (Day, 1994g).

Certain foundation types should not be used for expansive soil. For example, a common type of foundation in southern California is the raised wood floor foundation. This type of foundation consists of perimeter wall footings and interior pad footings that support a raised wood floor. This type of foundation creates a crawl space below the wood floor, such as shown in Fig. 9.14. This foundation type tends to be easily damaged by expansive soil, as shown in Fig. 9.14, where expansive soil has uplifted one of the interior pad footings. Note the wet condition of the soil in the crawl space.

9.5 PAVEMENTS AND FLATWORK

Pavements

1. *Flexible pavements.* Flexible pavements can be damaged by heave of compacted clay subgrade. This is often caused by water seeping through the joints in the concrete gutters or cracks in the asphalt concrete pavement. The water then percolates downward through the base and is absorbed by the compacted clay subgrade, resulting is expansion of the clay. The asphalt pavement will have more heave relative to the curbs resulting in cracks parallel to the curbs and gutters. Once cracked, the asphalt pavement allows for

TABLE 9.3 Expansive Soil Treatment Alternatives

Method (1)	Salient points (2)
Removal and replacement	Nonexpansive, impermeable fill must be available and economical. Nonexpansive soils can be compacted at higher densities than expansive clay, producing high bearing capacities. If granular fill is used, special precautions must be taken to control drainage away from the fill, so water does not collect in the pervious material. Replacement can provide safe slab-on-grade construction. Expansive material may be subexcavated to a reasonable depth, then protected by a vertical and/or a horizontal membrane. Sprayed asphalt membranes are effectively used in highway construction.
Remolding and compaction	Beneficial for soils having low potential for expansion, high dry density, low natural water content, and in a fractured condition. Soils having a high swell potential may be treated with hydrated lime, thoroughly broken up, and compacted—if they are lime reactive. If lime is not used, the bearing capacity of the remolded soil is usually lower since the soil is generally compacted wet of optimum at a moderate density. Quality control is essential. If the active zone is deep, drainage control is especially important. The specific moisture-density conditions should be maintained until construction begins and checked prior to construction.
Surcharge loading	If swell pressures are low and some deformation can be tolerated, a surcharge load may be effective. A program of soil testing is necessary to determine the depth of the active zone and the maximum swell pressures to be counteracted. Drainage control is important when using a surcharge. Moisture migration can be both vertical and horizontal.
Prewetting	Time periods up to as long as a year or more may be necessary to increase moisture contents in the active zone. Vertical sand drains drilled in a grid pattern can decrease the wetting time. Highly fissured, desiccated soils respond more favorably to prewetting. Moisture contents should be increased to at least 2–3% above the plastic limit. Surfactants may increase the percolation rate. The time needed to produce the expected swelling may be significantly longer than the time to increase moisture contents. It is almost impossible to adequately prewet dense unfissured clays. Excess water left in the upper soil can cause swelling in deeper layers at a later date. Economics of prewetting can compare favorably to other methods, but funds must be available at an early date in the project. Lime treatment of the surface soil following prewetting can provide a working table for equipment and increase soil strength. Without lime treatment soil strength can be significantly reduced, and the wet surface may make equipment operation difficult. The surface should be protected against evaporation and surface slaking. Quality control improves performance.

TABLE 9.3 Expansive Soil Treatment Alternatives (Continued)

Method (1)	Salient points (2)
Lime treatment	Sustained temperatures over 70°F for a minimum of 10 to 14 days is necessary for the soil to gain strength. Higher temperatures over a longer time produce higher strength gains. Organics, sulfates, and some iron compounds retard the pozzolanic reaction of lime. Gypsum and ammonium fertilizers can increase the soil's lime requirements. Calcareous and alkaline soil have good reactivity. Poorly drained soils have higher reactivities than well-drained soils. Usually 2–10% lime stabilizes reactive soil. Soil should be tested for lime reactivity and percentage of lime needed. The mixing depth is usually limited to 12–18 in., but large tractors with ripper blades have successfully allowed in-place mixing of 2 ft of soil. Lime can be applied dry or in a slurry, but excess water must be present. Some delay between application and final mixing improves workability and compaction. Quality control is especially important during pulverization, mixing, and compaction. Lime-treated soils should be protected from surface and groundwater. The lime can be leached out and the soil can lose strength if saturated. Dispersion of the lime from drill holes is generally ineffective unless the soil has an extensive network of fissures. Stress relief from drill holes may be a factor in reducing heave. Smaller-diameter drill holes provide less surface area to contact the slurry. Penetration of pressure-injected lime is limited by the slow diffusion rate of the lime, the amount of fracturing in the soil, and the small pore size of clay. Pressure injection of lime may be useful to treat layers deeper than possible with the mixed-in-place technique.
Cement treatment	Portland cement (4–6%) reduces the potential for volume change. Results are similar to lime, but shrinkage may be less with cement. Method of application is similar to mix-in-place lime treatment, but there is significantly less time delay between application and final placement. Portland cement may not be as effective as lime in treating highly plastic clays. Portland cement may be more effective in treating soils that are not lime-reactive. Higher strength gains may result with cement. Cement-stabilized material may be prone to cracking and should be evaluated prior to use.
Salt treatment	There is no evidence that use of salts other than NaCl or $CaCl_2$ is economically justified. Salts may be leached easily. Lack of permanence of treatment may make salt treatment uneconomical. The relative humidity must be a least 30% before $CaCl_2$ can be used. Calcium and sodium chloride can reduce frost heave by lowering the freezing point of water. $CaCl_2$ may be useful to stabilize soils having a high sulfur content.
Fly ash	Fly ash can increase the pozzolanic reaction of silty soils. The gradation of granular soils can be improved.

TABLE 9.3 Expansive Soil Treatment Alternatives (Continued)

Method (1)	Salient points (2)
Organic compounds	Spraying and injection are not very effective because of the low rate of diffusion in expansive soil. Many compounds are not water-soluble and react quickly and irreversibly. Organic compounds do not appear to be more effective than lime. None is as economical and effective as lime.
Horizontal barriers	Barrier should extend far enough from the roadway or foundation to prevent horizontal moisture movement into the foundation soils. Extreme care should be taken to securely attach barrier to foundation, seal the joints, and slope the barrier down and away from the structure. Barrier material must be durable and nondegradable. Seams and joints attaching the membrane to a structure should be carefully secured and made waterproof. Shrubbery and large plants should be planted well away from the barrier. Adequate slope should be provided to direct surface drainage away from the edges of the membranes.
Asphalt	When used in highway construction, a continuous membrane should be placed over subgrade and ditches. Remedial repair may be less complex than for concrete pavement. Strength of pavement is improved over untreated granular base. Can be effective when used in slab-on-grade construction.
Rigid barrier	Concrete sidewalks should be reinforced. A flexible joint should connect sidewalk and foundation. Barriers should be regularly inspected to check for cracks and leaks.
Vertical barrier	Placement should extend as deep as possible, but equipment limitations often restrict the depth. A minimum of half of the active zone should be used. Backfill material in the trench should be impervious. Types of barriers that have provided control of moisture content are capillary barrier (coarse limestone), lean concrete, asphalt and ground-up tires, polyethylene, and semihardening slurries. A trenching machine is more effective than a backhoe for digging the trench.
Membrane-encapsulated soil layers	Joints must be carefully sealed. Barrier material must be durable to withstand placement. Placement of the first layer of soil over the bottom barrier must be controlled to prevent barrier damage.

Source: Nelson and Miller (1992).

further infiltration of water, accelerating the heave process. There can be a variety of heave and crack patterns in pavements on expansive clay subgrade (Van der Merwe and Ahronovitz, 1973; Day, 1995c).

Another effect of moisture infiltration into the compacted clay subgrade is a reduction in the undrained shear strength of the soil. Moisture causes a softening of the compacted clay. The process involves an increase in moisture content and a reduction in negative pore pressures, which result in a decrease in the undrained shear strength. This makes the subgrade weaker and more susceptible to pavement deterioration, such as alligator cracking and rutting.

FIGURE 9.14 Expansive soil damage of a raised wood floor foundation.

Equation (8.12) does not have a factor to account for the expansiveness of the subgrade. For example, it is not unusual for a medium expansive clay subgrade and a very highly expansive clay subgrade to both have an R-value of less than 10. As a result, for both the medium and very highly expansive clay subgrade, the thickness of asphalt concrete and base material are similar, given the same traffic index. But because the percent swell is much greater for the very highly expansive soil (see Table 9.1), this pavement will have more uplift and cracking as moisture infiltrates into the clay subgrade.

In some cases, project specifications require compaction of the subgrade to a high density, for example, requiring the subgrade to be compacted to a minimum of 95 percent relative compaction (Modified Proctor). This can lead to problems, especially if the clays are wet, because it may be difficult to dry out and effectively compact the clays to a high density. Also, once the clays are dried out and densified, there is a much greater potential for additional expansion when water infiltrates the base. A better solution would be to keep the clay subgrade in a wet condition and increase the base thickness to compensate for the lower undrained shear strength of the subgrade.

2. *Concrete pavements.* The same expansive soil mechanisms that affect flexible pavement can also impact concrete pavements. The damage caused by expansive soil may be more severe for concrete pavements than asphalt pavements, because they are usually more brittle. For example, Fig. 9.15 shows two views of concrete pavement damage due to expansive soil.

3. *Mitigation measures.* Methods to mitigate the effects of clay expansion include compacting the clay subgrade wet of optimum or adding lime or cement to the clay subgrade (see Table 9.3). When compacting the clay subgrade wet of optimum, a geofabric can be used to prevent the aggregate base from being pushed down into and contaminated by the soft clay subgrade.

FIGURE 9.15 Two views of concrete pavement damage due to expansive soil.

Contractors often believe that for construction on expansive soil, the subgrade should be hard and dense. Thus they allow the clay to dry out and then fill in the desiccation cracks and tamp the soil to create a firm, level subgrade. This then creates a solid and rock-like subgrade of dense clay having a low moisture content. This method of subgrade preparation is probably the worst possible method for expansive soil. Once the flatwork is constructed, moisture inevitably accumulates beneath the flatwork, with resulting heave and extensive damage.

As previously mentioned, the best approach is to get the subgrade in a wet condition prior to construction, so that when water does seep into the base after the construction is complete, there will not be damaging expansion of the clay subgrade. For construction atop expansive soil, asphalt paving is usually preferred because it is more flexible and can be more easily repaired than concrete.

One construction method for concrete pavements is to use open-graded gravel as base material over the subgrade material. Extra base can be added to compensate for some gravel being pushed into the clay subgrade or a permeable geofabric can be placed atop the clay subgrade. Once the base (open-graded gravel) is in place, the gravel is flooded with water. The water will seep down through the gravel and be absorbed by the underlying clay subgrade. Cylindrical holes can be excavated into the subgrade to facilitate the wetting of the subgrade. Periodically, soil samples can be taken from the subgrade to check the moisture condition and swell of the subgrade. After the subgrade has been sufficiently moistened, the pavement section can be constructed. Usually extra steel reinforcement and an increased thickness of concrete pavement are needed to compensate for the poor bearing conditions of the subgrade. Figures 9.16 and 9.17 show this construction process for a concrete cul-de-sac. In Fig. 9.16, the open-graded gravel is in place, the subgrade has been wetted, steel rebars (reinforcing bars atop concrete chairs) are in place, and some segments of the concrete have been placed. Note that individual segments were placed in order to create concrete joints. Figure 9.17 shows the placement of the concrete with the reinforcement bars positioned about mid-height in the concrete section.

Flatwork

1. *Upward movement.* Flatwork can be defined as appurtenant structures that surround a building or house such as concrete walkways, patios, driveways, and pool decks. It is the lightly loaded structures, such as pavements or lightly loaded foundations, that are commonly damaged by expansive soil. Because flatwork usually only supports its own weight, it can be especially susceptible to expansive–soil related damage such as shown in Fig. 9.18. The arrows in Fig. 9.18 indicate the amount of uplift of the lightly loaded exterior sidewalk relative to the heavily loaded exterior wall of the tilt-up building. The sidewalk heaved to such an extent that the door could not be opened and the door had to be rehung so that it opened into the building.

Another example is Fig. 9.19, which shows cracking to a concrete driveway due to expansive soil uplift. The expansive soil uplift tends to produce a distinct crack pattern, which has been termed a *spider* or *x-type* crack pattern.

Besides the concrete itself, utilities can also be damaged due to the upward movement of the flatwork. For example, Fig. 9.20 shows a photograph of concrete flatwork that was uplifted due to expansive soil heave. The upward movement of the flatwork bent the utility line and chipped off the stucco as shown in Fig. 9.20.

2. *Walking.* Besides differential movement of the flatwork, there can also be progressive movement of flatwork away from the structure. This lateral movement is known as "walking."

9.30 ANALYSIS OF GEOTECHNICAL DATA AND ENGINEERING COMPUTATIONS

FIGURE 9.16 Construction of concrete cul-de-sac on highly expansive clay.

FIGURE 9.17 Placement of concrete for cul-de-sac shown in Fig. 9.16. (Note steel reinforcement is near center of fresh concrete.)

EXPANSIVE SOIL 9.31

FIGURE 9.18 Sidewalk uplift due to expansive soil.

FIGURE 9.19 Driveway cracking due to expansive soil.

FIGURE 9.20 Concrete flatwork uplifted by expansive soil.

Figure 9.21 shows an example of walking of flatwork where a backyard concrete patio slab was built atop highly expansive soils. The patio slab originally abutted the wall of the house, but is now separated from the house by about 4 cm (1.5 in.). At one time, the gap was filled in with concrete, but as Fig. 9.21 shows, the patio slab has continued to walk away from the house. In many cases, there will be appurtenant structures (such as patio shade cover) that are both attached to the house and also derive support from the flatwork. As the flatwork walks away from the house, these appurtenant structures are pulled laterally and frequently damaged.

The results of a field experiment indicated that most walking occurs during the wet period (Day, 1992c). The expansion of the clay causes the flatwork to move up and away from the structure. During the dry period, the flatwork does not return to its original position. Then during the next wet period, the expansion of the clay again causes an upward and outward movement of the flatwork. The cycles of wetting and drying cause a progressive movement of flatwork away from the building. An important factor in the amount of walking is the moisture condition of the clay prior to construction of the flatwork. If the clay is dry, then more initial upward and outward movement can occur during the first wet cycle.

9.6 CONSTRUCTION ON EXPANSIVE ROCK

There are several different mechanisms that can cause the expansion of rock. Some rock types, such as shale, slate, mudstone, siltstone, and claystone, can be especially susceptible to expansion. Some common mechanisms that can cause rock to expand are as follows:

1. *Rebound.*   For cut areas, where the overburden has been removed by erosion or by mass-grading operations, the rock will rebound because of the release in overburden pressure. The rebound can cause the opening of cracks and joints. Usually rebound of

rock occurs during the rock excavation, and, because it is a relatively rapid process, it is not included in the engineering analyses.

2. *Expansion due to physical factors.* Rock, especially soft sedimentary and fractured rock, can expand because of the physical growth of plant roots or the freezing of water in the fractures. Studies have also shown that the precipitation of gypsum in rock pores, cracks, and joints can cause rock expansion and disintegration. Such conditions occur in arid climates where subsurface moisture evaporates at ground surface, precipitating the minerals in the rock pores. Gypsum crystals have been observed to grow in rock fractures and are believed to exert the most force at their growing end (Hawkins and Pinches, 1987). Gypsum growth has even been observed in massive sandstone, and resulted in significant heave of the rock (Hollingsworth and Grover, 1992).

3. *Expansion due to weathering.* Probably the most frequent cause of heave of rock and resulting damage to foundations is weathering. Weathering of rock can occur by physical and chemical methods. Typical types of chemical weathering include oxidation, hydration of clay minerals, and the chemical alteration of the silt-size particles to clay. Factors affecting oxidation include the presence of moisture and oxygen (aerobic conditions), biological activity, acidic environment, and temperature (Hollingsworth and Grover, 1992). As indicated in Table 1.1, pyritic shale often expands on exposure to air and moisture. Another example is bentonite, which is a rock that is composed of montmorillonite clay minerals. This type of rock will also rapidly weather and greatly expand when exposed to air and moisture.

It is often difficult to predict the amount of heave that will occur because of expansive rock. One approach is similar to the testing of clay specimens, where a rock specimen is placed in an oedometer, subjected to the anticipated overburden pressure, submerged in distilled water, and then the expansion of the rock is measured (ASTM D 4546-96, 1998). However, this approach could considerably underestimate the expansion potential of the rock because in an intact (unweathered) state, it is much less expansive than in a fractured and weathered state.

FIGURE 9.21 Walking of flatwork on expansive soil.

Another approach is to break apart the rock and then subject the fragments to wetting and drying cycles. The cycles of wetting and drying are often very effective in rapidly weathering expansive rock (Day, 1994e). Once the rock fragments have sufficiently weathered, an expansion index test (see Sec. 9.1.2) could be performed on this material. On the basis of the Expansion Index Test results, the type of foundation could then be selected (by using Table 9.2, for example).

PROBLEMS

The problems have been divided into basic categories as indicated below:

Expansive Soil Potential

1. A clay has a liquid limit = 80, a plastic limit = 20, and on a dry mass basis, 60 percent of the soil particles are finer than 0.002 mm. Using Fig. 9.1, determine the expansion potential of this clay. *Answer:* Very high expansion potential.
2. For Prob. 1, determine the expansion potential of the clay using Table 9.1. *Answer:* On the basis of the plasticity index, the clay has a very high expansion potential.
3. A clayey gravel has 33 percent particles (dry mass basis) passing the No. 4 sieve, and the expansion index (EI) for the soil passing the No. 4 sieve is 135. What is the estimated expansion potential of this clayey gravel? *Answer:* Low expansion potential.

Rate of Swell

4. Assume the upper 2 m of a clay deposit has the swell behavior shown in Fig. 9.2. Assume the ground surface is deliberately flooded. How long will it take for 90 percent of primary swell to occur? *Answer:* 1.2 years.
5. For Prob. 4, assume that in addition to the ground surface being flooded, the groundwater table rises to a depth of 2 m below ground surface. How long will it take for 90 percent of primary swell to occur? *Answer:* 0.3 year.
6. Using the data shown in Fig. 9.2, determine the secondary swell ratio $C_{\alpha s}$. *Answer:* The equation used to calculate the secondary compression ratio (Sec. 7.4.3) can be used to calculate $C_{\alpha s} = 0.003$.

Computation of Foundation Heave

7. Use the data shown in Fig. 9.11. If the allowable foundation heave = 0.9 in., what depth of undercut and replacement with nonexpansive soil would be required? *Answer:* Depth of undercut and replacement = 2 ft.
8. Use the data shown in Fig. 9.11, except assume that the swell tests performed on undisturbed soil specimens indicate the following:

Depth below ground surface of soil specimen, ft	Percent swell
2	4
4	3
6	2
8	1
10	0

Using the above data, calculate the total heave of the foundation. *Answer:* Total heave = 2.4 in.

9. Use the data shown in Fig. 9.12. Assume that the swell pressure (P'_s) = 100 kPa. Determine the total heave of the foundation, assuming the foundation is constructed at ground surface and its weight is neglected. *Answer:* Total heave of the foundation = 84 mm.

10. Use the data shown in Fig. 9.12. Assume that the swell index (C_s) = 0.20. Determine the total heave of the foundation assuming the foundation is constructed at ground surface and its weight is neglected. *Answer:* Total heave of the foundation = 228 mm.

11. Use the data shown in Fig. 9.12. Assume that the final condition (P_f) will be a groundwater table at a depth of 0.5 m with hydrostatic pore water pressures below the groundwater table and zero pore water pressures above the groundwater table. Also assume the total unit weight γ_t of 18 kN/m³ can be used for the clay above and below the groundwater table. Determine the total heave of the foundation, assuming the foundation is constructed at ground surface and its weight is neglected. *Answer:* Total heave of the foundation = 126 mm.

12. Use the data from Prob. 11. Also assume that there is a mat foundation with the bottom of the mat located at a depth of 0.5 m and that the mat foundation exerts a vertical pressure = 25 kPa on the soil at this level. If the mat foundation is large enough so that one-dimensional conditions exist beneath the center of the mat foundation, determine the total heave of the center of the mat foundation. *Answer:* Total heave of the center of the mat foundation = 61 mm.

13. Use the data from Prob. 11. Also assume that there is a square footing (1.2 m by 1.2 m) with the bottom of the square footing located at a depth of 0.5 m and that the square footing exerts a vertical pressure = 50 kPa on the soil at this level. Using the 2:1 approximation, determine the total heave of the square footing. *Answer:* Total heave of the square footing = 59 mm.

14. Solve Prob. 13, but assume that an undisturbed soil specimen was obtained from layer 3 (see Fig. 9.12) and testing in the oedometer apparatus indicates a swell pressure P'_s of 100 kPa. Using the 2:1 approximation, determine the total heave of the square footing. *Answer:* Total heave of the square footing = 44 mm.

Pier and Grade Beam Support

15. Use the data shown in Fig. 9.12. Assume the 2 m of swelling clay has an undrained shear strength s_u of 50 kPa. It is proposed to construct a concrete pier and grade beam foundation such that the bottom of the grade beams is located at a depth of 0.4 m below ground surface. The piers will have a diameter of 0.3 m. Also assume that there is unweathered shale [adhesion value (c_A) = 80 kPa] below the 2 m of swelling clay. If each pier supports a dead load (including the weight of the pier) of 20 kN, determine the depth of the piers below ground surface (include a factor of safety = 2.0). Also determine the thickness of the air gap that should be provided beneath the grade beams (include a factor of safety = 1.5). *Answers:* Depth of piers below ground surface = 3.0 m, thickness of air gap = 110 mm.

CHAPTER 10
SLOPE STABILITY

The following notation is used in this chapter:

SYMBOL	DEFINITION
$a, \Delta x$	Width of a slice
c	Cohesion based on a total stress analysis
c'	Cohesion based on an effective stress analysis
d, D	Depth of the slip surface
F	Factor of safety for slope stability
h	Depth below the ground surface (for calculation of r_u)
L	Length of the slip surface
N	Normal force on the slip surface
N'	Effective normal force on the slip surface
r_u	Pore water pressure ratio
T	Shear force along the slip surface
u	Pore water pressure
U	Pore water force acting on the slip surface
W	Total weight of the failure wedge or failure slice
α	Slope inclination
β	Angular distortion as defined by Boscardin and Cording (1989)
ε_h	Horizontal strain of the foundation
ϕ	Friction angle based on a total stress analysis
ϕ'	Friction angle based on an effective stress analysis
γ_b	Buoyant unit weight of saturated soil below the groundwater table
γ_t	Wet unit weight of soil
γ_w	Unit weight of water
σ'_n	Effective normal stress on the slip surface
τ	Shear stress along the slip surface
τ_f	Shear strength of the soil

10.1 TYPICAL TYPES OF SLOPE MOVEMENT

Slope movement can be divided into six basic categories, as described below:

 1. *Rockfalls or topples.* This is usually an extremely rapid movement that includes the free fall of rocks, movement of rocks by leaps and bounds down the slope face, and/or

the rolling of rocks or fragments of rocks down the slope face (Varnes, 1978). A rock topple is similar to a rockfall, except that there is a turning moment about the center of gravity of the rock which results in an initial rotational-type movement and detachment from the slope face. Rockfalls are discussed in Sec. 10.3.

   2. *Surficial slope stability.* Surficial slope instability involves shear displacement along a distinct failure (or slip) surface. As the name implies, surficial slope stability analyses deal with the outer face of the slope, generally up to 1.2 m (4 ft) deep. The typical surficial slope stability analysis assumes the failure surface is parallel to the slope face. Surficial slope stability analysis is discussed in Sec. 10.4.

   3. *Gross slope stability.* As contrasted with a shallow (surficial) stability analysis, gross slope stability involves an analysis of the entire slope. In the stability analysis of slopes, curved or circular slip surfaces are often assumed. Terms such as *fill slope stability analysis* and *earth* or *rock slump analysis* have been used to identify similar processes. Gross slope stability is discussed in Sec. 10.5.

   4. *Landslides.* Gross slope stability could be referred to as *landslide analysis.* However, landslides in some cases may be so large that they involve several different slopes. Landslides are discussed in Sec. 10.6.

   5. *Debris flow.* Debris flow is commonly defined as soil with entrained water and air that moves readily as a fluid on low slopes. Debris flow can include a wide variety of soil-particle sizes (including boulders) as well as logs, branches, tires, and automobiles. Other terms, such as mud flow, debris slide, mud slide, and earth flow, have been used to identify similar processes. While categorizing flows based on rate of movement or the percentage of clay particles may be important, the mechanisms of all these flows are essentially the same (Johnson and Rodine, 1984). Debris flow is discussed in Sec. 10.7.

   6. *Creep.* Creep is generally defined as an imperceptibly slow and more or less continuous downward and outward movement of slope-forming soil or rock (Stokes and Varnes, 1955). Creep can effect both the near-surface (surficial) soil or deep-seated (gross) materials. The process of creep is frequently described as viscous shear that produces permanent deformations, but not failure as in landslide movement. Creep is discussed in Sec. 10.8.

   Table 10.1 presents a checklist for slope stability and landslide analysis (adapted from Sowers and Royster, 1978). This table provides a comprehensive list of the factors that may need to be considered by the geotechnical engineer when designing slopes.

   An important consideration for the design of dams is slope stability and therefore dams have been included in the chapter (Sec. 10.9).

10.2 ALLOWABLE LATERAL MOVEMENT

Compared to the settlement of buildings, less published work is available on the allowable lateral movement. To evaluate the lateral movement of buildings, a useful parameter is the horizontal strain ε_h, defined as the change in length divided by the original length of the foundation. Figure 10.1 shows a correlation between horizontal strain ε_h and severity of damage (Boone, 1996; originally from Boscardin and Cording, 1989). Assuming a 6-m- (20-ft-) wide zone of the foundation subjected to lateral movement, Fig. 10.1 indicates that a building can be damaged by as little as 3 mm (0.1 in.) of lateral movement. Figure 10.1 also indicates that a lateral movement of 25 mm (1 in.) would cause "severe" to "very severe" building damage.

   It should be mentioned that in Fig. 10.1, Boscardin and Cording (1989) used a *distortion factor* in their calculation of angular distortion β for foundations subjected to settlement

TABLE 10.1 Checklist for the Study of Slope Stability and Landslides

Main topic (1)	Relevant items (2)
Topography	Contour map, consider land form and anomalous patterns (jumbled, scarps, bulges). Surface drainage, evaluate conditions such as continuous or intermittent drainage. Profiles of slope, to be evaluated along with geology and the contour map. Topographic changes, such as the rate of change by time and correlate with groundwater, weather, and vibrations.
Geology	Formations at site, consider the sequence of formations, colluvium (bedrock contact and residual soil), formations with bad experience, and rock minerals susceptible to alteration. Structure: evaluate three-dimensional geometry, stratification, folding, strike and dip of bedding or foliation (changes in strike and dip and relation to slope and slide), and strike and dip of joints with relation to slope. Also investigate faults, breccia, and shear zones with relation to slope and slide. Weathering, consider the character (chemical, mechanical, and solution) and depth (uniform or variable).
Groundwater	Piezometric levels within slope, such as normal, perched levels, or artesian pressures with relation to formations and structure. Variations in piezometric levels due to weather, vibration, and history of slope changes. Other factors include response to rainfall, seasonal fluctuations, year-to-year changes, and effect of snowmelt. Ground surface indication of subsurface water, such as springs, seeps, damp areas, and vegetation differences. Effect of human activity on groundwater, such as groundwater utilization, groundwater flow restriction, impoundment, additions to groundwater, changes in ground cover, infiltration opportunity, and surface water changes. Groundwater chemistry, such as dissolved salts and gases and changes in radioactive gases.
Weather	Precipitation from rain or snow. Also consider hourly, daily, monthly, or annual rates. Temperature, such as hourly and daily means or extremes, cumulative degree-day deficit (freezing index), and sudden thaws. Barometric changes.
Vibration	Seismicity, such as seismic events, microseismic intensity, and microseismic changes. Human induced from blasting, heavy machinery, or transportation (trucks, trains, etc.).
History of Slope Changes	Natural processes, such as long-term geologic changes, erosion, evidence of past movement, submergence, or emergence. Human activities, including cutting, filling, clearing, excavation, cultivation, paving, flooding, and sudden drawdown of reservoirs. Also consider changes caused by human activities, such as changes in surface water, groundwater, and vegetation cover. Rate of movement from visual accounts, evidence in vegetation, evidence in topography, or photographs (oblique, aerial, stereoptical data, and spectral changes). Also consider instrumental data, such as vertical changes, horizontal changes, and internal strains and tilt, including time history. Correlate movements with groundwater, weather, vibration, and human activity.

Source: Adapted from Sowers and Royster, 1978. Reprinted with permission from *Landslides: Analysis and Control,* Special Report 176. Copyright 1978 by the National Academy of Sciences. Courtesy of the National Academy Press, Washington, D.C.

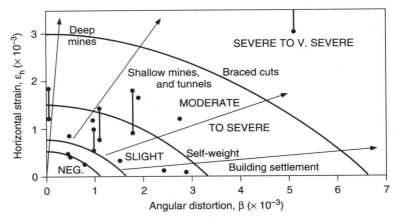

FIGURE 10.1 Relationship of damage to angular distortion and horizontal extension strain. (*From Boscardin and Cording, 1989; reprinted with permission from the American Society of Civil Engineers.*)

from mines, tunnels, and braced cuts. Because of this distortion factor, the angular distortion β by Boscardin and Cording (1989) in Fig. 10.1 is different from the definition (δ/L) as used in Chap. 7.

The ability of the foundation to resist lateral movement will depend on the tensile strength of the foundation. Those foundations that can not resist the tensile forces imposed by lateral movement will be the most severely damaged. For example, Figs. 10.2 and 10.3 show damage to a tilt-up building. For a tilt-up building, the exterior walls are cast in segments upon the concrete floor slab, and then, once they gain sufficient strength, they are tilted up into position. The severe damage shown in Figs. 10.2 and 10.3 was caused by slope movement, which affected the tilt-up building because it was constructed near the top of the slope. Figure 10.2 shows lateral separation of the concrete floor slab at the location of a floor joint. Figure 10.3 shows separation at the junction of two tilt-up panels. Because of the presence of joints between tilt-up panels and joints in the concrete floor slab, the building was especially susceptible to slope movement, which literally pulled apart the tilt-up building.

Those foundations that have joints or planes of weakness, such as the tilt-up building shown in Figs. 10.2 and 10.3, will be most susceptible to damage from lateral movement. Buildings having a mat foundation or a posttensioned slab would be less susceptible to damage because of the high tensile resistance of these foundations.

10.3 ROCKFALL

Definition. A rockfall is defined as a relatively free-falling rock or rocks that have detached themselves from a cliff, steep slope, cave, arch, or tunnel (Stokes and Varnes, 1955). The movement may be by the process of a vertical fall, by a series of bounces, or by rolling down the slope face. The free-fall nature of the rocks and the lack of movement along a well-defined slip surface differentiate a rockfall from a rockslide.

Basic Factors Governing a Rockfall. A rock slope or the rock exposed in a tunnel is characterized by a heterogeneous and discontinuous medium of solid rocks that are separated by discontinuities. The rocks composing a rockfall tend to detach themselves from these preexisting discontinuities in the slope or tunnel walls. The sizes of the individual rocks in a rockfall are governed by the attitude, geometry, and spatial distribution of the rock discontinuities. The basic factors governing the potential for a rockfall include (Piteau and Peckover, 1978):

1. The geometry of the slope or tunnel.
2. The system of joints and other discontinuities and the relation of these systems to possible failure surfaces.
3. The shear strength of the joints and discontinuities.
4. Destabilizing forces such as water pressure in the joints, freezing water, or vibrations.

Rockfall Examples. A rockfall can cause significant damage to structures in its path. For example, Fig. 10.4 shows a large rock that detached itself from the slope and landed on the roof of a house. The house is a one-story structure, having typical wood frame construction, and a stucco exterior. Figure 10.5 shows the location where the rock detached itself from the slope. The rock that impacted the house was part of the Santiago Peak volcanics, which consists of an elongated belt of mildly metamorphosed volcanic rock. The rock is predominately dacite and andesite and can be classified as hard and extremely resistant to weathering and erosion (Kennedy, 1975). The part of the house below the area of impact of the rock was severely damaged, as shown in Figs. 10.6 and 10.7. The oven, refrigerator, and other items are turned over at an angle in Fig. 10.6. The hanging baskets in the upper left corner of Fig. 10.6 provide a vertical plane of reference. The damage shown in Figs. 10.6 and 10.7 was a result of the force of impact which crushed the roof and interior walls that were beneath the rock. There was also damage in the house at other locations away from the area of rock impact. For example, Fig. 10.8 shows interior wallboard cracking and

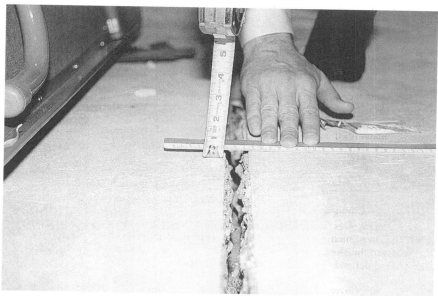

FIGURE 10.2 Damage due to lateral movement.

FIGURE 10.3 Joint separation between wall panels.

distortion of the door frames. Note that the door frames in Fig. 10.8 were originally rectangular, but are now highly distorted. Figure 10.9 shows how the roof literally opened up under the impact of the rock. Figure 10.10 shows damage to a roof beam in the house.

A second example of a rockfall is shown in Fig. 10.11. The top portion of the slope shown in Fig. 10.11 is unstable, and blocks of rock have toppled off and then tumbled down the slope face. As shown in Fig. 10.12, the major reason for the rockfall is a near vertical face at the top of slope. Contributing factors in the rockfall are adversely oriented joint planes. For example, Fig. 10.11 shows that one series of joint planes is near vertical. The Rose Canyon fault zone is only about 120 to 150 m (400 to 500 ft) from the site (Kennedy, 1975) and movement of this nearby fault may have contributed to the fracturing of the rock. Since the rockfall occurred in the winter rainy season, another contributing factor was probably water (from rainstorms) seeping along the joint planes. This water provided additional pressure and softening needed to dislodge the rocks.

Remedial Measures. The investigation and analysis of rockfall will often be jointly performed by the engineering geologist and geotechnical engineer. If the analysis indicates

FIGURE 10.4 Rockfall in Santiago Peak volcanics.

FIGURE 10.5 Location of rockfall.

that rockfalls are likely to impact the site, then remedial measures can be implemented. For slopes, the main measures to prevent damage from a rockfall are to alter the slope configuration, retain the rocks on the slope, intercept the falling rocks before they reach the structure, and direct the falling rocks around the structure. Altering the slope can include such measures as removing the unstable or potentially unstable rocks, flattening the slope, or incorporating benches into the slope (Piteau and Peckover, 1978). Measures to retain the rocks on the slope face include the use of anchoring systems (such as bolts, rods, or dowels), shotcrete applied to the rock slope face, and retaining walls. Intercepting or deflecting

10.8 ANALYSIS OF GEOTECHNICAL DATA AND ENGINEERING COMPUTATIONS

FIGURE 10.6 Damage to kitchen area.

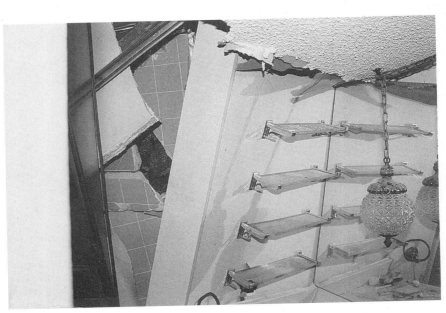

FIGURE 10.7 Damage to bathroom.

FIGURE 10.8 Damage to interior wallboard and distortion of door frames.

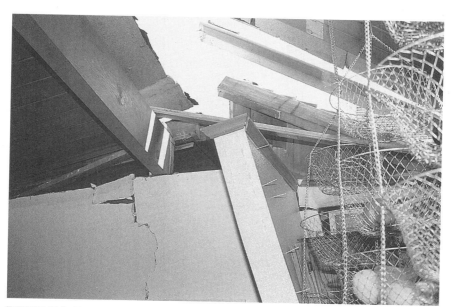

FIGURE 10.9 Damage to the roof.

10.10 ANALYSIS OF GEOTECHNICAL DATA AND ENGINEERING COMPUTATIONS

FIGURE 10.10 Damage to roof beam.

falling rocks around the structure can be accomplished by using toe-of-slope ditches, wire mesh catch fences, and catch walls (Peckover, 1975). Recommendations for the width and depth of toe-of-slope ditches have been presented by Ritchie (1963) and Piteau and Peckover (1978).

Rockfall in Tunnel Excavations. Of all engineering structures, tunnels are most vulnerable to small, troublesome, and large catastrophic failures (Feld and Carper, 1997). One reason is because the tunnel excavation allows for the sudden release of large confining pressures, resulting in rock strains that are not easily predicted. Another reason is the unpredictability of subsurface conditions, such as the possibility of encountering groundwater that can flood the tunnel. Because of the problems with tunnel excavations, there has been the development of tunnel coring or boring machines (for soft and hard rock) that can provide some protection during excavation of the heading. Other measures to help stabilize tunnel rock faces include anchoring systems (such as bolts, rods, and dowels) and shotcrete. Chemical grouts have also been used to fill in rock discontinuities and joints in order to stabilize the tunnel crown and walls. There are many references on tunnel excavation and construction details, such as Sandström, 1963a, 1963b; Szechy, 1973; Wahlstrom, 1973; Attewell et al., 1986; Sinha, 1989, 1991; and Mahtab and Grasso, 1992.

10.4 SURFICIAL SLOPE STABILITY

Definition and Failure Mechanism. Surficial failures of slopes are quite common throughout the United States (Day and Axten, 1989; Wu et al., 1993). In southern California, surficial failures usually occur during the winter rainy season, after a prolonged rainfall or during a heavy rainstorm, and are estimated to account for more than 95 percent of the problems associated with slope movement upon developed properties (Gill, 1967). Figure 10.13 illustrates a typical surficial slope failure. The surficial failure by definition is shallow with the failure surface usually at a depth of 1.2 m (4 ft) or less (Evans, 1972). In many cases, the failure surface is parallel to the slope face. The common surficial failure mechanism for clay slopes in southern California is as follows:

1. During the hot and dry summer period, the slope face can become desiccated and shrunken. The extent and depth of the shrinkage cracks depend on many factors, such as the temperature and humidity, the plasticity of the clay, and the extraction of moisture by plant roots.
2. When the winter rains occur, water percolates into the fissures causing the slope surface to swell and saturate with a corresponding reduction in shear strength. Initially, water percolates downward into the slope through desiccation cracks and in response to the suction pressures of the dried clay.

FIGURE 10.11 Location of rockfall (asterisk shows location of rockfall; arrow points to rockfall debris at toe of slope.).

10.12 ANALYSIS OF GEOTECHNICAL DATA AND ENGINEERING COMPUTATIONS

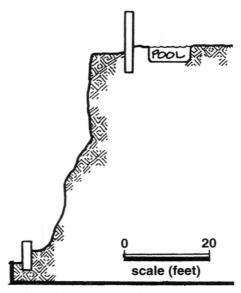

FIGURE 10.12 Cross section through slope.

3. As the outer face of the slope swells and saturates, the permeability parallel to the slope face increases. With continued rainfall, seepage develops parallel to the slope face.
4. Because of a reduction in shear strength due to saturation and swell coupled with the condition of seepage parallel to the slope face, failure occurs.

Surficial Stability Equation. The factor of safety is used to determine the stability of a slope. A factor of safety of 1.0 indicates a failure condition, while a factor of safety greater than 1.0 indicates that the slope is stable. The higher the factor of safety, the higher the stability of a slope.

To determine the factor of safety F for surficial stability, an effective stress slope stability analysis is performed and the following equation is used:

$$F = \frac{\text{shear strength of soil}}{\text{shear stress in the soil}} = \frac{c' + \sigma'_n \tan \phi'}{\tau} \tag{10.1}$$

where σ'_n = effective normal stress on the assumed failure surface and τ = shear stress in the soil along the assumed failure surface. Figure 10.14 (from Lambe and Whitman, 1969) shows the derivation of N' and T, assuming an infinite slope with seepage parallel to the slope face. Assuming that the failure surface will be at a depth = d, and that the width of the slip surface (C to D) is 1.0, then $a = \cos \alpha$ and $\sigma'_n = N'$ and $\tau = T$. Substituting these values into Eq. (10.1) gives

$$F = \frac{c' + \gamma_b d \cos^2 \alpha \tan \phi'}{\gamma_t d \cos \alpha \sin \alpha} \tag{10.2}$$

Since an effective stress analysis is being performed with steady-state flow conditions, effective shear strength soil parameters (ϕ' = effective friction angle; c' = effective cohesion) must be used in the analysis. In Eq. (10.2), α = slope inclination; γ_b = buoyant unit weight of the soil; and γ_t = total unit weight of the soil. As mentioned, the parameter d = depth of the failure surface where the factor of safety is computed. The factor of safety for surficial stability [(Eq. (10.2)] is highly dependent on the effective cohesion value of the soil. Automatically assuming $c' = 0$ for Eq. (10.2) may be overly conservative.

Because of the shallow nature of surficial failures, the normal stress on the failure surface is usually low. Studies have shown that the effective shear strength envelope for soil can be nonlinear at low effective stresses. For example, Fig. 10.15 (from Maksimovic, 1989b) shows the effective stress failure envelope for compacted London clay. Note the nonlinear nature of the shear strength envelope. For an effective stress of about 100 to 300 kPa (2100 to 6300 psf), the failure envelope is relatively linear and has effective shear strength parameters of $\phi' = 16°$ and $c' = 25$ kPa (520 psf). But below about 100 kPa (2100

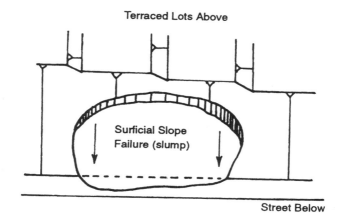

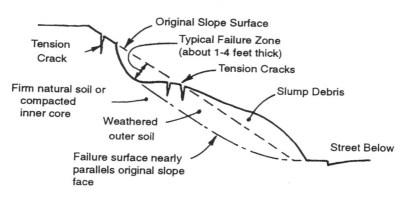

FIGURE 10.13 Illustration of typical surficial slope failure: (*a*) plan view; (*b*) cross-sectional view.

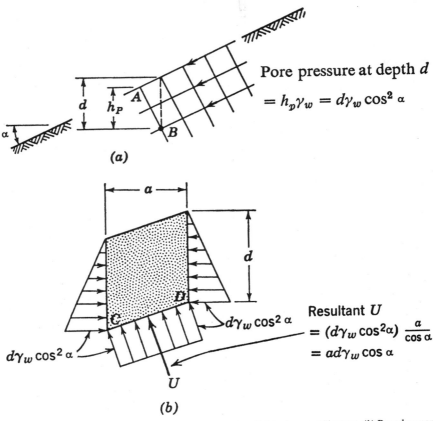

FIGURE 10.14 Analysis of infinite slope with seepage parallel to slope. (*a*) Flow net. (*b*) Boundary pore pressures. (*From Lambe and Whitman, 1969; reproduced with permission from John Wiley & Sons.*)

psf), the failure envelope is curved (see Fig. 10.15*b*) and the shear strength is less than the extrapolated line from high effective stresses. Because of the nonlinear nature of the shear strength envelope, the shear strength parameters (c' and ϕ') obtained at high normal stresses can overestimate the shear strength of the soil and should not be used in Eq. (10.2) (Day, 1994d). As an example, suppose a fill slope having an inclination of 1.5:1 ($\alpha = 33.7°$) was constructed using compacted London clay having the shear strength envelope shown in Fig. 10.15. Table 10.2 lists the parameters used for the calculations and shows that using the linear portion of the shear strength envelope (Fig. 10.15*a*), the factor of safety = 2.5; but for the actual nonlinear shear strength envelope (Fig. 10.15*b*), the factor of safety = 0.98, indicating a failure condition.

Surficial Failures

1. *Cut and natural slopes.* Figure 10.16 shows a picture of a surficial slope failure in a cut slope. The slope is located in Poway, California, and was created in 1991 by cutting down the hillside during the construction of the adjacent road. The cut slope has an area of

$$N' + U = \gamma_t a d \cos \alpha$$
$$\therefore \quad N' = \gamma_b a d \cos \alpha$$
$$T = \gamma_t a d \sin \alpha$$

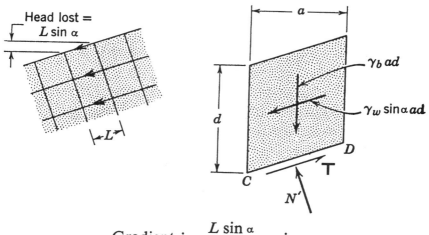

FIGURE 10.14 (*Continued*). (*c*) Analysis of force equilibrium with moments balanced by side forces. (*From Lambe and Whitman, 1969; reproduced with permission from John Wiley & Sons.*)

$$\text{Gradient } i = \frac{L \sin \alpha}{L} = \sin \alpha$$

Summing $\perp$ to CD:
$$N' = \gamma_b a d \cos \alpha$$
Summing $=$ to CD:
$$T = \gamma_b a d \sin \alpha + \gamma_w a d \sin \alpha = \gamma_t a d \sin \alpha$$

(d)

FIGURE 10.14 (*Continued*). (*d*) Alternative analysis using seepage forces and buoyant unit weight. (*From Lambe and Whitman, 1969; reproduced with permission from John Wiley & Sons.*)

10.16 ANALYSIS OF GEOTECHNICAL DATA AND ENGINEERING COMPUTATIONS

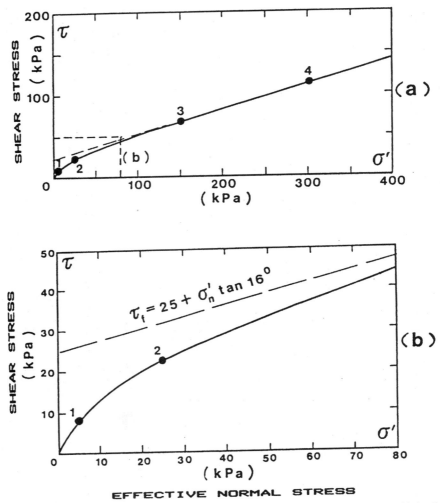

FIGURE 10.15 Failure envelope for compacted London clay. (*a*) Investigated stress range with detail; (*b*) low stress range. (*From Maksimovic, 1989b; reprinted with permission from the American Society of Civil Engineers.*)

about 400 m² (4000 ft²), a maximum height of 6 m (20 ft), and a slope inclination that varies from 1.5:1 (34°) to 1:1 (45°). These are rather steep slope inclinations, but are not uncommon for cut slopes in rock. The failure mechanism is a series of thin surficial failures, about 0.15 m (0.5 ft) thick. The depth to length ratio (*D/L*) of a single failure mass is around 5 to 6 percent. For this ratio, the slides shown in Fig. 10.16 fall within the classification range (3 to 6 percent) that Hansen (1984) has defined as "shallow surface slips."

The type of rock exposed in the cut slope is the Friars formation. This rock is of middle to late Eocene and has thick layers of nonmarine lagoonal sandstone and claystone

TABLE 10.2 Example of Surficial Stability Calculations

Shear strength (1)	Factor of safety (2)
Linear portion of shear strength envelope, $\phi' = 16°$ and $c' = 25$ kPa (520 psf); i.e., Fig. 10.15a	2.5
Nonlinear portion of shear strength envelope at low effective stress, i.e., Fig. 10.15b	0.98

Note: Both surficial stability analyses use Eq. (10.2), with slope inclination (α) = 33.7°, total unit weight (γ_t) = 19.8 kN/m³ (126 pcf), buoyant unit weight (γ_b) = 10.0 kN/m³ (63.6 pcf), and depth of seepage and failure plane (d) = 1.2 m (4 ft).

FIGURE 10.16 Surficial failure, cut slope for road.

10.18 ANALYSIS OF GEOTECHNICAL DATA AND ENGINEERING COMPUTATIONS

(Kennedy, 1975). The clay minerals are montmorillonite and kaolinite. The Friars formation is common in San Diego and Poway, California, and is a frequent source of geotechnical problems such as landslides and heave of foundations.

The cause of the surficial slope failures shown in Fig. 10.16 was weathering of the Friars formation. Weathering breaks down the rock and reduces the effective shear strength of the material. The weathering process also opens up fissures and cracks, which increases the permeability of the near-surface rock and promotes seepage of water parallel to the slope face. This weathering process is illustrated in Fig. 10.17 (Ortigao et al., 1997). Failure will eventually occur when the material has weathered to such a point that the effective cohesion approaches zero.

Surficial failures are most common for cut slopes in soft sedimentary rock, such as claystones or weakly cemented sandstones. As mentioned, the most common reason for the surficial failure is that a relatively steep slope (such as 1.5:1 or 1:1) is excavated into the sedimentary rock. Then, with time, as vegetation is established on the slope face and the face of the cut slope weathers, the probability of surficial failures increases. The surficial failure usually develops during or after a period of heavy and prolonged rainfall.

A similar process as described above can cause surficial failures in natural slopes. There is often an upper weathered zone of soil that slowly grades into solid rock with depth. The upper weathered zone of soil is often much more permeable because of its loose soil structure and hence higher porosity than the underlying dense rock. Water that seeps into the natural slope from rainstorms will tend to flow in the outer layer of the slope which has the much higher permeability than the deeper unweathered rock. Thus the seepage condition for natural slopes is often similar to that as shown in Fig. 10.17. Figure 10.18 shows a surficial failure in the weathered outer portion of a natural slope composed of sedimentary rock.

2. *Fill slope.* Figure 10.19 shows a surficial slope failure in a fill slope. Studies by Pradel and Raad (1993) indicate that in southern California, fill slopes made of clayey or

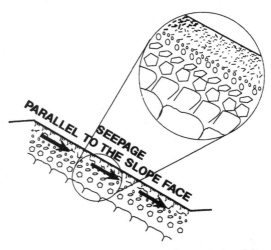

FIGURE 10.17 Illustration of near-surface weathering of claystone and zone of seepage parallel to slope face. (*Adapted from Ortigao et al., 1997; reprinted with permission from the American Society of Civil Engineers.*)

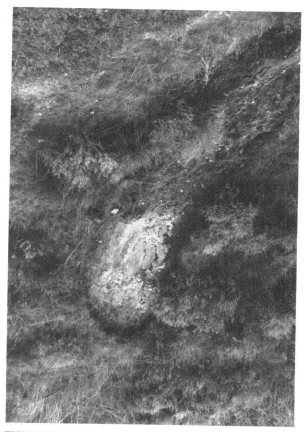

FIGURE 10.18 Surficial slope failure in the upper weathered zone of sedimentary rock.

silty soils are more prone to develop the conditions for surficial instability than slopes made of sand or gravel. This is probably because water tends to migrate downward in sand or gravel fill slopes, rather than parallel to the slope face.

As shown in Fig. 10.19, surficial failures cause extensive damage to landscaping. Surficial failures can even carry large trees downslope (see Fig. 10.19). Besides the landscaping, there can be damage to the irrigation and drainage lines. The surficial failure can also damage appurtenant structures, such as fences, walls, or patios.

A particularly dangerous condition occurs when the surficial failure mobilizes itself into a debris flow. In such cases, severe damage can occur to any structure located in the path of the debris flow. Figure 10.20 shows partial mobilization of the surficial failure, which flowed over the sidewalk and into the street. Surficial failures, such as the failure shown in Figs. 10.19 and 10.20, can be sudden and unexpected, without any warning of potential failure. Other surficial failures, especially in clays, may have characteristic signs of imminent failure. For example, Fig. 10.21 shows a clay slope having a series of nearly continuous

10.20 ANALYSIS OF GEOTECHNICAL DATA AND ENGINEERING COMPUTATIONS

FIGURE 10.19 Surficial slope failure in a fill slope.

semicircular ground cracks. After this picture was taken (during the rainy season), this slope failed in a surficial failure mode.

Surficial failures can also develop on the downstream face of earth dams. For example, Sherard et al. (1963) state:

> Shallow slides, most of which follow heavy rainstorms, do not as a rule extend into the embankment in a direction normal to the slope more than 4 or 5 feet [1.2–1.5 m]. Some take place soon after construction, while others occur after many years of reservoir operation.... Shallow surface slips involving only the upper few inches of the embankment have sometimes occurred when the embankment slopes have been poorly compacted. This is a frequent difficulty in small, cheaply constructed dams, where the construction forces often do not make the determined effort necessary to prevent a loose condition in the outer slope. The outer few feet then soften during the first rainy season, and shallow slides result.

Effect of Vegetation. A contributing factor in the instability of the slope shown in Fig. 10.21 was the loss of vegetation due to fire. Also notice in Fig. 10.19 that the surficial failure appeared to have developed just beneath the bottom of the grass roots. Roots can provide a large resistance to shearing. The shear resistance of root-permeated homogeneous and stratified soil has been studied by Waldron (1977) and Merfield (1992). The increase in shear strength due to plant roots is a direct result of mechanical reinforcement of the soil and an indirect result of removal of soil moisture by transpiration. Even grass roots can provide an increase in shear resistance of the soil equivalent to an effective cohesion of 3 to 5 kPa (60 to 100 psf) (Day, 1993b). When the vegetation is damaged or destroyed by fire, the slope can be much more susceptible to surficial failure such as shown in Fig. 10.21.

Surficial Stability Analysis. In order to calculate the factor of safety for surficial stability of cut, natural, or fill slopes, the method of analysis should be as follows:

1. For cut slopes, determine if the rock is likely to weather. Local experience can be of use in identifying sedimentary rocks, such as claystones or shales, that are known to quickly weather.
2. Obtain samples of the fill or weathered rock. Perform shear strength tests, such as drained direct shear tests, at confining pressures as low as possible to obtain the effective shear strength parameters ϕ' and c'.
3. The factor of safety for surficial stability [Eq. (10.2)] is very dependent on the value of effective cohesion c'. If the shear strength tests indicate a large effective cohesion intercept, then perform additional drained shear strength tests to verify this cohesion value.
4. Use Eq. (10.2) to calculate the factor of safety. The parameters in Eq. (10.2) can be determined as follows:
    a. Inclination α: The slope inclination can be measured for natural slopes or based on the anticipated constructed condition for cut and fill slopes.
    b. Total unit weight γ_t: For fill slopes, the total unit weight γ_t can be based on anticipated unit weight conditions at the end of grading. For cut or natural slopes, the total

FIGURE 10.20 Partial mobilization of surficial failure.

FIGURE 10.21 Ground cracks associated with incipient surficial instability.

unit weight γ_t can be obtained from the laboratory testing of undisturbed soil samples. Note that the total unit weight used in Eq. (10.2) must be based on a saturated ($S = 100\%$) condition.

  c. Buoyant unit weight γ_b: From the total unit weight for saturated soil, the buoyant unit weight γ_b can be obtained by Eq. (6.10) or (6.14).
  d. Depth of failure surface (d): In Eq. (10.2), the depth d of the failure surface is also the depth of seepage parallel to the slope face. In southern California, the value of d is typically assumed to equal 1.2 m (4 ft) for fill slopes. This may be overly conservative for cut or natural slopes, and different values of d may be appropriate given local rainfall and weathering conditions.
5. The acceptable minimum factor of safety for surficial stability is often 1.5. However, as previously mentioned, root reinforcement can significantly increase the surficial stability of a slope. It may be appropriate to accept a lower factor of safety in cases where deep-rooting plants will be quickly established on the slope face.

SLOPE STABILITY

Design and Construction. If the factor of safety for surficial stability is deemed to be too low, there are many different methods that can be used to increase the factor of safety. For example, the surficial stability can be increased by building a flatter slope (i.e., decreasing the slope inclination). It has been observed that 1.5:1 (horizontal:vertical) or steeper slope inclinations are often most susceptible to surficial instability, while 2:1 or flatter slope inclinations have significantly fewer surficial failures.

The surficial stability could also be increased by facing the slope with soil that has a higher shear strength. This process can be performed during the grading of the site and is similar to the installation of a stabilization fill (see Standard Detail No. 3, App. C).

Another option is shown in Figs. 10.22 and 10.23, where the face of the slope has been covered with gunite. The gunite facing will not allow rainwater to infiltrate the slope, which should prevent surficial instability from developing.

Instead of using gunite for steep slopes, the outer zone of the slope could be constructed with layers of geogrid, such as shown in Fig. 10.24. The typical construction steps for slopes reinforced with geogrid are as follows:

1. *Excavation of benches.* Benches are cut into the hillside as shown in Fig. 10.24. Benches provide favorable (i.e., not out-of-slope) frictional contact between the new fill mass and the horizontal portion of the bench. The benches are also used for the placement of the drainage system.

2. *Installation of drains.* After the benches have been excavated, drains are installed as shown in Fig. 10.24. The vertical drains are often placed at a 3-m (10-ft) spacing and are used to intercept seepage that may be migrating through the ground. The horizontal drains collect the water from the vertical drains and dispose of the water off-site. Figure 10.25 shows a photograph of the excavation of benches and installation of drains.

3. *Construction of the slope.* The slope is built with layers of geogrid and compacted fill. The usual design specification is to compact the fill to a minimum of 90 percent of the

FIGURE 10.22 Slope face covered with gunite.

10.24 ANALYSIS OF GEOTECHNICAL DATA AND ENGINEERING COMPUTATIONS

FIGURE 10.23 Another view of slope face covered with gunite.

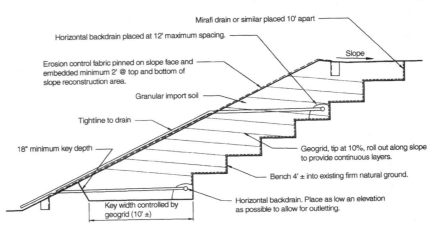

FIGURE 10.24 Construction of slope using geogrid.

FIGURE 10.25 Excavation of benches and installation of drains.

Modified Proctor maximum dry density. Figure 10.26 shows a photograph of the installation of the geogrid.

4. *Erosion control fabric.* At the end of the construction process, an erosion control fabric can be pinned to the face of the slope and then the slope face is planted. Figure 10.27 shows a photograph of the completed slope with an erosion control fabric on the slope surface.

In the soil, the geogrid acts as soil reinforcement, providing reinforcement effects similar to those of plant roots. Note in Fig. 10.24 that the geogrid is tipped back into the slope in order to get the geogrid as perpendicular to the potential failure surface as possible. The main design requirements are the type and vertical spacing of the geogrid. The design will depend on such factors as the shear strength of the soil, the slope inclination, and the depth d of the potential failure surface used in the analysis. The design could be based on Eq. (10.2), where a geogrid resisting force is included in the numerator of Eq. (10.2).

Soil Report Recommendations. Upon completion of the surficial stability analysis, the recommendations would normally be included in a soils report. An example of typical wording concerning the surficial stability of a site is as follows (see App. E):

> Permanent slopes in bedrock or fill should be constructed at a slope ratio not steeper than 2:1 (horizontal:vertical). In order to enhance the potential for further favorable performance of the permanent slopes, these areas should be landscaped at the completion of grading. In order to improve the surficial stability of the slope face, plants should consist of deep-rooted varieties requiring little watering. A landscape architect would be the best party to consult regarding actual types of plants and planting configuration. To retard the tendency of weathering of the slope face, irrigation should be planned to achieve uniform moisture conditions well below the saturation level. If automatic timing devices are utilized in conjunction with irrigation systems, provision should be made for interrupting normal watering during and

FIGURE 10.26 Installation of geogrid during construction of a slope.

FIGURE 10.27 Completion of slope shown in Fig. 10.26, with erosion control fabric on the slope surface.

following periods of rainfall. Property owners and/or maintenance personnel should be made aware that improper slope maintenance, altering site drainage, overwatering, and burrowing animals can be detrimental to surficial slope stability. It is recommended that, when a landscape irrigation system is installed, piping be anchored to the slope face instead of being placed in excavated trenches in the slope faces.

10.5 GROSS SLOPE STABILITY

In contrast to a shallow (surficial) slope failure discussed in the previous section, a gross slope failure often involves shear displacement of the entire slope. For example, Figs. 10.28 and 10.29 present two examples of gross slope failures. Figure 10.28 shows a rotational slope failure where the slide mass has moved downward and outward into the street. Figure 10.29 shows a translational slope failure that developed on steeply inclined out-of-slope rock bedding planes. Different terms, such as *slides* or *slumps,* are often used to identify gross slope instability.

As in surficial stability analysis, the objective of a gross slope stability analysis is to determine the factor of safety of an existing or proposed slope. For permanent slopes, the minimum factor of safety is 1.5. Gross slope stability analyses can be performed as either a total stress analysis (short-term condition using undrained shear strength) or an effective stress analysis (long-term condition using the drained shear strength). For a total stress slope stability analysis, the total unit weight of the soil is utilized and the groundwater table is not considered in the analysis. For an effective stress analysis, the pore water pressures (such as those based on a groundwater table) must be included in the analysis.

FIGURE 10.28   Rotational gross slope failure.

10.28 ANALYSIS OF GEOTECHNICAL DATA AND ENGINEERING COMPUTATIONS

FIGURE 10.29 Translational gross slope failure.

Wedge Method. The simplest type of gross slope stability analysis uses a free-body diagram such as illustrated in Fig. 10.30, where there is a planar slip surface inclined at an angle α to the horizontal. The wedge method is a two-dimensional analysis based on a unit length of slope. A wedge-type failure is similar to the translational failure shown in Fig. 10.29. The assumption in this slope stability analysis is that there will be a wedge-type failure of the slope along a planar slip surface. The factor of safety F of the slope can be derived by summing forces parallel to the slip surface, and is:

Total stress analysis:

$$F = \frac{\text{resisting force}}{\text{driving force}} = \frac{cL + N \tan \phi}{W \sin \alpha} = \frac{cL + W \cos \alpha \tan \phi}{W \sin \alpha} \quad (10.3a)$$

Effective stress analysis:

$$F = \frac{\text{resisting force}}{\text{driving force}} = \frac{c'L + N' \tan \phi'}{W \sin \alpha} = \frac{c'L + (W \cos \alpha - uL) \tan \phi'}{W \sin \alpha} \quad (10.3b)$$

where F = factor of safety for gross slope stability (dimensionless parameter)
 c, ϕ = undrained shear strength of the slide plane
 c', ϕ' = drained shear strength of the slide plane
 L = length of the slip surface (m or feet)
 N = normal force, i.e., the force acting perpendicular to the slip surface ($N = W \cos \alpha$)
 W = total weight of the failure wedge (i.e., $W = \gamma_t$ times the area of the wedge); units are kN (or lb) for a unit length of slope
 u = average pore water pressure along the slip surface (kPa or psf)
 α = slip surface inclination (degrees)

Because the wedge method is a two-dimensional analysis based on a unit length of slope (i.e., length = 1 m or 1 ft), the numerator and denominator of Eq. (10.3) are in kN (or lb). As in the surficial stability analysis [Eq. (10.1)], the resisting force in Eq. (10.3) is equal to the shear strength (in terms of total stress or effective stress) of the soil along the slip surface. The driving force [Eq. (10.3)] is caused by the pull of gravity, and it is equal to the component of the weight of the wedge parallel to the slip surface.

1. Total stress analysis. The total stress analysis would be applicable to the quick construction of the slope or a condition where the slope is loaded very quickly. A total stress analysis could be performed by using the consolidated undrained shear strength (c and ϕ) or the undrained shear strength (s_u) of the slip surface material. When the undrained shear strength is used, $s_u = c$ and $\phi = 0$ are substituted into Eq. (10.3a).

Example. A slope has a height of 9.1 m (30 ft), and the slope face is inclined at a 2:1 (horizontal:vertical) ratio. Assume a wedge-type analysis where the slip surface is planar through the toe of the slope and is inclined at a 3:1 (horizontal:vertical) ratio. The total unit weight of the slope material (γ_t) = 18.1 kN/m³ (115 pcf). Using the undrained shear strength parameters from Fig. 6.16 [i.e., c = 14.5 kPa (2.1 psi) and $\phi = 0$], calculate the factor of safety.

Solution. The area of the wedge is first determined from simple geometry and is equal to 41.4 m² (450 ft²). For a unit length of the slope, the total weight W of the wedge equals the area times total unit weight, or 750 kN per meter of slope length (52,000 pounds per foot of slope length). Using Eq. (10.3a) and the following values:

c = 14.5 kPa (300 psf)

$\phi = 0$

Length of slip surface (L) = 29 m (95 ft)

Slope inclination (α) = 18°

Total weight of wedge (W) = 750 kN/m (51,700 lb/ft)

we find the factor of safety F of the slope is 1.80.

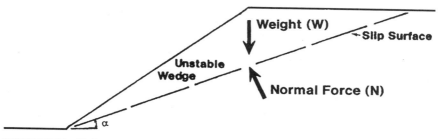

FIGURE 10.30 Wedge method.

2. Effective stress analysis. The purpose of the effective stress slope stability analysis is to model the long-term condition of the slope, and effective stresses must be utilized along the slip surface. The term $W \cos \alpha - uL$ [Eq. (10.3b)] is equal to the effective normal force on the slip surface. Because the analysis uses the boundary pore water pressures and total unit weight of the soil, the weight W in the numerator and denominator of Eq. (10.3b) is equal to the total weight of the wedge.

Example. Assume the same situation as the previous example except that the slip surface has the effective shear strength shown in Fig. 6.18 [i.e., $c' = 3.4$ kPa (70 psf), $\phi' = 29°$]. Also assume that piezometers have been installed along the slip surface and the average measured steady-state pore water pressure (u) = 2.4 kPa (50 psf). Calculate the factor of safety of the failure wedge based on an effective stress analysis.

Solution. Using Eq. (10.3b) and the following values:

$c' = 3.4$ kPa (70 psf)

$\phi' = 29°$

Length of slip surface (L) = 29 m (95 ft)

Slope inclination (α) = 18°

Average pore water pressure acting on the slip surface (u) = 2.4 kPa (50 psf)

Total weight of wedge (W) = 750 kN/m (51,700 lb/ft)

we find the factor of safety F of the slope is 1.95.

Method of Slices. The most commonly used method of gross slope stability analysis is the *method of slices*, where the failure mass is subdivided into vertical slices and the factor of safety is calculated from force equilibrium equations. A circular arc slip surface and rotational type of failure mode are often used for the method of slices, and, for homogeneous soil, a circular arc slip surface provides a lower factor of safety than assuming a planar slip surface. The slope failure shown in Fig. 10.28 is an example of a rotational failure.

Figure 10.31 shows an example of a slope stability analysis using a circular arc slip surface. The failure mass has been divided into 30 vertical slices. The calculations are similar to the wedge-type analysis, except that the resisting and driving forces are calculated for each slice and then summed up in order to obtain the factor of safety of the slope. For the *ordinary method of slices* (also known as the *Swedish circle method* or *Fellenius method*; Fellenius, 1936), the equation used to calculate the factor of safety is identical to Eq. (10.3), with the resisting and driving forces calculated for each slice and then summed up in order to obtain the factor of safety.

Commonly used methods of slices to obtain the factor of safety are listed in Table 10.3. The method of slices is not an exact method because there are more unknowns than equilibrium equations. This requires that an assumption be made concerning the interslice forces. Table 10.3 presents a summary of the assumptions for the various methods. For example, Fig. 10.32 shows that for the ordinary method of slices (Fellenius, 1936), it is assumed that the resultant of the interslice forces are parallel to the average inclination of the slice α. It has been determined that because of this interslice assumption for the ordinary method of slices, this method provides a factor of safety that is too low for some situations (Whitman and Bailey, 1967). As a result, the other methods listed in Table 10.3 are used more often than the ordinary method of slices.

Because of the tedious nature of the calculations, computer programs are routinely used to perform the analysis. Duncan (1996) states that the nearly universal availability of computers and much improved understanding of the mechanics of slope stability analyses have brought about considerable change in the computational aspects of slope stability analysis.

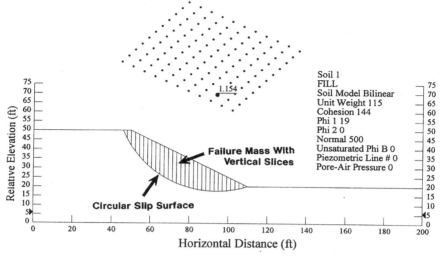

FIGURE 10.31 Example of a slope stability analysis using the Geo-Slope computer program.

TABLE 10.3 Assumptions Concerning Interslice Forces for Different Methods of Slices

Type of method of slices (1)	Assumption concerning interslice forces (2)	Reference (3)
Ordinary method of slices	Resultant of the interslice forces is parallel to the average inclination of the slice	Fellenius (1936)
Bishop simplified method	Resultant of the interslice forces is horizontal (no interslice shear forces)	Bishop (1955)
Janbu simplified method	Resultant of the interslice forces is horizontal (a correction factor is used to account for interslice shear forces)	Janbu (1968)
Janbu generalized method	Location of the interslice normal force is defined by an assumed line of thrust	Janbu (1957)
Spencer method	Resultant of the interslice forces is of constant slope throughout the sliding mass	Spencer (1967, 1968)
Morgenstern-Price method	Direction of the resultant of interslice forces is determined by using a selected function	Morgenstern and Price (1965)

Sources: Lambe and Whitman (1969); Geo-Slope (1991).

Analyses can be done much more thoroughly, and, from the point of view of mechanics, more accurately than was possible previously. However, problems can develop because of a lack of understanding of soil mechanics, soil strength, and the computer programs themselves, as well as the inability to analyze the results in order to avoid mistakes and misuse (Duncan, 1996).

10.32 ANALYSIS OF GEOTECHNICAL DATA AND ENGINEERING COMPUTATIONS

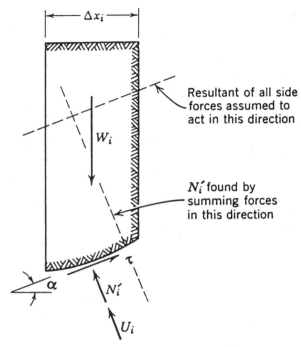

FIGURE 10.32 Forces acting on a vertical slice and the assumption concerning interslice forces for the ordinary method of slices. (*Adapted from Lambe and Whitman, 1969.*)

1. *Total stress analysis.* The slope stability analysis shown in Fig. 10.31 was performed by using the SLOPE/W (Geo-Slope, 1991) computer program. The following data was input into the computer program:

- *Slope cross section.* The slope has a height of 9.1 m (30 ft) and a 2:1 (horizontal:vertical) slope inclination.
- *Type of slope stability analysis* A total stress slope stability analysis was selected.
- *Shear strength parameters.* The shear strength parameters from Fig. 6.16 were used in the slope stability analysis.
- *Total unit weight of soil.* The total unit weight (γ_t) = 18.1 kN/m³ (115 pcf).
- *Critical slip surface.* For this analysis, the computer program was requested to perform a trial-and-error search for the critical slip surface (i.e., the slip surface having the lowest factor of safety). Note the grid of points that has been produced above the slope. Each one of these points represents the center of rotation of a circular arc slip surface passing through the toe of the slope. The computer program has actually performed 121 slope stability analyses. In Fig. 10.31, the dot with the number 1.154 indicates the center of rotation of the circular arc slip surface with the lowest factor of safety (i.e., lowest factor of safety = 1.15).

In a previous example, the factor of safety of this slope was determined on the assumption of a planar failure surface through the toe. That analysis indicated a factor of safety of

1.80. The analysis with a circular arc slip surface (Fig. 10.31) indicates a factor of safety of 1.15. This demonstrates that for a homogeneous slope, a slope stability analysis using circular arc slip surfaces will generate a lower factor of safety than using planar slip surfaces.

   2. *Effective stress analysis.* The slope stability program can also perform effective stress slope stability analyses. For this type of analysis, the effective shear strength parameters c' and ϕ' are input into the computer program. The pore water pressures must also be input into the computer program. In most cases, the pore water pressures have a significant impact on slope stability and they are often very difficult to estimate. There are several different options that can be used concerning the pore water pressures, as follows:

- *Zero pore water pressure.* A common assumption for slopes that have or will be constructed with drainage devices, such as the drainage systems shown in Standard Detail No. 4 (App. C), is to use a pore water pressure equal to zero.
- *Groundwater table.* A second option is to specify a groundwater table. For the soil above the groundwater table, it is common to assume zero pore water pressures. If the groundwater table is horizontal, then the pore water pressures below the groundwater table are typically assumed to be hydrostatic [i.e., Eq. (6.18)]. For the condition of seepage through the slope (i.e., a sloping groundwater table), the computer program can develop a flow net in order to estimate the pore water pressures below the groundwater table (Sec. 6.6). Figure 10.33 shows a rotational slide and Fig. 10.34 shows the pore water pressures along the failure surface due to a sloping groundwater table.
- *Pore water pressure ratio* r_u. The pore water pressure ratio $(r_u) = u/\gamma_t h$, where u = pore water pressure, γ_t = total unit weight of the soil, and h = depth below the ground surface. If a value of $r_u = 0$ is selected, then the pore water pressures u are assumed to be equal to zero in the slope. Suppose an r_u value is used for the entire slope. In many cases the total unit weight is about equal to 2 times the unit weight of water (i.e., $\gamma_t = 2\gamma_w$), and thus a value of $r_u = 0.25$ is similar to the effect of a groundwater table at midheight of the slope. A value of $r_u = 0.5$ would be similar to the effect of a groundwater table corresponding to the ground surface. The pore water pressure ratio r_u can be used for existing slopes where the pore water pressures have been measured in the field, or for the design of proposed slopes where it is desirable to obtain a quick estimate of the effect of pore water pressures on the stability of the slope.

FIGURE 10.33 Example of rotational slide movement.

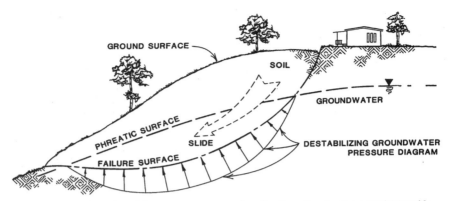

FIGURE 10.34 Illustration of the pore water pressure along the slip surface due to a groundwater table.

Other Slope Stability Considerations. The objective of the slope stability analysis is to accurately model the existing or design conditions of the slope. Some of the important factors that may need to be considered in a slope stability analysis are as follows:

1. *Different soil layers.* If a proposed slope or existing slope contains layers of different soil types with different engineering properties, then these layers must be input into the slope stability computer program. Most slope stability computer programs have this capability.

2. *Slip surfaces.* In some cases, a composite-type slip surface may have to be included in the analysis. This option is discussed in the next section.

3. *Tension cracks.* It has been stated that tension cracks at the top of the slope can reduce the factor of safety of a slope by as much as 20 percent. Such cracks are usually regarded as an early and important warning sign of impending failure in cohesive soil (Cernica, 1995a). Slope stability programs often have the capability to model or input tension crack zones. The destabilizing effects of water in tension cracks or even the expansive forces caused by freezing water can also be modeled by some slope stability computer programs.

4. *Surcharge loads.* There may be surcharge loads (such as a building load) at the top of the slope or even on the slope face. Most slope stability computer programs have the capability of including surcharge loads. In some computer programs, other types of loads, such as those due to tieback anchors, can also be included in the analysis.

5. *Nonlinear shear strength envelope.* In some cases, the shear strength envelope is nonlinear (for example, see Fig. 10.15). If the shear strength envelope is nonlinear, then a slope stability computer program that has the capability of using a nonlinear shear strength envelope in the analysis should be used.

6. *Plane strain condition.* Similar to strip footings, long uniform slopes will be in a plane strain condition. As discussed in Sec. 8.2, the friction angle ϕ is about 10 percent higher in the plane strain condition as compared to the friction angle ϕ measured in the triaxial apparatus (Meyerhof, 1961; Perloff and Baron, 1976). Since plane strain shear strength tests are not performed in practice, there will be an additional factor of safety associated with the plane strain condition. For uniform fill slopes that have a low factor of safety, it is often observed that the "end" slopes (slopes that make a 90° turn) are the first to show indications of slope movement. This is because the end slope is not sub-

jected to a plane strain condition and the shear strength is actually lower than in the center of a long, continuous slope.

7. *Progressive failure.* For the method of slices, the factor of safety is an average value of all the slices. Some slices, such as at the toe of the slope, may have a lower factor of safety that is balanced by other slices which have a higher factor of safety. For those slices that have a low factor of safety, the shear stress and strain may exceed the peak shear strength. For some soils, such as stiff-fissured clays, there may be a significant drop in shear strength as the soil deforms beyond the peak values. This reduction in shear strength will then transfer the load to an adjacent slice, which will cause it to experience the same condition. Thus the movement and reduction of shear strength will progress along the slip surface, eventually leading to failure of the slope. As mentioned in Sec. 5.2.3, the progressive nature of the failure may even reduce the shear strength of the soil to its residual value ϕ'_r.

8. *Pseudo-static analysis.* Slope stability analysis can also be adapted to perform earthquake analysis. This option will be discussed in Chap. 11.

9. *Other structures.* Slope stability analysis can be used for other types of engineering structures. For example, the stability of a retaining wall is often analyzed by considering a slip surface beneath the foundation of the wall.

Soil Report Recommendations. Upon completion of the gross slope stability analysis, the recommendations would normally be included in a soils report. An example of typical wording concerning the gross slope stability of a site is as follows (see App. E):

1. *Temporary slopes.* Gross slope stability analyses were performed for temporary slopes for the restrained basement walls (maximum height of 20 ft) in bedrock. Gross stability computations are included in the appendix. The results indicate that the temporary back-cuts should be grossly stable at a ratio of 0.5:1.0 (horizontal:vertical). Sloughing, however, of loose and/or fractured materials could occur and thus cleaning of temporary slopes of loose material should be conducted as excavation proceeds. Parking of equipment or stockpiling of materials at the top of the temporary slopes should be prohibited.

2. *Permanent slopes.* Gross slope stability analyses were performed for permanent slopes, not to exceed 40 ft (12 m) in height. Gross stability computations are included in the appendix. The results indicate that permanent slopes in bedrock or fill should be constructed at a slope ratio not steeper than 2:1 (horizontal:vertical). For slopes between 30 and 40 ft (9 to 12 m) in height, it is recommended that a gunite drainage swale (Fig. 4.5) should be provided at approximately midheight of the slope. Surface water should not be allowed to flow over the top of slope and the slope faces should be landscaped at the completion of grading.

3. *Foundations and slope areas.* Foundations planned within or adjacent to slope areas should be deepened to provide sufficient horizontal distance from the bottom, outer edge of the foundations to daylight. The distance should be equal to half the slope height or 5 ft (1.5 m), whichever is greater. Foundation details under the influence of this recommendation should be forwarded along with the structural load information to the geotechnical engineer for review.

10.6 LANDSLIDES

Landslides can be some of the most challenging projects worked on by geotechnical engineers and engineering geologists. As discussed in Sec. 4.7 (Figs. 4.24 to 4.34), landslides can cause extensive damage to structures and may be very expensive to stabilize when they impact developed property. In terms of damage, the National Research Council (1985) states:

Landsliding in the United States causes at least $1 to $2 billion in economic losses and 25 to 50 deaths each year. Despite a growing geologic understanding of landslide processes and a rapidly developing engineering capability for landslide control, losses from landslides are continuing to increase. This is largely a consequence of residential and commercial development that continues to expand onto the steeply sloping terrain that is most prone to landsliding.

Figure 10.35 shows an example of a landslide and Table 10.4 presents common nomenclature used to describe landslide features (Varnes, 1978). Landslides are described as mass movement of soil or rock that involves shear displacement along one or several rupture surfaces, which are either visible or may be reasonably inferred (Varnes, 1978). As previously mentioned, it is the shear displacement along a distinct rupture surface that distinguishes landslides from other types of soil or rock movement such as falls, topples, or flows.

Landslides are generally classified as either rotational or translational. Rotational landslides are due to forces that cause a turning movement about a point above the center of gravity of the failure mass, which results in a curved or circular surface of rupture. Translational landslides occur on a more or less planar or gently undulatory surface of rupture. Translational landslides are frequently controlled by weak layers, such as faults, joints, or bedding planes; examples include the variations in shear strength between layers of tilted bedded deposits or the contact between firm bedrock and weathered overlying material. For example, Fig. 10.36 shows a translational landslide that developed when the toe of slope was removed by stream erosion.

Active landslides are those that are either currently moving or that are only temporarily suspended, which means that they are not moving at present but have moved within the last cycle of seasons (Varnes, 1978). Active landslides have fresh features, such as a main scarp, transverse ridges and cracks, and a distinct main body of movement. The fresh features of an active landslide enable the limits of movement to be easily recognized.

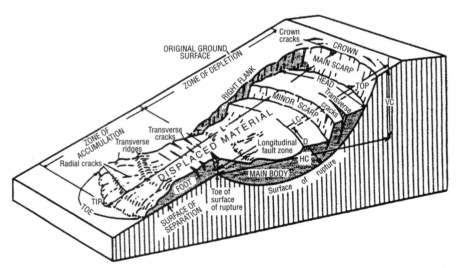

FIGURE 10.35 Landslide illustration. [*Reprinted with permission from Landslides: Analysis and Control, Special Report 176 (from Varnes, 1978). Copyright 1978 by the National Academy of Sciences, Courtesy of the National Academy Press, Washington, D.C.*]

TABLE 10.4 Common Landslide Nomenclature

Terms (1)	Definitions (2)
Main scarp	A steep surface on the undisturbed ground around the periphery of the slide, caused by the movement of slide material away from the undisturbed ground. The projection of the scarp surface under the displaced material becomes the surface of rupture.
Minor scarp	A steep surface on the displaced material produced by differential movements within the sliding mass.
Head	The upper parts of the slide material along the contact between the displaced material and the main scarp.
Top	The highest point of contact between the displaced material and the main scarp.
Toe, surface of rupture	The intersection (sometimes buried) between the lower part of the surface of rupture and the original ground surface.
Toe	The margin of displaced material most distant from the main scarp.
Tip	The point on the toe most distant from the top of the slide.
Foot	That portion of the displaced material that lies downslope from the toe of the surface of rupture.
Main body	That part of the displaced material that overlies the surface of rupture between the main scarp and toe of the surface of rupture.
Flank	The side of the landslide.
Crown	The material that is still in place, practically undisplaced and adjacent to the highest parts of the main scarp.
Original ground surface	The slope that existed before the movement which is being considered took place. If this is the surface of an older landslide, that fact should be stated.
Left and right	Compass directions are preferable in describing a slide, but if right and left are used they refer to the slide as viewed from the crown.
Surface of separation	The surface separating displaced material from stable material but not known to have been a surface on which failure occurred.
Displaced material	The material that has moved away from its original position on the slope. It may be in a deformed or undeformed state.
Zone of depletion	The area within which the displaced material lies below the original ground surface.
Zone of accumulation	The area within which the displaced material lies above the original ground surface.

Source: Reprinted with permission from *Landslides: Analysis and Control, Special Report 176* (from Varnes, 1978). Copyright 1978 by the National Academy of Sciences. Courtesy of the National Academy Press, Washington, D.C.

Generally active landslides are not significantly modified by the processes of weathering or erosion.

Landslides that have long since stopped moving are typically modified by erosion and weathering; or may be covered with vegetation so that the evidence of movement is obscure. The main scarp and transverse cracks will have been eroded or filled in with debris. Such landslides are generally referred to as *ancient* or *fossil landslides* (Zaruba and Mencl, 1969; Day, 1995d). These landslides have commonly developed under different climatic conditions thousands or more years ago.

Many different conditions can trigger a landslide. Landslides can be triggered by an increase in shear stress or a reduction in shear strength. The following factors contribute to an increase in shear stress:

1. Removing lateral support, such as erosion of the toe of the landslide by streams or rivers.
2. Applying a surcharge at the head of the landslide, such as the construction of a fill mass for a road.
3. Applying a lateral pressure, such as the raising of the groundwater table.
4. Applying vibration forces, such as an earthquake or construction activities.

Factors that result in a reduction in shear strength include:

1. Natural weathering of soil or rock.
2. Development of discontinuities, such as faults or bedding planes.
3. Increase in moisture content or pore water pressure of the slide plane material.

Landslide Example. The purpose of this subsection is to present an additional example of landslide movement. This landslide illustrates the typical mode and causes of failure. Figures 10.37 to 10.43 show pictures of a landslide which occurred in Olivenhain, California, in March 1998. A major contributing factor in this landslide failure was heavy El Niño winter rains. The figures show the following:

- *Figures 10.37 and 10.38.* An interesting feature of this landslide was how the toe of the landslide rolled down the hillside. The smaller arrow in Fig. 10.37 points to groundwater which was observed flowing out of the toe of the landslide. The two larger arrows in Fig. 10.37 point to areas where the toe of the landslide is overriding and rolling down the slope. Figure 10.38 is a close-up view that shows the toe of the landslide rolling down the slope.

- *Figures 10.39 and 10.40.* These two photographs show ground cracks and distress at the main body of the landslide. In Fig. 10.39, the force of the landslide has crushed a gunite drainage ditch. Figure 10.40 shows ground cracks and distortion of a gunite drainage ditch at another location.

- *Figures 10.41 to 10.43.* These three photographs show the head of the landslide and the main scarp. This area is at the top of the landslide and the mass movement of the landslide has caused the ground to drop down. In Fig. 10.41, the main scarp runs underneath the wood deck, which has been displaced downward and laterally. The arrow in Fig. 10.41 points to the main scarp and Fig. 10.42 is a close-up view of this area. Another view of the main scarp is shown in Fig. 10.43, where the ground surface has dropped downward by about 2 m (7 ft). The vertical distance between the two arrows in Fig. 10.43 is the amount of vertical ground displacement caused by movement of the landslide. Note in Fig. 10.43 the slick and grooved nature (i.e., slickensides) of the material exposed on the main scarp.

Method of Analysis

Method of Analysis for Active Landslide. Except in cases where the landslide fails because of a sudden loading or unloading (such as during an earthquake, Chap. 11), the usual procedure is to perform an effective stress slope stability analysis. Figure 10.44 shows an example of an effective stress analysis for the Laguna Niguel landslide discussed in Sec. 4.7. The actual slope stability analysis was performed by using the SLOPE/W (Geo-Slope, 1991) computer program. Although the method of slices was originally developed for circular slip surfaces, the analyses can be readily adapted to planar or composite slip surfaces.

In order to calculate the factor of safety for the landslide based on an effective stress analysis, the following parameters were input into the SLOPE/W (Geo-Slope, 1991) computer program:

1. *Landslide cross section.* The landslide cross section was developed by the engineering geologist (Fig. 4.23b). The different layers in Fig. 10.44 have been identified as "ef" (engineered fill), Q_{al} (alluvium), T_m (fractured formational rock), and slide plane material.

FIGURE 10.36 Landslide caused by erosion of the slope toe. Arrow indicates direction of landslide movement.

FIGURE 10.37 Olivenhain landslide: toe of the landslide. The smaller arrow points to groundwater observed coming out of the landslide and the larger arrows point to the toe of the landslide rolling downslope.

FIGURE 10.38 Olivenhain landslide: close-up view of the toe of the landslide rolling downslope.

FIGURE 10.39 Olivenhain landslide: main body of the landslide, which has crushed a gunite drainage ditch.

FIGURE 10.40 Olivenhain landslide: another view of the main body of the landslide, showing ground cracks and damage to another drainage ditch.

FIGURE 10.41 Olivenhain landslide: view of the head of the landslide. Arrow points to the main scarp.

FIGURE 10.42 Olivenhain landslide: close-up view of the main scarp indicated in Fig. 10.41.

FIGURE 10.43 Olivenhain landslide: another view of the main scarp. The vertical distance between arrows indicates the amount that the ground has dropped down.

2. *Failure mass.* As mentioned in Sec. 10.5, the computer program can efficiently search for the critical failure surface that has the lowest factor of safety. Another option is to specify the location of the slip surface. For the Laguna Niguel landslide, the location of the slip surface was determined from inclinometer monitoring. This location of the slip surface was then input into the computer program (i.e., the slip surface is "fully specified"). Note in Fig. 10.44 that for the stability analysis of the input failure mass, the computer program has divided the failure mass into 64 vertical slices.

3. *Groundwater table.* The location of the groundwater table was determined from piezometer readings and input into the slope stability program. The dashed line in Fig. 10.44 is the location of the groundwater table.

4. *Drained residual shear strength.* The drained residual shear strength (Secs. 5.2.3 and 6.5.2) of the slide plane material was obtained from laboratory tests using the modified Bromhead ring shear apparatus (Stark and Eid, 1994). Figure 10.45 presents the drained residual failure envelope and Fig. 10.46 shows the stress-displacement plots for the ring shear test on the slide plane specimen. It can be seen in Fig. 10.45 that the failure envelope is nonlinear. Also shown in Fig. 10.45 is a summary of residual shear strength and secant residual friction angles for various effective normal stresses. Given the great depth of the slide plane (i.e., a high effective normal stress), a drained residual friction angle ϕ'_r of 12° was selected for the slope stability analysis.

5. *Shear strength of other soil layers.* The effective shear strength parameters c' and ϕ' were determined from laboratory shear strength tests, and the data is summarized in Fig. 10.44.

6. *Total unit weight.* In laboratory testing of soil and rock specimens, total unit weights of the various strata were determined, and the values are summarized in Fig. 10.44.

10.44 ANALYSIS OF GEOTECHNICAL DATA AND ENGINEERING COMPUTATIONS

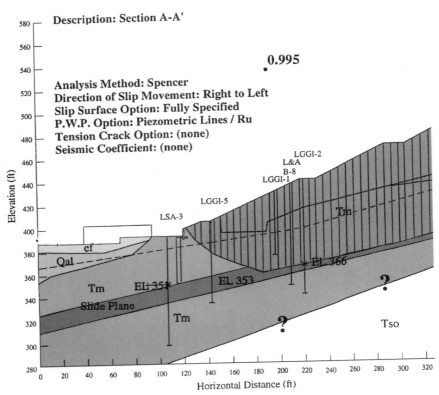

FIGURE 10.44 Slope stability analysis of the Laguna Niguel landslide.

Based on the input parameters, the factor of safety using the Spencer method of slices as calculated by the SLOPE/W (Geo-Slope, 1991) computer program is 0.995. This value is consistent with the actual failure (factor of safety = 1.0) of the slope.

Method of Analysis for an Ancient Landslide. If an ancient landslide is discovered during the design stage of a new project, the method of analysis would be similar to that as described above. Extensive subsurface exploration would be required so that the engineering geologist could develop a cross section of the ancient landslide based on subsurface exploration. The drained residual friction angle could be obtained from laboratory shear strength tests. The unit weight of the landslide could be estimated from laboratory testing of undisturbed specimens of the landslide materials.

A major unknown in the effective stress slope stability analysis of the ancient landslide would be the groundwater table. It is not uncommon that the groundwater table will rise once the project has been completed. This is because there will be additional infiltration of water into the landslide mass from irrigation or leaky pipes. The approximate location of the future (long-term) groundwater table could be based on the local topography and presence of drainage facilities that are to be installed during development of the site. Once the cross section of the landslide, location of the slip surface, estimated location of the long-term groundwater table, residual shear strength parameters, and unit weight are determined, the factor of safety of the ancient landslide mass would be calculated by a slope stability

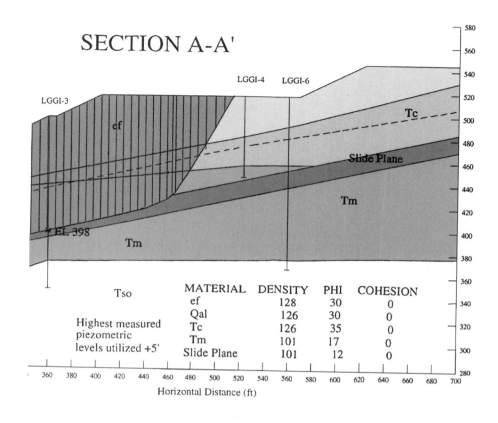

program such as shown in Fig. 10.44. The standard requirement is that a landslide must have a factor of safety of at least 1.5.

Stabilization of Slopes and Landslides. If the analysis shows that the landslide has too low a factor of safety, there are three basic approaches that can be used to increase the factor of safety of a slope or landslide: (1) increase the resisting forces, (2) decrease the driving forces, or (3) rebuild the slope.

1. *Increase the resisting forces.* Methods to increase the resisting forces include the construction of a buttress at the toe of the slope (see Standard Detail No. 3, App. C), or the installation of piles or reinforced concrete pier walls which provide added resistance to the slope. The design and construction of reinforced pier walls to stabilize landslides will be further discussed in Sec. 15.8.

Another technique is soil nailing, which is a practical and proved system used to stabilize slopes by reinforcing the slope with relatively short, fully bonded inclusions such as steel bars (Bruce and Jewell, 1987).

2. *Decrease the driving forces.* Methods to decrease the driving forces include the lowering of the groundwater table by improving surface drainage facilities, installing underground drains, or pumping groundwater from wells. Other methods to decrease the

Effective Normal Stress (psf)	Residual Shear Strength (psf)	Secant Residual Friction Angle (deg.)
1044	292	15.6
2088	529	14.2
4175	940	12.7
8351	1805	12.2
14614	2410	9.4

LIQUID LIMIT = 86%
PLASTIC LIMIT = 40%
CLAY-SIZE FRACTION = 40%

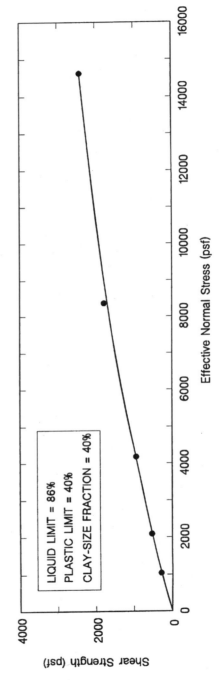

FIGURE 10.45 Drained residual failure envelope for Laguna Niguel slide plane material.

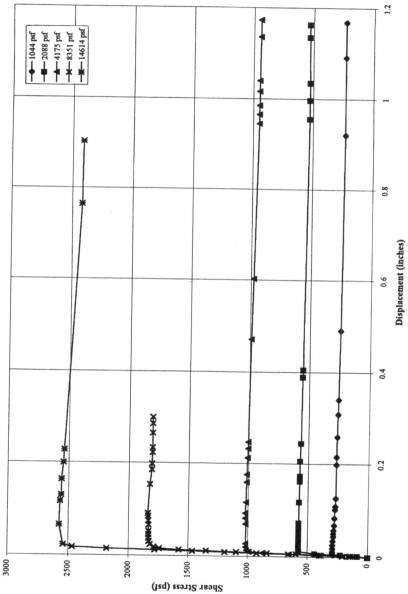

FIGURE 10.46 Shear stress-displacement relationships for the Laguna Niguel slide plane material.

driving forces consist of removing soil from the head of the landslide or regrading the slope in order to decrease its height or slope inclination.

3. *Rebuild the slope.* The slope failure could also be rebuilt and strengthened by using geogrids or other soil reinforcement techniques (Rogers, 1992). Other techniques could be employed during the grading of the site, such as the construction of a shear key. A shear key is defined as a deep and wide trench cut through the landslide mass and into intact material below the slide. The shear key is excavated from ground surface to below the basal rupture surface, and then backfilled will soil that provides a higher shear strength than the original rupture surface. By interrupting the original weak rupture surface with a higher-strength soil, the factor of safety of the landslide is increased. During construction, the shear key is also normally provided with a drainage system. It is generally recognized that during the excavation of a shear key, there is a risk of the failure of the landslide. To reduce this risk, a shear key is usually constructed in several sections with only a portion of the slide plane exposed at any given time. Figure 10.47 shows a cross section through an ancient landslide where a shear key was installed during grading of the site. At this project, the shear key was not successful in stabilizing the landslide and a portion of the landslide became reactivated (Day, 1999).

10.7 DEBRIS FLOW

Debris flows cause a tremendous amount of damage and loss of life throughout the world. An example is the loss of 6000 lives from the devastating flows that occurred in Leyte, Philippines, on November 5, 1991, due to deforestation and torrential rains from tropical storm Thelma. Because of continued population growth, deforestation, and poor land development practices, it is expected that debris flows will increase in frequency and devastation.

A debris flow is commonly defined as soil with entrained water and air that moves readily as a fluid on low slopes. As shown in Fig. 10.20, in many cases there is an initial surficial slope failure that transforms itself into a debris flow (Ellen and Fleming, 1987; Anderson and Sitar, 1995, 1996). Figure 10.48 shows two views of a debris flow. The upper photograph shows the source area of the debris flow and the lower photograph shows how the debris flow forced its way into the house.

Debris flow can include a wide variety of soil-particle sizes (including boulders) as well as logs, branches, tires, and automobiles. Other terms, such as *mud flow, debris slide, mud slide,* and *earth flow,* have been used to identify similar processes. While categorizing flows based on rate of movement or the percentage of clay particles may be important, the mechanisms of all these flows are essentially the same (Johnson and Rodine, 1984).

There are generally three segments of a debris flow: the source area, main track, and depositional area (Baldwin et al., 1987). The source area is the region where a soil mass becomes detached and transforms itself into a debris flow. The main track is the path over which the debris flow descends the slope and increases in velocity, depending on the slope steepness, obstructions, channel configuration, and the viscosity of the flowing mass. Where the debris flow encounters a marked decrease in slope gradient and deposition begins, this is called the depositional area.

Prediction of Debris Flow. It can be very difficult to predict the potential for a debris flow at a particular site. The historical method is one means of predicting debris-flow activity in a particular area. For example, as Johnson and Rodine (1984) indicate, many alluvial

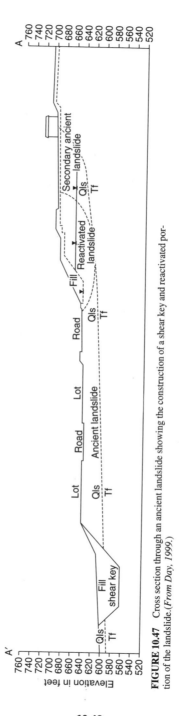

FIGURE 10.47 Cross section through an ancient landslide showing the construction of a shear key and reactivated portion of the landslide. *(From Day, 1999.)*

10.50 ANALYSIS OF GEOTECHNICAL DATA AND ENGINEERING COMPUTATIONS

FIGURE 10.48 Two views of a debris flow. Upper photograph shows the area of detachment and the lower photograph shows how the flow forced its way into the house.

fans in southern California contain previous debris-flow deposits, which in the future will likely again experience debris flow. However, using the historical method for predicting debris flow is not always reliable. For example, the residences of Los Altos Hills experienced an unexpected debris flow mobilization from a road fill after several days of intense rainfall (Johnson and Hampton, 1969). Using the historical method to predict debris flow is not always reliable, because the area can be changed, especially by society's activities.

Johnson and Rodine (1984) stated that a single parameter should not be used to predict either the potential or actual initiation of a debris flow. Several parameters appear to be of prime importance. Two such parameters, which numerous investigators have studied, are rainfall amount and rainfall intensity. For example, Neary and Swift (1987) stated that hourly rainfall intensity of 90 to 100 mm/h (3.5 to 4 in./h) was the key to triggering debris flows in the southern Appalachians. Other important factors include the type and thickness of soil in the source area, the steepness and length of the slope in the source area, the destruction of vegetation by fire or logging, and other society-induced factors such as the cutting of roads.

The engineering geologist is usually the best individual to investigate the possibility of a debris flow impacting the site. Based on this analysis, measures can be taken to prevent a debris flow from damaging the site. For example, grading could be performed to create a raised building pad so that the structure is elevated above the main debris flow path. Other measures include the construction of retention basins, deflection walls, or channels to control or direct the debris flow away from the structures.

Debris Flow Example. This section deals with a debris flow adjacent to the Pauma Indian Reservation, in San Diego County, California. The debris flow occurred in January 1980, during heavy and intense winter rains. A review of aerial photographs indicates that alluvial fans are being built at the mouths of the canyons in this area because of past debris flows.

The debris flow in January 1980 hit a house, which resulted in a lawsuit being filed. The author was retained in June of 1990 as an expert for one of the cross-defendants, a lumber company. The case settled out of court in July 1990.

The lumber company had cut down trees on the Pauma Indian Reservation and it was alleged that the construction of haul roads and the lack of tree cover contributed to the debris flow. These factors were observed to be contributing factors in the debris flow. For example, after the debris flow, deep erosional channels were discovered in the haul roads, indicating that at least a portion of the soil in the flow came from the road subgrade.

Figure 10.49 presents a topographic map of the area, showing the location of the house and the path of the debris flow. The debris flow traveled down a narrow canyon (top of Fig. 10.49) that had an estimated drainage basin of 0.8 km^2 (200 acres). The complete extent of the source area could not be determined because of restricted access. The source area and main track probably extended from an elevation of about 980 m (3200 ft) to 400 m (1300 ft), and the slope inclination of the canyon varied from about 34 to 20°. At an elevation of about 400 m (1300 ft), the slope inclination changes to about 7°, corresponding to the beginning of the depositional area.

Figure 10.50 presents a photograph of the house, which is a one-story single-family structure, having a typical wood frame construction and a stucco exterior. It was observed in the area of the site that the debris flow was uniform with a thickness of about 0.6 m (2 ft). There was no damage to the structural frame of the house for two reasons: (1) The debris flow impacted the house and moved around the sides of the house, rather then through the house, because of the lack of any opening on the impact side of the house (Fig. 10.50). (2) The debris flow had traveled about 370 m (1200 ft) in the depositional area before striking the house and then proceeded only about 15 m (50 ft) beyond the house. By the time the debris flow reached the house, its energy had been nearly spent.

10.52 ANALYSIS OF GEOTECHNICAL DATA AND ENGINEERING COMPUTATIONS

A sample of the debris-flow material was tested for its grain-size distribution. The soil composing the debris flow was classified as a nonplastic silty sand (SM), with 16 percent gravel, 69 percent sand, and 15 percent silt and clay. The material in the debris flow was derived from weathering of rock from the source area. Geologic maps of the area indicated that the watershed is composed of metamorphic rocks, probably the Julian schist.

On the basis of the historical method, for the site shown in Fig. 10.49, it seems probable that there will be future debris flows. Preventive solutions include the restriction of houses in the depositional area or the construction of houses on elevated building pads. Notice in the center of Fig. 10.49 that two houses were built very close to the path of the debris flow, but they are situated at an elevation about 15 m (50 ft) above the depositional area and, hence, were unaffected by the debris flow.

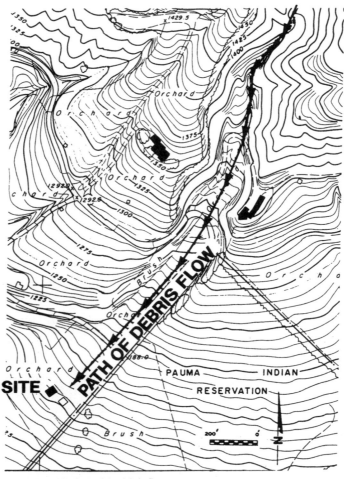

FIGURE 10.49 Path of the debris flow.

FIGURE 10.50 Photograph of site. Arrows indicate height of the debris flow.

10.8 SLOPE SOFTENING AND CREEP

This section describes the process of slope softening and creep. Both of these processes cause lateral movement of the slopes. As a practical matter, slope softening and creep need be considered only for plastic (cohesive) soil. Slopes composed of cohesionless soil, such as sands and gravels (Table 5.5), usually do not need to be evaluated for slope softening or creep.

Slope Softening. In many urban areas, there is a tendency toward small lot sizes because of the high cost of the land, with most of the lot occupied by building structures. A common soil engineering recommendation is to have the bottom edge of the footing at least 1.5 m (5 ft) horizontally from the face of any slope, irrespective of the slope height or soil type. This recommendation results in many buildings being constructed near the top of fill slopes.

Fill in slope areas is generally placed and compacted at optimum moisture content, which is often well below saturation. After construction of the slope, additional moisture is introduced into the fill by irrigation, rainfall, groundwater sources, and leaking water pipes. At optimum moisture content, a compacted clay fill can have a high shear strength because of negative pore water pressures. As water infiltrates the clay, the slope softens as the pore spaces fill with water and the pore water pressures tend toward zero. If a groundwater table then develops, the pore water pressures will become positive. The elimination of negative pore water pressure results in a decrease in effective stress and deformation of the slope in order to mobilize the needed shear stress to maintain stability. This process of moisture infiltration into a compacted clay slope which results in slope deformation has been termed *slope softening* (Day and Axten, 1990). The moisture migration into a cohesive fill slope which leads to slope deformation has also been termed *lateral fill extension*.

Some indications of slope softening include the rear patio pulling away from the structure, pool decking pulling away from the coping, tilting of improvements near the top of the slope, stair-step cracking in walls perpendicular to the slope, and downward deformation of that part of the building near the top of the slope. In addition to the slope movement caused by the slope softening process, there can be additional movement due to the process of creep.

Creep. Creep is generally defined as an imperceptibly slow and more or less continuous downward and outward movement of slope-forming soil or rock (Stokes and Varnes, 1955). Creep can affect both the near-surface soil or deep-seated materials. The process of creep is frequently described as viscous shear that produces permanent deformations, but not failure as in landslide movement. Typically the amount of movement is governed by the following factors: shear strength of the clay, slope angle, slope height, elapsed time, moisture conditions, and thickness of the active creep zone (Lytton and Dyke, 1980).

The process of creep is often divided into three different stages: primary or transient; steady state or secondary; and tertiary, which could lead to failure of the slope. These three stages of creep are illustrated in Fig. 10.51. Because of its relatively short duration, the primary (transient) creep is often ignored in slope deformation analysis. The secondary phase of creep often produces a relatively constant rate of strain, and depending on the slope conditions, it could continue for a considerable number of years. If the tertiary phase of creep is reached, the strain rate accelerates, and the slope could ultimately be subjected to a shear failure.

Example of Slope Creep. The purpose of this subsection is to present an example of a project having secondary creep (slope creep at a relatively constant rate). Figure 10.52 presents a cross section through the site, which was experiencing creep of a compacted clay slope. Typical damage included differential foundation displacement of the building located near the top of slope and cracking and separation of the patios located at the top of slope (Day, 1999). A boring was excavated at the top of slope (Boring B-1) and an inclinometer casing was installed in the boring (see Fig. 10.52 for location of Boring B-1).

An inclinometer is a device used to measure the amount of lateral movement of slopes. The procedure is to install a flexible casing into a vertical borehole. Then the inclinometer survey is performed by lowering an inclinometer probe into the flexible vertical casing. As the slope deforms, the flexible casing moves with the slope. By performing successive surveys, the shape and position of the flexible casing can be measured and the lateral deformation of the slope can be determined.

Figure 10.53 shows the results of the inclinometer monitoring. In Fig. 10.53, the vertical scale is depth below the ground surface and the horizontal scale is the lateral (out-of-slope) movement. The initial survey (base reading) was obtained on February 11, 1987. Additional successive surveys were performed from 1987 to 1995. As shown in Fig. 10.53, the inclinometer readings indicate a slow, continuous lateral movement (steady state creep) of the slope. The inclinometer data does not show movement on a discrete failure surface, but rather indicates a progressive creep of the entire fill slope. The inclinometer recorded the greatest amount of lateral movement [33 mm (1.3 in.)] at a depth of 0.3 m (1 ft). The amount of slope movement decreases with depth.

Figure 10.54 shows a plot of the amount of lateral slope movement versus time for readings at a depth of 0.3 m (1 ft) and 3.0 m (10 ft). This plot shows that the steady-state creep is about 4.1 mm/year (0.16 in./year) at a depth of 0.3 m (1 ft), and about 1.3 mm/year (0.05 in./year) at a depth of 3 m (10 ft).

An effective stress slope stability analysis was performed by using the SLOPE/W (Geo-Slope, 1991) computer program. In Fig. 10.52, the cross section was developed from the pre- and postgrading topographic maps and the results of the boring. The effective shear strength parameters of the compacted clay fill were obtained from drained direct shear strength tests,

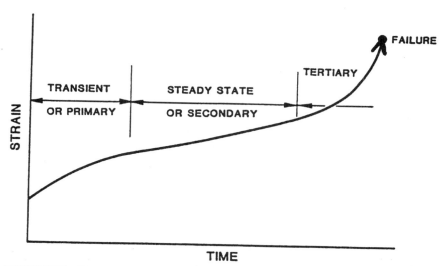

FIGURE 10.51 Three stages of creep. (*After Price, 1966.*)

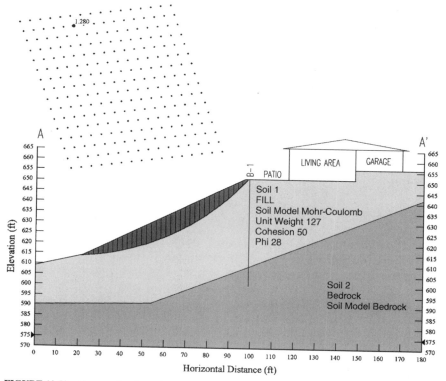

FIGURE 10.52 Cross section through a compacted clay slope subjected to creep.

and $c' = 2.4$ kPa (50 psf) and $\phi' = 28°$ were input into the computer program. Also used in the analysis was a total unit weight of the compacted clay fill (γ_t) of 20 kN/m³ (127 pcf). The pore water pressures were assumed to be equal to zero.

According to the SLOPE/W (Geo-Slope, 1991) computer program, the minimum factor of safety of the compacted fill slope, based on an effective stress analysis and using the Janbu method of slices, is 1.28. The results of the slope stability analysis are shown on Fig. 10.52. The stability analysis indicates that the fill closest to the slope face has the lowest factor of safety, which is consistent with the greatest amount of movement recorded from the inclinometer. In addition to the low factor of safety of the slope, there are apparently two additional factors that contributed to the creep of the slope:

1. *Loss of peak shear strength.* In the stability analysis, the peak effective shear strength parameters were used [$c' = 2.4$ kPa (50 psf) and $\phi' = 28$]. Figure 10.55 shows the shear strength versus horizontal deformation for the drained direct shear test specimen having a normal pressure during testing of 19 kPa (400 psf). The peak strength is identified in this figure. Note that with continued deformation, the shear strength decreases and reaches an ultimate value (which is less than the peak). It has been stated (Lambe and Whitman,

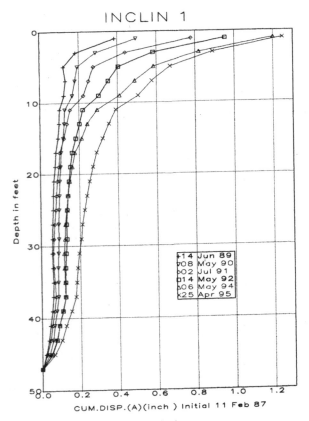

FIGURE 10.53 Inclinometer monitoring.

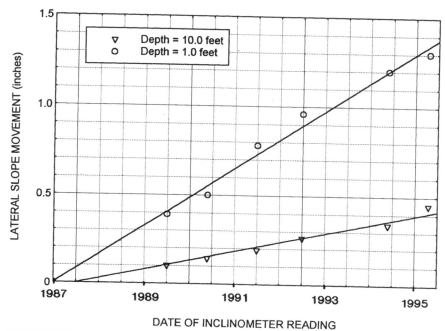

FIGURE 10.54 Lateral slope movement versus time.

1969) that overconsolidated soil, such as compacted clay, has a peak drained shear strength which it loses with further strain, and thus the overconsolidated and normally consolidated strengths approach each other at large strains. This reduction in shear strength of the compacted clay as it deforms during drained loading would result in an even lower factor of safety (lower than 1.28), which would further promote creep of the slope.

  2. *Seasonal moisture changes.* It has been stated that seasonal moisture changes can cause creep of fill slopes. For example, Bromhead (1984) states:

> These strains reveal themselves (usually at the surface) as deformations or ground movement. Some movement, particularly close to the surface, occurs even in slopes considered stable, as a result of seasonal moisture content variation; creep rates of a centimeter or so per annum are easily reached. Because of the shallow nature it is only likely to affect poorly-founded structures.

At this site, there would be seasonal moisture changes that probably contributed to the near-surface creep of the compacted clay fill slope.

Method of Analysis. The engineering geologist will often be involved with the studies of creep of natural slopes. In many cases, subsurface exploration (for example, test pits or trenches) excavated into the slope face will reveal the depth of active creep, which can be identified by the lateral offset of rock strata or soil layers.

  The zone of a compacted clay slope subjected to slope softening and creep is often difficult to estimate because it depends on many different factors. In general, the higher the

10.58 ANALYSIS OF GEOTECHNICAL DATA AND ENGINEERING COMPUTATIONS

plasticity index of the clay, the larger the zone of slope softening and creep. Also, the lower the factor of safety of the slope face, the more active the creep zone. In some cases grading options such as facing the slope with a cohesionless soil stabilization fill (see Standard Detail No. 3, App. C) can be used to mitigate the effects of slope creep.

For projects where clay slopes can not be avoided, the depth of slope softening and creep will be at least as deep as the depth of seasonal moisture changes in the clay (see Fig. 9.4). Structures should not be founded on the slope face or near the top of slope that corresponds to the zone of clay having seasonal moisture changes. Also, this zone should not be relied upon for support, such as passive pressure support for retaining wall footings (Chap. 15).

The clay slope could creep at a depth that is deeper than the depth of seasonal moisture changes. One design approach is to test the clay in the laboratory and model its slope softening and creep behavior. For example, a clay specimen could be prepared at anticipated field conditions, placed in the triaxial apparatus, sealed in a rubber membrane, and then subjected to the anticipated horizontal and vertical total stresses based on the slope configuration (stress path method). The measurement of the deformation upon saturation and deformation measurements versus time during the loading could be used to estimate the depth of the creep zone and the amount of anticipated slope movement.

10.9 DAMS

The design and construction of dams is complex and will be discussed only briefly in this section. Depending on the size and location of the dam, the design and construction

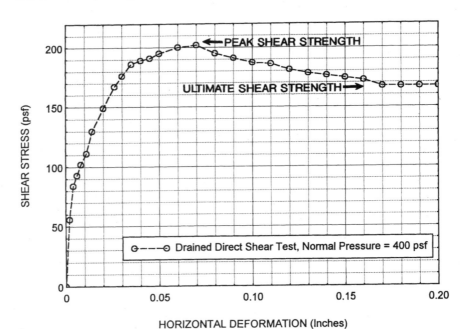

FIGURE 10.55 Drained direct shear test, silty clay fill.

may have to meet various state, local, and federal regulations. An excellent reference on the design and construction of earth dams is *Earth and Earth-Rock Dams*, by Sherard et al. (1963).

10.9.1 Large Dams

A dam failure has the potential to cause more damage and death than the failure of any other type of civil engineering structure. The worst type of failure is when the reservoir behind a large dam is full and the dam suddenly ruptures, causing a massive flood wave to surge downstream. When this type of sudden dam failure occurs without warning, the toll can be especially high. For example, the sudden collapse in 1928 of the St. Francis Dam in California killed about 450 people, which was California's second most destructive disaster, exceeded in loss of life and property only by the San Francisco earthquake of 1906. Like many dam failures, most of the dam was washed away and the exact cause of the failure is unknown. The consensus of forensic geologists and engineers is that the failure was due to adverse geologic conditions at the site. Three distinct geologic conditions could have led to the disaster: (1) slipping of the rock beneath the easterly side of the dam along weak geologic planes; (2) slumping of rocks on the westerly side of the dam as a result of water saturation; or (3) seepage of water under pressure along a fault beneath the dam (*Committee Report*, 1928; Association of Engineering Geologists, 1978; Schlager, 1994).

According to Middlebrooks' (1953) comprehensive study of earth dams, the most common causes of catastrophic failure are:

Overtopping. The most frequent cause of failure is water flowing over the top of the earth dam. This generally happens during heavy or record-breaking rainfall, which causes so much water to enter the reservoir that the spillway can not handle the flow or the spillway becomes clogged. Once the earth dam is overtopped, the erosive action of the water can quickly cut through the shells and core of the dam.

Piping. The second most common cause of earth dam failure is piping (Middlebrooks, 1953). Piping is defined as the progressive erosion of the dam at areas of concentrated leakage. As water seeps through the earth dam, seepage forces are generated that exert a viscous drag force on the soil particles. If the forces resisting erosion are less than the seepage forces, the soil particles are washed away and the process of piping commences. The forces resisting erosion include the cohesion of the soil, interlocking of individual soil particles, confining pressure from the overlying soil, and the action of any filters (Sherard et al., 1963). Figure 10.56 shows a series of photographs of the progressive piping failure of a small earth dam (from Sherard et al., 1963). The arrows in Fig. 10.56 point to the location of the piping failure.

There can be many different reasons for the development of piping in an earth dam. Some of the more common reasons are as follows (Sherard et al., 1963):

- *Poor construction control,* which can result in inadequately compacted or pervious layers in the embankment.
- *Inferior compaction* adjacent to concrete outlet pipes or other structures.
- *Poor compaction and bond* between the embankment and the foundation or abutments.
- *Leakage through cracks* that developed when portions of the dam were subjected to tensile strains caused by differential settlement of the dam.

- *Cracking* in outlet pipes, which is often caused by foundation settlement, spreading of the base of the dam, or deterioration of the pipe itself.
- *Leakage through the natural foundation soils* under the dam.

Leakage of the natural soils under the dam can be a result of the natural variation of the foundation material. Any seepage erupting on the downstream side of the dam is likely to cause "sand boils," which are circular mounds of soil deposited as the water exits the ground surface (see Fig. 10.57). Sand boils, if unobserved or unattended, can lead to complete failure by piping (Sherard et al., 1963).

Clean (uncemented) sand and nonplastic silt are probably the most susceptible to piping, with clay being the most resistant. There can be exceptions, such as dispersive clays, which Perry (1987) defines as follows:

> Dispersive clays are a particular type of soil in which the clay fraction erodes in the presence of water by a process of deflocculation. This occurs when the interparticle forces of repulsion exceed those of attraction so that the clay particles go into suspension and, if the water is flowing such as in a crack in an earth embankment, the detached particles are carried away and piping occurs.

According to Sherard (1972), one of the largest known areas of dispersive clays in the United States is north-central Mississippi. The cause of failure for several earth dams has been attributed to the piping of dispersive clays (Bourdeaux and Imaizumi, 1977; Stapledon and Casinader, 1977; Sherard et al., 1972). The pinhole test is a laboratory test that can be used to classify the clay as being either highly dispersive, moderately dispersive, slightly dispersive, or nondispersive. The test method consists of evaluating the erodibility of clay soils by causing water to flow through a small hole punched through the specimen (ASTM D 4647-93, 1998).

Slope Instability. Another common cause of dam failures is slope instability. There could be a gross slope failure of the upstream or downstream faces of the dam or sliding of the dam foundation. The development of these failures is similar to that of the gross slope failures and landslides discussed in Secs. 10.5 and 10.6, and slope stability analyses can be used to design the upstream and downstream faces of the dam. Design analyses can be grouped into three general categories (Sherard et al., 1963):

1. Analysis of stability during construction, where a failure usually occurs through the natural ground underlying the dam. A total and/or effective stress slope stability analysis could be performed, depending on the type of soil beneath the dam and the rate of construction.
2. Stability analyses of the downstream slope during reservoir operation. Usually an effective stress analysis is performed, assuming a full reservoir condition. A flow net can be used to estimate the groundwater level and pore water pressures within the earth dam.
3. Stability analyses of the upstream slope after reservoir drawdown. Like item 1 above, the type of slope stability analysis (total stress versus effective stress analysis) would depend on the soil type and assumed rate of drawdown.

Other Analyses. Besides the three most important design considerations (overtopping, piping, and slope instability), there can be many other analyses required for the design of earth dams (see Sherard et al., 1963). The checklist provided in Table 10.1 may also prove useful for the design of dams.

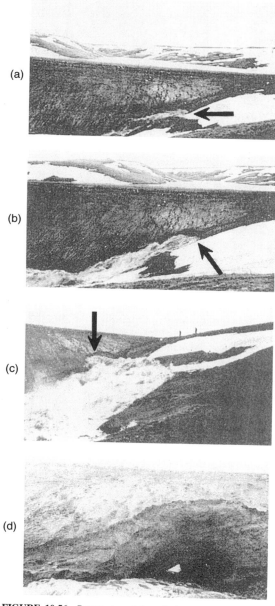

FIGURE 10.56 Progressive piping of abutment leak at small dam: (*a*) 3:30 P.M.; (*b*) 3:45 P.M.; (*c*) 4:30 P.M.; (*d*) 5:30 P.M. (*From Sherard et al., 1963.*)

FIGURE 10.57 Typical sand boil. (*From Sherard et al., 1963.*)

10.9.2 Small Dams

Small dams have been classified as those dams having a height of less than 12 m (40 ft) or those that impound a volume of water less than a million cubic meters (1000 acre-ft) (Corns, 1974). Sowers (1974) states that, although failures of large dams are generally more spectacular, failures of small dams occur far more frequently. Reasons for the higher frequency of failures include: (1) lack of appropriate design, (2) owners believe that the consequences of failure of a small dam will be minimal, (3) small dam owners frequently have no previous experience with dams, and (4) smaller dams are often not maintained (Sowers, 1974). On the basis of his experience, Griffin (1974) lists some of the common deficiencies with small dams:

1. Subsurface or geological investigations were minimal or nonexistent.
2. No provision was made for future maintenance, and maintenance was totally ignored for many years.
3. Slopes of many of the structures were constructed too steep for routine maintenance.
4. Construction supervision varied from inadequate to nonexistent.
5. Hydrological design was deficient in that the top elevation of the dam was set an arbitrary distance above the pool with no flood routings being made. These hydrological deficiencies are now amplified by development within the watershed.
6. Inadequate purchase of land to protect the investment in the project.

10.9.3 Landslide Dams

Landslide dams usually develop when there is a blockage of a valley by landslides, debris flow, or rockfall. Concerning landslide dams, Schuster (1986) states:

SLOPE STABILITY

Landslide dams have proved to be both interesting natural phenomena and significant hazards in many areas of the world. A few of these blockages attain heights and volumes that rival or exceed the world's largest man-made dams. Because landslide dams are natural phenomena and thus are not subject to engineering design (although engineering methods can be utilized to alter their geometries or add physical control measures), they are vulnerable to catastrophic failure by overtopping and breaching. Some of the world's largest and most catastrophic floods have occurred because of failure of these natural dams.

According to Schuster and Costa (1986), most landslide dams are short-lived. In their study of 63 landslide dams, 22 percent failed in less than 1 day after formation, and half failed within 10 days. According to Schuster and Costa (1986), overtopping was by far the most frequent cause of landslide-dam failure.

PROBLEMS

The problems have been divided into basic categories as indicated below:

Surficial Slope Stability

1. For the surficial stability analysis summarized in Table 10.2, calculate the pore water pressure u at the slip surface. *Answer:* $u = 8.1$ kPa.
2. For the surficial stability analysis summarized in Table 10.2, calculate the shear stress τ along the slip surface. *Answer:* $\tau = 11.0$ kPa.
3. For the surficial stability analysis summarized in Table 10.2, calculate the effective normal stress σ'_n on the slip surface. *Answer:* $\sigma'_n = 8.3$ kPa.
4. For the surficial stability analysis summarized in Table 10.2, calculate the shear strength τ_f along the slip surface. *Answer:* Using the linear shear strength envelope, $\tau_f = 27.4$ kPa. For the nonlinear portion of the shear strength envelop at low effective stress, $\tau_f = 10.8$ kPa.
5. Using the data from Probs. 2 and 4, calculate the ratio τ_f/τ and compare the values with the values for the factor of safety in Table 10.2. *Answer:* Values are the same because $\tau_f/\tau =$ factor of safety.
6. Perform the surficial stability calculations summarized in Table 10.2 for a slope inclination α of 2:1 (horizontal:vertical). *Answers:* For the linear portion of the shear strength envelope, $F = 2.9$. For the nonlinear portion of the shear strength envelope at low effective stress, $F = 1.37$.

Gross Slope Stability

7. Revise Eq. (10.3) so that it includes a vertical surcharge load Q applied to the failure wedge. *Answers:* Total stress analysis:

$$F = \frac{cL + (W + Q) \cos \alpha \tan \phi}{(W + Q) \sin \alpha}$$

Effective stress analysis:

$$F = \frac{c'L + [(W + Q) \cos \alpha - uL] \tan \phi'}{(W + Q) \sin \alpha}$$

8. Revise Eq. (10.3) so that it includes a horizontal earthquake force $= bW$, where $b = a$ constant and $W =$ weight of the wedge. *Answers:* Total stress analysis:

$$F = \frac{cL + (W \cos \alpha - bW \sin \alpha) \tan \phi}{W \sin \alpha + bW \cos \alpha}$$

Effective stress analysis:

$$F = \frac{c'L + (W \cos \alpha - bW \sin \alpha - uL) \tan \phi'}{W \sin \alpha + bW \cos \alpha}$$

9. For the two example problems in Sec. 10.5, assume there is a uniform vertical surcharge pressure of 10 kPa applied to the entire top of the slope. Calculate the factors of safety. *Answers:* For the total stress example problem, $F = 1.62$. For the effective stress example problem, $F = 1.94$.

10. For the two examples in Sec. 10.5, assume that an earthquake causes a horizontal force $= 0.1\ W$ (i.e., $b = 0.1$). Calculate the factors of safety (pseudo-static slope stability analysis). *Answers:* For the total stress example problem, $F = 1.39$. For the effective stress example problem, $F = 1.46$.

11. A computer program is used to determine the factor of safety of a slope. Assume that the computer program uses the ordinary method of slices. For a particular slice located within the assumed failure mass, the total weight W of the slice $= 100$ kN, the average slope inclination of the slice $(\alpha) = 26°$, the pore water pressure u at the bottom of the slice $= 20$ kPa, the width of the slice $(\Delta x) = 1.0$ m, and the peak effective shear strength parameters along the assumed slip surface are $\phi' = 25°$ and $c' = 2$ kPa. Assuming an effective stress slope stability analysis is being performed, calculate the shear stress τ and the shear strength τ_f along the base of the slice. *Answers:* $\tau = 39.4$ kPa and $\tau_f = 39.7$ kPa.

12. For Prob. 11, calculate the factor of safety of the slice. If the factor of safety for the entire failure mass $= 1.2$ and the slope is composed of stiff-fissured clay that has a substantial reduction in shear strength for strain past the peak values, is progressive failure likely for this slope? *Answers:* $F = \tau_f/\tau = 1.01$, and, yes, progressive failure is likely for this slope.

13. Figure 10.58 presents a plot of the lateral downslope movement of a slope. The data was obtained from inclinometer monitoring of a casing that was installed at the top of slope. The base reading for the inclinometer monitoring is August 1991, with readings taken through September 1992. Based on the inclinometer results (Fig. 10.58), what can be concluded concerning the stability of the slope? *Answer:* The slope is deforming laterally on a slip surface located about 7 m below ground surface.

Creep

14. Assume a 2:1 (horizontal:vertical) slope is composed of Soil B (Fig. 9.4). Also assume the slope height is 20 m. On the basis of the depth of seasonal moisture changes and assuming slope creep will occur in this zone, how far back from the top of slope should the structure be located? *Answer:* 9 m.

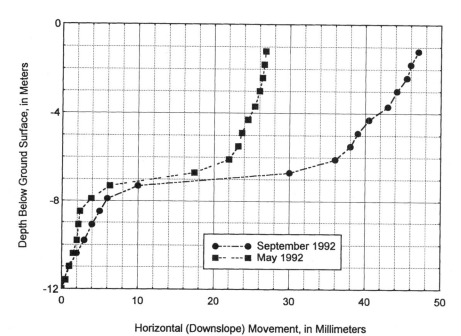

FIGURE 10.58 Results from inclinometer monitoring.

CHAPTER 11
EARTHQUAKES

The following notation is used in this chapter:

SYMBOL	DEFINITION
a_{max}	Peak acceleration at the ground surface
C_b	SPT correction factor to account for the borehole diameter
C_N	SPT correction factor to account for the vertical effective stress
C_r	SPT correction factor to account for the drilling rod length
E_m	SPT correction factor to account for the hammer efficiency
g	Acceleration of gravity
M	Earthquake magnitude
N_{60}	SPT N-value corrected for field testing procedures
$(N_1)_{60}$	SPT N-value corrected for both field testing procedures and σ'_{v0}
r_d	Depth reduction factor
W	Weight of the sliding mass
z	Depth below ground surface
σ_{v0}	Total vertical stress
σ'_{v0}	Vertical effective stress

11.1 INTRODUCTION

Chapter 11 presents a brief discussion of earthquakes. A structure could be damaged by the seismic energy of the earthquake or actual rupture of the ground due to fault movement. Other problems caused by the ground shaking from earthquakes include the settlement of loose soil, slope movement or failure, and liquefaction of loose granular soil below the groundwater table. Figures 11.1 and 11.2 show the collapse of an apartment building during the Northridge, California, earthquake caused by the weak shear resistance of the first floor garage area.

Earthquakes throughout the world cause a considerable amount of death and destruction. Much as diseases will attack the weak and infirm, earthquakes damage those structures that have inherent weaknesses or age-related deterioration. Those buildings that are nonreinforced, poorly constructed, weakened from age or rot, or underlain by soft or unstable soil are most susceptible to damage. For example, Fig. 11.3 shows the collapse of a brick chimney during the 1994 Northridge, California, earthquake. The chimney was poorly constructed and, as shown in Fig. 11.3, there were no connections between the chimney and house.

Besides buildings, earthquakes can damage other structures, such as earth dams. Sherard et al. (1963) indicate that in the majority of dams shaken by severe earthquakes,

11.2 ANALYSIS OF GEOTECHNICAL DATA AND ENGINEERING COMPUTATIONS

FIGURE 11.1 Building collapse caused by the 1994 Northridge, California, earthquake.

FIGURE 11.2 View inside the collapsed first floor parking garage (arrows point to columns).

FIGURE 11.3 Collapse of brick chimney, 1994 Northridge, California, earthquake.

two primary types of damage have developed: (1) longitudinal cracks at the top of the dam and (2) crest settlement. In the case of the Sheffield Dam, a complete failure did occur, probably due to liquefaction of the very loose and saturated lower portion of the embankment (Sherard et al., 1963).

11.2 EARTHQUAKES

A structure can be damaged by many different earthquake effects. Examples previously provided include a lack of shear resistance (Figs. 11.1 and 11.2) and poor construction practices (Fig. 11.3). The purpose of the following sections is to provide a brief summary of the common earthquake effects that need to be considered by the geotechnical engineer.

11.2.1 Fault and Ground Rupture Zone

Surface fault rupture caused by the earthquake is important because it has caused severe damage to buildings, bridges, dams, tunnels, canals, and underground utilities (Lawson et al., 1908; Ambraseys, 1960; Duke, 1960; California Department of Water Resources, 1967; Bonilla, 1970; Steinbrugge, 1970). Fault displacement is defined as the relative movement of the two sides of a fault, measured in a specific direction (Bonilla, 1970). Figure 11.4 shows the displacement of rock strata caused by the Carmel Valley Fault, located at Torry Pines, California.

Examples of very large surface fault rupture are the 11 m (35 ft) of vertical displacement in the Assam earthquake of 1897 (Oldham, 1899) and the 9 m (29 ft) of horizontal

11.4 ANALYSIS OF GEOTECHNICAL DATA AND ENGINEERING COMPUTATIONS

FIGURE 11.4 Carmel Valley Fault, located at Torry Pines, California.

movement during the Gobi-Altai earthquake of 1957 (Florensov and Solonenko, 1965). The length of the fault rupture can be quite significant. For example, the estimated length of surface faulting in the 1964 Alaskan earthquake varied from 600 to 720 km (Savage and Hastie, 1966; Housner, 1970).

A recent (geologically speaking) earthquake caused the fault rupture shown in Fig. 11.5. The fault is located at the base of the Black Mountains, in California. The vertical fault displacement caused by the earthquake is the vertical distance between the two arrows in Fig. 11.5. The fault displacement occurred in an alluvial fan being deposited at the base of the Black Mountains. Most structures would be unable to accommodate the huge vertical displacement shown in Fig. 11.5.

In addition to fault rupture, there can also be ground rupture away from the main trace of the fault. These ground cracks could be caused by many different factors, such as movement of subsidiary faults, auxiliary movement that branches off from the main fault trace, or ground rupture caused by the differential or lateral movement of underlying soil deposits. For example, Fig. 11.6 shows ground rupture during the 1994 Northridge,

California, earthquake. The direction of the ground shear movement shown in Fig. 11.6 is toward the northwest. The ground movement sheared both the concrete patio and adjacent pool and knocked the house off its foundation.

Since most structures will be unable to resist the shear movement associated with surface faulting and ground rupture, one design approach is to simply restrict construction in the active fault shear zone. The best individual to determine the location and width of the active fault shear zone is the engineering geologist. Seismic study maps, such as the *State of California Special Studies Zones Maps* (1982), which were developed as part of the Alquist-Priolo Special Studies Zones Act, delineate the approximate location of active fault zones that require special geologic studies. These maps also indicate the approximate locations of historic fault offsets, which are indicated by year of earthquake-associated event, as well as the locations of ongoing fault displacement due to fault creep. There are many other geologic references, such as the cross section shown in Fig. 4.2, that can be used to identify active shear fault zones. Trenches, such as shown in Figs. 4.20 and 4.21, can be excavated across the fault zone to more accurately identify the width of the active fault shear zone. Critical structures, such as essential transportation routes (see Sec. 2.2) that must cross the active shear fault zones, will need special designs to resist the earthquake forces induced by ground rupture.

11.2.2 Liquefaction

The typical subsurface condition that is susceptible to liquefaction is a loose or very loose sand, that has been newly deposited or placed, with a groundwater table near ground surface. During an earthquake, the ground shaking causes the loose sand to contract, resulting in an increase in pore water pressure. Because the seismic shaking occurs so quickly, the

FIGURE 11.5 Fault rupture at the base of the Black Mountains (arrows indicate the amount of vertical displacement caused by the earthquake).

FIGURE 11.6 Ground rupture, 1994 Northridge, California, earthquake.

cohesionless soil is subjected to an undrained loading (total stress analysis). The increase in pore water pressure causes an upward flow of water to the ground surface, where it emerges in the form of mud spouts or sand boils. The development of high pore water pressures due to the ground shaking (i.e., the effective stress becomes zero) and the upward flow of water may turn the sand into a liquefied condition, which has been termed *liquefaction*. Structures on top of the loose sand deposit that has liquefied during an earthquake will sink or fall over, and buried tanks will float to the surface when the loose sand liquefies (Seed, 1970).

Liquefaction can also cause lateral movement of slopes and create flow slides (Ishihara, 1993). Seed (1970) states:

If liquefaction occurs in or under a sloping soil mass, the entire mass will flow or translate laterally to the unsupported side in a phenomenon termed a flow slide. Such slides also develop in loose, saturated, cohesionless materials during earthquakes and are reported at Chile (1960), Alaska (1964), and Niigata (1964).

An example of lateral movement of liquefied sand is shown in Fig. 11.7 (from Kerwin and Stone, 1997). This damage occurred to a marine facility at Redondo Beach King Harbor during the 1994 Northridge, California, earthquake. The 5.5 m (18 ft) of horizontal displacement was caused by the liquefaction of an offshore sloping fill mass that was constructed as a part of the marine facility.

There can also be liquefaction of seams of loose saturated sands within a slope. This can cause the entire slope to move laterally along the liquefied layer at the base. These types of gross slope failures caused by liquefied seams of soil caused extensive damage during the 1964 Alaskan earthquake (Shannon and Wilson, Inc., 1964; Hansen, 1965). It has been observed that slope movement of this type typically results in little damage to structures located on the main slide mass, but buildings located in the graben area are subjected to large differential settlements and are often completely destroyed (Seed, 1970).

Main Factors for Liquefaction of Cohesionless Soil.   There are many factors that govern the liquefaction process. The most important factors are as follows:

1. *Earthquake intensity and duration.*   In order to have liquefaction of soil, there must be ground shaking. The character of the ground motion, such as acceleration and frequency content, determines the shear strains that cause the contraction of the soil particles and the development of excess pore water pressures leading to liquefaction. The most common

FIGURE 11.7   Damage to marine facility, 1994 Northridge, California, earthquake. (*From Kerwin and Stone, 1997; reprinted with permission from the American Society of Civil Engineers.*)

cause of liquefaction is the seismic energy released during an earthquake. The potential for liquefaction increases as the earthquake intensity and duration increase. Sites located near the epicenter of major earthquakes will be subjected to the largest intensity and duration of ground shaking (i.e., higher number of applications of cyclic shear strain). Besides earthquakes, other conditions can cause liquefaction, such as subsurface blasting.

2. *Groundwater table.* The condition most conducive to liquefaction is a near-surface groundwater table. Unsaturated soil located above the groundwater table will not liquefy.

3. *Soil type.* The soil types susceptible to liquefaction are nonplastic (cohesionless) soils. Seed et al. (1983) state that, on the basis of both laboratory testing and field performance, the great majority of clayey soils will not liquefy during earthquakes.

An approximate listing of cohesionless soils from least to most resistant to liquefaction are clean sands, nonplastic silty sands, nonplastic silt, and gravels. There could be numerous exceptions to this sequence. For example, Ishihara (1985, 1993) describes the case of tailings derived from the mining industry that were essentially composed of ground-up rocks and were classified as rock flour. Ishihara (1985, 1993) states that the rock flour in a water-saturated state did not possess significant cohesion and behaved as if it were a clean sand. These tailings were shown to exhibit as low a resistance to liquefaction as clean sand.

4. *Soil relative density D_r.* Cohesionless soils in a very loose relative density state are susceptible to liquefaction, while the same soil in a very dense relative density state will not liquefy. Very loose nonplastic soils will contract during the seismic shaking which will cause the development of excess pore water pressures. Very dense soils will dilate during seismic shaking and are not susceptible to liquefaction.

5. *Particle size gradation.* Poorly graded nonplastic soils tend to form more unstable particle arrangements and are more susceptible to liquefaction than well-graded soils.

6. *Placement conditions.* Hydraulic fills (fill placed under water) tend to be more susceptible to liquefaction because of the loose and segregated soil structure created by the soil particles falling through water.

7. *Drainage conditions.* If the excess pore water pressure can quickly dissipate, the soil may not liquefy. Thus gravel drains or gravel layers can reduce the liquefaction potential of adjacent soil.

8. *Confining pressures.* The greater the confining pressure, the less susceptible the soil is to liquefaction. Conditions that can create a higher confining pressure are a deeper groundwater table, soil that is located at a deeper depth below ground surface, and a surcharge pressure applied at ground surface. Case studies have shown that the possible zone of liquefaction usually extends from the ground surface to a maximum depth of about 15 m (50 ft). Deeper soils generally do not liquefy because of the higher confining pressures.

9. *Aging.* Newly deposited soils tend to be more susceptible to liquefaction than old deposits of soil. Older soil deposits may already have been subjected to seismic shaking or the soil particles may have deformed or been compressed into more stable arrangements.

Liquefaction Analysis. The most common type of analysis to determine the liquefaction potential is to use the standard penetration test (SPT) or the cone penetration test (CPT) (Seed et al., 1985; Stark and Olson, 1995). The analysis is based on the simplified method proposed by Seed and Idriss (1971). The method of analysis is as follows:

1. *Seismic shear stress ratio (SSR) caused by the earthquake.* The first step in the liquefaction analysis is to determine the seismic shear stress ratio (SSR). The seismic shear stress ratio induced by the earthquake at any point in the ground is estimated as follows (Seed and Idriss, 1971):

$$\text{SSR} = 0.65\, r_d \left(\frac{a_{max}}{g}\right)\left(\frac{\sigma_{v0}}{\sigma'_{v0}}\right) \quad (11.1)$$

where SSR = seismic shear stress ratio (dimensionless parameter)
a_{max} = peak acceleration measured or estimated at the ground surface of the site (m/s²)
g = acceleration of gravity (9.81 m/s²)*
σ_{v0} = total vertical stress at a particular depth where the liquefaction analysis is being performed (kPa) (in order to calculate the total vertical stress, the total unit weight γ_t of the soil layers must be known)
σ'_{v0} = vertical effective stress at that same depth in the soil deposit where σ_{v0} was calculated (kPa) (in order to calculate the vertical effective stress, the location of the groundwater table must be known).

The term r_d = depth reduction factor, which can be estimated in the upper 10 m of soil as (Kayan et al., 1992):

$$r_d = 1 - (0.012)(z) \quad (11.2)$$

where z = depth in meters below the ground surface where the liquefaction analysis is being performed (i.e., the same depth used to calculate σ_{v0} and σ'_{v0})

2. *Seismic shear stress ratio that will cause liquefaction of the soil.* The second step is to determine the seismic shear stress ratio (SSR) that will cause liquefaction of the *in situ* soil. Figure 11.8 presents a chart that can be used to determine the seismic shear stress ratio (SSR) that will cause liquefaction of the *in situ* soil. In order to use this chart, the results of the standard penetration test (SPT) must be expressed in terms of the SPT $(N_1)_{60}$ value. In liquefaction analysis, the SPT N_{60} value [Eq. (4.3)] is corrected for the overburden pressure. When a correction is applied to the SPT N_{60} value to account for the effect of overburden pressure, these values are referred to as SPT $(N_1)_{60}$ values. The procedure consists of multiplying the N_{60} value by a correction C_N in order to calculate the SPT $(N_1)_{60}$ value, or:

$$(N_1)_{60} = C_N N_{60} = (100/\sigma'_{v0})^{0.5} N_{60} \quad (11.3)$$

where $(N_1)_{60}$ = standard penetration N-value corrected for both field testing procedures and overburden pressure
C_N = correction factor to account for the overburden pressure [as indicated in Eq. (11.3), C_N is approximately equal to $(100/\sigma'_{v0})^{0.5}$, where σ'_{v0} is the vertical effective stress in kPa]
N_{60} = standard penetration N-value corrected for testing procedures [note that N_{60} is calculated by using Eq. (4.3)]

Once the corrected SPT $(N_1)_{60}$ has been calculated, Fig. 11.8 can be used to determine the seismic shear stress ratio (SSR) that will cause liquefaction of the *in situ* soil. Note that Fig. 11.8 is for a projected earthquake of 7.5 magnitude. The figure also has different curves that are to be used according to the percent fines in the soil. For a given $(N_1)_{60}$ value,

---

*Usually the engineering geologist will determine the peak acceleration at the ground surface at the site from fault, seismicity, and attenuation studies. Typically the engineering geologist provides a peak ground acceleration in the form of a_{max}/g = a constant. For example, the engineering geologist may determine that the peak ground surface acceleration at a site is $a_{max}/g = 0.1$, in which case the value of 0.1 (dimensionless) is substituted into Eq. (11.1) in place of a_{max}/g.

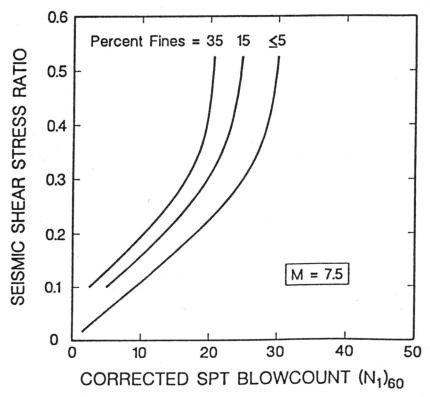

FIGURE 11.8 Relationship between seismic shear stress ratio (SSR) triggering liquefaction and $(N_1)_{60}$ values for clean and silty sands for $M = 7.5$ earthquakes. (*After Seed and DeAlba, 1986; reproduced from Stark and Olson, 1995; reprinted with permission of the American Society of Civil Engineers.*)

soils with more fines have a higher seismic shear stress ratio (SSR) that will cause liquefaction of the *in situ* soil.

Figure 11.9 presents a chart for clean sands (5 percent or less fines) and different magnitude earthquakes. The magnitude 7.5 curve in Fig. 11.9 is similar to the magnitude 7.5 curve for 5 percent or less fines in Fig. 11.8.

3. *Compare seismic shear stress ratios.* The final step in the liquefaction analysis is to compare the seismic shear stress ratio (SSR) values. If the SSR value from Eq. (11.1) is greater than the SSR value obtained from either Fig. 11.8 or Fig. 11.9, then liquefaction could occur during the earthquake, and vice versa.

Example. It is planned to construct a building on a cohesionless soil deposit (fines less than 5 percent). There is a nearby major active fault and the engineering geologist has determined that the peak ground acceleration $(a_{max}) = 0.4g$. Assume the site conditions are the same as stated in Prob. 18 (Chap. 6), i.e., a level ground surface with the groundwater table located 1.5 m below ground surface and the standard penetration test performed at a depth of 3 m. If the earthquake magnitude (M) = 7.5, will the saturated clean sand located at a depth of 3 m below ground surface liquefy during the anticipated earthquake?

Solution. From Prob. 18 (Chap. 6), σ'_{v0} = 43 kPa and N_{60} = 5. Using the total unit weights from Prob. 18 (Chap. 6), σ_{v0} = 58 kPa. Since z = 3 m, r_d = 0.96. Using the following values:

$r_d = 0.96$
$a_{max}/g = 0.4$
$(\sigma_{v0}/\sigma'_{v0}) = (58/43) = 1.35$

and inserting the above values into Eq. (11.1), we find the seismic shear stress ratio (SSR) caused by the earthquake is 0.34.

The next step is to determine the seismic shear stress ratio (SSR) that will cause liquefaction of the *in situ* soil. From Prob. 18 (Chap. 6), the N-value corrected for field testing procedures (N_{60}) = 5. Using Eq. (11.3) with σ'_{v0} = 43 kPa and N_{60} = 5, we find

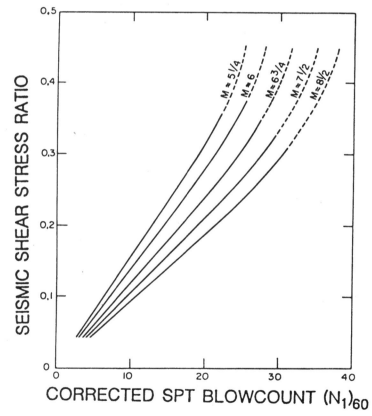

FIGURE 11.9 Relationship between seismic shear stress ratio (SSR) triggering liquefaction and $(N_1)_{60}$ values for clean sand for different magnitude earthquakes. (*After Seed et al., 1983; reprinted with permission of the American Society of Civil Engineers.*)

11.12 ANALYSIS OF GEOTECHNICAL DATA AND ENGINEERING COMPUTATIONS

$(N_1)_{60} = 8$. Entering Fig. 11.8 with $(N_1)_{60} = 8$ and intersecting the curve labeled ≤ 5% fines, the seismic shear stress ratio (SSR) that will cause liquefaction of the *in situ* soil at a depth of 3 m = 0.08.

The final step is to compare the SSR caused by the earthquake (SSR = 0.34) with the SSR that will cause liquefaction of the *in situ* soil (SSR = 0.08). From a comparison of the SSR values, it is probable that during the earthquake the *in situ* sand located at a depth of 3 m below ground surface will liquefy.

In the above liquefaction analysis, there are many different equations and corrections that are applied to the seismic shear stress ratio (SSR). For example, there are four different corrections (E_m, C_b, C_r, and σ'_{v0}) that are applied to the SPT N-value in order to calculate the $(N_1)_{60}$ value. All of these different equations and various corrections may provide the engineer with a sense of high accuracy, when in fact, the entire analysis is only a gross approximation. The analysis should be treated as such and engineering experience and judgment are essential in the final determination of whether or not a site has liquefaction potential.

11.2.3 Slope Movement and Settlement

Besides liquefaction of loose saturated sands, other soil conditions can result in slope movement or settlement during an earthquake. For example, Grantz et al. (1964) describe an interesting case of ground vibrations from the 1964 Alaskan earthquake that caused 0.8 m (2.6 ft) of alluvium settlement. Other loose soils, such as cohesionless sand and gravel, will also be susceptible to settlement due to the ground vibrations from earthquakes.

Slopes having a low factor of safety can experience large horizontal movement during an earthquake. Types of slopes most susceptible to movement during earthquakes include those slopes composed of soil that loses shear strength with strain (such as sensitive soil) and ancient landslides that can become reactivated by seismic forces (Day and Poland, 1996).

11.2.4 Translation and Rotation

An unusual effect caused by earthquakes is translation and rotation of objects. For example, Fig. 11.10 shows a photograph of a brick mailbox that rotated and translated (moved laterally) during the Northridge earthquake. The initial position in Fig. 11.10 refers to the pre-earthquake position of the mailbox.

Earthquakes have caused the rotation of other movable objects, such as grave markers (Athanasopoulos, 1995; Yegian et al., 1994). According to Athanasopoulos (1995), such objects will rotate in such a manner as to be aligned with the strong component of the earthquake. Besides rotation, translation (lateral movement) can also occur during an earthquake. The objects will tend to move in the same direction as the propagation of energy waves, i.e., in a direction away from the epicenter of the earthquake.

11.3 ESTIMATING EARTHQUAKE GROUND MOVEMENT

Often the geotechnical engineer will be required to estimate the amount of foundation displacement caused by earthquake-induced soil movement. For example, the *Uniform*

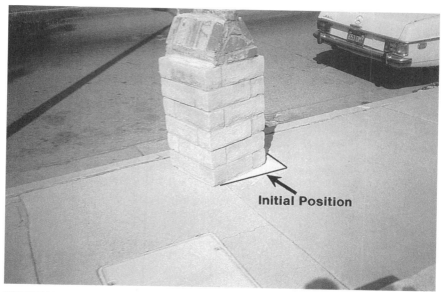

FIGURE 11.10 Rotation of brick mailbox, 1994 Northridge, California, earthquake.

Building Code (1997), which is the building code required for construction in California, states (code provision submitted by the author, adopted in May 1994):

> The potential for soil liquefaction and soil strength loss during earthquakes shall be evaluated during the geotechnical investigation. The geotechnical report shall assess potential consequences of any liquefaction and soil strength loss, including estimation of differential settlement, lateral movement or reduction in foundation soil-bearing capacity, and discuss mitigating measures. Such measures shall be given consideration in the design of the building and may include, but are not limited to, ground stabilization, selection of appropriate foundation type and depths, selection of appropriate structural systems to accommodate anticipated displacement or any combination of these measures.

The intent of this building code requirement is to obtain an approximate estimate of the foundation displacement caused by the earthquake induced soil movement. In terms of accuracy of the calculations used to determine the earthquake induced soil movement, Tokimatsu and Seed (1984) conclude:

> It should be recognized that, even under static loading conditions, the error associated with the estimation of settlement is on the order of ±25 to 50%. It is therefore reasonable to expect less accuracy in predicting settlements for the more complicated conditions associated with earthquake loadin.... In the application of the methods, it is essential to check that the final results are reasonable in light of available experience.

Pseudostatic Approach. A vast majority of foundation and earthwork designs are based on the pseudostatic approach (Coduto, 1994). This method ignores the cyclic nature of earthquakes and treats them as if they apply an additional static force upon the slope, retaining wall, or foundation element. For example, as will be discussed in Sec. 15.3, a common approach for the design of retaining walls is to subject the retaining wall to an additional static force [P_E, Eq. (15.11)] that represents the seismic energy of the earthquake.

For slope stability analysis, the pseudostatic approach is to apply a lateral force acting through the center of the sliding mass, where the lateral force = $(W)(a_{max}/g)$ and W = weight of the failure wedge or sliding mass of soil. Many local jurisdictions require specific values of a_{max}/g that must be used in the pseudostatic analysis (typically $a_{max}/g = 0.1$ to 0.2 in southern California). Most slope stability computer programs (Sec. 10.5) have the ability to perform pseudostatic slope stability analyses and the only additional data that needs to be input is the value of a_{max}/g. In southern California, an acceptable minimum factor of safety of the slope is 1.1 for pseudostatic slope stability analysis.

The major unknown in the pseudostatic method is the shear strength of the soil. In most cases, the dynamic shear resistance of the soil is assumed to be equal to the static shear strength (i.e., the shear strength of the soil that exists prior to the earthquake). This pseudostatic approach can be substantially in error for the following conditions (Coduto, 1994):

1. *Sensitive clay.* These soils will lose a portion of their shear strength during cyclic loading. The higher the sensitivity [Eq. (5.5)], the greater the loss of shear strength for a given shear strain.

2. *Saturated loose to medium sands.* Even if these cohesionless soils do not liquefy, they can still develop large excess pore water pressures during the earthquake and lose some or a significant portion of their shear strength.

3. *Immediate settlement.* There can be significant settlement for foundations on soft saturated clays because of undrained plastic flow when the foundations are overloaded during the seismic shaking.

4. *Uplift loads.* The foundation may be subjected to uplift loads during an earthquake. The foundation is often more susceptible to failure during an uplift load, as was demonstrated by uplift failures of deep foundations in Mexico City during the 1985 earthquake.

Coduto (1994) suggests that in most situations, it is probably best to use dynamic shear strengths equal to or less than the static shear strengths. Lower shear strengths would be used for sensitive clays or soils likely to experience a reduction in shear strength, such as loose sands that may not liquefy, but will lose shear strength during an earthquake.

Settlement. Estimating the settlement of a structure caused by an earthquake is usually very difficult. For example, if the upper layer of a sand deposit supports a heavy structure and this sand layer liquefies during an earthquake, the building will literally sink into the liquefied soil. In this instance, there would be a large amount of foundation settlement and extensive damage to the building.

Figure 11.11 shows one method that can be used to estimate ground surface subsidence of saturated clean sands. As in Figs. 11.8 and 11.9, the SPT $(N_1)_{60}$ value is first calculated. Also, the seismic shear stress ratio (SSR) is calculated from Eq. (11.1). By entering the chart with the SPT $(N_1)_{60}$ value and the seismic shear stress ratio (SSR) calculated from Eq. (11.1), the volumetric strain (%) can be estimated. Note that Fig. 11.11 can be used only to estimate the volumetric strain (%) of the soil, and it does not include the amount of settlement of the building due to its weight, which can cause the building to sink into the soil as described in the above example.

In the example in Sec. 11.2.2, SPT $(N_1)_{60} = 8$ and the earthquake-induced seismic shear stress ratio (SSR) = 0.34. Entering Fig. 11.11 with these two values, we find the volumetric strain (%) is about 2.8 percent. If the *in situ* sand layer is 1 m thick, then the estimated settlement of this 1-m-thick layer during this earthquake would be 2.8 cm (i.e., 1 m times 0.028 = 2.8 cm). Tokimatsu and Seed (1984) also present a simplified method of determining the settlement of *in situ* dry loose sands during seismic shaking.

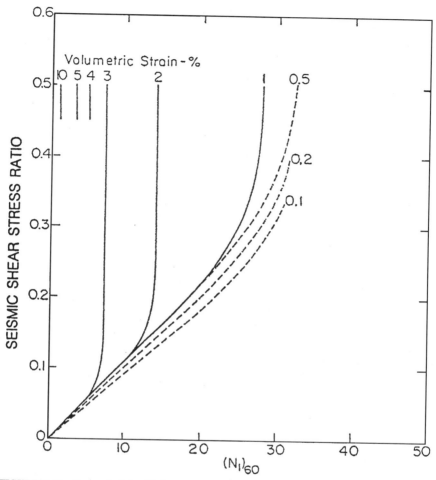

FIGURE 11.11 Proposed relationship between the earthquake-induced seismic shear stress ratio and $(N_1)_{60}$ to obtain volumetric strain for saturated clean sand. (*From Tokimatsu and Seed, 1984.*)

11.4 SELECTION OF FOUNDATION TYPE

Depending on the estimation of ground movement due to seismic shaking, there are many different options that can be used to reduce foundation displacement during the earthquake. For example, prior to construction of the project, the engineering properties of the soil or slope could be improved. Loose sands could be densified to reduce liquefaction susceptibility. The potential lateral deformation of slopes could be decreased by installing shear keys or other slope stabilization devices. For other projects, deep foundations can be installed that derive support from soil that is located below the depth of possible liquefaction.

The type of foundation can be very important in the performance of the structure during the earthquake. Foundations such as mats or posttensioned slabs may enable the building to remain intact, even with substantial movements. This section presents a discussion of the effect of foundation type on the damage to single-family houses during the Northridge, California, earthquake. The emphasis of this section is on earthquake-related foundation damage and not foundation damage caused by earthquake-induced soil movement.

The Northridge, California, earthquake occurred on January 17, 1994 at 4:31 A.M. The earthquake magnitude was 6.7. There was collapse of specially designed structures such as multistory buildings, parking garages, and freeways. In some areas, the most severe damage would indicate a Modified Mercalli intensity of IX, although MM VII to VIII was more widespread. Because the Northridge earthquake occurred in a suburban community, damage to single-family houses was common.

Damage estimates varied considerably. For both public and private facilities, the total cost of the Northridge earthquake was on the order of $20 to $25 billion. This makes the Northridge earthquake California's most expensive natural disaster. Given the significant damage caused by this earthquake, the number of deaths was relatively low. This was partly because most people were asleep at home at the time of the earthquake.

The Northridge earthquake data was obtained from site investigations performed by the author in 1994. This work was requested by several different insurance companies. Most of the residences investigated were single-family, detached houses having wood frame construction with exterior stucco and interior wallboard or lath and plaster.

For those locations with similar earthquake intensity and duration, the type of foundation was a factor in the severity of damage. For most single-family houses in southern California, the foundations are either raised wood floor or slab-on-grade. This is because, as a result of the climate, frost heave is not a concern and the foundations tend to be of shallow depth.

Raised Wood Floor with Isolated Interior Posts. This type of raised wood floor foundation consists of continuous concrete perimeter footings and interior (isolated) concrete pads. The floor beams span between the continuous perimeter footings and the isolated interior pads. The continuous concrete perimeter footings are typically constructed so that they protrude about 0.3 to 0.6 m (1 to 2 ft) above adjacent pad grade. The interior concrete pad footings are not as high as the perimeter footings, and short wood posts are used to support the floor beams. The perimeter footings and interior posts elevate the wood floor and provide for a crawl space below the floor.

In southern California, the raised wood floor foundation having isolated interior pads is common for houses 30 years or older. Most newer houses are not constructed with this foundation type. In some cases, the raised wood floor was not adequately bolted to the concrete foundation. Only a few bolts, or in some cases nails, were used to attach the wood sill plate to the concrete foundation.

In general, damage was more severe to houses having this type of raised wood floor foundation. There may be several different reasons for this behavior.

1. *Lack of shear resistance of wood posts.* As previously mentioned, in the interior, the raised wood floor beams are supported by short wood posts bearing on interior concrete pads. During the earthquake, these short posts were vulnerable to collapse or tilting.

2. *No bolts or inadequate bolted condition.* Because in many cases the house was not adequately bolted to the foundation, it slid or even fell off of the foundation during the earthquake. In other cases the bolts were spaced too far apart and the wood sill plate split, allowing the house to slide off the foundation.

3. *Age of residence.* The houses having this type of raised wood floor foundation were older. The wood was more brittle and in some cases weakened by rot or termite damage.

In some cases, the concrete perimeter footings were nonreinforced or had weakened due to prior soil movement, so they were more susceptible to cracking during the earthquake.

Slab-on-Grade. In southern California, the concrete slab-on-grade is the most common type of foundation for houses constructed within the past 20 years. It consists of perimeter and interior continuous footings, interconnected by a slab-on-grade. Construction of the slab-on-grade begins with the excavation of the interior and perimeter continuous footings. Steel reinforcing bars are commonly centered in the footing excavations, and a wire mesh is used as reinforcement for the slab. The concrete is usually placed at the same time for both the footings and the slab in order to create a monolithic foundation. Unlike the raised wood floor foundation, the slab-on-grade does not have a crawl space.

In general, for those houses with a slab-on-grade, the wood sill plate was observed to be bolted to the concrete foundation. In many cases, the earthquake caused the development of an exterior crack in the stucco at the location where the sill plate meets the concrete foundation. In some cases, the crack could be found on all four sides of the house. The crack developed when the house framing bent back and forth during the seismic shaking.

Of the two foundations subjected to similar earthquake intensity and duration, those houses having a slab-on-grade generally had the best performance. This is because the slab-on-grade is typically stronger due to steel reinforcement and monolithic construction, the houses are newer (less wood rot and concrete deterioration), there is greater frame resistance because of the construction of shear walls, and the wood sill plate is in continuous contact with the concrete foundation.

It should be mentioned that although the slab-on-grade generally had the best performance, there were such houses that were nevertheless severely damaged. In many cases, these houses did not have adequate shear walls, there were numerous wall openings, or there was poor construction. The construction of a slab-on-grade by itself was not enough to protect the structure from collapse if the structural frame above the slab did not have adequate shear resistance.

Besides determining the type of foundation to resist earthquake related effects, the geotechnical engineer could also be involved with the retrofitting of existing structures. For the two foundation types previously discussed, the raised wood floor with isolated posts is rarely used for new construction. But there are numerous older houses that have this foundation type and in many cases, the wood sill plate is inadequately bolted to the foundation. Bolts or tie-down anchors could be installed to securely attach the wood framing to the concrete foundation. Wood bracing or plywood could be added to the open areas between posts to give the foundation more shear resistance.

In the Northridge, California, earthquake, the observed foundation damage indicated the importance of tying together the various foundation elements. To resist damage during the earthquake, the foundation should be monolithic, with no gaps in the footings or planes of weakness due to free-floating slabs.

PROBLEMS

1. Use the data from the example in Sec. 11.2.2, but assume that there is a vertical surcharge pressure applied at ground surface that equals 20 kPa. Determine the seismic shear stress ratio (SSR) induced by the earthquake. *Answer:* SSR = 0.31.

2. Use the data from the example in Sec. 11.2.2, but assume that $a_{max}/g = 0.1$ and the sand contains 15 percent fines. Will there be liquefaction of the soil? *Answer:* SSR induced by the earthquake = 0.085 and the SSR that will cause liquefaction of the *in situ* soil = 0.13, therefore liquefaction will not occur.

11.18 ANALYSIS OF GEOTECHNICAL DATA AND ENGINEERING COMPUTATIONS

3. Use the data from the example in Sec. 11.2.2, but assume that $a_{max}/g = 0.2$ and the earthquake magnitude $(M) = 5\frac{1}{4}$. Will there be liquefaction of the soil? *Answer:* SSR induced by the earthquake $= 0.17$ and the SSR that will cause liquefaction of the *in situ* soil $= 0.13$, therefore liquefaction will probably occur.

4. Assume a site has a groundwater table near ground surface. The following data was determined for the site:

Layer depth	Earthquake-induced SSR	$(N_1)_{60}$
2 to 3 m	0.18	10
3 to 5 m	0.20	5
5 to 7 m	0.22	7

Estimate the total settlement of these layers caused by a magnitude (M) 7.5 earthquake. *Answer:* 17 cm.

5. Use the data from the second example in Sec. 10.5 (i.e., the slope with a 9.1-m height and a planar slip surface where $\phi' = 29°$ and $c' = 3.4$ kPa). Assume that during the design earthquake, it is anticipated that the pore water pressure along the slip surface will increase 100 percent over the steady-state pore water pressure (i.e., $u_e = 2.4$ kPa). Using a pseudostatic slope stability analysis with $a_{max}/g = 0.2$ (i.e., horizontal earthquake force $= bW = 0.2W$) and including both the static (u) and earthquake-induced excess pore water pressure (u_e), calculate the factor of safety of the slope for the design earthquake condition. *Answer:* $F = 1.04$.

CHAPTER 12
EROSION

The following notation is used in this chapter:

SYMBOL	DEFINITION
A	Soil loss (universal soil loss equation)
C	Vegetative cover factor (universal soil loss equation)
E	Storm energy
H	Height of the slope
I_{30}	30-minute rainfall intensity
K	Soil erodibility factor (universal soil loss equation)
L	Length of slope in feet
L'	Length factor [Eq. (12.3)]
m	Exponent for Eq. (12.3)
P	Erosion control practice factor (universal soil loss equation)
R	Rainfall erosion index (universal soil loss equation)
s	Slope steepness, expressed as a percentage
S	Combined slope length and steepness factor (universal soil loss equation)
α	Slope inclination

12.1 INTRODUCTION

The process of slope erosion can begin as slopewash. The term *slopewash* refers to a form of erosion in which water moves as a thin and relatively uniform film (Rice, 1988). With time, the flow of water may concentrate into a series of slightly deeper pathways (called rills), and this is known as *rillwash*. With continued erosion, gullies could emerge and ultimately there could be the formation of channeled stream flow through the slope. Table 12.1 lists five levels of slope erosion and Fig. 12.1 shows severe erosion of a slope.

The factors that affect sheet and rill erosion have been discussed by Smith and Wischmeier (1957). They indicate that there are two principal processes of sheet erosion: raindrop impact and transportation of soil particles by flowing water. According to Smith and Wischmeier (1957), the amount of soil loss is governed by six factors: length of slope, slope gradient, ground cover, soil type, management, and rainfall. Soil type would include such factors as type and degree of cementation of particles.

As previously mentioned in Sec. 10.9, clean (uncemented) sand and nonplastic silt are most susceptible to erosion, with clay being the most resistant with certain exceptions, such as dispersive clays. McElroy (1987) describes unique erosion features caused by dispersive clays. For example, the erosion of fill or cut slopes can take the form of vertical or near-vertical

12.2 ANALYSIS OF GEOTECHNICAL DATA AND ENGINEERING COMPUTATIONS

FIGURE 12.1 Example of severe erosion of a slope (note broken drainage ditch in lower left corner of photograph).

TABLE 12.1 Level of Erosion

Level of erosion (1)	Classification (2)	Description (3)
1	Very slight	Minor erosion; at the base of the slope, minor accumulation of debris
2	Slight	Erosion consists of rills, which may be up to about 8 cm deep; some debris at the base of the slope
3	Appreciable	Rills up to about 0.3 m deep; debris at the base of the slope
4	Severe	Rills from about 0.3 to 1 m deep and gullies are beginning to form; considerable debris at the base of the slope
5	Very severe	Deep erosion channels, consisting of rills and gullies; development of pipes causing underground erosion; very large accumulation of debris at the base of the slope

tunnels called "jugs." McElroy (1987) states that jugs may develop as a result of small drying cracks, rodent holes, openings created by decaying roots, animal and vehicle tracks, or other small surface depressions that permit rainfall or surface water to collect. The jugs frequently consist of a small hole (often less than 25 mm) on the surface of the cut or fill slope, and this hole can extend to a depth of 3 m (10 ft) with a bottom diameter of 1 m (3 ft). When these jugs collapse, the slope can erode to form severe rills and gullies.

Having the results of subsurface exploration and laboratory testing, the geotechnical engineer is often the best individual to evaluate the erosion potential of the on-site soils. For example, as indicated in App. B (Sec. F), preliminary geotechnical reports (i.e., feasibility reports) should provide recommendations on "special planting and irrigation measures, slope coverings and other erosion control measures which may be apparent from the preparation of the geotechnical report."

Underground erosion, such as the piping of soil from or underneath an earth dam, has been discussed in Sec. 10.9.1. The purpose of this chapter is to focus solely on ground surface erosion.

12.2 PRINCIPLES OF EROSION CONTROL

It has been stated that although agriculture produces the largest percentage of the total sediment load, construction activities cause the most concentrated form of erosion (Goldman et al., 1986). The disturbance of the land surface during construction can increase the soil erosion by a factor of 2 to 40,000 times the preconstruction erosion rate (Wolman and Schick, 1967; Virginia Soil and Water Conservation Commission, 1980). The most damaging effect of construction is the removal of ground cover, which normally means vegetation. Ground cover can also include ground surface stabilization treatments, such as open-graded gravel, mulch, and jute netting. According to Goldman et al. (1986), vegetation performs the following functions:

- Shields the soil surface from the impact of falling rain
- Slows the velocity of runoff
- Holds soil particles in place, and
- Maintains the soil's capacity to absorb water.

Three common types of ground surface erosion due to the loss of vegetation during site construction are shown in Figs. 12.2 to 12.4, as follows:

FIGURE 12.2 Formation of a gully at a mass grading project.

FIGURE 12.3 Slope erosion at a recently graded site (arrows point to locations where surface runoff has flowed over the top of slope).

FIGURE 12.4 At a mass graded project, erosion has undermined the concrete sidewalk.

1. *Gully erosion.* Figure 12.2 shows the formation of a gully during the mass grading of a project. The gully has formed where the surface water has become concentrated. Note in Fig. 12.2 that the gully makes a right angle turn and is enlarging and progressing along the drainage path.
2. *Slope erosion.* Figure 12.3 shows slope erosion at a recently graded site. The water flows off of the graded pad and over the top of a fill slope. As shown in Fig. 12.3, the water has become concentrated and has eroded gullies into the face of the fill slope.
3. *Undermining of structures.* The erosion of soil can undermine structures, such as foundations, bridge piers, and gunite drainage ditches. Figure 12.4 shows an example of erosion that has undermined a concrete sidewalk.

The purpose of the remainder of this section is to discuss the basic principles of erosion control. The items that will be discussed include the universal soil loss equation, methods that can be used to reduce erosion during construction, and the design and construction of systems that can be used to trap sediment on site.

Universal Soil Loss Equation. The basic form of the universal soil loss equation is to calculate the soil loss A by multiplying together five different factors, as follows (Goldman et al., 1986):

$$A = RKSCP \tag{12.1}$$

where A = soil loss, tons per acre per year [tons/(acre·year)]
R = rainfall erosion index [(100 ft-tons/acre)(in./h)]
K = soil erodibility (tons/acre per unit of R)
S = combined slope length and steepness factor (dimensionless)
C = vegetative cover factor (dimensionless)
P = erosion control practice factor (dimensionless)

Note that Eq. (12.1) is an empirical equation, and the units of the equation are tons per acre per year. In order to determine the amount of soil loss (in tons) per year, the value of A would be multiplied by the area (in acres) under consideration. To convert A from tons per acre to SI units (tonnes per hectare), multiply A by 2.24. The individual terms in Eq. (12.1) are discussed below.

1. *Rainfall erosion index R.* The rainfall erosion index combines both a rainfall energy component and a rainfall intensity component into one number. There is a two-step process in calculating the rainfall erosion index R. First, an average rainfall year is selected and the total energy E times the maximum 30-minute rainfall intensity I_{30} is calculated for a given storm. Second, the E times I_{30} values for all the major storms in an area are added up for this average rainfall year.

Figure 12.5a and b presents rainfall erosion index R values for the continental United States. Those values west of 104° are only rough approximations because the rainfall erosion index values can vary significantly between adjacent areas because of the differing topography. In Fig. 12.5, the highest values are $R = 550$ for the general area of New Orleans, which receives numerous storms of high intensity in an average year. The lower values in the west are a result of the much lower amount of rainfall in an average year. The rainfall erosion index R also does not consider erosion due to snowmelt runoff.

As previously mentioned, the universal soil loss equation determines the soil loss (in tons) per acre for an average year. It is possible to determine A for a shorter time period. For

12.6 ANALYSIS OF GEOTECHNICAL DATA AND ENGINEERING COMPUTATIONS

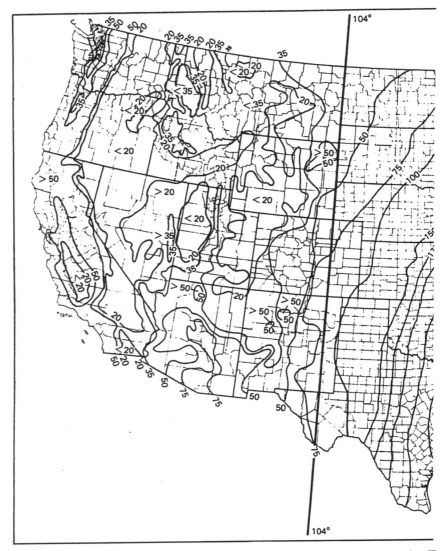

FIGURE 12.5a Rainfall erosion index R values for the continental United States, western portion. The units of R are [(100 ft-tons/acre)(in./h)]. (*Reproduced from Goldman et al., 1986; reprinted with permission from McGraw-Hill, Inc.*)

example, suppose grading work will be performed during January and the average rainfall in January is 2 in. If the annual average rainfall is 20 in., then the value of A for only January would be 10 percent (i.e., 2/20 = 10 percent) of the value calculated from Eq. (12.1).

2. *Soil erodibility factor K.* The soil erodibility factor K is a measure of the ability of rainfall and runoff to detach and transport the soil particles. There are two main factors that

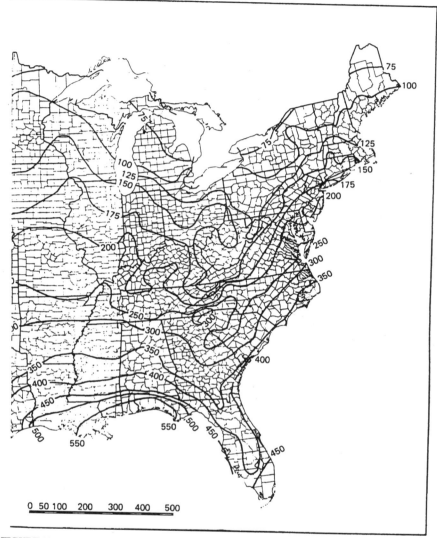

FIGURE 12.5b Rainfall erosion index R values for the continental United States, eastern portion. The units of R are [(100 ft-tons/acre)(in./h)]. *(Reproduced from Goldman et al., 1986; reprinted with permission from McGraw-Hill, Inc.)*

govern a soil's susceptibility to erosion: particle size and soil plasticity. Values of K range from a low of 0.02 to a high of 0.7.

Table 12.2 presents recommended values of the soil erodibility factor K. Note that nonplastic silt (MN) has the highest susceptibility to erosion ($K = 0.7$). This is because of the small size of the silt-size particles; without any soil plasticity to hold the soil particles together, they are easily eroded. At the other extreme is a highly plastic clay (CH), where the plasticity is so great (PI > 30) that rainwater is unable to detach the clay particles ($K = 0.02$). Other soils with relatively low soil erodibility factors are gravels, which are large-size particles that resist erosion due to their weight.

In Table 12.2, ranges of K are provided. Lower values would be used if the soil contains a higher gravel content or has a higher plasticity index, and vice versa. Note that some soils, such as a silt of low plasticity (ML), have a wide range in values for K. For silt of low plasticity (ML), the value of K increases as the plasticity index decreases. As the PI approaches zero, the soil becomes nonplastic and the value of K approaches 0.7 (value for nonplastic silt). Likewise, for clayey sands (SC), K is very low when the clayey sand has a high plasticity index ($K = 0.1$), but as the plasticity index approaches zero, the value of K increases to the nonplastic sand value ($K = 0.7$).

There are other factors that may need to be considered in the selection of the soil erodibility factor. Some of these factors are as follows:

- *Organic matter.* A small amount of organic matter in the soil tends to decrease the soil erodibility factor K. This is probably because of the absorptive and possibly binding capability of organic matter.
- *Cementing agents.* In some cases there may be cementing agents that bind the soil particles together. If these cementing agents have a low solubility in water, then the soil erodibility factor K listed in Table 12.2 would be too high.
- *Dispersive clays.* Dispersive clays may have a high plasticity index, but as previously discussed, they can be susceptible to erosion because of the process of deflocculation of the clay-size particles (Perry, 1987). Values listed in Table 12.2 would be too low for dispersive clays.

3. *Combined slope length and steepness factor S.* This factor combines both the effects of the slope length and the slope gradient. The combined slope length and slope gradient factor S is the ratio of soil loss per unit area on a site to the corresponding loss from a 73-ft-long experimental plot with a 9 percent slope (Goldman et al., 1986). The following equation (Wischmeier and Smith, 1965, 1978) can be used to calculate the combined slope length and steepness factor S:

$$S = \frac{65s^2 L'}{s^2 + 10{,}000} + \frac{4.6sL'}{(s^2 + 10{,}000)^{0.5}} + 0.065L' \qquad (12.2)$$

where s = slope steepness expressed as a percentage and L' = length factor defined as follows:

$$L' = (L/73)^m \qquad (12.3)$$

where L = length of the slope in feet and m has the following values:

$m = 0.2$ for $s < 1\%$

$m = 0.3$ for $1\% \leq s < 3\%$

$m = 0.4$ for $3\% \leq s < 5\%$

$m = 0.5$ for $s \geq 5\%$

In Eq. (12.2), the slope steepness must be inserted as a percentage. For example, a 2:1 (horizontal:vertical) slope has a slope steepness of 50 percent, and $s = 50$ for Eq. (12.2). Note

TABLE 12.2 Recommended Values of the Soil Erodibility Factor K

Major divisions (1)	Subdivisions (2)	ISBP symbol (3)	Typical names (4)	Laboratory classification criteria (5)	Soil erodibility factor K (6)
Nonplastic soils	Gravels (greater fraction of total sample is retained on No. 4 sieve)	GW	Well-graded gravels, sandy gravels, silty-sandy gravels	$C_u \geq 4$ and $1 \leq C_c \leq 3$	0.05 to 0.1
		GP	Poorly graded gravels, gravel-sand-silt mixtures	Does not meet above criteria, % passing No. 200 sieve < 15%	0.1 to 0.15
		GM	Poorly graded, nonplastic silty gravels	Does not meet above criteria, % passing No. 200 $\geq$ 15%	0.15 to 0.2
	Sands (greater fraction of total sample is between No. 4 and No. 200 sieves)	SW	Well-graded sands and gravelly sands	$C_u \geq 6$ and $1 \leq C_c \leq 3$	0.5 to 0.6
		SP	Poorly graded sands or sand-gravel-silt mixtures	Does not meet above criteria, % passing No. 200 sieve > 15%	0.6 to 0.7
		SM	Poorly graded, nonplastic silty sands	Does not meet above criteria, % passing No. 200 sieve $\geq$ 15%	0.6 to 0.7
	NP silt	MN	Nonplastic silts, rock flour	Greater fraction of the total sample passes the No. 200 sieve.	0.7
Plastic soils	Plastic silts (minus No. 40 fraction plots below A-line)	GM$^{(1)}$	Plastic silty gravels, gravel-silt mixtures	$\geq$ 50% retained on the No. 200 sieve with the greater fraction of gravel size	0.1 to 0.2
		SM$^{(1)}$	Plastic silty sands, sand-silt mixtures	$\geq$ 50% retained on the No. 200 sieve with the greater fraction of sand size	0.3 to 0.7
		ML MI MH	Plastic silts, sandy silts, and clayey silts	Silt of low plasticity (ML) PI $\leq$ 10 Intermediate plasticity (MI) $10 < PI \leq 30$ Silt of high plasticity (MH) PI > 30	0.3 to 0.7 0.2 to 0.3 0.1 to 0.2
	Clays (minus No. 40 fraction plots on or above A-line)	GC$^{(1)}$	Clayey gravels, gravel-clay mixtures	$\geq$ 50% retained on the No. 200 sieve with the greater fraction of gravel size	0.05 to 0.2
		SC$^{(1)}$	Clayey sands, sand-clay mixtures	$\geq$ 50% retained on the No. 200 sieve with the greater fraction of sand size	0.1 to 0.7
		CL CI CH	Clay, sandy clays, and silty clays	For clay of low plasticity (CL) PI $\leq$ 10 Intermediate plasticity (CI) $10 < PI \leq 30$ For clay of high plasticity (CH) PI > 30	0.1 to 0.6 0.02 to 0.1 0.02

Note: (1) High, intermediate, or low plasticity based on Eq. (5.17).

that by plugging the values from the experimental plot ($s = 9\%$ and $L = 73$ ft) into the above equations, the value of the combined slope length and steepness factor S is equal to 1.0.

Example 1. As an example of the use of Eqs. (12.2) and (12.3), suppose that a building pad will be constructed at a slope gradient of 100:1 (horizontal:vertical) and the slope length L of the graded pad is equal to 100 ft. For this example, $s = 1$, $m = 0.3$, and $L = 100$ ft. Substituting L and m into Eq. (12.3) gives $L' = 1.1$. Substituting $L' = 1.1$ and $s = 1$ into Eq. (12.2) gives a value of the combined slope length and steepness factor S of 0.13. This value is less than 1 because the slope gradient ($s = 1\%$) is much less than the slope gradient used for the experimental plot ($s = 9\%$).

Example 2. Suppose a slope will be constructed such that it has a height of 70 ft and will have a slope gradient of 1:1 (horizontal:vertical). In this case, with a slope height of 70 ft, the length of the slope surface (L) will be 100 ft (i.e., $L = H/\sin \alpha$, where $H =$ height of slope and $\alpha =$ slope inclination). For a 1:1 slope, the slope gradient s equals 100. Inserting $L = 100$ and $m = 0.5$ into Eq. (12.3) gives a value of L' of 1.2. Substituting $s = 100$ and $L' = 1.2$ into Eq. (12.2) gives a value of the combined slope length and steepness factor S of 43. This value is greater than 1 because the slope gradient ($s = 100\%$) is much greater than the slope gradient used for the experimental plot ($s = 9\%$).

In these two examples, the values of the combined slope length and steepness factor S was 0.13 for the first example and 43 for the second example. The second example has an S value that is 330 times greater than the first example. As these two examples illustrate, the S value can have a wider variation than any other factor in the universal soil loss equation.

4. *Vegetative cover factor C.* The vegetative cover factor accounts for the type and extent of vegetation at the site. Table 12.3 (from Goldman et al., 1986) presents C values for different ground cover conditions. Note in Table 12.3 that bare soil, which is the typical condition for grading or construction projects, has a C value of 1.0. For a site where the native vegetation is undisturbed, the C value is only 0.01. Thus the erosion for a bare site will be 100 times greater than the same site having native vegetation in an undisturbed state. For sites where there is little native vegetation, such as badlands, a C value of 1.0 should be used.

TABLE 12.3 Vegetative Cover Factor C for the Universal Soil Loss Equation

Type of cover (1)	C factor (2)
No vegetative cover (bare ground)	1.0
Native vegetation (undisturbed)	0.01
Temporary seeding:	
90% cover, annual grasses, no mulch	0.1
Wood fiber mulch, 3/4 ton/acre, with seed*	0.5
Excelsior mat, jute*	0.3
Straw mulch:*	
1.5 tons/acre, tacked down	0.2
4 tons/acre, tacked down	0.05

*For slopes up to 2:1 (horizontal:vertical).
Source: This table was developed by Goldman et al. (1986), on the basis of the work by Kay (1983), U.S. Department of Agriculture (1977), and Wischmeier and Smith (1965).

TABLE 12.4 Erosion Control Practice Factor P for the Universal Soil Loss Equation

Surface condition (1)	P value (2)
Compacted and smooth	1.3
Trackwalked along contour*	1.2
Trackwalked up and down slope†	0.9
Punched straw	0.9
Rough, irregular cut	0.9
Loose to 12 in. (30 cm) depth	0.8

*Tread marks oriented up and down slope.
†Tread marks oriented parallel to contours.
Source: This table was developed by Goldman et al. (1986), on the basis of the work by the U.S. Department of Agriculture (1977).

5. *Erosion control practice factor P.* The final term in the universal soil loss equation is the erosion control practice factor P. It is included in the equation to account for the management of the site. For example, Table 12.4 (from Goldman et al., 1986) presents typical P values for the construction of a site. Higher P values apply to sites where the graded surface is smooth and uniform, while lower P values are applicable for surfaces that are rough and irregular. As indicated in Table 12.4, changing the graded surface condition does not have much effect on the P value and, for the five factors in the universal soil loss equation, the P value has the least effect on soil loss at a construction site.

In using the universal soil loss equation, the final result should be considered only an estimate of the soil loss at a site. This is because of the many limitations of the universal soil loss equation, such as (Goldman et al., 1986):

- *Empirical equation.* The equation is empirical, and it obtains the soil loss by simply multiplying together five different factors that have been observed to have a significant impact on the erosion process.
- *Average conditions.* The equation is based on average rainfall conditions. Unusual weather patterns, such as El Niño, which caused much heavier rainfall in California during the winter of 1997–1998, can produce significantly more soil loss.
- *Sheet and rill erosion.* The equation was developed to predict the soil loss for sheet and rill erosion. Concentrated flow of water, such as that in gullies, will result in a much larger volume of eroded soil than as predicted by the universal soil loss equation.
- *Sediment deposition.* The universal soil loss equation predicts soil loss and not soil deposition. Gravel-size particles and coarse sand may only be eroded to the toe of a slope and may not need to be considered in the analysis of sediment storage facilities.

Example of the Use of the Universal Soil Loss Equation. The following presents an example of the use of the universal soil loss equation:

Site:	Earth dam, 800 ft long and 40 ft high
Location:	Northwest corner of South Carolina
Slope inclination:	Downstream face will be constructed at a 2.5:1 (horizontal:vertical) slope ratio

12.12 ANALYSIS OF GEOTECHNICAL DATA AND ENGINEERING COMPUTATIONS

Soil type: Red clay fill (CI), having LL = 38 and PI = 14

Graded condition: Downstream slope face will be trackwalked up and down slope

Required: Sediment loss in 1 year for a bare downstream face of the dam

Solution

R = rainfall erosion index [(100 ft-tons/acre)(in./h)]. From Fig. 12.5b, $R = 300$

K = soil erodibility (tons/acre per unit of R). From Table 12.2, for a CI soil, the PI is closer to the lower limit, therefore use $K = 0.1$

S = combined slope length and steepness factor (dimensionless). Slope height = 40 ft, therefore slope length (L) = 108 ft. Slope steepness (s) = 40 (2.5:1). From Eqs. (12.2) and (12.3), $S = 13$.

C = vegetative cover factor (dimensionless). Bare face, therefore $C = 1.0$ (Table 12.3).

P = erosion control practice factor (dimensionless). Trackwalked, therefore $P = 0.9$ (Table 12.4).

Therefore:

$$A = RKSCP = (300)(0.1)(13)(1.0)(0.9) = 350 \text{ tons per acre per year}$$

$$\text{Area} = (800 \text{ ft})(108 \text{ ft})/(43{,}560 \text{ ft}^2) = 2 \text{ acres}$$

Total soil loss per year = (350 tons/acre)(2 acres) = 700 tons (1.4 million pounds)

Design and Construction. The advantage of the universal soil loss equation is that it can provide an estimate of the soil loss for a given construction project. Obviously, for the above example of the construction of an earth dam, a total soil loss of 1.4 million lb would be unacceptable. In terms of design and construction, there are 10 basic principles that can be used to reduce the amount of soil erosion and avoid sediment transport onto adjacent property, as follows (Goldman et al., 1986):

1. *Fit the development to the terrain.* The best method to mitigate the risk of creating erosion and sedimentation problems is to disturb as little of the land surface as possible. For example, the development could be tailored to the natural contours of the land, which would then require little grading, with less ground surface erosion.

2. *Time grading and construction to minimize soil exposure.* Rainfall intensity is a key factor in slope erosion. If possible, it would be best to not perform the grading and construction activities during a time of year subjected to intense rainstorms. For example, in many parts of the country, there is a dry season when the amount of erosion will be much less than during the wet season. In California, the dry season is during the summer and the wet season is during the winter. The best situation would be to plan the grading operation during the summer months and have the grading completed in time to have the slopes seeded prior to the rainy season.

3. *Retain existing vegetation whenever feasible.* As previously mentioned, the vegetative cover factor C can vary from 0.01 for native vegetation (undisturbed state) to 1.0 for a bare slope (Table 12.3). Thus there will be 100 times more soil loss for the denuded slope condition. Whenever possible, the existing trees and other natural vegetation should be incorporated into the site development plan.

4. *Vegetate and mulch denuded areas.* The areas that have been denuded by the site grading or construction process should be planted as soon as possible. In many cases, it may

be possible to perform the site grading in phases, with the denuded areas planted at the completion of each phase of grading.

5. *Divert runoff away from denuded areas.* Runoff should always be diverted away from denuded areas and not allowed to flow over the tops of slopes. If surface runoff is not diverted, there could be the formation of gullies such as shown in Fig. 12.3.

6. *Minimize length and steepness of slopes.* Slope length and slope steepness are critical factors in the amount of soil loss caused by erosion. This is because the slope length and slope steepness primarily determine the velocity of runoff. The energy (hence erosive potential) of flowing water increases as the square of the velocity. Thus long, continuous slopes allow for the surface runoff to gain momentum, which can result in the runoff becoming concentrated in gullies.

The combined length and steepness factor S can be significantly decreased by installing gunite drainage ditches (see Fig. 4.5) on the face of the slope. As an example of the use of drainage ditches, suppose that for the example of the earth dam that a midslope drainage ditch will be installed. In this case, the slope length will be cut in half ($L = 54$ ft). The combined length and steepness factor S from Eq. (12.2) is 9.2. Therefore the amount of soil loss would be reduced by about 30 percent by simply constructing a midslope drainage ditch.

7. *Keep runoff velocities low.* Soil erosion can be reduced by keeping the runoff velocities low. For example, as indicated in Table 12.4, the erosion control practice factor P can be reduced by constructing a rough and irregular slope face instead of a smooth slope surface.

8. *Prepare drainage ways and outlets to handle concentrated or increased runoff.* The grading and construction process will increase the amount of runoff from the project. The project plans should include proper drainage ways and outlets to handle this increased runoff.

9. *Trap sediment on site.* There are many different devices that can be used to trap sediment on site. The most basic are devices that filter the sediment from the water, such as silt fences. Figure 12.6 shows another example of a commonly used device used to trap sediment on site. At this project, straw bales were placed along the toe of a recently graded hillside. As Fig. 12.6 shows, the straw bales have not been totally effective because sediment has been deposited on both the sidewalk and in the adjacent street.

More reliable devices that can be used to trap sediment on site are shown in Figs. 12.7 and 12.8. Figure 12.7 shows the typical features of a sediment trap and Fig. 12.8 shows an elevated sediment trap connected to a storm drain line.

In a sediment basin (Fig. 12.7), loose and porous soil will accumulate. At one project, fill was placed atop the basin sediment and a structure was built on top of the fill. The structure was subjected to substantial settlement and distress as the loose and porous sediment collapsed upon moisture infiltration. It is important to remove and recompact all sediment located in the basin prior to placing a structural fill at the basin area.

10. *Inspect and maintain control measures.* Inspection and maintenance of erosion control systems are essential to the reduction in volume of soil loss. Without cleaning and maintenance, the systems shown in Figs. 12.7 and 12.8 will fill with sediment and no longer trap the eroded soil.

12.3 EROSION-PRONE LANDFORMS

As mentioned in the previous section, the disturbance of the land surface during construction can increase the soil erosion by a factor of 2 to 40,000 times the preconstruction erosion rate (Wolman and Schick, 1967; Virginia Soil and Water Conservation Commission,

FIGURE 12.6 Straw bales being used to trap sediment at the toe of a graded slope. (Note that the straw bales have not been totally effective, since there is sediment on the sidewalk and in the street).

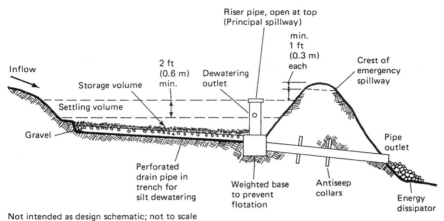

FIGURE 12.7 Typical features of a sediment basin. (*From Goldman et al. (1986); reprinted with permission from McGraw-Hill, Inc.*)

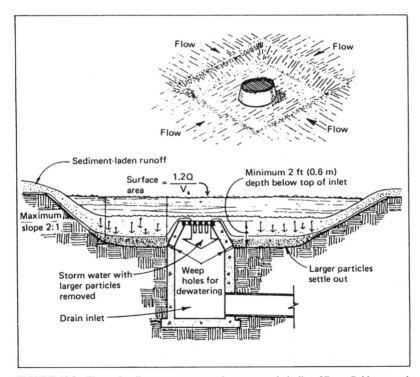

FIGURE 12.8 Elevated sediment trap connected to a storm drain line. [*From Goldman et al. (1986); based on the Virginia Erosion and Sediment Control Handbook (1980). Reprinted with permission from McGraw-Hill, Inc.*]

12.16 ANALYSIS OF GEOTECHNICAL DATA AND ENGINEERING COMPUTATIONS

1980). There are other types of landforms that are especially susceptible to erosion, such as sea cliffs and badlands.

12.3.1 Sea Cliffs

Factors Affecting Erosion of Sea Cliffs. Sea cliffs can be more susceptible to erosion because of a lack of vegetation caused by salt accumulation in the soil from sea spray. A reduction in vegetation will provide less rain impact protection and less root reinforcement to hold the soil in place. Another factor that causes sea cliff erosion, unique to the ocean environment, is toe erosion by ocean wave impact and scour.

Figure 12.9 shows a photograph of various sea cliff erosion control measures, in Solana Beach, California. The sea cliff shown in Fig. 12.9 consists of weakly cemented to moderately cemented sandstone. The left side of the photograph shows the natural slope with no irrigation, except perhaps at the top-of-slope. Notice that the erosion of the face of the slope has created numerous rills. The sea cliff protection consists predominantly of boulder riprap placed at the toe of the slope.

In the middle of Fig. 12.9, a second (more rigorous) system of sea cliff protection has been constructed. At the toe of the slope, there is a concrete sea wall to protect the slope from ocean waves. Above the sea wall, a crib wall has been constructed which supports a cement-treated slope. The entire sea cliff face has been essentially rebuilt with a more resistant, man-made erosion control system.

At the right side of Fig. 12.9, there is the sea wall to again protect the toe of the slope from ocean waves. But above the sea wall, there is extensive planting and irrigation to promote the growth of vegetation. The slope face is essentially completely vegetated.

FIGURE 12.9 Sea cliff erosion protection.

FIGURE 12.10 Sea cliff erosion protection.

Figure 12.10 shows other commonly used erosion control measures for sea cliffs in Solana Beach, California. The left side of Fig. 12.10 shows a gunite retaining structure. Typically the gunite is anchored to the rock by tieback anchors. On the right side of Fig. 12.10, segmented multilevel retaining structures, with reinforced earth backfill, have been constructed to protect the toe and slope face from erosion.

As a localized stabilization measure, the caves or gullies can be filled with grout or gunite. Figure 12.11 shows an example of this erosion control measure. Note in Fig. 12.11 that a lowering of the beach has exposed the bottom of the grout. Undermining of the grout is one disadvantage of this erosion control measure.

12.3.2 Badlands

Badlands have been defined as follows (Stokes and Varnes, 1955):

> An area, large or small, characterized by extremely intricate and sharp erosional sculpture. Badlands usually develop in areas of soft sedimentary rocks such as shale, but may also occur in decomposed igneous rocks, loess, etc. The divides are sharp and the slopes are scored by intricate systems of ravines and furrows. Fantastic erosional forms are commonly developed through the unequal erosion of hard and soft layers. Vegetation is scanty or lacking and there is a notable lack of coarse detritus. Badlands occur chiefly in arid or semiarid climates where the rainfall is concentrated in sudden heavy showers. They may, however, occur in humid regions where vegetation has been destroyed, or where soil and coarse detritus are lacking.

Figures 12.12 and 12.13 present two views of badlands near Death Valley, California. The arrows in Figs. 12.12 and 12.13 point to the remains of an asphalt roadway located in

12.18 ANALYSIS OF GEOTECHNICAL DATA AND ENGINEERING COMPUTATIONS

FIGURE 12.11 Sea cliff erosion protection.

Golden Canyon. According to the National Park Service, a 4-day rainstorm in February 1976 dropped 5.8 cm (2.3 in.) of rain in this area. There was tremendous runoff that undermined and eroded away the pavement, as shown in Figs. 12.12 and 12.13. Such intense erosive action can easily wash away any vegetation that may have gained a foothold in the badlands.

Figure 12.14 shows erosion of a debris flow. The cap rock has protected the underlying debris flow material from erosion and created this unusual landform.

PROBLEMS

In the following problems, use the data from "Example of the Use of the Universal Soil Loss Equation" in Sec. 12.2 (i.e., the earth dam in South Carolina).

1. Assuming uniform soil loss of soil from the slope, calculate the thickness of soil (parallel to the slope face) lost during 1 year's time if the dry unit weight = 108 pcf. *Answer:* Thickness = 1.8 in.

2. Assume that the slope will be bare for only 2 months, when there is 20 percent of the annual rainfall amount. The slope will then be covered with excelsior mat (jute) and it is anticipated that it will take an additional 3 months to establish ground cover (30 percent of the rainfall occurs in these three months). Calculate the soil loss for the 5 months it takes to establish ground cover at the site. *Answer:* Soil loss = 203 tons.

3. Assume the slope has been planted with native vegetation that is fully established. Calculate the soil loss per year (A) for the vegetated slope. *Answer:* A = 7 tons per year.

FIGURE 12.12 Badlands (arrow points to remains of the asphalt pavement).

4. Assume that the slope face is constructed with sand that has a $K = 0.7$. Calculate the soil loss per year (A). *Answer:* $A = 4900$ tons per year.
5. Assume that the slope is constructed at a 3:1 (horizontal:vertical) ratio (i.e., $s = 33.3$). Calculate the percent decrease in soil loss. *Answer:* For a 3:1 slope versus a 2.5:1 slope, there is a 25 percent reduction in soil loss.

FIGURE 12.13 Badlands (arrow points to remains of the asphalt pavement).

FIGURE 12.14 Badlands (a cap rock has prevented erosion of the underlying debris flow).

CHAPTER 13
DETERIORATION

13.1 INTRODUCTION

All man-made and natural materials are susceptible to deterioration. This topic is so broad that it is not possible to cover every type of geotechnical and foundation element susceptible to deterioration. Instead, this chapter will discuss four of the more common types of deterioration: sulfate attack of concrete, pavement distress, frost-related damage, and damage caused by tree roots.

In terms of deterioration, the National Science Foundation (NSF, 1992) states:

> The infrastructure deteriorates with time, due to aging of the materials, excessive use, overloading, climatic conditions, lack of sufficient maintenance, and difficulties encountered in proper inspection methods. All of these factors contribute to the obsolescence of the structural system as a whole. As a result, repair, retrofit, rehabilitation, and replacement become necessary actions to be taken to insure the safety of the public.

13.2 SULFATE ATTACK OF CONCRETE

Sulfate attack of concrete is defined as a chemical and/or physical reaction between sulfates (usually in the soil or groundwater) and concrete or mortar, primarily with hydrated calcium aluminate in the cement-paste matrix, often causing deterioration (ACI, 1990). Sulfate attack of concrete occurs throughout the world, especially in arid areas, such as the southwestern United States. In arid regions, the salts are drawn up into the concrete and then deposited on the concrete surface as the groundwater evaporates, as shown in Figs. 13.1 to 13.3. Sulfate attack of concrete can cause a physical loss of concrete (Fig. 13.4) or the unusual cracking and discoloration of concrete such as shown in Fig. 13.5.

Typically the geotechnical engineer obtains the representative soil or groundwater samples to be tested for sulfate content. The geotechnical engineer can do an in-house analysis of the soil samples or groundwater, but a more common situation is to send the samples to a chemical laboratory for testing. There are different methods to determine the soluble sulfate content in soil or groundwater. One method is to precipitate out and then weigh the sulfate compounds. A faster and easier method is to add barium chloride to the solution and then compare the turbidity (relative cloudiness) of barium sulfate with known concentration standards.

13.2 ANALYSIS OF GEOTECHNICAL DATA AND ENGINEERING COMPUTATIONS

FIGURE 13.1 Concrete sidewalk deterioration, Mojave Desert (salts were deposited on concrete surface from evaporating groundwater).

Once the soluble sulfate content has been determined, the geotechnical or foundation engineer can then recommend measures, such as using type V cement, to mitigate the effects of sulfate on the concrete. As indicated in App. B, "Technical Guidelines for Soil and Geology Reports," the geotechnical engineer is often required to provide "soluble sulfate specifications and recommendations" for concrete foundations.

Mechanisms of Sulfate Attack of Concrete. There has been considerable research, testing, and chemical analysis of sulfate attack. Two different mechanisms of sulfate attack have been discovered: chemical reactions and the physical growth of crystals.

1. *Chemical reactions.* The chemical reactions involved in sulfate attack of concrete are complex. Studies (Lea, 1971; Mehta, 1976) have discovered two main chemical reactions. The first is a chemical reaction of sulfate and calcium hydroxide (which was generated during the hydration of the cement) to form calcium sulfate, commonly known as *gypsum*. The second is a chemical reaction of gypsum and hydrated calcium aluminate to form calcium sulfoaluminate, commonly called *ettringite* (ACI, 1990). As with many chemical reactions, the final product of ettringite causes an increase in volume of the concrete.

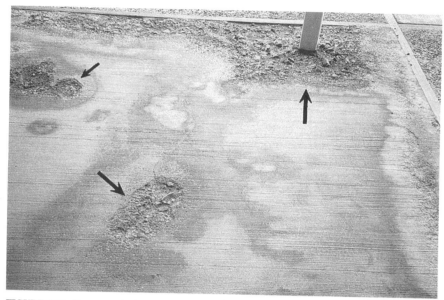

FIGURE 13.2 Concrete patio deterioration, Mojave Desert (arrows point to deterioration of concrete surface).

FIGURE 13.3 Concrete patio deterioration, Mojave Desert (note salt deposits and deterioration of concrete surface).

FIGURE 13.4 Physical loss of concrete due to sulfate attack.

FIGURE 13.5 Cracking and discoloration of concrete due to sulfate attack.

Hurst (1968) indicates that the chemical reactions produce a compound of twice the volume of the original tricalcium aluminate compound. Concrete has a low tensile strength and thus the increase in volume fractures the concrete, allowing for more sulfates to penetrate the concrete, resulting in accelerated deterioration.

2. *Physical growth of crystals.* The physical reaction of sulfate has been studied by Tuthill (1966) and Reading (1975). They conclude that there can be crystallization of the sulfate salts in the pores of the concrete. The growth of crystals exerts expansive forces within the concrete, causing flaking and spalling of the outer concrete surface. Besides sulfate, the concrete, if porous enough, can be disintegrated by the expansive force exerted by the crystallization of almost any salt in its pores (Tuthill, 1966, Reading, 1975). Damage due to crystallization of salt is commonly observed in areas where water is migrating through the concrete and then evaporating at the concrete surface. Examples include the surfaces of concrete dams, basement and retaining walls that lack proper waterproofing, and concrete structures that are partially immersed in salt-bearing water (such as seawater) or soils.

Sulfate Resistance of Concrete. The geotechnical engineer should be aware of the factors that increase the sulfate resistance of concrete. In general, the degree of sulfate attack of concrete will depend on the type of cement used, quality of the concrete, soluble sulfate concentration that is in contact with the concrete, and the surface preparation of the concrete (Mather, 1968).

1. *Type of cement.* There is a correlation between the sulfate resistance of cement and its tricalcium aluminate content. As previously discussed, it is the chemical reaction of hydrated calcium aluminate and gypsum that forms ettringite. Therefore, limiting the tricalcium aluminate content of cement reduces the potential for the formation of ettringite. It has been stated that the tricalcium aluminate content of the cement is the greatest single factor that influences the resistance of concrete to sulfate attack, where in general, the lower the tricalcium aluminate content, the greater the sulfate resistance (Bellport, 1968).

As indicated in Table 13.1, of the types of Portland cements, the most resistant is type V, in which the tricalcium aluminate content must be less than 5 percent. The ACI (ACI, 1990), Portland Cement Association (*Design*, 1994), and *Uniform Building Code* (Table 19-A-4, 1997) have essentially the same requirements (e.g., see Table 13.1) for normal-weight concrete subjected to sulfate attack. Depending on the percentage of soluble sulfate (SO_4) in the soil or groundwater, a certain cement type is required as indicated in Table 13.1.

TABLE 13.1 Typical Requirements for Concrete Foundations Exposed to Soluble Sulfate

Classification of sulfate exposure (1)	Soluble sulfate (SO_4) in the soil, %, based on dry weight (2)	Dissolved sulfate (SO_4) in the groundwater, parts per million (3)	Required Portland cement type (4)	Maximum water-cement ratio (5)
Negligible	0.0 to 0.1	0 to 150	Any type	No requirement
Moderate	0.1 to 0.2	150 to 1500	Type II	0.50
Severe	0.2 to 2.0	1500 to 10,000	Type V	0.45
Very severe	Over 2.0	Over 10,000	Type V with pozzolan	0.45

2. *Quality of concrete.* In general, the more impermeable the concrete, the more difficult for the waterborne sulfate to penetrate the concrete surface. To have a low permeability, the concrete must be dense, have a high cement content, and a low water-cement ratio. Using a low water-cement ratio decreases the permeability of mature concrete (*Design*, 1994). A low water-cement ratio is a requirement of ACI (1990) for concrete subjected to soluble sulfate in the soil or groundwater. For example, the water-cement ratio must be equal to or less than 0.45 for concrete exposed to severe or very severe sulfate exposure. There are many other conditions that can affect the quality of the concrete. For example, a lack of proper consolidation of the concrete can result in excessive voids. Another condition is the corrosion of reinforcement, which may crack the concrete and increase its permeability. Cracking of concrete may also occur when structural members are subjected to bending stresses. For example, the tensile stress due to a bending moment in a footing may cause the development of microcracks, which increase the permeability of the concrete.

3. *Concentration of sulfates.* As previously mentioned, it is often the geotechnical engineer who obtains the soil or water specimens that will be tested for soluble sulfate concentration. In some cases, after construction is complete, the sulfate may become concentrated on crack faces. For example, water evaporating through cracks in concrete flatwork will deposit the sulfate on the crack faces. This concentration of sulfate may cause accelerated deterioration of the concrete.

4. *Surface preparation of concrete.* An important factor in concrete resistance is the surface preparation, such as the amount of curing of the concrete. Curing results in a stronger and more impermeable concrete (*Design*, 1994), which is better able to resist the effects of salt intrusion.

Design and Construction. The soil or groundwater samples should be obtained after the site has been graded and the location of the proposed building is known. For the planned construction of shallow foundations, near-surface soil samples should be obtained for the sulfate testing. For the planned construction of deep foundations that consist of concrete piles or piers, soil and groundwater samples for sulfate testing should be taken at various depths that encompass the entire length of the concrete foundation elements.

The geotechnical engineer should select representative soil samples for testing. If the site contains clean sand or gravel, then these types of soil often have low sulfate contents because of their low capillary rise and high permeability, which enable any sulfate to be washed from the soil. But clays can have a higher sulfate content because of their high capillary rise, which enables them to draw up sulfate-bearing groundwater, and their low permeability, which prevents the sulfate from being washed from the soil.

Sulfate concentrations in the soil and groundwater can vary throughout the year. For example, the highest concentration of sulfates in the soil and groundwater will tend to occur at the end of the dry season or after a long dry spell. Likewise, the lowest concentration of sulfates will occur at the end of the rainy season, when the sulfate has been diluted or partially flushed out of the soil. The geotechnical engineer should recognize that sampling of soil or groundwater at the end of a heavy rainfall season will most likely be unrepresentative of the most severe condition.

Soil that contains gypsum should always be tested for sulfate content. This is because soluble sulfate (SO_4) is released as gypsum ($CaSO_4 \cdot 2H_2O$) weathers. Gypsum is an evaporite and can rapidly weather upon exposure to air and water.

There are many environments that could lead to the chemical attack of concrete. For example, if a site has been previously used as a farm, there may be fertilizers or animal wastes in the soil that will have a detrimental effect on concrete. A corrosive environment, such as drainage of acid mine water, will also lead to the deterioration of concrete. Another corrosive environment is seawater, which has both a moderate soluble sulfate content and a high salt content which can attack concrete through the process of chemical reactions involving sulfate attack and deterioration of concrete by the physical growth of salt crystals.

Soil Report Recommendations. Upon completion of the sulfate testing and analysis, the recommendations would normally be included in a soils report. An example of typical wording concerning sulfate testing and analysis at a proposed building site is as follows (see App. E):

> Type of Cement for Construction: Evaluation of soluble sulfate content of samples considered representative of the predominate material types on site suggests that type V cement is not a requirement for use in construction. Type I or II should be utilized.

13.3 PAVEMENT DETERIORATION

There can be many types of pavement deterioration or failure, including rutting, alligator cracking, bleeding, block cracking, raveling, corrugation, and the development of potholes or depressions. Descriptions and photographs of these types of pavement deterioration are presented by ASTM (e.g., ASTM D 5340-93 and ASTM E 1778-96a, 1997).

The damage and deterioration of pavements caused by expansion of a clay subgrade has been discussed in Sec. 9.5. The detrimental effects of water trapped in slow-draining pavement and base will be discussed in Sec. 16.2.1, "Groundwater." Besides expansive soil and groundwater, there can be other factors that contribute to pavement deterioration. Probably the most common causes of premature pavement deterioration or failure are from heavier-than-expected traffic loads or an unanticipated higher volume of traffic.

Factors that should be considered during the design and construction phases of the pavement in order to prevent deterioration include (NAVFAC DM-21.3, 1978):

- Characteristics, strength, and in-place density of the subgrade, base, and asphalt or concrete surface.
- Seasonal fluctuations of groundwater and effectiveness of pavement drainage.
- Frost susceptibility of the pavement section and the effect of freeze-thaw conditions on the subgrade.
- Presence of weak or compressible layers in the subgrade.
- Variability of the subgrade, which may cause differential surface movements.

Other important factors in preventing deterioration of pavements include:

- *Thickness of pavement section.* The construction of the pavement surface and base course must be at least as thick as calculated during the design phase.
- *Strength parameters.* If the *R*-value or California bearing ratio (CBR) values were assumed during the design phase, then the geotechnical engineer should check these values prior to construction at the site.
- *Proper construction.* It is important to have proper compaction of the subgrade and base and to use high-quality aggregate that does not degrade during compaction and from traffic loads. The pavement may need a drainage system to prevent the buildup of water pressures in the subgrade and base materials.

13.4 FROST

There have been extensive studies on the detrimental effects of frost (Casagrande, 1931; Kaplar, 1970; Young and Warkentin, 1975; Reed et al., 1979). Two common types of damage related to frost are: (1) freezing of water in cracks and (2) formation of ice lenses. In many cases, deterioration or damage is not evident until the frost has melted. In these instances, it may be difficult for the geotechnical engineer to conclude that frost was the primary cause of the deterioration.

Freezing of Water in Cracks. There is about a 10 percent increase in volume of water when it freezes, and this volumetric expansion of water can cause deterioration or damage to many different types of materials. Common examples include rock slopes and concrete, as discussed below.

- *Rock slopes.* The expansive forces of freezing water result in a deterioration of the rock mass, additional fractures, and added driving (destabilizing) forces. Feld and Carper (1997) describe several rock slope failures caused by freezing water, such as the February 1957 failure where 900 Mg (1000 tons) of rock fell out of the slope along the New York State Thruway, closing all three southbound lanes north of Yonkers.
- *Concrete.* Durability is defined by the American Concrete Institute as the ability to resist weathering, chemical attack, abrasion, or any other type of deterioration (ACI, 1982). Durability is affected by strength, but also by density, permeability, air entrainment, dimensional stability, characteristics and proportions of constituent materials, and construction quality (Feld and Carper, 1997). Durability is harmed by freezing and thawing, sulfate attack, corrosion of reinforcing steel, and reactions between the various constituents of the cements and aggregates. Damage to concrete caused by freezing could occur during the original placement of the concrete or after it has hardened. To prevent damage during placement, it is important that the fresh concrete not be allowed to freeze. Air-entraining admixtures can be added to the concrete mixture to help protect the hardened concrete from freeze-thaw deterioration.

Formation of Ice Lenses. Frost penetration and the formation of ice lenses in the soil can damage shallow foundations and pavements. The frost penetration will cause heave of the structure if moisture is available to form ice lenses in the underlying soil. The spring thaw will then melt the ice resulting in settlement of the foundation or a weakened subgrade that will make the pavement surface susceptible to deterioration or failure. Damage to highways in the United States and Canada because of frost action is estimated to amount to millions of dollars annually (Holtz and Kovacs, 1981). It is well known that silty soils are more likely to form ice lenses because of their high capillarity and sufficient permeability that enable them to draw up moisture to the ice lenses.

Feld and Carper (1997) describe several interesting cases of damage due to frost action. At Fredonia, New York, the frost from a deep-freeze storage facility froze the soil and heaved the foundations upward 100 mm (4 in.). A system of electrical wire heating was installed to maintain soil volume stability.

Another case involved an extremely cold winter in Chicago, where frost penetrated below an underground garage and broke a buried sprinkler line. This caused an ice buildup which heaved the structure above the street level and sheared off several supporting columns.

As these cases show, it is important that the foundation be constructed below the depth of frost action. Usually there will be local building requirements on the minimum depth of the foundation to prevent damage caused by the formation of ice lenses. There have also been studies to determine the annual maximum frost depths for various site conditions and 50-year or 100-year return periods (e.g., DeGaetano et al., 1997).

Permafrost. Permafrost is often defined as perennially frozen soil. Another frequently used definition of permafrost is that portion of the ground that remains below freezing temperatures for 2 or more years. The bottom of permafrost lies at depths ranging from a few feet to over a thousand feet. The *active layer* is defined as the upper few inches to several feet of ground that is frozen in winter but thawed in summer.

Permafrost often requires special design and construction measures, which are beyond the scope of this book.

13.5 TREE ROOTS

Section 9.3.3 describes how damage can occur to foundations on expansive clay due to the extraction of moisture by the tree roots. A root barrier can be constructed around the structure to prevent roots from gaining access to the area below the foundation.

Damage can also be caused by the physical growth of the tree roots. Deterioration usually occurs to lightly loaded structures such as sidewalks, patios, roads, and block walls, where the physical increase in size of growing roots cause uplift and differential movement. There has been a considerable amount written on the destructive effects of tree roots. For example, it has been stated (Perry and Merschel, 1987):

> The most destructive weapon in a plant's arsenal is its roots. These find their way into microscopic cracks where the long process of leveling cities begins. Roots, even when they are small, develop tremendous hydraulic forces that enlarge cracks, separate bricks and lift concrete and blacktop.

Figure 13.6 presents a photograph of the most common type of damage due to tree roots. In many areas of southern California, land is expensive and most of the building pad is occupied by houses and associated roads and flatwork. The space allocated to plants can be small, such as designated planter areas around the perimeter of the house. As the trees mature, the root systems expand underneath the flatwork. As shown in Fig. 13.6, the tree root has grown beneath the concrete sidewalk, causing cracking and differential movement of the concrete sidewalk.

Figures 13.7 and 13.8 show damage to a road caused by growing tree roots. Subsurface exploration revealed that the road was constructed by placing asphalt concrete atop the outer edge of the paved shoulder. This allowed the base to be in direct contact with the soil from the planter areas. The curbs in this case did not act as a barrier to restrict tree roots from gaining direct access to the roadway base. As shown in Figs. 13.7 and 13.8, the roots penetrated the base, and as they grew, the asphalt curb and the roadway were uplifted and cracked. Note in Fig. 13.7 that the damage was done by a eucalyptus tree of a relatively young age.

For these examples of tree-root uplift, it was observed that damage more frequently occurs to lightly loaded structures, such as sidewalks, roads, and block walls. Damage seems to be less severe for soft or loose soils that deform as the roots grow, rather than the dense or hard soils such as the base in Fig. 13.8.

In order to mitigate root intrusion underneath structures, root barriers can be constructed or a landscape architect can be consulted concerning the best tree planting locations and varieties to prevent intrusive root growth.

FIGURE 13.6 Sidewalk uplift due to tree root.

13.6 HISTORIC STRUCTURES

The repair or maintenance of historic structures presents unique challenges to the geotechnical and foundation engineer. The geotechnical engineer could be involved in many different types of problems with historic structures. Common problems include structural weakening and deterioration due to age or environmental conditions, original poor construction practices, inadequate design, or faulty maintenance. For example, Fig. 13.9 shows the exposed foundation of the Bunker Hill Monument, which was erected at the top of Bunker Hill in 1825 in tribute to the courageous Minutemen who died at Breed's Hill.

Figure 13.10 shows the mortar between the hornblende granite blocks that compose the foundation. In some cases the mortar was completely disintegrated, while in other cases, the mortar could be easily penetrated, as shown in Fig. 13.10, where a pen has been used to pierce the mortar. The causes of the disintegration were chemical and physical weathering, such as cycles of freezing and thawing of the foundation.

In contrast to New England where many of the historic structures are made of brick or stone, in the southwestern United States the historic structures are commonly constructed of adobe. Adobe is a sun-dried brick made of soil mixed with straw. The preferred soil type is a clayey material that shrinks upon drying to form a hard, rocklike brick. Straw is added to the adobe to provide tensile reinforcement. The concept is similar to the addition of steel fibers to modern concrete.

Adobe construction was developed thousands of years ago by the indigenous people, and traditional adobe brick construction was commonly used prior to the twentieth century. The use of adobe as a building material resulted in part from the lack of abundant alternate construction materials, such as trees. The arid climate of the southwestern United States also helped to preserve the adobe bricks.

Adobe structures in the form of multifamily dwellings were typically constructed in low-lying valleys, adjacent to streams, or near springs. In an arid environment, a year-round water source was essential for survival. Mud was scooped out of streambeds and used to

FIGURE 13.7 Condition prior to test pit excavation.

13.12 ANALYSIS OF GEOTECHNICAL DATA AND ENGINEERING COMPUTATIONS

FIGURE 13.8 Condition after test pit excavation.

make the adobe bricks. As a matter of practicality, many structures were constructed close to the source of the mud. These locations were frequently in floodplains, which were periodically flooded, or in areas underlain by a shallow groundwater table.

Adobe construction does not last forever; many historic adobe structures have simply "melted" away. The reasons for deterioration of adobe with time include the uncemented or weakly cemented nature of adobe that makes it susceptible to erosion or disintegration from rainfall, periodic flooding, or water infiltration due to the presence of a shallow groundwater table. For many historic adobe structures and typical archeological sites throughout the world, all that remains is a mound of clay.

Case Study. The purpose of this section is to describe the performance of a historic adobe structure known as the Guajome Ranch house, located in Vista, California. The Guajome Ranch house, a one-story adobe structure, is considered to be one of the finest large Mexican colonial ranch houses remaining in southern California ("Guajome," 1986). The name Guajome means *frog pond* (Engstrand and Ward, unpublished report, 1991).

DETERIORATION

The main adobe structure was built during the period 1852–1854. It was constructed with the rooms surrounding and enclosing a main inner courtyard, which is typical Mexican architecture (Fig. 13.11). The main living quarters were situated in the front of the house. Sleeping quarters were located in another wing, and the kitchen and bakehouse in a third wing. At the time of initial construction, the house was sited in a vast rural territory, and was self-sufficient with water from a nearby stream and a food supply for its habitants provided by farming and livestock.

Additional rooms were added around 1855, enclosing a second courtyard (Fig. 13.11). Several additional rooms were added in 1887 as indicated in Fig. 13.11. Contemporary additions (not shown in Fig. 13.11) include a garage and gable-roof sewing room (built near the parlor).

The Guajome Ranch was purchased by the County of San Diego from the Couts family in 1973. The Guajome Ranch has been declared a National Historic Landmark; restoration was performed by the County of San Diego. The author was a member of the restoration team and investigated the damage caused by poor surface drainage at the site.

1. *Deteriorated condition.* The original exterior adobe walls, which are 0.6–1.2 m (2–4 ft) thick, have been covered with stucco as a twentieth-century modification. Portions of the original low-pitched tile roof have been replaced with corrugated metal.

The Guajome Ranch house was built on gently sloping topography. In the 1855 courtyard, water drained toward one corner and then passed beneath the building. The arrows in Fig. 13.11 indicate the path of the surface water. At the location where the water comes in contact with the adobe, there was considerable deterioration, as shown in Fig. 13.12. Note in this photograph that the individual adobe blocks are visible.

Both the interior and exterior of the structure had considerable deterioration due to a lack of maintenance. The adobe was further eroded when the gable-roofed sewing room

FIGURE 13.9 Exposed foundation of the Bunker Hill Monument.

13.14 ANALYSIS OF GEOTECHNICAL DATA AND ENGINEERING COMPUTATIONS

FIGURE 13.10 Deteriorated mortar between the granite blocks of the Bunker Hill Monument foundation.

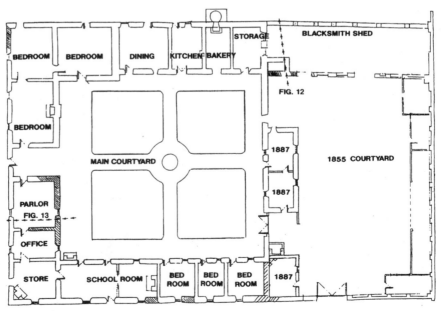

FIGURE 13.11 Site plan.

burned in 1974 and water from the fire hoses severely eroded the adobe. The hatched walls in Fig. 13.11 indicate those adobe walls having the most severe damage.

The main courtyard drains toward the front of the house, as indicated by the arrows in Fig. 13.11. There are no drains located underneath the front of the house; water is removed by simply letting it flow in ditches beneath the floorboards. Figure 13.13 shows the drainage beneath the front of the house. The moisture has contributed to the deterioration

FIGURE 13.12 Deterioration of adobe. The lower photograph is a close-up view of the upper photograph. Note that the individual adobe blocks are visible in the lower photograph.

of the wood floorboards. Because the water is in contact with the adobe, the parlor had some of the most badly damaged adobe walls in the structure.

2. *Laboratory testing of original adobe materials.* Classification tests performed on remolded samples of the original adobe bricks indicate the soil can be classified as a silty sand to clayey sand (SM-SC), the plasticity index varies from about 2 to 4, and the liquid limit is about 20. Based on dry weight, the original adobe bricks contain about 50 percent sand-size particles, 40 percent silt-size particles, and 10 percent clay-size particles smaller than 0.002 mm.

The resistance of the original adobe bricks to moisture is an important factor in their preservation. To determine the resistance of the adobe to moisture infiltration, an index test for the erosion potential of an adobe brick was performed (Day, 1990b). The test consisted of trimming an original adobe brick to a diameter of 6.35 cm (2.5 in.) and a height of 2.54 cm (1.0 in.). Porous plates having a diameter of 6.35 cm (2.5 in.) were placed on the top and bottom of the specimen. The specimen of adobe was then subjected to a vertical stress of 2.9 kPa (60 psf) and was unconfined in the horizontal direction. A dial gauge measured vertical deformation.

After obtaining an initial dial gauge reading, the adobe specimen was submerged in distilled water. Time-versus-dial readings were then recorded. The dial readings were converted to percent strain and plotted versus time, as shown in Fig. 13.14. When initially submerged in distilled water, some soil particles sloughed off in the horizontal (unconfined) direction, but the specimen remained essentially intact. This indicates that there is a weak bond between the soil particles.

After submergence in distilled water for 8 days, the water was removed from the apparatus and the adobe specimen was placed outside and allowed to dry in the summer sun. After 7 days of drying, the adobe specimen showed some shrinkage and corresponding cracking.

The adobe specimen was again submerged in distilled water and after only 18 minutes, the specimen had completely disintegrated as soil particles sloughed off in the horizontal (unconfined) direction. The experiment demonstrated the rapid disintegration of the adobe when it is subjected to wetting-drying cycles. Initially, the adobe is weakly cemented and resistant to submergence, but when the soaked adobe is dried, it shrinks and cracks, allowing for an accelerated deterioration when again submerged.

3. *Drainage repair.* Drainage was probably not a major design consideration when the Guajome Ranch house was built. However, the deterioration of the adobe where water is present indicates the importance of proper drainage. The drainage repair consisted of regrading of the courtyards so that surface water flows to box and grate inlets connected to storm-water piping. Surface drainage water is removed from the site through underground pipes. The new drainage system prevents surface water from coming in contact with the adobe foundation.

In summary, the interior and exterior of the Guajome Ranch house had severe adobe and wood floor deterioration in areas where water from surface drainage came in contact with the structure (Figs. 13.12 and 13.13). The results of laboratory testing demonstrated the rapid disintegration of the adobe when it is subjected to cycles of wetting and drying. The repair for damage related to surface drainage consisted of a new drainage system that prevents surface water from coming in contact with the adobe foundation.

DETERIORATION 13.17

FIGURE 13.13 Drainage beneath parlor.

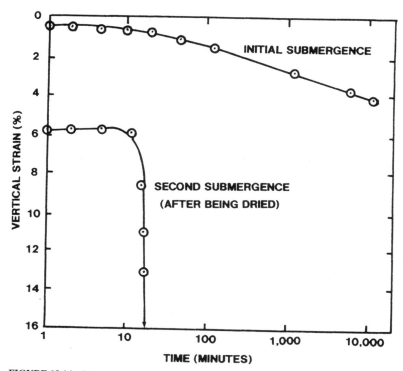

FIGURE 13.14 Laboratory test results.

CHAPTER 14
UNUSUAL SOIL

14.1 INTRODUCTION

An unusual soil can be defined as a soil that has rare or unconventional engineering behavior. Such soils are frequent causes of damage and distress or result in construction cost overruns. This is often because of their unusual properties that were not identified during the subsurface exploration or properly evaluated during the design phase. Some unusual soils are discussed below:

14.2 EXAMPLES OF UNUSUAL SOIL

Rock Flour (or Bull's Liver). This soil consists predominantly of silt-size particles, but has little or no plasticity. Nonplastic rock flour contains particles of quartz, ground to a very fine state by the abrasive action of glaciers. Terzaghi and Peck (1967) state that because of its fine particle size, this soil is often mistaken as clay.

In describing this soil, the term "bull's liver" apparently comes from its *in situ* appearance. It has been observed that in a saturated state, it quakes like jelly from shock or vibration and can even flow like a liquid (Sowers and Sowers, 1970).

Peat. Peat is composed of partially decayed organic matter (humic and nonhumic substances), where the remains of leaves, stems, twigs, and roots can be identified. The places where peat accumulates are known as peat bogs or peat moors. Its color ranges from light brown to black. Peat is unusual because it has a very high water content which makes it extremely compressible. This almost always makes it unsuitable for supporting foundations (Terzaghi and Peck, 1967).

Nonwelded Tuff and Volcanic Ash. Tuff is a pyroclastic rock, originating as airborne debris from explosive volcanic eruptions. The largest fragments (in excess of 64mm) are called *blocks* and *bombs*, fragments between 4 mm and 64 mm are called *lapilli*, fragments between 4 mm and 0.25 mm are *ash*, and the finest fragments (less than 0.25 mm) are *volcanic dust* (Compton, 1962).

An important aspect of tuff is the degree of welding, which can be described as either nonwelded, partially welded to varying degrees, or densely welded. Welding is generally caused by fragments that are hot when deposited, and because of this heat, the sticky glassy fragments may actually fuse together (Best, 1982). There are distinct changes in the original shards and pumice fragments, such as the union and elongation of the glassy shards and flattening of the

FIGURE 14.1 Deposit of oversize particles.

pumice fragments, which are characteristic of completely welded tuff (Ross and Smith, 1961). The degree of welding depends on many factors, such as type of fragments, plasticity of the fragments (which depends on the emplacement temperature and chemical composition), thickness of the resulting deposit, and rate of cooling (Smith, 1960).

Deposition of volcanic ash directly from the air may result in an unconsolidated (geologically speaking) deposit, which would then be called *ash*; but indurated deposits are called *tuff*. Nonwelded tuff has an engineering behavior similar to volcanic ash. These materials have been used as mineral filler in highways and other earth-rock construction. Some types of volcanic ash have been used as pozzolanic cement and as admixtures in concrete to retard undesired reactions between cement alkalies and aggregates.

Natural deposits of nonwelded tuff and volcanic ash are unusual because they have very low dry densities (e.g., 1 Mg/m^3) due to the presence of lightweight glass and pumice. The materials are also highly susceptible to erosion, which can cause the development of unusual eroded landforms known as *pinnacles*.

Loess. Loess is widespread in the central portion of the United States. It consists of uniform cohesive wind-blown silt, commonly light brown, yellow, or gray in color, with most of the particle sizes between 0.01 and 0.05 mm (Terzaghi and Peck, 1967). The cohesion is commonly a result of calcareous cement which binds the particles together. An unusual feature of loess is the presence of vertical root holes and fractures that make it much more permeable in the vertical direction than the horizontal direction. Another unusual feature of loess is that it can form near-vertical slopes, but when saturated, the cohesion is lost and the slope will fail or the ground surface settle.

Caliche. This type of material is common in arid or semiarid parts of the southwestern United States. It consists of soil that is normally cemented together by calcium carbonate.

When water evaporates near or at ground surface, the calcium carbonate is deposited in the void spaces between soil particles. Caliche is generally strong and stable in an undisturbed state, but it can become unstable if the cementing agents are leached away by water from leaky pipes or sewers or from the infiltration of irrigation water.

Debris Flow and Alluvial Fan Deposits. As previously mentioned, a debris flow can transport a wide variety of soil particle sizes, including boulders and cobbles. Boulders, cobbles, and coarse gravel are typically described as *oversize particles,* and the finer soil particles are described as the *soil matrix.* Figure 14.1 shows a deposit of oversize particles. The oversize particles can be deposited in alluvial fans or from debris flow with the finer soil particles filling in the void spaces. The debris flow and alluvial fan deposits often consist of oversize particles that primarily carry the overburden pressure, such as shown in Fig. 14.2, where the four identified cobbles are in direct contact and are carrying the overburden pressure. Such debris flow and alluvial fan deposits are often unstable because of the erratic and unsteady arrangement of oversize particles and loose matrix soil.

Varved Clay. Varved clays ordinarily form as lake deposits. They consist of alternating layers of soil. Each varve represents the deposition during a year, where the lower sandy part is deposited during the summer, and the upper clayey part is then deposited during the winter when the surface of the lake is frozen and the water is tranquil. This causes an unusual variation in shear strength in the soil, where the horizontal shear strength along the clay portion of the varve is much less than the vertical shear strength. This can cause the stability of structures founded on varved clay to be overestimated, resulting in a bearing-capacity-type failure.

Bentonite. Bentonite is a deposit consisting mainly of montmorillonite clay particles. It is derived from the alteration of volcanic tuff or ash. Bentonite is mined to make products that are used as impermeable barriers such as geosynthetic clay liners (GCL), which are bentonite/geosynthetic composites. Because bentonite consists almost exclusively of montmorillonite, it will swell, shrink, and cause more expansive-soil-related damage than any other type of soil.

Sensitive or Quick Clays. As mentioned in Sec. 5.2.3, the sensitivity S_t of a clay is defined as the undisturbed or natural undrained shear strength divided by its remolded shear strength. On the basis of this value, the clay can be rated as having a sensitivity from "low" to "quick" (Holtz and Kovacs, 1981). An unusual feature of highly sensitive or quick clays is that the *in situ* water content is often greater than the liquid limit (liquidity index greater than one). Figure 4.42 shows sensitive Canadian clay that has a water content greater than its liquid limit.

Sensitive clays have unstable bonds between particles. As long as these unstable bonds are not broken, the clay can support a heavy load. But once remolded, the bonding is destroyed and the shear strength is substantially reduced. For example, sensitive Leda Clay, from Ottawa, Ontario, has a high shear strength in the undisturbed state, but once remolded, the clay is essentially a fluid (no shear strength). There are reports of entire hillsides of quick clays becoming unstable and then simply flowing away (Lambe and Whitman, 1969).

Diatomaceous Earth. Diatoms are defined as microscopic, single-celled plants of the class Bacillariophyceae, which grow in both marine and fresh water (Bates and Jackson, 1980). Diatoms secrete outer shells of silica, called frustules, in a great variety of forms which can accumulate in sediments in enormous amounts (Bates and Jackson, 1980). Deposits of diatoms have low dry density and high moisture content because the structure of the diatom is an outer shell of silica that can contain water. Common shapes of diatoms

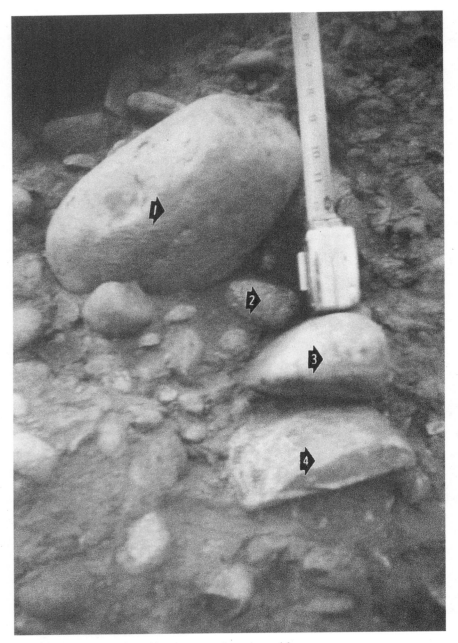

FIGURE 14.2 Erratic and unstable arrangement of oversize particles.

are rodlike, spherical, or circular disks having a typical length or diameter of about 0.03 to 0.11 mm (Fig. 14.3; from Spencer, 1972). Diatoms typically have rough surface features, such as protrusions and indentations.

A natural deposit of diatoms is commonly referred to as diatomaceous earth or diatomite. Diatomaceous earth usually consists of fine, white, siliceous powder, composed mainly of diatoms or their remains (Terzaghi and Peck, 1967; Stokes and Varnes, 1955). Diatomite is a organogenetic sedimentary rock containing frustules of diatoms and sometimes mixed with shells of radiolarians, spicules of sponges, and foraminifera (Mottana et al., 1978). Industrial uses of diatomaceous earth or diatomite are as filters to remove impurities, as abrasives to polish soft metals, and when mixed with nitroglycerin, as an absorbent in the production of dynamite (Mottana et al., 1978).

Diatomaceous earth can be very compressible when used as fill. For example, Fig. 14.4 (Day, 1995e) shows the one-dimensional vertical settlement of diatomaceous earth measured by the oedometer apparatus. The initial dry density of the diatomaceous earth was 0.87 Mg/m^3 (54.3 pcf). During the testing of the diatomaceous earth, there was a distinct popping sound at high vertical pressures (i.e., 1600 kPa), a result of the diatoms (which are essentially hollow shells of silica) being crushed together. The crushing together of the diatoms is the reason for the high compressibility at high vertical pressures.

14.3 CASE STUDY OF UNUSUAL SOIL

The purpose of this section is to present a case study of the construction of a site using unusual soil. In particular, the project deals with the deformation of fill slopes at Canyon Estates, located in Mission Viejo, California. The author was retained as an expert for the plaintiffs. The case settled out of court in early 1997 for over $10 million. The purpose of this case study is to describe the slope movement, present data on observed damages, and discuss the causes of the slope movement. A discussion of the legal issues of the project, which are important for geotechnical engineers practicing in the United States, will also be presented. As of the date of preparation of this case study, no repairs had been performed at the site.

The Canyon Estates project is an approximately 350-acre parcel of land. Earth moving operations were utilized to fill canyons in order to develop roads and terraced level building pads in 1983 to 1988. Approximately 750 one- and two-story single-family residences had been built on these pads. At the time of the study, there were an additional 200 houses in various stages of construction. Structures appurtenant to all the homes include side yard property line block

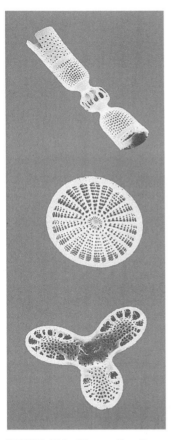

FIGURE 14.3 Diatoms. Top diatom enlarged approximately 1200 times; middle and bottom diatoms enlarged approximately 300 times. (*Reprinted from The Dynamics of the Earth, copyright 1972 by the Thomas Y. Crowell Company, Inc.*)

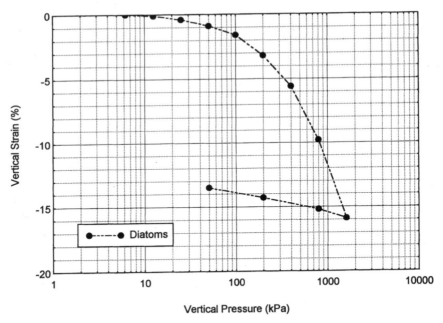

FIGURE 14.4 Vertical strain versus vertical pressure. (*From Day 1995e, reprinted with permission of American Society of Civil Engineering*)

walls, which are typically connected to top-of-slope iron-post-and-block walls. Homeowner improvements include patio flatwork, decks, and pools.

Because of the filled-in canyons and buttressed slopes, most building pads contained rear yard descending fill slopes with some building pads also containing a side yard descending fill slope. The fill slopes were constructed at approximately 2:1 (horizontal:vertical) inclinations and commonly range between 6 to 12 m (20 to 40 ft) in height throughout the site. In some areas, the fill slopes were constructed to approximately 20 m (65 ft) in height. Because of the presence of adverse (out-of-slope) bedding in cut slope areas, the cut slopes were overexcavated and buttressed with fill. These buttressed slopes were generally constructed with an equipment-width (4.6 m, 15 ft) key and extended from the slope toe to the top.

Monterey Formation. The fill used to create the fill slopes at Canyon Estates was derived from the Monterey formation. The Monterey formation is generally classified as a siltstone or claystone. An unusual feature of the Monterey formation is that it contains diatoms or broken fragments of diatoms. As previously discussed, deposits of diatoms are unusual because they have a low dry density and high moisture content due to the structure of the diatom which is an outer shell of silica that can contain water (Fig. 14.3).

Index Properties of Fill. According to the Unified Soil Classification System, the fill at Canyon Estates was classified as a silty clay (CH) to clayey silt (MH). The liquid limit was generally between 60 to 70. Based on 24 particle size analyses of fill throughout the site, the percentage of clay particles (finer than 0.002 mm) varied from 28 to 52 percent, based on dry weight. X-ray diffraction tests indicated that the clay particles are predominantly montmorillonite. The majority of the remaining particles are of silt size with a significant portion being the remains of diatoms. The diatoms do affect the index properties because in some cases the limits plot well below the A-line, while inorganic soil containing montmorillonite usually plots just below the U-line (see Fig. 5.18). In some areas, the process of compaction did not completely pulverize the fragments of Monterey formation that were ripped from cut areas, and the fill did contain oversize particles, but this was usually less than 10 percent, based on dry weight.

New Method of Compaction (ADC). Because of the high moisture content of the diatomaceous material encountered during site grading and the difficulties in compacting the soil to the industry standard of a minimum of 90 percent relative compaction (based on Modified Proctor), an alternative method of determining fill compaction was utilized. This alternative method required compacting the fill to 95 percent of the maximum achievable density (ADC). The maximum achievable density is the soil density obtained by compacting a sample (one point) at an *in situ* moisture content using the Modified Proctor compaction energy. In essence, the unusual nature of the diatomaceous earth resulted in the development of a new method for the placement and testing of the soil at the site.

As a result of using the ADC test method, fill was compacted at a higher than optimum moisture content and to a relative compaction that was less than the industry standard. Average relative compaction measured during the investigation was about 82 percent. Because of the lower density achieved by the ADC method, there was a corresponding reduction in shear strength which allowed the slopes to deform more than typically expected. For example, Figs. 14.5 through 14.7 present unconsolidated undrained (UU) triaxial compression tests performed on typical fill specimens remolded to relative compaction of either 70 percent (Fig. 14.5), 80 percent (Fig. 14.6), or 90 percent (Fig. 14.7) and tested at different confining pressures. Note that in these figures, the Mohr circles from the unconsolidated undrained (UU) triaxial tests are shown, but the failure envelope has not been drawn. This data shows a substantial reduction in the undrained shear strength as the relative compaction decreases.

Basis for Filing of Lawsuit. The complaint was initially filed by the Canyon Estates Homeowners Association in Orange County Superior Court on April 3, 1992. The first and preeminent basis for filing the lawsuit was "strict liability." The concept of strict liability means that developers of mass-produced housing will be found liable for any defects at the project, regardless of whether the defect resulted from a failure to comply with the standard of practice. In California, it is not necessary to prove negligence on the part of the builder, but rather all that needs to be shown is that the project has a defect. A defect exists when a product or some component of it fails to perform normally when used for its intended purpose by the consumer.

At Canyon Estates, it was contended that the movement of the slopes with consequential failure of lateral support to hundreds of homes atop those slopes was a defect, in that the slopes were failing to perform normally when used for their intended purpose by the consumer.

Damage at the Site. Visual inspections of fill slope areas and top-of-slope improvements such as walls, flatwork, or pools revealed a variety of distress. Typically, damage consisted of cracking and separation of block walls near the slope tops. In most cases, the pilasters at the slope top pulled away from the side yard property line walls (Figs. 14.8 and 14.9). Some of these separations were on the order of 20 cm (8 in.) in width. Another common type of damage to the side yard property walls was stair-step cracking which occurred up to 4.5 m to 6 m (15 to 20 ft) back from the slope top (Figs. 14.10 and 14.11). Often, cracks extended

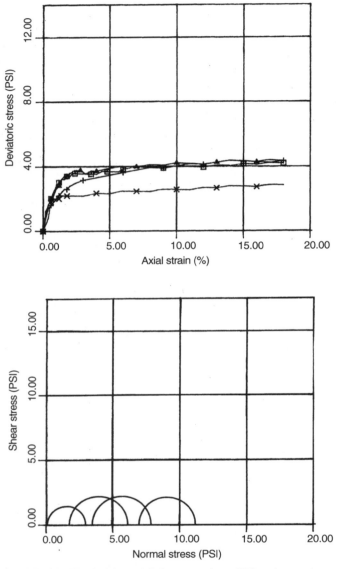

FIGURE 14.5 UU triaxial tests (relative compaction = 70%, moisture content = 49%).

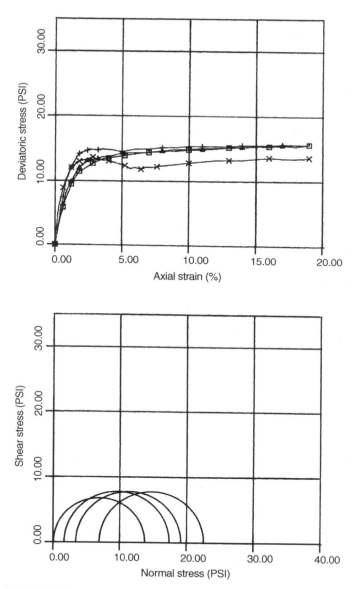

FIGURE 14.6 UU triaxial tests (relative compaction = 80%, moisture content = 39%).

14.10 ANALYSIS OF GEOTECHNICAL DATA AND ENGINEERING COMPUTATIONS

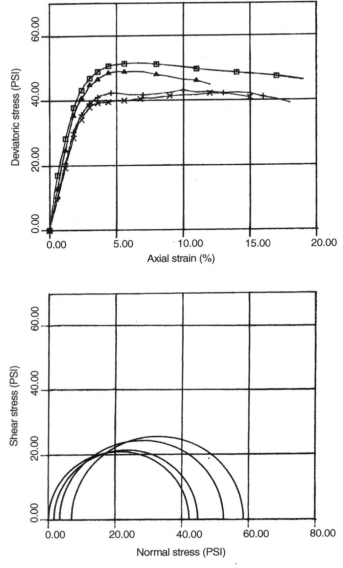

FIGURE 14.7 UU triaxial tests (relative compaction = 90%, moisture content = 28%).

FIGURE 14.8 Eight inches of separation of top-of-slope wall.

through the wall footings as shown in Fig. 14.12. The damage indicated predominately a lateral component of movement as shown in Figs. 14.8 to 14.12.

In general, the higher the slope, the greater the observed magnitude of cracking and separations. Also, at lot corners where slopes descended in both the back and side yards, distress was greater.

Observations of backyard improvements, such as patio slabs, planter walls, and pool areas have exhibited similar cracks and separations (Figs. 14.13 and 14.14). These features generally align parallel to the slope and in some areas were noted as far as 6 m (20 ft) back from the slope tops, indicating fairly extensive slope movement. In some cases, it was observed that a gap opened up between the concrete driveway and the concrete garage slab. In some cases, the gap exceeded 2.5 cm (1 in.) and the gap was believed to be an effect of slope deformation, which was pulling the posttensioned house foundations downslope.

During the investigation of site conditions, it was noted that some repairs had been made to the damage. Repairs were noted to consist of localized patching with grout or extending of wrought iron fences to connect pilaster/wall separations. Both types of repair were

FIGURE 14.9 Four inches of separation of top-of-slope wall.

observed to be ineffective. In most cases, the patches reopened and wrought iron extensions disconnected, indicating ongoing slope movement. Some of the reopening of cracks exceeded 5 cm (2 in.).

Inclinometer Monitoring. Thirteen inclinometer casings were installed in the rear yards (near the top of slope) throughout the Canyon Estates site in order to monitor the rate and magnitude of lateral movement. The 13 inclinometers were monitored from mid-1994 up to the time the case settled out of court (early 1997). About half the inclinometers displayed inconclusive or low-level slope movement during the monitoring period. The other half did show continuous lateral movement during the monitoring period. These inclinometers displaying ongoing movement were generally located at the highest-fill slopes or in areas where there were both rear and side yard descending slopes.

Figures 14.15 to 14.18 show the downslope (A direction) movement of two of the inclinometers at Canyon Estates as drawn by the Geo-Slope inclinometer plot program. In each figure, the plot on the left is the lateral (downslope) deformation versus depth, while the plot on the right is the lateral deformation versus time.

Figures 14.15 to 14.17 are the plots for one of the inclinometers. The right side of Fig. 14.15 shows the lateral deformation versus time for data at a depth of 0.6 m (2 ft), Fig. 14.16 shows the lateral deformation versus time for data at a depth of 1.8 m (6 ft), and Fig. 14.17 shows the lateral deformation versus time for data at a depth of 3.6 m (12 ft). Note the cyclic behavior of the lateral deformation versus time for the data at a depth of 0.6 m (2 ft). Since the inclinometer was not installed in the slope, but rather on the flat part of the building pad near the top of slope, this cyclic inclinometer behavior was attributed to wetting and drying cycles. When the building pad dried out, there was a pulling of the inclinometer in the direction away from the slope face. Note that the lateral deformation versus time at depths of 1.8 m (6 ft, Fig. 14.16) and 3.6 m (12 ft, Fig. 14.17) do not show the cyclic lateral deformation, indicating that this phenomenon is a near-surface condition.

Figure 14.18 presents the plot for a second inclinometer. Note in this figure that the depth of movement was rather deep, on the order 4.9 to 5.5 m (16 to 18 ft). Also, the inclinometers did not indicate a discrete plane of failure, but rather a progressive decrease in lateral movement with depth.

FIGURE 14.10 Cracking and separation of wall near top of slope.

14.14 ANALYSIS OF GEOTECHNICAL DATA AND ENGINEERING COMPUTATIONS

FIGURE 14.11 Stair-step cracking of wall near top of slope.

FIGURE 14.12 Cracking through wall and footing.

FIGURE 14.13 Lateral movement of pool deck.

Slope Stability. Slope stability total stress (undrained) analysis was performed by using the SLOPE/W (Geo-Slope, 1991) computer program. By inputting the slope configuration (2:1 horizontal:vertical slope inclination), undrained shear strength parameters from Figs. 14.5 to 14.7, and total unit weight of soil (18.1 kN/m^3, 115 pcf), the computer program calculated the factor of safety using the Janbu simplified method of slices. The slope stability analyses are summarized in Table 14.1.

The slope stability analyses show that at a relative compaction of 70 percent, the fill slopes would fail. At 80 percent relative compaction, which is about the average for the site, the taller slopes (18 m, 60 ft) have a low factor of safety of 1.17. Figure 14.19 presents the slope stability analysis for the condition of 80 percent relative compaction for a slope of 18 m (60 ft) height. For a condition of 90 percent relative compaction, the slopes have a very high factor of safety because of the high undrained shear strength.

Slope Deformation. The results of slope stability analyses show a low factor of safety for total stress (undrained) conditions for the higher fill slopes (Table 14.1). These results indicate that one cause of the lateral movement at the site was undrained creep of the compacted slopes. The slope stability analysis shows that the taller slopes would be most susceptible to the undrained creep.

A second factor in the movement was slope softening. On the basis of a comparison of initial as-compacted moisture contents and those obtained from this investigation, some of the fill slopes were subjected to an increase in moisture content with time. It was concluded that a second probable cause of the slope deformation was slope softening.

As previously mentioned, the near-surface wetting and drying of the building pad did affect the inclinometer (Fig. 14.15). It was concluded that a third probable cause of the near-surface slope deformation was seasonal moisture changes.

FIGURE 14.14 Separation between patio and pool bond beam.

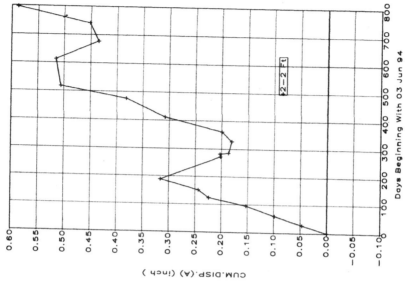

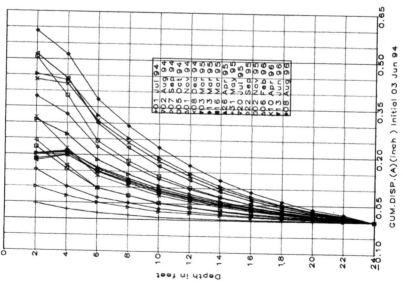

FIGURE 14.15 Inclinometer no. 7 (displacement plot for depth = 2 ft).

14.17

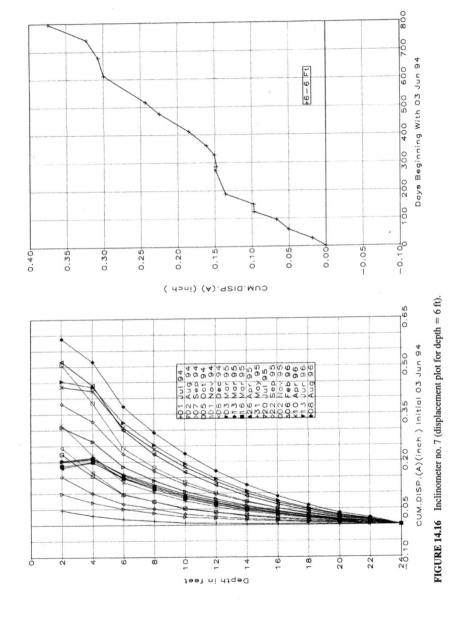

FIGURE 14.16 Inclinometer no. 7 (displacement plot for depth = 6 ft).

14.18

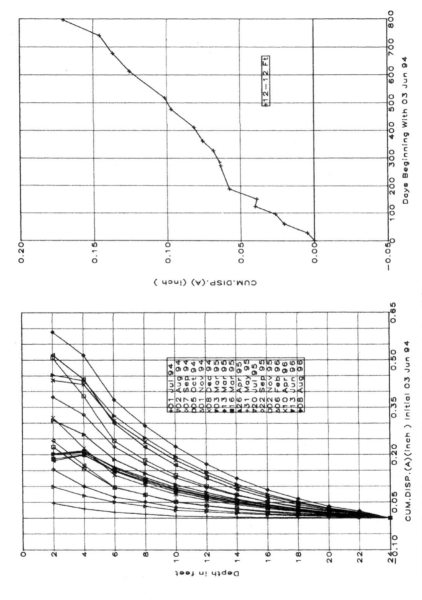

FIGURE 14.17 Inclinometer no. 7 (displacement plot for depth = 12 ft).

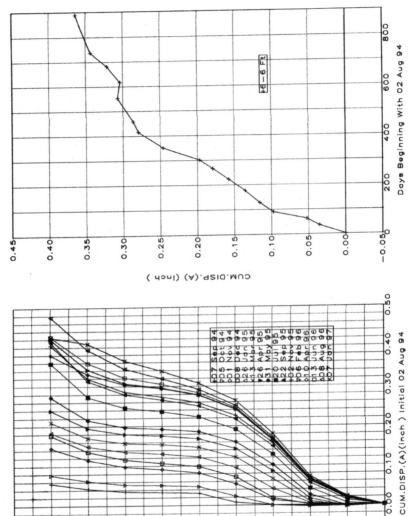

FIGURE 14.18 Inclinometer no. 12 (displacement plot for depth = 6 ft).

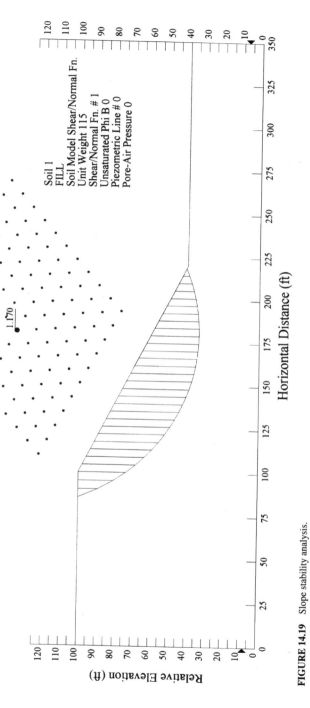

FIGURE 14.19 Slope stability analysis.

TABLE 14.1 Summary of Slope Stability Analysis

Relative compaction, % (1)	Undrained shear strength (2)	Slope height (3)	Factor of safety (4)
70	Fig. 14.5	9 m (30 ft)	1.15
		18 m (60 ft)	0.97
80	Fig. 14.6	9 m (30 ft)	1.89
		18 m (60 ft)	1.17
90	Fig. 14.7	9 m (30 ft)	5.11
		18 m (60 ft)	3.65

Note: All slope inclinations are 2:1 (horizontal:vertical), and the total unit weight of soil = 18.1 kN/m^3 (115 pcf).

Summary. Figures 14.8 to 14.14 show pictures of typical severe damage at the Canyon Estates project. The cause of the damage was lateral deformation of the fill slopes. The data indicated that there were three separate mechanisms that caused the slope deformation: undrained creep, slope softening, and near-surface downslope movement due to seasonal moisture changes.

The results of slope stability analysis (Table 14.1) indicate the importance of relative compaction on the undrained shear strength of the soil. The project compaction specifications were based on the presence of unusual soil (diatomaceous earth) at the site. As a consequence, a new method of compaction (ADC method) was developed for the site. This compaction method resulted in an average degree of fill compaction of only 82 percent. This low compaction was considered a primary reason for the lateral deformation of the fill slopes.

Engineers must be aware of legal issues when working on specific projects. In California, the courts have adopted the concept of "strict liability." This means that in order to obtain a monetary judgment, negligence need not be proven. All that needs to be shown is that the project has a defect that is causing damage. At the Canyon Estates project, the defect was the lateral deformation of fill slopes that damaged block walls and homeowner improvements.

CHAPTER 15
RETAINING WALLS

The following notation is used in this chapter:

SYMBOL	DEFINITION
a	Horizontal distance from W to the toe of the footing
a_{max}	Peak acceleration at the ground surface
A_p	Anchor pull force (sheet pile wall)
c	Cohesion based on a total stress analysis
c'	Cohesion based on an effective stress analysis
c_a	Adhesion between the bottom of the footing and the underlying soil
d_1	Depth from ground surface to the groundwater table
d_2	Depth from the groundwater table to the bottom of the sheet pile wall
D	Depth of the retaining wall footing (Secs. 15.2 to 15.4)
D	Portion of the sheet pile wall anchored in soil (Fig. 15.29)
D	Diameter of the pier (Sec. 15.8)
e	Lateral distance from P_v to the toe of the retaining wall
F	Factor of safety
g	Acceleration of gravity
H	Height of the retaining wall (Secs. 15.2 to 15.5)
H	Unsupported face of the sheet pile wall (Fig. 15.29)
H	Distance from the slip surface to ground surface (Sec. 15.8)
k_A	Active earth pressure coefficient
k_0	Coefficient of earth pressure at rest
k_p	Passive earth pressure coefficient
L	Length of the slip surface (Sec. 15.8)
M_{max}	Maximum moment in the sheet pile wall
N	Sum of the wall weights W plus P_v (if applicable)
P_A	Active earth pressure resultant force
P_E	Earthquake-induced horizontal force on the retaining wall
P_H	Horizontal component of the active earth pressure resultant force
P_i	Required pier wall force that is inclined at an angle α
P_L	Lateral design force for each pier
P_p	Passive resultant force
P_Q	Resultant force due to a uniform surcharge on top of the wall backfill
P_v	Vertical component of the active earth pressure resultant force
P_1	Active earth pressure resultant force (i.e., $P_1 = P_A$, Fig. 15.23)
P_2	Resultant force due to a uniform surcharge (i.e., $P_2 = P_Q$, Fig. 15.23)

Q	Uniform vertical surcharge pressure acting on the wall backfill
s_u	Undrained shear strength of the soil
S	On-center spacing of the piers
u	Average pore water pressure along the slip surface
W	Resultant of the vertical retaining wall loads (Secs. 15.2 to 15.4)
W	Total weight of the failure wedge (Sec. 15.8)
$\bar{x}$	Location of N from the toe of the footing
Y	Horizontal displacement of the retaining wall
Z_T	Total length of the pier
Z_1	Depth to adequate bearing material
α	Groundwater location parameter (Fig. 15.30)
α	Angle of inclination of the slip surface (Sec. 15.8)
β	Slope inclination behind or in front of the retaining wall
δ, ϕ_{cv}	Friction angle between bottom of wall footing and underlying soil
δ, ϕ_w	Friction angle between the back face of the wall and the soil backfill
ϕ	Friction angle based on a total stress analysis
ϕ'	Friction angle based on an effective stress analysis
γ_b	Buoyant unit weight of the soil
γ_t	Total unit weight of the soil
θ	Back face inclination of the retaining wall
σ_{avg}	Average bearing pressure of the retaining wall foundation
σ_{mom}	That portion of the bearing pressure due to the eccentricity of N

15.1 INTRODUCTION

A retaining wall is defined as a structure whose primary purpose is to provide lateral support for soil or rock. In some cases, the retaining wall may also support vertical loads. Examples include basement walls and certain types of bridge abutments.

Cernica (1995a) lists and describes various types of retaining walls. Some of the more common types of retaining walls are gravity walls, counterfort walls, cantilevered walls, and crib walls. Gravity retaining walls are routinely built of plain concrete or stone, and the wall depends primarily on its massive weight to resist failure from overturning and sliding. Counterfort walls consist of a footing, a wall stem, and intermittent vertical ribs (called *counterforts*) which tie the footing and wall stem together. Crib walls consist of interlocking concrete members that form cells which are then filled with compacted soil.

Although mechanically stabilized earth retaining walls have become more popular in the past decade, cantilever retaining walls are still probably the most common type of retaining structure. There are many different types of cantilevered walls, with the common features being a footing that supports the vertical wall stem. Typical cantilevered walls are T-shaped, L-shaped, or reverse L-shaped (Cernica, 1995a).

Clean granular material (no silt or clay) is the standard recommendation for backfill material. There are several reasons for this recommendation:

1. *Predictable behavior.* Import granular backfill generally has a more predictable behavior in terms of earth pressure exerted on the wall. Also, expansive soil related forces (Chap. 9) will not be generated by clean granular soil.

2. *Drainage system.* To prevent the buildup of hydrostatic water pressure on the retaining wall, a drainage system is often constructed at the heel of the wall. The drainage system will be more effective if highly permeable soil, such as clean granular soil, is used as backfill.

3. *Frost action.* In cold climates, frost action has caused many retaining walls to move so much that they have become unusable. If freezing temperatures prevail, the backfill soil can be susceptible to frost action, where ice lenses will form parallel to the wall and cause horizontal movements of up to 0.6 to 0.9 m (2 to 3 ft) in a single season (Sowers and Sowers, 1970). Backfill soil consisting of clean granular soil and the installation of a drainage system at the heel of the wall will help to protect the wall from frost action.

Movement of retaining walls (i.e., active condition) involves the shear failure of the wall backfill, and the analysis will naturally include the shear strength of the backfill soil (Sec. 6.5). As in the analysis of strip footings (Sec. 8.2) and slope stability (Sec. 10.5), for most field situations involving retaining structures, the backfill soil is in a plane strain condition (i.e., the soil is confined along the long axis of the wall). As previously mentioned, the friction angle ϕ is about 10 percent higher in the plane strain condition compared to the friction angle ϕ measured in the triaxial apparatus. In practice, plane strain shear strength tests are not performed, which often results in an additional factor of safety for retaining wall analyses.

The next section (15.2) of this chapter discusses the basic retaining wall equations for a simple retaining wall. The following sections will then discuss in more detail the design and construction of retaining walls (15.3), restrained retaining walls (15.4), mechanically stabilized earth retaining walls (15.5), and sheet pile walls (15.6). The final two sections are devoted to temporary retaining systems used for construction purposes (15.7) and pier walls used to stabilize slopes (15.8).

15.2 BASIC RETAINING WALL ANALYSES

Figure 15.1 shows a reverse L-shaped cantilever retaining wall. This type of simple retaining wall will be used to introduce the basic types of retaining wall design analyses. The pressure exerted on the back side of the wall is the active earth pressure. The footing is supported by the vertical bearing pressure of the soil or rock. Lateral movement of the wall is resisted by passive earth pressure and slide friction between the footing and bearing material. The following discussion of the design analyses for retaining walls is divided into two categories: (1) simple retaining wall without wall friction and (2) simple retaining wall with wall friction.

15.2.1 Simple Retaining Wall without Wall Friction

Active Earth Pressure. As shown in Fig. 15.1, the active earth pressure is often assumed to be horizontal by neglecting the friction developed between the vertical wall stem and the backfill. This friction force has a stabilizing effect on the wall and therefore it is usually a safe assumption to ignore friction. However, if the wall should settle more than the backfill because, for example, of high vertical loads imposed on the top of the wall, then a negative skin friction can develop between the wall and backfill which has a destabilizing effect on the wall.

In the evaluation of the active earth pressure, it is common for the soil engineer to recommend clean granular soil as backfill material. In order to calculate the active earth pressure

15.4 ANALYSIS OF GEOTECHNICAL DATA AND ENGINEERING COMPUTATIONS

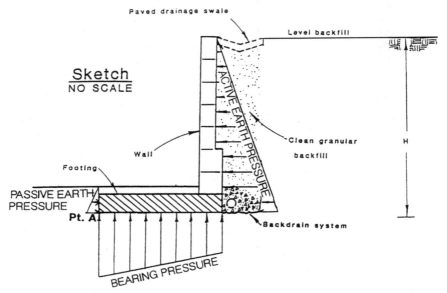

FIGURE 15.1 Retaining wall design pressures.

resultant force P_A, in kN per linear meter of wall or pounds per linear foot of wall, the following equation is used for clean granular backfill:

$$P_A = \tfrac{1}{2} k_A \gamma_t H^2 \tag{15.1}$$

where k_A = active earth pressure coefficient, γ_t = total unit weight of the backfill, and H = height over which the active earth pressure acts as defined in Fig. 15.1. The active earth pressure coefficient k_A is equal to:

$$k_A = \tan^2(45° - \tfrac{1}{2}\phi) \tag{15.2}$$

where ϕ = friction angle of the clean granular backfill. Equation (15.2) is known as the active Rankine state, after the British engineer Rankine who in 1857 obtained this relationship.

In Eq. (15.1), the product of k_A times γ_t is referred to as the equivalent fluid pressure (even though the product is actually a unit weight). In the design analysis, the soil engineer usually assumes a total unit weight γ_t of 18.9 kN/m³ (120 pcf) and a friction angle ϕ of 30° for the granular backfill. From Eq. (15.2) for ϕ = 30°, the active earth pressure coefficient k_A is 0.333. Multiplying 0.333 times the total unit weight γ_t of backfill results in an equivalent fluid pressure of 6.3 kN/m³ (40 pcf).

This is a common recommendation for equivalent fluid pressure from soil engineers. It is valid for the conditions of clean granular backfill, a level ground surface behind the wall, a backdrain system, and no surcharge loads. Note that this recommended value of equivalent fluid pressure of 6.3 kN/m³ (40 pcf) does not include a factor of safety and is the actual pressure that would be exerted on a smooth wall when the friction angle ϕ of the granular backfill equals 30°. When designing the vertical wall stem in terms of wall thickness and size and location of steel reinforcement, a factor of safety F can be applied to the active earth pressure in Eq. (15.1). A factor of safety may be prudent because higher wall pressures will most likely be generated during compaction of the backfill or when translation of the footing is restricted (Goh, 1993).

Additional important details concerning the active earth pressure are as follows:

1. *Sufficient movement.* There must be sufficient movement of the retaining wall in order to develop the active earth pressure of the backfill. For example, Table 15.1 (from NAVFAC DM-7.2, 1982) indicates the amount of wall rotation that must occur for different backfill soils in order to reach the active earth pressure state.

2. *Triangular distribution.* As shown in Fig. 15.1, the active earth pressure is a triangular distribution, and thus the active earth pressure resultant force P_A is located at a distance equal to $1/3\,H$ above the base of the wall.

3. *Surcharge pressure.* If there is a uniform surcharge pressure Q acting upon the entire ground surface behind the wall, then there would be an additional horizontal pressure exerted upon the retaining wall equal to the product of k_A times Q. Thus the resultant force P_Q, in kN per linear meter of wall or pounds per linear foot of wall, acting on the retaining wall due to the surcharge Q is equal to:

$$P_Q = Q H k_A \qquad (15.3)$$

where Q = uniform vertical surcharge (kPa or psf) acting upon the entire ground surface behind the retaining wall, k_A = active earth pressure coefficient [Eq. (15.2)], and H = height of the retaining wall. Because this pressure acting upon the retaining wall is uniform, the resultant force P_Q is located at midheight of the retaining wall.

4. *Active wedge.* The *active wedge* is defined as that zone of soil involved in the development of the active earth pressures upon the wall. This active wedge must move laterally in order to develop the active earth pressures. It is important that building footings or other load-carrying members are not supported by the active wedge, or else they will be subjected to lateral movement. Figure 15.2 shows an illustration of the active wedge of soil behind the retaining wall. As indicated in Fig. 15.2, the active wedge is inclined at an angle of $45° + \phi/2$ from the horizontal.

Passive Earth Pressure. As shown in Fig. 15.1, the passive earth pressure is developed along the front side of the footing. Passive pressure is developed when the wall footing moves laterally into the soil and a passive wedge is developed such as shown in Fig. 15.2. In order to calculate the passive resultant force P_p, the following equation is used, assuming that there is cohesionless soil in front of the wall footing:

TABLE 15.1 Magnitudes of Wall Rotation to Reach Active and Passive States

Soil type and condition (1)	Rotation (Y/H) for active state (2)	Rotation (Y/H) for passive state (3)
Dense Cohesionless	0.0005	0.002
Loose Cohesionless	0.002	0.006
Stiff Cohesive	0.01	0.02
Soft Cohesive	0.02	0.04

Note: Y = wall displacement and H = height of the wall.
Source: From NAVFAC DM-7.2, 1982.

15.6   ANALYSIS OF GEOTECHNICAL DATA AND ENGINEERING COMPUTATIONS

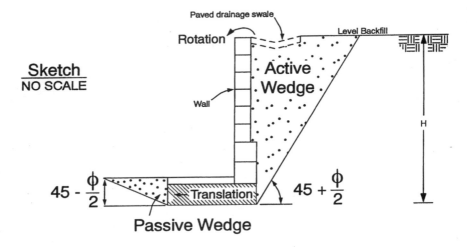

Note: For active and passive wedge development there must be movement of the retaining wall as illustrated above.

FIGURE 15.2   Active wedge behind retaining wall.

$$P_p = \tfrac{1}{2} k_p \gamma_t D^2 \tag{15.4}$$

where P_p = passive resultant force in kN per linear meter of wall or pounds per linear foot of wall, k_p = passive earth pressure coefficient, γ_t = total unit weight of the soil located in front of the wall footing, and D = depth of the wall footing (vertical distance from the ground surface in front of the retaining wall to the bottom of the footing). The passive earth pressure coefficient k_p is equal to:

$$k_p = \tan^2 (45° + \tfrac{1}{2}\phi) \tag{15.5}$$

where ϕ = friction angle of the soil in front of the wall footing. Equation (15.5) is known as the passive Rankine state.

In order to develop passive pressure, the wall footing must move laterally into the soil. As indicated in Table 15.1 (from NAVFAC DM-7.2, 1982), the wall translation to reach the passive state is at least twice that required to reach the active earth pressure state.

Usually it is desirable to limit the amount of wall translation by applying a reduction factor to the passive pressure. A commonly used reduction factor is 2.0 (Lambe and Whitman, 1969). The soil engineer routinely reduces the passive pressure by half (reduction factor = 2.0) and then refers to the value as the *allowable passive pressure*. To limit wall translation, the structural engineer should use the allowable passive pressure for design of the retaining wall. The passive pressure may also be limited by building codes. For example, the allowable passive soil pressure, in terms of equivalent fluid pressure, is 16 to 32 kN/m³ (100 to 200 pcf) according to the *Uniform Building Code* (1997).

If the soil in front of the retaining wall is a plastic (clayey) soil, then usually the long-term effective stress analysis will govern. For the effective stress analysis, the effective cohesion c' and effective friction angle ϕ' can be determined from laboratory tests. As a conservative approach, often the effective cohesion c' is ignored and the effective friction angle ϕ' is used in Eq. (15.5) in order to determine the allowable passive resistance.

Footing Bearing Pressure. In order to calculate the footing bearing pressure, the first step is to sum the vertical loads, such as the wall and footing weights. The vertical loads can be represented by a single resultant vertical force, per linear meter or foot of wall, that is offset by a distance from the toe of the footing. This can then be converted to a pressure distribution as shown in Fig. 15.1. The largest bearing pressure is routinely at the toe of the footing (point A, Fig. 15.1). The largest bearing pressure should not exceed the allowable bearing pressure (Chap. 8), which is usually provided by the soil engineer or by local building code specifications.

Sliding Analysis. The factor of safety F for sliding of the retaining wall is often defined as the resisting forces divided by the driving force. The forces are per linear meter or foot of wall, or:

$$F = \frac{W \tan \delta + P_p}{P_A} \qquad (15.6)$$

where δ = friction angle between the bottom of the concrete foundation and bearing soil, W = weight of the wall and footing, P_p = allowable passive resultant force [P_p from Eq. (15.4) divided by a reduction factor], and P_A = active earth pressure resultant force from Eq. (15.1). The typical recommendation for minimum factor of safety for sliding is 1.5 to 2.0 (Cernica, 1995a).

In some situations, there may be adhesion between the bottom of the footing and the bearing soil. This adhesion is often neglected because the wall is designed for active pressures, which typically develop when there is translation of the footing. Translation of the footing will break the adhesive forces between the bottom of the footing and the bearing soil and therefore adhesion is often neglected for the factor of safety of sliding.

Overturning Analysis. The factor of safety F for overturning of the retaining wall is calculated by taking moments about the toe of the footing (point A, Fig. 15.1), and is:

$$F = \frac{Wa}{\frac{1}{3} P_A H} \qquad (15.7)$$

where W = weight of the wall and footing, a = lateral distance from W to the toe of the footing, and P_A = active earth pressure resultant force [Eq. (15.1)]. Note in this equation that the moment due to passive pressure is neglected. The reason is that with a rotation-type failure mode, the wall may not move enough laterally to induce passive earth pressures. The typical recommendation for minimum factor of safety for overturning is 1.5 to 2.0 (Cernica, 1995a).

15.2.2 Simple Retaining Wall with Wall Friction

In some cases, the geotechnical engineer may want to include the friction between the soil and the rear side of the retaining wall. For this situation, the design analysis is more complicated, as discussed below.

Active Earth Pressure. A common equation that is used to calculate the active earth pressure coefficient k_A for the case of wall friction is the Coulomb equation, which is shown in Fig. 15.3 (from NAVFAC DM-7.2, 1982). The Coulomb equation can also be used if the back face of the wall is sloping or if there is a sloping backfill behind the wall. Once the active earth pressure coefficient k_A is calculated, the active earth pressure resultant force P_A can be calculated by using Eq. (15.1).

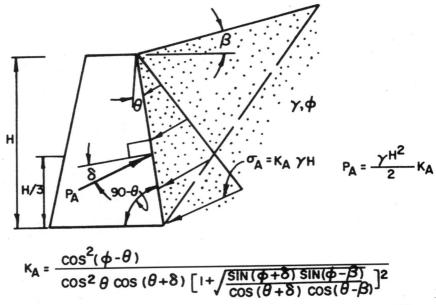

FIGURE 15.3 Coulomb's earth pressure (k_A) equation. (*Reproduced from NAVFAC DM-7.2, 1982.*)

Figure 15.4a, b, and c (from Lambe and Whitman, 1969) presents an example of a proposed concrete retaining wall that will have a height of 20 ft (6.1 m) and a base width of 7 ft (2.1 m). The wall will be backfilled with sand that has a total unit weight γ_t of 110 pcf (17.3 kN/m³), friction angle ϕ of 30°, and an assumed wall friction ϕ_w of 30° Although $\phi_w = 30°$ will be used for this example problem, more typical values of wall friction are $\phi_w = \frac{3}{4}\phi$ for the wall friction between granular soil and wood or concrete walls, and $\phi_w = 20°$ for the wall friction between granular soil and steel walls such as sheet pile walls.

For the example problem shown in Fig. 15.4, the value of the active earth pressure coefficient k_A can be calculated by using Coulomb's equation (Fig. 15.3) and inserting the following values:

- Slope inclination: $\beta = 0$ (no slope inclination)
- Back face of the retaining wall: $\theta = 0$ (vertical back face of the wall)
- Friction between the back face of the wall and the soil backfill: $\delta = \phi_w = 30°$
- Friction angle of backfill sand: $\phi = 30°$

For the above values in Coulomb's equation (Fig. 15.3), the value of the active earth pressure coefficient (k_A) = 0.297.

By using Eq. (15.1) with $k_A = 0.297$, total unit weight (γ_t) = 110 pcf (17.3 kN/m³), and the height of the retaining wall (H) = 20 ft (6.1 m, see Fig. 15.4a), the active earth pressure resultant force (P_A) = 6540 pounds per linear foot of wall (95.4 kN per linear meter of wall). As indicated Fig. 15.4a, the active earth pressure resultant force ($P_A = 6540$ lb/ft) is inclined at an angle of 30° because of the wall friction assumptions. The vertical ($P_v = 3270$ lb/ft) and horizontal ($P_H = 5660$ lb/ft) resultants of P_A are also shown in Fig. 15.4a.

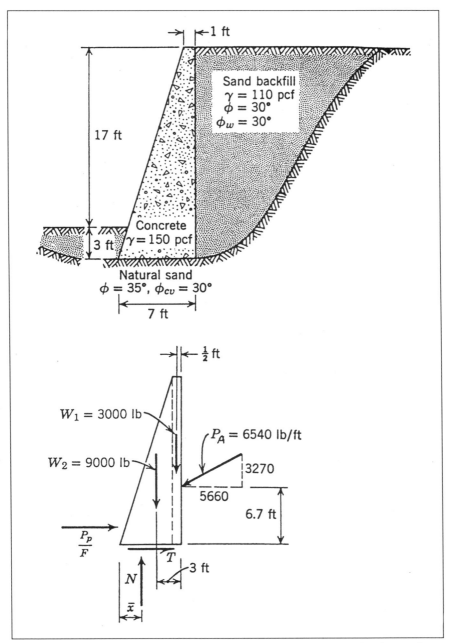

FIGURE 15.4a Example problem: Cross section of proposed retaining wall and resultant forces acting on the retaining wall. (*From Lambe and Whitman, 1969; reproduced with permission of John Wiley & Sons.*)

Note in Fig. 15.3 that even with wall friction, the active earth pressure is still a triangular distribution acting upon the retaining wall, and thus the location of the active earth pressure resultant force P_A is at a distance of ⅓H above the base of the wall, or 6.7 ft (2.0 m).

Passive Earth Pressure. As shown in Fig. 15.4a, the passive earth pressure is developed by the soil located at the front of the retaining wall. Usually wall friction is ignored for the passive earth pressure calculations. For the example problem shown in Fig. 15.4, the passive resultant force (P_p) was calculated by using Eqs. (15.4) and (15.5) and neglecting wall friction and the slight slope of the front of the retaining wall (see Fig. 15.4c for passive earth pressure calculations).

Footing Bearing Pressure. The procedure for the calculation of the footing bearing pressure is as follows:

1. *Calculate N.* As indicated in Fig. 15.4b, the first step is to calculate N (15,270 lb/ft), which equals the sum of the weight of the wall, footing, and vertical component of the active earth pressure resultant force (i.e., $N = W + P_A \sin \phi_w$).
2. *Determine $\bar{x}$.* The value of $\bar{x}$ (2.66 ft) is calculated as shown in Fig. 15.4b. The moments are determined about the toe of the retaining wall. Then $\bar{x}$ equals the difference in the opposing moments divided by N.
3. *Determine average bearing pressure.* The average bearing pressure (2180 psf) is calculated in Fig. 15.4c as N divided by the width of the footing (7 ft).
4. *Calculate moment about the centerline of the footing.* The moment about the centerline of the footing is calculated as N times the eccentricity (0.84 ft).
5. *Find section modulus.* The section modulus of the footing is calculated as shown in Fig. 15.4c.
6. *Determine portion of bearing stress due to moment.* The portion of the bearing stress due to the moment (σ_{mom}) is determined as the moment divided by the section modulus.
7. *Calculate maximum bearing stress.* The maximum bearing stress is then calculated as the sum of the average stress ($\sigma_{avg} = 2180$ psf) plus the bearing stress due to the moment ($\sigma_{mom} = 1570$ psf).

As indicated in Fig. 15.4c, the maximum bearing stress is 3750 psf (180 kPa). This maximum bearing stress must be less than the allowable bearing pressure (Chap. 8). It is also a standard requirement that the resultant normal force N be located within the middle third of the footing as illustrated in Fig. 15.4b.

Sliding Analysis. The factor of safety F for sliding of the retaining wall is often defined as the resisting forces divided by the driving force. The forces are per linear meter or foot of wall, or:

$$F = \frac{N \tan \delta + P_p}{P_H} \quad (15.8)$$

where $\delta = \phi_{cv}$ = friction angle between the bottom of the concrete foundation and bearing soil; N = sum of the weight of the wall, footing, and vertical component of the active earth pressure resultant force (i.e., $N = W + P_A \sin \phi_w$); P_p = allowable passive resultant force [P_p from Eq. (15.4) divided by a reduction factor]; and P_H = horizontal component of the active earth pressure resultant force (i.e., $P_H = P_A \cos \phi_w$).

Find. Adequacy of wall.
Solution. The first step is to determine the active thrust;
The next step is to compute the weights:

$$W_1 = (1)(20)(150) = 3000 \text{ lb/ft}$$
$$W_2 = \tfrac{1}{2}(6)(20)(150) = 9000 \text{ lb/ft}$$

Next N and $\bar{x}$ are computed:

$$N = 9000 + 3000 + 3270 = 15{,}270 \text{ lb/ft}$$
$$\text{Overturning moment} = 5660(6.67) - 3270(7) = 37{,}800 - 22{,}900 = 14{,}900$$
$$\text{Moment of weight} = (6.5)(3000) + (4)(9000) = 19{,}500 + 36{,}000 = 55{,}500$$
$$\text{Ratio} = 3.73 \quad \underline{\underline{\text{OK}}}$$

$$\bar{x} = \frac{55{,}500 - 14{,}900}{15{,}270} = \frac{40{,}600}{15{,}270} = 2.66 \text{ ft} \quad \underline{\underline{\text{OK}}}$$

The location of N

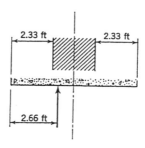

FIGURE 15.4b Example problem: Calculation of the factor of safety of overturning and the location of the resultant force N. (*From Lambe and Whitman, 1969; reproduced with permission of John Wiley & Sons.*)

There are variations of Eq. (15.8) that are used in practice. For example, as illustrated in Fig. 15.4c, the value of P_p is subtracted from P_H in the denominator of Eq. (15.8), instead of P_p being used in the numerator. For the example problem shown in Fig. 15.4, the factor of safety for sliding $(F) = 1.79$ when passive pressure is included and $F = 1.55$ when passive pressure is excluded. As previously mentioned, the typical recommendation for minimum factor of safety for sliding is 1.5 to 2.0 (Cernica, 1995a).

Overturning Analysis. The factor of safety F for overturning of the retaining wall is calculated by taking moments about the toe of the footing (point A, Fig. 15.1), and is:

$$F = \frac{Wa}{\tfrac{1}{3}P_H H - P_v e} \quad (15.9)$$

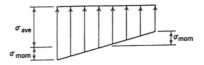

Next the bearing stress is computed. The average bearing stress is $15{,}270/7 = 2180$ psf. Assuming that the bearing stress is distributed linearly, the maximum stress can be found

$$\sigma_{mom} = \frac{M}{S}$$

where

M = moment about $\mathcal{L} = 15{,}270(3.5 - 2.66) = 12{,}820$ lb-ft/ft

S = section modulus $= \frac{1}{6}B^2 = \frac{1}{6}(7)^2 = 8.17$ ft^2

where B is width of base

$$\sigma_{mom} = \frac{12{,}820}{8.17} = 1570 \text{ psf}$$

Maximum stress = $2180 + 1570 = 3750$ psf

Finally, the resistance to horizontal sliding is checked. Assuming passive resistance without wall friction,

$$K_p = 3$$

$$P_p = \tfrac{1}{2}(110)(3^2)(3) = 1500 \text{ lb/ft}$$

With reduction factor of 2,

$$\frac{P_p}{F} = 750 \text{ lb/ft}$$

$$T = 5660 - 750 = 4910 \text{ lb/ft}$$

$$N \tan 30° = 8810 \text{ lb/ft}$$

$$\frac{N \tan \phi_{cv}}{T} = 1.79 < 2 \quad \text{not OK}$$

Ignoring passive resistance

$$T = 5660 \text{ lb/ft}$$

$$\frac{N \tan \phi_{cv}}{T} = 1.55 > 1.5 \quad \text{OK}$$

FIGURE 15.4c Example problem: Calculation of the maximum bearing stress and the factor of safety of sliding. (*From Lambe and Whitman 1969; reproduced with permission of John Wiley & Sons.*)

where a = lateral distance from the resultant weight of the wall and footing (W) to the toe of the footing, P_H = horizontal component of the active earth pressure resultant force, P_v = vertical component of the active earth pressure resultant force, and e = lateral distance from the location of P_v to the toe of the wall. In Fig. 15.4b, the factor of safety (ratio) for overturning is calculated to be 3.73. As previously mentioned, the typical recommendation for minimum factor of safety for overturning is 1.5 to 2.0 (Cernica, 1995a).

15.3 DESIGN AND CONSTRUCTION OF RETAINING WALLS

The previous section dealt with simple retaining walls, such as those shown in Figs. 15.1 and 15.4a, b, and c. This section will provide an additional discussion of the design and construction of retaining walls.

Figure 15.5a, b, and c (from NAVFAC DM-7.2, 1982) shows several examples of different types of retaining walls. Figure 15.5a shows gravity and semigravity retaining walls, Fig. 15.5b shows cantilever and counterfort retaining walls, and the design analyses are shown in Fig. 15.5c. Although the equations in Fig. 15.5c include an adhesion value c_a, as previously mentioned, the adhesion is often neglected. This is because active pressures develop when there is translation (movement) of the footing which would tend to break the adhesive resistance.

Special Design Cases: Retaining Walls at the Top of Slopes. Retaining walls are sometimes constructed at the top of slopes. In this case, there is a descending ground surface in front of the retaining wall, and Eq. (15.5) can not be used to determine the passive earth pressure coefficient. For either a descending slope ($-\beta$) or ascending slope ($+\beta$) in front of the retaining wall, the following equation can be used to determine the passive earth pressure coefficient k_p:

$$k_p = \frac{\cos^2 \phi}{[1 - (\sin^2 \phi + \sin \phi \cos \phi \tan \beta)^{0.5}]^2} \qquad (15.10)$$

where ϕ = friction angle of the soil in front of the retaining wall and β = slope inclination measured from a horizontal plane, where a descending slope in front of the retaining wall has a negative β value. Although not readily apparent, if $\beta = 0$, Eq. (15.10) will give exactly the same values of k_p as Eq. (15.5). For example, substituting $\beta = 0$ and $\phi = 30°$ into Eq. (15.10) gives $k_p = 3$, which is exactly the value obtained from Eq. (15.5).

Suppose that a retaining wall is constructed at the top of a 2:1 (horizontal:vertical) slope. In this case, the slope inclination is $\beta = -26.6°$. Assuming $\phi = 30°$ for the soil composing the slope, then according to Eq. (15.10), the value of $k_p = 1.12$. Thus for the case of a 2:1 descending slope in front of the retaining wall, the passive resistance will be significantly reduced (k_p decreases from 3 to 1.12). Many retaining walls tilt or deform downslope because the condition of a sloping ground surface in front of the retaining wall significantly reduces the passive resistance for the wall foundation.

There can be additional factors that contribute to the failure of retaining walls constructed at the top of slopes. In some cases, there may be a low factor of safety of the entire slope which is exacerbated by the construction of a top of slope wall. In other cases, where the soil in front of the retaining wall is clayey, there can be slope creep which causes a soil gap to open up at the front of the retaining wall footing. Figure 15.6 shows a retaining wall at the top of a clayey fill slope. The arrow points to a soil gap that has opened up in front of the retaining wall because of creep of the clayey slope. In these cases of a descending clayey slope in front of the retaining wall, the depth of creep of the outer face of the slope must be estimated (see Sec. 10.8) and then the passive pressure must be utilized only below the depth of surface slope creep.

Earthquake Analysis. It is difficult to accurately predict the additional lateral forces that will be generated on a retaining wall during an earthquake. Some factors affecting the magnitude of earthquake forces on the wall are the size and duration of the earthquake,

15.14 ANALYSIS OF GEOTECHNICAL DATA AND ENGINEERING COMPUTATIONS

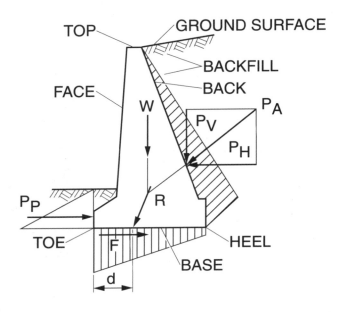

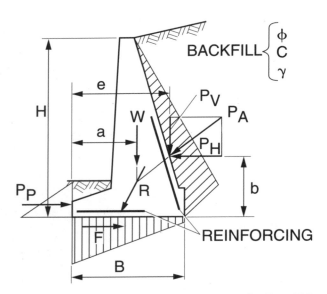

FIGURE 15.5a Gravity and semigravity retaining walls. (*From NAVFAC DM-7.2, 1982.*)

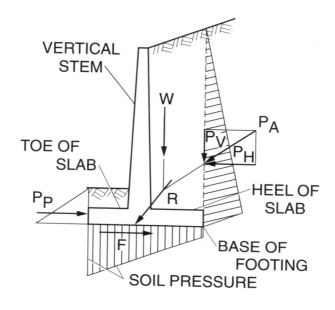

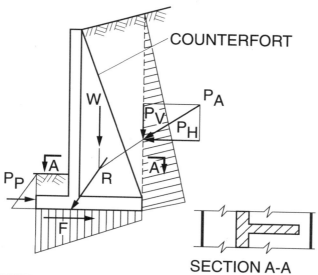

FIGURE 15.5b Cantilever and counterfort retaining walls. (*From NAVFAC DM-7.2, 1982.*)

LOCATION OF RESULTANT

MOMENTS ABOUT TOE:

$$d = \frac{Wa + P_V e - P_H b}{W + P_V}$$

ASSUMING $P_P = 0$

OVERTURNING

MOMENTS ABOUT TOE:

$$F = \frac{Wa}{P_H b - P_V e} \geq 1.5$$

IGNORE OVERTURNING IF R IS WITHIN MIDDLE THIRD (SOIL), MIDDLE HALF (ROCK).
CHECK R AT DIFFERENT HORIZONTAL PLANES FOR GRAVITY WALLS.

RESISTANCE AGAINST SLIDING

$$F = \frac{(W + P_V) \tan \delta + C_a B}{P_H} \geq 1.5$$

$$F = \frac{(W + P_V) \tan \delta + C_a B + P_P}{P_H} \geq 2.0$$

$$F = (W + P_V) \tan \delta + C_a B$$

C_a = ADHESION BETWEEN SOIL AND BASE

$\tan \delta$ = FRICTION FACTOR BETWEEN SOIL AND BASE

W = INCLUDES WEIGHT OF WALL AND SOIL IN FRONT FOR GRAVITY AND SEMIGRAVITY WALLS. INCLUDES WEIGHT OF WALL AND SOIL ABOVE FOOTING, FOR CANTILEVER AND COUNTERFORT WALLS.

FIGURE 15.5c Design analysis for retaining walls shown in Figs. 15.5a and 15.5b. *(From NAVFAC DM-7.2, 1982.)*

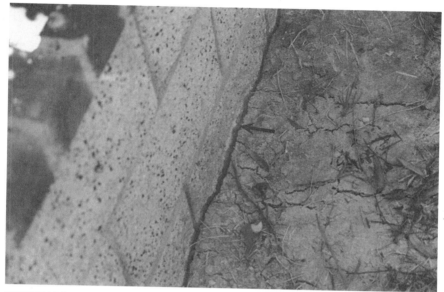

FIGURE 15.6 Retaining wall located at the top of a clayey slope. The arrow points to a soil gap that has opened up in front of the wall because of creep of the slope.

the distance from the earthquake epicenter to the site, and the mass of soil retained by the wall. Many retaining walls are designed for only the active earth pressure and then fail when additional forces are generated by the earthquake.

A simple approach, based on the work of Seed and Whitman (1970), is to include in the design analysis an additional horizontal force P_E to account for the additional loads imposed on the retaining wall by the earthquake, as follows:

$$P_E = \tfrac{3}{8}(a_{max}/g)\gamma_t H^2 \quad (15.11)$$

where a_{max} = peak acceleration at the ground surface, γ_t = total unit weight of the backfill soil, H = height of the retaining wall, and g = earth's gravity (9.81 m/s^2). As an example of the use of Eq. (15.11), assume that the maximum horizontal ground acceleration for the retaining wall shown in Fig. 15.4 will be 0.20g. Then substituting a_{max} = 0.2g, γ_t = 110 pcf, and H = 20 ft into Eq. (15.11), we find the additional horizontal force acting on the retaining wall due to the earthquake (P_E) = 3300 lb per linear foot of wall (48.1 kN per meter of wall). The location of this earthquake-induced force can be assumed to act at a distance of 0.6 H above the base of the wall.

Because P_E is a short-term loading that may never occur during the life of the retaining wall, it is common to allow a one-third increase in the bearing pressure and passive resistance for earthquake analysis. Also, for the analysis of sliding and overturning of the retaining wall, it is common to accept a lower factor of safety (1.1 to 1.2) under the combined static and earthquake loads.

Construction of the Retaining Wall. There are many construction factors that can result in excessive lateral movement, bearing capacity failures, sliding failures, or failure by overturning of the retaining wall. Common causes can include inadequate subsurface exploration or laboratory testing, incorrect design, improper construction, or unanticipated loadings. Typical construction-related problems are discussed below:

1. *Clay backfill.* A frequent cause of failure is backfilling the wall with clay. As previously mentioned, clean granular sand or gravel is usually recommended as backfill material. This is because of the undesirable effects of using clay or silt as a backfill material (Sec. 9.3.1). When clay is used as backfill material, the clay backfill can exert swelling pressures on the wall (Fourie, 1989; Marsh and Walsh, 1996). The highest swelling pressures develop when water infiltrates a backfill consisting of a clay that was compacted to a high dry density at a low moisture content. The type of clay particles that will exert the highest swelling pressures are montmorillonite. Because the clay backfill is not free draining, there could also be additional hydrostatic forces or ice-related forces that substantially increase the thrust on the wall.

2. *Inferior backfill soil.* To reduce construction costs, soil available on site is sometimes used for backfill. This soil may not have the properties, such as being a clean granular soil with a high shear strength, that were assumed during the design stage. Using on-site available soil, rather than importing granular material, is probably the most common reason for retaining wall failures.

3. *Compaction-induced pressures.* As previously mentioned, one reason for applying a factor of safety F to the active earth pressure [Eq. (15.1)] is that larger wall pressures will typically be generated during compaction of the backfill. By using heavy compaction equipment in close proximity to the wall, excessive pressures can be developed that damage the wall. The best compaction equipment, in terms of exerting the least compaction induced pressures on the wall, are small vibrator plate (hand-operated) compactors such as models VPG 160B and BP 19/75 (Duncan et al., 1991). The vibrator plates effectively densify the granular backfill, but do not induce high lateral loads because of their light weight. Besides hand-operated compactors, other types of relatively lightweight equipment (such as a bobcat) can be used to compact the backfill.

4. *Failure of the back-cut.* There could also be the failure of the back-cut for the retaining wall. A vertical back-cut is often used when the retaining wall is less than 1.5 m (5 ft) high. In other cases, the back-cut is usually sloped. The back-cut can fail if it is excavated too steeply and does not have an adequate factor of safety.

Retaining wall movement is gradual and the wall deforms by intermittently tilting or moving laterally. It is possible that a failure can occur suddenly, such as when there is a slope-type failure beneath the wall or when the foundation of the wall fails because of inadequate bearing capacity. These rapid failure conditions could develop if the wall is supported by soft clay and there is an undrained shear failure beneath the foundation.

The construction of a cantilever retaining wall is shown in Figs. 15.7 to 15.16, as follows:

- *Back-cut slope (Fig. 15.7).* This photograph shows the first step in the construction of the retaining wall, which was the excavation of the back-cut slope.
- *Construction of the retaining wall (Figs. 15.8 to 15.10).* These three photographs show the construction of the retaining wall. In Figs. 15.8 to 15.10, the foundation has already been installed and the retaining wall is in the process of being constructed.
- *Backdrainage system (Fig. 15.11).* In this photograph, the retaining wall is finished and the drainage system has been installed. The top of the black filter fabric, which contains a perforated pipe surrounded by open-graded gravel, is visible in Fig. 15.11.
- *Compaction of backfill (Figs. 15.12 and 15.13).* These two photographs show the compaction of import granular soil behind the retaining wall. The fill should not be com-

FIGURE 15.7 Excavation of back-cut slope.

pacted until the retaining wall concrete has gained sufficient strength. Note that a bobcat (lightweight equipment) is being used to compact the backfill.

- *Construction of inclined slope (Figs. 15.14 and 15.15).* These photographs show the construction of a fill slope above the retaining wall. For the compaction of this fill, heavier compaction equipment is being utilized.

- *Final construction condition (Fig. 15.16).* This last photograph shows the final constructed condition of the retaining wall. The last operation will be to plant and irrigate the slope above the retaining wall.

15.4 RESTRAINED RETAINING WALLS

As mentioned in the previous section, in order for the active wedge (Fig. 15.2) to be developed, there must be sufficient movement of the retaining wall (Table 15.1). There are many cases where movement of the retaining wall is restricted. Examples include massive bridge abutments, rigid basement walls, and retaining walls that are anchored in nonyielding rock. These cases are often described as *restrained retaining walls*.

In order to determine the earth pressure acting on a restrained retaining wall, Eq. (15.1) can be utilized where the coefficient of earth pressure at rest k_0 is substituted for k_A. The value of k_0 can be estimated from Eqs. (6.21) and (6.22). A common value of k_0 that is used for restrained retaining walls is 0.5. Restrained retaining walls are especially susceptible to higher earth pressures induced by heavy compaction equipment, and extra care must be taken during the compaction of backfill for restrained retaining walls.

15.20 ANALYSIS OF GEOTECHNICAL DATA AND ENGINEERING COMPUTATIONS

FIGURE 15.8 Construction of the retaining wall.

Soil Report Recommendations. Upon the completion of the retaining wall analyses (Secs. 15.2 to 15.4), the recommendations would normally be included in a soils report. An example of typical wording concerning the design of retaining walls is as follows (see App. E):

> Where free unrestrained walls are planned to retain level, predominantly granular, nonexpansive, imported backfill, an active lateral soil pressure of 45 pounds per cubic foot (7 kN/m^3) equivalent fluid pressure may be utilized for retaining wall design. Appropriate allowances should be made for anticipated surcharge loadings. Unless walls incorporate appropriately designed backdrainage systems, allowances should be made for seepage and/or hydrostatic forces. For retaining walls restrained against rotation, the above value should be increased to 70 pounds per cubic foot (11 kN/m^3) equivalent fluid pressure. The restrained basement walls should be backfilled with predominately granular, nonexpansive granular import material and

FIGURE 15.9 Construction of the retaining wall.

provided with a backdrainage system. Neither spread nor wall footings should be founded in the granular backfill.

It should be noted that the use of heavy compaction equipment in close proximity to retaining walls can result in excess wall movement (strains greater than those normally associated with the construction of retaining walls) and/or soil pressures exceeding design values. In this regard, care should be taken during backfilling operations, and lightweight equipment, such as hand-operated tampers, should be used for compaction.

The dynamic increment in lateral earth pressure due to earthquakes should be considered in retaining wall design. Using the Seed and Whitman (1970) recommendation and the peak ground accelerations previously presented in this report, retaining walls should be designed to resist an additional lateral soil pressure of 20 pounds per cubic foot (3 kN/m^3) equivalent fluid pressure for unrestrained walls and 30 pcf (5 kN/m^3) equivalent fluid pressure for restrained walls. For earthquake conditions, the pressure resultant force should be assumed to act at a distance of $0.6H$ above the base of the wall.

FIGURE 15.10 Construction of the retaining wall.

15.5 MECHANICALLY STABILIZED EARTH RETAINING WALLS

Mechanically stabilized earth retaining walls (also known as MSE retaining walls) are typically composed of strip- or grid-type (geosynthetic) reinforcement (Koerner, 1990). Because they are often more economical to construct than conventional concrete retaining walls, mechanically stabilized earth retaining walls have become very popular in the past decade.

Construction of a Mechanically Stabilized Earth Retaining Wall. The construction of a mechanically stabilized earth retaining wall proceeds as follows:

1. *Excavation of key and installation of drainage system (Fig. 15.17).* This photograph shows the first step in the construction of a mechanically stabilized earth retaining wall. A key (slot) has been excavated into the natural ground and a drainage system is being constructed at the back of the key. The drainage system consists of a perforated drain pipe, surrounded by open graded gravel, which is in turn surrounded by the geofabric (see "Geofabric Alternative," Standard Detail No. 1, App. C).
2. *Construction of mechanically stabilized earth retaining wall (Figs. 15.18 to 15.20).* A mechanically stabilized earth retaining wall is composed of three elements: (*a*) wall facing material, (*b*) soil reinforcement, such as strip- or grid-type reinforcement, and (*c*) compacted fill between the soil reinforcement.
    a. *Wall facing element.* Installation of the wall facing element is shown in Fig. 15.18. The arrows in Fig. 15.18 point to the wall facing elements, which are precast concrete members. Other types of wall facing elements are often utilized, such as wood

planks or concrete interlocking panels. Note in Fig. 15.18 that one worker is using a crowbar to align the wall facing members. The wall facing members are often connected together by steel dowels.
b. *Geogrid.* Figure 15.19 shows the black geogrid, which is a type of geosynthetic. Geogrids are usually composed of a high-strength polymer used to create an open grid pattern. The large arrows in Fig. 15.19 indicate the width of the geogrid, which

FIGURE 15.11 Installation of backdrainage system.

15.24 ANALYSIS OF GEOTECHNICAL DATA AND ENGINEERING COMPUTATIONS

FIGURE 15.12 Compaction of backfill.

is considered to be the zone of mechanically stabilized earth. Geogrid is transported to the site in rolls, which are then spread out and cut as needed. Geogrid has different tensile strengths along the roll, versus perpendicular to the roll, and it is important that the highest strength of the geogrid be placed in a direction that is perpendicular to the wall face. The small arrow in Fig. 15.19 points to a splice, where the geogrid is overlapped. A smaller width of geogrid is used at the end of the wall

FIGURE 15.13 Compaction of backfill.

FIGURE 15.14 Construction of fill slope above retaining wall.

15.26 ANALYSIS OF GEOTECHNICAL DATA AND ENGINEERING COMPUTATIONS

FIGURE 15.15 Construction of fill slope above retaining wall.

FIGURE 15.16 Final constructed condition.

FIGURE 15.17 Excavation of key and installation of drainage system for a mechanically stabilized earth retaining wall.

(location of small arrow) because the required resistance is less at the wall perimeter. The geogrid is often attached to the wall facing elements and the vertical spacing of the geogrid is equal to the thickness of the wall facing elements.

c. *Compaction of fill.* The arrows in Fig. 15.20 point to granular (nonplastic) fill which is being placed and will be compacted once the section of wall elements are in place. Since this mechanically stabilized earth retaining wall has a drain only in the

15.28 ANALYSIS OF GEOTECHNICAL DATA AND ENGINEERING COMPUTATIONS

FIGURE 15.18 Two views of the installation of the wall facing elements for a mechanically stabilized earth retaining wall. The arrows point to the wall facing elements.

FIGURE 15.19 Two views of the installation of the geogrid for a mechanically stabilized earth retaining wall. The large arrows indicate the width of the mechanically stabilized zone and the small arrow points to a geogrid splice.

FIGURE 15.20 Two views of the placement of fill on top of the geogrid. The arrows point to the granular fill placed on top of the geogrid.

key, granular (permeable) soil is being used to prevent the buildup of pore water pressure behind or in the mechanically stabilized earth zone. For geogrid installed in cohesive (plastic) soil, a backdrain system is often installed as shown in Standard Detail No. 8 (App. C). The geogrid and compacted fill derive frictional resistance and interlocking resistance between each other. When the mechanically stabilized soil mass is subjected to shear stress, the soil tends to transfer the shear stress to the stronger geogrid. Also, if high stress concentrations develop in the mechanically stabilized soil mass, such as along a potential slip surface, then the geogrid tends to redistribute stresses away from areas of high stress. Lightweight equipment must be used to compact the fill that is placed on top of each layer of geogrid. Heavy compaction equipment could push the wall facing elements out of alignment.

3. *Final constructed condition (Figs. 15.21 and 15.22).* These two photographs show the final constructed condition of the mechanically stabilized earth retaining wall. The arrow in Fig. 15.21 points to a gap in the wall facing element, which can be used to install plants which will grow over the wall facing elements and eventually blend the wall into the surrounding vegetation.

Design Analysis for Mechanically Stabilized Earth Retaining Wall. The design analysis for a mechanically stabilized earth retaining wall is more complex than for a cantilevered retaining wall. For a mechanically stabilized retaining wall, both the internal and external stability must be checked.

1. *External stability.* The analysis for the external stability is similar to a gravity retaining wall. For example, Figs. 15.23 and 15.24 (adapted from AASHTO, 1996) present the design analysis for external stability for a level backfill condition and a sloping backfill condition.

    In both Figs. 15.23 and 15.24, the zone of mechanically stabilized earth mass is treated in the same fashion as a massive gravity retaining wall. The following analyses must be performed:

    a. *Allowable bearing pressure.* The bearing pressure due to the reinforced soil mass must not exceed the allowable bearing pressure.
    b. *Factor of safety of sliding.* The reinforced soil mass must have an adequate factor of safety for sliding ($F = 1.5$ to 2).
    c. *Factor of safety of overturning.* The reinforced soil mass must have an adequate factor of safety ($F = 2$) for overturning about point O.
    d. *Resultant of vertical forces (N).* The resultant of the vertical forces (N) must be within the middle third of the base of the reinforced soil mass.
    e. *Stability of reinforced soil mass.* The stability of the entire reinforced soil mass (i.e., shear failure below the bottom of the wall) would have to be checked.

    Note in Fig. 15.23 that two forces (P_1 and P_2) are shown acting on the reinforced soil mass. The first force (P_1) is determined from the standard active earth pressure resultant equation [i.e., Eq. (15.1)]. The second force (P_2) is due to a uniform surcharge Q applied to the entire ground surface behind the mechanically stabilized earth retaining wall [i.e., Eq. (15.3)]. If the wall does not have a surcharge, then P_2 is equal to zero.

    Figure 15.24 presents the active earth pressure force for an inclined slope behind the retaining wall. Note in Fig. 15.24 that the friction δ of the soil along the back side of the reinforced soil mass has been included in the analysis. The value of k_A would be obtained from Coulomb's earth pressure equation (Fig. 15.3). As a conservative approach, the friction angle δ can be assumed to be equal to zero and then $P_H = P_A$. Note in both Figs. 15.23 and 15.24 that the minimum width of the reinforced soil mass must be at least seven-tenths the height of the reinforced soil mass.

FIGURE 15.21 Final constructed condition of the mechanically stabilized earth retaining wall. The arrow points to one of the openings in the wall facing element which can be used to establish vegetation on the wall face.

2. *Internal stability.* In terms of the internal stability, Fig. 15.25 shows the predicted (from the method of slices) and measured slip surfaces in a mechanically stabilized zone consisting of geotextile (Zornberg et al., 1998). As Fig. 15.25 shows, there is very good agreement between the actual location of the slip surface and the location of the critical slip surface predicted by slope stability analysis (i.e., method of slices).

RETAINING WALLS   15.33

FIGURE 15.22 Final constructed condition of the mechanically stabilized earth retaining wall.

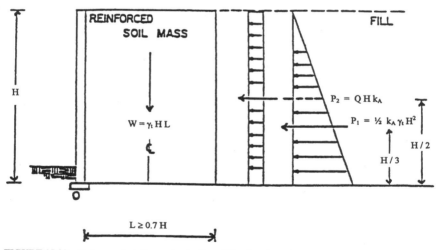

FIGURE 15.23 Design analysis for mechanically stabilized earth retaining wall having horizontal backfill. (*Adapted from AASHTO, 1996.*)

To check the stability of the mechanically stabilized zone, a slope stability analysis can be performed where the soil reinforcement is modeled as horizontal forces equivalent to the allowable tensile resistance of the geotextile. In addition to calculating the factor of safety, the pull-out resistance of the reinforcement along the slip surface should also be checked.

15.34 ANALYSIS OF GEOTECHNICAL DATA AND ENGINEERING COMPUTATIONS

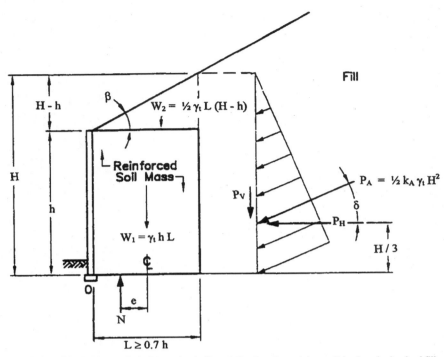

FIGURE 15.24 Design analysis for mechanically stabilized earth retaining wall having sloping backfill. (*Adapted from AASHTO, 1996.*)

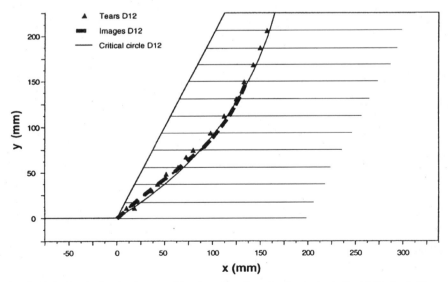

FIGURE 15.25 Predicted and measured location of the slip surface for mechanically stabilized soil. (*From Zornberg et al., 1998; reprinted with permission of American Society of Civil Engineers.*)

The analysis of mechanically stabilized earth retaining walls is based on active earth pressures. It is assumed that the wall will move enough to develop the active wedge. As with concrete retaining walls, it is important that building footings or other load-carrying members are not supported by the mechanically stabilized earth retaining wall and the active wedge, or else they will be subjected to lateral movement.

15.6 SHEET PILE WALLS

Sheet pile retaining walls are widely used for waterfront construction and consist of interlocking members that are driven into place. Individual sheet piles come in different sizes and shapes, such as shown in Fig. 15.26. Sheet piles have an interlocking joint (see Fig. 15.26) that enables the individual segments to be connected together to form a solid wall.

Figures 15.27 and 15.28 show two views of the placement of a sheet pile wall at the beach. The arrow in Fig. 15.27 points to an individual sheet pile, which is designated a "PZ" type profile (see Fig. 15.26). The arrow in Fig. 15.28 points to the top of the sheet pile which has been cut off at ground surface.

Earth Pressures Acting on Sheet Pile Wall. There are many different types of design methods that are used for sheet pile walls. Figure 15.29 (from NAVFAC DM-7.2, 1982) shows the most common type of design method. In Fig. 15.29, the term H represents the unsupported face of the sheet pile wall. As indicated in Fig. 15.29, this sheet pile wall is being used as a waterfront retaining structure, and the level of water in front of the wall is at the same elevation as the groundwater table elevation behind the wall. For highly permeable soil, such as clean sand and gravel, this often occurs because the water can quickly flow underneath the wall in order to equalize the water levels.

In Fig. 15.29, the term D represents that portion of the sheet pile wall that is anchored in soil. Also shown in Fig. 15.29 is a force designated as A_p. This represents a restraining force on the sheet pile wall due to the construction of a tieback, such as a rod that has a grouted end or is attached to an anchor block. Tieback anchors are often used in sheet pile wall construction in order to reduce the bending moments in the sheet pile. When tieback anchors are used, the sheet pile wall is typically referred to as an *anchored bulkhead,* while if no tiebacks are utilized, the wall is called a *cantilevered sheet pile wall.*

Sheet pile walls tend to be relatively flexible. Thus, as indicated in Fig. 15.29, the design is based on active and passive earth pressures. The soil behind the wall is assumed to exert an active earth pressure on the sheet pile wall. At the groundwater table (point A), the active earth pressure is equal to:

$$\text{Active earth pressure at point } A \text{ (kPa or psf)} = k_A \gamma_t d_1 \qquad (15.12)$$

where k_A = active earth pressure coefficient from Eq. (15.2) (dimensionless parameter); the friction between the sheet pile wall and the soil is usually neglected in the design analysis
γ_t = total unit weight of the soil above the groundwater table (kN/m³ or pcf)
d_1 = depth from the ground surface to the groundwater table (m or ft)

In using Eq. (15.12), a unit length (1 m or 1 ft) of sheet pile wall is assumed. At point B in Fig. 15.29, the active earth pressure equals:

$$\text{Active earth pressure at point } B \text{ (kPa or psf)} = k_A \gamma_t d_1 + k_A \gamma_b d_2 \qquad (15.13)$$

where γ_b = buoyant unit weight of the soil below the groundwater table and d_2 = depth from the groundwater table to the bottom of the sheet pile wall. For a sheet pile wall having

Steel Sheet Piling Sections

Profile	Section Index	District Rolled	Driving Distance per Pile (In.)	Weight Per Foot (Lbs.)	Weight Per Square Foot of Wall (Lbs.)	Web Thickness (In.)	Section Modulus Per Pile (In.³)	Section Modulus Per Foot of Wall (In.³)	Area Per Pile (In.²)	Moment of Inertia Per Pile (In.⁴)	Moment of Inertia Per Foot of Wall (In.⁴)
Interlock with Each Other	PSX32	H.	16½	44.0	32.0	29/64	3.3	2.4	12.94	5.1	3.7
	PS32*	H.S.	15	40.0	32.0	½	2.4	1.9	11.77	3.6	2.9
	PS28	H.S.	15	35.0	28.0	3/8	2.4	1.9	10.30	3.5	2.8
Interlock with Each Other	PSA28*	H.	16	37.3	28.0	½	3.3	2.5	10.98	6.0	4.5
	PSA23	H.S.	16	30.7	23.0	3/8	3.2	2.4	8.99	5.5	4.1
	PDA27	H.S.	16	36.0	27.0	3/8	14.3	10.7	10.59	53.0	39.8
	PMA22	H.S.	19⅝	36.0	22.0	3/8	8.8	5.4	10.59	22.4	13.7
	PZ38	H.	18	57.0	38.0	3/8	70.2	46.8	16.77	421.2	280.8

15.36

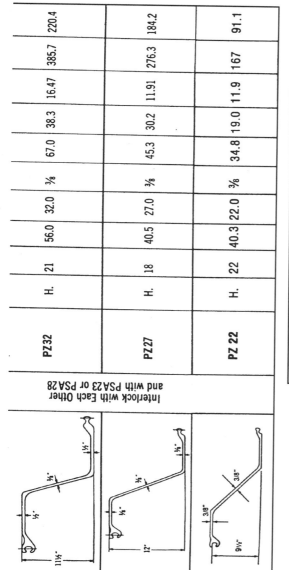

FIGURE 15.26 Steel sheet piling sections. (*From USS Steel Sheet Piling Design Manual, 1984.*)

FIGURE 15.27 Installation of sheet pile wall. Arrow points to an individual sheet pile member.

FIGURE 15.28 Installation of sheet pile wall. Arrow points to sheet pile that has been cut off at ground surface.

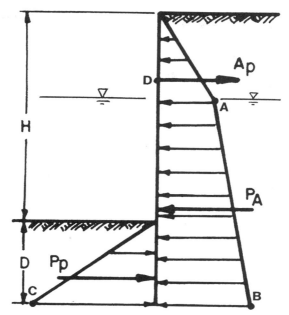

FIGURE 15.29 Earth pressure diagram for design of sheet pile wall. (*Reproduced from NAVFAC DM-7.2, 1982.*)

assumed values of H and D (see Fig. 15.29), and for the calculated values of active earth pressure at points A and B, the active earth pressure resultant force P_A, in kN per linear meter of wall or pounds per linear foot of wall, can be calculated.

The soil in front of the wall is assumed to exert a passive earth pressure on the sheet pile wall. The passive earth pressure at point C in Fig. 15.29 is equal to:

$$\text{Passive earth pressure at point } C \text{ (kPa or psf)} = k_p \gamma_b D \qquad (15.14)$$

where the passive earth pressure coefficient k_p can be calculated from Eq. (15.5). Similar to the analysis of cantilever retaining walls, it is desirable to limit the amount of sheet pile wall translation by applying a reduction factor (such as reduction factor = 2) to this passive pressure. Once the allowable passive pressure is known at point C, the passive resultant force P_p can be readily calculated.

As an alternative solution for the passive pressure, Eq. (15.4) can be used to calculate P_p with the buoyant unit weight γ_b substituted for the total unit weight γ_t and then a reduction factor is applied to P_p.

Note that a water pressure has not been included in the analysis. This is because the water level is the same on both sides of the wall, and water pressure cancels out and thus should not be included in the analysis.

Design Analysis for Anchored Bulkhead. The design of sheet pile walls requires the following analyses: (1) evaluation of the earth pressures that act on the wall, such as shown in Fig. 15.29; (2) determination of the required depth D of piling penetration;

15.40 ANALYSIS OF GEOTECHNICAL DATA AND ENGINEERING COMPUTATIONS

(3) calculation of the maximum bending moment M_{max} which is used to determine the maximum stress in the sheet pile; and (4) selection of the appropriate piling type, size, and construction details.

A typical design process is to assume a depth D (Fig. 15.29) and then calculate the factor of safety for toe failure (i.e., toe kick-out) by the summation of moments at the tieback anchor (point D). The factor of safety is defined as the moment due to the passive force divided by the moment due to the active force. Values of acceptable factor of safety for toe failure are 2 to 3. An alternative solution is to first select the factor of safety and then develop the active and passive resultant forces and moment arms in terms of D. By solving the equation, the value of D for a specific factor of safety can be directly calculated.

Once the depth D of the sheet pile wall is known, the anchor pull A_p must be calculated. The anchor pull is determined by the summation of forces in the horizontal direction, or:

$$A_p = P_A - P_p/F \qquad (15.15)$$

where P_A and P_p are the resultant active and passive forces (see Fig. 15.29) and F is the factor of safety that was obtained from the toe failure analysis. Based on the earth pressure diagram (Fig. 15.29) and the calculated value of A_p, elementary structural mechanics can be used to determine the maximum moment in the sheet pile wall. The maximum moment divided by the section modulus can then be compared with the allowable design stresses listed in Fig. 15.26.

Design Analysis for Cantilevered Sheet Pile Wall. In the case of cantilevered sheet pile walls, the sheet piling is driven to an adequate depth to become fixed as a vertical cantilever in resisting the lateral active earth pressure. Cantilevered walls usually undergo large lateral deflections and are readily affected by scour and erosion in front of the wall. Because of these factors, penetration depths can be quite high, resulting in excess stresses and severe yield. According to the *USS Steel Sheet Piling Design Manual* (1984), cantilevered walls using steel sheet piling are restricted to a maximum height H of approximately 4.6 m (15 ft).

Design charts have been developed for the analysis of cantilevered sheet pile walls. For example, Fig. 15.30 is a design chart reproduced from the *USS Steel Sheet Piling Design Manual* (1984). The value of α is based on the location of the water level as indicated in Fig. 15.30. The chart is entered with the ratio of the passive earth pressure coefficient k_p divided by the active earth pressure coefficient k_A. By intersecting either the depth ratio curves or moment ratio curves, the depth of the sheet pile wall D or maximum moment M_{max} can be calculated. Note that the depth D from Fig. 15.30 corresponds to a factor of safety of 1.0. The *USS Steel Sheet Piling Design Manual* (1984) states that increasing the depth of embedment D by 20 to 40 percent will provide for a factor of safety of toe failure of approximately 1.5 to 2.0.

Example. Assume a cantilevered sheet pile wall with $k_A = 0.2$, $k_p = 8.65$, $\alpha = 1.0$, $H = 13.8$ ft (4.2 m), and $\gamma_b = 57$ pcf (9 kN/m³). Determine the total length and the maximum moment in the sheet pile wall.

Solution.
1. *Total length* $(H + D)$. Entering Fig. 15.30 at $k_p/k_A = 43$, and intersecting the depth ratio curve of $\alpha = 1$, find $D/H = 0.6$. Therefore, using $H = 13.8$ ft (4.2 m), the value of D for a factor of safety of 1 is equal to 8.3 ft (2.5 m). Using a 30 percent increase in embedment depth, the required embedment depth $(D) = 11$ ft (3.3 m). Thus, the total length of the sheet pile wall is 25 ft (7.5 m).

RETAINING WALLS

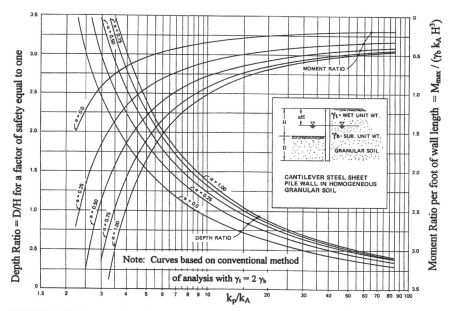

FIGURE 15.30 Design chart for cantilever sheet pile wall. (*From USS Steel Sheet Piling Design Manual, 1984.*)

2. *Maximum moment* (M_{max}). Entering Fig. 15.30 at $k_p/k_A = 43$, and intersecting the moment ratio curve of $\alpha = 1$, $M_{max}/(\gamma_b\, k_A\, H^3) = 0.53$. Thus the maximum moment in the sheet pile wall = 15,900 ft-lb/ft (70.7 kN·m/m).

Other Considerations. Some other important design considerations for sheet pile walls include the following:

1. *Soil layers.* The active and passive earth pressures should be adjusted for soil layers having different engineering properties.

2. *Penetration depth.* The penetration depth D of the sheet pile wall should be increased by at least an additional 20 percent to allow for the possibility of dredging and scour. Deeper penetration depths may be required, depending on the results of a scour analysis.

3. *Surcharge loads.* The ground surface behind the sheet pile wall is often subjected to surcharge loads. Equation (15.3) can be used to determine the active earth pressure resultant force due to a uniform surcharge pressure applied to the ground surface behind the wall. Note in Eq. (15.3) that the entire height of the sheet pile wall (i.e., $H + D$; see Fig. 15.29) must be used in place of H. Typical surcharge pressures exerted on sheet pile walls are caused by railroads, highways, dock loading facilities and merchandise, ore piles, and cranes.

4. *Unbalanced hydrostatic and seepage forces.* The previous discussion has assumed that the water level on both sides of the sheet pile wall are at the same elevation. Depending on factors such as the watertightness of the sheet pile wall and the backfill

permeability, it is possible that the groundwater level could be higher than the water level in front of the wall, in which case the wall would be subjected to water pressures. This condition could develop when there is a receding tide or a heavy rainstorm that causes a high groundwater table. A flow net can be used to determine the unbalanced hydrostatic and upward seepage forces in the soil in front of the sheet pile wall.

5. *Other loading conditions.* The sheet pile wall may have to be designed to resist the lateral loads due to ice thrust, wave forces, ship impact, mooring pull, and earthquake forces. If granular soil behind or in front of the sheet pile wall is in a loose state, it could be susceptible to liquefaction during an earthquake.

6. *Factors increasing the stability.* A factor usually not considered in the design analysis is the densification of loose sand during driving of the sheet piles. However, since sheet piles are relatively thin, the densification effect would be less for a sheet pile than for a comparable round pile, because the sheet pile displaces less soil.

Another beneficial effect in the design analysis is that many sheet pile walls used for waterfront construction are relatively long and hence the soil is in a plane strain condition. As previously discussed, the plane strain shear strength is higher than the shear strength determined from conventional shear strength tests.

Cohesive Soil. For cohesive soil, the long-term condition often governs the design. In this case, an effective stress analysis can be performed (c' and ϕ') using the estimated location of the groundwater table. For convenience, the effective cohesion is neglected and the analysis is performed by using only ϕ'. Therefore, the long-term effective stress condition for cohesive soil is analyzed as described in the preceding subsections for sheet pile walls having granular soil. Since ϕ' for a cohesive soil is usually less than ϕ' for granular soil, the active earth pressure will be higher and the passive resistance lower for cohesive soil.

Cofferdams. Steel sheet piling is widely used for the construction of a cofferdam, which is defined as a temporary structure designed to support the sides of an excavation and to exclude water from the excavation. Tomlinson (1986) presents an in-depth discussion of the different types of cofferdams and construction techniques. For example, the first step in the construction of a bridge foundation could be the installation of a large circular cofferdam. Then the river water would be pumped out from inside the cofferdam and the river bottom would be excavated to the desired bearing stratum for the bridge foundation.

In general, the following loads could be exerted on the cofferdam:

- Hydrostatic groundwater pressures located outside the cofferdam. If the cofferdam is constructed in a river or other body of water, then the hydrostatic water pressure should be based on the anticipated high-water level.
- Hydrostatic pressure of water inside the cofferdam if the water level outside falls below the interior level during initial pumping of water from the cofferdam. This condition could cause bursting of the cofferdam by interlock tension.
- Earth pressure outside the cofferdam.
- Other loads, such as surcharge loads, wave pressures, and earthquake loading.

On the basis of these loading conditions, the cofferdam must have an adequate factor of safety for the following conditions:

1. Heave of the soil located at the bottom of the cofferdam.
2. Sliding of the cofferdam along its base.
3. Overturning and tilting of the cofferdam.

4. The inward or outward yielding of the cofferdam (i.e., maximum moment and shear in the sheet piling and interlock tension).

5. Piping of soil which could undermine the cofferdam or lead to sudden flooding of the interior of the cofferdam.

An important consideration in the design of cofferdams is the ability of the sheet pile wall to deflect. If struts are used that brace the opposite sides of the excavation (i.e., braced excavation), the deflection of the sheet pile wall will be restricted. The next section will discuss the earth pressures exerted on braced excavations. If the sheet pile wall is able to deform—if raking braces are used, for example—then the deflections of the wall during construction will permit mobilization of active pressures in accordance with the previous discussion. The *USS Steel Sheet Piling Design Manual* (1984) and NAVFAC DM-7.2 (1982) present a further discussion of the design of cofferdams.

15.7 TEMPORARY RETAINING WALLS

Temporary retaining walls are often used during construction, such as for the support of the sides of an excavation that is made below grade in order to construct the building foundation. If the temporary retaining wall has the ability to develop the active wedge (Fig. 15.2), then the basic active earth pressure principles described in the previous sections can be used for the design of the temporary retaining walls.

Especially in urban areas, movement of the temporary retaining wall may have to be restricted to prevent damage to adjacent property. If movement of the retaining wall is restricted, the earth pressures will typically be between the active k_A and at-rest k_0 values.

15.7.1 Braced

For some projects, the temporary retaining wall may be constructed of sheeting (such as sheet piles) that are supported by horizontal braces, also known as *struts*. Near or at the top of the temporary retaining wall, the struts restrict movement of the retaining wall and prevent the development of the active wedge. Because of this inability of the retaining wall to deform at the top, earth pressures near the top of the wall are in excess of the active (k_A) pressures (see Fig. 15.31). At the bottom of the wall, the soil is usually able to deform into the excavation which results in a reduction in earth pressure, and the earth pressures at the bottom of the excavation tend to be constant or even decrease as shown in Fig. 15.31.

The earth pressure distributions shown in Fig. 15.31 were developed from actual measurements of the forces in struts during the construction of braced excavations (Terzaghi and Peck, 1967). In Fig. 15.31, case *a* shows the earth pressure distribution for braced excavations in sand and cases *b* and *c* show the earth pressure distribution for clays. In Fig. 15.31, the distance H represents the depth of the excavation (i.e., the height of the exposed wall surface). The earth pressure distribution is applied over the exposed height H of the wall surface, with the earth pressures transferred from the wall sheeting to the struts (the struts are labeled with the forces F_1, F_2, etc.).

Any surcharge pressures, such as surcharge pressures on the ground surface adjacent the excavation, must be added to the pressure distributions shown in Fig. 15.31. In addition, if the sand deposit has a groundwater table that is above the level of the bottom of the excavation, then water pressures must be added to the case *a* pressure distribution shown in Fig. 15.31.

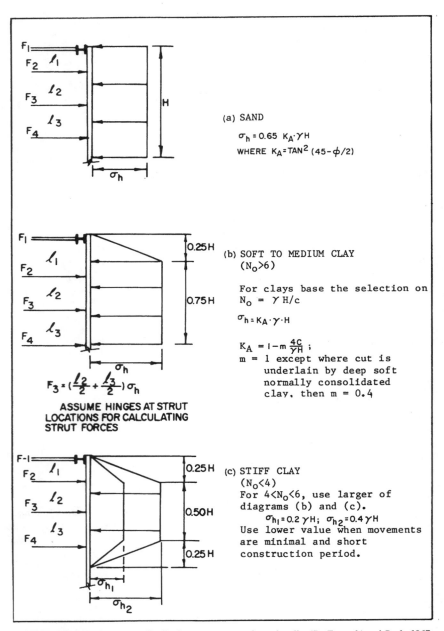

FIGURE 15.31 Earth pressure distribution on temporary braced walls. (*By Terzaghi and Peck, 1967; reproduced from NAVFAC DM-7.2 1982.*)

Because the excavations are temporary (i.e., short-term condition), the undrained shear strength ($s_u = c$) is used for the analysis of the earth pressure distributions for clay. The earth pressure distributions for clay (i.e., cases *b* and *c*) are not valid for permanent walls or for walls where the groundwater table is above the bottom of the excavation.

15.7.2 Steel I Beam and Wood Lagging

Figure 15.32 shows a picture of a temporary cantilevered steel I beam and wood lagging retaining wall. This type of temporary retaining wall consists of steel I beams that are either driven into place or installed in predrilled holes with the bottom portion cemented into place. During the main excavation of the interior area, wood lagging is installed between the steel I beams. The earth pressure exerted on the wood lagging is transferred to the steel I-beam flanges.

In some cases, the steel I beams will be braced with struts and the earth pressure distributions would be similar to those as shown in Fig. 15.31. For cantilevered steel I-beam and wood lagging retaining walls, such as shown in Fig. 15.32, there will be sufficient movement of the top of the wall to develop the active earth pressure. Tieback anchors, such as shown in Fig. 15.33, are often installed to reduce the bending moments of the steel I beams for deep excavations.

15.7.3 Utility Trench Shoring

A trench is generally defined as a narrow excavation made below the ground surface that may or may not be shored. Figure 15.34 shows an example of a trench excavation for the construction of a storm drain line.

FIGURE 15.32 Temporary steel I-beam and wood lagging cantilevered retaining wall.

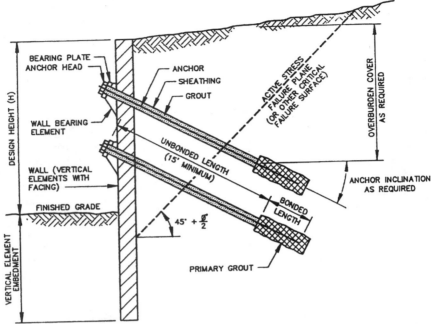

FIGURE 15.33 Cross section showing tieback anchors for retaining walls. (*Reproduced with permission from AASHTO 1996.*)

Each year, numerous workers are killed or injured in trench or excavation cave-ins. These accidents typically occur because the trench was not shored or the shoring system was inadequate (Thompson and Tanenbaum, 1977). In the United States, fatalities in trench cave-ins represent a substantial portion of the construction fatalities. For example, the Ohio Bureau of Workers' Compensation indicated a total of 271 cave ins during the period of 1986–1990. The cause of death in many cases is suffocation. Other workers die from the force of the cave in, which causes a crushing injury to the chest.

Petersen (1963) described the death of Peter Reimer, a construction worker, who decided that a trench needed no shoring because, he said, "it looks as solid as a brick wall." Reimer was alone in the trench when one side caved-in, but he was only buried up to his chest. Three fellow workers started digging Reimer out, but they had to scramble to safety when the other side of the trench caved-in. Reimer's body was uncovered half an hour later. As Petersen states, the first rule of a cave-in is to brace the excavation to prevent further cave-ins, then proceed with the rescue.

On large construction projects or deep excavations for buildings, money is generally available to perform an exploration program that defines both the soil and groundwater conditions over the full extent of the project. This results in an engineered shoring design that is commonly adjusted to satisfy the varying site conditions. However, utility trench excavations are smaller projects with typically less money for subsurface exploration. Also, utility trench excavations frequently extend over longer distances than large projects or deep building excavations. In many situations for utility trench excavations, it is the contractor who determines which shoring system to install on the basis of the superintendent's experience and the observed soil and groundwater conditions. When trench walls are loose or sandy, the danger is apparent to the contractor and shoring is installed. As in the Reimer case (Petersen, 1963), many problems develop when the trench looks hard and compact, and the contractor will take chances and install minimal or no shoring on the assumption that it will be stable for a few hours.

A common form of shoring for unstable soil is close sheeting. Figure 15.35 presents a diagram of close sheeting ("Trench," 1984). The main components of the shoring system are continuous sheeting from the top to the bottom of the trench, which is held in place by

FIGURE 15.34 Trench excavation for a storm drain line.

15.48 ANALYSIS OF GEOTECHNICAL DATA AND ENGINEERING COMPUTATIONS

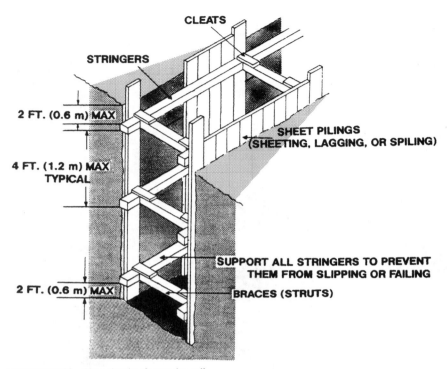

FIGURE 15.35 Close sheeting for running soil.

stringers and cross braces (struts). In many cases, the cross braces are hydraulic jacks rather than wood members. Another common type of shoring uses a sliding trench metal shield (NAVFAC DM-7.2, 1982).

Many injuries occur after the utility lines have been placed in the bottom of the trench and it is time to backfill. If the trench has remained open for a long time, there can be much more pressure on the shoring system than when it was installed. A worker who is in the trench and knocks out a cross brace will release this built-up pressure and perhaps cause failure of other members or collapse. According to Petersen (1963), the following is the correct procedure to backfill a trench.

1. Backfill and compact the trench to a point just below the bottom cross brace.
2. Enter the trench and remove only the bottom cross brace.
3. Backfill and compact to the next level of cross braces.
4. Remove these cross braces and continue the procedure until only the uprights or sheeting remain in the ground.
5. Pull out the uprights or sheeting using either a backhoe or front-end loader.

When designing trench shoring, the engineer could use the earth pressures for the condition of braced excavations (i.e., Fig. 15.31). According to structural engineering principles, this earth pressure diagram could be used to determine the maximum moments and maximum shear in the sheeting and the compressive force in the cross braces. Based on the structural analysis and including an appropriate factor of safety, the size of individual members could then be selected.

After obtaining the size of the individual shoring members, the design engineer should check the requirements versus the minimum state or local requirements. For example, the California Occupational Safety and Health Administration (Cal/OSHA) has developed a system to design shoring based on four different material types. The previous federal OSHA classification system ("Excavations," 1989) was to design shoring for either one of two soil conditions: hard compact soil or running soil. Running soil is defined as earth material where the angle of repose is approximately zero, as in the case of soil in a nearly liquid state, or dry, unpacked sand that flows freely under slight pressure. Running material also includes loose or disturbed earth that can only be contained with close sheeting (*Standards*, 1991). Hard compact soil was defined as all earth material not classified as running soil.

The new Cal/OSHA shoring design system (effective September 25, 1991) categorizes soil and rock deposits in a hierarchy of stable rock, type A, type B, and type C, in decreasing order of stability (*Standards*, 1991). Stable rock is defined as solid rock that can be excavated with vertical sides and remain intact while exposed. The definitions of type A, B, and C materials are rather extensive, but in general, the classifications are based on the strength of the material. For example, if the trench were excavated in cohesive soil, then it would be type A if the unconfined compressive strength was greater than 150 kPa (1.5 tsf), type B if the unconfined compressive strength was between 50 and 150 kPa (0.5 and 1.5 tsf), and type C if the unconfined compressive strength was less than 50 kPa (0.5 tsf). California OSHA has tables (*Standards*, 1991) that provide the minimum timber shoring requirements for each soil type and a specific depth of trench. A type C soil would require a more substantial shoring system than a type A soil, given the same depth of the trench.

In designing the shoring system, other important factors to be considered by the design engineer include the following:

1. The possibility of a temporary increase in the depth, length, or width of the trench.
2. The excavation of a trench below the groundwater table.
3. The presence of surcharge loads from adjacent soil or material stockpiles, buildings, and other loads.
4. The presence of vibrations from passing traffic, earthquakes, jackhammers, and other sources.
5. The possibility that the trench will be left open a long time.
6. The possibility of a change in climate, such as frequent rainstorms or thawing of frozen ground.
7. The possibility that surface-water flow may not be diverted away from the top of the trench.

15.8 PIER WALLS

As previously mentioned in Sec. 10.6, drilled piers are frequently used as a restraining system to stabilize slopes (Zaruba and Mencl, 1969). The terms *piers, shafts,* and *caissons* are sometimes used interchangeably by engineers. The common feature is that a cylindrical hole is drilled into the ground and then the hole is filled with concrete. The hole may be cased with a metal shell (a casing) in order to keep the hole from collapsing and sometimes to facilitate the cleaning of the bottom of the hole. The lower part of the hole may be belled out to develop a larger end-bearing area, thereby increasing the vertical load-carrying capacity of the drilled pier for a given allowable end-bearing pressure (Cernica, 1995b). General construction details for drilled piers are presented by Reese et al. (1981, 1985).

There are many different uses for drilled piers. Contiguous drilled piers are sometimes constructed to retain large open cuts more than 30 m (100 ft) deep (Abramson et al., 1996). When used to stabilize slopes, the drilled piers must pass through the slip surface and be embedded into a bearing stratum that can provide passive resistance to the destabilizing force transmitted by the unstable slope. Because soil arching can transfer lateral loads to the drilled piers, the piers are typically spaced a distance of two or three pier diameters apart.

In order to stabilize slopes, the required resistance of the pier wall may be very large. This can result in deep pier walls, having large pier diameters and substantial reinforcement to resist the overturning moment (Abramson et al., 1996). The construction cost can be reduced when the pier wall is combined with tiebacks. A tieback is normally constructed at the top of the pier by drilling a hole into the slope, installing a tieback into the hole, and then filling the anchoring portion of the hole with high-strength grout. The purpose of the tieback is to transfer a portion of the destabilizing force to a zone behind the slip surface. The tiebacks consist of high-tensile-strength steel cables, tendons, or rods.

Three examples of the installation of pier walls to stabilize slopes are presented below.

Portuguese Bend Landslide. A famous case of the failure of a reinforced-concrete pier wall involves the Portuguese Bend landslide, located in Palos Verdes Peninsula, California. The landslide is about 105 ha (260 acres) in size with a typical thickness between 30 to 45 m (100 to 150 ft) (Watry and Ehlig, 1995). The Portuguese Bend landslide started to move in mid-1956 after the placement of a fill embankment for the extension of a roadway across the top of the landslide (Ehlig, 1992). The slip surface of the Portuguese Bend landslide occurs in bentonite layers within the Miocene Monterey formation. Movement of the Portuguese Bend landslide has destroyed about 160 homes.

At the toe of the landslide, a cantilevered pier wall was constructed over a period of several months in 1957. A total of 23 reinforced-concrete piers were installed. The concrete piers were 1.2 m (4 ft) in diameter and embedded 3 m (10 ft) below the slip surface. It was reported that the rate of movement of the landslide decreased 50 percent after installation of the concrete piers, although it has been argued that the slowing was a result of other factors such as seasonal drying (Ehlig, 1986). Shortly after construction of the concrete pier wall, all of the piers failed by being either plucked out of the ground, tilted downslope, or sheared off by the landslide movement (Watry and Ehlig, 1995).

Fill Slope Failure. A second example of the construction of a pier wall to stabilize a slope involves a site in San Diego, California. The site contains a single-family house with an 18-m- (60-ft-) high descending rear yard fill slope inclined at approximately 1.5:1 (horizontal:vertical). In 1990, the homeowner discovered a pipe leak beneath the house. Water from the pipe leak infiltrated the fill mass and caused a slope failure. The slope failure resulted in 0.3 to 0.5 m (1 to 1.5 ft) of ground surface subsidence and the opening of tension ground cracks to approximately 100 mm (4 in.) in width. Figure 15.36 shows the ground cracks and foundation damage in the crawl space below the house.

The stabilization of the fill slope was achieved by the construction of a pier wall. Figures 15.37 and 15.38 show two views of the construction of the pier wall. Nineteen 0.9-m- (3-ft-) diameter piers, spaced 2.4 to 2.7 m (8 to 9 ft) on center, were installed through the fill mass and anchored in the underlying bedrock. Tiebacks were used to reduce the bending moments for the ten most heavily loaded piers. The arrow in Fig. 15.37 points to the location of one of the inclined tieback anchors. The pier wall stabilized the fill slope by providing the needed lateral resistance. A retaining wall was included at the top of the piers, and the final constructed condition is shown in Fig. 15.39.

FIGURE 15.36 Ground cracks and foundation damage. Arrows point to ground cracks.

FIGURE 15.37 Construction of pier wall. Arrow points to tieback anchor.

FIGURE 15.38 Construction of pier wall.

FIGURE 15.39 Final constructed condition of the pier wall.

RETAINING WALLS   **15.53**

Laguna Niguel Landslide.   The Laguna Niguel landslide has been discussed in Sec. 4.7. As shown in the cross section (Fig. 4.23b), the landslide dropped downward and outward, creating a very high unsupported head scarp (Fig. 4.29). This head scarp was unstable and with time, it was anticipated that the landslide movement would progress upslope. In order to stabilize the ground surface above the main scarp (i.e., the crown of the landslide), a reinforced-concrete pier wall was installed. After excavation of the pier hole, a steel reinforcement cage was installed into the borehole as shown in Figs. 15.40 and 15.41. After

FIGURE 15.40   Construction of the pier wall at the crown of the Laguna Niguel landslide. The photograph shows the installation of the steel reinforcement cage in the pier hole.

FIGURE 15.41 Construction of the pier wall at the crown of the Laguna Niguel landslide. The photograph shows a close-up view of the installation of the steel reinforcement cage in the pier hole.

installation of the steel reinforcement cage, the holes were filled with concrete. It was proposed that once all the piers were installed, the ground in front of the reinforced concrete pier wall would be excavated, and tieback anchors would be installed.

Design of Pier Walls. One approach for the design of pier walls is to use slope stability analysis (method A). The factor of safety of the slope, when stabilized by the pier wall, is first selected. Depending on such factors as the size of the slope failure, the proximity of critical facilities, or the nature of the pier wall stabilization (temporary versus permanent), a factor of safety of 1.2 to 1.5 is routinely selected. The next step is to determine the lateral design force P_L that each pier must resist in order to increase the factor of safety of the slope up to the selected value.

Figure 15.42 shows an unstable slope having a planar slip surface inclined at an angle α to the horizontal (wedge method; see Sec. 10.5). The most common approach is to perform an effective stress analysis (long-term condition). For an effective stress analysis, the shear strength of the slip surface can be defined by the effective friction angle ϕ' and effective cohesion c'. The factor of safety of the slope with the pier wall is derived by summing forces parallel to the slip surface, and is:

$$F = \frac{\text{resisting forces}}{\text{driving force}} = \frac{c'L + (W \cos \alpha - uL) \tan \phi' + P_i}{W \sin \alpha} \qquad (15.16)$$

where L = length of the slip surface, W = total weight of the failure wedge, u = average pore water pressure along the slip surface, and P_i = required pier wall force that is inclined at an angle α as shown in Fig. 15.42.

The elements of Eq. (15.16) are determined as follows:

α and L The angle of inclination α and the length of the slip surface L are based on the geometry of the failure wedge.

W Samples of the failure wedge material can be used to obtain the total unit weight. By knowing the total unit weight, we can calculate the total weight W of the failure wedge (Fig. 15.42).

ϕ' and c' The shear strength parameters ϕ' and c' can be determined from laboratory shear testing of slip surface specimens.

u By installing piezometers in the slope, the pore water pressure u can be measured.

F As previously mentioned, a factor of safety F of 1.2 to 1.5 is routinely selected for the slope stabilized with a pier wall.

The only unknown in Eq. (15.16) is P_i (inclined pier wall force; Fig. 15.42). The following equation can be used to calculate the lateral design force P_L for each pier having an on-center spacing of S:

$$P_L = SP_i \cos \alpha \qquad (15.17)$$

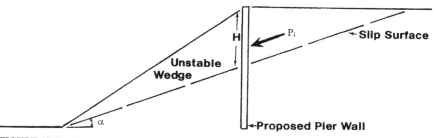

FIGURE 15.42 Design of pier wall for wedge slope failure.

15.56 ANALYSIS OF GEOTECHNICAL DATA AND ENGINEERING COMPUTATIONS

The location of the lateral design force P_L is ordinarily assumed to be at a distance of $1/3\, H$ above the slip surface, where H is as defined in Fig. 15.42. The lateral design force P_L would be resisted by passive pressure exerted on that portion of the pier below the slip surface.

The above analysis to determine the lateral design force P_L for each pier was based on a wedge type of slope failure as illustrated in Fig. 15.42. If the slip surface is a circular arc, then the method of slices can be used to calculate the lateral design force P_L. Circular arc slip surfaces may develop for slope failures in uniform soil.

A second approach for the design of pier walls is to use earth pressure theory (method B), as illustrated in Fig. 15.43. The destabilizing pressure acting on the pier wall is assumed to be the active earth pressure. The active pressure is applied in a tributary fashion as if a continuous wall exists. The active pressure can be applied over the entire length Z_T of the pier.

The resistance to movement of the pier wall is from passive pressure. To account for arching, the passive pressure can be applied over 2 pier diameters (for pier spacing of $2D$ or greater). Note that the passive pressure begins at a depth of Z_1, which is defined as the depth to adequate lateral bearing material, determined from subsurface investigation.

To calculate the lateral design force P_L for each pier, the following equation is used:

$$P_L = 0.5\, k_A \gamma_t Z_T^2 S \qquad (15.18)$$

where k_A = active earth pressure coefficient and γ_t = total unit weight. Because the active earth pressure is the actual driving pressure exerted on the pier wall, a factor of safety F can be included in Eq. (15.18). The factor of safety F could be the same value (1.2 to 1.5) as used in method A (slope stability method).

As mentioned in Sec. 15.2.1, the value of k_A times γ_t is ordinarily referred to as the *equivalent fluid pressure*. The value of k_A can be calculated for a given shear strength of unstable slope material. In Fig. 15.43, the value of 60 psf/ft is equal to the equivalent fluid pressure recommended for the design of a pier wall to stabilize fill and landslide debris that moved during the Northridge earthquake (Day and Poland, 1996).

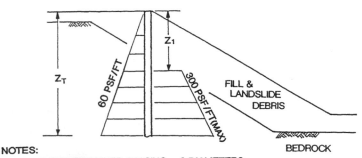

NOTES:
1) MAXIMUM PIER SPACING = 3 DIAMETERS

2) APPLY ACTIVE PRESSURE IN TRIBUTARY FASHION AS IF A WALL EXISTS. APPLY PASSIVE RESISTANCE OVER 2 PIER DIAMETERS

3) FOR TYPICAL CANTILEVER DESIGN, STRUCTURAL ENGINEER SHOULD ASSUME ABOUT 0.5 TO 1% ROTATION OF WALL IS REQUIRED TO MOBILIZE MAXIMUM PASSIVE RESISTANCE. TOTAL DEFLECTION SHOULD BE CALCULATED BY ADDING THIS ROTATION TO THE STRUCTURAL DEFLECTION FROM BENDING STRESS.

FIGURE 15.43 Design of pier walls by earth pressure theory.

Earth pressure theory (method B) can be used to calculate the lateral design force P_L when the slope deformation varies with depth (such as creep of clayey slopes), rather than failure on a distinct slip surface (method A).

As previously mentioned, because soil arching can transfer lateral loads to the drilled piers, the piers are typically spaced a distance of 2 or 3 diameters apart. However, for some soil, this spacing may be too great. For example, Figs. 15.44 and 15.45 show pictures of soil movement around concrete piers. The soil shown in Figs. 15.44 and 15.45 was classified as a silty clay, having a liquid limit of 56 and a plasticity index of 32 (high plasticity). At this site, the clay was plastic enough to simply flow around the piers.

Table 15.2 presents recommendations for maximum spacing of piers versus soil or rock type. This table is based on the performance of existing pier walls. In general, the pier spacing should be decreased as the rock becomes more fractured or the soil becomes more plastic.

Construction of Pier Walls. There are many references on the construction of piers (e.g., Reese et al., 1981, 1985). A major difficulty with the excavation of piers is access for drilling equipment. Because of the normal large size of the pier hole, great depth of excavation, and high resistance of the lateral bearing strata, a truck-mounted drill rig is required. This can make access a serious limitation for the construction of pier walls. If there is not

FIGURE 15.44 Soil movement around concrete piers.

FIGURE 15.45 Soil movement around concrete piers.

TABLE 15.2 Allowable Pier Spacing

Material type (1)	Largest on-center pier spacing (2)
Intact rock	No limit
Fractured rock	$4D$
Clean sand or gravel	$3D$
Clayey sand or silt	$2D$
Plastic clay	$1.5D$

Note: D = diameter of pier.

enough room for the drill rig or if the topography is too steep, the option of using a pier wall for slope stabilization may have to be abandoned.

Vertical loads can be imposed upon the pier by the tieback anchor or from the slope movement (Fig. 15.37). These vertical loads must be resisted by end bearing and/or skin friction in the bearing strata. Loose soil or debris is frequently knocked into the hole during the construction of the pier. This generally occurs when the prefabricated steel cage is lowered into the pier hole. When the steel cage strikes the side wall of the pier, soil is inevitably knocked to the bottom of the pier hole. Once the steel cage is in place, it is nearly impossible to remove the accumulation of loose soil at the bottom of the pier. One solution to this problem is to design the piers to resist vertical loads only through skin friction between the concrete and the bearing strata. In this case, end-bearing resistance is neglected. Since lateral loads usually govern the depth of embedment, it may be economical to simply neglect end bearing and design the piers to resist vertical loads through skin friction.

A small-diameter inclined hole is ordinarily used for the tieback anchor. This means that it will not be possible to visually observe the installation conditions of the tieback. Common problems are tiebacks not centered in the hole or inadequately grouted tieback zones which are unable to generate the required frictional resistance. Because of the numerous difficulties in construction of tieback anchors, field testing of the anchors is essential in order to be sure of an acceptable capacity. Table 15.3 presents test procedures and acceptance criteria for tieback anchors.

PROBLEMS

The problems have been divided into basic categories as indicated below.

Basic Retaining Wall Analyses

1. Using the retaining wall shown in Fig. 15.1, assume $H = 4$ m, the thickness of the reinforced concrete wall stem = 0.4 m, the reinforced concrete wall footing is 3-m wide by 0.5-m-thick, the ground surface in front of the wall is level with the top of the wall footing, and the unit weight of concrete = 23.5 kN/m³. The wall backfill will consist of sand having $\phi = 32°$ and $\gamma_t = 20$ kN/m³. Also assume that there is sand in front of the footing with these same soil properties. The friction angle between the bottom of the footing and the bearing soil (δ) = 24°. For the condition of a level backfill and neglecting the wall friction on the backside of the wall and the front side of the footing, calculate the active earth pressure coefficient k_A, the passive earth pressure coefficient k_p, the active earth pressure resultant force P_A, and the passive resultant force P_p using a reduction factor = 2.0. Answers: $k_A = 0.307$, $k_p = 3.25$, $P_A = 49.2$ kN/m, and $P_p = 4.07$ kN/m.

2. For Prob. 1, determine the amount of wall displacement needed to develop the active wedge if the sand will be compacted into a dense state. Answer: 0.2 cm.

3. For Prob. 1, determine the width of the active wedge at the top of the retaining wall. Answer: 2.2 m.

4. For Prob. 1, determine the resultant normal force N and the distance of N from the toe of the footing $\bar{x}$. Answers: $N = 68.2$ kN/m and $\bar{x} = 1.16$ m.

5. Using the data from Probs. 1 and 4, determine the maximum bearing pressure q' and the minimum bearing pressure (q'') exerted by the retaining wall foundation. Answer: $q' = 37.9$ kPa and $q'' = 7.5$ kPa.

6. For Prob. 1, determine the allowable bearing capacity q_{all} of the retaining wall foundation using Eq. (8.1), assuming the soil beneath the footing is cohesionless sand (ϕ = 32°) and using a factor of safety = 3. Assume the groundwater table is well below the bottom of the retaining wall foundation. On the basis of a comparison of the allowable bearing pressure q_{all} and the largest bearing pressure exerted by the retaining wall foundation q', is the design acceptable in terms of the foundation bearing pressures? Answers: $q_{all} = 210$ kPa, and, since $q_{all} > q'$, the design is acceptable in terms of bearing pressures.

7. Using the data from Prob. 1, calculate the factor of safety F for sliding of the retaining wall. Answer: $F = 0.70$.

8. Using the data from Probs. 1 and 4, calculate the factor of safety F for overturning of the retaining wall. Answer: $F = 2.2$.

TABLE 15.3 Test Procedures and Acceptance Criteria for Each Tieback

A. Stages and Observation Periods for 24-Hour Test (Performance Test)

Load level	Recommended period of observations (in minutes)
Seating/alignment load = 0.1DL	None
0.25DL	10
0.50DL	10
0.75DL	10
1.00DL	10
1.25DL	10
1.50DL	10
1.75DL	30
2.00DL	480

DL = design load

After each stage of loading, the load should be reduced to seating/alignment load and replaced after 2 minutes. Deflection should be measured immediately after application of each load level and at 5-minute intervals thereafter. Measurements should also be taken immediately prior to any unloading and reloading steps.

B. Load stages and observation periods for proof test

Load Level	Recommended period of observations (in minutes)
Seating/alignment load = 0.1/DL	None
0.50DL	10
1.00DL	10
1.25DL	10
1.50DL	30

DL = design load

After each stage of loading, the load should be reduced to seating/alignment load and replaced after 2 minutes. Deflection should be measured before and after application of each loading increment.

Acceptance criteria:

1. Total deflection during the long term testing should not exceed 12 inches.
2. Creep deflection at 200% design load should not exceed 0.1 in. during the 4-hour period.
3. Total deflection during short-term testing should not exceed 12 inches.
4. Creep deflection at 150% design load should not exceed 0.1 in. during the 15-minute period.
5. All the tiebacks should be locked off at 120% of the design load.
6. Following final tensioning, the nongrouted portion of the anchor boring should be grouted.

9. For the retaining wall described in Prob. 1, it is proposed to increase the factor of safety for sliding by installing a concrete key at the bottom of the retaining wall foundation. Determine the depth of the key (measured from the ground surface in front of the retaining wall) that is needed to increase the factor of safety of sliding to 1.5. Neglect the weight of the concrete key in the analysis and include a reduction factor of 2 for P_p. Answer: $D = 1.64$ m.

Retaining Walls with Wall Friction

10. Solve Prob. 1, but include wall friction in the analysis (use Coulomb's earth pressure equation, Fig. 15.3). Assume the friction angle between the back side of the retaining wall and the backfill is equal to $3/4\ \phi$ (i.e., $\phi_w = 3/4\phi = 24°$). Answers: $k_A = 0.275$, $P_A = 43.9$ kN/m, $P_H = 40.1$ kN/m, and $P_v = 17.9$ kN/m.

11. For Prob. 10, determine the resultant normal force N and the distance of N from the toe of the footing $\bar{x}$. Answers: $N = 86.1$ kN/m and $\bar{x} = 1.69$ m.

12. Using the data from Probs. 10 and 11, determine the maximum bearing pressure q' and the minimum bearing pressure q'' exerted by the retaining wall foundation. Answer: $q' = 39.6$ kPa and $q'' = 17.8$ kPa.

13. For Prob. 10, determine the allowable bearing capacity q_{all} of the retaining wall foundation using Eq. (8.1), assuming the soil beneath the footing is cohesionless sand ($\phi = 32°$) and using a factor of safety = 3. Assume the groundwater table is well below the bottom of the retaining wall foundation. On the basis of a comparison of the allowable bearing pressure q_{all} and the largest bearing pressure exerted by the retaining wall foundation q', is the design acceptable in terms of the foundation bearing pressures? Answers: $q_{all} = 210$ kPa and, since $q_{all} > q'$, the design is acceptable in terms of bearing pressures.

14. Using the data from Prob. 10, calculate the factor of safety F for sliding of the retaining wall. Answer: $F = 1.06$.

15. Using the data from Probs. 10 and 13, calculate the factor of safety F for overturning of the retaining wall. Answer: $F = \infty$.

16. For the retaining wall described in Prob. 10, it is proposed to increase the factor of safety for sliding by installing a concrete key at the bottom of the retaining wall foundation. Determine the depth of the key (measured from the ground surface in front of the retaining wall) that is needed to increase the factor of safety of sliding to 1.5. Neglect the weight of the concrete key in the analysis and include a reduction factor of 2 for P_p. Answer: $D = 1.16$ m.

17. Using the retaining wall shown at the top of Fig. 15.5b (i.e., a cantilevered retaining wall), assume $H = 4$ m, the thickness of the reinforced concrete wall stem = 0.4 m and the wall stem is located at the centerline of the footing, the reinforced concrete wall footing is 2-m wide by 0.5-m thick, the ground surface in front of the wall is level with the top of the wall footing, and the unit weight of concrete = 23.5 kN/m³. The wall backfill will consist of sand having $\phi = 32°$ and $\gamma_t = 20$ kN/m³. Also assume that there is sand in front of the footing with these same soil properties. The friction angle between the bottom of the footing and the bearing soil (δ) = 24°. For the condition of a level backfill and assuming total mobilization of the shear strength along the vertical plane at the heel of the wall, calculate the active earth pressure coefficient k_A, the passive earth pressure coefficient k_p, the active earth pressure resultant force P_A, the vertical P_v and horizontal components P_H of P_A, and the passive resultant force P_p using a reduction factor = 2.0. Answers: $k_A = 0.277$, $k_p = 3.25$, $P_A = 44.3$ kN/m, $P_v = 23.5$ kN/m, $P_H = 37.6$ kN/m, and $P_p = 4.07$ kN/m.

18. For Prob. 17, determine the resultant normal force N and the distance of N from the toe of the footing $\bar{x}$. *Answers:* $N = 135.9$ kN/m and $\bar{x} = 1.05$ m.

19. Using the data from Probs. 17 and 18, determine the maximum bearing pressure q' and the minimum bearing pressure q'' exerted by the retaining wall foundation. *Answer:* $q' = 78.1$ kPa and $q'' = 57.8$ kPa.

20. For Prob. 17, determine the allowable bearing capacity q_{all} of the retaining wall foundation using Eq. (8.1), assuming the soil beneath the footing is cohesionless sand ($\phi = 32°$) and using a factor of safety = 3. Assume the groundwater table is well below the bottom of the retaining wall foundation. On the basis of a comparison of the allowable bearing pressure q_{all} and the largest bearing pressure exerted by the retaining wall foundation q', is the design acceptable in terms of the foundation bearing pressures? *Answers:* $q_{all} = 160$ kPa and, since $q_{all} > q'$, the design is acceptable in terms of bearing pressures.

21. Using the data from Prob. 17, calculate the factor of safety F for sliding of the retaining wall. *Answer:* $F = 1.72$.

22. Using the data from Probs. 17 and 18, calculate the factor of safety F for overturning of the retaining wall. *Answer:* $F = 47$.

23. For the retaining wall described in Prob. 17, it is proposed to increase the factor of safety for sliding by installing a concrete key at the bottom of the retaining wall foundation. Determine the depth of the key (measured from the ground surface in front of the retaining wall) that is needed to increase the factor of safety of sliding to 2.0. Neglect the weight of the concrete key in the analysis and include a reduction factor of 2 for P_p. *Answer:* $D = 0.95$ m.

Additional Design Features for Retaining Walls

24. For the example problem shown in Fig. 15.4a, b, and c, assume that there is a vertical surcharge pressure of 200 psf located at ground surface behind the retaining wall. Calculate the factor of safety for sliding and factor of safety for overturning, and determine if N is within the middle third of the retaining wall foundation. *Answers:* Factor of safety for sliding = 1.48, factor of safety for overturning = 2.64, and N is not within the middle third of the retaining wall foundation.

25. For the example problem shown in Fig. 15.4a, b, and c, assume that the ground surface behind the retaining wall slopes upward at a 3:1 (horizontal:vertical) slope inclination. Calculate the factor of safety for sliding, factor of safety for overturning, and determine if N is within the middle third of the retaining wall foundation. *Answers:* Factor of safety for sliding = 1.32, factor of safety for overturning = 2.73, and N is not within the middle third of the retaining wall foundation.

26. For the example problem shown in Fig. 15.4a, b, and c, assume that the retaining wall will be subjected to a peak acceleration at the ground surface $(a_{max}) = 0.20g$ [i.e., use Eq. (15.11)]. Calculate the factor of safety for sliding and the factor of safety for overturning. *Answers:* Factor of safety for sliding = 1.07 and factor of safety for overturning = 1.02.

27. Assume a 2.5:1 (horizontal:vertical) slope is composed of Soil B (Fig. 9.4) and the slope height is 20 m. Also assume that the depth of slope creep will correspond to the depth of seasonal moisture changes and that the effective shear strength parameters for this clay are $\phi' = 28°$ and $c' = 4$ kPa and $\gamma_t = 19$ kN/m³. It is proposed to construct a retaining wall at the top of the fill slope. If the required passive resistance force P_p for the proposed retaining wall is equal to 10 kN/m, determine the required depth of the retaining foundation (measured from the top of the fill slope). Use a reduction factor = 2 for P_p. *Answer:* $D = 5.6$ m.

Restrained Retaining Wall

28. For the example problem shown in Fig. 15.4a, b, and c, assume that the retaining wall is part of a bridge abutment and that the lateral deformation of the wall will be restricted. Considering the retaining wall to be a restrained wall and, using Eq. (6.21) to calculate the coefficient of earth pressure at rest, determine the resultant of the lateral pressure exerted on the retaining wall by the backfill soil. *Answer:* Resultant force = 11,000 lb/ft.

Mechanically Stabilized Earth Retaining Walls

29. Using the mechanically stabilized earth retaining wall shown in Fig. 15.23, assume $H = 20$ ft, the width of the mechanically stabilized earth retaining wall = 14 ft, the depth of embedment at the front of the mechanically stabilized zone = 3 ft, and the soil behind the mechanically stabilized zone is a clean sand with a friction angle (ϕ) = 30° and a total unit weight of (γ_t) = 110 pcf. Also assume that there is sand in front of the wall with these same properties, and there is a level backfill with no surcharge pressures (i.e., $P_2 = 0$). Further assume that the mechanically stabilized zone will have a total unit weight (γ_t) = 120 pcf, there will be no shear stress (i.e., $\delta = 0°$) along the vertical back side of the mechanically stabilized zone, and $\delta = 23°$ along the bottom of the mechanically stabilized zone. Calculate the active earth pressure coefficient k_A, the passive earth pressure coefficient k_p, the active earth pressure resultant force P_A, and the passive resultant force P_p using a reduction factor = 2.0. *Answers:* $k_A = 0.333$, $k_p = 3.0$, $P_A = 7330$ lb/ft, and $P_p = 740$ lb/ft.

30. For Prob. 29, determine the width of the active wedge at the top of the mechanically stabilized earth retaining wall. *Answer:* 25.5 ft, measured from the front left upper corner of the mechanically stabilized earth retaining wall (Fig. 15.23).

31. For Prob. 29, determine the resultant normal force N and the distance of N from the toe of the mechanically stabilized zone, i.e., point O. *Answers:* $N = 33,600$ lb and $\bar{x} = 5.55$ ft.

32. Using the data from Probs. 29 and 31, determine the maximum bearing pressure q' and the minimum bearing pressure q'' exerted by the mechanically stabilized earth retaining wall. *Answer:* $q' = 3890$ psf and $q'' = 910$ psf.

33. For Prob. 29, determine the allowable bearing capacity q_{all} of the base of the mechanically stabilized earth retaining wall using Eq. (8.1), assuming the soil beneath the wall is cohesionless sand ($\phi = 30°$ and $\gamma_t = 120$ pcf) and using a factor of safety = 3. Assume the groundwater table is well below the bottom of the mechanically stabilized zone. On the basis of a comparison of the allowable bearing pressure q_{all} and the largest bearing pressure exerted at the base of the mechanically stabilized earth retaining wall q', is the design acceptable in terms of the bearing pressures? *Answers:* $q_{all} = 3340$ psf, and since $q' > q_{all}$, the design is unacceptable in terms of bearing pressures.

34. Using the data from Prob. 29, calculate the factor of safety F for sliding of the mechanically stabilized earth retaining wall. *Answer:* $F = 2.05$.

35. Using the data from Probs. 29 and 31, calculate the factor of safety F for overturning of the mechanically stabilized earth retaining wall. *Answer:* $F = 4.81$.

36. Use the data from Prob. 29 and assume that there is a vertical surcharge pressure of 200 psf located at ground surface behind the mechanically stabilized earth retaining wall. Calculate the factor of safety for sliding, factor of safety for overturning, and the maximum pressure exerted by the base of the mechanically stabilized earth retaining wall. *Answers:* Factor of safety for sliding = 1.73, factor of safety for overturning = 3.78, and maximum pressure (q') = 4300 psf.

15.64 ANALYSIS OF GEOTECHNICAL DATA AND ENGINEERING COMPUTATIONS

37. Use the data from Prob. 29 and assume that the ground surface behind the mechanically stabilized earth retaining wall slopes upward at a 3:1 (horizontal:vertical) slope inclination. Also assume that the 3:1 slope does not start at the upper front corner of the rectangular reinforced soil mass (such as shown in Fig. 15.24), but instead the 3:1 slope starts at the upper back corner of the rectangular reinforced soil mass. Calculate the factor of safety for sliding, factor of safety for overturning, and the maximum pressure exerted by the retaining wall foundation. *Answers:* Factor of safety for sliding = 1.60, factor of safety for overturning = 3.76, and maximum pressure (q') = 4310 psf.

38. For Prob. 29, the internal stability of the mechanically stabilized zone is to be checked by using the wedge analysis. Assume a planar slip surface that is inclined at an angle of 61° (i.e., $\alpha = 61°$) and passes through the toe of the mechanically stabilized zone. Also assume that the mechanically stabilized zone contains 40 horizontal layers of Tensar SS2 geogrid, which has an allowable tensile strength = 300 lb per foot of wall length for each geogrid. In the wedge analysis, these 40 layers of geogrid can be represented as an allowable horizontal resistance force = 12,000 lb per foot of wall length (i.e., 40 layers times 300 lb). If the friction angle ϕ of the sand = 32° in the mechanically stabilized zone, calculate the factor of safety for internal stability of the mechanically stabilized zone using the wedge analysis. *Answer: F = 1.82.*

Sheet Pile Walls with Tieback Anchors

39. Using the sheet pile wall diagram shown in Fig. 15.29, assume that the soil behind and in front of the sheet wall is uniform sand with a friction angle (ϕ') = 33°, buoyant unit weight (γ_b) = 64 pcf, and above the groundwater table, the total unit weight (γ_t) = 120 pcf. Also assume that the sheet pile wall has $H = 30$ ft and $D = 20$ ft, the water level in front of the wall is at the same elevation as the groundwater table which is located 5 ft below the ground surface, and the tieback anchor is located 4 ft below the ground surface. Neglecting wall friction, calculate the active earth pressure coefficient k_A, the passive earth pressure coefficient k_p, the active earth pressure resultant force P_A, and the passive resultant force P_p using no reduction factor. *Answers: $k_A = 0.295$, $k_p = 3.39$, $P_A = 27,500$ lb/ft, and $P_p = 43,400$ lb/ft.*

40. For Prob. 39, calculate the factor of safety for toe failure (i.e., toe kick-out). *Answer: F = 2.19.*

41. For Prob. 39, calculate the anchor pull force A_p assuming that the anchors will be spaced 10 ft on center. *Answer: $A_p = 76.8$ kips.*

42. For Prob. 39, calculate the depth D that will provide a factor of safety for toe failure (i.e., toe kick-out) equal to 1.5. *Answer: D = 14.6 ft.*

43. For Prob. 39, assume that there is a uniform vertical surcharge pressure = 200 psf applied to the ground surface behind the sheet pile wall. Calculate the factor of safety for toe failure. *Answer: F = 2.03.*

44. For Prob. 39, assume that the ground surface slopes upward at a 2:1 (horizontal:vertical) slope ratio behind the sheet pile wall. Calculate the factor of safety for toe failure. *Answer: F = 1.46.*

45. For Prob. 39, assume that the ground in front of the sheet pile wall (i.e., the passive earth zone) slopes downward at a 3:1 (horizontal:vertical) slope ratio. Calculate the factor of safety for toe failure. *Answer: F = 1.18.*

46. For Prob. 39, assume that there is a nearby active fault and the engineering geologist has indicated that the peak ground acceleration at the site (a_{max}) = 0.2g for the design earthquake. Using Eq. (15.11) with H = total height of the sheet pile wall (i.e., $H + D$; Fig. 15.29) and substituting γ_b for γ_t, calculate the factor of safety for toe failure for the design earthquake condition. *Answer: F = 1.76.*

47. For Prob. 39, assume that there are two different horizontal layers of sand at the site. The first layer of sand is located only behind the sheet pile wall and extends from the ground surface to a depth of 30 ft. This sand layer has the engineering properties stated in Prob. 39. The second horizontal layer of sand is located on both sides of the sheet pile wall. All the sand in front of the sheet pile wall is composed of this second layer, and the soil behind the sheet pile wall below a depth of 30 ft also consists of this second layer. The second sand layer is a clean sand, in a loose state, and has a friction angle (ϕ') = 30° and a buoyant unit weight (γ_b) = 60 pcf. Neglecting wall friction, calculate the active earth pressure coefficients k_A, the passive earth pressure coefficient k_p, the active earth pressure resultant force P_A, and the passive resultant force P_p using no reduction factor. *Answers:* k_A = 0.295 (upper sand layer), k_A = 0.333 (lower sand layer), k_p = 3.0, P_A = 29,400 lb/ft, and P_p = 36,000 lb/ft.

48. For Prob. 47, calculate the factor of safety for toe failure (i.e., toe kick-out). *Answer:* F = 1.67.

49. For Prob. 47, calculate the anchor pull force A_p assuming that the anchors will be spaced 10 ft on center. *Answer:* A_p = 78.4 kips.

50. For Prob. 47, assume that there is a uniform vertical surcharge pressure = 200 psf applied to the ground surface behind the sheet pile wall. Calculate the factor of safety for toe failure. *Answer:* F = 1.55.

51. For Prob. 47, assume that the clean sand located 10 ft below the ground in front of the sheet pile wall (passive zone) has an $(N_1)_{60}$ value = 5. Also assume that the clean sand located 40 ft below the ground surface behind the sheet pile wall (active zone) has an $(N_1)_{60}$ value = 25. For a magnitude (M) 7.5 design earthquake with a_{max} = 0.2g, will the soil in front and behind the sheet pile wall be subjected to liquefaction during the design earthquake? If so, what is the likely consequence of liquefaction? *Answer:* At a depth of 10 ft in front of the sheet pile wall, the SSR induced by the earthquake = 0.26 (assuming the groundwater table is at ground surface) and the SSR that will cause liquefaction of the *in situ* soil = 0.06, therefore liquefaction is likely in the passive earth pressure zone. For the sand located at a depth of 40 ft below the ground surface in the active zone, the SSR induced by the earthquake = 0.20 and the SSR that will cause liquefaction of the *in situ* soil = 0.30, therefore liquefaction is unlikely in the active earth pressure zone. For liquefaction in the passive zone, toe failure of the sheet pile wall is likely.

52. Use the data from Prob. 47, except assume that the second soil layer is a clay that has effective shear strength parameters of c' = 2 kPa and ϕ' = 25°. Neglecting the effective cohesion value and using an effective stress analysis, calculate the factor of safety for toe failure. *Answer:* F = 1.16.

Cantilevered Sheet Pile Walls

53. For the example problem in Sec. 15.6, calculate the depth D of the sheet pile wall and the maximum moment M_{max} in the sheet pile wall if the sand has a friction angle ϕ' of 30°. Neglect friction between the sheet pile wall and the sand. *Answers:* D = 27 ft and M_{max} = 55,000 ft-lb/ft.

54. For the example problem in Sec. 15.6, calculate the depth D of the sheet pile wall and the maximum moment M_{max} in the sheet pile wall if the water level on both sides of wall is located at the top of the sheet pile wall (i.e., α = 0). *Answers:* D = 8 ft and M_{max} = 6000 ft-lb/ft.

Braced Excavations

55. A braced excavation will be used to support the vertical sides of a 20-ft-deep excavation (i.e., H = 20 ft in Fig. 15.31). If the site consists of a sand with a friction angle (ϕ)

= 32° and a total unit weight (γ_t) = 120 pcf, calculate the earth pressure σ_h and the resultant earth pressure force acting on the braced excavation. Assume the groundwater table is well below the bottom of the excavation. *Answers:* σ_h = 480 psf, and the resultant force = 9600 lb per linear foot of wall length.

56. Solve Prob. 55, but assume the site consists of a soft clay having an undrained shear strength (s_u) = 300 psf (i.e., $c = s_u$ = 300 psf). *Answers:* σ_h = 1200 psf, and the resultant force = 21,000 lb per linear foot of wall length.

57. Solve Prob. 55, but assume the site consists of a stiff clay having an undrained shear strength (s_u) = 1200 psf and use the higher earth pressure condition (i.e., σ_{h2}). *Answers:* σ_{h2} = 960 psf, and resultant force = 14,400 lb per linear foot of wall length.

Steel I Beam and Wood Lagging

58. A cantilevered steel I-beam and wood lagging system is installed for a 20-ft-deep excavation. Assume the site consists of a sand with a friction angle (ϕ) = 32° and a total unit weight (γ_t) = 120 pcf, and the temporary retaining wall deforms enough to develop active earth pressures. Also assume the groundwater table is well below the bottom of the excavation. If the steel I beams are spaced 7 ft on center, calculate the active earth pressure resultant force acting on each steel I beam. *Answer:* The resultant force = 51,600 lb acting on each steel I beam.

59. For Prob. 58, assume that each steel I beam is braced (i.e., a braced excavation). For the braced excavation, calculate the earth pressure resultant force acting on each steel I beam. *Answer:* The resultant force = 67,200 lb acting on each steel I beam.

60. Assume that for a steel I beam and wood lagging temporary retaining wall, tieback anchors will be used to reduce the bending moments in the steel I beams. Each tieback will have a diameter of 6 in., and the allowable frictional resistance between the grouted tieback anchor and the soil is 10 psi. If each tieback must resist a load of 20 kips, determine the bonded length. *Answer:* Bonded length = 8.8 ft.

Pier Walls

61. A slope will have a height of 9.1 m and the slope face will be inclined at a 2:1 (horizontal:vertical) ratio. Assume a wedge-type analysis where the slip surface is planar through the toe of the slope and is inclined at a 3:1 (horizontal:vertical) ratio. The total unit weight of the slope material (γ_t) = 18.1 kN/m³ and the slip surface has the drained residual shear strength parameters that are shown in Fig. 5.6 (i.e., $c'_r = 0$ and $\phi'_r = 7°$). Also assume that piezometers have been installed along the slip surface and the average measured steady-state pore water pressure (u) = 2.4 kPa. Calculate the lateral pier resistance force (P_L) for each pier (pier diameter = 1 m, pier spacing = 4 m) in order to increase the factor of safety of the failure wedge to 1.5. *Answer:* P_L = 1050 kN per pier.

CHAPTER 16
GROUNDWATER AND MOISTURE MIGRATION

16.1 INTRODUCTION

This final chapter of Part 3 (Analysis of Geotechnical Data and Engineering Computations) deals with groundwater and moisture migration. The groundwater table (also known as the *phreatic surface*) is the top surface of underground water, the location of which is often determined from piezometers, such as an open standpipe. A perched groundwater table refers to groundwater occurring in an upper zone separated from the main body of groundwater by underlying unsaturated rock or soil.

Groundwater and moisture migration can affect all types of civil engineering projects. Probably more failures in geotechnical and foundation engineering are either directly or indirectly related to groundwater and moisture migration than any other factor. In the previous chapters, the discussion of groundwater and moisture migration has included the following:

- *Subsurface exploration (Sec. 4.3).* One of the main purposes of subsurface exploration is to determine the location of the groundwater table. The groundwater table (▼ symbol) is often shown on the subsoil profiles (see Figs. 4.39 to 4.41).

- *Laboratory testing (Sec. 5.2.4).* The rate of flow through soil is dependent on its permeability. The constant-head permeameter (Fig. 5.8) or the falling-head permeameter (Fig. 5.9) can be used to determine the coefficient of permeability (also known as the hydraulic conductivity). Equations to calculate the coefficient of permeability in the laboratory are based on Darcy's law ($v = ki$).

- *Pore water pressure (Sec. 6.3).* The pore water pressure u is calculated by using Eq. (6.18) for soil below the groundwater table and Eq. (6.19) for soil above the groundwater table that is saturated because of capillary rise.

- *Shear strength (Sec. 6.5).* The shear strength of saturated soil can be influenced by the water that fills the soil pores. For example, normally consolidated plastic soils have an increase in pore water pressure because the soil structure wants to contract during undrained shear, while heavily overconsolidated plastic soils have a decrease in pore water pressure because the soil structure wants to dilate during undrained shear.

- *Permeability and seepage (Sec. 6.6).* The basic principles of permeability and seepage are presented in Sec. 6.6. The primary method to estimate the seepage quantities and the distribution of pore water pressure is the two-dimensional flow net. Flow nets have numerous applications, such as the study of earth dams, effect of cutoff walls, consequences of dewatering, drawdown due to pumping of groundwater from wells, effect of groundwater on foundations, and the study of slope stability.

- *Collapsible soil (Sec. 7.3).* Collapsible soil is broadly defined as a soil with an unstable particle structure, often of low density and low moisture content, that collapses when there is moisture migration into the soil.
- *Consolidation (Sec. 7.4).* The increase in vertical pressure due to the weight of a structure constructed on top of a soft saturated clay or organic soil will initially be carried by excess pore water pressure u_e. As water slowly drains from the clay, the excess pore water pressure slowly dissipates. Drainage of water from the clay layer causes the soil particles to compress together (i.e., consolidate).
- *Subsidence due to extraction of groundwater (Sec. 7.6.3).* Pumping can cause a lowering of the groundwater table, which then increases the overburden pressure on underlying sediments. This can cause consolidation of soft clay deposits or compression of the loose soil or porous rock structure, which then results in ground surface subsidence.
- *Bearing capacity (Sec. 8.2).* The bearing capacity of soil with a high groundwater table is lower than the same soil having a groundwater table at a great depth. Groundwater can also exert buoyancy effects on submerged foundation elements and hydrostatic uplift forces on underground basements, storage tanks, etc.
- *Expansive soil (Chap. 9).* Moisture migration from the clay, such as that due to surface drying, creates a desiccated clay deposit. If a structure is then constructed at ground surface, moisture will migrate by capillarity and thermo-osmosis to the clay located underneath the structure, resulting in an increase in water content of the clay and heave of the center of the structure (center lift).
- *Surficial slope stability (Sec. 10.4).* The development of surficial instability is often due to saturation and groundwater seepage in the outer face of the slope. Equation (10.2) was derived by assuming seepage of groundwater parallel to the slope surface to a depth d.
- *Debris flow (Sec. 10.7).* Debris flow is commonly defined as soil with entrained water and air that moves readily as a fluid on low slopes. As shown in Fig. 10.20, in many cases there is an initial surficial slope failure that transforms itself into a debris flow.
- *Slope softening (Sec. 10.8).* As water infiltrates a compacted clay slope, the pore spaces fill with water and the pore water pressure tends to zero. If a groundwater table then develops, the pore water pressures will become positive. The elimination of negative pore water pressure results in a decrease in effective stress and deformation of the slope (i.e., slope softening) in order to mobilize the needed shear stress to maintain stability.
- *Dams (Sec. 10.9).* Design analysis of dams can be grouped into three general categories: (1) stability during construction, usually involving a failure through the natural ground that often has a groundwater table, (2) stability of the downstream slope during reservoir operation, where there is a steady-state groundwater flow through the dam, and (3) stability analysis of the upstream slope after rapid drawdown, which causes seepage of groundwater down and through the slope surface.
- *Frost (Sec. 13.4).* Near-surface silty soils can form ice lenses because of their high capillarity and sufficient permeability that enables them to draw up moisture from the groundwater table. The development of ice lenses can cause heave of the structure. The spring thaw then will melt the ice, resulting in settlement of the foundation or a weakened subgrade that will make the pavement surface susceptible to deterioration or failure.

16.2 GROUNDWATER

As described above, groundwater can cause or contribute to failure because of excess saturation, seepage pressures, or uplift forces. It has been stated that uncontrolled saturation and seepage causes many billions of dollars a year in damage (Cedergren, 1989). Other geotechnical and foundation problems due to groundwater are as follows (Harr, 1962; Collins and Johnson, 1988; Cedergren, 1989):

- Piping failures of dams, levees, and reservoirs.
- Seepage pressures that cause or contribute to slope failures and landslides.
- Deterioration and failure of roads due to the presence of groundwater in the base or subgrade.
- Highway and other fill foundation failures caused by perched groundwater.
- Earth embankment and foundation failures caused by excess pore water pressures.
- Retaining wall failures caused by hydrostatic water pressures.
- Canal linings, drydocks, and basement or spillway slabs uplifted by groundwater pressures.
- Soil liquefaction, caused by earthquake shocks, because of the presence of loose granular soil that is below the groundwater table.
- Transportation of contaminants by the groundwater.

Proper drainage design and construction of drainage facilities can mitigate many of these groundwater problems. For example, for canyon and drainage channels where fill is to be placed, a canyon subdrain system such as shown in Standard Detail No. 1 (App. C) should be installed to prevent the buildup of groundwater in the canyon fill. Figure 16.1 shows a photograph of the installation of a canyon subdrain system. The drain consists of a perforated pipe (perforations on the underside of the pipe) and open graded gravel around the pipe, with the gravel wrapped in a geofabric that is used to prevent the gravel and pipe from being clogged with soil particles.

16.2.1 Pavements

Probably the most common engineering facilities damaged by groundwater are pavements. Table 4.8 presents the drainage properties of different compacted soils. It has been stated that groundwater in pavements accelerates the damage rates by hundreds of times over the damage rates of pavements with no groundwater. This premature failure of thousands of miles of pavements and billions of dollars in losses a year could be avoided by good pavement drainage practices (Cedergren, 1989). The key element in a good drainage system is a layer of highly permeable material (such as open-graded gravel) protected by filters or geofabric so that the permeable material will not become clogged by the intrusion of soil fines. A drainage system, to remove the water from the base, is also required.

There are several ways that groundwater can enter the base material. In areas having a high groundwater table or artesian conditions, water can be forced upward into the base material. Water can also flow downward through pavement cracks or joints. There can also be the development of a perched groundwater condition, where water moves laterally through the base from adjacent planter areas, medians, or shoulders.

Figure 16.2 shows an example of the effect of groundwater on pavement deterioration. At this site, water was observed to be coming up through the pavement as shown in Fig. 16.2. A test pit was excavated in a utility trench, with the result that groundwater bubbled up from the ground as shown in Fig. 16.3. It was observed that the native soil was clayey, but the utility trenches had been backfilled with granular (permeable) soil. At the low points in the streets, the water exited the utility trenches and flowed up through the pavement surface as shown in Fig. 16.2. The continuous flow of groundwater from the utility trenches led to premature deterioration of the pavements.

One construction method to prevent the flow of groundwater through utility trenches is to use a grout (such as a cement slurry) in the utility line or storm drain bedding zone. The remainder of the trench could then be backfilled and compacted with on-site native soil.

FIGURE 16.1 Construction of a canyon subdrain.

This should provide the trench with a permeability equal to or less than the surrounding native soil.

16.2.2 Slopes

Groundwater can affect slopes in different ways. Table 16.1 presents common examples of the influence of groundwater on slope failures. The main destabilizing factors of groundwater on slope stability are as follows (Cedergren, 1989):

FIGURE 16.2 Flow of groundwater through top of pavement.

FIGURE 16.3 Groundwater exiting a test pit excavated into a utility trench.

16.6 ANALYSIS OF GEOTECHNICAL DATA AND ENGINEERING COMPUTATIONS

TABLE 16.1 Common Examples of Slope Failures

Kind of slope (1)	Conditions leading to failure (2)	Type of failure and its consequences (3)
Natural earth slopes above developed land areas (homes, industrial)	Earthquake shocks, heavy rains, snow, freezing and thawing, undercutting at toe, mining excavations	Mud flows, avalanches, landslides; destroying property, burying villages, damming rivers
Natural earth slopes within developed land areas	Undercutting of slopes, heaping fill on unstable slopes, leaky sewers and water lines, lawn sprinkling	Usually slow creep type of failure; breaking water mains, sewers, destroying buildings, roads
Reservoir slopes	Increased soil and rock saturation, raised water table, increased buoyancy, rapid drawdown	Rapid or slow landslides; damaging highways, railways, blocking spillways, leading to overtopping of dams, causing flood damage with serious loss of life
Highway or railway cut or fill slopes	Excessive rain, snow, freezing, thawing, heaping fill on unstable slopes, undercutting, trapping groundwater	Cut slope failures blocking roadways, foundation slipouts removing roadbeds or tracks; property damage, some loss of life
Earth dams and levees, reservoir ridges	High seepage levels, earthquake shocks; poor drainage	Sudden slumps leading to total failure and floods downstream; much loss of life, property damage
Excavations	High groundwater level, insufficient groundwater control, breakdown of dewatering systems	Slope failures or heave of bottoms of excavations; largely delays in construction, equipment loss, property damage

Source: From Cedergren, 1989; reprinted with permission from John Wiley & Sons.

1. Reducing or eliminating cohesive strength.
2. Producing pore water pressures which reduce effective stresses, thereby lowering shear strength.
3. Causing horizontally inclined seepage forces which increase the driving forces and reduce the factor of safety of the slope.
4. Providing for the lubrication of slip surfaces.
5. Trapping of groundwater in soil pores during earthquakes or other severe shocks, which leads to liquefaction failures.

There are many different construction methods to mitigate the effects of groundwater on slopes. During construction of slopes, built-in drainage systems can be installed. For example, Standard Detail No. 4 in App. C shows a drainage detail for the construction of a stabilization fill slope. Figure 16.4 shows the actual installation of the drainage system.

FIGURE 16.4 Construction of drainage system during construction of a fill slope for a mass grading project. Note that the pipe is surrounded by open-graded gravel. The gravel is surrounded by a geofabric to prevent soil from clogging the gravel and drain line.

For existing slopes, drainage devices such as trenches or galleries, relief wells, or horizontal drains can be installed. Another common slope stabilization method is the construction of a drainage buttress at the toe of a slope. In its simplest form, a drainage buttress can consist of cobbles or crushed rock placed at the toe of a slope. The objective of the drainage buttress is to be as heavy as possible to stabilize the toe of the slope and also have a high permeability so that seepage is not trapped in the underlying soil.

16.8 ANALYSIS OF GEOTECHNICAL DATA AND ENGINEERING COMPUTATIONS

There can also be other indirect effects of groundwater on slopes. For example, when the groundwater evaporates at the toe of the slope, salt deposits can form. Both the high groundwater table and the surface salt deposits can kill or stunt the growth of plants and trees. Deposits of salts due to evaporation of groundwater are known as *evaporites*. In arid or semiarid regions where moisture is evaporating at the ground surface, they can form on or just beneath the ground surface. The three most common evaporites are gypsum, anhydrite, and sodium chloride (rock salt). Figure 16.5 shows salt deposited at ground surface due to the evaporation of groundwater at Death Valley Junction, California.

Examples of the effects of groundwater on slope vegetation are shown in Figs. 16.6 and 16.7. Note the relatively sparse vegetation near the toe of the slope. Most of the vegetation at the toe of the slope died as a result of the high groundwater table and salt deposits. As shown in Fig. 16.7, the salt deposits have formed a crust on top of the ground surface.

At the site shown in Figs. 16.6 and 16.7, there was formational material consisting of alternating beds of sandstone, siltstone, and claystone of the Eocene Santiago formation. Because of the adverse out-of-slope dip of the bedding, the slope was faced with fill (i.e., a stabilization fill) and a key was constructed at the toe of the slope. This condition is illustrated in Fig. 16.8, which presents a cross section through the slope. A subdrain was reportedly placed at the back of the fill key, although it may have been improperly installed or become clogged during construction. The arrows in Fig. 16.8 show the path of groundwater through the stabilization fill and up through the toe of the slope where it created the conditions shown in Figs. 16.6 and 16.7. These conditions could have been prevented by the installation of an appropriate drainage system, such as shown in Standard Detail No. 4 (App. C).

FIGURE 16.5 Salts deposited at ground surface from evaporating groundwater (Death Valley Junction, California).

FIGURE 16.6 Toe of slope.

16.3 MOISTURE MIGRATION THROUGH FLOOR SLABS

Moisture migration into buildings is one of the major challenges faced by engineers, architects, and contractors. Many times, the project architect provides waterproofing recommendations. But in other instances, the geotechnical engineer may need to provide recommendations. For example, the *Technical Guidelines for Soil and Geology Reports* (App. B) requires the geotechnical engineer to provide "drainage and backfill requirements including waterproofing of living areas and suitable drains."

Moisture can migrate into the structure through the foundation, exterior walls, and through the roof. Four ways that moisture can penetrate a concrete floor slab are by water vapor, hydrostatic pressure, leakage, and capillary action (WFCA, 1984). Water vapor acts in accordance with the physical laws of gases, traveling from one area to another area whenever there is a difference in vapor pressure between the two areas. Hydrostatic pressure is the buildup of water pressure beneath the floor slab, which can force large quantities of water through slab cracks or joints. Leakage refers to water traveling from a higher to a lower elevation solely by the force of gravity, and such water can surround or flood the area below the slab. Capillary action is different from leakage in that water can travel from a lower to a higher elevation. The controlling factor in the height of capillary rise in soils is the pore size (Holtz and Kovacs, 1981). Open-graded gravel has large pore spaces and hence very low capillary rise. This is why open-graded gravel is frequently placed below the floor slab to act as a capillary break (Butt, 1992).

Moisture that travels through the concrete floor slab can damage such floor coverings as carpet, hardwood, and vinyl. When a concrete floor slab has floor coverings, the moisture can collect at the top of the slab, where it weakens the floor-covering adhesive. Hardwood floors can be severely affected by moisture migration through slabs because

16.10 ANALYSIS OF GEOTECHNICAL DATA AND ENGINEERING COMPUTATIONS

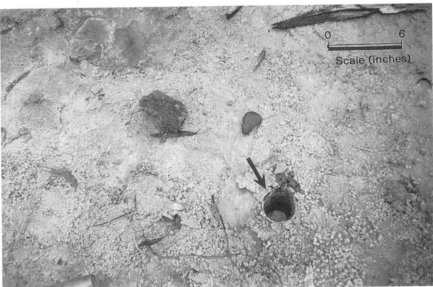

FIGURE 16.7 Toe of slope. The arrow in the upper photograph points to the location of the lower photograph.

they can warp or swell from the moisture. Moisture that penetrates the floor slab can also cause musty odors or mildew growth in the space above the slab. Some people are allergic to mold and mildew spores. They can develop health problems from the continuous exposure to such allergens. In most cases, the moisture that passes through the concrete slab contains dissolved salts. As the water evaporates at the slab surface, the salts form a white crystalline deposit, commonly called efflorescence. The salt can build up underneath the floor covering, where it attacks the adhesive as well as the flooring material itself. Figure 16.9 shows a photograph of salt deposits (white areas) and the growth of mold (dark areas) caused by severe moisture migration through a concrete slab-on-grade foundation.

Oliver (1988) believes that it is the shrinkage cracks in concrete that provide the major pathways for rising dampness. Sealing of slab cracks may be necessary in situations of rising dampness affecting sensitive floor coverings.

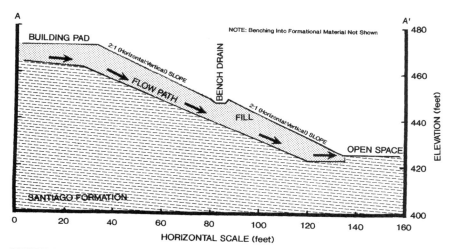

FIGURE 16.8 Cross section A-A'.

FIGURE 16.9 Salt deposits and the growth of mold caused by moisture migration through a concrete floor slab (carpets have been removed).

16.12 ANALYSIS OF GEOTECHNICAL DATA AND ENGINEERING COMPUTATIONS

As an example of damage due to moisture migration through a concrete floor slab, Fig. 16.10 shows damage to wood flooring installed on top of a concrete slab. Only 6 months after completion of the house, the wood flooring developed surface moisture stains and warped upward in some places as much as 150 mm (0.5 ft). Most of the moisture stains developed at the joints where the wood planking had been spliced together. The joints would be the locations where most of the moisture penetrates the wood flooring. In Fig. 16.10, the arrow points to one of the moisture stains. The asterisk in Fig. 16.10 shows the location of the upward warping of the wood floor.

16.3.1 Design and Construction Details

As previously mentioned, the geotechnical engineer often provides the design and construction specifications that are used to prevent both water vapor and capillary rise through floor slabs. An example of below-slab recommendations (WFCA, 1984) is as follows:

> Over the subgrade, place 4 inches (10 cm) to 8 inches (20 cm) of washed and graded gravel. Place a leveling bed of 1 to 2 inches (2.5 to 5 cm) of sand over the gravel to prevent moisture barrier puncture. Place a moisture barrier over the sand leveling bed and seal the joints to prevent moisture penetration. Place a 2 inch (5 cm) sand layer over the moisture barrier.

The gravel layer should consist of open-graded gravel. This means that the gravel should not contain any fines and that all the soil particles are retained on the gravel size sieves. This will provide for large void spaces between the gravel particles. Large void spaces in the gravel will help prevent capillary rise of water through the gravel (Day, 1992d). In addition

FIGURE 16.10 Damage to wood flooring. Arrow points to moisture stain; asterisk indicates upward warping of wood floor.

to a gravel layer, the installation of a moisture barrier (such as visqueen) will further reduce the moisture migration through concrete (Brewer, 1965).

In some areas, there may be local building requirements for the construction of moisture barriers. For example, Fig. 16.11 indicates the County of San Diego requirements for below-slab moisture barriers when constructing on clays. Note in this figure that 4 in. (10 cm) of open-graded gravel or rock is required below the concrete slabs. A 6-mil. visqueen moisture barrier is required below the open-graded gravel. These recommendations are similar to the example listed above, except that the sand layers have been omitted. The purpose of the sand layer is to prevent puncture of the visqueen moisture barrier. When using specifications similar to Fig. 16.11, it is best to use rounded gravel so that the visqueen is not punctured during placement and densification of the gravel.

The below-slab design and construction details presented above will not be able to resist a high groundwater table or artesian groundwater pressure. In these cases, a more extensive waterproofing system will need to be installed. One common approach is to provide a below-foundation waterproofing system and a sump equipped with a pump to catch and dispose of any water that seeps through the waterproofing system. A sump is defined as a small pit excavated in the ground or through the basement floor to serve as a collection basin for surface runoff or groundwater. A sump pump is used to periodically drain the pit when it fills with water. For example, Fig. 16.12 shows a collector box (sump) located below a concrete floor slab. When the water level reaches a certain level in the collector box, the submerged pump will be activated, and the collection box will be emptied of water.

Another possible drainage system to deal with a high groundwater table is shown in Fig. 16.13. This below-grade garage was constructed with a waterproofing system. As an additional measure, a pressure relief system was installed as shown in Fig. 16.13. When groundwater rises, the box (large arrow) is forced upward by any groundwater that seeps through the waterproofing system, and the water then flows to the drain (small arrow) and

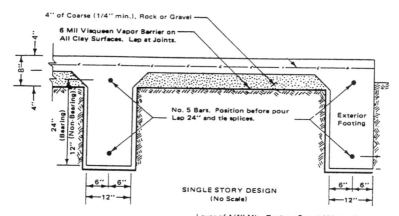

FIGURE 16.11 Moisture barrier specifications according to the County of San Diego (1983).

16.14 ANALYSIS OF GEOTECHNICAL DATA AND ENGINEERING COMPUTATIONS

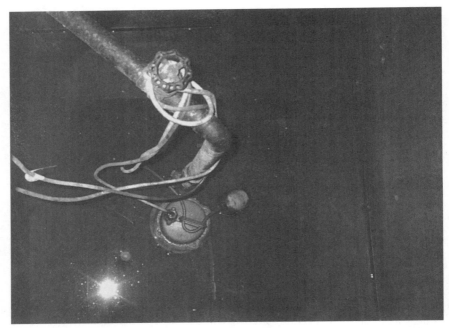

FIGURE 16.12 A sump (collector box) containing a submerged pump for the disposal of groundwater seeping through the waterproofing system located below the foundation.

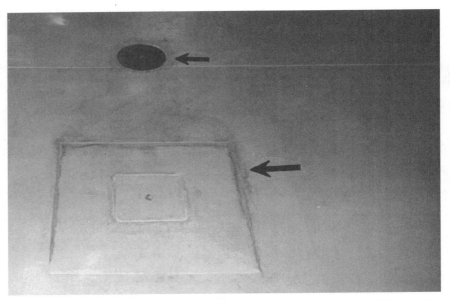

FIGURE 16.13 Groundwater pressure relief system. The large arrow points to a pressure relief cover which is uplifted by the groundwater pressure. The water then flows to the drain indicated by the small arrow.

ultimately to a sump pump for disposal. Note the water stains on the concrete surface, which indicate that the system is functioning.

16.4 MOISTURE MIGRATION THROUGH BASEMENT WALLS

As with concrete floor slabs, water can penetrate basement walls by hydrostatic pressure, capillary action, and water vapor. If a groundwater table exists behind the basement wall, then the wall will be subjected to hydrostatic pressure, which can force large quantities of water through wall cracks or joints. A subdrain is usually placed behind the basement wall to prevent the buildup of hydrostatic pressure. Such drains will be more effective if the wall is backfilled with granular (permeable) soil rather than clay.

Another way for moisture to penetrate basement walls is by capillary action in the soil or the wall itself. By capillary action, water can travel from a lower to higher elevation in the soil or wall. Capillary rise in walls is related to the porosity of the wall and the fine cracks in both the masonry and, especially, the mortar. To prevent moisture migration through basement walls, an internal or surface waterproofing agent is used. Chemicals can be added to cement mixes to act as internal waterproofs. More common are the exterior applied waterproof membranes. Oliver (1988) lists and describes various types of surface-applied waterproof membranes.

A third way that moisture can penetrate through basement walls is by water vapor. As with concrete floor slabs, water vapor can penetrate a basement wall whenever there is a difference in vapor pressure between the two areas.

16.4.1 Types of Damage

Moisture that travels through basement walls can damage wall coverings, such as wood paneling. Moisture traveling through the basement walls can also cause musty odors or mildew growth in the basement areas. If the wall should freeze, then the expansion of freezing water in any cracks or joints may cause deterioration of the wall. As with concrete floor slabs, the moisture that is passing through the basement wall will usually contain dissolved salts. The penetrating water may often contain salts originating from the ground or mineral salts naturally present in the wall materials. As the water evaporates at the interior wall surface, the salts form white crystalline deposits (efflorescence) on the basement walls. Figures 16.14 and 16.15 show photographs of the build-up of salt deposits on the interior surface of basement walls.

The salt crystals can accumulate in cracks or wall pores, where they can cause erosion, flaking, or ultimate deterioration of the contaminated surface. This is because the process of crystallization often involves swelling and considerable forces are generated. Another problem is if the penetrating water contains dissolved sulfates, because they can accumulate and thus increase their concentration on the exposed wall surface, resulting in chemical deterioration of the concrete (ACI, 1990).

The effects of dampness, freezing, and salt deposition are major contributors to the weathering and deterioration of basement walls. Some of the other common deficiencies that contribute to moisture migration through basement walls are as follows (Diaz et al., 1994; Day, 1994f):

1. The wall is poorly constructed (for example, joints are not constructed to be watertight), or poor-quality concrete that is highly porous or shrinks excessively is used.
2. There is no waterproofing membrane or there is a lack of waterproofing on the basement wall exterior.

16.16 ANALYSIS OF GEOTECHNICAL DATA AND ENGINEERING COMPUTATIONS

FIGURE 16.14 Two views of groundwater migration through basement walls at a condominium complex in San Diego, California.

FIGURE 16.15 Two views of groundwater migration through basement walls at a condominium complex in Los Angeles, California.

16.18 ANALYSIS OF GEOTECHNICAL DATA AND ENGINEERING COMPUTATIONS

3. There is improper installation, such as a lack of bond between the membrane and the wall, or deterioration with time of the waterproofing membrane.
4. There is no drain, the drain is clogged, or there is improper installation of the drain behind the basement wall. Clay, rather than granular backfill, is used.
5. There is settlement of the wall, which causes cracking or opening of joints in the basement wall.
6. There is no protection board over the waterproofing membrane. During compaction of the backfill, the waterproofing membrane is damaged.
7. The waterproofing membrane is compromised. This happens, for example, when a hole is drilled through the basement wall.
8. There is poor surface drainage, or downspouts empty adjacent the basement wall.

16.4.2 Design and Construction Details

The main structural design and construction details to prevent moisture migration through basement walls is a drainage system at the base of the wall to prevent the buildup of hydrostatic water pressure and a waterproofing system applied to the exterior wall surface.

A typical drainage system for basement walls is shown in Fig. 16.16. A perforated drain is installed at the bottom of the basement wall footing. Open-graded gravel, wrapped in a geofabric, is used to convey water down to the perforated drain. The drain outlet should be tied to a storm drain system.

The waterproofing system frequently consists of a high-strength membrane, a primer for wall preparation, a liquid membrane for difficult to reach areas, and a mastic to seal holes in the wall. The primer is used to prepare the concrete wall surface for the initial application of the membrane and to provide long-term adhesion of the membrane. A protection board is commonly placed on top of the waterproofing membrane to protect it from damage during compaction of the backfill soil. Self-adhering waterproofing systems have been developed to make the membrane easier and quicker to install.

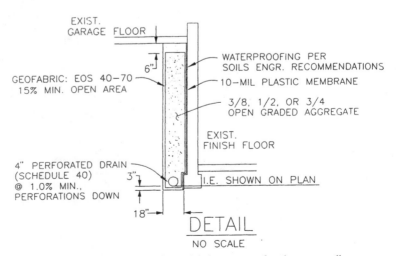

FIGURE 16.16 Typical waterproofing and drainage system for a basement wall.

GROUNDWATER AND MOISTURE MIGRATION **16.19**

16.5 SURFACE DRAINAGE

Inadequate surface drainage can be an important factor in triggering soils problems. For example, as previously discussed, water ponding adjacent to a foundation can contribute to expansive soil edge lift, the infiltration of ponding water can raise the groundwater table, and inadequate surface drainage can contribute to moisture intrusion problems through floor slabs or basement walls.

Figure 16.17 shows an example of poor surface drainage and ponding of water at a condominium complex. In Fig. 16.17, the condominium is visible along the right side of the photograph. In order to prevent such ponding of water at the site, it is important to have proper drainage gradients around the structure. In addition, the surface runoff must be transferred to suitable disposal systems, such as storm drain lines. Standard Detail No. 9 (App. C) shows two alternatives for lot surface drainage. Figure 4.5 shows typical construction specifications for drainage ditches that are used to drain surface runoff from slopes.

16.6 PIPE BREAKS AND CLOGS

Pipes can be classified as either pressurized or nonpressurized. Common pressurized pipes are the water lines that provide potable water to the building occupants. An example of a nonpressurized pipe is a sewer line, which may only be intermittently filled with effluent.

A pressurized pipe break can introduce large volumes of water into the ground. This water can trigger collapse of loose soil, heave of expansive soil, or a raising of the groundwater table which may lead to slope instability. The large volume of water from a broken pressurized pipe

FIGURE 16.17 Poor surface drainage at a condominium complex.

16.20 ANALYSIS OF GEOTECHNICAL DATA AND ENGINEERING COMPUTATIONS

can also erode and transport soil particles, leading to the development of voids below the structure. For example, Figs. 16.18 and 16.19 show two views of the collapse of a street due to the erosion of the base and subgrade caused by a pressurized pipe break.

Structural damage can also develop because of a nonpressurized break. At one building, there was substantial damage to the front bearing wall caused by the sudden subsidence of the ground surface. Subsurface exploration discovered that there was a sewer line that ran underneath the bearing wall. The top of the sewer line was broken and soil had filtered down into the sewer line. Periodic cleaning of the sewer line probably helped to enlarge the void that was developing above the sewer line. Eventually the void collapsed, causing the ground surface subsidence and damage to the overlying bearing wall.

Besides pipe breaks, there can be damage due to pipe clogs. There are many different ways that a pipe can become clogged. For example, the pipe can become clogged with debris or the overburden pressure can crush the pipe. Figure 16.20 shows an example of a clogged storm drain. The storm drain was a critical drainage facility and when it became clogged during a heavy rainstorm, there was extensive flooding and damage to the adjacent area. As shown in Fig. 16.20, the cause of the clog was a sewer main line that was constructed right through the center of the storm drain. The area below the sewer line became clogged with rocks, while the area above the sewer line was plugged by a plastic bottle (see Fig. 16.20). During the heavy rainstorm, the plastic bottle may have floated on top of the storm water and then become lodged in place above the sewer line. The plastic bottle shown in Fig. 16.20 was probably the final plug for the storm drain which then led to the flooding and damage of the surrounding area.

In summary, the geotechnical engineer should determine if the site contains soil conditions, such as expansive soil or collapsible soil, that may be especially susceptible to pipe leaks or breaks. If this is the case, special pipes, connections, or shutoff valves that are triggered by a reduction in water pressure may be required.

FIGURE 16.18 Street collapse due to water line break.

FIGURE 16.19 Close-up view of the street collapse.

FIGURE 16.20 Sewer main constructed through storm drain.

16.7 PERCOLATION TESTS FOR SEWAGE DISPOSAL SYSTEM

The geotechnical engineer is often involved with the design of subsurface sewage disposal systems for building construction in rural areas. This is because the design involves the determination of key geotechnical parameters, such as the location of groundwater and the permeability of the soil. The permeability of the soil is assessed by performing percolation tests.

The procedure for performing percolation tests varies in different localities. Because the specifications vary so much from place to place, the geotechnical engineer should always obtain the governing applicable standards. Appendix D (from the County of San Diego, 1991) presents an example of a percolation test procedure for the design of subsurface sewage disposal systems for single-family residences in rural areas.

The first step in the analysis is to determine the approximate location of the sewage disposal system. Figure 16.21 shows an example of the proposed construction of a single-family dwelling in a rural area. The area delineated as "prop. 23.5′ × 50′ disposal field" in Fig. 16.21 shows the location of the proposed subsurface sewage disposal system.

Sewage from the residence first flows into the septic tank. As indicated in Table D-1 (App. D), the septic tank size is based on the number of bedrooms (hence number of occupants) in the house, with a 1000-gallon septic tank required for a two- to three-bedroom

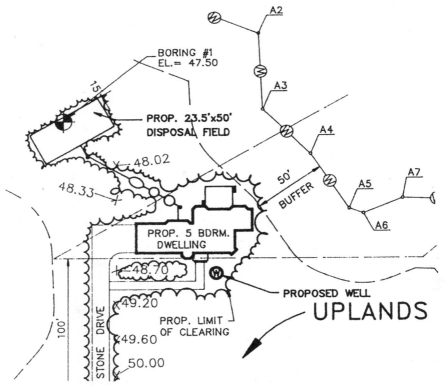

FIGURE 16.21 Plot plan showing the location of the proposed sewage disposal system for a single-family residence.

house and a 1500-gallon septic tank required for a five- to six-bedroom house. After settling of solids and partial decomposition in the septic tank, the effluent drains through a tight line pipe to the leach field. The leach field is constructed by first excavating trenches, then placing perforated pipes that are surrounded by open-graded gravel in the trench, and finally backfilling the trench. In the leach field, the effluent seeps from the perforated pipes and into the ground.

As indicated in App. D, the purpose of the percolation test is as follows:

> Determine the area necessary to properly treat and maintain sewage underground, to size the disposal system with adequate infiltration surface on the basis of expected hydraulic conductivity of the soil and the rate of loading, and to provide for a system with long term expectation of satisfactory performance.

The steps in the percolation test as indicated in App. D are as follows:

1. A cylindrical hole is excavated into the ground. The hole should have a minimum diameter of 6 in. (15 cm) and a maximum diameter of 10 in. (25 cm). The depth of the hole must be at least 14 in. (36 cm) and should correspond to the approximate depth that the leach lines will be installed. Often the hole can be excavated by using a hand auger.

2. A 2-in.- (5-cm-) thick layer of open-graded gravel is placed in the bottom of the hole. Clear water is then added to the hole and the depth of water is maintained at 12 to 14 in. (30 to 36 cm) in order to allow the soil to presoak (see App. D for details on presoaking). Presoaking is important because it allows desiccation cracks in clayey soil to close up and the suction pressures to decrease. If the test were performed without presoaking a clayey soil, a high and unrepresentative percolation rate could be obtained as the water quickly flows through the desiccation cracks.

3. After presoaking is complete, the percolation test is performed. Depending on the results of the presoaking, there are three procedures (Cases No. 1 through 3) that are used to perform the percolation test (see App. D). In general, the percolation test consists of providing an initial height of 6 in. (15 cm) of water above the open-graded gravel and recording the drop of this water level over a time period of 30 minutes.

4. The percolation rate is calculated as 30 minutes divided by the drop in elevation (inches) of the water level over the 30-minute time period. The units of the percolation rate are minutes per inch (mpi). As an example of the calculation of the percolation rate, suppose the water level in the hole drops $3/4$ in. in the 30-minute time period. Then the percolation rate would then be 30 minutes divided by 0.75 in., or 40 mpi.

The average percolation rate is calculated as the sum of the percolation rate at each test hole divided by the number of test holes. From the average percolation rate, two design parameters are obtained:

- *Primary disposal trench length.* For an average percolation rate, the total primary disposal trench length (which contains the perforated pipe for the leach field) can be determined by using Table D-1 (App. D).
- *Lot size.* As indicated in Table D-2 (App. D), the average percolation rate determines the minimum required lot size. For example, if the average percolation rate is 61 to 90 mpi, then the minimum lot size is 3 acres. If the average percolation rate is 91 to 120 mpi, then the minimum lot size is 5 acres.

There are many site factors that may need to be considered in the design of the sewage disposal system. For example, site factors that could impact the design include a high groundwater table, placement of the leach field on a steep slope, nitrate impacts on vegetation growth, and a confined soil layer having a high horizontal permeability which could transport the effluent onto adjacent property.

In summary, the percolation test is a straightforward and simple test that is widely used in the United States to design sewage disposal systems for rural properties. Besides its simplicity, the percolation test has an advantage in that it is a field test that measures the actual performance of the *in situ* soil in terms of water infiltration. Disadvantages are that the design is empirical and there are many site factors besides soil percolation that can impact a sewage disposal system.

PROBLEMS

1. A proposed highway project is planned to consist of a porous asphalt concrete surface layer underlain by an 8-in-thick open-graded gravel base layer. The base material will be in contact with a drain system located in the shoulder of the highway. During heavy rainstorms, it is anticipated that the infiltration of water into the base will be 0.5 ft^3 per minute per linear foot length of the highway. Assume the open graded gravel has a coefficient of permeability k of 20 ft/min and the slope of the base to the drain pipe is 2 percent (i.e., $i = 2\%$). Will the base be able to absorb this rate of water infiltration? *Answer:* No, because Q for the base when flowing full = 0.27 ft^3/min.

2. In Fig. 16.8, assume the Santiago formation is relatively impervious compared to the fill. Also assume that the fill has a coefficient of permeability (k) = 0.3 ft/day, the fill-bedrock contact is at a 2:1 (horizontal:vertical) ratio, the perched groundwater table extends from the Santiago formation to a height of 7 ft (vertical distance) above the Santiago formation, and that the slope is 200 ft long. Determine the quantity of groundwater that will flow out of the toe of the slope per day assuming the drainage system at the back of the key is inoperable. *Answer:* 168 ft^3/day.

3. A cantilevered retaining wall (3 m in height) has a granular backfill with $\phi = 30°$ and $\gamma_t = 20$ kN/m^3. Neglect wall friction and assume the drainage system fails and the water level rises 1.5 m above the bottom of the retaining wall. Determine the initial active earth pressure resultant force P_A, the resultant force (due to earth plus water pressure) on the wall due to the rise in water level, and the percent increase in force against the wall due to the rise in water level. *Answers:* $P_A = 30$ kN/m (initial condition). With a rise in water level, the force acting on the wall = 37.3 kN/m, representing a 24 percent increase in force acting on the retaining wall.

4. A single-family dwelling will be constructed in rural San Diego county. The house will contain five bedrooms, and the average percolation test recorded a $^1/_4$-in. drop in water level in a 30-minute time interval. Determine the average percolation rate, septic tank size, required minimum lot size, and primary disposal line length. *Answers:* average percolation rate = 120 mpi, required septic tank size = 1500 gallons, required minimum lot size = 5 acres, and primary disposal leach line length = 1420 ft.

PART · 4

PERFORMANCE OR ENGINEERING EVALUATION OF CONSTRUCTION

A scraper, used for both the excavation and compaction of fill.

CHAPTER 17
GRADING

The following notation is used in this chapter:

SYMBOL	DEFINITION
G	Bulk specific gravity of oversize particles
M_{ds}	Dry mass of the matrix soil
M_o	Mass of the oversize particles
RC	Relative compaction
s	Standard deviation
V	Total volume of the excavated hole (sand cone test)
w	Water content (also known as the moisture content)
w_{opt}	Optimum water content, also known as the optimum moisture content
ρ_a	Average value of the laboratory maximum dry density
ρ_d	Field dry density
ρ_{dm}	Dry density of the matrix material
$\rho_{d\,max}$	Laboratory maximum dry density
ρ_s	Density of soil particles
ρ_w	Density of water

17.1 INTRODUCTION

Chapters 6 through 16 (Part 3 of the book) deal with the analysis of geotechnical data and engineering computations. As indicated in Fig. 1.11, Part 4 of the book (Chaps. 17 and 18) deals with the performance or engineering evaluation of construction. At this stage of the project, the subsurface exploration, laboratory testing, and engineering analysis for the project are complete. Normally, the geotechnical engineer and engineering geologist will prepare a report presenting the results of the subsurface investigation and design recommendations as required for the project. These types of reports are often known as preliminary or feasibility reports. In many cases, the report would follow a format similar to that as indicated in App. E. Report preparation will be discussed in the final chapter (Chap. 19).

Since most building sites start out as raw land, the first step in site construction work usually involves the grading of the site. Grading is defined as any operation consisting of excavation, filling, or combination thereof. The Glossary (App. A, Glossary 4) presents a list of common construction and grading terms and their definitions.

Most projects involve grading and it is an important part of geotechnical engineering. For example, the *Orange County Grading Manual* (1993) states:

The soil engineer's area of responsibility shall include, but need not be limited to, the professional inspection and approval concerning the preparation of ground to receive fills, testing for required compaction, stability of all finish slopes, design of buttress fills, subdrain installation and incorporation of data supplied by the engineering geologist.

17.2 GRADING SPECIFICATIONS

It is important to prepare a set of grading specifications for the project. These specifications are often used to develop the grading plans, which are basically a series of maps that indicate the type and extent of grading work to be performed at the site. Appendix C presents an example of grading specifications. Included at the end of the grading specifications in App. C are Standard Details Nos. 1 through 9, which illustrate the proper grading procedure for various activities. Often the grading specifications will be included as an appendix in the preliminary or feasibility report prepared by the geotechnical engineer and engineering geologist.

A typical grading process would include the following:

1. *Easements.* The first step in the grading operation is to determine the location of any on-site utilities and easements. The on-site utilities and easements often need protection so that they are not damaged during the grading operation.
2. *Clearing, brushing, and grubbing.* Clearing, brushing, and grubbing are defined as the removal of vegetation (grass, brush, trees, and similar plant types) by mechanical means. This debris is often stockpiled at the site, such as shown in Fig. 17.1. It is important that this debris be removed from the site and not accidentally placed within the structural fill mass.
3. *Cleanouts.* This grading process deals with the removal of unsuitable bearing material at the site, such as loose or porous alluvium, colluvium, and uncompacted fill.
4. *Benching (hillside areas).* Benching is defined as the excavation of relatively level steps into earth material on which fill is to be placed. Standard Details Nos. 1 through 6 (App. C) show benching operations required for various grading operations.
5. *Canyon subdrain.* A subdrain is defined as a pipe-and-gravel or similar drainage system placed in the alignment of canyons or former drainage channels. After placement of the subdrain, structural fill is placed on top of the subdrain. See App. C, Standard Detail No. 1, for details on the construction of a canyon subdrain.
6. *Scarifying and recompaction.* In flat areas that have not been benched, scarifying and recompaction of the ground surface is performed by compaction equipment in order to get a good bond between the in-place material and compacted fill.
7. *Cut and fill rough grading operations.* Rough grading operations involve the cutting of earth materials from high areas and compaction of fill in low areas, in conformance with grading plans. Other activities could be performed during rough grading operations, such as:
    a. *Ripping or blasting of rock.* Large rock fragments can be removed from the site or disposed of in windrows. Standard Detail No. 7 (App. C) illustrates the construction of a windrow.
    b. *Cut/fill transition.* Figure 7.5 illustrates a cut/fill transition. It is the location in a building pad where on one side the pad has been cut down exposing natural or rock material, while on the other side, fill has been placed. Standard Detail No. 6 (App. C) presents one method to deal with cut/fill transitions.
    c. *Slope stabilization.* Examples of slope stabilization using earth materials include stabilization fill (Standard Detail No. 3, App. C), buttress fill (Standard Detail No. 3, App. C), drainage buttress, and shear keys. Such devices should be equipped with backdrain systems.
    d. *Fill slopes.* In creating a fill slope, it is often difficult to compact the outer edge of the fill mass. For example, Fig. 17.2 shows the attempted compaction at the slope face.

FIGURE 17.1 Brush stockpile from grading operation.

Because there is no confining pressure, the soil deforms downslope without increasing in density, as shown in Fig. 17.2. To deal with this situation, the slope can be overbuilt and then cut back to the compacted core as illustrated in Fig. 17.3. The second best alternative is to use conventional construction procedures such as back-rolling techniques or by using a bulldozer to track-walk the slope (see Fig. 17.4).
  e. *Revision of grading operations.* Every grading job is different and there could be a change in grading operations based on field conditions.
8. *Fine grading (also known as precise grading).* At the completion of the rough grading operations, fine grading is performed to obtain the finish elevations in accordance with the precise grading plan.
9. *Slope protection and erosion control.* Although usually not the responsibility of the grading contractor, upon completion of the fine grading, slope protection and permanent erosion control devices are installed.
10. *Trench excavations.* Utility trenches are excavated in the proposed road alignments and building pads for the installation of the on-site utilities. The excavation and compaction of utility trenches is often part of the grading process. Once the utility lines are installed, scarifying and recompaction of the road subgrade is performed and base material is placed and compacted.
11. *Footing and foundation excavations.* Although usually not part of the grading operation, the footing and foundation elements can be excavated as shown in Fig. 1.1.

As previously mentioned, the geotechnical engineer and engineering geologist are actively involved during the grading process. Typical activities include observing the grading operation as well as performing compaction tests during the placement of fill. Some of the key grading activities that should be attended or observed by the geotechnical engineer and engineering geologist are as follows:

17.6 PERFORMANCE OR ENGINEERING EVALUATION OF CONSTRUCTION

FIGURE 17.2 Lack of compaction at the slope face.

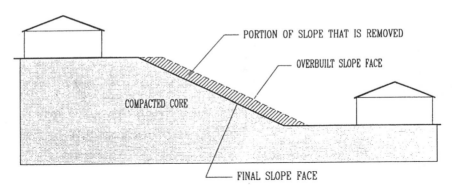

FIGURE 17.3 Illustration of overbuilding and cutting back slope face.

- *Pregrading meeting.* A pregrading meeting should be arranged between the geotechnical engineer, engineering geologist, grading contractor, client, and local building inspector. The purpose of the pregrading meeting is to review the grading plans and specifications, discuss the construction schedule such as the dates when key activities will be started and finished, and evaluate potential problems that may develop during the grading.
- *Initial observation.* Usually the pregrading meeting is held at the site. This will give the geotechnical engineer and engineering geologist an opportunity to view the site and note any changes since the subsurface exploration, such as recently dumped debris, erosion, etc.
- *Clearing, grubbing, and canyon cleanout.* A site visit should be planned as a final check that all brush and unsuitable material has been removed from the site. This site visit should be performed before fill placement operations are started.

- *Rough grading observations.* During rough grading of the site, the geotechnical engineer and engineering geologist should observe key grading activities, such as ripping, benching, keying of slopes, construction of earth structures (buttresses, fill stabilization, etc.), treatment of cut/fill transitions, and installation of subsurface drainage systems. The compaction process must also be reviewed and field density tests taken to assess the quality of the fill.
- *Final grade observation.* The final condition of the site should be observed at the completion of grading.

17.3 COMPACTION FUNDAMENTALS

An important part of the grading of the site is the compaction of fill. Compaction is defined as the densification of a fill by mechanical means. This physical process of getting the soil into a dense state can increase the shear strength, decrease the compressibility, and decrease the permeability of the soil. There are four basic factors that affect compaction of fill, as follows:

1. *Soil type.* Nonplastic (i.e., cohesionless) soil, such as sands and gravels, can be effectively compacted by using a vibrating or shaking type of compaction operation. Plastic (i.e., cohesive) soil, such as silts and clays, is more difficult to compact and requires a kneading or manipulation type of compaction operation. If the soil contains oversize particles, such as coarse gravel and cobbles, they tend to interfere with the compaction process and reduce the effectiveness of compaction for the finer soil particles. Typical values of dry density for different types of compacted soil are listed in Table 4.8.

2. *Material gradation.* Those soils that have a well-graded grain size distribution can generally be compacted into a denser state than a poorly graded soil that is composed of soil particles of about the same size. For example, a well-graded decomposed granite (DG) can have a maximum dry density of 2.2 Mg/m^3 (137 pcf), while a poorly graded sand can have a maximum dry density of only 1.6 Mg/m^3 (100 pcf, Modified Proctor).

3. *Water content.* The water content is an important parameter in the compaction of soil. Water tends to lubricate the soil particles, thus helping them slide into dense arrangements. However, too much water and the soil becomes saturated and often difficult to compact. There is an optimum water content at which the soil can be compacted into its densest state for a given compaction energy. Typical optimum moisture contents (Modified Proctor) for different soil types are as follows:

  a. Clay of high plasticity (CH): optimum moisture content $\geq 18\%$
  b. Clay of low plasticity (CL): optimum moisture content = 12 to 18%
  c. Well-graded sand (SW): optimum moisture content = 10%
  d. Well-graded gravel (GW): optimum moisture content = 7%

Some soils may be relatively insensitive to compaction water content. For example, open-graded gravels and clean coarse sands are so permeable that water simply drains out of the soil or is forced out of the soil during the compaction process. These types of soil can often be placed in a dry state and then vibrated into dense particle arrangements.

4. *Compaction effort (or energy).* The compactive effort is a measure of the mechanical energy applied to the soil. Usually the greater the amount of compaction energy applied to a soil, the denser the soil will become. There are exceptions, such as pumping soils (discussed in

Sec. 17.6) which can not be densified by an increased compaction effort. Compactors are designed to use one or a combination of the following types of compaction effort:

   a. Static weight or pressure
   b. Kneading action or manipulation
   c. Impact or a sharp blow
   d. Vibration or shaking

17.3.1 Laboratory Compaction Testing

The laboratory compaction test consists of compacting a soil at a known water content into a mold of specific dimensions using a certain compaction energy. The procedure is repeated for various water contents to establish the compaction curve. The most common testing procedures (compaction energy, number of soil layers in the mold, etc.) are the Modified Proctor (ASTM D 1557-91, 1998) and the Standard Proctor (ASTM D 698-91, 1998). The term *Proctor* is in honor of R. R. Proctor, who in 1933 showed that the dry density of a soil for a given compactive effort depends on the amount of water the soil contains during compaction.

In California, there is almost exclusive use of the Modified Proctor compaction specifications. For the Modified Proctor (ASTM D 1557-91, 1998, procedure A), the soil is compacted into a 10.2-cm- (4-in.-) diameter mold that has a volume of 944 cm^3 (^{1}/$_{30}$ ft^3), where five layers of soil are compacted into the mold with each layer receiving 25 blows from a 44.5-N (10-lbf) hammer that has a 0.46-m (18-in.) drop. The Modified Proctor has a compaction energy of 2,700 kN-m/m^3 (56,000 ft-lbf/ft^3). The test procedure is to prepare soil at a certain water content, compact the soil into the mold; then, by recording the mass of soil within the mold, the wet density of the compacted soil is obtained. From the water content of the compacted soil, the dry density can be calculated (i.e., divide the wet density by $1 + w$, where w is in a decimal form). This compaction procedure is repeated for the soil at different water contents, and then the data is plotted on a graph in order to obtain the compaction curve.

Figure 17.5 shows the compaction curve for a nonplastic silty sand. The compaction curve shows the relationship between the dry density and water content for a given compaction effort. The compaction data presented in Fig. 17.5 was obtained by using the Modified Proctor specifications. Note in Fig. 17.5 that the vertical axis is dry density and the horizontal axis is water content. The density of the dry soil particles is of interest, which is why the dry density, and not wet density, is the vertical axis.

Four data points, representing four individual soil specimens compacted into the Proctor mold, are shown in Fig. 17.5. The four data points are connected in order to create the compaction curve. The three lines to the right of the compaction curve are each known as a *zero air voids curve*. These curves represent the relationship between water content and dry density for a condition of saturation ($S = 100$ percent) for a specified specific gravity. Note how the right side of the compaction curve is approximately parallel to the zero air voids curve. This is often the case for many soil types and can be used as a check on the laboratory test results.

The peak point of the compaction curve is the laboratory maximum dry density. In Fig. 17.5, the laboratory maximum dry density is 122.5 pcf (1.96 Mg/m^3). The water content corresponding to the laboratory maximum dry density is known as the *optimum moisture content*. In Fig. 17.5, the optimum moisture content w_{opt} for this soil is 11 percent. This laboratory data is important because it tells the grading contractor that the moisture content of the soil should be about 11 percent for the most efficient compaction of the soil. Also, as indicated in App. C, the grading specifications often require compaction "near optimum," which means that fill should be compacted at a water content that is from 1 percent below to 3 percent above optimum (i.e., for the data in Fig. 17.5, $w = 10\%$ to 14%).

FIGURE 17.4 Slope face compacted by track-walking techniques.

17.3.2 Field Compaction

For mass grading operations, usually heavy construction equipment, such as shown in Figs. 17.6 to 17.10, is used to excavate and compact fill. The three most common types of equipment used for grading operations are the bulldozer, scraper, and trucks:

1. *Bulldozer (Figs. 17.6 and 17.7).* The bulldozer is used to clear the land of debris and vegetation (clearing, brushing, and grubbing), excavate soil from the borrow area, cut haul roads, spread out dumped fill, and compact the soil.
2. *Scraper (Figs. 17.8 and 17.9).* The scraper is used to excavate ("scrape up") soil from the borrow area, transport it to the site, and dump it at the site, and the rubber tires of the scraper can be used to compact the soil. Push-pull scrapers can be used in tandem to provide additional energy to excavate hard soil or soft rock.
3. *Trucks.* If the borrow area is quite a distance from the site, then dump trucks may be required to transport the borrow soil to the site. Dump trucks are also needed to transport soil on public roads or to import select material. Especially in the arid climate of the southwestern United States, soil taken from the borrow area may be in a dry state and water will need to be added to the soil to bring it up to optimum moisture content. Water trucks are typically used for this operation.

The *Caterpillar Performance Handbook* (1997), which is available at Caterpillar dealerships, is a valuable reference because it not only lists rippability versus types of equipment (see Sec. 4.7), but also indicates types and models of compaction equipment, equipment sizes and dimensions, and performance specifications. Compaction equipment can generally be grouped into five main categories, as follows:

17.10 PERFORMANCE OR ENGINEERING EVALUATION OF CONSTRUCTION

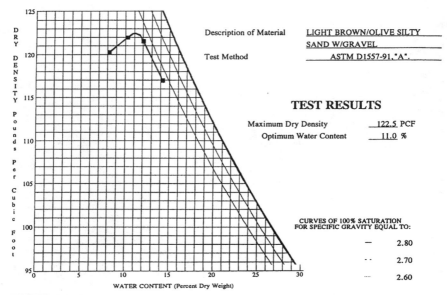

FIGURE 17.5 Compaction curve for a silty sand based on the Modified Proctor test specifications.

1. *Static weight or pressure.* This type of compaction equipment applies a static or relatively uniform pressure to the soil. Examples include the compaction by the rubber tires of a scraper, the tracks of a bulldozer, and a smooth drum roller.
2. *Kneading action or manipulation.* The sheepsfoot roller, which has round or rectangular-shaped protrusions or "feet," is ideally suited to applying a kneading action to the soil. This has proven to be effective in compacting silts and clays.
3. *Impact or a sharp blow.* There are compaction devices, such as the high-speed tamping foot and the Caterpillar tamping foot, that compact the soil by imparting an impact or sharp blow to the soil.
4. *Vibration or shaking.* Nonplastic sands and gravels can be effectively compacted by vibrations or shaking. An example is the smooth drum vibratory soil compactor.
5. *Chopper wheels.* This type of compaction equipment has been specially developed for the compaction of waste products at municipal landfills.

Table 4.8 presents a summary of different types of compaction equipment best suited to compact different types of soil.

A common objective of the grading operations is to balance the volume of cut and fill. This means that just enough earth material is cut from the high areas to fill in the low areas. A balanced cut and fill operation means that no soil needs to be imported or exported from the site, leading to a reduced cost of the grading operation.

Typical specifications for fill placement and compaction are indicated in App. C. Fill placement should proceed in thin lifts, i.e., 15 to 20 cm (6 to 8 in.) loose thickness. Each lift should be moisture-conditioned and thoroughly compacted. The desired moisture condition should be maintained or reestablished, where necessary, during the period between successive lifts. Selected lifts should be tested to ascertain that the desired compaction is being achieved.

FIGURE 17.6 Bulldozer spreading out fill for compaction.

FIGURE 17.7 Bulldozer in the process of compacting fill.

17.12 PERFORMANCE OR ENGINEERING EVALUATION OF CONSTRUCTION

FIGURE 17.8 Scraper in operation.

FIGURE 17.9 Close-up view of the portion of the scraper where the soil is stored during transport.

FIGURE 17.10 An excavator and soil compactor being used on a mass grading project.

There are many excellent publications on field compaction. For example, *Moving the Earth* (Nichols and Day, 1999) presents an in-depth discussion of the practical aspects of earth moving equipment and earthwork operations.

17.3.3 Relative Compaction

The most common method of assessing the quality of the field compaction is to calculate the relative compaction (RC) of the fill, defined as:

$$RC = \frac{100 \, \rho_d}{\rho_{d\,max}} \qquad (17.1)$$

where $\rho_{d\,max}$ = laboratory maximum dry density and ρ_d = field dry density.

In California, typical mass grading specifications require a minimum relative compaction of 90 percent. For example, the common fill compaction specifications adopted by city and county agencies are from the *Uniform Building Code* (1997), which states: "All fills shall be compacted to a minimum of 90 percent of maximum density." In some cases, such as for the compaction of roadway base or for the lower portions of deep fill, a higher compaction standard of a minimum relative compaction of 95 percent is specified.

As discussed in Sec. 17.3.1, the maximum dry density ($\rho_{d\,max}$) is obtained from the laboratory compaction curve. In order to determine ρ_d for Eq. (17.1), a field density test must be performed. Field density tests can be classified as either destructive or nondestructive tests (Holtz and Kovacs, 1981). Probably the most common destructive method of determining the field dry density is through the use of the sand cone apparatus (ASTM D 1556-96, 1998). The test procedure consists of excavating a hole in the ground, filling the hole with sand using the sand cone apparatus, and then determining the volume of the hole from the amount of sand required to fill the hole. By dividing the wet mass of soil removed from the hole by the volume of the hole, the wet density of the fill can be calculated. The water content w of the soil extracted from the hole can be determined and thus the dry density ρ_d can then be calculated.

Another type of destructive test for determining the field dry density is the drive cylinder (ASTM D 2937-94, 1998). This method involves the driving of a steel cylinder of known volume into the soil. From the mass of soil within the cylinder, the wet density can be calculated. Once the water content w of the soil is obtained, the dry density ρ_d of the fill can be calculated.

Probably the most common type of nondestructive field test is the nuclear method (ASTM D 2922-96, 1998). In this method, the wet density is determined by the attenuation of gamma radiation. The nuclear method can give inaccurate results (density too high) when oversize particles, such as coarse gravel and cobbles, are present. Likewise, if there is a large void in the source-detector path, then unusually low density values may be recorded.

NAVFAC DM-7.2 (1982) presents guidelines on the number of field density tests for different types of grading projects, as follows:

- One test for every 380 m³ (500 yd³) of material placed for embankment construction.
- One test for every 380 to 750 m³ (500 to 1000 yd³) of material for canal or reservoir linings or other relatively thin fill sections.
- One test for every 75 to 150 m³ (100 to 200 yd³) of backfill in trenches or around structures, depending upon total quantity of material involved.
- At least one test for every full shift of compaction operations on mass earthwork.
- One test whenever there is a definite suspicion of a change in the quality of moisture control or effectiveness of compaction.

There are many different guidelines concerning the number of field density tests for specific grading activities. For example, the Grading Specifications presented in App. C indicate that field density tests for mass graded structural fill should be taken at about every 2 vertical feet or 750 m³ (1000 yd³) of fill placed and that the actual test interval may vary as field conditions dictate.

It is rare for the licensed geotechnical engineer to perform field density testing on a daily basis because of the repetitive and time-consuming nature of such work. For large mass grading operations, it is common to have technicians performing the field density testing. The technician will have to be able to perform the field density tests, classify different soil types (by visual and tactile methods), and insist on remedial measures when compaction falls below the specifications.

17.4 TYPES OF FILL

As previously mentioned, fill is defined as a deposit of earth material placed by artificial means. There are different types of fill, as follows (Monahan, 1986; Greenfield and Shen, 1992):

1. *Engineered (or structural) fill.* This refers to a fill in which the geotechnical engineer has, during grading, made sufficient tests to enable the conclusion that the fill has been placed in substantial compliance with the recommendations of the geotechnical engineer and the governing agency requirements. Standard Detail No. 5 in App. C shows typical canyon fill placement specifications. Structural fills are used to support all types of structures.
2. *Hydraulic fill* This refers to a fill placed by transporting soils through a pipe using large quantities of water. These fills are generally loose because they have little or no mechanical compaction during construction.

3. *Dumped or uncontrolled fill.* This refers to fill that was not documented with compaction testing as they were placed or fill that may have been compacted but there is no documentation of testing or the amount of effort that was used to perform the compaction. Dumped or uncontrolled fill should not be used to support structures.

In placing structural fill, it is usually not economical or possible from a time standpoint to perform a laboratory maximum dry density test for every field density test. NAVFAC DM-7.2 (1982) recommends a laboratory maximum dry density test for every 10 to 20 field density tests, depending on the variability of the materials. For mass grading operations in southern California, it is common to have a much higher ratio, such as one laboratory maximum dry density test for every 60 to 70 field density tests.

The typical situation in placing structural fill is that the field technician will have a family of compaction curves corresponding to different soils. It is then the technician's responsibility to select the appropriate laboratory maximum dry density corresponding to the field tested soil. The technician must use experience and judgment to match up the soil types from the laboratory maximum dry density tests with the field soil. When several different laboratory maximum dry density tests are given for the same general soil type, it is common for the technician to select a laboratory maximum dry density that will provide a passing result. For example, if the field dry density is low, a low maximum dry density is selected. If the field dry density is high, a high maximum dry density is selected. The result is fill that may not meet project specifications because of the uncertainty in matching the laboratory maximum dry density with the fill soil type. The purpose of the remainder of this section is to discuss types of fill and the selection of the appropriate laboratory maximum dry density for each fill type.

Structural fill can generally be divided into four basic types, as follows: select (processed) import, uniform borrow, mixed borrow, and borrow having oversize particles. These four types of structural fill are individually discussed below.

Select Import. Select import refers to a processed material. The material may be derived from several different sources, then screened and mixed to provide a material of specified gradation. For example, a common select import is granular base material, which may have to meet specifications for gradation, wear resistance, and shear strength (*Standard Specifications for Public Works Construction,* 1997). Other uses for select import include backfill for retaining walls and utilities, and even for mass graded fill.

The main characteristics of select import are a well-graded granular soil, which has a high laboratory maximum dry density, typically in the range of 2.0 to 2.2 Mg/m^3 (125 to 135 pcf). As a processed material, the particle size gradation for each batch of fill should be similar. Usually an import material will have all laboratory maximum dry density values within 0.05 Mg/m^3 (3 pcf) and a standard deviation of 0.02 Mg/m^3 (1 pcf) or less. Since results of the laboratory maximum dry density are within a narrow band, a field technician could use either the average or highest value of the laboratory maximum dry density without much effect on the relative compaction.

Uniform Borrow. Uniform borrow typically refers to a natural material that will consistently have the same soil classification and similar grain size distribution. An example of a possible uniform borrow could be a natural deposit of beach sand. Other uniform borrow could be formational rock, such as deposits of sandstone or siltstone.

Figure 17.11 presents laboratory test results on a uniform borrow material. The fill was derived from a formational rock, classified as a weakly cemented shale. When used as fill, the material is classified as a silty clay, having a liquid limit between 41 and 50. The laboratory maximum dry density varies from 1.84 to 1.97 Mg/m^3 (115 to 123 pcf), with an average value of 1.92 Mg/m^3 (120 pcf) and a standard deviation of 0.035 Mg/m^3 (2.2 pcf).

As the name implies, the main characteristic of the material is its uniformity. Usually a uniform borrow material has consistently the same soil classification, with all laboratory maximum dry density values within 0.13 Mg/m³ (8 pcf) and a standard deviation of 0.05 Mg/m³ (3 pcf) or less.

For uniform borrow soil, such as that represented by the data shown in Fig. 17.11, it is often not possible to match a specific laboratory maximum dry density to the soil in the field. This is because the grain size distribution curves and the plasticity characteristics are too similar to distinguish the soil based on visual or tactile methods. Unless the maximum dry density happens to be performed on exactly the same soil tested in the field, it

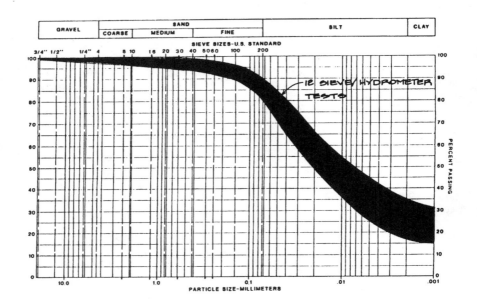

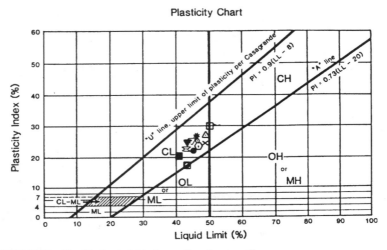

FIGURE 17.11 Classification test data for a uniform borrow soil.

is not possible to accurately match the laboratory maximum dry density to the soil tested in the field.

The maximum error in relative compaction for a uniform borrow is about 6 percent, or a difference in relative compaction from 84 percent to 90 percent. In order to have an error on the safe side, one of the higher maximum dry density values should be selected. For example, the following value of laboratory maximum dry density $\rho_{d\,max}$ could be used:

$$\rho_{d\,max} = \rho_a + s \qquad (17.2)$$

where ρ_a = average value of laboratory maximum dry density and s = standard deviation. When Eq. (17.2) is used, the maximum error on the unsafe side for uniform borrow is about 1 percent, or a difference in relative compaction from 89 percent to 90 percent.

Mixed Borrow. Mixed borrow contains material of different classifications. For example, mixed borrow could be a deposit of alluvium that contains alternating layers of sand, silt, and clay. Mixed borrow could also be formational rock that contains thin alternating layers of sandstone and claystone. The main characteristic of mixed borrow is that each load of fill could have soils with significantly different grain size distributions and soil classifications. For example, Fig. 17.12 shows a photograph of a fill that was derived from a mixed borrow. The fill contains many different soil types, all jumbled up and mixed together.

One method to deal with mixed borrow material is to thoroughly mix each load of import and then perform a laboratory maximum dry density test on that batch of import soil. Another option is to thoroughly mix each batch of mixed borrow material and then perform a one-point Proctor test (Holtz and Kovacs, 1981). This method consists of performing a single-point maximum dry density test (in the field) on soil that has a water content that is dry of optimum moisture content. The maximum dry density is then estimated on the basis of the observation that compaction curves on similar soil types have the same basic shape. This procedure is illustrated in Fig. 17.13 where point A is the one-point Proctor test performed on the soil that has a water content dry of optimum and then the compaction curve is drawn. Note in Fig. 17.13 that the laboratory maximum dry density (point B) is obtained by using the line of optimums, which is a line drawn through the peak point of the compaction curves.

Borrow with Oversize Particles. The last basic type of fill is borrow having oversize particles, which are typically defined as those particles retained on the 3/4-in. (19-mm) U.S. standard sieve, i.e., coarse gravel and cobble-size particles. The soil matrix is defined as those soil particles that pass the 3/4-in. (19-mm) U.S. standard sieve. When a field density test (such as a sand cone test) is performed, the soil excavated for the test can be sieved on the 3/4-in. (19-mm) sieve in order to determine the mass of oversize particles. The elimination method (Day, 1989) can then be used to mathematically eliminate the volume of oversize particles in order to calculate the dry density of the matrix material (ρ_{dm}), by using the following equation:

$$\rho_{dm} = \frac{M_{ds}}{V - M_o/(G\,\rho_w)} \qquad (17.3)$$

where

M_{ds} = dry mass of the matrix soil
V = total volume of the excavated hole
M_o = mass of oversize particles
G = bulk specific gravity of the oversize particles
ρ_w = density of water

The relative compaction is calculated by dividing the dry density of the matrix material [ρ_{dm}, Eq. (17.3)] by the laboratory maximum dry density, where the laboratory compaction test is performed on the matrix material. By using the elimination method to calculate the rel-

17.18 PERFORMANCE OR ENGINEERING EVALUATION OF CONSTRUCTION

FIGURE 17.12 Fill that was derived from a mixed borrow source.

ative compaction of the matrix material, the compaction state of the matrix soil is controlled. This is desirable because it is the matrix soil (not the oversize particles) that usually governs the compressibility, shear strength, and permeability of the soil mass.

If the matrix soil can be considered to be a uniform borrow material, then the procedure for selecting the laboratory maximum dry density in the field is the same as previously discussed for uniform borrow material. If the matrix material is a mixed borrow material, then the procedure for selecting the laboratory maximum dry density in the field is the same as previously discussed for a mixed borrow material. Other methods have been developed to deal with fill containing oversize particles (Saxena et al., 1984; Houston and Walsh, 1993).

Checking Field Compaction. As previously mentioned, for mass grading projects, the number of field density tests per volume of compacted fill is often very low (i.e., one field density test per 1000 yd^3 of fill). It is important that the field technician perform the density tests on areas where compaction is suspect. For example, the technician should not perform field compaction tests in the haul road area, because this path receives continuous traffic and will usually be in a dense compacted state. Likewise, testing in the wheel paths of the compaction equipment will yield high values. Often the field technician uses a metal rod to probe for possible poorly compacted fill zones. Field density tests would then be performed in these areas of possible poor compaction.

17.5 UTILITY TRENCH AND BACKFILL COMPACTION

As discussed in Sec. 17.3, heavy compaction equipment is commonly used to compact fill for mass grading operations. But many projects require backfill compaction where the construction space is too small to allow for such heavy equipment. For example, Fig. 17.14 shows a bulldozer pushing fill into a drain line trench during the mass grading of a project. As shown in Fig. 17.14, there is not enough space for the bulldozer to effectively compact the soil in the trench.

Hand-operated vibratory plate compactors, such as shown in Fig. 17.15, are ideally suited for compacting fill in small or tight spaces. The hand-operated compactors can be used for all types of restricted access areas, such as the compaction of fill in utility trenches, behind basement walls, or around storm drain access vaults such as shown in Figure 17.16. Recommendations for the compaction of storm drain backfill, as well as the width of the storm drain trench, the size of the pipe bedding zone, and soil type for the pipe bedding, have been presented in Fig. 4.4.

For many projects having small or tight spaces, the backfill is simply dumped in place or compacted with minimal compaction effort. These factors of limited access, poor compaction process, and lack of compaction testing frequently lead to backfill settlement. At the site shown in Fig. 17.17, gravel was simply dumped behind the garage walls and then stairs were constructed atop the gravel backfill. Vibrations due to subsequent construction and the migration of water down through the gravel caused the loose gravel to settle, resulting in damage to the stairs, as shown in Fig. 17.17. The settlement of the

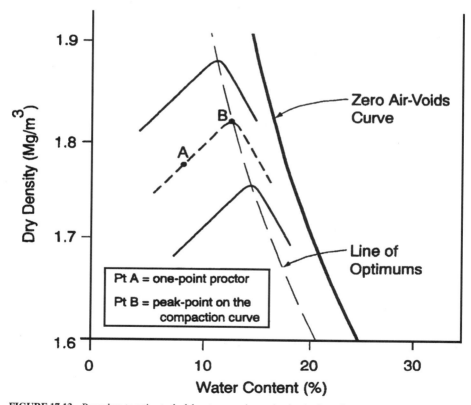

FIGURE 17.13 Procedure to estimate the laboratory maximum dry density from the one-point Proctor test.

FIGURE 17.14 Bulldozer pushing fill into a drain line trench.

FIGURE 17.15 Hand-operated vibratory plate compactor used for compacting fill in tight spaces.

FIGURE 17.16 Storm drain access vault. Note that wood forms are being used to confine the fluid concrete. After removal of forms and compaction of fill around the concrete vault, asphalt will be placed on top of the concrete vault and flush with the top of the round metal ring.

FIGURE 17.17 Damage to stairs caused by backfill settlement.

gravel backfill could have been avoided if the gravel had been placed in layers and then compacted by using a vibrating hand tamper.

17.6 PUMPING

Pumping is a form of bearing capacity failure that occurs during compaction of fill. A commonly used definition of pumping is the softening and squeezing of clay from underneath the compaction equipment. Continual passes of the compaction equipment can cause a decrease in the undrained shear strength of the wet clay, and the pumping may progressively worsen. Figures 17.18 and 17.19 show the pumping of a wet clay subgrade.

Pumping is dependent on the penetration resistance of the compacted clay. Figure 17.20 (from Turnbull and Foster, 1956) presents data on the California bearing ratio (CBR) of compacted clay and shows that the penetration resistance approaches zero (i.e., the clay can exhibit pumping) when the clay has a water content that is wet of optimum. Also note in the lower part of Fig. 17.20 that the laboratory maximum dry density increases and the optimum moisture content decreases as the compaction energy increases (more blows per layer).

Besides clay, pumping can occur in all types of plastic soils. Near- or at-surface plastic soils will absorb water from rainstorms. As shown in Figs. 17.21 to 17.24, the absorbed water makes the plastic soil soft and fluid-like and very difficult to compact. The rainwater can also cause construction slopes to slump, as shown in Fig. 17.25. Any attempt to compact such soil would result in pumping, rather than densification, of the plastic soil.

There are many different methods to stabilize pumping soil. The most commonly used method is to simply allow the plastic soil to dry out. Other methods include adding a chem-

FIGURE 17.18 Pumping of clay subgrade during construction.

FIGURE 17.19 Pumping of clay subgrade during construction.

ical agent (such as lime) to the clay or placing a geotextile on top of the pumping clay to stabilize its surface (Winterkorn and Fang, 1975).

Another common procedure to stabilize pumping clay is to add gravel to the clay. The typical procedure is to dump angular gravel at ground surface and then work it in from the surface. The angular gravel produces a granular skeleton, which then increases both the undrained shear strength and penetration resistance of the mixture (Day, 1996b).

17.7 ADJACENT PROPERTY DAMAGE

This final section of Chap. 17 briefly discusses damage to adjacent property due to grading operations. If the grading operations will be adjacent to developed property, the geotechnical engineer and engineering geologist should evaluate the possibility of damage to off-site facilities. Many lawsuits are initiated when off-site property is damaged during the grading operations.

Figures 17.26 and 17.27 show an example of a grading operation that damaged off-site property. Two views of the grading process, which consisted of the cutting and filling of the ground, are shown in Fig. 17.26. When the toe of a slope was cut into during the grading operation, there was a resulting slope failure as shown in Fig. 17.27. The arrow at the top of the photograph indicates the original level of the backyard of a homeowner's property. The lower arrow points to the ground that was originally located at the uppermost

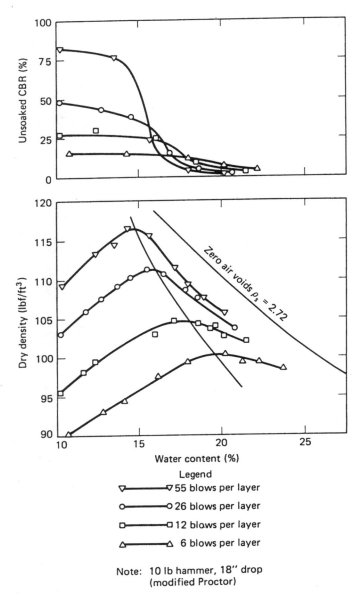

FIGURE 17.20 California bearing ratio (CBR) versus water content for a compacted clay. (*From Turnbull and Foster, 1956; reprinted with permission from the American Society of Civil Engineers.*)

FIGURE 17.21 Rainwater ponded at mass grading project. Note idle compaction equipment in the distance.

FIGURE 17.22 Rainwater ponded at a mass grading project. Note the idle water truck in the upper right corner of the photograph.

FIGURE 17.23 Soft and fluid-like state of plastic soil caused by rainwater at a mass grading project.

FIGURE 17.24 Compaction wheel ruts caused by pumping of plastic soil at a mass grading project.

FIGURE 17.25 Slump of a construction slope caused by rainwater.

arrow. The vertical distance between the two arrows is the large distance that the ground surface dropped down. In such cases, the stability of the off-site land must be investigated prior to grading, and mitigation measures must be adopted to prevent damage to off-site property such as shown in Fig. 17.27.

PROBLEMS

The problems have been divided into basic categories as indicated below:

Laboratory Compaction Test

1. Using the laboratory compaction test data shown in Fig. 17.5, calculate the degree of saturation for soil having a dry density and water content that correspond to the peak point of the compaction curve. Assume the specific gravity of solids (G) = 2.65. *Answer: S* = 83%.

2. The following data was obtained from a laboratory compaction test performed in accordance with the standard Proctor compaction specifications ($1/30$-ft^3 mold):

Test no.	Compacted wet soil (from mold), lb	Water content of soil, %
1	4.14	11.0
2	4.26	12.5
3	4.37	14.0
4	4.33	15.5

17.28 PERFORMANCE OR ENGINEERING EVALUATION OF CONSTRUCTION

FIGURE 17.26 Two views of the grading operation.

FIGURE 17.27 Off-site damage caused by the grading operation. The vertical distance between the two arrows is the distance that the ground surface dropped down.

Determine the laboratory maximum dry density and optimum moisture content. *Answers:* $\rho_{d\,max} = 115$ pcf (1.84 Mg/m³), $w_{opt} = 14.0\%$.

3. If a higher compaction energy is to be used for soil pertaining to Prob. 2, what will happen to the laboratory maximum dry density and optimum moisture content? *Answer:* Higher compaction energy results in a higher laboratory maximum dry density and a lower optimum moisture content.

4. Use the data shown in Fig. 17.5 and assume that the line of optimums is parallel to the zero air voids curve. On a slightly different soil, a one-point Proctor test (Fig. 17.13) dry of optimum has a dry density of 117 pcf at a water content of 8.0%. Estimate the laboratory maximum dry density of this slightly different soil. *Answer:* $\rho_{d\,max} = 120$ pcf (1.92 Mg/m³).

Field Density Tests

5. During grading, a sand cone test was performed on fill. The following data was obtained:

   Volume of hole: 2000 cm³

   Mass of soil removed from hole: 4.0 kg

   Water content of soil: 8.3%

   Determine the field dry density ρ_d. *Answer:* $\rho_d = 1.85$ Mg/m³ (115 pcf).

6. During grading, a 12-in.-diameter sand cone apparatus was used to perform a field density test on a fill that contains oversize particles. The following data was obtained:

Volume of hole: 11,200 cm³
Mass of wet matrix soil plus oversize particles removed from hole: 22.4 kg
Mass of oversize particles: 8.6 kg
Water content of matrix soil: 9.3%

Assume that the oversize particles are dense and hard rock with negligible absorption and have a bulk specific gravity = 2.67. Determine the percent oversize particles, based on dry weight, and the field dry density of the matrix soil (ρ_{dm}). *Answer:* Percent oversize particles = 40.5%, dry density of matrix soil (ρ_{dm}) = 1.58 Mg/m³ (98.8 pcf).

Relative Compaction

7. Using the results from Prob. 5 and the laboratory compaction curve shown in Fig. 17.5, calculate the relative compaction. *Answer:* RC = 94%.
8. Using the results from Prob. 6 and assuming the compaction curve shown in Fig. 17.5 was performed on the matrix soil, calculate the relative compaction of the matrix soil. *Answer:* RC = 81%.

Borrow Computations

9. Project specifications require a relative compaction of 95 percent (modified Proctor). Construction of a highway embankment requires 10,000 yd³ of fill. Assume the borrow soil has an *in situ* dry density of 94 pcf. Also assume that this borrow soil has the laboratory compaction curve as shown in Fig. 17.5. Determine the total volume of soil that must be excavated from the borrow area. *Answer:* 12,380 yd³.
10. Project specifications require a relative compaction of 90 percent (modified Proctor). Construction of a building pad requires 5000 yd³ of fill. Assume the borrow soil has an *in situ* wet density of 128 pcf and an *in situ* water content of 6.5 percent. Also assume that this borrow soil has the laboratory compaction curve shown in Fig. 17.5. Determine the total volume of soil that must be excavated from the borrow area. *Answer:* 4590 yd³.
11. Use the data from Prob. 9 and assume that the water content of the *in situ* borrow soil is 8.0 percent and that the embankment fill must be compacted at optimum moisture content (w_{opt} = 11.0%). How much water (gallons) must be added to the soil during compaction? *Answer:* 113,000 gal.

CHAPTER 18
CONSTRUCTION SERVICES

18.1 INTRODUCTION

Chapter 17 deals with site grading work, which includes the testing of fill to determine if the compaction meets the project specifications. Site grading work is a major construction activity that involves geotechnical engineers and engineering geologists. At the end of grading, there is often additional field work, such as tests performed on the compacted road subgrade (CBR, R-value tests, etc.) or the testing of the rough-graded building pads to determine the presence of expansive soils (expansion index tests).

Usually there are additional services by the geotechnical engineer during the actual construction of the project. Examples of these types of services are as follows:

1. *Field observations concerning load or performance tests.* There are numerous types of field load or performance tests. For example, load tests are common for pile foundations and are used to determine their load-carrying capacity. This process involves driving or installing the pile to the desired depth. Then the load is applied to the pile and the deformation behavior of the pile is measured. ASTM standards for the field testing of piles have been developed. Typical pile load tests and the applicable ASTM standard are listed below:

- Static Axial Compressive Load Test (ASTM D 1143-94, 1998)
- Static Axial Compressive Load Test for Piles in Permafrost (ASTM D 5780-95, 1998)
- Static Axial Tension Load Test (ASTM D 3689-90, 1998)
- Lateral Load Test (ASTM D 3966-90, 1998)
- Dynamic Testing of Piles (ASTM D 4945-89, 1998)

Besides load tests, there can be all types of performance tests during construction that will need to be observed by the geotechnical engineer. For example, percolation tests (Sec. 16.7) may be performed for the design of septic-tank disposal fields. Another example is the field testing of tieback anchors as discussed in Sec. 15.8.

2. *Field observations concerning unusual or unanticipated conditions.* Unusual or unanticipated conditions during construction often lead to delays and additional expenses. If the construction personnel observe any signs of impending earth failure, such as ground cracks, ground subsidence, or unanticipated seepage, then it is important that the geotechnical engineer and engineering geologist review the site conditions. Unanticipated earth movement can affect all types of construction projects. For example, Fig. 18.1 shows a slope failure during the grading of a project. The process of grading can undermine the toe of a slope or surcharge the head of a slope, leading to a failure such as shown in Fig. 18.1.

FIGURE 18.1 Slope failure during the grading of a site. The backpack located in the middle of the photograph provides a scale for the size of the ground cracks.

3. *Field observations to confirm subsurface conditions.* During the construction of the project, the geotechnical engineer will often be asked or required to review the soil or rock bearing conditions of the foundation. For example, Fig. 18.2 shows the excavation of a pier where a socket has been excavated into the rock. A light has been lowered into the pier to observe the socket portion of the pier. Figure 18.3 shows another example of the type of field observations that would be required to confirm bearing conditions. At the site shown in Fig. 18.3, rain water has flooded the footing trenches. This water and loose soil or mud at the bottom of the footing trench must be cleaned out prior to placement of steel and concrete. Note also in Fig. 18.3 that the steel reinforcement has been prefabricated at ground surface and once the footings are dry and clean, the prefabricated steel reinforcement sections will be lifted and lowered into the footing trench.

4. *Field observations to check dimensions of geotechnical elements.* This type of service involves measuring the dimensions of geotechnical elements (such as the depth and width of footings) to make sure that they conform to the requirements of the construction plans. This service is often performed at the same time as the field observation to confirm bearing conditions.

5. *Field observations for essential structures.* For an essential structure (Sec. 2.2), the client or governing authorities may require that the geotechnical engineer be on site the entire time that the geotechnical elements of the project are being constructed. This is to allow documentation of the entire construction process and visual observation of all geotechnical conditions that could impact the project.

In many cases, field observations to confirm bearing conditions and check foundation dimensions will be required by the local building department. In addition, a letter indicating the outcome of the observations must be prepared by the engineer to satisfy the local building department. Building departments often refer to these types of reports as *final inspection*

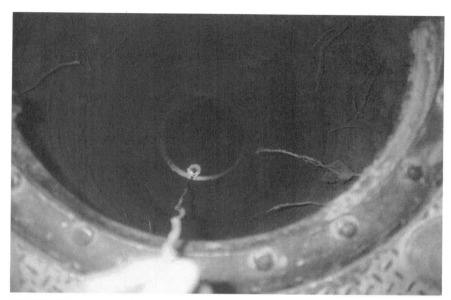

FIGURE 18.2 Pier excavation. A light has been lowered into the hole to observe the socket portion of the pier.

FIGURE 18.3 Footing excavations. Note the flooded condition of the footing trenches.

18.4 PERFORMANCE OR ENGINEERING EVALUATION OF CONSTRUCTION

reports. The local building department often considers these reports to be so important that they may not issue a certificate of occupancy until the reports have been submitted and accepted. An example of such a report for the construction of a foundation is as follows:

> *Footing Observations:* The footings at the site have been observed and are generally in conformance with the approved building plans. Additionally, the footings have been approved for installing steel reinforcement and the soil conditions are substantially in conformance with those observed during the subsurface exploration. Furthermore, the footing excavations extend to the proper depth and bearing strata. Care should be taken to keep all loose soil and debris out of the footing excavations prior to placement of concrete.

18.2 MONITORING

Another broad category of construction services performed by geotechnical engineers is the installation of monitoring devices. There are many types of monitoring devices used by geotechnical engineers. The usual purpose of the installation of monitoring devices is to measure the performance of the structure as it is being built. Monitoring devices could also be installed to monitor existing adjacent structures, groundwater conditions, or slopes that may be impacted by the new construction.

Monitoring devices are especially important in urban areas where there are often adjacent structures that could be damaged by the construction activities. Damage to an adjacent structure can result in a lawsuit. Frequent causes of damage to an adjacent structure include the lowering of the groundwater table or lateral movement of temporary underground shoring systems. Monitoring devices are essential for adjacent historic structures, which tend to be brittle and easily damaged, and can be very expensive to repair. For example, Feld and Carper (1997) describe the construction of the John Hancock Tower that damaged the adjacent historic Trinity Church in Boston. According to court records, the retaining walls (used for the construction of the John Hancock basement) moved 84 cm (33 in.) as the foundation was under construction in 1969. This movement of the retaining walls caused the adjacent street to sink 46 cm (18 in.) and caused the foundation of the adjacent Trinity Church to shift, which resulted in structural damage and 13 cm (5 in.) of tilting of the central tower. The resulting lawsuit was settled in 1984 for about $12 million. An interesting feature of this lawsuit was that the Trinity Church was irreparably damaged, and the damage award was based on the cost of completely demolishing and reconstructing the historic masonry building (ASCE, 1987; *Engineering News Record,* 1987).

Some of the more common monitoring devices are as follows:

Inclinometers. The horizontal movement preceding or during the movement of slopes can be investigated by successive surveys of the shape and position of flexible vertical casings installed in the ground (Terzaghi and Peck, 1967). The surveys are performed by lowering an inclinometer probe into the flexible vertical casing. The inclinometer probe is capable of measuring its deviation from the vertical. An initial survey (base reading) is performed, and then successive surveys are compared to the base reading to obtain the horizontal movement of the slope.

Figure 18.4 (from Slope Indicator, 1998) shows a sketch of the inclinometer probe in the casing and the calculations used to obtain the lateral deformation. Inclinometers are often installed to monitor the performance of earth dams and during the excavation and grading of slopes where lateral movement might affect off-site structures. Inclinometers are also routinely installed to monitor the lateral ground movement due to the excavation of building basements and underground tunnels.

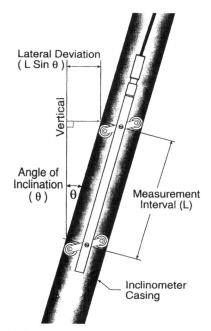

FIGURE 18.4 Inclinometer probe in a casing. (*Reprinted with permission from the Slope Indicator Company.*)

Piezometers. Piezometers are installed in order to monitor pore water pressures in the ground. Several different types are commercially available, including borehole, embankment, and push-in piezometers. Figure 18.5 (from Slope Indicator, 1998) shows an example of a borehole piezometer.

In their simplest form, piezometers can consist of a standpipe that can be used to monitor groundwater levels and obtain groundwater samples. Figure 18.6 (from Slope Indicator, 1998) shows an example of a standpipe piezometer. It is standard procedure to install piezometers when an urban project requires dewatering in order to make excavations below the groundwater table. Piezometers are also used to monitor the performance of earth dams and dissipation of excess pore water pressure associated with the consolidation of soft clay deposits.

Settlement Monuments or Cells. Settlement monuments or settlement cells can be used to monitor settlement or heave. Figure 18.7 (from Slope Indicator, 1998) shows a diagram of the installation of a pneumatic settlement cell and plate. More advanced equipment includes settlement systems installed in borings that can not only measure total settlement, but also the incremental settlement at different depths. Settlement monuments or cells are often installed to measure the deformation of the foundation or embankments during construction or to monitor the movement of existing structures that are located adjacent the area of construction.

Pressure and Load Cells. A total pressure cell measures the sum of the effective stress and pore water pressure. The total pressure cell can be manufactured from two circular plates of stainless steel. The edges of the plates are welded together to form a sealed cavity, which is filled with fluid. Then a pressure transducer is attached to the cell. The total pressure acting on the sensitive surface is transmitted to the fluid inside the cell and measured by the pressure transducer (Slope Indicator, 1998).

Total pressure cells are often used to monitor total pressure exerted on a structure to verify design assumptions and to monitor the magnitude, distribution, and orientation of stresses. For example, load cells are commonly installed during the construction of earth dams to monitor the stresses within the dam core. During construction of the earth dam, the total pressure cells are often installed in arrays with each cell being placed in a different orientation and then covered with compacted fill. For the monitoring of earth pressure on retaining walls, the total pressure cell is typically placed into a recess so that the sensitive side is flush with the retaining wall surface.

Load cells are similar in principle to total pressure cells. They can be used for many different types of geotechnical engineering projects. For example, Fig. 18.8 (from Slope

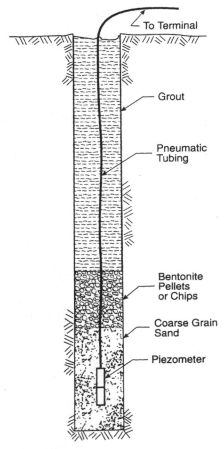

FIGURE 18.5 Pneumatic piezometer installed in a borehole. (*Reprinted with permission from the Slope Indicator Company.*)

Indicator, 1998) shows a center-hole load cell which is designed to measure loads in tiebacks. This center-hole load cell can also be used to measure loads in rock bolts and cables. As shown in Fig. 18.8, for best results, the load cell is centered on the tieback bar and bearing plates are placed above and below the cell. The bearing plates must be able to distribute the load without bending or yielding.

Crack Monitoring Devices. For construction in congested urban areas, it is essential to monitor the performance of adjacent buildings, especially if they already have cracking. This can often be the case in historic districts of cities, where old buildings may be in a weakened or cracked state. Monitoring of existing cracks in adjacent buildings should be performed where there is pile driving or blasting at the construction site. The blasting of rock could be for the construction of an underground basement or for the construction of road cuts. People are often upset by the noise and vibrations from pile driving and blasting and will claim damage due to the vibrations from these construction activities. By monitoring the width of existing cracks, the geotechnical engineer will be able to evaluate these claims of damage.

A simple method to measure the widening of cracks in concrete or brickwork is to install crack pins on both sides of the crack. By periodically measuring the distance between the pins, the amount of opening or closing of the crack can be determined.

Other crack monitoring devices are commercially available. For example, Fig. 18.9 shows two types of crack monitoring devices. For the Avongard crack monitoring device, there are two installation procedures: (1) the ends of the device are anchored by the use of bolts or screws or (2) the ends of the device are anchored with epoxy adhesive. The center of the Avongard crack monitoring device is held together with clear tape, which is cut once the ends of the monitoring device have been securely fastened with bolts, screws, or epoxy adhesive.

Other Monitoring Devices. There are many other types of monitoring devices that can be used by the geotechnical engineer. Some commercially available devices include borehole and tape extensometers, soil strainmeters, beam sensors and tiltmeters, and strain gauges.

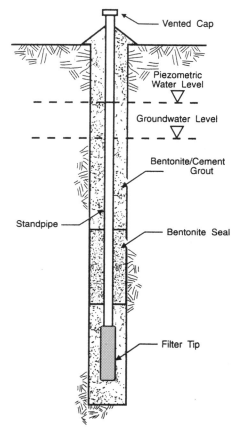

FIGURE 18.6 Standpipe (Casagrande) piezometer. (*Reprinted with permission from the Slope Indicator Company.*)

18.3 OBSERVATIONAL METHOD

Concerning the observational method, Terzaghi and Peck (1967) state:

> Design on the basis of the most unfavorable assumptions is inevitably uneconomical, but no other procedure provides the designer in advance of construction with the assurance that the soil-supported structure will not develop unanticipated defects. However, if the project permits modifications of the design during construction, important savings can be made by designing on the basis of the most probable rather than the most unfavorable possibilities. The gaps in the available information are filled by observations during construction, and the design is modified in accordance with the findings. This basis of design may be called the observational procedure.

18.8 PERFORMANCE OR ENGINEERING EVALUATION OF CONSTRUCTION

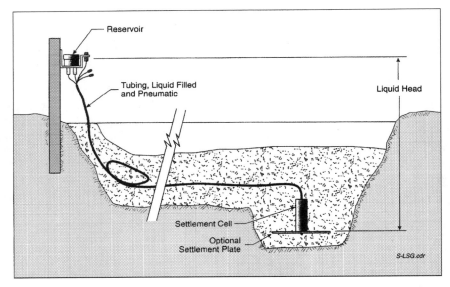

FIGURE 18.7 Pneumatic settlement cell installation. (*Reprinted with permission from the Slope Indicator Company.*)

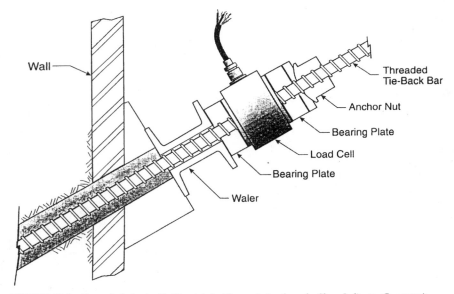

FIGURE 18.8 Center-hole load cell. (*Reprinted with permission from the Slope Indicator Company.*)

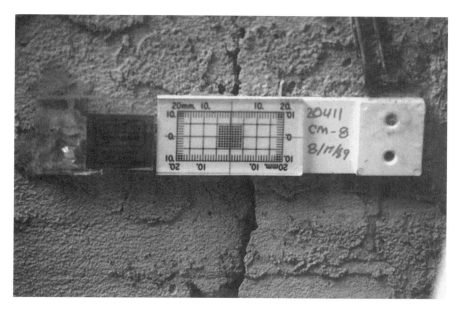

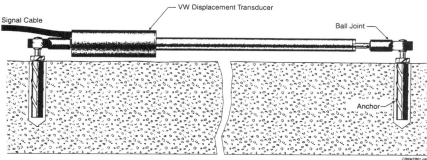

FIGURE 18.9 Crack monitoring devices. The upper photograph shows the Avongard crack monitoring device. The lower diagram shows the VW Crackmeter. (*Reprinted with permission from the Slope Indicator Company.*)

. . . In order to use the observational procedure in earthwork engineering, two requirements must be satisfied. First of all, the presence and general characteristics of the weak zones must be disclosed by the results of the subsoil exploration in advance of construction. Secondly, special provisions must be made to secure quantitative information concerning the undesirable characteristics of these zones during construction before it is too late to modify the design in accordance with the findings.

As mentioned above, the observational method is used during the construction of the project. It is a valuable technique because it allows the geotechnical engineer, on the basis of observations and testing during construction, to revise the design and provide a more economical foundation or earth structure. The method is often used during the installation of deep foundations, where field performance testing or observations are essential in confirming that the foundation is bearing on the appropriate strata.

In some cases, the observational method can be misunderstood or misused. For example, at one project the geotechnical engineer discovered the presence of a shallow groundwater table and indicated in the feasibility report that the best approach would be to use the observational method, where the available information on the groundwater table would be supplemented by observations during construction. When the excavation for the underground garage was made, extensive groundwater control was required and an expensive dewatering system was installed. But the cost of the expensive dewatering system had not been anticipated by the client. The client was very upset at the high cost of the dewatering system because a simple design change of using the first floor of the building as the garage (i.e., above-grade garage) and adding an extra floor to the building would have been much less expensive than dealing with the groundwater. As the client stated: "If I had known that the below-grade garage was going to cost this much, I would never have attempted to construct it."

As this case illustrates, the observational method must not be used in place of a plan of action, but rather to make the plan more economical in keeping with observed subsurface conditions during construction. The client must understand that the savings associated with the observational method may be offset by construction delays resulting from the redesign of the project.

18.4 UNDERPINNING

This final section of Part 4 deals with underpinning. There are many different situations where a structure may need to be underpinned. Tomlinson (1986) lists several underpinning possibilities:

- To support a structure that is sinking or tilting due to ground subsidence or instability of the superstructure.
- As a safeguard against possible settlement of a structure when excavating close to and below its foundation level.
- To support a structure while making alterations to its foundation or main supporting members.
- To enable the foundations to be deepened for structural reasons, for example to construct a basement beneath a building.
- To increase the width of a foundation to permit heavier loads to be carried, for example, when increasing the story height of a building.
- To enable a building to be moved bodily to a new site.

Tomlinson (1986) also indicates that each underpinning project is unique and requires highly skilled personnel, and therefore it should be attempted only by experienced firms. Because each job is different, individual consideration of the most economical and safest scheme is required for each project. Common methods of underpinning include the construction of continuous strip foundations, piers, and piles. To facilitate the underpinning process, the ground can be temporarily stabilized by freezing the ground or by injecting grout or chemicals into the soil (Tomlinson, 1986). Further discussion is presented in *Underpinning* by Prentis and White (1950) and *Underpinning* by Thorburn and Hutchison (1985).

Underpinning by Using a Reinforced Mat or Reinforced Mat Supported by Piers. The remainder of this section will discuss underpinning to support a structure that is sinking or tilting because of ground subsidence or instability of the superstructure. The most expensive and rigorous method of underpinning would be to entirely remove the existing foundation and install a new foundation. This method of repair is usually reserved for cases involving a large magnitude of soil movement. For example, Fig. 18.10 shows the manometer survey (i.e., a floor-level survey) of a building containing two condominium units at a project called Timberlane in Scripps Ranch, California. The building shown in Fig. 18.10 was constructed in 1977 and was underlain by poorly compacted fill that increased in depth toward the front of the building. In 1987, the amount of fill settlement was estimated to be 100 mm (4 in.) at the rear of the building and 200 mm (8 in.) at the front of the building. As shown in Fig. 18.10, the fill settlement caused 80 mm (3.2 in.) of differential settlement for the conventional slab-on-grade and 99 mm (3.9 in.) for the second floor. The reason the second floor had more differential settlement was because it extended out over the garage. Note in Fig. 18.10 that the foundation tilts downward from the rear to the front of the building, or in the direction of deepening fill. Typical damage consisted of cracks in the slab-on-grade, exterior stucco cracks, interior wallboard damage, ceiling cracks, and racked door frames. From Table 7.2, the damage was classified as severe. Because of ongoing fill settlement, the future (additional) differential settlement of the foundation was estimated to be 100 mm (4 in.).

In order to reduce the potential for future damage due to the anticipated fill settlement, it was decided to underpin the building by installing a new foundation. The type of new foundation for the building was a reinforced mat, 380 mm (15 in.) thick, and reinforced with No. 7 bars, 305 mm (12 in.) on center, each way, top and bottom. In order to install the reinforced mat, the connections between the building and the existing slab-on-grade were severed and the entire building was raised about 2.4 m (8 ft). Figure 18.11 shows the building in its raised condition. Steel beams, passing through the entire building, were used

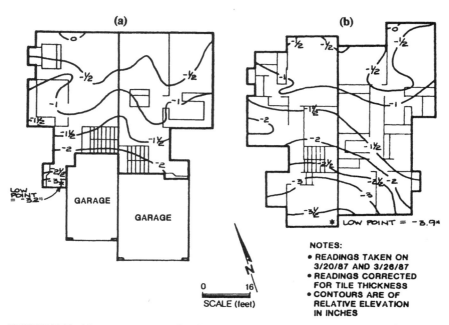

FIGURE 18.10 Manometer survey: (*a*) first floor; (*b*) second floor.

FIGURE 18.11 Raised building.

to lift the building during the jacking process. After the building was raised, the existing slab-on-grade foundation was demolished. The formwork for the construction of the reinforced mat is shown in Fig. 18.12. The mat was designed and constructed so that it sloped 50 mm (2 in.) upward from the back to the front of the building. It was anticipated that with future settlement, the front of the building would settle 100 mm (4 in.) such that the mat would eventually slope 50 mm (2 in.) downward from the back to the front of the building.

After placement and hardening of the new concrete for the mat, the building was lowered onto its new foundation. The building was then attached to the mat and the interior and exterior damages were repaired. Flexible utility connections were used to accommodate the difference in movement between the building and settling fill.

Another underpinning option is to remove the existing foundation and install a mat supported by piers. The mat transfers building loads to the piers, which are embedded in a firm bearing material. For a condition of soil settlement, the piers will usually be subjected to down-drag loads from the settling soil.

The piers are usually at least 0.6 m (2 ft) in diameter to enable downhole logging to confirm end-bearing conditions. The piers can either be built within the building, or the piers can be constructed outside the building with grade beams used to transfer loads to the piers. Given the height of a drill rig, it is usually difficult to drill within the building (unless it is raised). The advantages of constructing the piers outside the building are that the height restriction is no longer a concern and a large, powerful drill rig can be used to quickly and economically drill the holes for the piers.

Figure 18.13 shows a photograph of the conditions at an adjacent building at Timberlane. Given the very large magnitude of the estimated future differential settlement for this building, it was decided to remove the existing foundation and then construct a mat supported by 0.76 m (2.5 ft) diameter piers. The arrow in Fig. 18.13 points to one of the piers.

In order to construct the mat supported by piers, the building was raised and then the slab-on-grade was demolished. With the building in a raised condition, a drill rig was used

FIGURE 18.12 Construction of mat foundation.

FIGURE 18.13 Construction of mat supported by piers. The arrow points to one of the piers.

to excavate the piers. The piers were drilled through the poorly compacted fill and into the underlying bedrock. The piers were belled at the bottom in order to develop additional end-bearing resistance. After drilling and installation of the steel reinforcement consisting of eight No. 6 bars with No. 4 ties at 0.3-m (1-ft) spacing, the piers were filled with concrete to near ground surface. Figure 18.14 shows a close-up of the pier indicated in Fig. 18.13. Note the bent steel reinforcement (No. 6 bars) at the top of the pier which is connected to the steel reinforcement in the mat.

Besides piers, the foundation can also be underpinned with piles or screw or earth anchors (Brown, 1992). Greenfield and Shen (1992) presents a list of the advantages and disadvantages of pier and pile underpinning installations.

Underpinning by Using Continuous Strip Foundations. Another type of underpinning consists of the construction of continuous strip foundations. This is usually a less expensive and rigorous method of underpinning than the reinforced mat described above. The underpinning by using continuous strip footings is often used when the amount of soil movement is less or the depth of problem soil is shallower than the conditions described above.

In addition to underpinning the foundation, this option can be used to fix the damaged foundation and then strengthen the foundation so that the damage does not reoccur. Underpinning by using continuous strip footings is a common type of repair for damage caused by expansive soil (Chen, 1988).

Figure 18.15 shows a cross section of a typical design for underpinning using a continuous strip footing. The construction of the footing starts with the excavation of slots in order to install the hydraulic jacks. The hydraulic jacks are used to temporarily support the foundation until the entire footing is exposed. Steel reinforcement is then tied to the existing foundation by using dowels. The final step is to fill the excavation with concrete. The jacks are left in place during the placement of the concrete. Figure 18.16 shows the installation of the underpinning of a structure using a continuous strip footing.

FIGURE 18.14 Close-up view of Fig. 18.13.

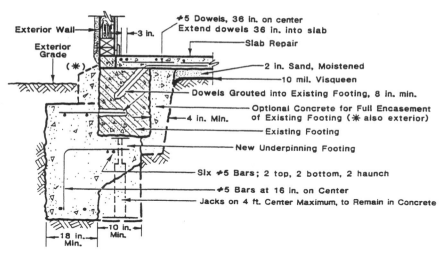

FIGURE 18.15 Underpinning by using continuous strip foundations.

FIGURE 18.16 Underpinning of a structure using continuous strip foundations.

In addition to the underpinning by using continuous strip footings, cracks in the foundation slabs will often need repair. Figure 18.17 shows the strip replacement method, which is one type of repair for concrete slab-on-grade cracks. The construction of the strip replacement starts by saw cutting out the area containing the concrete crack. Figure 18.17 indicates that a distance of 0.3 m (1 ft) on both sides of the concrete crack be saw cut. This is to provide enough working space to install reinforcement and the dowels. After the new reinforcement (No. 3 bars) and dowels are installed, the area is filled with a new portion of concrete.

18.16 PERFORMANCE OR ENGINEERING EVALUATION OF CONSTRUCTION

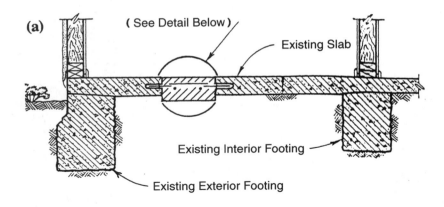

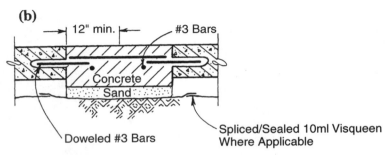

Saw-Cut 12" Each Side of Crack, Dowel #3 Bars 6"(min.) Into Existing Slab. Provide 5" Concrete Section with #3 Bars, 12" O.C. Both Ways. Underlay with 2" Moist Sand. Where Visqueen Exists, Splice/Seal in a Replacement Section

FIGURE 18.17 Concrete crack repair: (*a*) strip replacement of floor cracks; (*b*) strip replacement detail.

Another option is to patch the existing concrete cracks. The objective is to return the concrete slab to a satisfactory appearance and provide structural strength at the cracked areas. It has been stated (Transportation Research Board, 1977) that a patching material must meet the following requirements:

1. Be at least as durable as the surrounding concrete.
2. Require a minimum of site preparation.
3. Be tolerant of a wide range of temperature and moisture conditions.
4. Be noninjurious to the concrete through chemical incompatibility.
5. Preferably be similar in color and surface texture to the surrounding concrete.

Figure 18.18 shows a typical detail for concrete crack repair. If there is differential movement at the crack, then the concrete may require grinding or chipping to provide a smooth transition across the crack. The material commonly used to fill the concrete crack is epoxy. Epoxy compounds consist of a resin, a curing agent or hardener, and modifiers that make them suitable for specific uses. The typical range (3400 to 35,000 kPa, 500 to 5000 psi) in tensile strength of epoxy is similar to its range in compressive strength (Schutz, 1984). Performance specifications for epoxy have been developed (e.g., C 881-90 "Standard Specification for Epoxy-Resin-Base Bonding Systems for Concrete"; ASTM, 1997). In order for the epoxy to be effective, it is important that the crack faces be free of contaminants (such as dirt) that could prevent bonding. In many cases, the epoxy is injected under pressure so that it can penetrate the full depth of the concrete crack. Figure 18.19 shows the installation of pressure injected epoxy into concrete slab cracks.

Underpinning Alternatives. The previous discussion has dealt with the strengthening and underpinning of the foundation in order to resist soil movement or bypass the problem soil. There are many other types of underpinning and soil treatment alternatives (Brown, 1990, 1992; Greenfield and Shen, 1992; Lawton, 1996). In some cases, the magnitude of soil movement may be so large that the only alternative is to demolish the structure. For example, movement of the Portuguese Bend landslide in Palos Verdes, California, has destroyed about 160 homes. But a few homeowners refuse to abandon their homes as they slowly slide

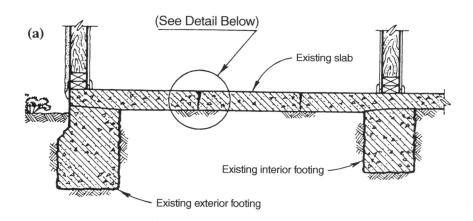

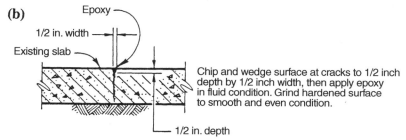

FIGURE 18.18 Concrete crack repair: (*a*) epoxy repair of floor cracks; (*b*) detail of crack repair with epoxy.

FIGURE 18.19 Pressure injection of epoxy into concrete slab cracks.

downslope. Some owners have underpinned their house foundations with steel beams supported by hydraulic jacks that are periodically used to relevel the house. Other owners have tried bizarre underpinning methods, such as supporting the house on huge steel drums.

An alternative method to underpinning is to treat the problem soil. For example, fluid grout can be injected into the ground to fill in joints, fractures, or underground voids in order to stabilize settling structures (Graf, 1969; Mitchell, 1970). Another option is mudjacking, which has been defined as a process whereby a water and soil-cement or soil-lime-cement grout is pumped beneath the slab, under pressure, to produce a lifting force which literally floats the slab to the desired position (Brown, 1992).

A commonly used alternative method to underpinning is compaction grouting, which consists of intruding a mass of very thick consistency grout into the soil, which both displaces and compacts the loose soil (Brown and Warner, 1973; Warner, 1982). Compaction grouting has proved successful in increasing the density of poorly compacted fill, alluvium, and compressible or collapsible soil. The advantages of compaction grouting are less expense and disturbance to the structure than foundation underpinning, and it can be used to relevel the structure. The disadvantages of compaction grouting are that it is difficult to analyze the results, it is usually ineffective near slopes or for near-surface soils because of the lack of confining pressure, and there is the danger of filling underground pipes with grout (Brown and Warner, 1973).

For expansive soil, underpinning alternatives can include horizontal or vertical moisture barriers to reduce the cyclic wetting and drying around the perimeter of the structure (Nadjer and Werno, 1973; Snethen, 1979; Williams, 1965). Drainage improvements and the repair of leaky water lines are also performed in conjunction with the construction of the moisture barriers. Other expansive soil stabilization options include chemical injection (such as a lime slurry) into the soil below the structure. The goal of such mitigation measures is to induce a chemical mineralogical change of the clay particles which will reduce the soil's tendency to swell.

PART 5

PREPARATION OR ENGINEERING EVALUATION OF GEOTECHNICAL REPORTS

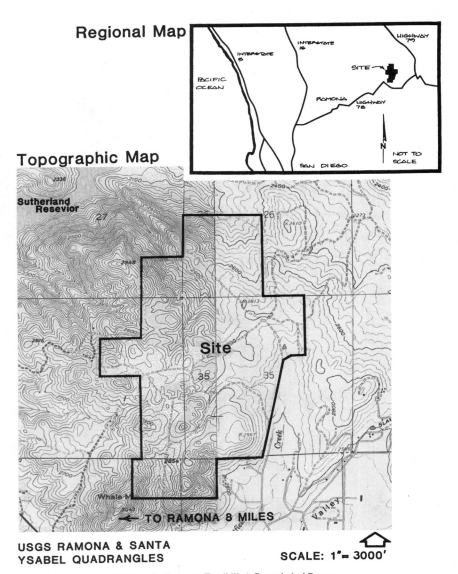

Example of a Location Map for a Preliminary (Feasibility) Geotechnical Report

CHAPTER 19
REPORTS

19.1 INTRODUCTION

The final chapter of this book deals with the preparation of reports and file management, evaluation of reports, and strategies to avoid civil liability. There are numerous types of geotechnical reports that are prepared for private clients or public agencies. As indicated in Figure 1.11, two common types of reports are preliminary (also known as feasibility) reports, required at the completion of the site investigation, and final grading reports, required on completion of the grading operations.

When preparing reports, it is best to follow standard guidelines. For example, App. B presents the *Technical Guidelines for Soil and Geology Reports* (from the *Orange County Grading Manual*). This document provides a detailed description of the required elements of preliminary and grading reports for Orange County, California. This document can be used as a guide during the preparation of the geotechnical report. As indicated in App. B, in California the reports usually have to be prepared jointly by the geotechnical engineer and engineering geologist. The guidelines for report preparation in App. B have been divided into six different parts, as follows:

Part I   Single-Family Dwellings (flatland).   Identifies the report content for precise grading permits on single-family dwellings in flatland areas (see Sec. D, App. B).

Part II   Single-Family Dwellings (hillside).   Identifies the report content for precise grading permits on single-family dwellings in hillside areas (additive to the requirements of Part I; see Sec. E, App. B).

Part III   Single-Family Dwellings (supplemental information).   Identifies additional report content which may be needed with Part I and Part II depending on the site conditions and proposed development (additive to the requirements of Parts I and II; see Sec. F, App. B).

Part IV   Commercial and Industrial Sites.   Identifies the report content for precise grading permits on commercial and industrial sites including apartment complexes (additive to the requirements of Part I and applicable items of Part III, see Sec. G, App. B).

Part V   Residential, Commercial, and Industrial Subdivisions (tracts and parcels).   Identifies the report content for preliminary grading permits of large commercial and industrial subdivisions and preliminary and precise grading permits of residential subdivisions in flatland and hillside areas (additive to the requirements of Part I and applicable items of Parts II and III; see Sec. H, App. B).

Part VI   Rough Grade Compaction Reports.   Identifies the report content for preliminary and rough grade compaction reports (see Sec. I, App. B).

19.4 PREPARATION OR ENGINEERING EVALUATION OF GEOTECHNICAL REPORTS

These guidelines presented in App. B may not be applicable for all types of projects, and the geotechnical engineer should always inquire about local building department report requirements. For the preparation of an engineering report, Wallace (1981) states that the purpose should be:

1. To tell the client what we did for the money paid us.
2. To summarize the results of our work, providing a written record of the activities that have just concluded.
3. To satisfy the requirements of some agency or agencies ranging from local building departments to state or federal agencies.
4. When monitoring construction, to offer an engineering opinion, based upon our work, regarding the contractor's compliance with the applicable portions of the job plans and specifications.

And the purpose of a written report should *not* be:

1. To guarantee, certify, or even accept the completed work. Guaranteeing is the business of contractors and manufactures. Certifying is the business of banks and notaries. Accepting is the business of owners and architects.
2. To provide more information or opinions than contracted for between your firm and your client.
3. To rehash resolved disputes or arguments between the various parties involved in the project.

19.2 PREPARATION OF REPORTS AND FILE MANAGEMENT

The purpose of Secs. 19.2 and 19.3 is to discuss the preparation of reports and file management.

19.2.1 Report Preparation

For preliminary (feasibility) reports, the written report should summarize the findings of the geotechnical investigation. Appendix E presents an example of a feasibility report. Specific items in the written report may include the following:

- Purpose of the investigation (including type of planned construction).
- Scope of services.
- Location map showing size of the project and existing development on the site.
- Findings from the document review.
- Summary of the site investigation including the data from the subsurface exploration (Chap. 4) and the laboratory test results (Chap. 5). The soil profile (see Figs. 4.39 to 4.42) should be included in the report to provide a summary of the subsoil conditions at the site.
- Engineering analyses, including the results of calculations or computer analyses.
- Geologic analyses, including the results of geologic studies such as geophysical techniques.
- Presentation of required geotechnical and geologic parameters, such as foundation design values, expected settlement of the structure, lateral pressures for retaining walls, etc.

- Conclusions and recommendations, which can vary depending on the type of project (see App. B). Common conclusions include an opinion on the geotechnical and geologic feasibility of the project, recommendations on ground preparation, fill placement and support, utility trenches, slope stability, etc.
- Appendixes, which can contain references, exploratory logs (such as boring, test pit, and trench logs), and the results of laboratory tests. The appendixes could also include detailed calculations or computer print-outs.
- Standard grading specifications (such as the document contained in App. C) which provides specifications for the grading to be performed on the site. The grading specifications are often included as an appendix to the report.

Feasibility reports are often prepared for clients who do not have an engineering background. In these cases, the report should be simply worded without resorting to engineering jargon. Report graphics can make it easier for the client to understand critical geotechnical or geologic features that could impact the proposed development.

An important aspect of preliminary or feasibility reports is to present the geotechnical and geologic conditions that will have the most impact on the proposed project. This is so important that it is often presented as the first section under the "Conclusions and Recommendations" part of the feasibility report. An example of typical wording in a soils report (see App. E) is as follows:

> The proposed construction is feasible from a geotechnical aspect. The recommendations herein address Phase I construction. Recommendations for future phases should be provided at the time of future development. Conventional spread and wall footings for Phase I can support the structural dead and live loads. Two constraints on the proposed construction are:
>
> 1. Nonuniform bearing conditions at the Administration Building, Projects Building, and Hardware Building consisting of both fill and bedrock within the building footprints, and
> 2. Results of expansion tests indicate that the soil and bedrock have a medium to high expansion potential. Influx of water from local irrigation, runoff, or leakage from the proposed artificial ponds and streams could cause expansive soil problems.

A common method of preparing a report is to use standardized formats. Reports or letters are prepared by filling in the blanks. Wallace (1981) states this is a dangerous practice, because it tends to discourage the thought process. The danger is that the writer will make the conditions of the job fit the standardized form, instead of modifying the report or letter. There is also the tendency to forget those items not included on the standardized form. A better approach is to use a report outline and write the individual sections. Only a few sections should be standardized, such as the closure or limitation of liability clause.

19.2.2 Daily Field Reports

Daily field reports are common on construction projects. For example, daily field reports are used during the placement and compaction of fill. The report could be prepared by a field technician, engineer, or geologist. The daily field report would typically list the type of compaction equipment, the area of fill compaction, the results of field compaction tests, and observations during construction operations.

Daily field reports are generally recorded on duplicate forms, which are commonly titled "Report of Field Observations and Testing." The form usually has the company's name, address, and telephone number as well as spaces for the date, writer, project name and number, and client. The form usually has a main section to document field observations

19.6 PREPARATION OR ENGINEERING EVALUATION OF GEOTECHNICAL REPORTS

and testing. Some common items listed on this section of the daily field report are the contractors' operations, test results, and sketches or diagrams.

In many cases, the daily field report will be an in-house document that is used to prepare a construction progress report or final project completion report. Frequently a copy of the daily field report will be given to the contractor, supervisor, or client. Wallace (1981) states:

> The degree of care which goes into a daily field report should not vary depending upon who the reader will be, i.e., if the client gets a copy, do a good job; if it's only for the front office, do a sloppy job. It should be remembered that, in the event of a lawsuit, all information in all files related to the problem is available to all parties concerned. Your angrily scribbled blast at the grading contractor complete with swearwords—may eventually be read aloud to a judge and jury as well as a whole courtroom full of spectators.

The writer of a daily field report should prepare the document with the same care as a client's final report. Several rough drafts may be needed before the completion of the final report. A daily field report should not be a hasty summary of events, but a well-thought-out document that reflects field observations accurately.

19.2.3 File Management

Examples of the contents of files include copies of reports sent to the client, daily field reports, correspondence, calculations, notes, and maps.

Some engineers believe that all documents, no matter how trivial or redundant, should be saved. Problems could occur if confidential, personal, or unrelated material is discovered in the file. At the other extreme are files devoid of all documents, except for reports issued to the client. This could cause problems because important photographs or testing of data may have been discarded. Neither extreme is desirable.

A file should be divided into separate compartments, such as confidential material, correspondence, testing, photographs, analysis, reports, and maps. Each individual compartment can be scrutinized for inappropriate material. Items that should be discarded include prior drafts of reports and letters, personal documents, rough notes or calculations (but keep the finished draft), and inappropriate or unrelated material. When a project is complete, the project manager should review the file and prepare it for the archives.

19.2.4 Examples of Poor File Management

Two examples of poor file management (involving lawsuits where the geotechnical engineer was sued) are as follows:

In one defendant geotechnical engineer's file, a cartoon was discovered. The cartoon was titled "West Corner of Cribwall" and depicted a cross section through a house and rear-yard cribwall slope. A person was drawn looking out the rear window and saying, "I hope I don't lose my back yard!" The cartoon may demonstrate to a jury that this defendant was indifferent to the failing condition of the cribwall slope. Since juries may be unable to understand complex stability analyses or factors of safety, they may consider the cartoon as evidence of the engineer's negligence. The file should have been thoroughly reviewed on completion of the project and the cartoon discarded.

The second example concerns a large residential development with numerous fill slopes. These fill slopes experienced surficial instability as well as slope deformation, which required the demolition of one house. During the ensuing lawsuit the following daily field report, prepared by the geotechnical engineer during the construction of the site, was discovered in the file:

[The owner] was still concerned about the fact that the material in the slope was not what was indicated in the plans, and I finally assured him that we would write a letter with our opinion that the material was adequate, and had adequate strength, such that deep-seated slope failure was not likely. This seemed to satisfy him at this time. [The city inspector] still appeared to be very concerned about the material, and could not be convinced that there was not something drastically wrong done during construction of the slope.

The memo is a candid admission of the fact that the slope was not built in accordance with the approved plans. Such self-incriminating memos could increase potential liability. Perhaps a better approach would have been to accurately document field conditions by describing the reasons, limits, and remedial measures for the area not being constructed in accordance with the plans.

19.3 ENGINEERING JARGON, SUPERLATIVES, AND TECHNICAL WORDS

Geotechnical and foundation engineers, like most professionals, have words or phrases that have special meaning. This engineering jargon can cause problems. For example, some words have legal implications, others may be misunderstood by the layman. This can result in increased liability on the part of the geotechnical engineer.

Attorneys have said that a report speaks for itself. Years after a project is complete, the official record will be the written report. The words in the report can be scrutinized and interpreted. It is essential that the report is clear and concise. Careful selection of words and phrases can be important in avoiding ambiguity. For example, two words, *inspect* and *observe* may mean exactly the same to an author, but could have significantly different meaning in a court of law. The purpose of the following section is to help the engineer avoid those words that can increase legal liability.

19.3.1 Engineering Jargon

Certification, Guarantee, Ensure, or Warrant. Certification reports are common in engineering practice. One example is a certification report for the compaction of fill. A typical report title and text are as follows:

> *Title:* Report of Certification of Compacted Fill
> *Text:* This report was prepared to certify the degree of compaction of fill materials placed and compacted at the site, and in conformance with the attached grading specifications.

As the fill at the development is being certified, the geotechnical engineer will most likely be sued if damage occurs because of fill problems. This is because the words *certification, guarantee, ensure,* and *warrant* can have legal definitions that increase the liability of the geotechnical engineer. For example, Narver (1993) states:

> In many states, a certification is synonymous with a warranty. If you certify that a project was constructed in accordance with the approved plans and specifications, you could be guaranteeing contractor compliance with all specifications. If it is later shown that the contractor was less than 100% in compliance, you might be held liable for the consequences on the theory that other parties may have relied upon your certification, or that you warranted another party's work.

19.8 PREPARATION OR ENGINEERING EVALUATION OF GEOTECHNICAL REPORTS

Instead of certifying the fill, the report title and text could be as follows:

Title: Report of Observations of Compacted Fill
Text: This report presents the results of fill compaction tests and observations during the placement and compaction of fill at the site.

Because of the potential increased liability associated with *certification, guarantee, ensure,* and *warrant,* geotechnical engineers should avoid the use of these words.

Controlling, Inspecting, Supervising. A primary responsibility of engineers is to review the work of junior engineers, builders, and contractors. In fact, registration as a civil engineer can not be obtained unless the engineer-in-training has had responsible charge. This may cause engineers-in-training to use words such as *inspect* or *supervise* in reports to comply with the responsible charge requirement. But such words have legal meaning. For example, Narver (1993) states:

> Observe means "to notice with care; to be on the watch respecting." By contrast, inspection has been determined by several courts to have a much broader meaning. A Missouri court found the word inspection to mean "to examine carefully or critically, investigate and test officially." A Pennsylvania court went a step further, finding that there was no substantial difference between the services of "inspection" and "supervision." This last definition is significant because it places upon a design professional who inspects a project the responsibilities of a supervisor. Consequently, by agreeing to perform "inspection services" or using the term "inspection" in conjunction with construction observation, the design professionals can potentially subject themselves to additional claims from the owner, contractors, workers, and the government for problems relating to compliance with plans and site safety.

Geotechnical engineers should avoid the use of *control, inspect,* or *supervise,* unless they carefully or critically investigate each step of the work. Alternatives include words such as *observe, review, study, look over* (the job), or *perform tests.*

19.3.2 Superlatives

Warriner (1957) states that comparison is the name given to the change in the form of adjectives and adverbs when they are used to compare the qualities of words they modify. There are three degrees of comparison: positive, comparative, and superlative. Table 19.1 lists three examples.

Commonly used superlatives are *must, always* (or *all*), *never,* and *none.* Engineers use superlatives because of their education and training. They are taught to be exact, with the result that superlatives are routinely used in everyday correspondence.

An example from the writer's experience involved the construction of a house on expansive soil. The geotechnical engineering firm involved with the construction stated in their final compaction report:

TABLE 19.1 Degrees of Comparison

Positive (1)	Comparative (2)	Superlative (3)
Good	Better	Best
Long	Longer	Longest
Many, much	More	Most

> In addition to capping lots as described above, the cut portion of cut/fill transition lots was undercut at least 3 feet and replaced with compacted fill so that all lots are either entirely cut or entirely fill lots.

In terms of expansive soil, the soil engineering firm stated that all lots have very low to low expansion potential. In both statements, the geotechnical engineering firm used superlatives, such as *all, at least,* and *entirely.*

Shortly after construction, the house began to experience cracking. The same geotechnical engineering firm investigated the damage to the house. They prepared a report on the damage to the house and stated the following:

> In our opinion, the distress is likely related to a combination of minor soil settlement within the less densely compacted fill soils, the presence of highly expansive soils near grade within the remaining portion of the building pad and the differing soil settlement characteristics of the fill soils and formation soils due to the cut portion of the lot not being undercut.

The geotechnical engineering firm was candid. They admitted that the cut portion of the lot was not entirely undercut and there was highly expansive soil near grade in the building pad. Superlatives in the final compaction report had an impact on the legal liability of the geotechnical engineering firm. Instead of using superlatives, the geotechnical engineering firm could have stated that based on observations during grading, the lots were graded as either cut or fill lots.

19.3.3 Technical Words

As indicated in the Glossary (App. A), geotechnical engineering has numerous technical words that have limited or unusual meanings. Sweet (1970) states that technical words will be interpreted as usually understood by persons in the profession or business to which they relate, unless clearly used in a different sense. But there may not be a consensus of opinion as to the exact meaning of a technical word. Words such as *critically expansive* or *collapsible soil* can be defined by specific parameters, but to a client such words may imply serious deficiencies with the project. Such technical words should be clearly defined or avoided so that false impressions are not formed.

19.4 ENGINEERING EVALUATION OF GEOTECHNICAL REPORTS

There could be many different situations where the geotechnical engineer reviews geotechnical reports prepared by others. The most common situation is where the geotechnical engineer reviews and signs an engineering report prepared by someone else in the same company, such as a staff engineer who is being supervised by the geotechnical engineer. In these cases, the geotechnical engineer must make sure that the report is technically correct and complete, addresses all the issues of the project (such as the items listed in App. B), is clear and concise, and the engineering jargon and superlatives have been removed.

Another common situation is where a geotechnical engineer prepares a preliminary (feasibility) report, but several years pass before the project is developed. At the time of project development, a different geotechnical engineer may be asked to work on the project and prepare the final "as-built" grading report. In this situation, the geotechnical engineer must review the original soils report and consider the following items:

1. *Change in site conditions.* If considerable time has elapsed since the preliminary (feasibility) report, the site may be significantly altered. For example, the site could contain piles of debris, old fill, or erosional scarps that were not present at the site during the feasibility study. Both the groundwater condition and site topography may be significantly different and a new feasibility study may be required.
2. *Report contents.* The geotechnical engineer should check the report contents. Some items may have been forgotten, or there may be mistakes or discrepancies with the data.
3. *Up-to-date.* The standard of practice is constantly evolving. An old preliminary (feasibility) report may not meet today's standards. Additional site investigation work (subsurface exploration and laboratory testing) as well as engineering and geologic analyses may be required to update the report.

19.5 STRATEGIES TO AVOID CIVIL LIABILITY

The final section of this book deals with strategies to avoid civil liability. Geotechnical engineering is a risky profession. Many engineers believe that if they practice perfect engineering or conform to the standard of care, they will be immune to civil liability. Unfortunately, if a project develops problems, all engineers may be named in the lawsuit, regardless of their innocence. In the United States, there are attorneys who sue everyone involved with a damaged project to secure as much money for their clients as possible (Meehan et al., 1993).

Today, a practicing engineer can not perform services without the fear of a lawsuit. A lawsuit that could result in substantial financial loss is every engineer's nightmare. Even if the engineer is vindicated, the drain on resources and the stress of deposition and trial can be debilitating (Day, 1992e).

One well-documented case history involves L'Ambiance Plaza, which collapsed during construction in Bridgeport, Connecticut, killing 28 workers. Fairweather (1992) dealt with the personal effect of the collapse on James O'Kon of O'Kon and Co., who was the engineer of record for L'Ambiance Plaza. James O'Kon summarized his experience:

> No suits or claims were filed against my firm. The contractors did not survive, the developer did not survive. O'Kon and Co. survived. I was declared guilty by the press, investigated for murder, almost lost my professional license, and threatened with business failure by a federal judge.

In their paper titled "Claims Analysis from Risk-Retention Professional Liability Group," Janney et al. (1996) conclude that of all alleged errors or omissions, the highest percentage of claims was for design errors leading to delays or extras. Common reasons for claims due to delays or extra work on construction projects are as follows:

1. The client may request changes or additions.
2. There can be a misinterpretation of the statement of work, by the engineer, architect, contractor, or even the owner.
3. There can be omissions due to an improperly defined scope of work.
4. The requirements for the project may be poorly defined, overly optimistic, or unrealistic in terms of scheduling.
5. There may be a failure to take into account all of the risks of construction, especially with new or innovative techniques.
6. There may be outside influences (such as natural disasters) that affect the project.

But perhaps the most frequent cause of claims for delays or extra work is due to the bidding system (Gould, 1995). The goal of an owner or public agency is to obtain the lowest possible cost to construct the project. It is usually the low bidder who wins the contract. In order to prepare a low bid, contractors may not use a sufficient markup for overhead or profit. Once the job is started, contractors will hope to make a profit by claiming expenses due to delays or extra work. Owens (1993) states:

> All plans have errors and omissions. Contractors are experts at mining these hidden gold nuggets for extra profits. Public works–type projects seem especially vulnerable. Owners, whether airport authorities or waste-water treatment-plant officials, micro-manage. There is always some deadline that must be met. Contractors will stall, then, when reminded about the schedule, accelerate and claim extra costs.

The different objective of the owner or public agency trying to construct a low-bid project and the contractor trying to make a profit through delays or extra work frequently results in litigation.

To reduce the number of claims due to design errors leading to delays or extras, Janney et al. (1996) have suggested the following:

1. Improved contracts between owners and constructors
2. Improved contracts between owners and design professionals and between design professionals
3. Partnering and dispute review boards (prompt resolution of emergency disputes)
4. Better control of realistic schedule for production of design documents
5. Better control of realistic turnaround time of all submittals
6. Quality control of design including regular "peer" checking, especially of details
7. Total quality management

There are other ways to limit one's potential liability. The following discussion is primarily focused toward practicing geotechnical and foundation engineers who provide consulting services in small to midsize companies.

19.5.1 Assessing Risk

Geotechnical engineers who have practiced in a locality for a long time can develop consulting services that have a low risk. For example, a geotechnical engineer may provide consulting service to the local or county government, which may have set procedures of trying to negotiate a fix for problems, rather than sue. Other examples of low-risk activities could be consulting work for longtime clients, where a personal relationship has been developed. Such a relationship results in solving problems, rather than trying to fix blame in a court of law. Often the best approach to avoiding a lawsuit is to work with the client to solve the problems as they develop.

Byer (1992) defines clients as low-budget speculators, high-budget speculators, remodelers, mass builders, and the professional builder. He assesses the risk of working for each of these clients and concludes that the client in a project can affect your liability on the project. He suggests that the speculators are high-risk clients because they will tend to cut corners, while the professional builder generally has a low risk of failure or problems because of the insistence on quality work.

Besides the type of client, the type of project can also affect liability. Some types of projects with a very high potential for liability are as follows:

19.12 PREPARATION OR ENGINEERING EVALUATION OF GEOTECHNICAL REPORTS

1. *Toxic waste.* The failure to find and treat all of the toxic waste is always a possibility.
2. *Condominium complexes.* Performing geotechnical or foundation design engineering services for these developments is very risky, especially in California, because if damage develops, the developer could be held strictly liable for that damage, regardless of negligence. When a developer is sued, cross-complaints against the design engineers are inevitable.
3. *Trench excavations.* There are probably more injuries and death due to trench cave-ins than any other construction activity. Frequently, trenches are excavated for long distances and encounter many different soil and groundwater conditions, making the design or observation of shoring difficult.

19.5.2 Insurance

One way to deal with potential liability is to purchase errors-and-omissions insurance. If sued, the case can be turned over to the insurance company. After the deductible is exhausted, the insurance company will have to pay the attorney's fees, provide the legal defense, and pay judgments up to the limits of the policy.

Insurance, however, can be very expensive, especially for high-risk activities such as geotechnical engineering. There is also the problem that if an individual or company is sued, the rates could significantly escalate. Some engineers believe that having insurance invites lawsuits. For example, a developer actually stated that he was more interested in the adequacy of the engineering consultant's insurance policy rather than professional abilities. He was quoted as saying, "It's your [the engineer's] insurance policy I want" (Meehan et al., 1993).

For uninsured defendants, Patton (1992) has stated:

> Since the financial condition and insurance policy limits are factors to be considered in determining whether or not a proposed settlement is made in "good faith," consideration should be made in keeping the operating corporation as empty as possible and without insurance coverage. Even where the plaintiff's damages are accepted as great and the liability certain, a disproportionately low settlement figure is often reasonable in the case of a relatively insolvent and uninsured defendant. This requires strong wills, steady nerves and determination on the part of the uninsured defendants.

19.5.3 Limitation of Liability Clauses

Another common strategy for avoiding lawsuits is limitation of liability clauses. These are usually included in the fine print of the contract between the consulting engineer and the client. An example of a limitation of liability clause is as follows:

> *Limitation of liability.* This contract contains a limitation of liability clause as follows: In the event of a legal proceeding or other action arising from or related to this agreement or the consultant's performance hereunder, the consultant's total liability to the client, whatever form the action, shall be limited to 1.4 times the professional fees collected by the consultant for the project.

Some of the problems with limitation of liability clauses are:

1. The limitation of liability clause may not be judgement-proof. Case law continually evolves, and it is possible that a limitation of liability clause will be ruled invalid by a judge or court of appeals.

2. The client may be simply unwilling to sign the contract with a limitation of liability clause. Then the choice is whether to accept the increased risk of the job or refuse the assignment.
3. There is always the possibility of a lawsuit by a third party, such as an injured bystander, who is not a party to the contract.

In summary, a practicing geotechnical engineer can not perform services without the fear of a lawsuit. Some strategies for avoiding civil liability include assessing the risk before accepting the assignment, purchasing errors-and-omissions insurance, and using limitation of liability clauses in contracts.

APPENDIX A
GLOSSARY

INTRODUCTION

The following is a list of commonly used geotechnical engineering and engineering geology terms and definitions. The glossary has been divided into four main categories:

1. Engineering Geology and Subsurface Exploration Terminology
2. Laboratory Testing Terminology
3. Terminology for Engineering Analysis and Computations
4. Construction and Grading Terminology

References.   References used for the Glossary include:

- *Glossary of Selected Geologic Terms With Special Reference to Their Use in Engineering* (1955)
- *Soil Mechanics in Engineering Practice* (1967)
- *Soil Mechanics* (1969)
- *Essentials of Soil Mechanics and Foundations* (1977)
- *An Introduction to Geotechnical Engineering* (1981)
- NAVFAC DM-7.1 (1982), NAVFAC DM-7.2 (1982), and NAVFAC DM-7.3 (1983)
- *Thickness Design—Asphalt Pavements for Highways and Streets* (1984)
- *Orange County Grading and Excavation Code* (1993)
- *Foundation Design, Principles and Practice* (1994)
- *Standard Specifications for Highway Bridges* (1996)
- *Caterpillar Performance Handbook* (1997)
- *Uniform Building Code* (1997)
- *American Society for Testing and Materials* (ASTM D 653-97 and ASTM D 4439-97, 1998)

Basic Terms

Civil Engineer   A professional engineer who is registered to practice in the field of civil works.

Civil Engineering   The application of the knowledge of the forces of nature, principles of mechanics, and the properties of materials for the evaluation, design, and construction of civil works for the beneficial uses of mankind.

Engineering Geologist   A geologist who is experienced and knowledgeable in the field of engineering geology.

Engineering Geology   The application of geologic knowledge and principles in the investigation and evaluation of naturally occurring rock and soil for use in the design of civil works.

Geologist   An individual educated and trained in the field of geology.

Geotechnical Engineer   A licensed individual who performs an engineering evaluation of earth materials including soil, rock, groundwater, and man-made materials and their interaction with earth retention systems, structural foundations, and other civil engineering works.

Geotechnical Engineering   A subdiscipline of civil engineering. Geotechnical engineering requires a knowledge of engineering laws, formulas, construction techniques, and the performance of civil engineering works influenced by earth materials. Geotechnical engineering encompasses many of the engineering aspects of soil mechanics, rock mechanics, foundation engineering, geology, geophysics, hydrology, and related sciences.

Rock Mechanics   The application of the knowledge of the mechanical behavior of rock to engineering problems dealing with rock. Rock mechanics overlaps with engineering geology.

Soil Mechanics   The application of the laws and principles of mechanics and hydraulics to engineering problems dealing with soil as an engineering material.

Soils Engineer   Synonymous with geotechnical engineer (see Geotechnical Engineer).

Soils Engineering   Synonymous with geotechnical engineering (see Geotechnical Engineering).

GLOSSARY 1
ENGINEERING GEOLOGY AND SUBSURFACE EXPLORATION TERMINOLOGY

Adobe Sun-dried bricks composed of mud and straw. Abode is commonly used for construction in the southwestern United States and in Mexico.

Aeolian (or Eolian) Particles of soil that have been deposited by the wind. Aeolian deposits include dune sands and loess.

Alluvium Detrital deposits resulting from the flow of water, including sediments deposited in river beds, canyons, flood plains, lakes, fans at the foot of slopes, and estuaries.

Aquiclude A relatively impervious rock or soil strata that will not transmit groundwater fast enough to furnish an appreciable supply of water to a well or spring.

Aquifer A relatively pervious rock or soil strata that will transmit groundwater fast enough to furnish an appreciable supply of water to a well or spring.

Artesian Groundwater that is under pressure and is confined by impervious material. If the trapped pressurized water is released, such as by drilling a well, the water will rise above the groundwater table and may even rise above the ground surface.

Ash Fine fragments of rock, between 4 mm and 0.25 mm in size, that originated as airborne debris from explosive volcanic eruptions.

Badlands An area, large or small, characterized by extremely intricate and sharp erosional sculpture. Badlands occur chiefly in arid or semiarid climates where the rainfall is concentrated in sudden heavy showers. They may, however, occur in humid regions where vegetation has been destroyed, or where soil and coarse detritus are lacking.

Bedding The arrangement of rock in layers, strata, or beds.

Bedrock A more or less solid, relatively undisturbed rock in place either at the surface or beneath deposits of soil.

Bentonite A soil or formational material that has a high concentration of the clay mineral montmorillonite. Bentonite is usually characterized by high swelling upon wetting. The term bentonite also refers to manufactured products that have a high concentration of montmorillonite, for example, bentonite pellets.

Bit A device that is attached to the end of the drill stem and is used as a cutting tool to bore into soil and rock.

Bog A peat covered area with a high groundwater table. The surface is often covered with moss and it tends to be nutrient poor and acidic.

Boring A method of investigating subsurface conditions by drilling a hole into the earth materials. Usually soil and rock samples are extracted from the boring. Field tests, such as the standard penetration test (SPT) and the vane shear test (VST), can also be performed in the boring.

Boring Log A written record of the materials penetrated during the subsurface exploration. See Table 4.9 for other types of information that should be recorded on the boring log.

Boulder A large detached rock fragment with an average dimension greater than 300 mm (12 in.).

California Bearing Ratio (CBR) The CBR can be determined for soil in the field or soil compacted in the laboratory. See Sec. 4.5 for specific details on the procedure for determining the CBR. The CBR is frequently used for the design of roads and airfields.

Casing A steel pipe that is temporarily inserted into a boring or drilled shaft in order to prevent the adjacent soil from caving.

Cobble A rock fragment, usually rounded or semirounded, with an average dimension between 75 and 300 mm (3 and 12 in.).

Cohesionless Soil A soil, such as a clean gravel or sand, that when unconfined, falls apart in either a wet or dry state.

Cohesive Soil A soil, such as a silt or clay, that when unconfined, has considerable shear strength when dried, and will not fall apart in a saturated state. Cohesive soil is also known as a *plastic soil,* or a soil that has a plasticity index.

Colluvium Generally loose deposits usually found near the base of slopes and brought there chiefly by gravity through slow continuous downhill creep.

Cone Penetration Test (CPT) A field test used to identify and determine the *in situ* properties of soil deposits and soft rock.

> **Electric Cone** A cone penetrometer that uses electric force transducers built into the apparatus for measuring cone resistance and friction resistance.
>
> **Mechanical Cone** A cone penetrometer that uses a set of inner rods to operate a telescoping penetrometer tip and to transmit the resistance force to the surface for measurement.
>
> **Mechanical-Friction Cone** A cone penetrometer with the additional capability of measuring the local side friction component of penetration resistance.
>
> **Piezocone** A cone penetrometer with the additional capability of measuring pore water pressure generated during the penetration of the cone.

Core Drilling Also known as *diamond drilling*; the process of cutting out cylindrical rock samples in the field.

Core Recovery (RQD) The RQD is computed by summing the lengths of all pieces of the rock core (NX size) equal to or longer than 10 cm (4 in.) and dividing by the total length of the core run. The RQD is multiplied by 100 to express it as a percentage.

Deposition The geologic process of laying down or accumulating natural material into beds, veins, or irregular masses. Deposition includes mechanical settling (such as sedimentation in lakes), precipitation (such as the evaporation of surface water to form halite), and the accumulation of dead plants (such as in a peat bog).

Detritus Any material worn or broken down from rocks by mechanical means.

Diatomaceous Earth Diatomaceous earth usually consists of fine, white, siliceous powder, composed mainly of diatoms and their remains.

Erosion The wearing away of the ground surface as a result of the movement of wind, water, and/or ice.

Fault A fracture in the earth's crust along which movement has occurred. A fault is considered active if movement has occurred within the last 11,000 years (Holocene geologic time).

Fines Refers to the silt- and clay-size particles in the soil.

Fold Bending or flexure of a layer or layers of rock. Examples of folded rock include anticlines and synclines. Usually folds are created by the massive compression of rock layers.

Fracture A visible break in a rock mass. Examples includes joints, faults, and fissures.

Geophysical Techniques Various methods of determining subsurface soil and rock conditions without performing subsurface exploration. A common geophysical technique is to induce a shock wave into the earth and then measure the seismic velocity of the wave's travel through the earth material. The seismic velocity has been correlated with the rippability of the earth material.

Groundwater Table Also known as *phreatic surface*; the top surface of underground water, the location of which is often determined from piezometers, such as an open standpipe. A perched groundwater table refers to groundwater occurring in an upper zone separated from the main body of groundwater by underlying unsaturated rock or soil.

Horizon One of the layers of a soil profile that can be distinguished by its texture, color, and structure.

   A Horizon The uppermost layer of a soil profile which often contains remnants of organic life. Inorganic colloids and soluble materials are often leached from this horizon.

   B Horizon The layer of a soil profile in which material leached from the overlying "A" horizon is accumulated.

   C Horizon Undisturbed parent material from which the overlying soil profile has been developed.

Inclinometer The horizontal movement preceding or during the movement of slopes can be investigated by successive surveys of the shape and position of flexible vertical casings installed in the ground. The surveys are performed by lowering an inclinometer probe into the flexible vertical casing.

In Situ Used in reference to the original in-place (or *in situ*) condition of the soil or rock.

Iowa Borehole Shear Test (BST) A field test where the device is lowered into an uncased borehole and then expanded against the sidewalls. The force required to pull the device toward ground surface is measured, and, much like a direct shear test, the shear strength properties of the *in situ* soil can then be determined.

Karst Topography A type of landform developed in a region of easily soluble limestone. It is characterized by vast numbers of depressions of all sizes, sometimes by great outcrops of limestone ledges, sinks, and other solution passages, an almost total lack of surface streams, and large springs in the deeper valleys.

Kelly A heavy tube or pipe, usually square or rectangular in cross section, that is used to provide a downward load when excavating an auger borehole.

Landslide Mass movement of soil or rock that involves shear displacement along one or several rupture surfaces, which are either visible or may be reasonably inferred.

Landslide Debris Material, generally porous and of low density, produced from instability of natural or man-made slopes.

Leaching The removal of soluble materials in soil or rock caused by percolating or moving groundwater.

Loess A wind deposited silt often having a high porosity and low density which is often susceptible to collapse of its soil structure upon wetting.

Mineral An inorganic substance that has a definite chemical composition and distinctive physical properties. Most minerals are crystalline solids.

Overburden The soil that overlies bedrock. In other cases, it refers to all material overlying a point of interest in the ground, such as the overburden pressure exerted on a clay layer.

Peat A naturally occurring highly organic deposit derived primarily from plant materials.

Penetration Resistance See Standard Penetration Test.

Percussion Drilling A drilling process in which a borehole is advanced by using a series of impacts to the drill rods and attached bit.

Permafrost Perennially frozen soil. Also defined as ground that remains below freezing temperatures for 2 or more years. The bottom of permafrost lies at depths ranging from a few feet to over a thousand feet. The *active layer* is defined as the upper few inches to several feet of ground that is frozen in winter but thawed in summer.

Piezometer A device installed for measuring the pore water pressure (or pressure head) at a specific point within the soil mass.

Pit (or Test Pit) An excavation made for the purpose of observing subsurface conditions, performing field tests, and obtaining soil samples. A pit also refers to an excavation in the surface of the earth from which ore is extracted, such as an open pit mine.

Pressuremeter Test (PMT) A field test that involves the expansion of a cylindrical probe within an uncased borehole.

Refusal During subsurface exploration, refusal means an inability to excavate any deeper with the boring equipment. Refusal could be due to many different factors, such as hard rock, boulders, or a layer of cobbles.

Residual Soil Soil derived by in-place weathering of the underlying material.

Rock Rock is a relatively solid mass that has permanent and strong bonds between the minerals. Rock can be classified as being sedimentary, igneous, or metamorphic.

Rotary Drilling A drilling process in which a borehole is advanced by rotation of a drill bit under constant pressure without impact.

Rubble Rough stones of irregular shape and sizes that are naturally or artificially broken from larger masses of rock. Rubble is often created during quarrying, stone cutting, and blasting.

Screw Plate Compressometer (SPC) A field test that involves a plate that is screwed down to the desired depth, and then, as pressure is applied, the settlement of the plate is measured.

Seep A small area where water oozes from the soil or rock.

Slaking The crumbling and disintegration of earth materials when exposed to air or moisture. Slaking can also refer to the breaking up of dried clay when submerged in water, due either to compression of entrapped air by inwardly migrating water or to the progressive swelling and sloughing off of the outer layers.

Slickensides Surfaces within a soil mass which have been smoothed and striated by shear movements on these surfaces.

Slope Wash Soil and/or rock material that has been transported down a slope by mass wasting assisted by runoff water not confined by channels (also see Colluvium).

Soil Sediments or other accumulations of mineral particles produced by the physical and chemical disintegration of rocks. Inorganic soil does not contain organic matter, while organic soil contains organic matter.

Soil Sampler A device used to obtain soil samples during subsurface exploration. Based on the inside clearance ratio and the area ratio, soil samples can be either disturbed or undisturbed.

Standard Penetration Test (SPT) A field test that consists of driving a thick-walled sampler (ID = 1.5 in., OD = 2 in.) into the soil by using a 140-lb hammer falling 30 in. The number of blows to drive the sampler 18 in. is recorded. The N-value (penetration resistance) is defined as the number of blows required to drive the sampler from a depth interval of 6 to 18 in.

Strike and Dip Strike and dip refer to a planar structure, such as a shear surface, fault, bed, etc. The strike is the compass direction of a level line drawn on the planar structure. The dip angle is measured between the planar structure and a horizontal surface.

Subgrade Modulus Also known as the *modulus of subgrade reaction*; this value is often obtained from field plate load tests and is used in the design of pavements and airfields.

Subsoil Profile Developed from subsurface exploration, a cross section of the ground that shows the soil and rock layers. A summary of field and laboratory tests could also be added to the subsoil profile.

Till Material created directly by glaciers, without transportation or sorting by water. Till often consists of a wide range of particle sizes, including boulders, gravel, sand, and clay.

Topsoil The fertile upper zone of soil which contains organic matter and is usually darker in color and loose.

Vane Shear Test (VST) An *in situ* field test that consists of inserting a four-bladed vane into the borehole and then pushing the vane into the clay deposit located at the bottom of the borehole. Once inserted into the clay, the maximum torque required to rotate the vane and shear the clay is measured. From the dimensions of the vane and the maximum torque, the undrained shear strength s_u of the clay can be calculated.

Varved Silt or Varved Clay A lake deposit with alternating thin layers of sand and silt (varved silt) or sand and clay (varved clay). It is formed by the process of sedimentation from the summer to winter months. The sand is deposited during the summer and the silt or clay is deposited in the winter when the lake surface is covered with ice and the water is tranquil.

Wetland Land which has a groundwater table at or near the ground surface, or land that is periodically under water, and supports various types of vegetation that are adapted to a wet environment.

GLOSSARY 2
LABORATORY TESTING TERMINOLOGY

Absorption Defined as the mass of water in the aggregate divided by the dry mass of the aggregate. Absorption is used in soil mechanics for the study of oversize particles or in concrete mix design.

Activity of Clay The ratio of plasticity index to percent dry mass of the total sample that is smaller than 0.002 mm in grain size. This property is related to the types of clay minerals in the soil.

Atterberg Limits Water contents corresponding to different behavior conditions of plastic soil.

  Liquid Limit The water content corresponding to the behavior change between the liquid and plastic state of a soil. The liquid limit is arbitrarily defined as the water content at which a pat of soil, cut by a groove of standard dimensions, will flow together for a distance of 12.7 mm (0.5 in.) under the impact of 25 blows in a standard liquid limit device.

  Plastic Limit The water content corresponding to the behavior change between the plastic and semisolid state of a soil. The plastic limit is arbitrarily defined as the water content at which the soil will just begin to crumble when rolled into a tread approximately 3.2 mm (⅛ in.) in diameter.

  Shrinkage Limit The water content corresponding to the behavior change between the semisolid to solid state of a soil. The shrinkage limit is also defined as the water content at which any further reduction in water content will not result in a decrease in volume of the soil mass.

Capillarity Also known as *capillary action*, the rise of water through a soil due to the fluid property known as surface tension. Due to capillarity, the pore water pressures are less than atmospheric because of the surface tension of pore water acting on the meniscus formed in void spaces between the soil particles. The height of capillary rise is inversely proportional to the pore size of the soil.

Clay-Size Particles Clay-size particles are finer than 0.002 mm. Most clay particles are flat or plate-like in shape, and as such they have a large surface area. The most common clay minerals belong to the kaolin, montmorillonite, and illite groups.

Coarse-Grained Soil According to the Unified Soil Classification System, coarse-grained soils have more than 50 percent soil particles (by dry mass) retained on the No. 200 U.S. standard sieve.

Coefficient of Consolidation A coefficient used in the theory of consolidation. It is obtained from laboratory consolidation tests and is used to predict the time-settlement behavior of field loading of fine-grained soil.

Cohesion There are two types of cohesion: (1) cohesion in terms of total stress and (2) cohesion in terms of effective stress. For total cohesion c, the soil particles are predominately held together by capillary tension. For effective stress cohesion c', there must be actual bonding or attraction forces between the soil particles.

Colloidal Soil Particles Generally refers to clay-size particles (finer than 0.002 mm) where the surface activity of the particle has an appreciable influence on the properties of the soil.

Compaction (Laboratory)

  Compaction Curve A curve showing the relationship between the dry density and the water content of a soil for a given compaction energy.

Compaction Test   A laboratory compaction procedure whereby a soil at a known water content is compacted into a mold of specific dimensions. The procedure is repeated for various water contents to establish the compaction curve. The most common testing procedures (compaction energy, number of soil layers in the mold, etc.) are the Modified Proctor (ASTM D 1557) or Standard Proctor (ASTM D 698).

Compression Index   For a consolidation test, the slope of the linear portion of the vertical pressure versus void ratio curve on a semilog plot.

Consistency of Clay   Generally refers to the firmness of a cohesive soil. For example, a cohesive soil can have a consistency that varies from "very soft" up to "hard."

Consolidation Test   A laboratory test used to measure the consolidation properties of cohesive soil. The specimen is laterally confined in a ring and is compressed between porous plates (oedometer apparatus).

Contraction (during Shear)   During the shearing of soil, the tendency of loose soil to decrease in volume (or contract).

Density   Defined as mass per unit volume. In the International System of Units (SI), typical units for the density of soil are Mg/m^3.

Deviator Stress   Difference between the major and minor principal stress in a triaxial test.

Dilation (during Shear)   During the shearing of soil, the tendency of dense soil to increase in volume (or dilate).

Double Layer   A grossly simplified interpretation of the positively charged water layer, together with the negatively charged surface of the particle itself. Two reasons for the attraction of water to the clay particle are: (1) dipolar structure of water molecule which causes it to be electrostatically attracted to the surface of the clay particle, and (2) the clay particles attract cations which contribute to the attraction of water by the hydration process. The "absorbed water layer" consists of water molecules that are tightly held to the clay particle face, such as by the process of hydrogen bonding.

Fabric (of Soil)   Definitions vary, but in general the fabric of soil often refers only to the geometric arrangement of the soil particles. In contrast, the soil structure refers to both the geometric arrangement of soil particles and the interparticle forces which may act between them.

Fine-Grained Soil   According to the Unified Soil Classification System, a fine-grained soil contains more than 50 percent (by dry mass) of particles finer than the No. 200 sieve.

Flocculation   When in suspension in water, the process of fines attracting each other to form a larger particle. In the hydrometer test, a dispersing agent is added to prevent flocculation of fines.

Friction Angle   In terms of effective shear stress, the soil friction is usually considered to be a result of the interlocking of the soil or rock grains and the resistance to sliding between the grains. A relative measure of a soil's frictional shear strength is the friction angle.

Gravel-Size Fragments   Rock fragments and soil particles that will pass the 3-in. (76-mm) sieve and be retained on a No. 4 (4.75 mm) U.S. standard sieve.

Hydraulic Conductivity (or Coefficient of Permeability)   For laminar flow of water in soil, both terms are synonymous and indicate a measure of the soil's ability to allow water to flow through its soil pores. The hydraulic conductivity is often measured in a constant-head or falling-head permeameter.

Laboratory Maximum Dry Density   The peak point of the compaction curve (see Compaction).

Moisture Content (or Water Content)   Moisture content and water content are synonymous. The definition of moisture content is the ratio of the mass of water in the soil divided by the dry mass of the soil, usually expressed as a percentage.

Optimum Moisture Content   The moisture content, determined from a laboratory compaction test, at which the maximum dry density of a soil is obtained using a specific compaction energy.

Overconsolidation Ratio (OCR)   The ratio of the preconsolidation vertical effective stress to the current vertical effective stress.

Peak Shear Strength The maximum shear strength along a shear failure surface.

Permeability The ability of water (or other fluid) to flow through a soil by traveling through the void spaces. A high permeability indicates flow occurs rapidly, and vice versa. A measure of the soil's permeability is the hydraulic conductivity, also known as the coefficient of permeability.

Plasticity Term applied to silt and clay, to indicate the soil's ability to be rolled and molded without breaking apart. A measure of a soil's plasticity is the plasticity index.

Plasticity Index The plasticity index is defined as the liquid limit minus the plastic limit, often expressed as a whole number (also see Atterberg Limits).

Sand Equivalent (SE) A measure of the amount of silt or clay contamination in fine aggregate as determined by ASTM D 2419 test procedures.

Sand-Size Particles Soil particles that will pass the No. 4 (4.75 mm) sieve and be retained on the No. 200 (0.075 mm) U.S. standard sieve.

Shear Strength The maximum shear stress that a soil or rock can sustain. Shear strength of soil is based on total stresses (i.e., undrained shear strength) or effective stresses (i.e., effective shear strength).

> **Effective Shear Strength** Shear strength of soil based on effective stresses. The effective shear strength of soil could be expressed in terms of the failure envelope, that is defined by effective cohesion c' and effective friction angle ϕ'.
>
> **Shear Strength in Terms of Total Stress** Shear strength of soil based on total stresses. The undrained shear strength of soil could be expressed in terms of the undrained shear strength s_u, or by using the failure envelope that is defined by total cohesion c and total friction angle ϕ.

Shear Strength Tests (Laboratory) There are many types of shear strength tests that can be performed in the laboratory. The objective is to obtain the shear strength of the soil. Laboratory tests can generally be divided into two categories:

> **Shear Strength Tests Based on Effective Stress** The purpose of these laboratory tests is to obtain the effective shear strength of the soil based on the failure envelope in terms of effective stress. An example is a direct shear test where the saturated, submerged, and consolidated soil specimen is sheared slowly enough that excess pore water pressures do not develop (this test is known as a consolidated-drained test).
>
> **Shear Strength Tests Based on Total Stress** The purpose of these laboratory tests is to obtain the undrained shear strength of the soil or the failure envelope in terms of total stresses. Examples are the unconfined compression test and the unconsolidated-undrained triaxial compression test.

Sieve Laboratory equipment consisting of a pan with a screen at the bottom. U.S. standard sieves are used to separate particles of a soil sample into their various sizes.

Silt-Size Particles That portion of a soil that is finer than the No. 200 sieve (0.075 mm) and coarser than 0.002 mm. Silt- and clay-size particles are considered to be "fines."

Soil Structure Definitions vary, but in general, the soil structure refers to both the geometric arrangement of the soil particles and the interparticle forces which may act between them. Common soil structures are as follows:

> **Cluster Structure** Soil grains that consist of densely packed silt- or clay-size particles.
>
> **Dispersed Structure** The clay-size particles are oriented parallel to each other.
>
> **Flocculated (or Cardhouse) Structure** The clay-size particles are oriented in edge-to-face arrangements.
>
> **Honeycomb Structure** Loosely arranged bundles of soil particles, having a structure that resembles a honeycomb.
>
> **Single-Grained Structure** An arrangement composed of individual soil particles. This is a common structure of sands.

Skeleton Structure An arrangement where coarser soil grains form a skeleton with the void spaces partly filled by a relatively loose arrangement of soil fines.

Specific Gravity The specific gravity of soil or oversize particles can be determined in the laboratory. Specific gravity is generally defined as the ratio of the density of the soil particles divided by the density of water.

Tensile Test For a geosynthetic, a laboratory test in which the geosynthetic is stretched in one direction to determine the force-elongation characteristics, breaking force, and the breaking elongation.

Texture (of Soil) The term texture refers to the degree of fineness of the soil, such as smooth, gritty, or sharp, when the soil is rubbed between the fingers.

Thixotropy The property of a remolded clay that enables it to stiffen (gain shear strength) in a relatively short time.

Triaxial Test A laboratory test in which a cylindrical specimen of soil or rock encased in an impervious membrane is subjected to a confining pressure and then loaded axially to failure.

Unconfined Compressive Strength The vertical stress which causes the shear failure of a cylindrical specimen of a plastic soil or rock in a simple compression test. For the simple compression test, the undrained shear strength s_u of the plastic soil is defined as one-half the unconfined compressive strength.

Unit Weight Unit weight is defined as weight per unit volume. In the International System of Units (SI), unit weight has units of kN/m^3. In the United States Customary System, unit weight has units of pcf (pounds-force per cubic foot).

Water Content (or Moisture Content) See Moisture Content.

Zero Air Voids Curve On the laboratory compaction curve, the zero air voids curve is often included. It is the relationship between water content and dry density for a condition of saturation ($S = 100\%$) for a specified specific gravity.

GLOSSARY 3
TERMINOLOGY FOR ENGINEERING ANALYSIS AND COMPUTATIONS

Adhesion Shearing resistance between two different materials. For example, for piles driven into clay deposits, there is adhesion between the surface of the pile and the surrounding clay.

Allowable Bearing Pressure Allowable bearing pressure is the maximum pressure that can be imposed by a foundation onto soil or rock supporting the foundation. It is derived from experience and general usage, and provides an adequate factor of safety against shear failure and excessive settlement.

Anisotropic Soil A soil mass having different properties in different directions at any given point, referring primarily to stress-strain or permeability characteristics.

Arching The transfer of stress from an unconfined area to a less-yielding or restrained structure. Arching is important in the design of pile or pier walls that have open gaps between the members.

Bearing Capacity

> **Allowable Bearing Capacity** The maximum allowable bearing pressure for the design of foundations.
>
> **Ultimate Bearing Capacity** The bearing pressure that causes failure of the soil or rock supporting the foundation.

Bearing Capacity Failure A foundation failure that occurs when the shear stresses in the adjacent soil exceed the shear strength.

Bell The enlarged portion of the bottom of a drilled shaft foundation. A bell is used to increase the end-bearing resistance. Not all drilled shafts have bells.

Collapsible Formations Examples of collapsible formations include limestone formations and deep mining of coal beds. Limestone can form underground caves and caverns which can gradually enlarge resulting in a collapse of the ground surface and the formation of a sinkhole. Sites that are underlain by coal or salt mines could also experience ground surface settlement when the underground mine collapses.

Collapsible Soil Broadly classified as soil that is susceptible to a large and sudden reduction in volume upon wetting. Collapsible soil usually has a low dry density and low moisture content. Such soil can withstand a large applied vertical stress with a small compression, but then experience much larger settlements after wetting, with no increase in vertical pressure. Collapsible soil can include fill compacted dry of optimum and natural collapsible soil, such as alluvium, colluvium, or loess.

Compressibility A decrease in volume that occurs in the soil mass when it is subjected to an increase in loading. Some highly compressible soils are loose sands, organic clays, sensitive clays, highly plastic and soft clays, uncompacted fills, municipal landfills, and permafrost soils.

Consolidation The consolidation of a saturated clay deposit is generally divided into three separate categories:

Initial or Immediate Settlement The initial settlement of the structure caused by undrained shear deformations, or in some cases contained plastic flow, due to two- or three-dimensional loading.

Primary Consolidation The compression of clays under load that occurs as excess pore water pressures slowly dissipate with time.

Secondary Compression The final component of settlement, which is that part of the settlement that occurs after essentially all of the excess pore water pressures have dissipated.

Creep An imperceptibly slow and more or less continuous movement of slope-forming soil or rock debris.

Critical Height or Critical Slope Critical height refers to the maximum height at which a vertical excavation or slope will stand unsupported. Critical slope refers to the maximum angle at which a sloped bank of soil or rock of given height will stand unsupported.

Crown Generally, the highest point. For tunnels, the crown is the arched roof. For landslides, the crown is the area above the main scarp of the landslide.

Dead Load Structural loads due to the weight of beams, columns, floors, roofs and other fixed members. Does not include nonstructural items such as furniture, snow, occupants, or inventory.

Debris Flow An initial shear failure of a soil mass which then transforms itself into a fluid mass which can move rapidly over the ground surface.

Depth of Seasonal Moisture Change Also known as the *active zone*; the layer of expansive soil subjected to shrinkage during the dry season and swelling during the wet season. This zone extends from ground surface to the depth of significant moisture fluctuation.

Desiccation The process of shrinkage of clays. The process involves a reduction in volume of the grain skeleton and subsequent cracking of the clay caused by the development of capillary stresses in the pore water as the soil dries.

Design Load All forces and moments that are used to proportion a foundation. The design load includes the dead weight of a structure, and in most cases, can include live loads. Considerable judgment and experience are required to determine the design load that is to be used to proportion a foundation.

Downdrag Force induced on deep foundation resulting from downward movement of adjacent soil relative to foundation element. Also referred to as *negative skin friction*.

Earth Pressure Usually used in reference to the lateral pressure imposed by a soil mass against an earth-supporting structure such as a retaining wall or basement wall.

    Active Earth Pressure k_A Horizontal pressure for a condition where the retaining wall has yielded sufficiently to allow the backfill to mobilize its shear strength.

    At-Rest Earth Pressure k_0 Horizontal pressure for a condition where the retaining wall has not yielded or compressed into the soil. This would also be applicable to a soil mass in its natural state.

    Passive Earth Pressure k_p Horizontal pressure for a condition such as a retaining wall footing that has moved into and compressed the soil sufficiently to develop its maximum lateral resistance.

Effective Stress Defined as the total stress minus the pore water pressure.

Equipotential Line A line connecting points of equal total head.

Equivalent Fluid Pressure Horizontal pressures of soil, or soil and water in combination, which increase linearly with depth and are equivalent to those that would be produced by a soil of a given density. Equivalent fluid pressure is often used in the design of retaining walls.

Exit Gradient The hydraulic gradient near the toe of a dam or the bottom of an excavation through which groundwater seepage is exiting the ground surface.

Finite Element Regular geometrical shape into which a soil and structure profile is subdivided for the purpose of numerical stress analysis.

Flow Line The path of travel traced by moving groundwater as it flows through a soil mass.

Flow Net A graphical representation used to study the flow of groundwater through a soil. A flow net is composed of flow lines and equipotential lines.

Head From Bernoulli's energy equation, the total head is defined as the sum of the velocity head, pressure head, and elevation head. Head has units of length. For seepage problems in soil, the velocity head is usually small enough to be neglected and thus for laminar flow in soil, the total head h is equal to the sum of the pressure head h_p and elevation head h_e.

Heave The upward movement of foundations or other structures caused by frost heave or expansive soil and rock. Frost heave refers to the development of ice layers or lenses within the soil that causes the ground surface to heave upward. Heave due to expansive soil and rock is caused by an increase in water content of clays or rocks, such as shale or slate.

Homogeneous Soil Soil that exhibits essentially the same physical properties at every point throughout the soil mass.

Hydraulic Gradient Difference in total head at two points divided by the distance between them. Hydraulic gradient is used in seepage analyses.

Isotropic Soil A soil mass having essentially the same properties in all directions at any given point, referring primarily to stress-strain or permeability characteristics.

Laminar Flow Groundwater seepage in which the total head loss is proportional to the velocity.

Liquefaction The sudden, large decrease of shear strength of a cohesionless soil caused by collapse of the soil structure, produced by shock or earthquake-induced shear strains, associated with a sudden but temporary increase of pore water pressures. Liquefaction causes the cohesionless soil to behave as a fluid.

Live Load Structural loads due to nonstructural members, such as furniture, occupants, inventory, and snow.

Mohr Circle A graphical representation of the stresses acting on the various planes at a given point in the soil.

Normally Consolidated The condition that exists if a soil deposit has never been subjected to an effective stress greater than the existing overburden pressure and if the deposit is completely consolidated under the existing overburden pressure.

Overconsolidated The condition that exists if a soil deposit has been subjected to an effective stress greater than the existing overburden pressure.

Piping The movement of soil particles as a result of unbalanced seepage forces produced by percolating water, leading to the development of ground surface boils or underground erosion voids and channels.

Plastic Equilibrium The state of stress of a soil mass that has been loaded and deformed to such an extent that its ultimate shearing resistance is mobilized at one or more points.

Pore Water Pressure The water pressure that exists in the soil void spaces.

    Excess Pore Water Pressure The increment of pore water pressures greater than hydrostatic values, produced by consolidation stress in compressible materials or by shear strain.

    Hydrostatic Pore Water Pressure Pore water pressure or groundwater pressures exerted under conditions of no flow where the magnitude of pore pressures increase linearly with depth below the groundwater table.

Porosity The ratio, usually expressed as a percentage, of the volume of voids divided by the total volume of the soil or rock.

Preconsolidation Pressure The greatest vertical effective stress to which a soil, such as a clay layer, has been subjected.

Pressure (or Stress) The load divided by the area over which it acts.

Principal Planes Each of three mutually perpendicular planes through a point in the soil mass on which the shearing stress is zero. For soil mechanics, compressive stresses are positive.

Intermediate Principal Plane The plane normal to the direction of the intermediate principal stress.

Major Principal Plane The plane normal to the direction of the major principal stress (highest stress in the soil).

Minor Principal Plane The plane normal to the direction of the minor principal stress (lowest stress in the soil).

Principal Stresses The stresses that occur on the principal planes. Also see Mohr Circle.

Progressive Failure Formation and development of localized stresses which lead to fracturing of the soil, which spreads and eventually forms a continuous rupture surface and a failure condition. Stiff fissured clay slopes are especially susceptible to progressive failure.

Quick Clay A clay that has a sensitivity greater than 16. Upon remolding, such clays can exhibit a fluid (or quick) condition.

Quick Condition A condition in which groundwater is flowing upward with a sufficient hydraulic gradient to produce zero effective stress in the sand deposit. Soil in this condition is called *quicksand*.

Relative Density Term applied to a sand deposit; defined as the ratio of (1) the difference between the void ratio in the loosest state and the *in situ* void ratio, to (2) the difference between the void ratios in the loosest and in the densest states.

Saturation (Degree of) Calculated as the volume of water in the void space divided by the total volume of voids. It is usually expressed as a percentage. A completely dry soil has a degree of saturation of 0 percent and a saturated soil has a degree of saturation of 100 percent.

Seepage The infiltration or percolation of water through soil and rock.

  Seepage Analysis An analysis to determine the quantity of groundwater flowing through a soil deposit. For example, by using a flow net, the quantity of groundwater flowing through or underneath an earth dam can be determined.

  Seepage Force The frictional drag of water flowing through the soil voids.

  Seepage Velocity The seepage velocity is the velocity of flow of water in the soil, while the superficial velocity is the velocity of flow into or out of the soil.

Sensitivity The ratio of the undrained shear strength of the undisturbed plastic soil to the remolded shear strength of the same plastic soil.

Settlement The permanent downward vertical movement experienced by structures as the underlying soil consolidates, compresses, or collapses due to the structural load or secondary influences.

  Differential Settlement The difference in settlement between two foundation elements or between two points on a single foundation.

  Total Settlement The absolute vertical movement of the foundation.

Shear Failure A failure in a soil or rock mass caused by shearing strain along one or more slip (rupture) surfaces.

  General Shear Failure Failure in which the shear strength of the soil or rock is mobilized along the entire slip surface.

  Local Shear Failure Failure in which the shear strength of the soil or rock is mobilized only locally along the slip surface.

  Progressive Shear Failure See Progressive Failure.

  Punching Shear Failure Shear failure where the foundation pushes (or punches) into the soil because of the compression of soil directly below the footing as well as vertical shearing around the footing perimeter.

Shear Plane (or Slip Surface) A plane along which failure of soil or rock occurs by shearing.

Shear Stress Stress that acts parallel to the surface element.

Slope Stability Analyses

Gross Slope Stability The stability of slope material below a plane approximately 0.9 to 1.2 m (3 to 4 ft) deep measured from and perpendicular to the slope face.

Surficial Slope Stability The stability of the outer 0.9 to 1.2 m (3 to 4 ft) of slope material measured from and perpendicular to the slope face. See App. C, Standard Detail No. 8, for typical repair of surficial failures.

Strain The change in shape of soil when it is acted upon by stress.

Normal Strain A measure of compressive or tensile deformations; defined as the change in length divided by the initial length. In geotechnical engineering, strain is positive when it results in compression of the soil.

Shear Strain A measure of the shear deformation of soil.

Subsidence Settlement of the ground surface over a very large area, such as caused by the extraction of oil from the ground or the pumping of groundwater from wells.

Swell Increase in soil volume, typically referring to volumetric expansion of clay due to an increase in water content.

Time Factor T A dimensionless factor, used in the Terzaghi theory of consolidation or swelling of cohesive soil.

Total Stress Defined as the effective stress plus the pore water pressure. The vertical total stress for uniform soil and a level ground surface can be calculated by multiplying the total unit weight of the soil by the depth below ground surface.

Underconsolidation The condition that exists if a soil deposit is not fully consolidated under the existing overburden pressure and excess pore water pressures exist within the soil. Underconsolidation occurs in areas where a cohesive soil is being deposited very rapidly and not enough time has elapsed for the soil to consolidate under its own weight.

Void Ratio The void ratio is defined as the volume of voids divided by the volume of soil solids.

GLOSSARY 4
CONSTRUCTION AND GRADING TERMINOLOGY

Aggregate   A granular material used for a pavement base, wall backfill, etc.

   Coarse Aggregate   Gravel or crushed rock that is retained on the No. 4 sieve (4.75 mm).

   Fine Aggregate   Often refers to sand (passes the No. 4 sieve and is retained on the No. 200 U.S. standard sieve).

   Open-Graded Aggregate   Generally refers to a gravel that does not contain any soil particles finer than the No. 4 sieve.

Apparent Opening Size   For a geotextile, a property which indicates the approximate largest particle that would effectively pass through the geotextile.

Approval   A written engineering or geologic opinion by the responsible engineer, geologist of record, or responsible principal of the engineering company concerning the process and completion of the work, unless it specifically refers to the building official.

Approved Plans   The current grading plans which bear the stamp of approval of the building official.

Approved Testing Agency   A facility whose testing operations are controlled and monitored by a registered civil engineer and which is equipped to perform and verify the tests as required by the local building code or building official.

As-Graded (or As-Built)   The surface conditions at the completion of grading.

Asphalt   A dark brown to black cementitious material where the main ingredient is bitumen (high-molecular-weight hydrocarbons) that occurs in nature or is obtained from petroleum processing.

Asphalt Concrete (AC)   A mixture of asphalt and aggregate that is compacted into a dense pavement surface. Asphalt concrete is often prepared in a batch plant.

Backdrain   Generally a pipe and gravel or similar drainage system placed behind earth retaining structures such as buttresses, stabilization fills, and retaining walls. See App. C, Standard Detail No. 4.

Backfill   Soil material placed behind or on top of an area that has been excavated. For example, backfill is placed behind retaining walls and in utility trench excavations.

Base Course or Base   A layer of specified or selected material of planned thickness constructed on the subgrade or subbase for the purpose of providing support to the overlying concrete or asphalt concrete surface of roads and airfields.

Bench   A relatively level step excavated into earth material on which fill is to be placed.

Berm   A raised bank or path of soil. For example, a berm is often constructed at the top of slopes to prevent water from flowing over the top of the slope.

Borrow   Earth material acquired from an off-site location for use in grading on a site.

Brooming   The crushing or separation of wood fibers at the butt (top of the pile) of a timber pile while it is being driven.

APPENDIX A

Building Official The city engineer, director of the local building department, or a duly delegated representative.

Bulking The increase in volume of soil or rock caused by its excavation. For example, rock or dense soil will increase in volume upon excavation or by being dumped into a truck for transportation.

Buttress Fill A fill mass, the configuration of which is designed by engineering calculations to stabilize a slope exhibiting adverse geologic features. A buttress is generally specified by minimum key width and depth and by maximum backcut angle. A buttress normally contains a backdrainage system. See App. C, Standard Detail No. 3.

Caisson Sometimes large-diameter piers are referred to as caissons. Another definition is a large structural chamber utilized to keep soil and water from entering into a deep excavation or construction area. Caissons may be installed by being sunk in place or by excavating the bottom of the unit as it slowly sinks to the desired depth.

Clearing, Brushing, and Grubbing The removal of vegetation (grass, brush, trees, and similar plant types) by mechanical means.

Clogging For a geotextile, a decrease in permeability due to soil particles that have either logged in the geotextile openings or have built up a restrictive layer on the surface of the geotextile.

Compaction The densification of a fill by mechanical means.

Compaction Equipment Compaction equipment can be grouped generally into five different types or classifications: sheepsfoot, vibratory, pneumatic, high-speed tamping foot, and chopper wheels (for municipal landfill). Combinations of these types are also available.

Compaction Production Compaction production is expressed in compacted cubic meters (m^3) or compacted cubic yards (yd^3) per hour.

Concrete A mixture of aggregates (sand and gravel) and paste (Portland cement and water). The paste binds the aggregates together into a rocklike mass as the paste hardens because of the chemical reactions between the cement and the water.

Contractor A person or company under contract or otherwise retained by the client to perform demolition, grading, and other site improvements.

Cut/Fill Transition The location in a building pad where on one side the pad has been cut down exposing natural or rock material, while on the other side, fill has been placed. See App. C, Standard Detail No. 6, for one method to deal with cut/fill transition lots.

Dam A structure built to impound water or other fluid products such as tailing waste and wastewater effluent.

>**Homogeneous Earth Dam** An earth dam whose embankment is formed of one soil type without a systematic zoning of fill materials.

>**Zoned Earth Dam** An earth dam embankment zoned by the systematic distribution of soil types according to their strength and permeability characteristics, usually with a central impervious core and shells of coarser materials.

Debris All products of clearing, grubbing, demolition, or contaminated soil material that is unsuitable for reuse as compacted fill and/or any other material so designated by the geotechnical engineer or building official.

Dewatering The process used to remove water from a construction site, such as pumping from wells in order to lower the groundwater table during a foundation excavation.

Dozer Slang for bulldozer construction equipment.

Drainage The removal of surface water from the site. See App. C, Standard Detail No. 9, for typical lot drainage specifications.

Drawdown The lowering of the groundwater table that occurs in the vicinity of a well that is in the process of being pumped.

Earth Material Any rock, natural soil, or fill, or any combination thereof.

Electroosmosis A method of dewatering, applicable for silts and clays, in which an electric field is established in the soil mass to cause the movement by electroosmotic forces of pore water to wellpoint cathodes.

Erosion Control Devices (Temporary) Devices which are removable and can rarely be salvaged for subsequent reuse. In most cases they will last no longer than one rainy season. They include sandbags, gravel bags, plastic sheeting (visqueen), silt fencing, straw bales, and similar items.

Erosion Control System A combination of desilting facilities and erosion protection, including effective planting to protect adjacent private property, watercourses, public facilities, and receiving waters from any abnormal deposition of sediment or dust.

Excavation The mechanical removal of earth material.

Fill A deposit of earth material placed by artificial means. An engineered (or structural) fill refers to a fill in which the geotechnical engineer has, during grading, made sufficient tests to enable the conclusion that the fill has been placed in substantial compliance with the recommendations of the geotechnical engineer and the governing agency requirements. See App. C, Standard Detail No. 5, for typical canyon fill placement specifications.

   Hydraulic Fill A fill placed by transporting soils through a pipe using large quantities of water. These fills are generally loose because they have little or no mechanical compaction during construction.

Footing A structural member typically installed at a shallow depth that is used to transmit structural loads to the soil or rock strata. Common types of footings include combined footings, spread (or pad) footings, and strip (or wall) footings.

Forms Usually made of wood, forms are used during the placement of concrete. Forms confine and support the fluid concrete as it hardens.

Foundation That part of the structure that supports the weight of the structure and transmits the load to underlying soil or rock.

   Deep Foundation A foundation that derives its support by transferring loads to soil or rock at some depth below the structure.

   Shallow Foundation A foundation that derives its support by transferring load directly to soil or rock at a shallow depth.

Freeze Also known as *setup*, an increase in the load capacity of a pile after it has been driven. Freeze is caused primarily by the dissipation of excess pore water pressures.

Geosynthetic A planar product manufactured from polymeric material and typically placed in soil to form an integral part of a drainage, reinforcement, or stabilization system. Types include geotextiles, geogrids, geonets, and geomembranes.

Geotextile A permeable geosynthetic composed solely of textiles.

Grade The vertical location of the ground surface.

   Existing Grade The ground surface prior to grading.

   Finished Grade The final grade of the site, which conforms to the approved plan.

   Lowest Adjacent Grade Adjacent the structure, the lowest point of elevation of the finished surface of the ground, paving, or sidewalk.

   Natural Grade The ground surface unaltered by artificial means.

   Rough Grade The stage at which the grade approximately conforms to the approved plan.

Grading Any operation consisting of excavation, filling, or combination thereof.

Grading Contractor A contractor licensed and regulated who specializes in grading work or is otherwise licensed to do grading work.

Grading Permit An official document or certificate issued by the building official authorizing grading activity as specified by approved plans and specifications.

Grouting The process of injecting grout into soil or rock formations to change their physical characteristics. Common examples include grouting to decrease the permeability of a soil or rock strata, or compaction grouting to densify loose soil or fill.

Hillside Site A site which entails cut and/or fill grading of a slope which may be adversely affected by drainage and/or stability conditions within or outside the site, or which may cause an adverse affect on adjacent property.

Jetting The use of a water jet to facilitate the installation of a pile. It can also refer to the fluid placement of soil, such as jetting in the soil for a utility trench.

Key A designed compacted fill placed in a trench excavated in earth material beneath the toe of a proposed fill slope.

Keyway An excavated trench into competent earth material beneath the toe of a proposed fill slope.

Lift During compaction operations, a lift is a layer of soil that is dumped by the construction equipment and then subsequently compacted as structural fill.

Necking A reduction in cross-sectional area of a drilled shaft as a result of the inward movement of the adjacent soils.

Owner Any person, agency, firm, or corporation having a legal or equitable interest in a given real property.

Permanent Erosion Control Devices Improvements which remain throughout the life of the development. They include terrace drains, down-drains, slope landscaping, channels, storm drains, etc.

Permit An official document or certificate issued by the building official authorizing performance of a specified activity.

Pier A deep foundation system, similar to a cast-in-place pile, that consists of column-like reinforced concrete members. Piers are often of large enough diameter to enable downhole inspection. Piers are also commonly referred to as *drilled shafts, bored piles,* or *drilled caissons.*

Pile A deep foundation system, consisting of relatively long, slender, columnlike members that are often driven into the ground.

>**Batter Pile** A pile driven in at an angle inclined to the vertical to provide higher resistance to lateral loads.
>
>**Combination End-Bearing and Friction Pile** A pile that derives its capacity from combined end-bearing resistance developed at the pile tip and frictional and/or adhesion resistance on the pile perimeter.
>
>**End-Bearing Pile** A pile whose support capacity is derived principally from the resistance of the foundation material on which the pile tip rests.
>
>**Friction Pile** A pile whose support capacity is derived principally from the resistance of the soil friction and/or adhesion mobilized along the side of the embedded pile.

Pozzolan For concrete mix design, a siliceous or siliceous and aluminous material which will chemically react with calcium hydroxide within the cement paste to form compounds having cementitious properties.

Precise Grading Permit A permit that is issued on the basis of approved plans which show the precise structure location, finish elevations, and all on-site improvements.

Relative Compaction The degree of compaction (expressed as a percentage) defined as the field dry density divided by the laboratory maximum dry density.

Ripping or Rippability The characteristic of rock or dense and rocky soils that can be excavated without blasting. Ripping is accomplished by using equipment such as a Caterpillar ripper, ripper-scarifier, tractor-ripper, or impact ripper. Ripper performance has been correlated with the seismic wave velocity of the soil or rock (see *Caterpillar Performance Handbook,* 1997).

Riprap Rocks that are generally less than 1800 kg (2 tons) in mass that are placed on the ground surface, on slopes or at the toe of slopes, or on top of structures to prevent erosion by wave action or strong currents.

Running Soil or Running Ground In tunneling or trench excavations, a granular material that tends to flow or "run" into the excavation.

Sand Boil The ejection of sand at ground surface, usually forming a cone shape, caused by underground piping.

Shear Key Similar to a buttress, but generally constructed by excavating a slot within a natural slope in order to stabilize the upper portion of the slope without grading encroachment into the lower portion of the slope. A shear key is also often used to increase the factor of safety of an ancient landslide.

Shotcrete Mortar or concrete pumped through a hose and projected at high velocity onto a surface. Shotcrete can be applied by a "wet" or "dry" mix method.

Shrinkage Factor When the loose material is worked into a compacted state, the shrinkage factor (SF) is the ratio of the volume of compacted material to the volume of borrow material.

Site The particular lot or parcel of land where grading or other development is performed.

Slope An inclined ground surface. For graded slopes, the steepness is generally specified as a ratio of horizontal:vertical (e.g., 2:1 slope). Common types of slopes include natural (unaltered) slopes, cut slopes, false slopes (temporary slopes generated during fill compaction operations), and fill slopes.

Slough Loose, noncompacted fill material generated during grading operations. Slough can also refer to a shallow slope failure, such as sloughing of the slope face.

Slump In the placement of concrete, the slump is a measure of consistency of freshly mixed concrete as measured by the slump test. In geotechnical engineering, a slump could also refer to a slope failure.

Slurry Seal In the construction of asphalt pavements, a slurry seal is a fluid mixture of bituminous emulsion, fine aggregate, mineral filler, and water. A slurry seal is applied to the top surface of an asphalt pavement in order to seal its surface and prolong its wearing life.

Soil Stabilization The treatment of soil to improve its properties. There are many methods of soil stabilization such as adding gravel, cement, or lime to the soil. The soil could also be stabilized by using geotextiles, by drainage, or by compaction.

Specification A precise statement in the form of specific requirements. The requirements could be applicable to a material, product, system, or engineering service.

Stabilization Fill Similar to a buttress fill. The configuration is typically related to slope height and is specified by the standards of practice for enhancing the stability of locally adverse conditions. A stabilization fill is normally specified by minimum key width and depth and by maximum backcut angle. A stabilization fill usually has a backdrainage system. See App. C, Standard Detail No. 3.

Staking During grading, staking is the process where a land surveyor places wood stakes that indicate the elevation of existing ground surface and the final proposed elevation according to the grading plans.

Structure That which is built or constructed, an edifice or building of any kind, or any piece of work artificially built up or composed of parts joined together in some definite manner.

Subdrain (for Canyons) A pipe and gravel or similar drainage system placed in the alignment of canyons or former drainage channels. After placement of the subdrain, structural fill is placed on top of the subdrain (see App. C, Standard Detail No. 1).

Subgrade For roads and airfields, the subgrade is defined as the underlying soil or rock that supports the pavement section (subbase, base, and wearing surface). The subgrade is also referred to as the *basement soil* or *foundation soil*.

Substructure and Superstructure The substructure is the foundation and the superstructure is the portion of the structure located above the foundation (includes beams, columns, floors, and other structural and architectural members).

Sulfate (SO_4) A chemical compound occurring in some soils which, at above certain levels of concentration, has a corrosive effect on ordinary Portland cement concrete and some metals.

Sump A small pit excavated in the ground or through the basement floor to serve as a collection basin for surface runoff or groundwater. A sump pump is used to periodically drain the pit when it fills with water.

Tack Coat In the construction of asphalt pavements, the tack coat is a bituminous material that is applied to an existing surface to provide a bond between different layers of the asphalt concrete.

Tailings In terms of grading, tailings are nonengineered fill which accumulates on or adjacent to equipment haul roads.

Terrace A relatively level step constructed in the face of a graded slope surface for drainage control and maintenance purposes.

Underpinning Piles or other types of foundations built to provide new support for an existing foundation. Underpinning is often used as a remedial measure.

Vibrodensification The densification or compaction of cohesionless soils by imparting vibrations into the soil mass so as to rearrange soil particles resulting in fewer voids in the overall mass.

Walls

Bearing Wall Any metal or wood stud walls that support more than 100 pounds per linear foot of superimposed load. Any masonry or concrete wall that supports more than 200 pounds per linear foot of superimposed load or is more than one story (UBC, 1997).

Cutoff Wall The construction of tight sheeting or a barrier of impervious material extending downward to an essentially impervious lower boundary to intercept and block the path of groundwater seepage. Cutoff walls are often used in dam construction.

Retaining Wall A wall designed to resist the lateral displacement of soil or other materials.

Water-Cement Ratio For concrete mix design, the ratio of the mass of water (exclusive of that part absorbed by the aggregates) to the mass of cement.

Well Point During the pumping of groundwater, the well point is the perforated end section of a well pipe where the groundwater is drawn into the pipe.

Windrow A string of large rock buried within engineered fill in accordance with guidelines set forth by the geotechnical engineer or governing agency requirements. See Appendix C, Standard Detail No. 7.

Workability of Concrete The ability to manipulate a freshly mixed quantity of concrete with a minimum loss of homogeneity.

APPENDIX B
TECHNICAL GUIDELINES FOR SOIL AND GEOLOGY REPORTS*

A. PREFACE

The ultimate responsibility for a safe design, construction, and maintenance of any grading project rests with the consulting engineers, geologists, contractors, and the owner. Since site conditions and the proposed development plan varies so greatly between projects, the Environmental Management Agency (EMA) recognizes that discretion and judgments must be used by the consulting professionals. It is, therefore, essential to enhance the general understanding between the permit applicants, consultants, and the EMA.

The purpose of these technical guidelines is to inform grading permit applicants and their professional consultants of the basic information looked for by the Environmental Management Agency (EMA) in reviewing preliminary (initial) soil and geology reports for grading permit applications and rough grade compaction reports.

The guidelines used for the preparation of this document are: the *Orange County Grading and Excavation Code,* the *Uniform Building Code,* the California State Board of Registration policy statement on adequacy of professional geological work as represented by the guidelines for standards of practice issued by the California Division of Mines and Geology, the Orange County Planning commission and Subdivision Committee conditions of approval, the *Orange County Subdivision Code,* and presently accepted geotechnical engineering and engineering geologic practices.

B. DESCRIPTION

The technical guidelines are divided into six parts to distinguish report content for different project types and topographic areas to be developed by grading. The more involved grading projects will encompass, but not be limited to, several parts listed below:

Part I *Single Family Dwellings (flatland)*: Identifies the report content for precise grading permits on single family dwellings in flatland areas.

---

*From Orange County Grading Manual, part of Orange County Grading and Excavation Code, prepared by Orange County, California.

Part II *Single Family Dwellings (hillside):* Identifies the report content for precise grading permits on single family dwellings in hillside areas (additive to the requirements of Part I).

Part III *Single Family Dwellings (supplemental information):* Identifies additional report content which may be needed with Part I and Part II depending on the site conditions and proposed development (additive to the requirements of Parts I and II).

Part IV *Commercial and Industrial Sites:* Identifies the report content for precise grading permits on commercial and industrial sites including apartment complexes (additive to the requirements of Part I and applicable items of Part III).

Part V *Residential, Commercial and Industrial Subdivisions (tracts and parcels):* Identifies the report content for preliminary grading permits of large commercial and industrial subdivisions and preliminary and precise grading permits of residential subdivisions in flatland and hillside areas (additive to the requirements of Part I and applicable items of Parts II and III).

Part VI *Rough Grade Compaction Reports:* Identifies the report content for preliminary and precise grading compaction reports.

Due to particular site conditions, proposed improvement or the policies of testing firms or project consultants, some of these items may be included in subsequent reports on the same project with the conditional approval of EMA.

C. GRADING PLAN REVIEW REPORT

A grading plan review report is an evaluation of the conclusions and recommendations in the preliminary soil and geology report as they relate to the proposed grading plan. It is usually required when there are changes in the proposed developments, consulting firms, soil engineer or engineering geologist, an update of the preliminary report or signatures are needed, or the project is a conversion to precise permit application. The grading plan review reports are supplements to the preliminary reports and are an opportunity for the consultants to review the planned development. The purpose is to determine if the preliminary reports are adequate and complete for the presently planned grading and construction on the site and if the conclusions and recommendations still apply to the proposed operations. It is not intended that the soil engineer or engineering geologist approve or disapprove the grading plan, but provides them an opportunity to update the preliminary reports and include additions or qualifications as necessary. The date and name of the person preparing the latest grading plan review should be identified for reference purposes.

D. PART I: TECHNICAL GUIDELINES FOR PRELIMINARY REPORTS (SOIL REPORTS) ON SINGLE FAMILY DWELLINGS IN FLATLAND AREAS

a. General
  1. Signature and RCE number of project soil engineer.
  2. Job address.
  3. Location, description, and/or location index map with reference north, scale, etc.

GUIDELINES FOR SOIL AND GEOLOGY REPORTS **B.3**

    4. Description of site conditions (topography, relief, vegetation, man-made features, drainage, and watershed).

    5. Proposed grading (general scope, amount, special equipment and/or methods if applicable).

    6. Planned construction (type of structure and use, type of construction and foundation/floor system, number of stories, and estimated structural loads).

b. Field Investigations

    1. Scope (date of work done, investigative methods, sampling methods, logs of borings/test pits, elevations of borings/test pits. Reference materials and samples to finished grade or footing elevations, and identify real or assumed elevations).

    2. Plan with legend showing: site limits, terrain features, man-made features, boring/test pit locations, and proposed improvements (including slope with ratios, soil limits, daylight lines, paving areas, retaining walls, subdrains, and overexcavation/cleanout/uncertified fill areas).

    3. Location of all samples taken at the surface or subsurface.

    4. Groundwater conditions and potential (future natural and artificial seepage effects).

c. Engineering /Material Characteristics and Testing

    1. Test methods used, type or condition of samples, applicable engineering, graphics and calculations, results of all tests, and sample locations of all test samples.

    2. Unified Soil Classification of materials.

    3. Material competency and strength.[1]
   - Field densities (and relative compaction where pertinent) and moisture content.
   - Shear strength of foundation material (drained or undrained conditions, effective stress or total stress analysis, and *in situ* or remolded samples must be identified).
   - Consolidation or settlement potential.
   - Expansion potential

    4. Maximum dry density and optimum moisture parameters of proposed fill material (if available) by ASTM D 1557 or approved equivalent.

    5. Shrinkage and/or bulking factors.

d. Foundation Design Criteria

    1. Footing depth and width.[1]

    2. Criteria for foundation material preparation.[1]

    3. Allowable bearing values based on testing.[1]

    4. Lateral pressures (active, passive, or at-rest conditions) and coefficient of friction.[1]

    5. Settlement (total, differential, and rate of settlement).

e. Reference

    1. In supplemental or grading plan review reports referencing earlier reports, supply copies of those referenced reports or applicable portions as required by the building official.

f. Conclusions and Recommendations

---

[1] Uniform Building Code requirements may be used as an alternative: soil classification of founding materials by UBC Standard 18.1 (1997) and use of minimums and maximums based on UBC Tables 18-I-A and 18-I-C (1997) or approved equivalent.

1. Ground preparation (clearing, unsuitable material removal, scarification, and moisture preparation).
2. Fill support:
    - Suitability and precompaction of *in situ* materials (describe test results and other pertinent data to be used to determine suitability).
    - Densification and moisture preparation or dewatering measures (equipment, surcharge, and settlement monitoring if applicable).
3. Placement of fill:
    - Material approval (on-site, imported).
    - Methods and standard (ASTM D 1557 or approved equivalent).
    - Testing (minimum 90% relative compaction by ASTM D 1556 or equivalent) and frequency of field density testing by vertical intervals and/or volume of fill.
4. Elimination of cut/fill or other differential transitions beneath improvements.
5. Utility trenches:
    - Backfill specification and recommendations under structures, pavements, and slopes (minimum 90% relative compaction using native materials) versus landscape and other areas.
6. Provisions for approval inspections and necessary testing during and on completion of grading.
7. Opinion as to adequacy of site for the proposed development. This opinion should also be summarized in the first page of the report.
8. Other pertinent geotechnical information for the safe development of the site.

E. PART II: TECHNICAL GUIDELINES FOR PRELIMINARY REPORTS (SOIL AND GEOLOGY REPORTS) ON SINGLE FAMILY DWELLINGS IN HILLSIDE AREAS

All guidelines listed in Part I for preliminary reports are applicable in addition to the following:

a. General
1. Engineering geology report with signature and CEG number of project engineering geologist (generally needed depending on site conditions and proposed developments).
2. Source of base map with date.
3. Geologist performing mapping (if different than signing CEG).
4. Geological setting including general description, index of site on portion of recent large scale geologic map (if available) and references to previous reports (or published papers) and aerial photo data on site area.
5. Topographic features and relationship to site geology (outcrop distribution, slope height and angles and/or ratios, dip slopes, cliffs, faults, contacts, erosion pattern, etc.).

b. Field Investigations
1. Geologic map showing site geology, approximate location of proposed keyways, proposed buttresses, proposed or existing subdrains, seeps or springs, etc., and be

suitable for the general purpose in its size, scale, and manifestation, and contains an adequate legend. The map should have highlighted representative geologic data of sufficient amount and location for evaluation of: general rock and soil unit distribution, geologic structure, downslope movement features (including soil/rock creep), groundwater conditions, subsidence/settlement features or potential, and other pertinent site characteristics.

   2. Substantiation of any known gross differences of opinion with recently available geologic reports or published data or maps on site area.
c. Earth Materials (Bedrock and Surficial Units)
   1. Unit classification, general lithologic type, geologic age, and origin.
   2. Unit description and characteristics (in sequence for relative age) including:
      - Composition, texture, fabric, lithification, moisture, etc.
      - Pertinent engineering geologic attributes (clayey, weak, loose; alignments, fissility, planar boundaries; pervious or water-bearing parts; susceptibility to mass wasting, erosion, piping, or compressibility).
      - Distribution, dimensions, or occurrence (supplemental to data furnished on illustrations).
      - Suitability as construction and foundation material.
      - Effects and extent of weathering (existing and relationship to project design and future site stability, material strength, etc.).
d. Geologic Structure
   1. General structure.
   2. Distribution of structural features including position, attitude, pattern and frequency of:
      - Fissures, joints, shears, faults, and other features of discontinuity.
      - Bedding, folds, and other planar features.
   3. Character of structural features including: continuity, width of zone and activity, dominant versus subordinate, planar nature, plunge, depth, open versus closed (degree of cementation or infilling), and gouge.
   4. Structural or cross sections (one or more appropriately positioned and referenced on map; especially through critical areas, slopes, and slides) of suitable size and engineering scale; with labeled units, features and structures; and a legend. These sections should correlate with surface and subsurface data showing representative dip components, projections, and stratigraphic/structural relationships.
e. Stability Features and Conditions
   1. Adequate mapping, sections and description showing position, dimensions and type of existing downslope movement features including soil/rock creep, flows, falls, and slides, if any.
   2. Activity, cause or contributing factors of downslope movement features.
   3. Recent erosion, deposition, or flooding features.
   4. Subsidence/settlement, piping, solution or other void features or conditions.
   5. Groundwater and surface drainage characteristics or features.
      - Surface expression (past and present) and permeability/porosity of near surface materials.
      - Actual or potential aquifers or conduits, perching situations, barriers, or other controls to percolation and groundwater movement and fluctuation of groundwater levels at the site.
f. Conclusions and Recommendations (Including Slope and Site Stability).
   1. Unsuitable material removal (canyon cleanout, overexcavation, etc.)

2. Keyways and benching for existing slopes steeper than 5:1.
3. Specifications for the method of placement and compaction of soil within the zone of the slope face.
4. Slope stability, i.e. susceptibility to mass-wasting (creep to rapid failure potential).
   - Favorable or unfavorable interrelationships of fractures (joints, shear, faults, or fault zones) to planar structure (bedding, contacts, folds, plunges, weathered zones, etc.) and to each other, that forms potential failure planes, veneers, masses, or blocks.
   - Favorable or unfavorable interrelationships of geologic structures, conditions, and potential failure planes to natural and/or man-made topography, that forms actual or potential adverse dips and contacts, adverse fractures (jointing, shearing, and faulting), adverse fold limbs or synclinal axes, and adverse earth masses or blocks.
   - Favorable or unfavorable interrelationships of height of existing or proposed slopes to present and future strength of earth materials (i.e. weathering effects, such as rate, depth, etc.).
   - Slope stability effects onto or from developed, natural, or proposed slopes of adjacent properties.
5. Statement of site stability and summary of actual and potential unstable situations relative to the proposed site configuration and necessary stabilization or remedial measures for downslope movement, erosion, groundwater, or settlement/subsidence effects. Opinion and recommendations of surficial and gross stability of natural and manufactured slopes.
6. Provisions for necessary inspections of excavations to competent material by the project engineering geologist and/or soil engineer and their approval and/or testing of material competency.
7. Geologic feasibility of the site for proposed development. This opinion should also be summarized in the first part of the report.

F. PART III: TECHNICAL GUIDELINES FOR PRELIMINARY REPORTS (SOIL AND GEOLOGY REPORTS) ON SINGLE FAMILY DWELLINGS: SUPPLEMENT TO PARTS I AND II

This section includes additional report content that may be necessary depending on project site conditions or proposed developments for either flatland or hillside locations.

a. General
1. Site conditions: Indicate distress to existing improvements in area (i.e. expansive, settlement, subsidence, or creep affected areas).
2. Proposed grading: Special grading equipment or methods needed for resistant, saturated, or other unusual materials or situations.
3. Proposed rock disposal methods (for clasts and residuals larger than 12 inches) and disposal areas (include on geotechnical plan if disposal area is on-site).
4. References to publications and other reports cited.

b. Engineering/Material Characteristics and Testing
1. Shear strength evaluations and results (drained or undrained conditions, effective stress or total stress analysis, and *in situ* or remolded samples).

2. Expansive analyses of foundation material (test by UBC Standard 18-2, 1997, or approved equivalent, and classify expansion potential by UBC Table 18-I-B, 1997).

3. Material density and/or penetration tests (Standard Penetration or other methods of known correlation to material density).

4. Soluble sulfate content of soils in contact with concrete.

5. Gradation and particle size analysis, if appropriate.

6. Atterberg limits analysis and parameters, if appropriate.

7. Geophysical survey (with graphics and results), if appropriate.

8. Include all test methods used, type or condition of sample used, applicable engineering graphics and calculations, results of all tests, and sample locations of all test samples.

c. Slope Stability Analysis. Slope stability analysis is dependent on slope height and ratios, strength of earth materials, internal structure, susceptibility to weathering, actual or potential groundwater, surficial covering, proximity to site improvements or structures, and proposed landscaping and maintenance.

1. Gross stability or natural man-made slopes with calculations, graphics, supporting data, and applicable parameters.

2. Surficial stability of slopes with calculations, graphics, supporting data, and applicable parameters.

d. Seismic Evaluation. Seismic evaluation should include regional seismicity; potential for strong shaking, ground rupture, and liquefaction; and applicable parameters (peak and/or design ground acceleration, duration of strong shaking, site period) or reference to UBC standards for earthquake design (Chapter 16, UBC, 1997).

e. Foundation Design Criteria. Foundation design criteria includes special provision for expansive earth materials.

1. Footing design and placement criteria.

2. Slab thickness, reinforcement, separation and expansion joints, construction joints, doweling, and ties.

3. Bridging and grade beam specifications and recommendations, when applicable.

4. Prestressed (or posttensioned) floatation slab specifications and recommendations if this system is proposed.

5. Exterior flatwork recommendations.

6. Moisture barriers and/or selective grading (aggregate or sand base or other subbase).

7. Other moisture measures:
   - Treatment prior to concrete placement: "prepour moistening," "presoaking," or "presaturation."
   - Drainage/irrigation controls to maintain moisture content in foundation materials (including increased positive drainage, paving, cut-off walls, sealed planters, gutters and downspouts, etc.).

f. Foundation Design Criteria and Other Special Provisions

1. Soluble sulfate content specifications and recommendations based on Table 19-A-4 (UBC, 1997).

2. Footing setback from base of slopes and other setbacks (faults, fracture zones, contacts, etc.).

3. Effects of adjacent loads when footings are at differing elevations.

4. Deep foundation systems:

- Allowable bearing values.
- Foundation design criteria, parameters, and calculations when applicable.
- Additional loads or potential load caused by geologic conditions (parameters and calculations).

  5. Engineering calculations with supporting data and applicable parameters used as a basis for recommended values. These will be needed depending on the values presented relative to the foundation materials, groundwater table, proposed improvements, and imposed loads.

g. Retaining Walls. Retaining wall design criteria based on proposed walls that are surcharged or greater than 0.9 m (3 ft) in height above the base.

  1. Slope surcharge and geologic surcharge factors, parameters, and calculations.
  2. Drainage and backfill requirements including waterproofing of living areas and suitable drains.
  3. Allowable bearing values, lateral bearing resistance, and coefficient of friction based on testing or UBC (Chapter 18, UBC, 1997).
  4. Active, passive, or at-rest lateral pressure.
  5. Footing setback from base of slopes.

h. Conclusions and Recommendations

  1. Corrective or selective grading.
  2. Subgrade specifications and recommendations.
  3. Soil cement or lime stabilization.
  4. Rock clast disposal.
  5. Blasting.
  6. Irrigation and drainage controls, dewatering, and surface and subsurface drains or subdrains.
  7. Special planting and irrigation measures, slope coverings, and other erosion control measures which may be apparent from the preparation of the geotechnical report.
  8. Slough walls (including free board on retaining walls).
  9. Protection of existing structures during grading.
  10. Foundation and wall excavation inspections and approval by engineering geologist and/or soil engineer.
  11. Shoring requirements.
  12. Actual or potential effects extending into site from adjacent areas or from the site into adjacent areas and recommendations pertaining to stability, erosion, sedimentation, groundwater, etc.
  13. Stabilization measures:
      - Fill blankets for pads or stabilization blankets for slopes.
      - Stabilization fills: specifications (including subdrains and landscape) and parameters (include stability analysis and calculations if geologically surcharged).
      - Buttress fills: specifications (including landscape), subdrains, stability analyses with calculations and supporting test data and parameters.
  14. Fill over cut slope specifications and recommendations.
  15. Subsidence, collapsible soil, and piping potential (include contributing factors, time frame, and recommendations).

G. PART IV: TECHNICAL GUIDELINES FOR PRELIMINARY SOIL AND GEOLOGY REPORTS ON PRECISE COMMERCIAL/INDUSTRIAL GRADING APPLICATIONS

This section includes the necessary report content in addition to Part I and applicable items of Parts II and III for the proposed commercial/industrial development.

a. Pavement Design. For pavement design, indicate areas and type of pavement on the geotechnical plan.

   1. AC pavement design criteria:
   - R-value testing: method (ASTM D 2844-94, 1998 or equivalent), results, sample locations, and provide minimum AC sections per excavation and grading code.
   - Traffic index or projected loading conditions.
   - AC structural sections for parking areas, access areas, service areas, and heavy vehicle areas.
   - Untreated base compaction recommendations (minimum 95% relative compaction).
   - Subgrade recommendations such as minimum depth, compaction (minimum 90% relative compaction), special recommendations for bridging, or soil stabilization (such as soil cement or lime treatment, overexcavation, selective grading, etc.).

   2. Concrete pavement:
   - Minimum thickness and reinforcement.
   - Size of placed or sawed sections and expansion joints.
   - Untreated base specifications and recommendations.
   - Subgrade recommendations.

b. Seismic Evaluation. Perform a seismic evaluation if the site involves a critical or major structure or is in close proximity to an active fault. See Part III for description of necessary content.

H. PART V: TECHNICAL GUIDELINES FOR PRELIMINARY SOIL AND GEOLOGY REPORTS ON RESIDENTIAL OR COMMERCIAL/INDUSTRIAL SUBDIVISIONS (TRACTS AND PARCELS); FLATLAND OR HILLSIDE AREAS

This section includes necessary report content in addition to Part I and the applicable items of Parts II and III.

a. Seismic Evaluation: For seismic evaluation of the site, see Part III for description of necessary content.
b. Evaluate Expansion Potential of Site.
c. Stability Evaluation of Site (Slopes, Tract Boundary Areas, etc.).

I. PART VI: TECHNICAL GUIDELINES FOR ROUGH GRADE COMPACTION REPORTS

a. General

     1. Signature and RCE number of the project soil engineer.
     2. Job address, lot number, and tract number.
     3. Grading permit number.
  b. Placement of Fill
     1. Purpose for which fill was placed.
     2. Preparation of natural grade to receive fill.
     3. Placement of fill (depth of layers, watering, etc.).
     4. Equipment used for compaction.
     5. Method of compacting outer slope area.
  c. Compaction Testing
     1. Test procedure in the field and laboratory.
     2. Plot plan with the location of all density tests.
     3. Summary of test results. Include test identification number, date tests were performed, maximum dry density, optimum moisture, field dry density, field moisture, relative compaction, approximate elevation of test, and approximate finish grade elevation of test site.
  d. Utility Trench Compaction Testing
     1. Location of tests.
     2. Depth of trench and tests.
     3. Method of backfill and compaction equipment.
     4. Summary of test results.
  e. Other Testing
     1. Summary of expansion test results. Identify lots or areas with swelling potential and plot test locations on a plot plan.
     2. Summary of soluble sulfate test results.
     3. Summary of R-value tests for asphalt concrete design, if applicable.
  f. As-Built Conditions
     1. Plot plan showing limits of the approved compacted fill area. Include approximate pad elevation, depth of fill, areas of overexcavation, canyon cleanout, keys, and subdrains.
     2. Treatment of "daylight" or cut/fill transition zones. Include extent of overexcavation outside of footings.
     3. Type of soil encountered during grading, such as fill, *in situ,* or imported borrow.
     4. Groundwater conditions identified and subdrains or other methods used to mitigate adverse effects.
     5. Geologic conditions encountered.
     6. Comments on changes made during grading and their effect on the recommendations made in the geotechnical report.
  g. Recommendations and Opinions
     1. Footing recommendations and bearing value on compacted fill.
     2. Footing and floor slab recommendations based on results of expansion and soluble sulfate tests. Include construction details of footings, if applicable.

3. Pavement structural section design recommendations and specifications if applicable.
4. Approval as to the adequacy of the site for the intended use, as affected by soil engineering and/or geologic factors.
5. Opinion as to the gross and surficial stability of all slopes.
6. Opinion as to the suitability of utility trench and retaining wall backfill.
7. A statement that the soil engineering and engineering geologic aspects of the grading have been inspected and are in compliance with the applicable conditions of the grading permit and the soil engineer's and engineering geologist's recommendations.

APPENDIX C
EXAMPLE OF GRADING SPECIFICATIONS

A. GENERAL

A1. The enclosed document consists of grading recommendations and standard details. This information should be considered to be a part of the project specifications.

A2. The contractor should not vary from these specifications without prior recommendation by the geotechnical engineer and the approval of the client or the authorized representative. Recommendations by the geotechnical engineer and/or client should not be considered to preclude requirements issued by the local building department.

A3. These grading specifications may be modified and/or superseded by recommendations contained in the text of the preliminary geotechnical report and/or subsequent reports.

A4. If disputes arise out of the interpretation of these grading specifications, the geotechnical engineer shall provide the governing interpretation.

B. OBLIGATIONS OF PARTIES

B1. The geotechnical engineer should provide observation and testing services and should make evaluations to advise the client on geotechnical matters. The geotechnical engineer should report the findings and recommendations to the client or the authorized representative.

B2. The client should be chiefly responsible for all aspects of the project. The client or authorized representative has the responsibility of reviewing the findings and recommendations of the geotechnical engineer. The client shall authorize or cause to have authorized the contractor and/or other consultants to perform work and/or provide services. During grading the client or the authorized representative should remain on-site or should remain reasonably accessible to all concerned parties in order to make decisions necessary to maintain the flow of the project.

B3. The contractor should be responsible for the safety of the project and satisfactory completion of all grading and other associated operations on construction projects, including, but not limited to, earth work in accordance with the project plans, specifications, and controlling agency requirements. During grading, the contractor or the authorized representative should remain on-site. Overnight and on days off, the contractor should remain accessible.

C. SITE PREPARATION

C1. The client, prior to any site preparation or grading, should arrange and attend a meeting among the grading contractor, the design structural engineer, the geotechnical engineer, representatives of the local building department, as well as any other concerned parties. All parties should be given at least 48 hours' notice.

C2. Clearing and grubbing should consist of the removal of vegetation such as brush, grass, woods, stumps, trees, roots of trees, and otherwise deleterious natural materials from the areas to be graded. Clearing and grubbing should extend to the outside of all proposed excavation and fill areas.

C3. Demolition should include removal of buildings, structures, foundations, reservoirs, utilities (including underground pipelines, septic tanks, leach fields, seepage pits, cisterns, mining shafts, tunnels, etc.) and other man-made surface and subsurface improvements from the areas to be graded. Demolition of utilities should include proper capping and/or rerouting pipelines at the project perimeter and cutoff and capping of wells in accordance with the requirements of the local building department and the recommendations of the geotechnical engineer at the time of demolition.

C4. Trees, plants, or man-made improvements not planned to be removed or demolished should be protected by the contractor from damage or injury.

C5. Debris generated during clearing, grubbing, and/or demolition operations should be wasted from areas to be graded and disposed off-site. Clearing, grubbing, and demolition operations should be performed under the observation of the geotechnical engineer.

C6. The client or contractor should obtain the required approvals from the local building department for the project prior to, during, and/or after demolition, site preparation and removals. The appropriate approvals should be obtained prior to proceeding with grading operations.

D. SITE PROTECTION

D1. Protection of the site during the period of grading should be the responsibility of the contractor. Unless other provisions are made in writing and agreed upon among the concerned parties, completion of a portion of the project should not be considered to preclude that portion or adjacent areas from the requirements for site protection until such time as the entire project is complete as identified by the geotechnical engineer, the client, and the local building department.

D2. The contractor should be responsible for the stability of all temporary excavations. Recommendations by the geotechnical engineer pertaining to temporary excavations (e.g., back-cuts) are made in consideration of stability of the completed project and, therefore, should not be considered to preclude the responsibilities of the contractor.

D3. Precautions should be taken during the performance of site clearing, excavations, and grading to protect the work site from flooding, ponding, or inundation by poor or improper surface drainage. Temporary provisions should be made during the rainy season to adequately direct surface drainage away from and off the work site. Where low areas cannot be avoided, pumps should be kept on hand to continually remove water during periods of rainfall.

D4. During periods of rainfall, plastic sheeting should be kept reasonably accessible to prevent unprotected slopes from becoming saturated. Where necessary during periods of rainfall, the contractor should install check-dams, desilting basins, riprap, sand bags, or other devices or methods necessary to control erosion and provide safe conditions.

D5. During periods of rainfall, the geotechnical engineer should be kept informed by the contractor as to the nature of remedial or preventative work being performed (e.g., pumping, placement of sandbags or plastic sheeting, other labor, dozing, etc.).

D6. Following periods of rainfall, the contractor should contact the geotechnical engineer and arrange a walkover of the site in order to visually assess rain-related damage. The geotechnical engineer may also recommend excavations and testing in order to aid in the assessments. At the request of the geotechnical engineer, the contractor shall make excavations in order to evaluate the extent of rain-related damage.

D7. Rain-related damage should be considered to include, but may not be limited to, erosion, silting, saturation, swelling, structural distress and other adverse conditions identified by the geotechnical engineer. Soil adversely affected should be classified as unsuitable materials and should be subject to overexcavation and replacement with compacted fill or other remedial grading as recommended by the geotechnical engineer.

D8. Relatively level areas, where saturated soils and/or erosion gullies exist to depths of greater than 1.0 foot, should be over-excavated to unaffected, competent material. Where less than 1.0 foot in depth, unsuitable materials may be processed in-place to achieve near-optimum moisture conditions, then thoroughly recompacted in accordance with the applicable specifications. If the desired results are not achieved, the affected materials should be over-excavated, then replaced in accordance with the applicable specifications.

D9. In slope areas, where saturated soil and/or erosion gullies exist to depths of greater than 1.0 foot, they should be over-excavated and replaced as compacted fill in accordance with the applicable specifications. Where affected materials exist to depths of 1.0 foot or less below proposed finished grade, remedial grading by moisture conditioning in-place, followed by thorough recompaction in accordance with these grading specifications, may be attempted. If the desired results are not achieved, all affected materials should be over-excavated and replaced as compacted fill in accordance with the slope repair recommendations herein. As field conditions dictate, other slope repair procedures may be recommended by the geotechnical engineer.

E. EXCAVATIONS

E1. *Unsuitable Materials*
E1.1. Materials which are unsuitable should be excavated under observation and recommendations of the geotechnical engineer. Unsuitable materials include, but may not be

limited to: (1) dry, loose, soft, wet, organic, or compressible natural soils, (2) fractured, weathered, or soft bedrock, (3) nonengineered fill, and (4) other deleterious fill materials.

E1.2. Material identified by the geotechnical engineer as unsatisfactory due to its moisture conditions should be over-excavated, watered, or dried, as needed, and thoroughly blended to a uniform near-optimum moisture condition prior to placement as compacted fill.

E2. Cut Slopes

E2.1. Unless otherwise recommended by the geotechnical engineer and approved by the local building department, permanent cut slopes should not be steeper than 2:1 (horizontal:vertical).

E2.2. If excavations for cut slopes expose loose, cohesionless, significantly fractured or otherwise unsuitable material, overexcavation and replacement of the unsuitable materials with a compacted stabilization fill should be accomplished as recommended by the geotechnical engineer. Unless otherwise specified by the geotechnical engineer, stabilization fill construction should conform to the requirements of Standard Detail No. 3.

E2.3. The geotechnical engineer should review cut slopes during excavation. The geotechnical engineer should be notified by the contractor prior to beginning slope excavations.

E2.4. If during the course of grading, adverse or potentially adverse geotechnical or geologic conditions are encountered which were not anticipated in the preliminary report, the geotechnical engineer or engineering geologist should explore, analyze, and make recommendations to treat these problems.

E2.5. When cut slopes are made in the direction of the prevailing drainage, a non-erodible diversion swale (brow ditch) should be provided at the top-of-cut.

E3. Pad Areas

E3.1. All lot pad areas having cut/fill transitions in the building footprint should be over-excavated to provide for a minimum of 3 feet of compacted fill over the entire pad area (see Standard Detail No. 6). Cut areas exposing significantly varying material types should also be over-excavated to provide for at least a 3-foot thick compacted fill blanket. Geotechnical conditions may require greater depth of over-excavation. The actual depth should be determined by the geotechnical engineer during grading.

E3.2. For pad areas created above cut or natural slopes, positive drainage should be established away from the top-of-slope. This may be accomplished utilizing a berm and/or an appropriate pad gradient. A gradient in soil areas away from the top-of-slopes of 2 percent or greater is recommended.

F. COMPACTED FILL

All fill materials should be compacted as specified below or by other methods specifically recommended by the geotechnical engineer. Unless otherwise specified, the minimum degree of compaction (relative compaction) should be 90 percent of the laboratory maximum density (Modified Proctor).

F1. Placement

F1.1. Prior to placement of compacted fill, the contractor should request a review by the geotechnical engineer of the exposed ground surface. Unless otherwise recommended, the exposed ground surface should then be scarified (six inches minimum), watered, or dried as needed, thoroughly blended to achieve near-optimum moisture conditions, then thoroughly compacted to a minimum of 90 percent of the maximum density (Modified

Proctor). The review by the geotechnical engineer should not be considered to preclude requirement of inspection and approval by the local building department.

F1.2. Compacted fill should be placed in thin horizontal lifts not exceeding eight inches in loose thickness prior to compaction. Each lift should be watered or dried as needed, thoroughly blended to achieve near-optimum moisture conditions then thoroughly compacted by mechanical methods to a minimum of 90 percent of laboratory maximum dry density (Modified Proctor). Each lift should be treated in a like manner until the desired finished grades are achieved.

F1.3. The contractor should have suitable and sufficient mechanical compaction equipment and watering apparatus on the job site to handle the amount of fill being placed in consideration of moisture retention properties of the materials. If necessary, excavation equipment should be "shut down" temporarily in order to permit proper compaction of fills. Earth moving equipment should only be considered a supplement and not substituted for conventional compaction equipment.

F1.4. When placing fill in horizontal lifts adjacent to areas sloping steeper than 5:1 (horizontal:vertical), horizontal keys and vertical benches should be excavated into the adjacent slope area. Keying and benching should be sufficient to provide at least six-foot wide benches and a minimum of four feet of vertical bench height within the firm natural ground, firm bedrock, or engineered compacted fill. No compacted fill should be placed in an area subsequent to keying and benching until the area has been reviewed by the geotechnical engineer. Material generated by the benching operation should be moved sufficiently away from the bench area to allow for the recommended review of the horizontal bench prior to placement of fill. Typical keying and benching details have been included in Standard Detail Nos. 1 and 2.

F1.5. Within a single fill area where grading procedures dictate two or more separate fills, temporary slopes (false slopes) may be created. When placing fill adjacent to a false slope, benching should be conducted in the same manner as above described. At least a 3-foot vertical bench should be established within the firm core of adjacent approved compacted fill prior to placement of additional fill. Benching should proceed in at least 3-foot vertical increments until the desired finished grades are achieved.

F1.6. Fill should be tested for compliance with the recommended relative compaction and moisture conditions. Field density testing should conform to ASTM Method of Test D 1556 (Sand Cone), D 2922 (Nuclear Method), and D 2937 (Drive-Cylinder). Tests should be provided for about every two vertical feet or 1,000 cubic yards of fill placed. Actual test interval may vary as field conditions dictate. Fill found not to be in conformance with the grading recommendations should be removed or otherwise handled as recommended by the geotechnical engineer.

F1.7. The contractor should assist the geotechnical engineer or field technician by digging test pits for removal determinations or for testing compacted fill.

F1.8. As recommended by the geotechnical engineer, the contractor should "shut down" or remove grading equipment from an area being tested.

F1.9. The geotechnical engineer should maintain a plan with estimated locations of field tests. Unless the client provides for actual surveying of test locations, the estimated locations by the geotechnical engineer should only be considered rough estimates and should not be utilized for the purpose of preparing cross sections showing test locations or in any case for the purpose of after-the-fact evaluating of the sequence of fill placement.

F2. Moisture

F2.1. For field testing purposes, "near-optimum" moisture will vary with material type and other factors including compaction procedure. "Near optimum" may be specifically recommended in Preliminary Investigation Reports and/or may be evaluated during grading. As a preliminary guideline "near optimum" should be considered from one percent below to three percent above optimum.

F2.2. Prior to placement of additional compacted fill following an overnight or other grading delay, the exposed surface or previously compacted fill should be processed by scarification, watered or dried as needed, thoroughly blended to near-optimum moisture conditions, then recompacted to a minimum of 90 percent of laboratory maximum dry density (Modified Proctor). Where wet or other dry or other unsuitable materials exist to depths of greater than one foot, the unsuitable materials should be over-excavated.

F2.3. Following a period of flooding, rainfall, or over-watering by other means, no additional fill should be placed until damage assessments have been made and remedial grading has been performed.

F3. Fill Material

F3.1. Excavated on-site materials which are acceptable to the geotechnical engineer may be utilized as compacted fill, provided trash, vegetation, and other deleterious materials are removed prior to placement.

F3.2. Where import materials are required for use on-site, the geotechnical engineer should be notified at least 72 hours in advance of importing, in order to sample and test materials from proposed borrow sites. No import materials should be delivered for use on-site without prior sampling and testing by the geotechnical engineer.

F3.3. Where oversized rock or similar irreducible material is generated during grading, it is recommended, where practical, to waste such material off-site or on-site in areas designated as "nonstructural rock disposal areas." Rock placed in disposal areas should be placed with sufficient fines to fill voids. The rock should be compacted in lifts to an unyielding condition. The disposal area should be covered with at least three feet of compacted fill which is free of oversized material. The upper three feet should be placed in accordance with these specifications for compacted fill.

F3.4. Rocks 12 inches in maximum dimension and smaller may be utilized within the compacted fill, provided they are placed in such manner that nesting of the rock is avoided. Fill should be placed and thoroughly compacted over and around all rock. The amount of rock should not exceed 40 percent by dry weight passing the $3/4$-inch sieve size. The 12-inch and 40 percent recommendations herein may vary as field conditions dictate.

F3.5. During the course of grading operations, rocks or similar irreducible materials greater than 12 inches maximum dimension (oversized material), may be generated. These rocks should not be placed within the compacted fill unless placed as recommended by the geotechnical engineer.

F3.6. Where rocks or similar irreducible materials of greater than 12 inches but less than four feet of maximum dimension are generated during grading, or otherwise desired to be placed within an engineered fill, special handling in accordance with Standard Detail No. 7 is recommended. Rocks greater than four feet should be broken down or disposed off-site. Rocks up to four feet maximum dimension should be placed below the upper 10 feet of any fill and should not be closer than 20 feet to any slope face. These recommendations could vary as locations of improvements dictate.

Where practical, the rocks should not be placed below areas where structures or deep utilities are proposed. The rocks should be placed in windrows on a clean, over-excavated or unyielding compacted fill or firm natural ground surface. Select native or imported granular soil (SE = 30 or higher) should be placed and thoroughly flooded over and around all windrowed rock, such that voids are filled. Windrows of large rocks should be staggered so that successive strata of the large rocks are not in the same vertical plane.

The contractor should be aware that the placement of rock in windrows will significantly slow the grading operation and may require additional equipment or special equipment.

F3.7. It may be possible to dispose of individual larger rock as field conditions dictate and as recommended by the geotechnical engineer at the time of placement.

F3.8. Material that is considered unsuitable by the geotechnical engineer should not be utilized in the compacted fill.

F3.9. During grading operations, placing and mixing the materials from the cut or borrow areas may result in soil mixtures which possess unique physical properties. Testing may be required of samples obtained directly from the fill areas in order to verify conformance with the specifications. Processing of these additional samples may take two or more working days. The contractor may elect to move the operation to other areas within the project, or may continue placing compacted fill, pending laboratory and field test results. Should the contractor use the second alternative, the fill may need to be removed and recompacted depending on the outcome of the laboratory and field tests.

F3.10. Any fill placed in areas not previously reviewed and evaluated by the geotechnical engineer may require removal and recompaction at the contractor's expense. Determination of over-excavation should be made upon review of field conditions by the geotechnical engineer.

F4. Fill Slopes

F4.1. Unless otherwise recommended by the geotechnical engineer and approved by the local building department, permanent fill slopes should not be steeper than 2:1 (horizontal:vertical).

F4.2. Except as specifically recommended otherwise or as otherwise provided for in these grading specifications, compacted fill slopes should be overbuilt and cut back to grade, exposing the firm, compacted fill inner core. The actual amount of overbuilding may vary as field conditions dictate. If the desired results are not achieved, the existing slopes should be over-excavated and reconstructed per the recommendation of the geotechnical engineer. The degree of overbuilding shall be increased until the desired compacted slope surface condition is achieved. Care should be taken by the contractor to provide thorough mechanical compaction to the outer edge of the overbuilt slope surface.

F4.3. Although no construction procedure produces a slope free from risk of future movement, overfilling and cutting back of slope to a compacted inner core is, given no other constraints, the most desirable procedure. Other constraints, however, must often be considered. These constraints may include property line situations, access, the critical nature of the development, and cost. Where such constraints are identified, slope face compaction on slopes of 2:1 or flatter may be attempted as a second-best alternative by conventional construction procedures including back-rolling techniques upon specific recommendation by the geotechnical engineer.

Fill placement should proceed in thin lifts (i.e., six to eight inch loose thickness). Each lift should be moisture conditioned and thoroughly compacted. The desired moisture condition should be maintained or reestablished, where necessary, during the period between successive lifts. Selected lifts should be tested to ascertain that desired compaction is being achieved. Care should be taken to extend compactive effort to the outer edge of the slope.

Each lift should extend horizontally to the desired finished slope surface or more as needed to ultimately establish desired grades. Grade during construction should not be allowed to roll off at the edge of the slope. It may be helpful to elevate slightly the outer edge of the slope. Slough resulting from the placement of individual lifts should not be allowed to drift down over previous lifts. At intervals not exceeding four feet in vertical slope height or the capability of available equipment, whichever is less, fill slopes should be thoroughly back-rolled utilizing a conventional sheepsfoot-type roller. Care should be taken to maintain the desired moisture conditions and/or reestablishing same as needed prior to back-rolling. Upon achieving final grade, the slopes should again be moisture conditioned and thoroughly back-rolled. The use of a side-boom roller will probably be necessary and vibratory methods are strongly recommended. Without delay, so as to avoid (if possible) further moisture conditioning, the slopes should then be grid-rolled to achieve a relatively smooth surface and uniformly compact condition.

In order to monitor slope construction procedures, moisture and density tests should be taken at regular intervals. Failure to achieve the desired results will likely result in a recommendation by the geotechnical engineer to over-excavate the slope surfaces followed by reconstruction of the slopes utilizing over-filling and cutting back procedures or further attempts at the conventional back-rolling approach. Other recommendations may also be provided which would be commensurate with field conditions.

F4.4. Where placement of fill above a natural slope or above a cut slope is proposed, the fill slope configuration as presented in Standard Detail No. 2 should be adopted.

F4.5. For pad areas above fill slopes, positive drainage should be established away from the top-of-slope. This may be accomplished utilizing a berm and pad gradients of at least 2 percent in soil areas. See Standard Detail No. 9.

F5. *Off-Site Fill*

F5.1. Off-site fill should be treated in the same manner as recommended in these specifications for site preparation, excavation, and compaction.

F5.2. Off-site canyon fill should be placed in preparation for future additional fill, as shown in the accompanying Standard Detail No. 5.

F5.3. Off-site fill subdrains temporarily terminated (up canyon) should be surveyed for future relocation and connection.

G. DRAINAGE

G1. Canyon subdrain systems specified by the geotechnical engineer should be installed in accordance with Standard Detail No. 1.

G2. Typical subdrains for a compacted fill buttress or slope stabilization fill should be installed in accordance with the specifications of Standard Detail Nos. 3 or 4.

G3. Roof, pad, and slope drainage should be directed away from slopes and areas of structures to suitable disposal areas via nonerodible devices (i.e., gutters, downspouts, concrete swales).

G4. For drainage over soil areas immediately away from structures (i.e., within four feet), a minimum of 4 percent gradient should be maintained. Pad drainage of at least 2 percent should be maintained over soil areas as shown in Standard Detail No. 9. Pad drainage may be reduced to at least 1 percent for flatland projects. Flatland projects are defined as those projects where no natural or man-made slopes exist that are greater than 10 feet in height or steeper than 2:1 (horizontal:vertical) slope ratio.

G5. Drainage patterns established at the time of fine grading should be maintained throughout the life of the project. Property owners should be made aware that altering drainage patterns can be detrimental to slope stability and foundation performance.

H. STAKING

H1. In all fill areas, the fill should be compacted prior to the placement of the stakes. This particularly is important on fill slopes. Slope stakes should not be placed until the slope is thoroughly compacted (back-rolled). If stakes must be placed prior to the completion of compaction procedures, it must be recognized that they will be removed and/or demolished at such time as compaction procedures resume.

H2. In order to allow for remedial grading operations, which could include over-excavations or slope stabilization, appropriate staking offsets should be provided. For finished slope and stabilization backcut areas, at least a 10-foot setback is recommended from proposed toes and tops-of-cut.

I. MAINTENANCE

I1. Landscape Plants. In order to enhance surficial slope stability, slope planting should be accomplished at the completion of grading. Slope planting should consist of deep-rooting vegetation. Plants native to the area of grading are generally desirable. A landscape architect would be the best party to consult regarding actual types of plants and planting configuration.

I2. Irrigation.
   I2.1. Irrigation pipes should be anchored to slope faces, not placed in trenches excavated into slope faces.
   I2.2. Slope irrigation should be minimized. If automatic timing devices are utilized on irrigation systems, provisions should be made for interrupting normal irrigation during periods of rainfall.
   I2.3. Though not a requirement, consideration should be given to the installation of near-surface moisture monitoring control devices. Such devices can aid in the maintenance of relatively uniform and reasonably constant moisture conditions.
   I2.4. Property owners should be made aware that over-watering of slopes is detrimental to slope stability.

I3. Maintenance
   I3.1. Periodic inspections of landscaped slope areas should be planned and appropriate measures should be taken to control weeds and enhance growth of the landscape plants. Some areas may require occasional replanting or reseeding.
   I3.2. Terrace drains and downdrains should be periodically inspected and maintained free of debris. Damage to drainage improvements should be repaired immediately.
   I3.3. Property owners should be made aware that burrowing animals can be detrimental to slope stability. A preventative program should be established to control burrowing animals.
   I3.4. As a precautionary measure, plastic sheeting should be readily available, or kept on hand, to protect all slope areas from saturation by periods of heavy or prolonged rainfall. This measure is strongly recommended, beginning with the period of time prior to landscape planting.

I4. Repairs
   I4.1. If slope failures occur, the geotechnical engineer should be contacted for a field review of site conditions and development of recommendations for evaluation and repair.
   I4.2. If slope failures occur as a result of exposure to periods of heavy rainfall, the failure area and currently unaffected areas should be covered with plastic sheeting to protect against additional saturation.
   I4.3. The accompanying Standard Detail No. 8 provides appropriate repair procedures for surficial slope failures.

J. TRENCH BACKFILL

J1. Utility trench backfill should, unless otherwise recommended, be compacted by mechanical means. Unless otherwise recommended, the degree of compaction should be a minimum of 90 percent of the laboratory maximum density (Modified Proctor).

J2. As an alternative, granular material (sand equivalent greater than 30) may be thoroughly jetted in-place. Jetting should only be considered to apply to trenches no greater than two feet in width and four feet in depth. Following jetting operations, trench backfill should be thoroughly mechanically compacted and wheel-rolled from the surface.

J3. Backfill of exterior and interior trenches extending below a 1:1 projection from the outer edge of foundations should be mechanically compacted to a minimum of 90 percent of the laboratory maximum density (Modified Proctor).

J4. Within slab areas, but outside the influence of foundations, trenches up to one foot wide and two feet deep may be backfilled with sand and densified by jetting, flooding or by mechanical means. If on-site materials are utilized, they should be wheel-rolled, tamped or otherwise compacted to a firm condition. For minor interior trenches, density testing may be deleted or spot testing may be elected if deemed necessary, based on review of backfill operations during construction.

J5. If utility contractors indicate that it is undesirable to use compaction equipment in close proximity to a buried conduit, the contractor may elect the utilization of lightweight mechanical compaction equipment or shading of the conduit with clean, granular material, which should be thoroughly jetted in-place above the conduit, prior to initiating mechanical compaction procedures. Other methods of utility trench compaction may also be appropriate, upon review by the geotechnical engineer at the time of construction.

J6. In cases where clean granular materials are proposed for use in lieu of native materials or where flooding or jetting is proposed, the procedures should be considered subject to review by the geotechnical engineer.

J7. Clean granular backfill and/or bedding are not recommended in slope areas unless provisions are made for a drainage system to mitigate the potential build-up of seepage forces.

K. STATUS OF GRADING

Prior to proceeding with any grading operation, the geotechnical engineer should be notified at least two working days in advance in order to schedule the necessary observation and testing services.

K1. Prior to any significant expansion or cutback in the grading operation, the geotechnical engineer should be provided with adequate notice (i.e., two days) in order to make appropriate adjustments in observation and testing services.

K2. Following completion of grading operations or between phases of a grading operation, the geotechnical engineer should be provided with at least two working days' notice in advance of commencement of additional grading operations.

EXAMPLE OF GRADING SPECIFICATIONS   C.11

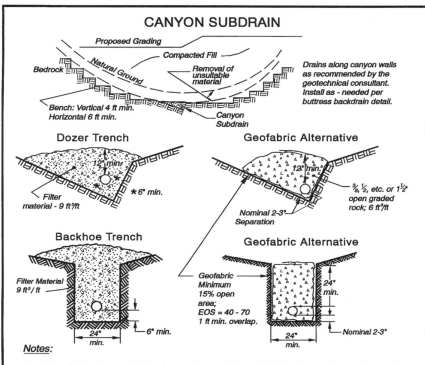

Notes:

1- Pipe be 4" min. diameter, 6" min. for runs of 500 ft to 1000 ft, 8" min. for runs of 1000ft. or greater.

2- Pipe should be schedule 40 PVC or similiar. Upstream ends should be capped.

3- Pipe should have 8 uniformly spaced 3/8" perforations per foot placed at 90° offset on underside of pipe. Final 20 feet of pipe should be nonperforated.

4- Filter material should be Calif. Class 2 Permable Material.

5- Appropriate gradiant should be provided for drainage; 2% minimum is recommended.

6- For the Geofabric Alternatives and gradients of 4% or greater, pipe may be omitted from the upper 500 ft. For runs of 500, 1000, and 1500 ft or greater 4", 6", and 8" pipe, respectively, should be provided.

STANDARD DETAIL NO. 1

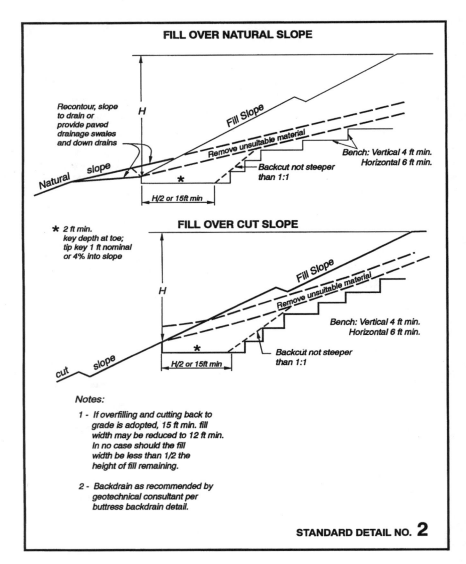

EXAMPLE OF GRADING SPECIFICATIONS

STABILIZATION FILL

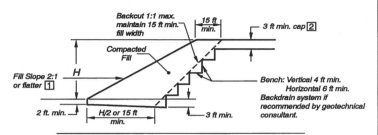

BUTTRESS FILL

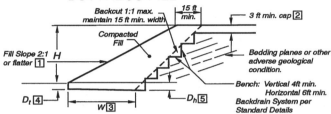

Notes:

[1] - If overfilling and cutting back to grade is adopted, 15 ft may be reduced to 12 ft. In no case should the fill width be less than 1/2 the fill height remaining.

[2] - A 3 ft blanket fill shall be provided above stabilization and buttress fills. The thickness may be greater as recommended by the geotechnical engineer.

[3] - W = designed width of key.

[4] - D_t = designed depth of key at toe

[5] - D_h = depth of key at heel; unless otherwise specified, $D_h = D_t + 1$ft

STANDARD DETAIL NO. 3

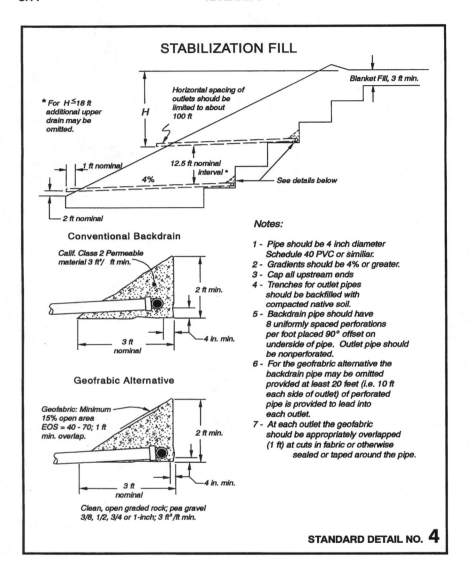

EXAMPLE OF GRADING SPECIFICATIONS C.15

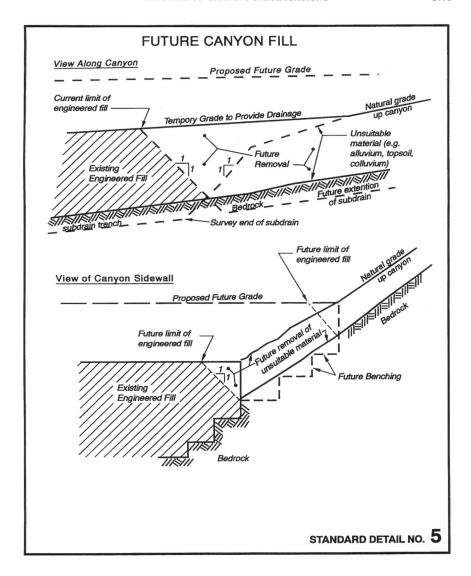

STANDARD DETAIL NO. 5

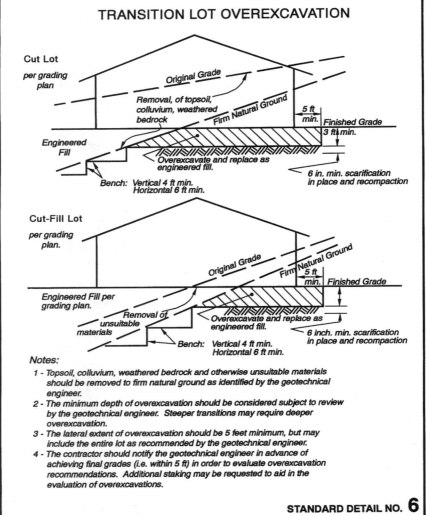

EXAMPLE OF GRADING SPECIFICATIONS C.17

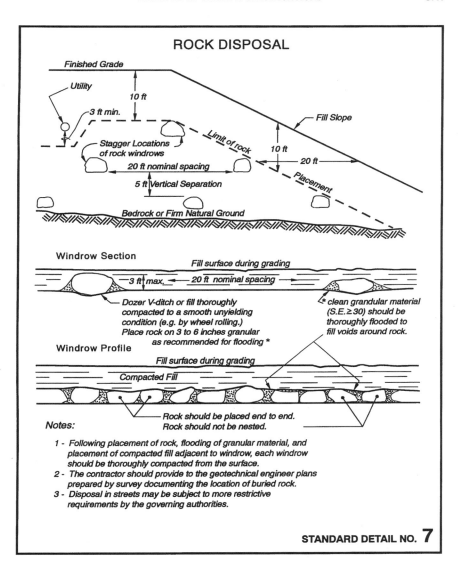

STANDARD DETAIL NO. 7

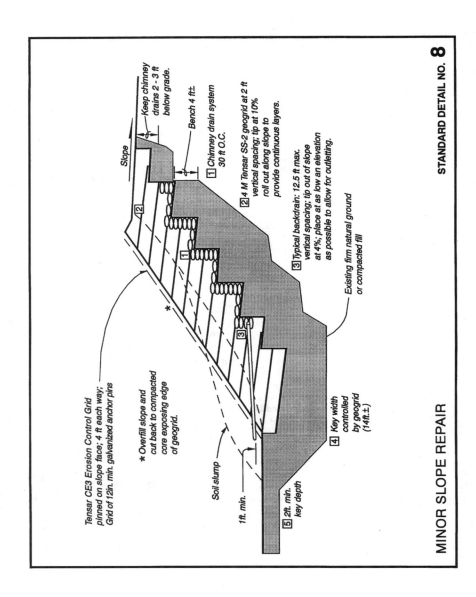

EXAMPLE OF GRADING SPECIFICATIONS C.19

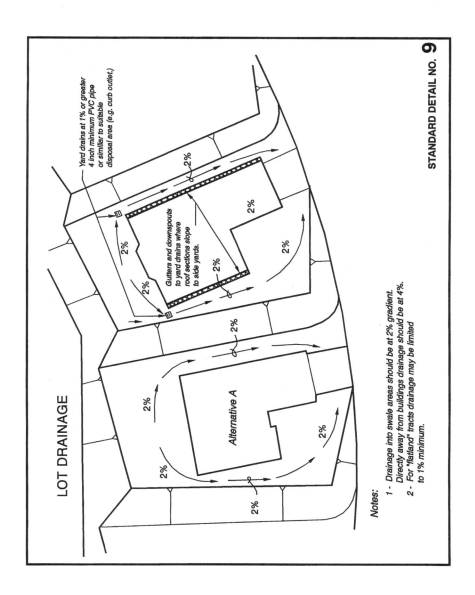

APPENDIX D
PERCOLATION TEST PROCEDURE*

Purpose: To establish clear direction and methodology for percolation testing in San Diego County.

Background: "Hybridization" of percolation test procedures, concerns expressed by industry regarding "perc" tests, and inconsistency in observed test methods have created a need to standardize a method that can be performed easily and will have results that can be duplicated on the basis of a single, county-wide test.

Objective of the Test: Determine the area necessary to properly treat and maintain sewage underground, to size the disposal system with adequate infiltration surface on the basis of expected hydraulic conductivity of the soil and the rate of loading, and to provide for a system with long term expectation of satisfactory performance.

Policy: All percolation testing in the county of San Diego shall follow procedures described herein. Any deviation shall be authorized only after receiving written approval by the department of environmental health services.

San Diego County Percolation Test Procedures:

1. **TEST HOLES**
   A. Number of Test Holes
      - A minimum of four test holes are required when percolation rates are less than 60 min./inch.
      - A minimum of six test holes are required when the average percolation rate is more that 60 min./inch. For those soils having an average percolation rate greater than 60 min./inch, the requirements of Table D.2 must be met.
      - Additional test holes may be necessary for adequate and sufficient information on a site specific basis for reasons that include, but are not limited to, unacceptable or failed tests, areas of the disposal field requiring defined limits for exclusion, a disposal system located outside the concentrated area, and soil conditions that are variable or inconsistent.

---

*Reproduced from "Percolation Test Procedure," issued by the Department of Health Services and Land Use Division, County of San Diego, California, 1991. Document has been edited for content.

B. Depth of Testing
- Test holes shall be representative of the leach line installation depth.
- Conditions which may require testing deeper than leach line depth include shallow consolidated rock, impervious layers, shallow groundwater, slope gradient that exceeds 25%, and other factors as might be determined by sound engineering practices.

C. Soil Classification
- All test holes and deep borings shall have adequate definition of soil types, consolidation, and rock.
- All borings are to be reported, including any which encountered groundwater or refusal.

D. Location of Test Holes
- Test holes shall be representative of the disposal area, demonstrating site conditions throughout the entire leach field, equal consideration of primary and reserve, and distances between test holes that is adequate to describe the disposal area.

E. Identification of Test Holes
- Stake and flag test holes, as needed, so holes can be located. Identify test holes with a test hole number or letter and depth of the test boring.

F. Drilling or Boring of Test Holes
- Diameter of each test hole shall be a minimum of six inches and a maximum of ten inches.
- If a backhoe excavation is used, a test hole that is 12 to 14 inches in depth should be excavated into the bottom of the trench.

G. Preparation of Test Holes
- The sides and bottom of the holes shall be scarified so as to remove the areas that became smeared by the auger or other tool used to develop the hole. Remove all loose material from the hole.
- Place 2 inches of open-graded gravel in the bottom of the hole, prior to running the percolation test (maximum size of open-graded gravel is $1/2$ inch).

2. PRESOAKING

A. Filling the Test Hole
- Carefully fill the test hole with 12 to 14 inches of clear water over the two inches of open-graded gravel.

B. Time and Duration of Presoaking
- Maintain 12 to 14 inches of clear water over the two inches of open-graded gravel for a minimum of four hours. After four hours, allow the water column to drop overnight. Testing must be done within 15 to 30 hours after the initial four hour presoak.
- Overnight option: If clay soils are present, it is recommended to maintain the 12 to 14 inches water column overnight. A siphon can be used to maintain the supply at a constant level.
- If overnight presoak option is used, it will eliminate the normal 1-hour presoak on the following day.
- In highly permeable sandy soils with no clay and/or silt, the presoak procedure may be modified. If after filling the hole twice with 12 to 14 inches of clear water, the water seeps completely away in less than 30 minutes, proceed immediately to Case 2 (Item 3) and refill to 6 inches above the open-graded gravel. If the test is done the following day, a presoak will be necessary.

C. Saturation and Swelling
- Saturation means that the void spaces between the soil particles are full of water. This can be accomplished in a short period of time.
- Swelling is caused by the intrusion of water into the void spaces between the soil particles. This is a slow process, especially in clay-type soil and is the reason for requiring a prolonged soaking.

D. Use of Inserts
- If side walls are not stable or sloughing results in changing depth, the test hole may be abandoned or retested after means are taken to shore up the sides. The holes shall be re-cleaned prior to restart of the test.
- Options for shoring or maintaining test hole stability include using a hardware cloth ($1/8$ inch grid) or perforated pipe or containers.

3. DETERMINATION OF THE PERCOLATION RATE

Case 1, the water remains overnight following the four hour presoak.

Case 2, fast soil with two columns of 12 to 14 inches of water percolating in less than 30 minutes during second presoak period.

Case 3, no water remains 15 to 30 hours after four hour presoak.

A. Case 1 Testing Procedure
- Adjust depth of water to six inches over the gravel.
- Take two readings at thirty minute intervals and report percolation rate as the slowest of the two readings. If a minimum amount of water remains due to a damaged hole or silting, the hole may be cleaned out and tested under Case 3, starting with the presoak.

B. Case 2 Testing Procedure
- After filling the hole twice with 12 to 14 inches of clear water, observe to see if the water will seep away in less than 30 minutes. If so, then proceed with this test procedure. If not, go to Case 3.
- Refill to 6 inches above the open-graded gravel.
- Measure from a fixed reference point at ten minute intervals over a period of one hour to the nearest $1/16$ inch. Add water at each ten minute time interval.
- Continue 10 minute readings as long as necessary to obtain a "stabilized" rate with the last two rate readings not varying more than $1/16$ inch. The last water drop will be considered the percolation rate.

C. Case 3 Testing Procedure
- Maintain a static head of 12 to 14 inches of clean water over the open-graded gravel for a period of one hour.
- Adjust water depth to 6 inches above the two inches of open-graded gravel and measure from a fixed reference point at thirty minute intervals to the nearest $1/16$ inch.
- Refill the hole as necessary to maintain a 6 inch column of water over the 2 inches of open-graded gravel. A fall of one inch can be allowed before refilling, for example; if fall is less than one inch, allow test to continue to next thirty minute reading interval.
- Continue 30 minute readings as long as necessary to obtain a "stabilized" rate with the last two rate readings not varying more than $1/16$ inch. The test shall run a minimum of three hours after the one hour presoak.
- The last water level drop is used to calculate the percolation rate.

4. CALCULATIONS AND MEASUREMENTS

A. Calculation Example
- Percolation rate is reported in minutes per inch, for example, 30 minute time interval with a $3/4$ inch fall would be 30 minutes divided by $3/4$ inch, or 40 minutes per inch (mpi).

B. Measurement Principles
- The time interval for readings are to reflect the actual times and are to be maintained as near as possible to the intervals outlined for the test (i.e., 10 or 30 minutes).
- Measurements to the nearest $1/16$ inch should be adjusted to the slowest rate. For example, a reading observed between $3/8$ inch and $5/16$ inch (80 mpi and 96 mpi) would be reported as 96 mpi.
- Measurements on an engineering scale (tenths of an inch) should follow the same principle. For example, a reading observed between 0.4 inch and 0.3 inch (75 and 100 mpi) would be reported as 100 mpi.

C. Measurement and Special Considerations
- Measurement from a fixed reference point shall be from a platform that is stable and represents the center of the hole.
- Accurate measuring devices are encouraged for the water level measurements, especially when the test depth is greater than 60 inches. A description of the measurement and clean-out methodology may be required.

5. REPORTS. All test data and required information shall be submitted on approved Environmental Health Service forms (four copies). Reports shall be signed with an original signature by the consultant. San Diego County Code requires all percolation testing to be done by a civil engineer, geologist, or environmental health specialist, registered in the state of California. These consultants are required to be on the approved list on file with the Environmental Health Services.

The percolation test is only one critical factor in siting an on-site disposal system. Site conditions may require special evaluation by a consultant qualified to technically address issues such as high groundwater, steep slopes, nitrate impacts, cumulative impacts, mounding, and horizontal transmissibility.

Companies whose consultants employ a technician are responsible for the work performed by the technician. It is incumbent upon the consultant to properly train, equip, and supervise anyone performing work under their direction and license.

TABLE D.1 Primary Disposal Trench Length (feet) Based on the Average Percolation Rate (minutes per inch)

Average Percolation Rate (minutes per inch)	Primary Disposal Trench Length (feet) Columns Below Indicate the Number of Bedrooms in the House				
	2	3	4	5	6
<3	200	240	270	280	300
5	240	290	320	320	340
10	275	330	370	410	420
15	300	360	400	450	470
20	315	380	430	470	530
25	330	400	450	500	600
30	350	420	470	540	650
35	365	440	490	590	710
40	380	460	520	630	750
45	400	480	540	670	800
50	415	500	560	700	840
55	430	520	580	740	890
60	450	540	610	770	930
65	500	590	660	780	940
70	550	640	710	790	950
75	600	690	760	810	960
80	650	740	810	860	970
85	700	790	860	910	980
90	755	845	915	965	1005
95	830	920	990	1040	1080
100	905	995	1065	1115	1155
105	980	1070	1140	1190	1230
110	1055	1145	1215	1265	1305
115	1130	1220	1290	1340	1380
120	1210	1300	1370	1420	1460
125	1310	1400	1470	1520	1560
130	1410	1500	1570	1620	1660
135	1510	1600	1670	1720	1760
140	1610	1700	1770	1820	1860
145	1710	1800	1870	1920	1960
150	1810	1900	1970	2020	2060

Notes: An 18-inch trench is the maximum allowable width to be used in determining the total linear drain line footage. Septic tank sizes: 2–3 bedrooms requires 1000 gallon, 4 bedrooms requires 1200 gallon, and 5–6 bedrooms requires 1500 gallon.

TABLE D.2 Standards and Requirements for Design and Installation of Subsurface Sewage Disposal Systems for Soils Having Poor Percolation (i.e., over 60 mpi)

Average Percolation Rate (minutes per inch)	Required Minimum Lot Size (acres)	Available Expansion Area (percent)
61 to 90	3	200
91 to 120	5	300
121 to 150	7	400
over 150	10	500

Notes: This table is for soils having an average percolation rate over 60 mpi. Percolation rates must be made at six different locations on the site of the proposed subsurface sewage disposal field. There shall be a minimum of 10 feet of soil above any impervious formation such as rock, clay, adobe, and/or groundwater table. Fractured and hard rock will not be considered as soil. Deep testing can be required to ensure uniform conditions exist below the disposal area. The land on which the subsurface sewage disposal system will have to be installed shall not have a slope gradient greater than 25%.

APPENDIX E
EXAMPLE OF A PRELIMINARY GEOTECHNICAL REPORT

A. INTRODUCTION

This report presents the results of the geotechnical and geologic investigation for a proposed 40-acre research facility. The research facility will include the construction of six large buildings, parking lots, and artificial ponds. The purpose of the investigation was to assess the feasibility of the project and provide geotechnical and geologic parameters for the design of the project.

B. SCOPE OF SERVICES

The scope of services for the project included the following:

- Review of published and unpublished geologic, seismicity, and soil engineering maps and reports pertinent to the project.
- Subsurface exploration consisting of 14 bucket auger borings to a maximum depth of 47 ft.
- Logging and sampling of exploratory borings to evaluate the geologic structure and to obtain samples for laboratory testing.
- Laboratory testing of samples representative of those obtained during the field investigation.
- Geologic and soil engineering evaluations of field and laboratory data which provide the basis for the conclusions and recommendations.
- Preparation of this report and other graphics presenting the findings, conclusions, and recommendations.

C. PROPOSED DEVELOPMENT AND SITE DESCRIPTION

The proposed development consists of six large buildings: Administration Building, Functional Building, Management Module, Projects Building, Hardware Building, and Central Plant. Future expansion is proposed for the Administration, Functional, Projects, and Hardware Buildings. The proposed buildings are to be clustered in the central section of the site with parking lots planned around the buildings. The location of the proposed buildings and parking lots are shown on the geotechnical map.

Anticipated loading conditions on a building by building basis were obtained from the structural engineer.

The site is located in gently sloping terrain. Topographically, the site has been rough-graded to the existing contours indicated on the geotechnical map. Cut and fill regrading is planned to achieve the final pad and pavement elevations.

D. FIELD INVESTIGATION AND LABORATORY TESTING

The field investigation performed during the course of this investigation consisted of geologic reconnaissance, mapping, and subsurface exploration consisting of 14 bucket auger borings. The field work was conducted under the direction of the engineering geologist or geotechnical engineer. A truck-mounted bucket auger drill rig was used to drill the borings to a maximum depth of 47 ft. The borings were logged by visual and tactile methods, selectively sampled, and backfilled. The logs of borings are included in the appendix.* The locations of the borings are shown on the geotechnical map. Soil samples recovered from the borings were placed in moisture-resistant containers and transported to the laboratory for testing.

The laboratory testing program consisted of moisture-density determinations, direct shear testing of undisturbed and remolded specimens, modified Proctor compaction tests, Atterberg limits, particle-size analyses, one-dimensional compression tests, expansion index tests, and soluble sulfate determinations. Results and descriptions of the laboratory tests performed are included in the appendix.*

E. GEOLOGY AND SEISMICITY

1. Geology

The site is located in the Peninsular Ranges Geomorphic Province of California near the western limits of the southern California batholith. The topography at the edge of the batholith changes from the typically rugged landforms developed over the granitic rocks to the more subdued landforms characteristic of the sedimentary bedrock of the local embayment.

The site is underlain by the Eocene-aged Mission Valley formation. The Mission Valley formation is a marine lagoonal and nonmarine sandstone with interbedded siltstone and claystone. Cemented layers of sandstone and siltstone were observed on site. The sandstone is typically off-white to gray, medium dense, friable, and fine-grained. The siltstone is poorly bedded to massive. The Mission Valley formation appears to be weathered to considerable depth. Heavy caliche deposits were noted to 15 to 20 ft below ground surface. Topsoil, alluvium, and colluvium have generally been removed by grading from the site.

The Mission Valley formation is considered to be a non-water-bearing material, although several zones of seepage were encountered during our investigation. These zones appear to be unrelated to each other and are the result of infiltrating rainwater.

---

*Appendix is not included in this example.

2. Seismicity

The site can be considered a seismically active area, as can all of southern California. There are, however, no active faults on or adjacent to the site. Seismic risk is considered low to moderate compared to other areas of southern California.

Seismic hazards within the site can be attributed to ground shaking resulting from events on distant active faults. There are several active and potentially active faults which can significantly affect the site. Based on an analysis of possible earthquake accelerations at the site, the most significant event is a 7.0 magnitude event on the Elsinore fault zone. The ground surface accelerations produced at the site by such an event would exceed those events on any other known fault.

A magnitude 7.0 earthquake on the Elsinore fault zone could produce a peak horizontal bedrock acceleration of $0.22g$ at the subject site with the duration of strong shaking exceeding 30 seconds. The material underlying the site consists primarily of sandstone and siltstone bedrock which will not be subjected to seismically induced settlement or liquefaction.

F. ENGINEERING EVALUATIONS

The results of the subsurface exploration indicate that the predominant material at the site will consist of Mission Valley formation bedrock with or without an upper layer of fill. Fill derived from the Mission Valley formation bedrock will be the predominant material encountered during regrading to obtain final pad and parking lot elevations. The Mission Valley formation bedrock underlying the site consists of generally horizontal layers of sandstone and siltstone. The intact bedrock is generally in a moist, medium-dense to dense state with numerous near-horizontal, moderately cemented layers. At fill/bedrock contact zones, the bedrock is weathered and fractured in a moist-to-wet and soft-to-firm state. Brief discussions of the engineering properties of the Mission Valley formation sandstone and siltstone and fill derived from this material are presented below:

1. Engineering Properties of the Mission Valley Formation Sandstone

Results of particle size analysis performed on remolded samples of the Mission Valley formation sandstone indicate that it contains approximately 60 percent fine-sand-size particles and 40 percent fines (silt- and clay-size particles) based on dry weight. The remolded sandstone has a medium to low expansion potential (expansion index test), depending on the amount of fines. Typical index and engineering properties of the Mission Valley formation sandstone are as follows:

a. Index Properties

- Water content $(w) = 16\% \pm 2\%$
- Total unit weight $(\gamma_t) = 132$ pcf
- Dry unit weight $(\gamma_d) = 114$ pcf
- Soil classification: plastic silty sand

b. Engineering Properties

- Peak shear strength of cemented sandstone (for effective normal stress range of 1000 to 4000 psf): friction angle $(\phi') = 15°$, cohesion intercept $(c') = 1200$ psf

- Ultimate shear strength of sandstone (for effective normal stress range of 1000 to 4000 psf): friction angle (ϕ') = 31°, cohesion intercept (c') = 150 psf
- Recompression ratio (RR) = 0.009
- Remolded expansion potential (expansion index test) = medium to low

2. Engineering Properties of Mission Valley Formation Siltstone

Results of particle size analysis performed on remolded samples of the Mission Valley formation siltstone indicate that it contains approximately 80 percent silt-size particles and 20 percent clay-size particles, on a dry weight basis. Atterberg limits performed on remolded siltstone indicate a plasticity index of 23 and a liquid limit of 48, which would classify the soil as a silty clay of intermediate plasticity (ISBP). Remolded siltstone typically has a high expansion potential (expansion index test). Typical index and engineering properties of the Mission Valley formation siltstone are as follows:

a. Index Properties

- Water content (w) = 17% ± 3%
- Total unit weight (γ_t) = 132 pcf
- Dry unit weight (γ_d) = 113 pcf
- Soil classification: silty clay of intermediate plasticity

b. Engineering Properties

- Shear strength of remolded siltstone (for effective normal stress range of 1000 to 4000 psf): friction angle (ϕ') = 21°, cohesion intercept (c') = 450 psf. (*Note:* The effective shear strength envelope for remolded siltstone is shown in Fig. E.1.)
- Recompression ratio (RR) of intact siltstone = 0.007.
- Compression ratio (CR) of remolded siltstone = 0.14.
- Remolded expansion potential (expansion index test) = high.

3. Engineering Properties of Fill Soils

At some locations, compacted fill presently overlies the bedrock. This fill is highly variable, consisting of sand-, silt-, and clay-size particles with differing proportions of each soil type at each different location. At some locations the fill is classified as a plastic sandy silt, while at other locations the fill is classified as a plastic silty clay. The fill thickness presently varies within the building areas.

4. General Description of Building Foundations and Loads

Foundation information was received from the structural engineer for the project and is summarized below on a building-by-building basis. The location of each building is shown on the geotechnical map.

a. *Administrative Building:* The proposed Administrative Building consists of an interconnected one and two-story structure. A two-story structure atop an 80-ft-wide

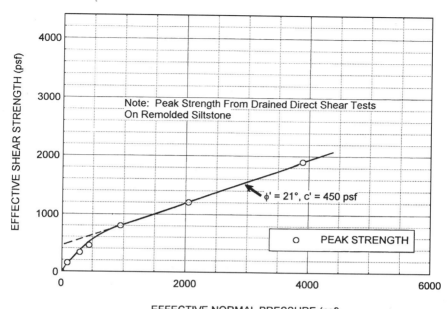

FIGURE E.1 Effective shear strength envelope for remolded siltstone from drained direct shear tests.

breezeway is planned for the central portion. Maximum column loads are expected to be 150 kips (dead load plus live load). Finished grade elevation is about 813 ft. A one-story structure is proposed at each end of the Administrative Building. Maximum column loads are expected to be 100 kips (dead load plus live load). Finished grade elevation is about 827 ft.

b. *Functional Building:* The proposed Functional Building consists of an interconnected two- and three-story structure. The two-story portion has a finished grade elevation of 814 ft and the three-story portion has a finished grade elevation of 800 ft. A restrained interior retaining wall (14 ft in height) is proposed to accommodate the differences in floor elevations. Column loads vary from 200 to 550 kips (dead load plus live load).

Six large (800 ft^2 in area) interior light wells are proposed for the Functional Building. Trees and other heavy vegetation will be contained within the light wells. The ground elevation in all light wells is expected to be 2 ft higher than the adjacent finished grade elevation.

c. *Management Module:* The proposed Management Module is expected to be supported on conventional spread footings. The maximum column loads are 350 kips (dead load plus live load). Finished grade elevation is about 795 ft with an elevated final floor elevation of 827 ft.

d. *Projects Building:* The proposed two-story Projects Building will consist of both cast-in-place concrete walls and tilt-up walls. Thus, the majority of the foundation will support exterior walls and consist of conventional wall footings with a maximum load of 8 kips per linear foot (dead load plus live load). Finished grade elevation is about 800 ft.

Four large (800 ft^2 in area) interior light wells containing vegetation are also proposed for the Projects Building. Columns with spread footings are located in each corner of the light wells.

e. *Hardware Building:* The proposed interconnected one and two-story Hardware Building has three sections with different final grade elevations (781, 786, and 799 ft). Restrained interior retaining walls (maximum height of 19 ft) are to be constructed to accommodate the differences in floor elevations. The upper level of the Hardware Building will have maximum column loads of 200 kips (dead load plus live load) while the lowest level has maximum column loads of 300 kips (dead load plus live load). A large (1600 ft^2 in area) interior light well containing vegetation is also proposed for the Hardware Building.

f. *Central Plant:* The proposed Central Plant will consist of a steel-framed building. Column loads were not available as of the date of this report. The finished grade elevation is 787 ft.

g. *Circulation Spine:* A proposed exterior walkway will connect all adjoining buildings.

5. General Discussion of Potential Soil-Foundation Problems

A discussion of the potential foundation problems is presented below.

a. **Cut/Fill Transitions.** In consideration of proposed cut/fill regrading to obtain final pad elevations, it is anticipated that some buildings will straddle cut/fill transitions created as a result of the regrading. Such transitions are considered generally undesirable because of nonuniform bearing conditions resulting from different bearing materials anticipated at final pad elevations after grading is complete. The following is a summary of the anticipated pad conditions upon completion of regrading:

- Administrative Building: mostly fill
- Functional Building: Mission Valley formation bedrock
- Management Module: Mission Valley formation bedrock
- Projects Building: mostly Mission Valley formation bedrock, some fill
- Hardware Building: mostly Mission Valley formation bedrock, some fill
- Central Plant: Mission Valley formation bedrock

It is anticipated that the Functional Building, Management Module, and Central Plant will have all footings founded directly on Mission Valley formation bedrock. Since the Projects Building and the Hardware Building may have some shallow fill at final pad elevations, deepened footings through this fill would be desirable so that all footings are founded in bedrock. Because of deeper fill, the foundation options for the Administration Building are (1) overexcavate the cut area, replace as compacted fill, and place all footings on fill; or (2) construct piers through the fill bearing on the Mission Valley formation bedrock.

The significant aspect of the first option would be the potential for differential settlement between the Administration Building and the adjacent Functional Building. The significant impact of the second option would be the additional cost for the deepened foundations.

b. **Expansive Soil.** Results of expansion index tests indicate that the Mission Valley formation siltstone and fill derived from this material will have a medium to high expansion potential. Figure E.2 presents laboratory results where the percent swell versus normal stress is plotted for remolded siltstone. Under low loads, the remolded siltstone can swell excessively (up to 33 percent swell at ground surface) if access to water is provided. From the data in Fig. E.2, the magnitude of ground heave for different depths of saturation is summarized below:

Depth of saturation = 1 ft, amount of ground heave = 1.6 in.
Depth of saturation = 2 ft, amount of ground heave = 2.7 in.
Depth of saturation = 3 ft, amount of ground heave = 3.6 in.
Depth of saturation = 4 ft, amount of ground heave = 4.4 in.

The amount of ground heave for only 1 ft of saturation could be enough to cause heaving and cracking of concrete walks, driveways, floor slabs, etc. On some projects, the potential adverse effects of expansive soils can be mitigated by selective grading. On the project, because of the predominance of Mission Valley siltstone, selective grading would require the importing of select soil. Some other measures to mitigate the effects of expansive soil are:

- Control seepage by utilizing drainage systems for planters, pavements, etc.
- Eliminate leakage from artificial ponds and streams by use of impervious liners.
- Provide additional steel reinforcement in slabs, footings, etc., to resist potential expansive forces.
- Beneath lightly loaded floor slabs, presaturate subgrade soils to allow them to swell and thus alleviate the majority of their potential expansion. Lime treatment of the subgrade is another option to mitigate potential expansion.

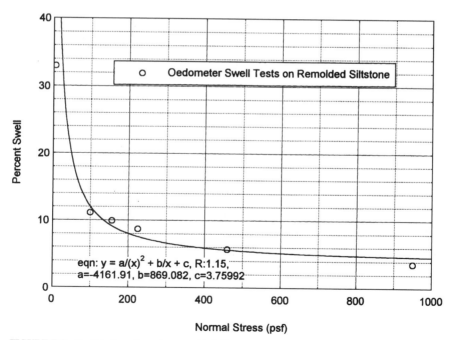

FIGURE E.2 Swell tests performed on remolded siltstone.

6. Groundwater

Groundwater, whether natural or that which may develop as a result of grading, drainage patterns, or irrigation, is often considered a major factor in structural distress due to expansive soils and undermining slopes, and a general nuisance where local seeps occur. To control water, subdrain systems and controlled irrigation are generally recommended.

7. Vibrating Machinery

It is our understanding that proposed development will contain vibrating machinery. Foundations for sensitive equipment may be required to be "vibration-free," which may, practically speaking, be impossible. It would be possible to design isolated foundation systems for vibrating machinery. Also, deepening the embedment and increasing the mass of the supporting foundation tends to decrease the amplitude of vibratory motion. Foundation requirements for vibrating machinery can best be evaluated when acceptable vibration criteria and machine parameters (i.e., mass, eccentricity, frequency, etc.) have been determined. The project mechanical engineer would be the best party to design the foundation systems. It is likely that the mechanical engineer would also utilize special vibration-isolating machine mounts.

G. CONCLUSIONS AND RECOMMENDATIONS

The proposed construction is feasible from a geotechnical aspect. The recommendations herein address Phase I construction. Recommendations for future phases should be provided at the time of future development. Conventional spread and wall footings for Phase I can support the structural dead and live loads.

Two constraints on the proposed construction are:

a. Nonuniform bearing conditions at the Administration Building, Projects Building, and Hardware Building consisting of both fill and bedrock within the building footprints, and

b. Results of expansion tests indicate that the soil and bedrock have a medium to high expansion potential. Influx of water from local irrigation, runoff, or leakage from the proposed artificial ponds and streams could cause expansive soil problems.

1. Grading

The site has been previously graded to the existing contours shown on the geotechnical map. Additional regrading is planned to achieve final pad and pavement grades. Regrading of the site should be done in accordance with the "Grading Specifications" (see App. C).

2. Recommended Type of Foundation

The recommended type of foundation to support dead and live loads is spread and continuous footings. To mitigate nonuniform bearing conditions due to the difference in material type, shear strength, and compressibility between fill and bedrock, it is recommended that footings for all structures be founded in Mission Valley formation bedrock.

It is anticipated that the proposed Functional Building, Management Module, and Central Plant will have conventional footings founded directly on bedrock. The Projects and Hardware Buildings may need some deepened footings to penetrate shallow fill. It may be possible to found the two-story section of the Administrative Building on bedrock using deepened footings; however, piers will be needed to reach bedrock for the one-story sections. By creating a uniform fill condition beneath the Administration Building, it would also be possible to found the footings in fill. This option, however, is not presently recommended because of differential settlement concerns.

Local deepening of foundations could also be required if soft, weathered bedrock materials are exposed at final grade. Care should be taken during grading so that foundation pads are not undercut.

3. Foundation and Slab Recommendations

Foundation and slab reinforcement recommendations provided herein are in consideration of expansive soil conditions. These minimum reinforcement levels have been empirically developed and are considered generally consistent with the Standards of Practice. The structural engineer should evaluate reinforcement conditions for structural loadings, shrinkage, and temperature stresses.

a. *Columns.* Columns should be founded on spread footings embedded in firm intact bedrock at a minimum depth of 24 in. below lowest adjacent grade. Reinforcement should be based on structural loading. As mentioned, some deepened footings to penetrate shallow fill will be required.

b. *Walls.* Exterior and interior wall footings should be continuous and founded on intact bedrock, with a minimum of 24 in. embedment below lowest adjacent grade. Reinforcement in exterior and interior footings should consist of a minimum of two No. 4 reinforcing bars placed one at the top and one at the bottom of the footing. At the perimeter of the buildings, a footing, grade beam, and/or cutoff wall should extend 24 in. below lowest adjacent grade.

c. *Footing excavations.* At the perimeter of significant exterior slabs, a 24-in. cutoff wall would also be desirable. Presoaking of footing excavations is not a requirement, but footings should not be allowed to dry out prior to the placement of concrete. Moisture maintenance by sprinkling may be needed. All footing excavations should be observed by the geotechnical engineer prior to placing reinforcing steel and concrete.

d. *Slabs.* Slabs should be a minimum of 5 in. thick, reinforced with a minimum of No. 3 bars (both ways) at 24 in. (on-center) spacing. At the perimeter of structures, interior and exterior slabs should be doweled to the perimeter footing or grade beam with No. 3 bars at 24 in. (on-center) spacing with the reinforcement connected to the slab reinforcement. Free-floating slabs would be a second-best alternative.

e. *Expansive soil treatment.* There are several measures that can be used to mitigate the effects of the on-site expansive soil. For example, there could be presoaking beneath the slabs or lime treatment of the subgrade. The preferred alternative is to remove the expansive soil and replace it with select import granular soil. It is considered likely that suitable material (i.e., decomposed granite, DG) can be located in relatively close proximity to the site. Capping with at least 3 ft of select material is recommended.

f. *Moisture barrier below slabs.* The slab subgrade should be topped with 4 in. of open-graded gravel (preferably rounded gravel), topped with a continuous plastic sheeting (10-mil visqueen or similar) that is sealed at all splices, then covered with 1 to 2 in. of clean sand.

4. Allowable Bearing Capacity

For intact Mission Valley formation (siltstone and sandstone) bedrock, the allowable bearing pressure for spread footings is 8000 pounds per square foot provided that the footing is at least 5 ft wide with a minimum of 2-ft embedment in firm, intact bedrock. For continuous wall footings, the allowable bearing pressure is 4000 pounds per square foot provided the footing is at least 2 ft wide with a minimum of 2-ft embedment in firm, intact bedrock. It is recommended that the structures be entirely supported by bedrock. It is anticipated that piers will be needed for the Administrative Building. Belled piers can be designed for an allowable end-bearing pressure of 12,000 pounds per square foot provided that the piers have a diameter of at least 2 ft, length of at least 10 ft, with a minimum embedment of 3 ft in firm, intact bedrock. It is recommended that the geotechnical engineer observe pier installation to confirm embedment requirements.

In designing to resist lateral loads, passive resistance of 1200 pounds per square foot per foot of depth to a maximum value of 6000 psf and a coefficient of friction equal to 0.35 may be utilized for embedment within firm bedrock.

For the analysis of earthquake loading, the above values of allowable bearing pressure and passive resistance may be increased by a factor of $1/3$.

5. Settlement

It is recommended that the buildings be supported by foundations anchored in bedrock. It is anticipated that piers will be needed for the Administrative Building. For foundations bearing on intact bedrock, the settlement for the anticipated maximum column load of 550 kips is estimated to be about 0.5 in. As indicated by the calculations presented in the appendix, the maximum differential settlement between columns is expected to be 0.25 in. or less. Settlement should consist of deformation of the bedrock and should occur during construction.

6. Coefficient of Subgrade Reaction

The coefficient of subgrade reaction depends on the material strength as well as the stress history of the bedrock material. Because plate load tests were not performed on the bedrock, an empirical estimation was used. The coefficient of subgrade reaction is estimated to be 350 pounds per cubic inch (pci) on the basis of material type.

7. Type of Cement for Construction

Evaluation of soluble sulfate content of samples considered representative of the predominate material types on site suggest that Type V cement is not a requirement for use in construction. Type I or II should be utilized.

8. Planters

Raised planters adjacent to structures should be tied to perimeter foundations, sealed at the bottom of the planters, and provided with appropriate drainage systems. Light wells containing plants and trees should be self-contained, sealed at the bottom, and provided with a drainage system that removes excess water to outside the building area. It is also recommended that adequate provisions be made to drain surface water away from buildings.

9. Retaining Structures

Where free unrestrained walls are planned to retain level, predominantly granular, nonexpansive, imported backfill, an active lateral soil pressure of 45 pounds per cubic foot (7 kN/m^3) equivalent fluid pressure may be utilized for retaining wall design. Appropriate allowances should be made for anticipated surcharge loadings. Unless walls incorporate appropriately designed backdrainage systems, allowances should be made for seepage and/or hydrostatic forces. For retaining walls restrained against rotation, the above value should be increased to 70 pounds per cubic foot (11 kN/m^3) equivalent fluid pressure. For the construction of temporary back-cut slopes for the retaining walls, see Sec. 13. The restrained basement walls should be backfilled with predominately granular, nonexpansive granular import material and provided with a backdrainage system. Neither spread nor wall footings should be founded in the granular backfill.

It should be noted that the use of heavy compaction equipment in close proximity to retaining walls can result in excess wall movement (strains greater than those normally associated with the construction of retaining walls) and/or soil pressures exceeding design values. In this regard, care should be taken during backfilling operations and lightweight equipment, such as hand-operated tampers, should be used for compaction.

The dynamic increment in lateral earth pressure due to earthquakes should be considered in retaining wall design. Using the Seed and Whitman (1970) recommendation and the peak ground accelerations previously presented in this report, retaining walls should be designed to resist an additional lateral soil pressure of 20 pounds per cubic foot equivalent fluid pressure for unrestrained walls and 30 pcf equivalent fluid pressure for restrained walls. For earthquake conditions, the pressure resultant force should be assumed to act at a distance of $0.6H$ above the base of the wall.

10. Pavements

Because of generally poor subgrade characteristics of the predominant soil types, generally heavy pavement sections can be anticipated. For traffic index (TI) values of 5.0 and 6.0, which are estimated for parking and truck driveway areas, the following pavement sections are recommended for planning purposes:

Traffic index	5.0	6.0
R-value	12	12
Pavement thickness	3 in.	3 in.
Aggregate base	9 in.	12 in.
Total thickness	12 in.	15 in.

At the completion of grading in the parking areas, R-values and pavement sections should be delineated on an area-by-area basis. Some variance in the above sections should be anticipated. It should be noted that if heavier traffic loadings are expected, this office should be informed.

Because of the presence of expansive soil on-site, consideration should be given to deepening planter area curb and gutter sections to the compacted subgrade. This recommendation is intended to provide for cutoff of irrigation water from free access to the subgrade via the granular base. Potential for saturation of the subgrade can be further reduced by providing irrigation overflow outlets through the curb faces. Because of the large area of parking and the potential for surface water infiltration, selectively located drainage trenches can also serve to enhance the potential for favorable performance of the pavement sections. Because of the nature of an expansive subgrade, some relatively minor future cracking of the surfacing could occur as a result of local heave. Such cracks should be treated by sanding and then sealing.

It is proposed to use concrete paving in areas where heavy truck traffic is anticipated (i.e., loading dock areas). For these areas, total truck traffic and truck loading information should be forwarded for use in evaluating section requirements.

11. Utility Trench Backfill

Utility trench backfill should be compacted by mechanical means. It is recommended that the degree of compaction be a minimum of 90 percent of the laboratory maximum dry density (Modified Proctor).

12. Artificial Ponds

Artificial ponds are proposed for the complex. Because the bedrock and fill are expansive, lining of the ponds and streams to prevent seepage is recommended. It is recommended that a reservoir specialist be consulted for further recommendations pertaining to the prevention of water seepage from artificial ponds and streams.

13. Slopes

a. Surficial Stability. Permanent slopes in bedrock or fill should be constructed at a slope ratio not steeper than 2:1 (horizontal:vertical). In order to enhance the potential for further favorable performance of the permanent slopes, these areas should be landscaped at the completion of grading. In order to improve the surficial stability of the slope face, plants should consist of deep-rooted varieties requiring little watering. A landscape architect would be the best party to consult regarding actual types of plants and planting configuration. To retard the tendency of weathering of the slope face, irrigation should be planned to achieve uniform moisture conditions well below the saturation level. If automatic timing devices are utilized in conjunction with irrigation systems, provision should be made for interrupting normal watering during and following periods of rainfall. Property owners and/or maintenance personnel should be made aware that improper slope maintenance, altering site drainage, overwatering, and burrowing animals can be detrimental to surficial slope stability. It is recommended that, when a landscape irrigation system is installed, piping be anchored to the slope face instead of excavating trenches into the slope faces.

b. Gross Stability, Temporary Slopes. Gross slope stability analyses were performed for temporary slopes for the restrained basement walls (maximum height of 20 ft) in bedrock. Gross stability computations are included in the appendix. The results indicate that the temporary back-cuts should be grossly stable at a ratio of 0.5:1.0 (horizontal:vertical). Sloughing, however, of loose and/or fractured materials could occur and thus cleaning of temporary slopes of loose material should be conducted as excavation proceeds. Parking of equipment or stockpiling of materials at the top of the temporary slopes should be prohibited.

c. Gross Stability, Permanent Slopes. Gross slope stability analyses were performed for permanent slopes, not to exceed 40 ft in height. Gross stability computations are included in the appendix.* The results indicate that permanent slopes in bedrock or fill should be constructed at a slope ratio not steeper than 2:1 (horizontal:vertical). For slopes between 30 to 40 ft in height, it is recommended that a gunite drainage swale should be provided at approximately midheight of the slope. Surface water should not be allowed to flow over the top of the slope and the slope faces should be landscaped at the completion of grading.

d. Foundations and Slope Areas. Foundations planned within or adjacent to slope areas should be deepened to provide sufficient horizontal distance from the bottom, outer edge of the foundations to daylight. The distance should be equal to half the slope height or 5 ft, whichever is greater. Foundation details under the influence of this recommendation should be forwarded along with the structural load information to the geotechnical engineer for review.

H. LIMITATIONS

The foundation investigation was performed using the degree of care and skill ordinarily exercised, under similar circumstances, by geotechnical engineers and geologists practicing in this or similar localities. No warranty, expressed or implied, is made as to the conclusions and professional advice included in this report.

The samples taken and used for testing and the observations are believed to be representative of the entire area. However, soil and geologic conditions can vary significantly between borings and surface outcrops. As in many developments, conditions revealed by excavations may be at variance with preliminary findings. If this occurs, the changed conditions must be evaluated by the geotechnical engineer and designs adjusted or alternative designs recommended.

---

*Appendix is not included in this example.

APPENDIX F
SOLUTION TO PROBLEMS

CHAPTER 4

Problem 1

$D_i = D_o - 2t$ (where t = wall thickness)

$D_i = 3.00 - 2(0.065) = 2.87$ in.

Inside clearance ratio = $100(D_i - D_e)/D_e = 100(2.87 - 2.84)/2.84 = 1.06\%$

Area ratio = $100(D_o^2 - D_i^2)/D_i^2 = 100(3^2 - 2.87^2)/2.87^2 = 9.26\%$

Problem 2

N-value = $8 + 9 = 17$ per Table 4.4, medium sand

Problem 3

$H = 4$ in. $= 0.333$ ft $\qquad D = 2$ in. $= 0.167$ ft

$s_u = T_{max}/[\pi(0.5\,D^2\,H + 0.167\,D^3)]$

$= 8.5/[\pi(0.5(0.167)^2\,0.333 + 0.167(0.167)^3)] = 500$ psf

Problem 4

$S_1 = 0.27$ in. $\qquad D_1 = 12$ in. $\qquad D = 5$ ft $= 60$ in.

$S = 4\,S_1/(1 + D_1/D)^2$

$= 4(0.27\text{ in.})/(1 + 12/60)^2 = 0.75$ in.

Problem 5

D10R tractor/ripper, therefore use Fig. 4.37. For granite with a seismic velocity = 12,000 to 15,000 feet per second, it is nonrippable.

CHAPTER 5

Problem 1

$P_f = 24.8$ lb $L_0 = 6.0$ in. $\Delta L = 0.8$ in. $D_o = 2.5$ in., therefore $A_o = 4.91$ in.^{2}

$\varepsilon_f = \Delta L/L_o = 0.8/6.0 = 0.133$

$A_f = A_o/(1 - \varepsilon_f) = 4.91/(1 - 0.133) = 5.66$ in.^{2}

$q_u = P_f/A_f = 24.8/5.66 = 4.38$ psi $= 630$ psf

$s_u = q_u/2 = 630/2 = 315$ psf and therefore "soft" consistency

Problem 2

$S_t = s_u$ undisturbed/s_u remolded $= 315/45 = 7$, therefore "medium" sensitivity

Problem 3

$Q = 782$ mL $= 782$ cm^3 $t = 31$ s $L = 2.54$ cm
$\Delta h = 2.0$ m $= 200$ cm

$D = 6.35$ cm and therefore $A = 31.67$ cm^2

$k = QL/(\Delta h A t) = [(782)(2.54)]/[(200)(31.67)(31)] = 0.01$ cm/s

Problem 4

For the standpipe, diameter = 0.635 cm, therefore $a = 0.317$ cm^2

For the specimen, diameter = 6.35 cm, therefore $A = 31.7$ cm^2

$h_0 = 1.58$ m $h_f = 1.35$ m $t = 11$ h $= 39,600$ s $L = 2.54$ cm

$k = 2.3 [(a L)/(A t)] \log (h_0/h_f)$

$\quad = 2.3 [(0.317)(2.54)/(31.7)(39,600)] \log (1.58/1.35) = 1.0 \times 10^{-7}$ cm/s

Problem 5

Plasticity index = liquid limit − plastic limit = 60 − 20 = 40

Entering Fig. 5.18 with liquid limit = 60 and plasticity index = 40, the predominant clay mineral in the soil is montmorillonite.

Problem 6

35% passes the No. 40 sieve, and therefore the clay-size fraction = 20/35 = 57%

Activity = PI/% clay = (93 − 18)/57 = 1.3

Problem 7

$C_u = D_{60}/D_{10} = 15/0.075 = 200$

$C_c = D_{30}^2/(D_{10} D_{60}) = (2.5)^2/[(0.075)(15)] = 5.6$

Problem 8

For the data from Prob. 7, and since 0.075 mm is the opening of the No. 200 sieve, the percent passing No. 200 sieve = 10%. Since D_{50} = 12 mm, which is a larger size than the No. 4 sieve, the majority of the soil particles are gravel. Since C_c does not meet the requirements for a well-graded gravel and the limits plot below the A line, per the USCS, there is a dual classification (because of the 10% fines) of GP-GM. For the ISBP, PI = (34) (0.18) = 6 and therefore the classification is a GM of low plasticity.

Problem 9

On the basis of the values of C_c and C_u, the sand is well-graded. Because of 4 percent nonplastic fines, the classification is SW for both the USCS and ISBP.

Problem 10

Since all the soil particles pass the No. 40 sieve, assume the soil is fine-grained. The limits plot below the A line and the LL is less than 50, therefore per the USCS system, the classification is ML. For PI = 16, the classification is MI per the ISBP.

Problem 11

The plasticity index = 40. For LL = 60 and PI = 40, the limits plot above the A line and the clay is classified as a CH for both the USCS and ISBP.

Problem 12

The LL (oven dry) divided by the LL (not dried) = 40/65 = 0.61 and therefore the soil is an organic soil. Since the LL is greater than 50, the classification per the USCS is OH.

Problem 13

Per the USCS, more than 50 percent of the soil particles are retained on the No. 200 sieve and therefore it is a coarse-grained soil. The majority of the soil particles are of sand size and since there are greater than 12 percent fines with the limits (LL = 85, PI = 67) plotting above the A line, the classification is SC per the USCS.

Per the ISBP, the soil is plastic and the PI = (67)(0.48) = 32. Since more than 50 percent is retained on the No. 200 sieve, with the majority being of sand size, then the classification is SC of high plasticity per the ISBP.

CHAPTER 6

Problem 1

For SI units, $V = 1$ m^3

$V_s = W/[(1 + w)(G)(\gamma_w)] = 19.0/[(1.18)(2.65)(9.81)] = 0.62$ m^3

$V_w = (\gamma_t - \gamma_d)/\gamma_w = [19.0 - (19/1.18)]/9.81 = 0.30$ m^3

$V_g = V - V_s - V_w = 1 - 0.62 - 0.30 = 0.08$ m^3

$M_s = \rho_d = 19/[(1 + 0.18)(9.81)] = 1.64$ Mg

$M_w = V_w \rho_w = (0.30)(1.0) = 0.3$ Mg

For United States Customary System Units, $V = 1$ ft^3

$V_s = W/[(1 + w)(G)(\gamma_w)] = 121/[(1.18)(2.65)(62.4)] = 0.62$ ft^3

$V_w = (\gamma_t - \gamma_d)/\gamma_w = [121 - (121/1.18)]/62.4 = 0.30$ ft^3

$V_g = V - V_s - V_w = 1 - 0.62 - 0.30 = 0.08$ ft^3

$M_s = \rho_d = 121/(1 + 0.18) = 103$ lb

$M_w = V_w \rho_w = (0.30)(62.4 \text{ pcf}) = 18$ lb

Problem 2

$\gamma_d = \gamma_t/(1 + w) = 19/(1 + 0.18) = 16.1$ kN/m^3 or 103 pcf

$e = V_v/V_s = (V_w + V_g)/V_s = (0.30 + 0.08)/0.62 = 0.61$

$n = V_v/V = (0.30 + 0.08)/1.0 = 0.38$ or 38%

$S = Gw/e = (2.65)(0.180)/0.61 = 0.78$ or 78%

Problem 3

$D_r = 100 (e_{max} - e)/(e_{max} - e_{min}) = 100 (0.85 - 0.61)/(0.85 - 0.30) = 44\%$

Problem 4

$\gamma_b = \gamma_t - \gamma_w = 19.5 - 9.81 = 9.7 \text{ kN/m}^3$ or 61.6 pcf

Problem 5

$\sigma_v = \gamma_t z = (19.5)(6) = 117$ kPa or 2480 psf

$u = \gamma_w z = (9.81)(6) = 59$ kPa or 1250 psf

$\sigma'_v = \sigma_v - u = 117 - 59 = 58$ kPa or 1230 psf

Problem 6

$\sigma_v = \gamma_t z = (19.5)(6) = 117$ kPa or 2480 psf

$u = \gamma_w z = (9.81)(4.5) = 44$ kPa or 940 psf

$\sigma'_v = \sigma_v - u = 117 - 44 = 73$ kPa or 1540 psf

Problem 7

Let z_1 = depth of the lake and z_2 = depth below lake bottom

$\sigma_v = \gamma_w z_1 + \gamma_t z_2 = (9.81)(3) + (19.5)(6) = 146$ kPa or 3100 psf

$u = \gamma_w(z_1 + z_2) = (9.81)(3 + 6) = 88$ kPa or 1870 psf

$\sigma'_v = \sigma_v - u = 146 - 88 = 58$ kPa or 1230 psf

Problem 8

Let z_1 = depth below ground surface and z_2 = distance above groundwater table

$\sigma_v = \gamma_t z_1 = (19.5)(1.5) = 29$ kPa or 620 psf

$u = -\gamma_w z_2 = (9.81)(3 - 1.5) = -14.7$ kPa or -310 psf

$\sigma'_v = \sigma_v - u = 29 - (-14.7) = 44$ kPa or 930 psf

Problem 9

$\sigma'_v = \sigma_v - u = 117 - (3)(9.81) = 88$ kPa or 1860 psf

Problem 10

In Fig. 4.40, the total unit weight values are shown on the left side of the figure. Using $\gamma_t = 122$ pcf for the sand-gravel layer, with z = thickness of each soil layer, gives:

$\sigma_v = \gamma_t z = (125)(21 - 12.5) + (101)(12.5 + 6) + (122)(4) + (119)(30) + (117)(25) + (123)(25) = 13{,}000$ psf or 6.3 kg/cm²

$u = \gamma_w d = (62.4)(12 + 90) = 6400$ psf or 3.1 kg/cm²

$\sigma'_v = \sigma_v - u = 13{,}000 - 6400 = 6600$ psf or 3.2 kg/cm²

Problem 11

Let H = thickness of the fill layer. Since the loading condition is one-dimensional,

$\Delta\sigma_v = \gamma_t H = (18.7)(3) = 56$ kPa or 1190 psf

Problem 12

For the loaded area, $B = 6$ m and $L = 10$ m. At a depth $z = 12$ m,

$P = B L \gamma_t H = (6)(10)(18.7)(3) = 3370$ kN

$\Delta\sigma_v = P/[(B + z)(L + z)]$ [Eq. (6.28)]

$= 3370/[(6 + 12)(10 + 12)] = 8.5$ kPa or 180 psf

Problem 13

$P = Q = 3370$ kN $r = 0$ $z = 12$ m

$\Delta\sigma_v = 3 Q z^3/[2\pi (r^2 + z^2)^{5/2}]$

$= 3 (3370)(12)^3/[2\pi (12^2)^{5/2}] = 11$ kPa or 250 psf

Problem 14

$q_0 = \gamma_t H = 56$ kPa $m = x/z = 3/12 = 0.25$ $n = y/z = 5/12 = 0.42$

From Fig. 6.4 and for the m and n values, $I = 0.044$

$\Delta\sigma_v = \sigma_z = 4 q_0 I = (4)(56)(0.044) = 9.9$ kPa or 210 psf

Problem 15

$q_0 = 56$ kPa $m = 0.25$ $n = 0.42$

From Fig. 6.5 and for the m and n values, $I = 0.027$

$\Delta\sigma_v = \sigma_z = 4 q_0 I = (4)(56)(0.027) = 6.1$ kPa or 130 psf

Problem 16

For Fig. 6.9 with the distance $AB = 12$ m, draw a rectangle with width $= 6/12 = 0.5 AB$ and length $= 10/12 = 0.83 AB$, with the center of the rectangle at the center of Fig. 6.9.

Counting the number of blocks within the rectangle = 34, and using Eq. (6.31) gives:
$$\Delta\sigma_v = \sigma_z = q_0 IN = (56)(0.005)(34) = 9.5 \text{ kPa or } 200 \text{ psf}$$

Summary of Values

Problem number (1)	Method of analysis (2)	$\Delta\sigma_v$ (3)
12	2:1 approximation	8.5 kPa (180 psf)
13	Concentrated load	11 kPa (250 psf)
14	Newmark chart (Fig. 6.4)	9.9 kPa (210 psf)
15	Westergaard chart (Fig. 6.5)	6.1 kPa (130 psf)
16	Newmark chart (Fig. 6.9)	9.5 kPa (200 psf)

Problem 17

For the 2:1 approximation, with $B = 6$ m, $L = 10$ m:

$\Delta\sigma_v = 0.1 \, q_0 = 5.6$ kPa

$\Delta\sigma_v = P/[(B + z)(L + z)]$ [Eq. (6.28)]

$5.6 = 3370/[(6 + z)(10 + z)]$; solving for $z = 16.6$ m

Problem 18

$\sigma'_v = \sigma_v - u = \gamma_t (1.5) + \gamma_b (1.5) = (18.9)(1.5) + (9.84)(1.5) = 43$ kPa

$N_{60} = 1.67 \, E_m C_b C_r N$ [Eq. (4.3)]

where $E_m = 0.45$ (donut hammer)

$\quad C_b = 1.0$ (100-mm-diameter hole)

$\quad C_r = 0.75$ (3-m length of drill rods)

$\quad N = 4 + 5 = 9$ (N-value)

Substituting these values into the equation gives

$N_{60} = 1.67 \, (0.45)(1.0)(0.75)(9) = 5$

In Fig. 4.13, for $N_{60} = 5$ and $\sigma'_v = 43$ kPa, $\phi' = 30°$

Problem 19

In Fig. 4.15, for $q_c = 40$ kg/cm² and $\sigma'_v = 43$ kPa, $\phi' = 40°$

Problem 20

Nonplastic cohesionless soil, therefore $c' = 0$

Specimen diameter = 6.35 cm, therefore area $(A) = 0.00317$ m^2

First test: $\sigma'_n = N/A = 150/0.00317 = 47{,}300$ Pa $= 47.3$ kPa

$\tau = T/A = 94/0.00317 = 29{,}700$ Pa $= 29.7$ kPa

$\tau = \sigma'_n \tan \phi'$ therefore $\tan \phi' = (29.7/47.3)$ or $\phi' = 32°$

Second test: $\sigma'_n = N/A = 300/0.00317 = 94{,}600$ Pa $= 94.6$ kPa

$\tau = T/A = 188/0.00317 = 59{,}300$ Pa $= 59.3$ kPa

$\tau = \sigma'_n \tan \phi'$ therefore $\tan \phi' = (59.3/94.6)$ or $\phi' = 32°$

Problem 21

$A_f = A_o/(1 - \varepsilon_f) = 9.68/[1 - (1.48/11.7)] = 11.1$ cm^2

$\sigma'_1 = \sigma'_v = \sigma_1 - u = [100 + (0.0484/0.00111)] - 45.6 = 98$ kPa

$\sigma'_3 = \sigma'_h = \sigma_3 - u = 100 - 45.6 = 54.4$ kPa

$p' = 0.5\,(\sigma'_1 + \sigma'_3) = (0.5)\,(98 + 54.4) = 76.2$ kPa

$q = 0.5\,(\sigma'_1 - \sigma'_3) = (0.5)(98 - 54.4) = 21.8$ kPa

For $c' = 0$, $a' = 0$, and $q = p' \tan \alpha'$ or $\tan \alpha' = q/p' = 21.8/76.2 = 0.286$, $\alpha' = 16°$

$\sin \phi' = \tan \alpha' = \tan 16°$

Solving for $\phi' = 17°$

$A_f = \Delta u/\Delta \sigma_1 = 45.6/(0.0484/0.00111) = 1.05$

Problem 22

On the basis of the low effective friction angle and the high A-value at failure, the most likely type of inorganic soil would be a normally consolidated clay of high plasticity (CH).

Problem 23

At point B, $E_u = \Delta \sigma_v/\varepsilon = 20/(0.13/10.67) = 1600$ kPa

Problem 24

At failure (point E): $\sigma_1 = \sigma_v = 50 + 80 = 130$ kPa

$\sigma_3 = \sigma_h = 50$ kPa

$p = 0.5\,(\sigma_1 + \sigma_3) = 0.5\,(130 + 50) = 90$ kPa

$q = 0.5\,(\sigma_1 - \sigma_3) = 0.5\,(130 - 50) = 40$ kPa

$q = a + p \tan \alpha$ and $a = c \cos \phi$ and $\tan \alpha = \sin \phi$

Therefore $q = c \cos \phi + p \sin \phi$

$40 = 5 \cos \phi + 90 \sin \phi$

$\phi = 23°$

Problem 25

At failure (point E): $\sigma'_1 = \sigma'_v = 130 - 6.7 = 123.3$ kPa

$\sigma'_3 = \sigma'_h = 50 - 6.7 = 43.3$ kPa

$p' = 0.5 (\sigma_1 + \sigma_3) = 0.5 (123.3 + 43.3) = 83.3$ kPa

$q = 0.5 (\sigma_1 - \sigma_3) = 0.5 (123.3 - 43.3) = 40$ kPa

$q = a' + p' \tan \alpha'$

Therefore $40 = 2 + 83.3 \tan \alpha'$ and $\alpha' = 24.5°$

Problem 26

$\tan \alpha' = \sin \phi'$ or $\tan 24.5° = \sin \phi'$ or $\phi' = 27°$

$a' = c' \cos \phi'$ or $2 = c' \cos 27°$ or $c' = 2.2$ kPa

Problem 27

$B = \Delta u / \Delta \sigma_c = 99.8/100 = 0.998$

$A_f = \Delta u / \Delta \sigma_1 = 6.7/80 = 0.08$

Problem 28

$d = D = 5.0$ cm $q = 1000/33 = 30.3$ cm³/s $H_c = 4$ m $= 400$ cm

For Case C (Fig. 6.23b) and a constant-head condition:

$k = q/[(2.75)(D)(H_c)] = 30.3/[(2.75)(5.0)(400)] = 0.006$ cm/s

Problem 29

$d = D = 0.5$ cm $H_1 = 4.0$ m $H_2 = 3.1$ m $t_2 - t_1 = 60$ s

For Case C (Fig. 6.23b) and a variable-head condition:

$k = \pi D/[11 (t_2 - t_1)] \ln (H_1/H_2)$

$= \pi (5.0)/[(11)(60)] \ln (4/3.1) = 0.006$ cm/s

Problem 30

A doubling of h_w (i.e., Δh), doubles Q per Eq. (6.46)

Problem 31

14.5 drops, therefore Δh lost = (14.5/18)(10) = 8.06 m

Pore water pressure (u) = 10 − 8.06 + 12 = 13.9 m or 137 kPa

Problem 32

$\sigma'_v = \sigma_v - u = (12)(19.8) - 137 = 101$ kPa

Problem 33

The length L of square labeled 18 is approximately = 4 m, therefore:

$i_e = h'/L = (10/18)/4 = 0.14$

$i_c = \gamma_b/\gamma_w = (19.8 - 9.81)/9.81 = 1.02$

$F = i_c/i_e = 1.02/0.14 = 7$

Problem 34

Since the stratum is sand, assume $G = 2.65$

$\gamma_b = \gamma_w (G - 1)/(1 + e)$ [Eq. (6.14)]

$(19.8 - 9.81) = 9.81 (2.65 - 1)/(1 + e)$

$e = 0.62$

$n = e/(1 + e) = 0.62/(1 + 0.62) = 0.38$

$v_s = k\,i/n = (0.1)(0.14)/0.38 = 0.037$ m/day with seepage in an approximately upward direction

Problem 35

Since the equipotential drops are all equal, the highest seepage velocity occurs where the length of the flow net squares are the smallest, or at the sheet pile tip.

Problem 36

$Q = k\,\Delta h\,t\,(n_f/n_d)$ [Eq. (6.46)]

$k = 1 \times 10^{-8}$ m/s $n_f = 10$ $n_d = 14$ $\Delta h = 20$ m

$Q = (1 \times 10^{-8})(20)(86{,}400)(10/14) = 0.012 \text{ m}^3$

Q of 0.012 m³ times L of 200 m $= 2.5$ m³/day

Problem 37

Number of equipotential drops $= 14$ (*Note:* for the uppermost flow channel, one of the equipotential drops occurs in the drainage filter.)

$h' = h_w/n_d = 20/14 = 1.43$ m

Problem 38

Since the equipotential drops are all equal, the highest seepage velocity occurs where the length of the flow net squares are the smallest, or in the soil that is located in front of the longitudinal drainage filter.

Problem 39

The soil located in front of the longitudinal drainage filter has the highest seepage velocity v_s, and this is the most likely location for piping of soil into the drainage filter.

Problem 40

At a point located at the centerline of the dam and 20 m below the top of the dam, the number of equipotential drops $= 6.2$.

6.2 drops, therefore Δh lost $= (6.2/14)(20) = 8.86$ m

Pore water pressure $(u) = 20 - 8.86 = 11.1$ m or 110 kPa

Problem 41

$\sigma'_v = \sigma_v - u = (20)(20) - 110 = 290$ kPa

Problem 42

$\tau_f = c' + \sigma'_n \tan \phi'$ [Eq. (6.32)]

In this case with a horizontal slip surface, $\sigma'_n = \sigma'_v = 290$ kPa

$\tau_f = 2 + (290) \tan 28° = 156$ kPa

Problem 43

$F =$ shear strength divided by shear stress or $F = \tau_f/\tau = 156/83 = 1.88$

CHAPTER 7

Problem 1

Perimeter footing: $(60)(30)(2) + (60)(42)(2) = 8640$ kN

Interior columns: 6 m spacing, therefore 24 interior columns, or

$24 (900) = 21,600$ kN

Floor slab: 6 kPa

Include live load from stored iems: 30 kPa

$\sigma_0 = (8640 + 21{,}600)/[(30)(42)] + 6 + 30 = 60$ kPa

Problem 2

Per Table 7.1, for a continuous steel frame, $\Delta = 0.002$, $L = 1.2$ cm (0.5 in.)

Problem 3

For sensitive machinery, per Fig. 7.3, $\delta/L = 1/750$

Assuming $\delta = \Delta$, $\Delta/L = 1/750$ and for $L = 6$ m, $\Delta = 0.8$ cm (0.3 in.)

Problem 4

Cracking occurs when $\Delta = 1.25$ in., therefore allowable $\Delta = 1.25/2.5 = 0.5$ in.

Problem 5

The maximum differential settlement (Δ) will occur between the center and edge of the tank. Thus $L = 15$ m and assuming $\delta = \Delta$, $\Delta/L = 1/200$ and $\Delta = 7.5$ cm (3 in.)

Problem 6

$\%C = 100\ \Delta H_c/H_0 = 100\ (2.8)/25.4 = 11\%$ (or "severe" collapse potential)

Problem 7

$\Delta = (5 - 1)\ (0.05) = 0.2$ ft $= 2.4$ in.

Problem 8

$\rho_{max} = 4\ (0.1/100 + 0.25/100 + 0.63/100 + 1.14/100) = 0.08$ ft $= 1.0$ in.

Problem 9

$$s_i = qBI_p(1-\mu^2)/E_u \quad \text{[Eq. (7.1)]}$$

For the square loaded area, assume a flexible loaded area on an elastic half-space of infinite depth with $E_u = 20{,}000$ kPa, $I_p = 1.12$ (center), $I_p = 0.56$ (corner), and $\mu = 0.5$ (saturated cohesive soil):

$s_i = (30)(20)(1.12)(1 - 0.5^2)/20{,}000 = 0.025$ m $= 2.5$ cm (center)

$s_i = (30)(20)(0.56)(1 - 0.5^2)/20{,}000 = 0.013$ m $= 1.3$ cm (corner)

Problem 10

$$s_i = qBI_p(1-\mu^2)/E_u \quad \text{[Eq. (7.1)]}$$

For the circular loaded area, assume a flexible loaded area on an elastic half-space of infinite depth with $E_u = 40{,}000$ kPa, $I_p = 1.0$ (center), and $\mu = 0.4$

$s_i = (50)(10)(1.0)(1 - 0.4^2)/40{,}000 = 0.010$ m $= 1.0$ cm (center)

Problem 11

Immediate settlement s_i due to undrained creep of the soft saturated clay was not included in the original settlement analysis by the design engineer.

Problem 12

According to the Casagrande construction technique (see Fig. 7.12), the maximum past pressure $(\sigma'_{vm}) = 20$ kPa (solid line) and $= 30$ kPa (dashed line).

Problem 13

$$C_c = \Delta e / \log(\sigma'_{vc2}/\sigma'_{vc1})$$

For the consolidation test with a solid line:

$e = 3.3$ at $\sigma'_{vc1} = 20$ kPa and $e = 1.6$ at $\sigma'_{vc2} = 80$ kPa

$C_c = (3.3 - 1.6)/\log(80/20) = 2.8$

For the consolidation test with a dashed line:

$e = 2.7$ at $\sigma'_{vc1} = 20$ kPa and $e = 1.6$ at $\sigma'_{vc2} = 80$ kPa

$C_c = (2.7 - 1.6)/\log(80/20) = 1.8$

Problem 14

$$C_{c\varepsilon} = \Delta\varepsilon/\log(\sigma'_{vc2}/\sigma'_{vc1})$$

For the consolidation test on the undisturbed soil specimen:

$\varepsilon = 0.02$ at $\sigma'_{vc1} = 2$ kg/cm² and $\varepsilon = 0.27$ at $\sigma'_{vc2} = 10$ kg/cm²

$C_{ce} = (0.27 - 0.02)/\log(10/2) = 0.36$

For the consolidation test on the disturbed soil specimen:

$\varepsilon = 0.09$ at $\sigma'_{vc1} = 2$ kg/cm² and $\varepsilon = 0.26$ at $\sigma'_{vc2} = 10$ kg/cm²

$C_{ce} = (0.26 - 0.09)/\log(10/2) = 0.24$

Problem 15

$C_{re} = \Delta\varepsilon/\log(\sigma'_{vc2}/\sigma'_{vc1})$ for recompression curve

$C_{ce} = \Delta\varepsilon/\log(\sigma'_{vc2}/\sigma'_{vc1})$ for virgin consolidation curve

For the recompression curve:

Over one log cycle, $\Delta\varepsilon = 0.04$

$C_{re} = (0.04)/\log(100/10) = 0.04$

For the virgin consolidation curve:

$\varepsilon = 0.03$ at $\sigma'_{vc1} = 70$ kPa and $\varepsilon = 0.1$ at $\sigma'_{vc2} = 100$ kPa

$C_{ce} = (0.1 - 0.03)/\log(100/70) = 0.45$

Problem 16

Divide the 2-m clay layer into two 1-m-thick layers as follows:
For layer 1:

Layer no. (1)	Layer thickness (2)	Depth to center of layer (3)	σ'_{v0} at the layer center of the layer (4)	$\Delta\sigma_v$ at the center of the layer (5)	Primary Consolidation Settlement (6)
1	1 m	10.5 m	147.9 kPa	24.3 kPa	2.6 cm
2	1 m	11.5 m	155.8 kPa	22.9 kPa	2.3 cm

$s_c = [C_c H_0/(1 + e_0)] \log[(\sigma'_{v0} + \Delta\sigma_v)/\sigma'_{v0}]$ [Eq. (7.4)]   (OCR = 1)

$= [(0.83)(1.0)/(1 + 1.1)] \log[(147.9 + 24.3)/147.9] = 0.026$ m $= 2.6$ cm

For layer 2:

$s_c = [C_c H_0/(1 + e_0)] \log[(\sigma'_{v0} + \Delta\sigma_v)/\sigma'_{v0}]$ [Eq. (7.4)]   (OCR = 1)

$= [(0.83)(1.0)/(1 + 1.1)] \log[(155.8 + 22.9)/155.8] = 0.023$ m $= 2.3$ cm

Total primary consolidation settlement $(s_c) = 2.6 + 2.3 = 4.9$ cm

Problem 17

Clay layer is underconsolidated (OCR < 1), therefore use Eq. (7.3)

$s_c = [C_c H_0/(1 + e_0)] \log [(\sigma'_{v0} + \Delta\sigma'_v + \Delta\sigma_v)/\sigma'_{v0}]$ [Eq. (7.3)]

$= [(0.83)(2.0)/(1 + 1.1)] \log [(125 + 27 + 24)/125] = 0.12$ m $= 12$ cm

Problem 18

$\sigma'_{v0} + \Delta\sigma_v = 152 + 24 = 176 > \sigma'_{vm} = 170$ kPa, therefore use Eq. (7.6)

$s_c = [C_r H_0/(1 + e_0)] \log (\sigma'_{vm}/\sigma'_{v0}) + [C_c H_0/(1 + e_0)] \log [(\sigma'_{v0} + \Delta\sigma_v)/\sigma'_{vm}]$

$= [(0.10)(2.0)/(1 + 1.1)] \log (170/152) + [(0.83)(2.0)/(1 + 1.1)] \log [(152 + 24)/170]$
$= 0.017$ m $= 1.7$ cm

Problem 19

$\Delta\sigma_v = 4\,\gamma_w - 4\,(\Delta\gamma_t) = (4)(9.81) - (4)(19.7 - 18.9) = 36$ kPa (*Note:* $\Delta\sigma_v$ can also be calculated as a the change from the initial σ'_{v0} to the final (after permanent lowering of the groundwater table) σ'_v.

$\sigma'_{v0} = 152 + 24 = 176$ kPa

$s_c = [C_c H_0/(1 + e_0)] \log [(\sigma'_{v0} + \Delta\sigma_v)/\sigma'_{v0}]$ [Eq. (7.4)]   (OCR = 1)

$= [(0.83)(2.0)/(1 + 1.1)] \log [(176 + 36)/176] = 0.064$ m $= 6.4$ cm

Problem 20

Divide the 2-m clay layer into two 1-m-thick layers as follows:

Layer no. (1)	Layer thickness (2)	Depth to center of layer (3)	σ'_{v0} at the center of the layer (4)	$\Delta\sigma_v$ at the center of the layer (5)	Primary consolidation settlement (6)
1	1 m	10.5 m	172.2 kPa	10 kPa	0.97 cm
2	1 m	11.5 m	178.7 kPa	30 kPa	2.66 cm

Note: In the above table, column 4 was obtained by adding together columns 4 and 5 from Prob. 16. Column 5 was obtained by using a linear distribution of pore water pressure decrease of zero at 10 m to 40 kPa at 12 m and then determining the corresponding pore water pressure decrease at the center of each layer.

For layer 1:

$s_c = [C_c H_0/(1 + e_0)] \log [(\sigma'_{v0} + \Delta\sigma_v)/\sigma'_{v0}]$ [Eq. (7.4)] (OCR = 1)

$= [(0.83)(1.0)/(1 + 1.1)] \log [(172.2 + 10)/172.2] = 0.097 \text{m} = 0.97 \text{ cm}$

For layer 2: $s_c = [C_c H_0/(1 + e_0)] \log [(\sigma'_{v0} + \Delta\sigma_v)/\sigma'_{v0}]$ [Eq. (7.4)] (OCR = 1)

$= [(0.83)(1.0)/(1 + 1.1)] \log [(178.7 + 30)/178.7] = 0.0266 \text{ m} = 2.66 \text{ cm}$

Total primary consolidation settlement $(s_c) = 0.97 + 2.66 = 3.6$ cm

Problem 21

From Fig. 7.28, $d_0 = 0$, $d_{100} = 2.02$, therefore $d_{50} = 1.01$ mm and $t_{50} = 50$ min

$H_{dr} = (10 - 1.01)/2 = 4.5$ mm (double drainage)

$c_v = (T)(H_{dr})^2/t_{50}$ [Eq. (7.14)]

From Table 7.4, for $U_{avg} = 50\%$, $T = 0.197$

$c_v = (0.197)(0.45 \text{ cm})^2/(3000 \text{ s}) = 1.3 \times 10^{-5} \text{ cm}^2/\text{s}$

Problem 22

$c_v = 0.32$ m^2/year and for single drainage, $H_{dr} = 2$ m

Use Eq. (7.14):

$t = (0.197)(2)^2/0.32 = 2.5$ years for $U_{avg} = 50\%$

$t = (0.848)(2)^2/0.32 = 10.6$ years for $U_{avg} = 90\%$

Problem 23

$U_{avg} = 2.5/5.0 = 50\%$ and $T = 0.197$ from Table 7.4

$t = 180$ days $= 1.6 \times 10^7$ s

$c_v = (0.197)(100 \text{ cm})^2/(1.6 \times 10^7 \text{ s}) = 1.3 \times 10^{-4} \text{ cm}^2/\text{s}$

Problem 24

$u_0 = 24$ kPa $u_e = (7.88 - 6)(9.81) = 18.4$ kPa

Equation (7.13): $U_z = 1 - (u_e/u_0) = 1 - (18.4/24) = 0.23$

Equation (7.12): $Z = z/H_{dr} = 0.5 H/0.5 H = 1.0$

Entering Fig. 7.14 with $U_z = 0.23$ and $Z = 1.0$, find $T = 0.20$

$c_v = (0.20)(100 \text{ cm})^2/(1.6 \times 10^7 \text{ s}) = 1.3 \times 10^{-4} \text{ cm}^2/\text{s}$

Problem 25

Since the initial specimen height = 10 mm, then the vertical axis in Fig. 7.28 can be converted to vertical strain ε_v by dividing the scale by 10.

$C_\alpha = \Delta\varepsilon_v/\Delta \log t$

At 1000 min, $\varepsilon_v = 0.21$, and at 100 min, $\varepsilon_v = 0.20$, therefore:

$C_\alpha = \Delta\varepsilon_v/\Delta \log t = (0.21 - 0.20)/[\log (1000) - \log (100)] = 0.01$

Problem 26

The time at the end of primary consolidation = $(1.0)(1)^2/0.32 = 3.1$ years

$s_s = C_\alpha H_0 \Delta \log t$ [Eq. (7.16)]

$= (0.01)(2)[\log (50) - \log (3.1)] = 0.024$ m $= 2.4$ cm

$\rho_{max} = s_i + s_c + s_s$ [Eq. (7.17)]

$= 0 + 5.0 + 2.4 = 7.4$ cm

Problem 27

$s = qBI_p (1 - \mu^2)/E_s$ [Eq. (7.1), with E_s substituted for E_u]

For N-value = 20 and uniform coarse sand, $E_s/N = 10$ or $E_s = 20,000$ kPa

For the square loaded area, assume a flexible loaded area on an elastic half-space of infinite depth with $E_s = 20,000$ kPa, $I_p = 1.12$ (center) and $I_p = 0.56$ (corner), and $\mu = 0.3$, therefore:

$s = (30)(20)(1.12)(1 - 0.3^2)/20,000 = 0.031$ m $= 3$ cm (center)

$s = (30)(20)(0.56)(1 - 0.3^2)/20,000 = 0.015$ m $= 1.5$ cm (corner)

ρ_{max} occurs at the center $= 3$ cm and $\Delta = 3 - 1.5 = 1.5$ cm

Problem 28

$s = q B I_p (1 - \mu^2)/E_s$ [Eq. (7.1), with E_s substituted for E_u]

For N-value = 40 and nonplastic silty sand, $E_s/N = 4$ or $E_s = 16,000$ kPa

For the circular loaded area, assume a flexible loaded area on an elastic half-space of infinite depth with $E_s = 16,000$ kPa, $I_p = 1.0$ (center), and $\mu = 0.3$

$s = \rho_{max} = (50)(10)(1.0)(1 - 0.3^2)/16,000 = 0.028$ m $= 2.8$ cm (center)

Problem 29

$s = qBI_p (1 - \mu^2)/E_s$ [Eq. (7.1), with E_s substituted for E_u]

For $q_c = 50$ kg/cm² and nonplastic silty sand, $E_s = 2\, q_c$ or $E_s = 10{,}000$ kPa

For the circular loaded area, assume a flexible loaded area on an elastic half-space of infinite depth with $E_s = 10{,}000$ kPa, $I_p = 1.0$ (center), and $\mu = 0.3$

$s = \rho_{max} = (50)(10)(1.0)(1 - 0.3^2)/10{,}000 = 0.046$ m $= 4.6$ cm (center)

Problem 30

$q = (230)(1000)/[(6)(6)] = 6400$ psf $= 3.2$ tsf

$s = q\, B\, I_p\, (1 - \mu^2)/E_s$ [Eq. (7.1), with E_s substituted for E_u]

For N-value $= 30$ and fine to medium sand, $E_s/N = 7$ or $E_s = 210$ tsf

For the square loaded area, assume a flexible loaded area on an elastic half-space of infinite depth with $E_s = 210$ tsf, $I_p = 1.12$ (center), and $\mu = 0.3$

$s = \rho_{max} = (3.2)(6)(1.12)(1 - 0.3^2)/(210) = 0.093$ ft $= 1.1$ in.

Problem 31

From Fig. 7.16, for $B = 6$ ft, $q = 3.4$ tsf for 1-in. settlement. Since actual $q = 3.2$ tsf, ρ_{max} will be slightly less than 1 in. (*Note:* For this problem, the theory of elasticity and Fig. 7.16 provide similar answers.)

Problem 32

Solve by trial and error:

Assuming $B = 8.5$ ft, from Fig. 7.16, $q = 3.2/2 = 1.6$ tsf

Maximum load $= (8.5)(8.5)(1.6) = 116$ tons $= 230$ kips

Problem 33

$\sigma_0 = 200 - 4\,(19.3) = 123$ kPa

Divide the sand below the foundation into two layers, as follows:

Layer no. (1)	Layer thickness (2)	Depth to center of layer (3)	σ'_{v0} at the center of the layer (4)	$\Delta\sigma_v$ at the center of the layer (5)	Settlement (6)
1	8 m	8 m	130 kPa	98.5 kPa	11 cm
2	8 m	16 m	220 kPa	67.5 kPa	1 cm

For layer 1:

$s = \Delta e \, H_0/(1 + e_0)$ [Eq. (7.18a)] and from Fig. 7.29 (8-m specimen):

$= (0.482 - 0.462)(8)/(1 + 0.5) = 0.11$ m $= 11$ cm

For layer 2:

$s = \Delta e \, H_0/(1 + e_0)$ [Eq. (7.18a)] and from Fig. 7.29 (16-m specimen):

$= (0.395 - 0.393)(8)/(1 + 0.4) = 0.01$ m $= 1$ cm

Total settlement $(\rho_{max}) = 11 + 1 = 12$ cm

Problem 34

A deep foundation system consisting of piles or piers embedded in the sandstone.

Problem 35

Assuming the weight of the soil excavated for the basement is approximately equal to the weight of the two-story structure, a floating foundation would be desirable.

Problem 36

Because the upper 16 ft of the site consists of overconsolidated clay, it would be desirable to use a shallow foundation system, assuming light loads.

Problem 37

No, because of the very high sensitivity of the clay, high-displacement piles will remold the clay and result in a loss of shear strength. The preferred option is to install low-displacement piles or use predrilled, cast-in-place concrete.

Problem 38

Divide the clay into four layers, as follows:

Layer no. (1)	Layer thickness (2)	Depth to center of layer (3)	σ'_{v0} at the center of the layer (4)	$\Delta\sigma_v$ at the center of the layer (5)	Primary consolidation settlement (6)
1	10 ft	76 ft	4600 psf	610 psf	0.10 ft
2	10 ft	86 ft	5100 psf	575 psf	0.085 ft
3	10 ft	96 ft	5700 psf	545 psf	0.073 ft
4	10 ft	106 ft	6300 psf	520 psf	0.063 ft

For layer 1:
$$s_c = [C_c H_0/(1 + e_0)] \log [(\sigma'_{v0} + \Delta\sigma_v)/\sigma'_{v0}] \quad [\text{Eq. (7.4)}] \quad (\text{OCR} = 1)$$
$$= [(0.35)(10)/(1 + 0.9)] \log [(4600 + 610)/4600] = 0.10 \text{ ft}$$

For layer 2:
$$s_c = [C_c H_0/(1 + e_0)] \log [(\sigma'_{v0} + \Delta\sigma_v)/\sigma'_{v0}] \quad [\text{Eq. (7.4)}] \quad (\text{OCR} = 1)$$
$$= [(0.35)(10)/(1 + 0.9)] \log [(5100 + 575)/5100] = 0.085 \text{ ft}$$

For layer 3:
$$s_c = [C_c H_0/(1 + e_0)] \log [(\sigma'_{v0} + \Delta\sigma_v)/\sigma'_{v0}] \quad [\text{Eq. (7.4)}] \quad (\text{OCR} = 1)$$
$$= [(0.35)(10)/(1 + 0.9)] \log [(5700 + 545)/5700] = 0.073 \text{ ft}$$

For layer 4:
$$s_c = [C_c H_0/(1 + e_0)] \log [(\sigma'_{v0} + \Delta\sigma_v)/\sigma'_{v0}] \quad [\text{Eq. (7.4)}] \quad (\text{OCR} = 1)$$
$$= [(0.35)(10)/(1 + 0.9)] \log [(6300 + 520)/6300] = 0.063 \text{ ft}$$

Total primary consolidation settlement $(s_c) = 0.10 + 0.085 + 0.073 + 0.063 = 0.32$ ft $= 4$ in.

CHAPTER 8

Problem 1

$$\gamma_a = \gamma_b + [(h' - D_f)/B](\gamma_t - \gamma_b) \quad [\text{Eq. (8.4)}]$$
Since $h' = D_f = 0.9$ m, $\gamma_a = \gamma_b$
$$q_{ult} = Q_{ult}/(BL) = \tfrac{1}{2} \gamma_a B N_\gamma + \gamma_t D_f N_q \quad [\text{Eq. (8.1)}]$$
With $c = 0$ then $q_{ult} = \tfrac{1}{2} (19 - 9.81)(1.2)(8) + (19)(0.9)(10) = 215$ kPa
$Q_{all} = (q_{ult})(B)/F = (215)(1.2)/3 = 86$ kN per linear meter of footing length

Problem 2

$$\gamma_a = \gamma_b + [(h' - D_f)/B](\gamma_t - \gamma_b) \quad [\text{Eq. (8.4)}]$$
For $h' = 1.5$ m, $D_f = 0.9$ m
$$\gamma_a = (19 - 9.81) + [(1.5 - 0.9)/1.2](19 - 9.19) = 14.1 \text{ kN/m}^3$$
$$q_{ult} = Q_{ult}/(BL) = \tfrac{1}{2} \gamma_a B N_\gamma + \gamma_t D_f N_q \quad [\text{Eq. (8.1)}]$$
With $c = 0$ then $q_{ult} = \tfrac{1}{2} (14.1)(1.2)(8) + (19)(0.9)(10) = 240$ kPa
$Q_{all} = (q_{ult})(B)/F = (240)(1.2)/3 = 95$ kN per linear meter of footing length

Problem 3

$q_{ult} = Q_{ult}/(BL) = \frac{1}{2} \gamma_t BN_\gamma + \gamma_t D_f N_q$ [Eq. (8.1) with $c = 0$]
$= \frac{1}{2}(19)(1.2)(8) + (15.7)(0.9)(10) = 230$ kPa

$Q_{all} = (q_{ult})(B)/F = (230)(1.2)/3 = 93$ kN per linear meter of footing length

Problem 4

Reference footing has a 1-ft depth and 1-ft width. Since actual footing width $= 4$ ft, there is a 60 percent increase allowed (i.e., 20 percent increase for 2-ft-wide footing, 40 percent increase for a 3-ft-wide footing, and 60 percent increase for a 4-ft-wide footing). For depth, there is a 40 percent increase allowed. Therefore:

$q_{all} = (75$ kPa$)(60\% + 40\% = 100\%$ increase$) = 150$ kPa

$Q_{all} = (q_{all})(B) = (150)(1.2) = 180$ kN per linear meter of footing width

Note that the UBC q_{all} is much greater than the q_{all} determined from the bearing capacity equation.

Problem 5

$q_{ult} = Q_{ult}/B^2 = 0.4 \gamma_t B N_\gamma + \gamma_t D_f N_q$ [Eq. (8.3) with $c = 0$]
$= (0.4)(19)(1.2)(8) + (19)(0.9)(10) = 244$ kPa

$Q_{all} = (q_{ult})(B^2)/F = (244)(1.2)^2/3 = 117$ kN

Problem 6

Same q_{all} as in Prob. 4, $q_{all} = 150$ kPa, therefore:

$Q_{all} = q_{all} B^2 = (150)(1.2)^2 = 216$ kN

Problem 7

For $\phi' = 40°$, from Table 8.2, $N_\gamma = 120$

$q_{ult} = Q_{ult}/B^2 = 0.4 \gamma_t B N_\gamma$ [Eq. (8.3) with $c = 0$ and $D_f = 0$]
$= (0.4)(125)(6)(120) = 36,000$ psf

$Q_{all} = (q_{ult})(B^2)/F = (36,000)(6)^2/3 = 430,000$ lb $= 430$ kips

The allowable load based on a bearing capacity analysis (430 kips) is much greater than the load (230 kips) that will cause 1-in. settlement. Therefore, settlement governs in the design of the footing.

Problem 8

For a 6-ft-wide footing, there is a 100% increase in allowable bearing capacity as compared to the 1-ft-wide footing. Therefore, $q_{\text{all}} = 3000$ psf

$$Q_{\text{all}} = (q_{\text{all}})(B^2) = (3000)(6)^2 = 108{,}000 \text{ lb} = 108 \text{ kips}$$

Note that for this dense sand, the UBC Q_{all} (108 kips) is much less than the Q_{all} that will cause about 1 in. of settlement (230 kips).

Problem 9

$$q_{\text{ult}} = (Q_{\text{all}})(F)/B = (150)(3)/B = 450/B$$

$$q_{\text{ult}} = {}^1\!/_2\, \gamma_t\, B\, N_\gamma + \gamma_t\, D_f\, N_q \qquad \text{[Eq. (8.1) with } c = 0\text{]}$$

$$450/B = {}^1\!/_2\,(19)(B)(8) + (19)(0.9)(10)$$

$$B = 1.56 \text{ m}$$

Problem 10

$$q_{\text{ult}} = 5\, s_u + \gamma_t\, D_f \qquad \text{[Eq. (8.5a)]}$$

$$= (5)(20) + (19)(0.9) = 117 \text{ kPa}$$

$$Q_{\text{all}} = q_{\text{ult}}\,(B)/F = (117)(1.2)/3 = 47 \text{ kN per linear meter of footing length}$$

Problem 11

At a depth of 0.9 m, $\sigma'_{vm} = 26$ kPa and $q_{\text{all}} = 39$ kPa, therefore settlement governs. It is recommended that the bearing pressure not exceed 26 kPa to prevent virgin consolidation.

Problem 12

$$q_{\text{ult}} = (Q_{\text{all}})(F)/B = (50)(3)/B = 150/B$$

$$q_{\text{ult}} = 5\, s_u + \gamma_t\, D_f \qquad \text{[Eq. (8.5a)]}$$

$$150/B = (5)(20) + (19)(0.9)$$

$$B = 1.28 \text{ m}$$

Problem 13

Total stress analysis [Eq. (8.5a)]

$$q_{\text{ult}} = 5 s_u + \gamma_t\, D_f = (5)(200) + (19)(0.9) = 1020 \text{ kPa}$$

Effective stress analysis [Eq. (8.1) using $c' = 5$ kPa, $\gamma_a = \gamma_b$, and from Table 8.2 by interpolation for $\phi' = 28°$, $N_c = 27$, $N_\gamma = 7$, and $N_q = 9$]

$q_{ult} = c' N_c + \frac{1}{2} \gamma_a B N_\gamma + \gamma_t D_f N_q$
$= (5)(27) + \frac{1}{2}(19 - 9.81)(1.2)(7) + (19)(0.9)(9) = 327$ kPa

Therefore the effective stress analysis governs:

$Q_{all} = (q_{ult})(B)/F = (327)(1.2)/3 = 131$ kN per linear meter of footing length

Problem 14

$q' = Q(B + 6e)/B^2 \qquad q'' = Q(B - 6e)/B^2$ [Eqs. (8.6a) and (8.6b)]
$q' = (100)[1.2 + (6)(0.15)]/(1.2)^2 = 146$ kPa
$q'' = (100)[1.2 - (6)(0.15)]/(1.2)^2 = 21$ kPa

And since $q' = 146$ kPa and $q_{all} = 90$ kPa, then $q' > q_{all}$ and q' is unacceptable

Problem 15

Q per linear meter of footing $= Q/B = 100$ kN/1.2 m $= 83.3$ kN/m
$q' = Q(B + 6e)/B^2 \qquad q'' = Q(B - 6e)/B^2$ [Eqs. (8.6a) and (8.6b)]
$q' = (83.3)[1.2 + (6)(0.15)]/(1.2)^2 = 122$ kPa
$q'' = (83.3)[1.2 - (6)(0.15)]/(1.2)^2 = 17$ kPa

And since $q' = 122$ kPa and $q_{all} = 80$ kPa, then $q' > q_{all}$ and q' is unacceptable

Problem 16

$q_{ult} = Q_p/(\pi r^2) = \sigma'_v N_q$ [Eq. (8.8b)]

Then

$N_q = Q_p/(\sigma'_v \pi r^2) = 250/[(87)(\pi)(0.3/2)^2] = 40.7$
$Q_p = \pi r^2 \sigma'_v N_q = \pi (0.4/2)^2 (87)(40.7) = 444$ kN and $Q_{all} = 148$ kN

Check: On an area basis (i.e., $0.4^2/0.3^2 = 1.78$), $Q_p = 1.78(250) = 444$ kN

Problem 17

Assuming the term $k \tan \phi'_w$ is the same for the larger- and smaller-diameter pile, the frictional resistance is proportional to the surface area (i.e., $0.4/0.3 = 1.33$), $Q_s = 1.33(250) = 333$ kN and $Q_{all} = 111$ kN

Problem 18

End bearing $= (0.6)(250) = 150$ kN

The end-bearing resistance is proportional to pile tip area (i.e., $0.4^2/0.3^2 = 1.78$), therefore

$Q_p = 1.78(150) = 267$ kN

Side friction $= (0.4)(250) = 100$ kN

The side friction resistance is proportional to a side area (i.e., 0.4/0.3 = 1.33), therefore

$Q_s = 1.33(100) = 133$ kN

$Q_p + Q_s = 267 + 133 = 400$ kN and $Q_{all} = 133$ kN

Problem 19

Because the length of pile = 20 ft, use column 3 of Table 8.1, or $q_{all} = 220$ kPa

Allowable pile tip resistance $= q_{all} \pi r^2 = (220)(\pi)(0.3/2)^2 = 16$ kN

Problem 20

From Table 8.2 for $\phi' = 41°$, $N_c = 84$, $N_\gamma = 140$, and $N_q = 110$

Use Eq. (8.3) with c', $B = L$, and γ_b:

$q_{ult} = 1.3\, c'\, N_c + 0.4\, \gamma_b\, B\, N_\gamma + \gamma_b\, D_f\, N_q$

Assume sandstone at -19 m:

$q_{ult} = (1.3)(50)(84) + (0.4)(11.7)(1)(140) + [(9.2)(9) + (11.7)(3)](110)$

$= 19{,}100$ kPa

Or $Q_{ult} = (19{,}100)(\pi)(1.0/2)^2 = 15{,}000$ kN or $Q_{all} = 5000$ kN

Problem 21

Because of the 9-m embedment in soil and 3-m embedment in sandstone, use column 3 of Table 8.1, or $q_{all} = 300$ kPa (sedimentary rock), therefore:

Allowable pier tip resistance $= q_{all} \pi r^2 = (300)(\pi)(1.0/2)^2 = 240$ kN

Problem 22

From Eq. (8.9b), down-drag load $= 2\pi r L\, \sigma'_v\, k \tan \phi_w$

For the sand layer, average $\sigma'_v = (0.25)(9.2) = 2.3$ kPa,

Down-drag load $= 2\pi (1.0/2)(0.5)(2.3)(0.5) \tan 20° = 0.7$ kN

For the 3-m-thick silt-peat layer, average $\sigma'_v = (0.5)(9.2) + (1.5)(9.2) = 18.4$ kPa

Down-drag load $= 2\pi (1.0/2)(3)(18.4)(0.4) \tan 15° = 18.6$ kN

Total down-drag load from both layers $= 0.7 + 18.6 = 19.3$ kN

Problem 23

Assume the pile cap is from elevation +20 to +21 ft and ignore its weight since it is approximately compensated by the weight of removed soil.

From Fig. 4.40, use an average $s_u = c = 0.6$ kg/cm² (1200 psf) from elevation -10 to -50 ft and an average $s_u = c = 0.4$ kg/cm² (800 psf) at elevation -50 ft.

From Fig. 8.6, for $c = 1200$ psf, $c_A/c = 0.62$ (average curve, all piles) and therefore $c_A = 740$ psf. From Eq. (8.10):

Q_{ult} = end bearing + side adhesion = $9 \pi c r^2 + 2 \pi c_A r z$

$= 9 \pi (800)(1.5/2)^2 + 2 \pi (740)(1.5/2)(40) = 13{,}000 + 140{,}000 = 153$ kips

For the pile group, use Fig. 8.7. The length $(L) = 70/1.5 = 47$ pile diameters, therefore use $L = 48$ curve. For spacing in pile diameters = 3, $G_e = 0.705$.

Ultimate load of group = $G_e \, n \, Q_{ult}$ (Fig. 8.7) = $(0.705)(81)(153) = 8700$ kips

For factor of safety = 3, $Q_{all} = 8700/3 = 2900$ kips

Problem 24

Divide the clay into four layers, as follows:

Layer no. (1)	Layer thickness (2)	Depth to center of layer (3)	σ'_{v0} at the center of the layer (4)	$\Delta\sigma_v$ at the center of the layer (5)	Primary consolidation settlement (6)
1	10 ft	76 ft	4600 psf	940 psf	0.149 ft
2	10 ft	86 ft	5100 psf	675 psf	0.099 ft
3	10 ft	96 ft	5700 psf	510 psf	0.069 ft
4	10 ft	106 ft	6300 psf	400 psf	0.049 ft

Calculation for the size of the pile group is shown in Fig. 8.7, or:

$B = L = (n - 1)(2 \, r)(s) + 2 \, r = (9 - 1)(1.5)(3) + 2 \, (1.5/2) = 37.5$ ft

$\Delta\sigma_v$ is calculated by using Eq. (6.28) with z = distance from elevation -37 ft to the center of the clay layer and $P = 2900$ kips.

For layer 1:

$s_c = [C_c \, H_0/(1 + e_0)] \log [(\sigma'_{v0} + \Delta\sigma_v)/\sigma'_{v0}]$ [Eq. (7.4)]   (OCR = 1)

$= [(0.35)(10)/(1 + 0.9)] \log [(4600 + 940)/4600] = 0.149$ ft

For layer 2:

$s_c = [C_c \, H_0/(1 + e_0)] \log [(\sigma'_{v0} + \Delta\sigma_v)/\sigma'_{v0}]$ Eq. (7.4)]   (OCR = 1)

$= [(0.35)(10)/(1 + 0.9)] \log [(5100 + 675)/5100] = 0.099$ ft

For layer 3:

$s_c = [C_c H_0/(1 + e_0)] \log [(\sigma'_{vo} + \Delta\sigma_v)/\sigma'_{vo}]$ [Eq. (7.4)] (OCR = 1)
$= [(0.35)(10)/(1 + 0.9)] \log [(5700 + 510)/5700] = 0.069$ ft

For layer 4:

$s_c = [C_c H_0/(1 + e_0)] \log [(\sigma'_{vo} + \Delta\sigma_v)/\sigma'_{vo}]$ [Eq. (7.4)] (OCR = 1)
$= [(0.35)(10)/(1 + 0.9)] \log [(6300 + 400)/6300] = 0.049$ ft

Total primary consolidation settlement $(s_c) = 0.149 + 0.099 + 0.069 + 0.049 = 0.37$ ft $= 4.4$ in.

Problem 25

From Fig. 4.41, for pile adhesion, use an average undrained shear strength $(s_u) = 350$ psf. For end bearing, use an average undrained shear strength $(s_u) = 400$ psf. From Fig. 8.6, use $c_A = c = s_u = 350$ psf. From Eq. (8.10):

Q_{ult} = end bearing + side adhesion = $9 \pi c r^2 + 2 \pi c_A r z$
$= 9 \pi (400)(1.5/2)^2 + 2 \pi (350)(1.5/2)(30) = 6400 + 49,500 = 56$ kips

$Q_{all} = Q_{ult}/F = 56/3 = 19$ kips

Problem 26

For the pile group, use Fig. 8.7. The length $(L) = 35/1.5 = 23$ pile diameters, therefore use $L = 24$ curve. For spacing in pile diameters $= 1.5$, $G_e = 0.41$

Ultimate load of group $= G_e n Q_{ult}$ (Fig. 8.7) $= (0.41)(81)(56) = 1900$ kips

For factor of safety = 3, $Q_{all} = 1900/3 = 630$ kips

Problem 27

From Fig. 4.42, for pile adhesion and end bearing, use an average undrained shear strength $(s_u) = 0.6$ kg/cm² (1200 psf). From Fig. 8.6, use $c_A/c = 0.67$ (average curve for concrete piles), and therefore $c_A = (0.67)(1200) = 800$ psf. From Eq. (8.10):

Q_{ult} = end bearing + side adhesion = $9 \pi c r^2 + 2 \pi c_A r z$
$= 9 \pi (1200)(1.0/2)^2 + 2 \pi (800)(1.0/2)(40) = 8000 + 100,000 = 108$ kips

$Q_{all} = Q_{ult}/F = 108/3 = 36$ kips

Problem 28

$T = 0.0032$ (TI)$(100 - R)/G_f$ [Eq. (8.12)] For TI = 6, AC $G_f = 2.32$

$T = 0.0032 (6)(100 - 65)/2.32 = 0.29$ ft $= 3.5$ in. for asphalt concrete

$TG_f = 0.0032$ (TI)(100 − R) = 0.0032 (6)(100 − 20) = 1.54

For the asphalt concrete, $TG_f = (3.5/12)(2.32) = 0.68$

Required = 1.54 − 0.68 = 0.86 ft or $T = 0.86/G_f = 0.86/1.1 = 0.78$ ft   Use 10 in.

Answer: 3.5 in. of asphalt concrete over 10 in. of base

Check: (3.5/12)(2.32) + (10/12)(1.1) = 1.59 > 1.54   OK

Problem 29

$T = 0.0032$ (TI)(100 − R)/G_f [Eq. (8.12)]   For TI = 5, AC $G_f = 2.50$

$T = 0.0032$ (5)(100 − 65)/2.50 = 0.22 ft = 2.7 in. (use 3 in. asphalt concrete)

$TG_f = 0.0032$ (TI)(100 − R) = 0.0032 (5)(100 − 20) = 1.28

For the asphalt concrete, $TG_f = (3/12)(2.50) = 0.62$

Required = 1.28 − 0.62 = 0.66 ft or $T = 0.66/G_f = 0.66/1.7 = 0.39$ ft   Use 5 in.

Answer: 3 in. of asphalt concrete over 5 in. of CTB base

Check: (3/12)(2.50) + (5/12)(1.7) = 1.33 > 1.28   OK

Problem 30

1. Asphalt concrete:

$T = 0.0032$ (TI)(100 − R)/G_f [Eq. (8.12)]   For TI = 5, AC $G_f = 2.50$

$T = 0.0032$ (5)(100 − 65)/2.50 = 0.22 ft = 2.7 in.   Use 3 in. asphalt concrete

2. Base:

$TG_f = 0.0032$ (TI)(100 − R) = 0.0032 (5)(100 − 45) = 0.88

For the asphalt concrete, $TG_f = (3/12)(2.50) = 0.62$

Required = 0.88 − 0.62 = 0.26 ft or $T = 0.26/G_f = 0.26/1.1 = 0.23$ ft or 3 in.

Use 4 in. of base because of minimum value requirement (see text).

3. Subbase:

$TG_f = 0.0032$ (TI)(100−R) = 0.0032 (5)(100 − 20) = 1.28

Asphalt concrete and base, $TG_f = (3/12)(2.50) + (4/12)(1.1) = 0.99$

Required = 1.28 − 0.99 = 0.29 ft or $T = 0.29/G_f = 0.29/1.0 = 0.29$ ft or 4 in.

Answer: 3 in. of asphalt concrete over 4 in. of base over 4 in. of subbase

4. *Check:*   (3/12)(2.50) + (4/12)(1.1) + (4/12)(1.0) = 1.33 > 1.28   OK

CHAPTER 9

Problem 1

LL = 80, PL = 20, therefore PI = 60 and activity $(A) = 60/60 = 1.0$

From Fig. 9.1, for $A = 1$ and clay fraction = 60%, "very high" expansion potential

Problem 2

For PI = 60, the clay has a "very high" expansion potential per Table 9.1

Problem 3

From Eq. (9.2), EI = (135)(0.33) = 45, therefore "low" expansion potential

Problem 4

From Eq. (9.3), $t = T H_{dr}^2/c_s$ where $c_s = 9 \times 10^{-4}$ cm²/s. For single boundary water infiltration, $H_{dr} = 2$ m = 200 cm, therefore:

$t = (0.848)(200)^2/9 \times 10^{-4} = 3.8 \times 10^7$ s = 1.2 years

Problem 5

For double boundary water infiltration, $H_{dr} = 100$ cm, therefore t = 0.3 year

Problem 6

The equation used to calculate the secondary compression ratio (Sec. 7.4.3) can be used to calculate the secondary swell ratio ($C_{\alpha s}$), or

$C_{\alpha s} = \Delta \varepsilon_v / \Delta \log t$

At $t = 100$ min, $\varepsilon_v = 0.063$ and at $t = 1000$ min, $\varepsilon_v = 0.066$, therefore:

$C_{\alpha s} = \Delta \varepsilon_v / \Delta \log t = (0.066 - 0.063)/(\log 1000 - \log 100) = 0.003$

Problem 7

0.9 in. = 0.075 ft. Entering the right side graph in Fig. 9.11 at total swell = 0.075 ft and intersecting the total swell curve, find the depth of undercut = 2 ft.

Problem 8

Plotting swell versus depth, the total swell is the area under the percent swell curve (from a depth of 1 to 10 ft), or:

Total swell = ½ (10 − 1)(4.5/100) = 0.203 ft = 2.4 in.

Problem 9

From Fig. 9.12, $\Delta h = [(h_0 \, C_s)/(1 + e_0)] \log (P_f/P_0)$, with $P_0 = 100$ kPa

Layer 1: $\Delta h = [(500)(0.1)/(1 + 1.0)] \log (4.5/100) = 33.7$ mm

Layer 2: $\Delta h = [(500)(0.1)/(1 + 1.0)] \log (13.5/100) = 21.7$ mm

Layer 3: $\Delta h = [(1000)(0.1)/(1 + 1.0)] \log (27/100) = 28.4$ mm

Total heave of foundation $= 33.7 + 21.7 + 28.4 = 84$ mm

Problem 10

From Fig. 9.12, $\Delta h = [(h_0 \, C_s)/(1 + e_0)] \log (P_f/P_0)$ with $C_s = 0.20$

Layer 1: $\Delta h = [(500)(0.2)/(1 + 1.0)] \log (4.5/200) = 82.4$ mm

Layer 2: $\Delta h = [(500)(0.2)/(1 + 1.0)] \log (13.5/200) = 58.6$ mm

Layer 3: $\Delta h = [(1000)(0.2)/(1 + 1.0)] \log (27/200) = 87.0$ mm

Total heave of foundation $= 82.4 + 58.6 + 87.0 = 228$ mm

Problem 11

The initial condition is a uniform $P'_s = 200$ kPa and the final condition is a groundwater table at a depth of 0.5 m.

From Fig. 9.12, $\Delta h = [(h_0 \, C_s)/(1 + e_0)] \log (P_f/P_0)$

Layer 1: $P_f = \sigma'_v = (0.25)(18) = 4.5$ kPa

$\Delta h = [(500)(0.1)/(1 + 1.0)] \log (4.5/200) = 41.2$ mm

Layer 2: $P_f = \sigma'_v = (0.5)(18) + (0.25)(18 - 9.81) = 11.0$ kPa

$\Delta h = [(500)(0.1)/(1 + 1.0)] \log (11.0/200) = 31.5$ mm

Layer 3: $P_f = \sigma'_v = (0.5)(18)+(1.0)(18 - 9.81) = 17.2$ kPa

$\Delta h = [(1000)(0.1)/(1 + 1.0)] \log (17.2/200) = 53.3$ mm

Total heave of foundation $= 41.2 + 31.5 + 53.3 = 126$ mm

Problem 12

The initial condition is a uniform P'_s condition $= 200$ kPa and the final condition is a groundwater table at a depth of 0.5 m.

From Fig. 9.12, $\Delta h = [(h_0 \, C_s)/(1 + e_0)] \log (P_f/P_0)$

Layer 2: $P_f = \sigma'_v = 25 + (0.25)(18 - 9.81) = 27.0$ kPa

$\Delta h = [(500)(0.1)/(1 + 1.0)] \log (27.0/200) = 21.7$ mm

Layer 3: $P_f = \sigma'_v = 25 + (1.0)(18 - 9.81) = 33.2$ kPa

$\Delta h = [(1000)(0.1)/(1 + 1.0)] \log (33.2/200) = 39.0$ mm

Total heave of center of mat foundation $= 21.7 + 39.0 = 61$ mm

Problem 13

The initial condition is a uniform $P'_s = 200$ kPa and the final condition is a groundwater table at a depth of 0.5 m. Note also that the square footing does not exert one-dimensional pressures, so the net pressure (net σ_0) must be used in the analysis, or net $\sigma_0 = 50 - (18)(0.5) = 41$ kPa

From Fig. 9.12, $\Delta h = [(h_0 \, C_s)/(1 + e_0)] \log (P_f/P_0)$

Layer 2: $P_f = \sigma'_v = [(41)(1.2)^2/(1.2 + 0.25)^2] + (18)(0.5) + (0.25)(18 - 9.81) = 39.1$ kPa

$\Delta h = [(500)(0.1)/(1 + 1.0)] \log (39.1/200) = 17.7$ mm

Layer 3: $P_f = \sigma'_v = [(41)(1.2)^2/(1.2 + 1.0)^2] + (18)(0.5) + (1.0)(18 - 9.81) = 29.4$ kPa

$\Delta h = [(1000)(0.1)/(1 + 1.0)] \log (29.4/200) = 41.6$ mm

Total heave of square footing $= 17.7 + 41.6 = 59$ mm

Problem 14

Layer 3: $P_f = \sigma'_v = [(41)(1.2)^2/(1.2 + 1.0)^2] + (18)(0.5) + (1.0)(18 - 9.81) = 29.4$ kPa

$\Delta h = [(1000)(0.1)/(1 + 1.0)] \log (29.4/100) = 26.6$ mm

Total heave of square footing $= 17.7 + 26.6 = 44$ mm

Problem 15

1. Depth to bottom of piers below ground surface:

From Fig. 8.6, for $s_u = c = 50$ kPa (1040 psf), the value of $c_A/c = 0.75$ (average curve for concrete piles), or $c_A = (0.75)(50) = 37.5$ kPa

$T_u = c_A \, 2 \pi R Z_a$ [Eq. (9.6)]   Piers start at bottom of grade beam (0.4 m)

$= (37.5)(2)(\pi)(0.3/2)(2 - 0.4) = 57$ kN

$T_r = P + c_A \, 2 \pi R Z_{na}$ [Eq. (9.7)]   Equating T_u and T_r gives:

$57 = 20 + (80)(2)(\pi)(0.3/2)(Z_{na})$

Therefore $Z_{na} = 0.48$ m

Use factor of safety $= 2$; therefore depth to bottom of piers below ground surface $= 2$ m $+ (2)(0.48$ m$) = 3$ m

2. Air gap below grade beams:

From Fig. 9.12, total heave of layers 2 and 3 = 73 mm. Using a factor of safety = 1.5, air gap = (1.5)(73) = 110 mm

CHAPTER 10

Problem 1

From Fig. 10.14, pore water pressure u at depth d (slip surface) is equal to:

$u = d\,\gamma_w \cos^2 \alpha = (1.2)(9.81) \cos^2 33.7° = 8.1$ kPa

Problem 2

From Fig. 10.14, shear stress (τ) along the slip surface is equal to:

$\tau = \gamma_t\, a\, d \sin \alpha = (19.8)(\cos 33.7°)(1.2)(\sin 33.7°) = 11.0$ kPa

Problem 3

Per Fig. 10.14, effective normal pressure σ'_n on the slip surface is equal to:

$\sigma'_n = \gamma_b\, a\, d \cos \alpha = (19.8 - 9.81)(\cos 33.7°)(1.2)(\cos 33.7°) = 8.3$ kPa

Problem 4

Use the linear extrapolated shear strength envelope in terms of effective stresses:
$\tau_f = c' + \sigma'_n \tan \phi' = 25 + (8.3) \tan 16° = 27.4$ kPa

For the actual nonlinear portion of the shear strength envelop at low effective stress (Fig. 10.15). $\sigma'_n = 8.3$ kPa and the shear strength (τ_f) = 10.8 kPa

Problem 5

For the linear extrapolated shear strength envelope, $\tau_f/\tau = 27.4/11.0 = 2.5$

For the actual nonlinear shear strength envelope, $\tau_f/\tau = 10.8/11.0 = 0.98$

Problem 6

1. For the linear extrapolated shear strength envelope:

$F = (c' + \sigma'_n \tan \phi')/(\gamma_t\, d \cos \alpha \sin \alpha)$ where $\sigma'_n = \gamma_b\, d \cos^2 \alpha$ [Eq. (10.2)]

$\sigma'_n = (10.0)(1.2)(\cos^2 26.6°) = 9.6$ kPa

$F = [25 + (9.6)(\tan 16°)]/[(19.8)(1.2)(\cos 26.6°)(\sin 26.6°)] = 2.9$

2. For the actual nonlinear shear strength envelope:

From Fig. 10.15, for $\sigma'_n = 9.6$ kPa, $\tau_f = 13.0$ kPa

$F = (13)/[(19.8)(1.2)(\cos 26.6°)(\sin 26.6°)] = 1.37$

Problem 7

Substitute $(W + Q)$ for W in Eqs. (10.3a) and (10.3b)

Problem 8

Component of the earthquake force normal to the slip surface $= -bW \sin \alpha$

Component of the earthquake force parallel to the slip surface $= bW \cos \alpha$

Adjust Eqs. (10.3a) and (10.3b) to include these two components.

Problem 9

1. Total stress example problem:

Width of wedge at the top of slope $= (3)(9.1) - (2)(9.1) = 9.1$ m

$Q = (9.1)(10) = 91$ kN per linear meter of slope length

$F = [(14.5)(29)]/[(750 + 91) \sin 18°] = 1.62$

2. Effective stress example problem: $W + Q = 750 + 91 = 841$ kN, $uL = 69.6$ kN

$F = [(3.4)(29) + (841 \cos 18° - 69.6) \tan 29°]/(841 \sin 18°) = 1.94$

Problem 10

1. Total stress example problem:

Horizontal earthquake force $= (0.1)(750) = 75$ kN per linear meter of slope length

$F = [(14.5)(29)]/[750 \sin 18° + 75 \cos 18°] = 1.39$

2. Effective stress example problem: $c'L = 98.6$ kN, $uL = 69.6$ kN

$F = [98.6 + (750 \cos 18° - 75 \sin 18° - 69.6) \tan 29°]/(750 \sin 18° + 75 \cos 18°) = 1.46$

Problem 11

From Fig. 10.32, length L of slip surface (for $\Delta x = 1$) $= 1/\cos 26°$

Shear stress: $\tau = W \sin \alpha/L = (100)(\sin 26°)(\cos 26°) = 39.4$ kPa

$\tau_f = c' + \sigma'_n \tan \phi'$ where $\sigma'_n = W \cos \alpha/L = (100) \cos^2 26° = 80.8$ kPa

Shear strength: $\tau_f = c' + \sigma'_n \tan \phi' = 2 + 80.8 \tan 25° = 39.7$ kPa

Problem 12

$F = \tau_f/\tau = 39.7/39.4 = 1.01$. Yes, progressive failure is likely for this slope.

Problem 13

The slope is deforming laterally on a slip surface located about 7 m below ground surface.

Problem 14

Depth of seasonal moisture changes = 4.5 m, therefore setback = (2)(4.5) = 9 m

CHAPTER 11

Problem 1

SSR induced by the earthquake = $0.65\, r_d\, (a_{max}/g)(\sigma'_{v0}/\sigma'_{v0})$ [Eq. (11.1)]

With surcharge; $\sigma_{v0} = 58 + 20 = 78$ kPa and $\sigma'_{v0} = 43 + 20 = 63$ kPa

SSR induced by the earthquake = 0.65 (0.96)(0.4)(78/63) = 0.31

Problem 2

SSR induced by the earthquake = $0.65\, r_d\, (a_{max}/g)(\sigma_{v0}/\sigma'_{v0})$ [Eq. (11.1)]

SSR = 0.65 (0.96)(0.1)(58/43) = 0.085 (or one-fourth of 0.34 = 0.085)

From Fig. 11.8 with $(N_1)_{60} = 8$ and the curve with 15 percent fines, SSR = 0.13

Liquefaction will not occur because SSR induced by the earthquake (0.085) is less than the SSR that will cause liquefaction of the *in situ* soil (0.13).

Problem 3

SSR induced by the earthquake = $0.65\, r_d\, (a_{max}/g)(\sigma_{v0}/\sigma'_{v0})$ [Eq. (11.1)]

SSR = 0.65 (0.96)(0.2)(58/43) = 0.17 (or one-half of 0.34 = 0.17)

From Fig. 11.9 with $(N_1)_{60} = 8$ and the $M = 5\frac{1}{4}$ curve, SSR = 0.13

Liquefaction will probably occur because SSR induced by the earthquake (0.17) is greater than the SSR that will cause liquefaction of the *in situ* soil (0.13).

Problem 4

Use Fig. 11.11 and enter the curve with the $(N_1)_{60}$ and SSR values:

For the 2- to 3-m layer: $\varepsilon_v = 2.6\%$ or settlement = $(0.026)(1.0) = 0.026$ m

For the 3- to 5-m layer: $\varepsilon_v = 4.0\%$ or settlement = $(0.04)(2.0) = 0.080$ m

For the 5- to 7-m layer: $\varepsilon_v = 3.1\%$ or settlement = $(0.031)(2.0) = 0.062$ m

Total settlement = $0.026 + 0.080 + 0.062 = 0.168$ m = 17 cm

Problem 5

$bW = 0.2\ W = (0.2)(750) = 150$ kN $c'L = 98.6$ kN

$uL = (u_s + u_e)L = (2.4 + 2.4)(29) = 139$ kN

$F = [98.6 + (750 \cos 18° - 150 \sin 18° - 139) \tan 29°]/(750 \sin 18° + 150 \cos 18°) = 1.04$

CHAPTER 12

Problem 1

Surface area = $(800)(108) = 86{,}400$ ft^2

Soil loss = 1,400,000 lb   Soil loss per unit area = $1{,}400{,}000/86{,}400 = 16.2$ psf

Depth = $16.2/108 = 0.15$ ft = 1.8 in.

Problem 2

First 2 months: soil loss = $(700)(0.20) = 140$ tons

Next 3 months, use Eq. (12.1) and $C = 0.3$ for excelsior mat (Table 12.3):

$A = RKSCP = (300)(0.1)(13)(0.3)(0.9) = 105$ tons/acre or 210 tons total

Soil loss = $(210)(0.30) = 63$ tons

Total soil loss for the 5 months = $140 + 63 = 203$ tons

Problem 3

Use Eq. (12.1) and $C = 0.01$ (Table 12.3):

$A = RKSCP = (300)(0.1)(13)(0.01)(0.9) = 3.5$ tons/acre or 7 tons per year

Problem 4

Use Eq. (12.1) and $K = 0.7$:

$A = RKSCP = (300)(0.7)(13)(1.0)(0.9) = 2450$ tons/acre or 4900 tons per year

Problem 5

For $s = 33.3$, $m = 0.5$ and therefore from Eq. (12.3):

$L' = (L/73)^{0.5} = (108/73)^{0.5} = 1.22$

From Eq. (12.2) with $L' = 1.22$ and $s = 33.3$,

$$S = \frac{65 s^2 L'}{s^2 + 10,000} + \frac{4.6 s L'}{(s^2 + 10,000)^{0.5}} + 0.065 L' = 9.75$$

Percent reduction $= 100 (13 - 9.75)/13 = 25\%$

CHAPTER 15

Problem 1

Equation (15.2): $k_A = \tan^2 (45° - \frac{1}{2}\phi) = \tan^2 [45° - (\frac{1}{2})(32°)] = 0.307$

Equation (15.5): $k_p = \tan^2 (45° + \frac{1}{2}\phi) = \tan^2 [45° + (\frac{1}{2})(32°)] = 3.25$

Equation (15.1): $P_A = \frac{1}{2} k_A \gamma_t H^2 = \frac{1}{2} (0.307)(20)(4)^2 = 49.2$ kN/m

Equation (15.4): $P_p = \frac{1}{2} k_p \gamma_t D^2 = \frac{1}{2} (3.25)(20)(0.5)^2 = 8.14$ kN/m

With reduction factor $= 2$, allowable $P_p = 4.07$ kN/m

Problem 2

From Table 15.1, $Y/H = 0.0005$ for dense sand, therefore

$Y = (0.0005)(400 \text{ cm}) = 0.2$ cm

Problem 3

From Fig. 15.2, the active wedge is inclined at $45° + \phi/2 = 45° + (32/2) = 61°$

Width of active wedge $= H/\tan 61° = (4)/\tan 61° = 2.2$ m

Problem 4

Footing weight $= (3)(0.5)(23.5) = 35.3$ kN/m

Stem weight $= (0.4)(3.5)(23.5) = 32.9$ kN/m

$N =$ weight of concrete wall $= 35.3 + 32.9 = 68.2$ kN/m

Take moments about the toe of the wall to determine x:

$N\bar{x} = -P_A (4/3) + (W)(\text{moment arms})$ or:

$(68.2)\bar{x} = -(49.2)(4/3) + (35.3)(3/2) + (32.9)(2.8)$

$\bar{x} = 79.5/68.2 = 1.165$ m

Problem 5

$q' = Q(B + 6e)/B^2 \qquad q'' = Q(B - 6e)/B^2$ [Eqs. (8.6a) and (8.6b)]

Note that Eqs. (8.6a) and (8.6b) are identical to the analysis presented in Fig. 15.4c

$e = $ eccentricity $= 1.5 - 1.165 = 0.335$ m

$q' = (68.2)[3 + (6)(0.335)]/(3)^2 = 37.9$ kPa

$q'' = (68.2)[3 - (6)(0.335)]/(3)^2 = 7.5$ kPa

Problem 6

For $\phi' = 32°$, from Table 8.2, $N_\gamma = 15$ and $N_q = 18$

$q_{ult} = \frac{1}{2} \gamma_t B N_\gamma + \gamma_t D_f N_q$ [Eq. (8.1) with $c = 0$]

$= \frac{1}{2}(20)(3)(15) + (20)(0.5)(18) = 630$ kPa

$q_{all} = 630/3 = 210$ kPa

Since $q' = 37.9$ kPa and $q_{all} = 210$ kPa, then $q_{all} > q'$ and q' is acceptable

Problem 7

$F = (N \tan \delta + P_p)/P_A$ [Eq. (15.6)]

$= (68.2 \tan 24° + 4.07)/49.2 = 0.70$

Problem 8

$F = (W)(a)/[(1/3)(P_A)(H)]$ [Eq. (15.7)]

$= [(35.3)(3/2) + (32.9)(2.8)]/[(49.2)(1/3)(4)] = 2.2$

Problem 9

$F = (N \tan \delta + P_p)/P_A$ [Eq. (15.6)]

$1.5 = (68.2 \tan 24° + P_p)/49.2$

$P_p = 43.5$ kN/m

Double P_p to account for reduction factor and use Eq. (15.4):

$P_p = \frac{1}{2} k_p \gamma_t D^2$

$(43.5)(2) = \frac{1}{2}(3.25)(20)(D)^2$

$D = 1.64$ m

Problem 10

$\delta = \phi_w = 24°$ $\phi = 32°$ $\theta = 0°$ $\beta = 0°$

Inserting the above values into Coulomb's equation (Fig. 15.3):

$k_A = 0.275$

Use Eq. (15.1): $P_A = \frac{1}{2} k_A \gamma_t H^2$

$P_A = \frac{1}{2} (0.275)(20)(4)^2 = 43.9$ kN/m

$P_H = P_A \cos 24° = (43.9)(\cos 24°) = 40.1$ kN/m

$P_v = P_A \sin 24° = (43.9)(\sin 24°) = 17.9$ kN/m

Problem 11

Footing weight = $(3)(0.5)(23.5) = 35.3$ kN/m

Stem weight = $(0.4)(3.5)(23.5) = 32.9$ kN/m

N = weight of concrete wall + P_v = 35.3 + 32.9 + 17.9 = 86.1 kN/m

Take moments about the toe of the wall to determine x:

$Nx = -P_H (4/3) + (W)(\text{moment arms}) + (P_v)(3)$

$(86.1) \bar{x} = -(40.1)(4/3) + (35.3)(3/2) + (32.9)(2.8) + (17.9)(3)$

$\bar{x} = 145/86.1 = 1.69$ m

Problem 12

$q' = Q (B + 6 e)/B^2$ $q'' = Q (B - 6 e)/B^2$ [Eqs. (8.6a) and (8.6b)]

$e = 1.5 - 1.69 = -0.19$ m

$q' = (86.1)[3 + (6)(0.19)]/(3)^2 = 39.6$ kPa

$q'' = (86.1)[3 - (6)(0.19)]/(3)^2 = 17.8$ kPa

Problem 13

The allowable bearing pressure is the same as in Prob. 6 ($q_\text{all} = 210$ kPa), and, since $q_\text{all} > q'$, the design is acceptable in terms of bearing pressures.

Problem 14

$F = (N \tan \delta + P_p)/P_H$ [Eq. (15.8)]   where $N = W + P_v = 86.1$ kN/m

$= (86.1 \tan 24° + 4.07)/40.1 = 1.06$

Problem 15

Overturning moment = $(P_H)(1/3)(H) - (P_v)(3) = (40.1)(1/3)(4) - (17.9)(3) = -0.23$

Therefore $F = \infty$

Problem 16

$F = (N \tan \delta + P_p)/P_H$ [Eq. (15.8)]   where $N = W + P_v = 86.1$ kN/m

$= (86.1 \tan 24° + P_p)/40.1$

$1.5 = (86.1 \tan 24° + P_p)/40.1$

$P_p = 21.8$ kN/m

Double P_p to account for reduction factor and use Eq. (15.4):

$P_p = \frac{1}{2} k_p \gamma_t D^2$

$(21.8)(2) = 1/2 (3.25)(20)(D)^2$

$D = 1.16$ m

Problem 17

$\delta = 32°$ $\phi = 32°$ $\theta = 0°$ $\beta = 0°$

Inserting the above values into Coulomb's equation (Fig. 15.3): $k_A = 0.277$

Equation (15.5): $k_p = \tan^2 (45° + 1/2 \phi) = \tan^2 [45° + (\frac{1}{2})(32°)] = 3.25$

From Eq. (15.1): $P_A = \frac{1}{2} k_A \gamma_t H^2$

$P_A = \frac{1}{2} (0.277)(20)(4)^2 = 44.3$ kN/m

$P_v = P_A \sin 32° = (44.3)(\sin 32°) = 23.5$ kN/m

$P_H = P_A \cos 32° = (44.3)(\cos 32°) = 37.6$ kN/m

Equation (15.4): $P_p = \frac{1}{2} k_p \gamma_t D^2 = \frac{1}{2} (3.25)(20)(0.5)^2 = 8.14$ kN/m

With reduction factor = 2, allowable $P_p = 4.07$ kN/m

Problem 18

Footing weight = $(2)(0.5)(23.5) = 23.5$ kN/m

Stem weight = $(0.4)(3.5)(23.5) = 32.9$ kN/m

Soil weight on top of footing = $(0.8)(3.5)(20) = 56$ kN/m

N = weights + $P_v = 23.5 + 32.9 + 56 + 23.5 = 135.9$ kN/m

Take moments about the toe of the wall to determine $\bar{x}$

$$N\bar{x} = -P_H(4/3) + (W)(\text{moment arms}) + (P_v)(2)$$
$$(135.9)\bar{x} = -(37.6)(4/3) + (56.4)(1) + (56)(1.6) + (23.5)(2)$$
$$\bar{x} = 143/135.9 = 1.05 \text{ m}$$

Problem 19

$q' = Q(B + 6e)/B^2$ $q'' = Q(B - 6e)/B^2$ [Eqs. (8.6a) and (8.6b)]

$e = $ eccentricity $= 1.0 - 1.05 = -0.05$ m

$q' = (135.9)[2 + (6)(0.05)]/(2)^2 = 78.1$ kPa

$q'' = (135.9)[2 - (6)(0.05)]/(2)^2 = 57.8$ kPa

Problem 20

For $\phi' = 32°$, from Table 8.2, $N_\gamma = 15$ and $N_q = 18$

$q_{ult} = \frac{1}{2}\gamma_t B N_\gamma + \gamma_t D_f N_q$ [Eq. (8.1) with $c = 0$]

$= \frac{1}{2}(20)(2)(15) + (20)(0.5)(18) = 480$ kPa

$q_{all} = 480/3 = 160$ kPa

Since $q' = 78.1$ kPa and $q_{all} = 160$ kPa, then $q_{all} > q'$ and q' is acceptable

Problem 21

$F = (N \tan \delta + P_p)/P_H$ [Eq. (15.8)]   where $N = W + P_v = 135.9$ kN/m

$= (135.9 \tan 24° + 4.07)/37.6 = 1.72$

Problem 22

Taking moments about the toe of the wall:

Overturning moment $= (P_H)(H/3) - (P_v)(2)$

$= (37.6)(4/3) - (23.5)(2) = 3.1$ kN·m/m

Moment of weights $= (56.4)(1) + (56)(1.6) = 146$ kN·m/m

$F = 146/3.1 = 47$

Problem 23

$F = (N \tan \delta + P_p)/P_H$ [Eq. (15.8)]   where $N = W + P_v = 135.9$ kN/m

$= (135.9 \tan 24° + P_p)/37.6$

$2.0 = (135.9 \tan 24° + P_p)/37.6$

$P_p = 14.7$ kN/m

Double P_p to account for reduction factor and use Eq. (15.4):

$P_p = \frac{1}{2} k_p \gamma_t D^2$

$(14.7)(2) = \frac{1}{2}(3.25)(20)(D)^2$

$D = 0.95$ m

Problem 24

$P_Q = Q H k_A = (200)(20)(0.297) = 1190$ lb/ft at 10 ft   [Eq. (15.3)]

$P_{QH} = 1190 \cos \phi_w = 1190 \cos 30° = 1030$ lb/ft

$P_{QV} = 1190 \sin \phi_w = 1190 \sin 30° = 595$ lb/ft

$N = 15{,}270 + 595 = 15{,}870$ lb/ft

$P_H = 5660 + 1030 = 6690$ lb/ft

1. Factor of safety for sliding:

$F = (N \tan \delta + P_p)/P_H$ [Eq. (15.8)]   where $\delta = \phi_{cv} = 30°$

$= (15{,}870 \tan 30° + 750)/6690 = 1.48$

2. Factor of safety for overturning:

Overturning moment $= 14{,}900 + (1030)(10) - 595 (7) = 21{,}030$

Moment of weight $= 55{,}500$

Therefore $F = 55{,}500/21{,}030 = 2.64$

3. Location of N:

$\bar{x} = (55{,}500 - 21{,}030)/15{,}870 = 2.17$ ft

Middle one-third of the foundation: $\bar{x} = 2.33$ to 4.67 ft

Therefore N is not within the middle one-third of the foundation.

Problem 25

$\delta = \phi_w = 30°$ $\phi = 30°$ $\theta = 0°$ $\beta = 18.4°$

Insert the above values into Coulomb's equation (Fig. 15.3): $k_A = 0.4065$

Use Eq. (15.1): $P_A = \frac{1}{2} k_A \gamma_t H^2$

$P_A = \frac{1}{2}(0.4065)(110)(20)^2 = 8940$ lb/ft

$P_v = P_A \sin 30° = (8940)(\sin 30°) = 4470$ lb/ft

$P_H = P_A \cos 30° = (8940)(\cos 30°) = 7740$ lb/ft

$N = 12,000 + 4470 = 16,470$ lb/ft

1. Factor of safety for sliding:

$F = (N \tan \delta + P_p)/P_H$ [Eq. (15.8)]   where $\delta = \phi_{cv} = 30°$

$= (16,470 \tan 30° + 750)/7740 = 1.32$

2. Factor of safety for overturning:

Overturning moment $= (7740)(20/3) - (4470)(7) = 20,310$

Moment of weight $= 55,500$

Therefore $F = 55,500/20,310 = 2.73$

3. Location of N:

$\bar{x} = (55,500 - 20,310)/16,470 = 2.14$ ft

For the middle one-third of the retaining wall foundation, $\bar{x} = 2.33$ to 4.67 ft.

Therefore, N is not within the middle one-third of the retaining wall foundation.

Problem 26

$P_E = (3/8)(a_{max}/g)\gamma_t H^2$ [Eq. (15.11)]

$= (3/8)(0.20)(110)(20)^2 = 3300$ lb/ft at 12 ft above wall base

$P_H = 5660 + 3300 = 8960$ lb/ft $N = 15,270$ lb/ft

1. Factor of safety for sliding:

$F = (N \tan \delta + P_p)/P_H$ [Eq. (15.8)]   where $\delta = \phi_{cv} = 30°$

$= (15,270 \tan 30° + 750)/8960 = 1.07$

2. Factor of safety for overturning:

Overturning moment $= 14,900 + (3300)(12) = 54,500$

Moment of weight $= 55,500$

Therefore $F = 55,500/54,500 = 1.02$

Problem 27

The depth of seasonal moisture change $= 4.5$ m. Assume that the depth of slope creep is along a plane that is at a depth of 4.5 m at the top of slope and this plane passes through the toe of the slope. Therefore the angle of inclination of this plane $= \tan^{-1} [(20 - 4.5)/50] = 17.2°$

Neglecting c' and using Eq. (15.10) with $\beta = -17.2°$ and $\phi' = 28°$, therefore $k_p = 1.61$

Equation (15.4): $P_p = 1/2\, k_p\, \gamma_t\, D^2 = \frac{1}{2}\,(1.61)(19)(D)^2$ with reduction factor = 2
$(10)(2) = \frac{1}{2}\,(1.61)(19)(D)^2$

$D = 1.1$ m

For a retaining wall to be constructed at the top of slope, the total depth to the bottom of the foundation = $4.5 + 1.1 = 5.6$ m

Problem 28

Equation (6.21) $k_0 = 1 - \sin \phi' = 1 - \sin 30° = 0.5$

$P = \frac{1}{2}\, k_0 \gamma_t H^2 = 1/2\,(0.5)(110)(20)^2 = 11{,}000$ lb/ft

Problem 29

Equation (15.2): $k_A = \tan^2(45° - 1/2\,\phi) = \tan^2[45° - (1/2)(30°)] = 0.333$

Equation (15.5): $k_p = \tan^2(45° + 1/2\,\phi) = \tan^2[45° + (1/2)(30°)] = 3.0$

Equation (15.1): $P_A = 1/2\, k_A\, \gamma_t\, H^2 = 1/2\,(0.333)(110)(20)^2 = 7330$ lb/ft

Equation (15.4): $P_p = 1/2\, k_p\, \gamma_t\, D^2 = 1/2\,(3.0)(110)(3)^2 = 1490$ lb/ft

With reduction factor = 2, allowable $P_p = 740$ lb/ft

Problem 30

From Fig. 15.2, the active wedge is inclined at $45° + \phi/2 = 45° + (30/2) = 60°$

Width of active wedge = $H/\tan 60° = (20)/\tan 60° = 11.5$ ft

As measured from the upper left corner: $11.5 + 14 = 25.5$ ft

Problem 31

N = weight of reinforced soil mass zone = $(H)(L)(\gamma_t)$

$= (20)(14)(120) = 33{,}600$ lb/ft

Take moments about the toe of the wall (point O) to determine $\bar{x}$:

$N\bar{x} = -P_A\,(20/3) + (W)(L/2)$

$(33{,}600)\bar{x} = -(7330)(20/3) + (33{,}600)(7)$

$\bar{x} = 186{,}000/33{,}600 = 5.55$ ft

Problem 32

$q' = Q(B + 6e)/B^2 \qquad q'' = Q(B - 6e)/B^2$ [Eqs. (8.6a) and (8.6b)]

e = eccentricity = $7 - 5.55 = 1.45$ ft

$q' = (33,600)[14 + (6)(1.45)]/(14)^2 = 3890$ psf

$q'' = (33,600)[14 - (6)(1.45)]/(14)^2 = 910$ psf

Problem 33

For $\phi' = 30°$, from Table 8.2, $N_\gamma = 8$ and $N_q = 10$

$q_{ult} = \frac{1}{2} \gamma_t B N_\gamma + \gamma_t D_f N_q$ [Eq. (8.1) with $c = 0$]

$= \frac{1}{2}(120)(14)(8) + (110)(3)(10) = 10,000$ psf

$q_{all} = 10,000/3 = 3340$ psf

Since $q' = 3890$ psf and $q_{all} = 3340$ psf, then $q' > q_{all}$ and the design is unacceptable in terms of bearing pressures

Problem 34

$F = (N \tan \delta + P_p)/P_A$ [Eq. (15.6)]

$= (33,600 \tan 23° + 740)/7330 = 2.05$

Problem 35

Take moments about the toe of the wall:

Overturning moment = $(P_A)(H/3) = (7330)(20/3) = 48,900$

Moment of weight = $(33,600)(14/2) = 235,000$

$F = 235,000/48,900 = 4.81$

Problem 36

$P_Q = P_2 = Q H k_A = (200)(20)(0.333) = 1330$ lb/ft at 10 ft [Eq. (15.3)]

$P_H = P_A + P_2 = 7330 + 1330 = 8660$ lb/ft $\qquad N = 33,600$ lb/ft

1. Factor of safety for sliding: $F = (N \tan \delta + P_p)/P_H$ [Eq. (15.8)]

$F = (33,600 \tan 23° + 740)/8660 = 1.73$

2. Factor of safety for overturning:

Overturning moment $= (7330)(20/3) + (1330)(10) = 62,200$

Moment of weight $= 235,000$

Therefore $F = 235,000/62,200 = 3.78$

3. Maximum pressure exerted by the bottom of the wall:

$\bar{x} = (235,000 - 62,200)/33,600 = 5.14$ $e =$ eccentricity $= 7 - 5.14 = 1.86$ ft

$q' = (33,600)[14 + (6)(1.86)]/(14)^2 = 4300$ psf

Problem 37

$\delta = 0°$ $\phi = 30°$ $\theta = 0°$ $\beta = 18.4°$

Insert the above values into Coulomb's equation (Fig. 15.3): $k_A = 0.427$

Use Eq. (15.1): $P_A = \frac{1}{2} k_A \gamma_t H^2 = \frac{1}{2}(0.427)(110)(20)^2 = 9390$ lb/ft

1. Factor of safety for sliding:

$F = (N \tan \delta + P_p)/P_A = (33,600 \tan 23° + 740)/9390 = 1.60$

2. Factor of safety for overturning:

Overturning moment $= (9390)(20/3) = 62,600$

Moment of weight $= 235,200$

Therefore $F = 235,200/62,600 = 3.76$

3. Maximum pressure exerted by the wall foundation:

$\bar{x} = (235,200 - 62,600)/33,600 = 5.14$ $e =$ eccentricity $= 7 - 5.14 = 1.86$ ft

$q' = (33,600)[14 + (6)(1.86)]/(14)^2 = 4310$ psf

Problem 38

For the failure wedge: $W = \frac{1}{2}(20)(11.1)(120) = 13,300$ lb/ft $R = 12,000$ lb/ft

$F = [(W \cos \alpha + R \sin \alpha) \tan \phi]/(W \sin \alpha - R \cos \alpha)$

For the failure wedge, $F = [(13,300 \cos 61° + 12,000 \sin 61°) \tan 32°]/(13,300 \sin 61° - 12,000 \cos 61°) = 1.82$

Problem 39

Equation (15.2): $k_A = \tan^2(45° - \frac{1}{2}\phi) = \tan^2[45° - (\frac{1}{2})(33°)] = 0.295$

Equation (15.5): $k_p = \tan^2(45° + \frac{1}{2}\phi) = \tan^2[45° + (\frac{1}{2})(33°)] = 3.39$

From 0 to 5 ft: $P_{1A} = \frac{1}{2} k_A \gamma_t (5)^2 = \frac{1}{2}(0.295)(120)(5)^2 = 400$ lb/ft

From 5 to 50 ft: $P_{2A} = k_A \gamma_t (5)(45) + \frac{1}{2} k_A \gamma_b (45)^2 = (0.295)(120)(5)(45) + \frac{1}{2}(0.295)(64)(45)^2 = 8000 + 19,100 = 27,100$

$P_A = P_{1A} + P_{2A} = 400 + 27,100 = 27,500$ lb/ft

Equation (15.4) with γ_b: $P_p = \frac{1}{2} k_p \gamma_b D^2 = \frac{1}{2}(3.39)(64)(20)^2 = 43,400$ lb/ft

Problem 40

Moment due to passive force $= (43,400)[26 + (2/3)(20)] = 1.71 \times 10^6$

Neglecting P_{1A}, moment due to active force $= (8000)[1 + (45/2)] + (19,100)[1 + (2/3)(45)] = 7.8 \times 10^5$

F = resisting moment/destabilizing moment $= (1.71 \times 10^6)/(7.8 \times 10^5) = 2.19$

Problem 41

Equation (15.15): $A_p = P_A - P_p/F = 27,500 - 43,400/2.19 = 7680$ lb/ft

For 10-ft spacing, therefore, $A_p = (10)(7680) = 76,800$ lb $= 76.8$ kips

Problem 42

Determine moments at the tie back anchor:

$P_p = \frac{1}{2} k_p \gamma_b D^2 = \frac{1}{2}(3.39)(64)(D)^2 = 108 D^2$ Moment arm $= 26 + 2/3 D$

$P_{2A} = k_A \gamma_t (5)(25 + D) + \frac{1}{2} k_A \gamma_b (25 + D)^2 = (177)(25 + D) + (9.44)(25 + D)^2$

Moment arms are:

First term: $1 + \frac{1}{2}(25 + D)$ Second term: $1 + (2/3)(25 + D)$

$$F = 1.5 = \frac{(108 D^2)[26 + (2/3)(D)]}{(177)(25 + D)[1 + \frac{1}{2}(25 + D)] + (9.44)(25 + D)^2 [1 + (2/3)(25 + D)]}$$

$D = 14.6$ ft

Problem 43

$P_Q = Q H k_A = (200)(50)(0.295) = 2950$ lb/ft at 25 ft   [Eq. (15.3)]

Moment due to active forces $= 7.8 \times 10^5 + (2950)(25 - 4) = 8.4 \times 10^5$

F = resisting moment/destabilizing moment $= (1.71 \times 10^6)/(8.4 \times 10^5) = 2.03$

Problem 44

$\delta = 0°$ $\phi = 33°$ $\theta = 0°$ $\beta = 26.6°$

Insert the above values into Coulomb's equation (Fig. 15.3): $k_A = 0.443$

Destabilizing moment = $(0.443/0.295)(7.8 \times 10^5) = 1.17 \times 10^6$

F = resisting moment/destabilizing moment = $(1.71 \times 10^6)/(1.17 \times 10^6) = 1.46$

Problem 45

Use Eq. (15.10) with $\beta = -18.4°$ and $\phi' = 33°$

Therefore $k_p = 1.83$

Resisting moment = $(1.83/3.39)(1.71 \times 10^6) = 9.23 \times 10^5$

F = resisting moment/destabilizing moment = $(9.23 \times 10^5)/(7.8 \times 10^5) = 1.18$

Problem 46

$P_E = (3/8)(a_{max}/g)\gamma_b H^2$ [Eq. (15.11) with γ_b substituted for γ_t]

$= (3/8)(0.20)(64)(50)^2 = 12{,}000$ lb/ft at 30 ft above wall bottom

Destabilizing moment = $7.8 \times 10^5 + (12{,}000)(16) = 9.7 \times 10^5$

F = resisting moment/destabilizing moment = $(1.71 \times 10^6)/(9.7 \times 10^5) = 1.76$

Problem 47

Upper sand: $k_A = \tan^2(45° - 1/2\,\phi) = \tan^2[45° - (1/2)(33°)] = 0.295$

Lower sand: $k_A = \tan^2(45° - 1/2\,\phi) = \tan^2[45° - (1/2)(30°)] = 0.333$

Equation (15.5): $k_p = \tan^2(45° + 1/2\,\phi) = \tan^2[45° + (1/2)(30°)] = 3.0$

From 0 to 5 ft: $P_{1A} = 1/2\,k_A\,\gamma_t(5)^2 = 1/2\,(0.295)(120)(5)^2 = 440$ lb/ft

From 5 to 30 ft: $P_{2A} = k_A\,\gamma_t\,(5)(25) + 1/2\,k_A\,\gamma_b\,(25)^2 = (0.295)(120)(5)(25) + \frac{1}{2}(0.295)(64)(25)^2 = 4400 + 5900 = 10{,}300$ lb/ft

From 30 to 50 ft: $P_{3A} = k_A\,[\gamma_t(5) + \gamma_b(25)](20) + 1/2\,k_A\,\gamma_b\,(20)^2 = (0.333)[(120)(5) + (64)(25)](20) + \frac{1}{2}(0.333)(60)(20)^2 = 14{,}700 + 4000 = 18{,}700$ lb/ft

$P_A = P_{1A} + P_{2A} + P_{3A} = 440 + 10{,}300 + 18{,}700 = 29{,}400$ lb/ft

Equation (15.4) with γ_b: $P_p = \frac{1}{2}k_p\,\gamma_b D^2 = \frac{1}{2}(3.0)(60)(20)^2 = 36{,}000$ lb/ft

Problem 48

Moment due to passive force = $(36{,}000)[26 + (2/3)(20)] = 1.42 \times 10^6$

Neglecting P_{1A}, moment due to active force = $(4400)[1 + (25/2)] + (5900)[1 + (2/3)(25)] + (14{,}700)[26 + \frac{1}{2}(20)] + (4000)[26 + (2/3)(20)] = 8.50 \times 10^5$

F = resisting moment/destabilizing moment = $(1.42 \times 10^6)/(8.50 \times 10^5) = 1.67$

Problem 49

Equation (15.15): $A_p = P_A - P_p/F = 29{,}400 - 36{,}000/1.67 = 7840$ lb/ft

For 10-ft spacing, therefore, $A_p = (10)(7840) = 78{,}400$ lb $= 78.4$ kips

Problem 50

$P_Q = QHk_A = (200)(30)(0.295) = 1770$ lb/ft at 35 ft above wall bottom

$P_Q = QHk_A = (200)(20)(0.333) = 1330$ lb/ft at 10 ft above wall bottom

Moment due to active forces $= 8.50 \times 10^5 + (1770)(11) + (1330)(36) = 9.17 \times 10^5$

$F =$ resisting moment/destabilizing moment $= (1.42 \times 10^6)/(9.17 \times 10^5) = 1.55$

Problem 51

1. Analysis for soil in front of sheet pile wall:

In front of the wall, assume the water level coincides with the ground surface.

$\text{SSR} = 0.65\, r_d\, (a_{max}/g)(\sigma_{v0}/\sigma'_{v0})$ [Eq. (11.1)]

$\sigma_{v0} = (10)(60 + 62.4) = 1224$ psf $\quad \sigma'_v = (10)(60) = 600$ psf

$r_d = 1 - (0.012)(z) = 1 - (0.012)(10/3.28) = 0.963$ [Eq. (11.2)]

$\text{SSR} = 0.65\, (0.963)(0.2)(1224/600) = 0.26$

From Fig. 11.8, for $(N_1)_{60} = 5$, $\text{SSR} = 0.06$

At a depth of 10 ft in front of the sheet pile wall, the SSR induced by the earthquake $= 0.26$ (assuming the groundwater table is at ground surface) and the SSR that will cause liquefaction of the *in situ* soil $= 0.06$, therefore liquefaction is likely in the passive earth pressure zone.

2. Analysis of soil behind sheet pile wall:

$\text{SSR} = 0.65\, r_d\, (a_{max}/g)(\sigma_{v0}/\sigma'_{v0})$ [Eq. (11.1)]

$\sigma_{v0} = (5)(120) + (25)(64 + 62.4) + (10)(60 + 62.4) = 4980$ psf

$\sigma'_v = (5)(120) + (25)(64) + (10)(60) = 2800$ psf

$r_d = 1 - (0.012)(z) = 1 - (0.012)(40/3.28) = 0.854$ [Eq. (11.2)]

$\text{SSR} = 0.65\, (0.854)(0.2)(4980/2800) = 0.20$

From Fig. 11.8, for $(N_1)_{60} = 25$, $\text{SSR} = 0.30$

For the sand located at a depth of 40 ft below the ground surface in the active zone, the SSR induced by the earthquake $= 0.20$ and the SSR that will cause liquefaction of the *in situ* soil $= 0.30$, therefore liquefaction is unlikely in the active earth pressure zone.

For liquefaction in the passive zone, toe failure of the sheet pile wall is likely.

Problem 52

Neglecting the effective cohesion c' in the analysis and taking moments about the tie back anchor:

Lower layer: $k_A = \tan^2(45° - \frac{1}{2}\phi) = \tan^2[45° - (\frac{1}{2})(25°)] = 0.406$

Equation (15.5): $k_p = \tan^2(45° + \frac{1}{2}\phi) = \tan^2[45° + (\frac{1}{2})(25°)] = 2.46$

Equation (15.4) with γ_b: $P_p = \frac{1}{2} k_p \gamma_b D^2 = \frac{1}{2}(2.46)(60)(20)^2 = 29{,}500$ lb/ft

Moment due to passive force $= (29{,}500)[26 + (2/3)(20)] = 1.16 \times 10^6$

From 30 to 50 ft: $P_{3A} = k_A [\gamma_t(5) + \gamma_b(25)](20) + \frac{1}{2} k_A \gamma_b (20)^2 = (0.406)[(120)(5) + (64)(25)](20) + \frac{1}{2}(0.406)(60)(20)^2 = 17{,}860 + 4870 = 22{,}700$ lb/ft

Neglecting P_{1A}, moment due to active force $= (4400)[1 + (25/2)] + (5900)[1 + (2/3)(25)] + (17{,}860)[26 + \frac{1}{2}(20)] + (4870)[26 + (2/3)(20)] = 9.98 \times 10^5$

F = resisting moment/destabilizing moment $= (1.16 \times 10^6)/(9.98 \times 10^5) = 1.16$

Problem 53

$k_A = 0.333$, $k_p = 3.0$, $k_p/k_A = 9$, $\alpha = 1.0$, $H = 13.8$ ft, and $\gamma_b = 57$ pcf

From Fig. 15.30, depth ratio $= 1.5$, moment ratio $= 1.1$

Therefore, for $H = 13.8$ ft, the value of D for a factor of safety of 1 is equal to $(1.5)(13.8) = 20.7$ ft. Using a 30 percent increase in embedment depth, the required embedment depth $(D) = 27$ ft

$M_{max}/(\gamma_b k_A H^3) = 1.1$ Thus the maximum moment in the sheet pile wall $=$
$(1.1)(57)(0.333)(13.8)^3 = 55{,}000$ ft-lb/ft

Problem 54

$k_A = 0.2$, $k_p = 8.65$, $k_p/k_A = 43$, $\alpha = 0$, $H = 13.8$ ft, and $\gamma_b = 57$ pcf

From Fig. 15.30, depth ratio $= 0.43$, moment ratio $= 0.20$

Therefore, for $H = 13.8$ ft, the value of D for a factor of safety of 1 is equal to $(0.43)(13.8) = 6$ ft. Using a 30 percent increase in embedment depth, the required embedment depth $(D) = 8$ ft

$M_{max}/(\gamma_b k_A H^3) = 0.20$ Thus the maximum moment in the sheet pile wall $=$
$(0.20)(57)(0.2)(13.8)^3 = 6000$ ft-lb/ft

Problem 55

$k_A = \tan^2(45° - \frac{1}{2}\phi) = \tan^2[45° - (\frac{1}{2})(32°)] = 0.307$

From Fig. 15.31, $\sigma_h = 0.65 k_A \gamma_t H = (0.65)(0.307)(120)(20) = 480$ psf

Resultant force $= \sigma_h H = (480)(20) = 9600$ lb/ft

Problem 56

$N_o = \gamma_t H/c = (120)(20)/300 = 8$, therefore use case (b) in Fig. 15.31

$k_A = 1 - m\,(4c)/(\gamma_t H)$

Use $m = 1$; therefore $k_A = 1 - [(4)(300)]/[(120)(20)] = 0.5$

From Fig. 15.31, $\sigma_h = k_A\,\gamma_t\,H = (0.5)(120)(20) = 1200$ psf

Use the earth pressure distribution shown in Fig. 15.31, i.e., case (b):

Resultant force $= \tfrac{1}{2}(1200)(0.25)(20) + (1200)(0.75)(20) = 21{,}000$ lb/ft

Problem 57

$N_o = \gamma_t H/c = (120)(20)/1200 = 2$

Therefore use case (c) in Fig. 15.31:

$\sigma_{h2} = 0.4\,\gamma_t H = (0.4)(120)(20) = 960$ psf

Use the earth pressure distribution shown in Fig. 15.31, i.e., case (c):

Resultant force $= \tfrac{1}{2}(960)(0.5)(20) + (960)(0.5)(20) = 14{,}400$ lb/ft

Problem 58

$k_A = \tan^2(45° - \tfrac{1}{2}\phi) = \tan^2[45° - (\tfrac{1}{2})(32°)] = 0.307$

$P_A = \tfrac{1}{2} k_A\,\gamma_t H^2 = \tfrac{1}{2}(0.307)(120)(20)^2 = 7370$ lb/ft

For 7-ft spacing, the resultant force $= (7370)(7) = 51{,}600$ lb on each steel I beam

Problem 59

$k_A = \tan^2(45° - \tfrac{1}{2}\phi) = \tan^2[45° - (\tfrac{1}{2})(32°)] = 0.307$

From Fig. 15.31, $\sigma_h = 0.65\,k_A\,\gamma_t\,H = (0.65)(0.307)(120)(20) = 480$ psf

Resultant force $= \sigma_h H = (480)(20) = 9600$ lb/ft, or for 7-ft spacing, 67,200 lb

Problem 60

Resistance force $= f_s \pi D L$ or $20{,}000 = (10)(\pi)(6)(L)$

Therefore $L = 106$ in. $= 8.8$ ft

Problem 61

$uL = (2.4)(29) = 69.6$ kN/m $\qquad W = 750$ kN/m $\qquad$ Use Eq. (15.16):

$1.5 = [(750 \cos 18° - 69.6) \tan 7° + P_i]/(750 \sin 18°)$ or $P_i = 270$ kN/m

Use Eq. (15.17): $P_L = S P_i \cos \alpha = (4)(270) \cos 18° = 1000$ kN per pier

CHAPTER 16

Problem 1

Equation (6.45): $Q = kiAt = (20)(0.02)(8/12)(1) = 0.27$ ft³ in 1 minute

The base will not be able to absorb a rate of infiltration of 0.5 ft³/min because when the base is flowing full, $Q = 0.27$ ft³.

Problem 2

From Fig. 10.14, $i = \sin \alpha = \sin 26.6°$

A = area perpendicular to the direction of flow = $LH \cos \alpha = (200)(7) \cos 26.6°$

$= 1250$ ft²

Equation (6.45): $Q = kiAt = (0.3)(\sin 26.6°)(1250)(1) = 168$ ft³/day

Problem 3

1. Initial active earth pressure resultant force:

Equation (15.2): $k_A = \tan^2 (45° - \frac{1}{2} \phi) = \tan^2 [45° - (\frac{1}{2})(30°)] = 0.333$

Equation (15.1): $P_A = \frac{1}{2} k_A \gamma_t H^2 = \frac{1}{2} (0.333)(20)(3)^2 = 30$ kN/m

2. Total force acting on wall due to rise in groundwater level:

Water pressure = $\frac{1}{2} \gamma_w H^2 = \frac{1}{2} (9.81)(1.5)^2 = 11.0$ kN/m

Above the groundwater table: $P_{1A} = \frac{1}{2}(0.333)(20)(1.5)^2 = 7.5$ kN/m

Below the groundwater table: $P_{2A} = (0.333)(20)(1.5)^2 + \frac{1}{2} k_A \gamma_b (1.5)^2 =$
$15.0 + \frac{1}{2}(0.333)(20 - 9.81)(1.5)^2 = 18.8$ kN/m

Total force = $11.0 + 7.5 + 18.8 = 37.3$ kN/m

3. Percent increase in force:

% increase = $100 (37.3 - 30)/30 = 24\%$

Problem 4

Average percolation rate = $(30 \text{ min})/(\frac{1}{4} \text{ in.}) = 120$ mpi

SOLUTION TO PROBLEMS   F.51

As indicated in the notes of Table D-1 (App. D), septic tank size = 1500 gallons for a five bedroom house. From Table D-2, the required minimum lot size for an average percolation rate = 120 mpi is 5 acres. From Table D-1, for an average percolation rate = 120 mpi and a house with five bedrooms, the primary disposal leach line length = 1420 ft.

CHAPTER 17

Problem 1

From Fig. 17.5: $\rho_d = 122.5$ pcf, $w = 11\%$

Assume $V = 1$ ft^3

$V_s = M_s/[(\rho_w)(G)] = (122.5)/[(62.4)(2.65)] = 0.74$ ft^3

$V_v = 1 - V_s = 1 - 0.74 = 0.26$ ft^3

$e = V_v/V_s = 0.26/0.74 = 0.35$

Equation (6.15): $Gw = Se$ or: $(2.65)(0.11) = S(0.35)$

Therefore $S = 83\%$

Problem 2

Test no. (1)	Dry soil (2)	Dry density (3)
1	4.14/(1 + 0.11) = 3.73 lb	3.73/(1/30) = 111.9 pcf
2	4.26/(1 + 0.125) = 3.79 lb	3.79/(1/30) = 113.7 pcf
3	4.37/(1 + 0.140) = 3.83 lb	3.83/(1/30) = 115.0 pcf
4	4.33/(1 + 0.155) = 3.75 lb	3.75/(1/30) = 112.5 pcf

Plot the dry density versus water content and find $\rho_{dmax} = 115$ pcf (1.84 Mg/m^3), $w_{opt} = 14.0\%$

Problem 3

Higher compaction energy results in a higher laboratory maximum dry density and a lower optimum moisture content.

Problem 4

In Fig. 17.5, the line of optimums can be drawn through the peak point of the compaction curve and parallel to the zero air voids curve. Then a second line is drawn through the point defined by dry density = 117 pcf and water content = 8.0% and parallel to the left side of the compaction curve. The intersection of this line and the line of optimums is at a dry density of 120 pcf. Therefore $\rho_{dmax} = 120$ pcf (1.92 Mg/m^3)

Problem 5

$$\text{Wet density of soil} = 4/2000 = 0.002 \text{ kg/cm}^3 = 2.0 \text{ Mg/m}^3$$

$$\rho_d = 2.0/(1 + 0.083) = 1.85 \text{ Mg/m}^3 \text{ (115 pcf)}$$

Problem 6

1. Percent oversize particles:

$$M_0 = 8.6 \text{ kg} \qquad M_{ds} = (22.4 - 8.6)/(1 + 0.093) = 12.6 \text{ kg}$$

$$\text{\% oversize particles} = M_0/(M_0 + M_{ds}) = (100)(8.6)/(8.6 + 12.6) = 40.5\%$$

2. Dry density of matrix soil:

Equation (17.3): $\rho_{dm} = M_{ds}/[V - M_0/(G \rho_w)]$ $\qquad V = 11{,}200 \text{ cm}^3 = 0.0112 \text{ m}^3$

$$M_0 = 8.6 \text{ kg} = 0.0086 \text{ Mg} \qquad M_{ds} = 12.6 \text{ kg} = 0.0126 \text{ Mg}$$

$$\rho_{dm} = 0.0126/[0.0112 - (0.0086/2.67)] = 1.58 \text{ Mg/m}^3 \text{ (98.8 pcf)}$$

Problem 7

$$\rho_d = 115 \text{ pcf} \qquad \rho_{d\max} = 122.5 \text{ pcf}$$

Therefore $RC = (100)(115)/122.5 = 94\%$

Problem 8

$$\rho_d = 98.8 \text{ pcf} \qquad \rho_{d\max} = 122.5 \text{ pcf}$$

Therefore $RC = (100)(98.8)/122.5 = 81\%$

Problem 9

Total vol. $= (10{,}000)(0.95)(122.5)/(94) = 12{,}380 \text{ yd}^3$

Problem 10

$$\rho_d = 128/(1 + 0.065) = 120.2 \text{ pcf}$$

Total vol. $= (5000)(0.90)(122.5)/(120.2) = 4590 \text{ yd}^3$

Problem 11

Total dry mass $= (12{,}380)(94)(27) = 3.14 \times 10^7 \text{ lb}$

Water to be added $= (3.14 \times 10^7)(0.11 - 0.08) = 943{,}000 \text{ lb}$

Number of gallons $= 943{,}000/8.34 = 113{,}000 \text{ gallons}$

APPENDIX G
REFERENCES

AASHTO (1996). *Standard Specifications for Highway Bridges,* 16th ed. Prepared by the American Association of State Highway and Transportation Officials (AASHTO), Washington, D.C.

ACI (1982). *Guide to Durable Concrete.* ACI Committee 201, American Concrete Institute, Detroit, Mich.

ACI (1990). *ACI Manual of Concrete Practice, Part 1, Materials and General Properties of Concrete.* American Concrete Institute, Detroit, Mich.

ASCE (1964). *Design of Foundations for Control of Settlement.* American Society of Civil Engineers, New York, 592 pp.

ASCE (1972). "Subsurface Investigation for Design and Construction of Foundations of Buildings." Task Committee for Foundation Design Manual. Part I, *Journal of Soil Mechanics,* ASCE, vol. 98, no. SM5, pp. 481–490; Part II, no. SM6, pp. 557–578; Parts III and IV, no. SM7, pp. 749–764.

ASCE (1976). *Subsurface Investigation for Design and Construction of Foundations of Buildings.* Manual No. 56. American Society of Civil Engineers, New York, 61 pp.

ASCE (1978). *Site Characterization and Exploration. Proceedings of the Specialty Workshop at Northwestern University,* C. H Dowding (ed.). New York, 395 pp.

ASCE (1987). *Civil Engineering Magazine,* American Society of Civil Engineers, New York, April.

ASTM (1970). "Special Procedures for Testing Soil and Rock for Engineering Purposes." ASTM Special Technical Publication 479, Philadelphia, 630 pp.

ASTM (1971). "Sampling of Soil and Rock." ASTM Special Technical Publication 483, Philadelphia, 193 pp.

ASTM (1998). *Annual Book of ASTM Standards: Concrete and Aggregates,* vol. 04.2. Standard No. C 127-93, "Standard Test Method for Specific Gravity and Absorption of Coarse Aggregate," West Conshohocken, Pa., pp. 64–68.

ASTM (1997). *Annual Book of ASTM Standards: Concrete and Aggregates,* vol. 04.2. Standard No. C 881-90, "Standard Specification for Epoxy-Resin-Base Bonding Systems for Concrete," West Conshohocken, Pa., pp. 436–440.

ASTM (1997). *Annual Book of ASTM Standards: Road and Paving Materials; Vehicle-Pavement Systems,* vol. 04.03. Standard No. D 5340-93, "Standard Test Method for Airport Pavement Condition Index Surveys," West Conshohocken, Pa, pp. 546–593.

ASTM (1997). *Annual Book of ASTM Standards: Road and Paving Materials; Vehicle-Pavement Systems,* vol. 04.03. Standard No. E 1778-96, "Standard Terminology Relating to Pavement Distress," West Conshohocken, Pa., pp. 843–849.

ASTM (1998). *Annual Book of ASTM Standards,* vol. 04.08, *Soil and Rock (I).* Standard No. D 420-93, "Standard Guide to Site Characterization for Engineering, Design, and Construction Purposes," West Conshohocken, Pa., pp. 1–7.

ASTM (1998). *Annual Book of ASTM Standards,* vol. 04.08, *Soil and Rock (II).* Standard No. D 422-90, "Standard Test Method for Particle-Size Analysis of Soils," West Conshohocken, Pa., pp. 10–20.

ASTM (1998). *Annual Book of ASTM Standards,* vol. 04.08, *Soil and Rock (II).* Standard No. D 427-93, "Standard Test Method for Shrinkage Factors of Soils by the Mercury Method," West Conshohocken, Pa., pp. 21–24.

ASTM (1998). *Annual Book of ASTM Standards,* vol. 04.08, *Soil and Rock* (*I*). Standard No. D 653-97, "Standard Terminology Relating to Soil, Rock, and Contained Fluids." Terms prepared jointly by the American Society of Civil Engineers and ASTM, West Conshohocken, Pa., pp. 42–76.

ASTM (1998). *Annual Book of ASTM Standards,* vol. 04.08, *Soil and Rock* (*I*). Standard No. D 698-91, "Test Method for Laboratory Compaction Characteristics of Soil Using Standard Effort," West Conshohocken, Pa., pp. 77–87.

ASTM (1998). *Annual Book of ASTM Standards,* vol. 04.08, *Soil and Rock* (*I*). Standard No. D 854-92, "Standard Test Method for Specific Gravity of Soils," West Conshohocken, Pa., pp. 88–91.

ASTM (1998). *Annual Book of ASTM Standards,* vol. 04.08, *Soil and Rock* (*I*). Standard No. D 1143-94, "Standard Test Method for Piles Under Static Axial Compressive Load," West Conshohocken, Pa., pp. 95–105.

ASTM (1998). *Annual Book of ASTM Standards,* vol. 04.08, *Soil and Rock* (*I*). Standard No. D 1196-97, "Standard Test Method for Nonrepetitive Static Plate Load Tests of Soils and Flexible Pavement Components, for Use in Evaluation and Design of Airport and Highway Pavements," West Conshohocken, Pa., pp. 111–112.

ASTM (1998). *Annual Book of ASTM Standards,* vol. 04.08, *Soil and Rock* (*I*). Standard No. D 1556-96, "Standard Test Method for Density and Unit Weight of Soil in Place by the Sand-Cone Method," West Conshohocken, Pa., pp. 120–125.

ASTM (1998). *Annual Book of ASTM Standards,* vol. 04.08, *Soil and Rock* (*I*). Standard No. D 1557-91, "Test Method for Laboratory Compaction Characteristics of Soil Using Modified Effort," West Conshohocken, Pa., pp. 126–133.

ASTM (1998). *Annual Book of ASTM Standards,* vol. 04.08, *Soil and Rock* (*I*). Standard No. D 1586-92, "Standard Test Method for Penetration Test and Split-Barrel Sampling of Soils," West Conshohocken, Pa., pp. 137–141.

ASTM (1998). *Annual Book of ASTM Standards,* vol. 04.08, *Soil and Rock* (*I*). Standard No. D 1587-94, "Standard Practice for Thin-Walled Tube Geotechnical Sampling of Soils," West Conshohocken, Pa., pp. 142–144.

ASTM (1998). *Annual Book of ASTM Standards,* vol. 04.08, *Soil and Rock* (*I*). Standard No. D 1883-94, "Standard Test Method for CBR (California Bearing Ratio) of Laboratory-Compacted Soil," West Conshohocken, Pa., pp. 159–165.

ASTM (1998). *Annual Book of ASTM Standards,* vol. 04.08, *Soil and Rock* (*I*). Standard No. D 2166-91, "Standard Test Method for Unconfined Compressive Strength of Cohesive Soil," West Conshohocken, Pa., pp. 172–176.

ASTM (1998). *Annual Book of ASTM Standards,* vol. 04.08, *Soil and Rock* (*I*). Standard No. D 2216-92, "Standard Test Method for Laboratory Determination of Water (Moisture) Content of Soil and Rock," West Conshohocken, Pa., pp. 188–191.

ASTM (1998). *Annual Book of ASTM Standards,* vol. 04.08, *Soil and Rock* (*I*). Standard No. D 2434-94, "Standard Test Method for Permeability of Granular Soils (Constant Head)," West Conshohocken, Pa., pp. 202–206.

ASTM (1998). *Annual Book of ASTM Standards,* vol. 04.08, *Soil and Rock* (*I*). Standard No. D 2435-96, "Standard Test Method for One-Dimensional Consolidation Properties of Soils," West Conshohocken, Pa., pp. 207–216.

ASTM (1998). *Annual Book of ASTM Standards,* vol. 04.08, *Soil and Rock* (*I*). Standard No. D 2487-93, "Standard Classification of Soils for Engineering Purposes (Unified Soil Classification System)," West Conshohocken, Pa., pp. 217–227.

ASTM (1998). *Annual Book of ASTM Standards,* vol. 04.08, *Soil and Rock* (*I*). Standard No. D 2573-94, "Standard Test Method for Field Vane Shear Test in Cohesive Soil," West Conshohocken, Pa., pp. 239–241.

ASTM (1998). *Annual Book of ASTM Standards,* vol. 04.08, *Soil and Rock* (*I*). Standard No. D 2844-94, "Standard Test Method for Resistance R-Value and Expansion Pressure of Compacted Soils," West Conshohocken, Pa., pp. 246–253.

ASTM (1998). *Annual Book of ASTM Standards,* vol. 04.08, *Soil and Rock* (*I*). Standard No. D 2850-95, "Standard Test Method for Unconsolidated-Undrained Triaxial Compression Test on Cohesive Soil," West Conshohocken, Pa., pp. 260–264.

REFERENCES

ASTM (1998). *Annual Book of ASTM Standards,* vol. 04.08, *Soil and Rock (I).* Standard No. D 2922-96, "Standard Test Methods for Density of Soil and Soil-Aggregate in Place by Nuclear Methods (Shallow Depth)," West Conshohocken, Pa., pp. 268–272.

ASTM (1998). *Annual Book of ASTM Standards,* vol. 04.08, *Soil and Rock (I).* Standard No. D 2937-94, "Standard Test Method for Density of Soil in Place by the Drive-Cylinder Method," West Conshohocken, Pa., pp. 275–278.

ASTM (1998). *Annual Book of ASTM Standards,* vol. 04.08, *Soil and Rock (I).* Standard No. D 2974-95, "Standard Test Methods for Moisture, Ash, and Organic Matter of Peat and Other Organic Soils," West Conshohocken, Pa., pp. 285–287.

ASTM (1998). *Annual Book of ASTM Standards,* vol. 04.08, *Soil and Rock (I).* Standard No. D 3080-90, "Standard Test Method for Direct Shear Test of Soils Under Consolidated Drained Conditions," West Conshohocken, Pa., pp. 300–305.

ASTM (1998). *Annual Book of ASTM Standards,* vol. 04.08, *Soil and Rock (I).* Standard No. D 3441-94, "Standard Test Method for Deep, Quasi-Static, Cone and Friction-Cone Penetration Tests," West Conshohocken, Pa., pp. 348–354.

ASTM (1998). *Annual Book of ASTM Standards,* vol. 04.08, *Soil and Rock (I).* Standard No. D 3689-90, "Standard Test Method for Individual Piles Under Static Axial Tensile Load," West Conshohocken, Pa., pp. 366–376.

ASTM (1998). *Annual Book of ASTM Standards,* vol. 04.08, *Soil and Rock (I).* Standard No. D 3966-90, "Standard Test Method for Piles Under Lateral Loads," West Conshohocken, Pa., pp. 389–403.

ASTM (1998). *Annual Book of ASTM Standards,* vol. 04.08, *Soil and Rock (I).* Standard No. D 4253-96, "Standard Test Methods for Maximum Index Density and Unit Weight of Soils Using a Vibratory Table," West Conshohocken, Pa., pp. 498–510.

ASTM (1998). *Annual Book of ASTM Standards,* vol. 04.08, *Soil and Rock (I).* Standard No. D 4254-96, "Standard Test Method for Minimum Index Density and Unit Weight of Soils and Calculation of Relative Density," West Conshohocken, Pa., pp. 511–518.

ASTM (1998). *Annual Book of ASTM Standards,* vol. 04.08, *Soil and Rock (I).* Standard No. D 4318-95, "Standard Test Method for Liquid Limit, Plastic Limit, and Plasticity Index of Soils," West Conshohocken, Pa., pp. 519–529.

ASTM (1998). *Annual Book of ASTM Standards,* vol. 04.08, *Soil and Rock (I).* Standard No. D 4429-93, "Standard Test Method for CBR (California Bearing Ratio) of Soils in Place," West Conshohocken, Pa., pp. 605–608.

ASTM (1998). *Annual Book of ASTM Standards,* vol. 04.08, *Soil and Rock (I).* Standard No. D 4452-95, "Standard Test Methods for X-Ray Radiography of Soil Samples," West Conshohocken, Pa., pp. 618–629.

ASTM (1998). *Annual Book of ASTM Standards,* vol. 04.08, *Soil and Rock (I).* Standard No. D 4546-96, "Standard Test Methods for One-Dimensional Swell or Settlement Potential of Cohesive Soils," West Conshohocken, Pa., pp. 663–669.

ASTM (1998). *Annual Book of ASTM Standards,* vol. 04.08, *Soil and Rock (I).* Standard No. D 4647-93, "Standard Test Method for Identification and Classification of Dispersive Clay Soils by the Pinhole Test," West Conshohocken, Pa., pp. 757–766.

ASTM (1998). *Annual Book of ASTM Standards,* vol. 04.08, *Soil and Rock (I).* Standard No. D 4648-94, "Standard Test Method for Laboratory Miniature Vane Shear Test for Saturated Fine-Grained Clayey Soil," West Conshohocken, Pa., pp. 767–772.

ASTM (1998). *Annual Book of ASTM Standards,* vol. 04.08, *Soil and Rock (I).* Standard No. D 4767-95, "Standard Test Method for Consolidated Undrained Triaxial Compression Test for Cohesive Soils," West Conshohocken, Pa., pp. 850–859.

ASTM (1998). *Annual Book of ASTM Standards,* vol. 04.08, *Soil and Rock (I).* Standard No. D 4829-95, "Standard Test Method for Expansion Index of Soils," West Conshohocken, Pa., pp. 860–863.

ASTM (1998). *Annual Book of ASTM Standards,* vol. 04.09, *Soil and Rock (II), Geosynthetics.* Standard No. D 4943-95, "Standard Test Method for Shrinkage Factors of Soils by the Wax Method," West Conshohocken, Pa., pp. 1–5.

ASTM (1998). *Annual Book of ASTM Standards,* vol. 04.09, *Soil and Rock (II), Geosynthetics.* Standard No. D 4945-96, "Standard Test Method for High-Strain Dynamic Testing of Piles," West Conshohocken, Pa., pp. 10–16.

ASTM (1998). *Annual Book of ASTM Standards,* vol. 04.09, *Soil and Rock (II), Geosynthetics.* Standard No. D 5084-97, "Standard Test Method for Measurement of Hydraulic Conductivity of Saturated Porous Materials Using a Flexible Wall Permeameter," West Conshohocken, Pa., pp. 62–69.

ASTM (1998). *Annual Book of ASTM Standards,* vol. 04.09, *Soil and Rock (II), Geosynthetics.* Standard No. D 5333-96, "Standard Test Method for Measurement of Collapse Potential of Soils," West Conshohocken, Pa., pp. 224–226.

ASTM (1998). *Annual Book of ASTM Standards,* vol. 04.09, *Soil and Rock (II), Geosynthetics.* Standard No. D 5780-95, "Standard Test Method for Individual Piles in Permafrost Under Static Axial Compressive Load," West Conshohocken, Pa., pp. 596–609.

ASTM (1998). *Annual Book of ASTM Standards,* vol. 04.09, *Soil and Rock (II), Geosynthetics.* Standard No. D 4439-97, "Standard Terminology for Geosynthetics," West Conshohocken, Pa., pp. 1219–1222.

Aberg, B. (1996). "Grain-Size Distribution for Smallest Possible Void Ratio." *Journal of Geotechnical Engineering,* ASCE, vol. 122, no. 1, pp. 74–77.

Abramson, L. W., Lee, T. S., Sharma, S., and Boyce, G. M. (1996). *Slope Stability and Stabilization Methods.* John Wiley & Sons, New York, 629 pp.

Al-Homoud, A. S., Basma, A. A., Husein Malkawi, A. I., and Al Bashabsheh, M. A. (1995). "Cyclic Swelling Behavior of Clays." *Journal of Geotechnical Engineering,* ASCE, vol. 121, no. 7, pp. 562–565.

Al-Homoud, A. S., Basma, A. A., Husein Malkawi, A. I., and Al Bashabsheh, M. A. (1997). Closure of "Cyclic Swelling Behavior of Clays." *Journal of Geotechnical and Geoenvironmental Engineering,* ASCE, vol. 123, no. 8, pp. 786–788.

Al-Khafaji, A. W. N., and Andersland, O. B. (1981). "Ignition Test for Soil Organic Content Measurement." *Journal of Geotechnical Engineering Division,* ASCE, vol. 107, no. 4, pp. 465–479.

Allen, L. R., Yen, B. C., and McNeill, R. L. (1978). "Stereoscopic X-ray Assessment of Offshore Soil Samples." *Offshore Technology Conference,* vol. 3, pp. 1391–1399.

Alphan, I. (1967). "The Empirical Evaluation of the Coefficient K_0 and K_{OR}." *Soil and Foundation,* vol. 7, no. 1, pp. 31–40.

Ambraseys, N. N. (1960). "On the Seismic Behavior of Earth Dams." *Proceedings of the Second World Conference on Earthquake Engineering,* vol. 1, Tokyo and Kyoto, pp. 331–358.

American Geological Institute (1982). *AGI Data Sheets for Geology in the Field, Laboratory, and Office.* Prepared by the American Geological Institute, Falls Church, Va., 61 data sheets.

Anderson, S. A., and Sitar, N. (1995). "Analysis of Rainfall-Induced Debris Flow." *Journal of Geotechnical Engineering,* ASCE, vol. 121, no. 7, pp. 544–552.

Anderson, S. A., and Sitar, N. (1996). Closure of "Analysis of Rainfall-Induced Debris Flow." *Journal of Geotechnical Engineering,* ASCE, vol. 122, no. 12, pp. 1025–1027.

Asphalt Institute (1984). *Thickness Design—Asphalt Pavements for Highways and Streets.* The Asphalt Institute, College Park, Md., 80 pp.

Association of Engineering Geologists (1978). *Failure of St. Francis Dam.* Southern California Section.

Athanasopoulos, G. A. (1995). Discussion of "1988 Armenia Earthquake II. Damage Statistics versus Geologic and Soil Profiles." *Journal of Geotechnical Engineering,* ASCE, vol. 121, no. 4, pp. 395–398.

Atkins, H. N. (1983). *Highway Materials, Soils, and Concretes,* 2d ed. Reston Publishing, Reston, Va., 377 pp.

Atterberg, A. (1911). "The Behavior of Clays with Water, Their Limits of Plasticity and Their Degrees of Plasticity." *Kungliga Lantbruksakademiens Handlingar och Tidskrift,* vol. 50, no. 2, pp. 132–158.

Attewell, P. B., Yeates, J., and Selby, A. R. (1986). *Soil Movements Induced by Tunneling and Their Effects on Pipelines and Structures.* Chapman and Hall, New York, 325 pp.

Baldwin, J. E., Donley, H. F., and Howard, T. R. (1987). "On Debris Flow/Avalanche Mitigation and Control, San Francisco Bay Area, California." *Debris Flow/Avalanches: Process, Recognition, and Mitigation,* The Geological Society of America, Boulder, Colo., pp. 223–226.

REFERENCES

Bates, R. L., and Jackson, J. A. (1980). *Glossary of Geology.* American Geological Institute, Falls Church, Va., 751 pp.

Bell, F. G. (1983). *Fundamentals of Engineering Geology.* Butterworths, London, 648 pp.

Bellport, B. P. (1968). "Combating Sulphate Attack on Concrete on Bureau of Reclamation Projects." *Performance of Concrete, Resistance of Concrete to Sulphate and Other Environmental Conditions,* University of Toronto Press, Toronto, pp. 77–92.

Best, M. G. (1982). *Igneous and Metamorphic Petrology.* W. H. Freeman and Company, San Francisco.

Biddle, P. G. (1979). "Tree Root Damage to Buildings—An Arboriculturist's Experience." *Arboricultural Journal,* vol. 3, no. 6, pp. 397–412.

Biddle, P. G. (1983). "Patterns of Soil Drying and Moisture Deficit in the Vicinity of Trees on Clay Soils." *Geotechnique,* London, vol. 33, no. 2, pp. 107–126.

Bishop, A. W. (1955). "The Use of the Slip Circle in the Stability Analysis of Slopes." *Geotechnique,* London, vol. 5, no. 1, pp. 7–17.

Bishop, A. W., and Henkel, D. J. (1962). *The Measurement of Soil Properties in the Triaxial Test,* 2d ed. Edward Arnold, London, 228 pp.

Bjerrum, L. (1963). "Allowable Settlements of Structures." *Proceedings of European Conference on Soil Mechanics and Foundation Engineering,* vol. 2, Wiesbaden, Germany, pp. 135–137.

Bjerrum, L. (1967a). "The Third Terzaghi Lecture: Progressive Failure in Slopes of Overconsolidated Plastic Clay and Clay Shales." *Journal of Soil Mechanics and Foundation Division,* ASCE, vol. 93, no. 5, Part 1, pp. 1–49.

Bjerrum, L. (1967b). "Engineering Geology of Norwegian Normally Consolidated Marine Clays as Related to Settlements of Buildings." Seventh Rankine Lecture, *Geotechnique,* vol. 17, no. 2, London, pp. 81–118.

Bjerrum, L. (1972). "Embankments on Soft Ground." *Proceedings of the ASCE Specialty Conference on Performance of Earth and Earth-Supported Structures.* Purdue University, West Lafayette, Ind., vol. 2, pp. 1–54.

Bjerrum, L. (1973). "Problems of Soil Mechanics and Construction on Soft Clays and Structurally Unstable Soils." Session 4, *Proceedings of the Eighth International Conference on Soil Mechanics and Foundation Engineering,* Moscow, vol. 3, pp. 111–159.

Blight, G. E. (1965). "The Time-Rate of Heave of Structures on Expansive Clays." *Moisture Equilibria and Moisture Changes in Soils Beneath Covered Areas,* G. D. Aitchison (ed.). Butterworth, Sydney, Australia, pp. 78–88.

Boardman, B. T., and Daniel, D. E. (1996). "Hydraulic Conductivity of Desiccated Geosynthetic Clay Liners." *Journal of Geotechnical Engineering,* ASCE, vol. 122, no. 3, pp. 204–215.

Bonilla, M. G. (1970). "Surface Faulting and Related Effects." Chapter 3 of *Earthquake Engineering,* Robert L. Wiegel coordinating editor. Prentice-Hall, Englewood Cliffs, N.J., pp. 47–74.

Boone, S. T. (1996). "Ground-Movement-Related Building Damage." *Journal of Geotechnical Engineering,* ASCE, vol. 122, no. 11, pp. 886–896.

Boone, S. T. (1998). Closure to "Ground-Movement-Related Building Damage." *Journal of Geotechnical and Geoenvironmental Engineering,* ASCE, vol. 124, no. 5, pp. 463–465.

Boscardin, M. D., and Cording, E. J. (1989). "Building Response to Excavation-Induced Settlement." *Journal of Geotechnical Engineering,* ASCE, vol. 115, no. 1, pp. 1–21.

Bourdeaux, G., and Imaizumi, H. (1977). "Dispersive Clay at Sabradinho Dam." *Dispersive Clays, Related Piping, and Erosion in Geotechnical Projects,* STP 625, American Society for Testing and Materials, Philadelphia, pp. 12–24.

Boussinesq, J. (1885). *Application des Potentiels à L'Ètude de L'Èquilibre et du Mouvement des Solides Èlastiques,* Gauthier-Villars, Paris.

Bowles, J. E. (1982). *Foundation Analysis and Design,* 3d ed. McGraw-Hill, New York, 816 pp.

Brewer, H. W. (1965). "Moisture Migration—Concrete Slab-on-Ground Construction." *Journal of the PCA Research and Development Laboratories,* May, pp. 2–17.

Bromhead, E. N. (1984). *Ground Movements and their Effects on Structures,* Chap. 3, "Slopes and Embankments," Attewell, P. B., and Taylor, R. K. (eds.). Surrey University Press, London, p. 63.

Brooker, E. W., and Ireland, H. O. (1965). "Earth Pressures at Rest Related to Stress History." *Canadian Geotechnical Journal,* vol. 2, no. 1, pp. 1–15.

Brown, D. R., and Warner, J. (1973). "Compaction Grouting." *Journal of the Soil Mechanics and Foundations Division,* ASCE, vol. 99, no. SM8, pp. 589–601.

Brown, R. W. (1990). *Design and Repair of Residential and Light Commercial Foundations.* McGraw-Hill, New York, 241 pp.

Brown, R. W. (1992). *Foundation Behavior and Repair, Residential and Light Commercial.* McGraw-Hill, New York, 271 pp.

Bruce, D. A., and Jewell, R. A. (1987). "Soil Nailing: The Second Decade." *International Conference on Foundations and Tunnels,* London, pp. 68–83.

Burland, J. B., Broms, B. B., and DeMello, V. F. B. (1977). "Behavior of Foundations and Structures: State of the Art Report." *Proceedings of the Ninth International Conference on Soil Mechanics and Foundation Engineering,* Japanese Geotechnical Society, Tokyo, vol. 2, pp. 495–546.

Butt, T. K. (1992). "Avoiding and Repairing Moisture Problems in Slabs on Grade." *The Construction Specifier.* December, pp. 17–27.

Byer, J. (1992). "Geocalamities—The Do's and Don'ts of Geologic Consulting in Southern California." *Engineering Geology Practice in Southern California,* B. W. Pipkin and R. J. Proctor (eds.). Star Publishing Company, Association of Engineering Geologists, Southern California section, Special Publication no. 4, pp. 327–337.

California Department of Water Resources (1967). "Earthquake Damage to Hydraulic Structures in California." California Department of Water Resources, Bulletin 116-3, Sacramento.

California Division of Highways (1973). *Flexible Pavement Structural Design Guide for California Cities and Counties,* Sacramento, 42 pp.

California Plain Language Pamphlet of the Professional Engineers Act and the Board Rules (1995). Prepared by the Board of Registration for Professional Engineers and Land Surveyors, Department of Consumer Affairs, State of California, Sacramento, 60 pp.

Casagrande, A. (1931). Discussion of "A New Theory of Frost Heaving" by A. C. Benkelman and F. R. Ohlmstead, *Proceedings of the Highway Research Board,* vol. 11, pp. 168–172.

Casagrande, A (1932). "Research on the Atterberg Limits of Soils." *Public Roads,* vol. 13, pp. 121–136.

Casagrande, A. (1936). "The Determination of the Pre-Consolidation Load and Its Practical Significance." Discussion D-34, *Proceedings of the First International Conference on Soil Mechanics and Foundation Engineering,* Cambridge, vol. 3, pp. 60–64.

Casagrande, A. (1940). "Seepage Through Dams." *Contributions to Soil Mechanics,* 1925–1940. Boston Society of Civil Engineers, Boston, pp. 295–336. Originally published in the *Journal of the New England Water Works Association,* June 1937.

Casagrande, A. (1948). "Classification and Identification of Soils." *Transactions,* ASCE, vol. 113, pp. 901–930.

Caterpillar Performance Handbook (1997), 28th edition. Caterpillar, Inc., Peoria, Ill., 1006 pp.

Cedergren, H. R. (1989). *Seepage, Drainage, and Flow Nets,* 3d ed. John Wiley & Sons, New York, 465 pp.

Cernica, J. N. (1995a). *Geotechnical Engineering: Soil Mechanics.* John Wiley & Sons, New York, 454 pp.

Cernica, J. N. (1995b). *Geotechnical Engineering: Foundation Design.* John Wiley & Sons, New York, 486 pp.

Cheeks, J. R. (1996). "Settlement of Shallow Foundations on Uncontrolled Mine Spoil Fill." *Journal of Performance of Constructed Facilities,* ASCE, vol. 10, no. 4, pp. 143–151.

Chen, F. H. (1988). *Foundations on Expansive Soil,* 2d ed. Elsevier Scientific, New York, 463 pp.

Cheney, J. E., and Burford, D. (1975). "Damaging Uplift to a Three-Story Office Block Constructed on a Clay Soil Following Removal of Trees." *Proceedings, Conference on Settlement of Structures,* Cambridge, Pentech Press, London, pp. 337–343.

City of San Diego Standard Drawings (1986). Document no. 769374, prepared by the City of San Diego, 285 pp.

Cleveland, G. B. (1960). "Geology of the Otay Clay Deposit, San Diego County, California." *California Division of Mines Special Report 64*, Sacramento, 16 pp.

Coduto, D. P. (1994). *Foundation Design, Principles and Practices.* Prentice-Hall, Englewood Cliffs, N.J., 796 pp.

Collins, A. G., and Johnson, A. I. (1988). *Ground-Water Contamination, Field Methods,* Symposium papers published by American Society for Testing and Materials, Philadelphia, 491 pp.

Committee Report for the State (1928). "Causes Leading to the Failure of the St. Francis Dam." California Printing Office.

Compton, R. R. (1962). *Manual of Field Geology.* John Wiley & Sons, New York, pp. 255–256.

Converse Consultants Southwest, Inc. (1990). *Soil and Foundation Investigation, Proposed 80-Acre Clayton-Alexander Parcel, North Las Vegas, Nevada.* CCSW Project no. 90-33264-01, Las Vegas, Nev., 46 pp.

Corns, C. F. (1974). "Inspection Guidelines—General Aspects." *Safety of Small Dams.* Proceedings of the Engineering Foundation Conference, Henniker, N.H., ASCE, New York, NY, pp. 16–21.

County of San Diego (1983). "Below Slab Moisture Barrier Specifications" (DPL no. 65B). Department of Planning and Land Use, County of San Diego, Calif.

County of San Diego (1991). "Percolation Test Procedure." Department of Environmental Health Services and Land Use Division, County of San Diego, Calif.

Cox, J. B. (1968). "A Review of the Engineering Characteristics of the Recent Marine Clays in South East Asia." Asian Institute of Technology, Research Report no. 6, Bangkok.

Cutler, D. F., and Richardson, I. B. (1989). *Tree Roots and Buildings,* 2d ed. Longman Scientific & Technical, England, pp. 1–67.

David, D., and Komornik, A. (1980). "Stable Embedment Depth of Piles in Swelling Clays." *Fourth International Conference on Expansive Soils,* Denver, ASCE, vol. 2, pp. 798–814.

Day, R. W. (1980). *Engineering Properties of the Orinoco Clay.* Thesis Submitted for the Master of Science in Civil Engineering (MCE) and Civil Engineer (CE) degrees. Massachusetts Institute of Technology, Cambridge, Mass., 140 pp.

Day, R. W. (1989). "Relative Compaction of Fill Having Oversize Particles." *Journal of Geotechnical Engineering,* ASCE, vol. 115, no. 10, pp. 1487–1491.

Day, R. W. (1990a). "Differential Movement of Slab-on-Grade Structures." *Journal of Performance of Constructed Facilities,* ASCE, vol. 4, no. 4, pp. 236–241.

Day, R. W. (1990b). "Index Test for Erosion Potential." *Bulletin of the Association of Engineering Geologists,* vol. 27, no. 1, pp. 116–117.

Day, R. W. (1991a). Discussion of "Collapse of Compacted Clayey Sand." *Journal of Geotechnical Engineering,* ASCE, vol. 117, no. 11, pp. 1818–1821.

Day, R. W. (1991b). "Expansion of Compacted Gravelly Clay." *Journal of Geotechnical Engineering,* ASCE, vol. 117, no. 6, pp. 968–972.

Day, R. W. (1992a). "Effective Cohesion for Compacted Clay." *Journal of Geotechnical Engineering,* ASCE, vol. 118, no. 4, pp. 611–619.

Day, R. W. (1992b). "Swell Versus Saturation for Compacted Clay." *Journal of Geotechnical Engineering,* ASCE, vol. 118, no. 8, pp. 1272–1278.

Day, R. W. (1992c). "Walking of Flatwork on Expansive Soils." *Journal of Performance of Constructed Facilities,* ASCE, vol. 6, no. 1, pp. 52–57.

Day, R. W. (1992d). "Moisture Migration Through Concrete Floor Slabs." *Journal of Performance of Constructed Facilities,* ASCE, vol. 6, no. 1, pp. 46–51.

Day, R. W. (1992e). "Depositions and Trial Testimony, a Positive Experience?" *Journal of Professional Issues in Engineering Education and Practice,* ASCE, vol. 118, no. 2, pp. 129–131.

Day, R. W. (1993a). "Expansion Potential According to the Uniform Building Code." *Journal of Geotechnical Engineering,* ASCE, vol. 119, no. 6, pp. 1067–1071.

Day, R. W. (1993b). "Surficial Slope Failure: A Case Study." *Journal of Performance of Constructed Facilities,* ASCE, vol. 7, no. 4, pp. 264–269.

Day, R. W. (1994a). "Inorganic Soil Classification System Based on Plasticity." *Bulletin of the Association of Engineering Geologists,* vol. 31, no. 4, pp. 521–527.

Day, R. W. (1994b). Discussion of "Evaluation and Control of Collapsible Soil." *Journal of Geotechnical Engineering,* ASCE, vol. 120, no. 5, pp. 924–925.

Day, R. W. (1994c). "Performance of Slab-on-Grade Foundations on Expansive Soil." *Journal of Performance of Constructed Facilities,* ASCE, vol. 8, no. 2, pp. 129–138.

Day, R. W. (1994d). "Surficial Stability of Compacted Clay: Case Study." *Journal of Geotechnical Engineering,* ASCE, vol. 120, no. 11, pp. 1980–1990.

Day, R. W. (1994e). "Weathering of Expansive Sedimentary Rock due to Cycles of Wetting and Drying." *Bulletin of the Association of Engineering Geologists,* vol. 31, no. 3, pp. 387–390.

Day, R. W. (1994f). "Moisture Migration Through Basement Walls." *Journal of Performance of Constructed Facilities,* ASCE, vol. 8, no. 1, pp. 82–86.

Day, R. W. (1994g). "Swell-Shrink Behavior of Compacted Clay." *Journal of Geotechnical Engineering,* ASCE, vol. 120, no. 3, pp. 618–623.

Day, R. W. (1995a). Discussion of "Numerical Analysis of Drained Direct and Simple Shear Tests." *Journal of Geotechnical Engineering,* ASCE, vol. 121, no. 2, pp. 223–227.

Day, R. W. (1995b). "Effect of Maximum Past Pressure on Two-Dimensional Immediate Settlement." *Journal of Environmental and Engineering Geoscience,* Joint Publication, AEG and GSA, vol. 1, no. 4, pp. 514–517.

Day, R. W. (1995c). "Pavement Deterioration: Case Study." *Journal of Performance of Constructed Facilities,* ASCE, vol. 9. no. 4, pp. 311–318.

Day, R. W. (1995d). "Reactivation of an Ancient Landslide." *Journal of Performance of Constructed Facilities,* ASCE, vol. 9, no. 1, pp. 49–56.

Day, R. W. (1995e). "Engineering Properties of Diatomaceous Fill." *Journal of Geotechnical Engineering,* ASCE, vol. 121, no. 12, pp. 908–910.

Day, R. W. (1996a). "Study of Capillary Rise and Thermal Osmosis." *Journal of Environmental and Engineering Geoscience,* Joint Publication, AEG and GSA, vol. 2, no. 2, pp. 249–254.

Day, R. W. (1996b). "Effect of Gravel on Pumping Behavior of Compacted Soil." *Journal of Geotechnical Engineering,* ASCE, vol. 122, no. 10, pp. 863–866.

Day, R. W. (1996c). Discussion of "Modified Oedometer for Arid Saline Soils." *Journal of Geotechnical Engineering,* ASCE, vol. 122, no. 1, pp. 83–84.

Day, R. W. (1997a). Discussion of "Grain-Size Distribution for Smallest Void Ratio." *Journal of Geotechnical and Geoenvironmental Engineering,* ASCE, vol. 123, no. 1, p. 78.

Day, R. W. (1997b). "Hydraulic Conductivity of a Desiccated Clay Upon Wetting." *Journal of Environmental and Engineering Geoscience,* Joint Publication, AEG and GSA, vol. 3, no. 2, pp. 308–311.

Day, R. W. (1998). Discussion of "Ground-Movement-Related Building Damage." *Journal of Geotechnical and Geoenvironmental Engineering,* ASCE, vol. 124, no. 5, pp. 462–463.

Day, R. W. (1999). *Forensic Geotechnical and Foundation Engineering.* McGraw-Hill, New York, 461 pp.

Day, R. W., and Axten, G. W. (1989). "Surficial Stability of Compacted Clay Slopes." *Journal of Geotechnical Engineering,* ASCE, vol. 115, no. 4, pp. 577–580.

Day, R. W., and Axten, G. W. (1990). "Softening of Fill Slopes due to Moisture Infiltration." *Journal of Geotechnical Engineering,* ASCE, vol. 116, no. 9, pp. 1424–1427.

Day, R. W., and Marsh, E. T. (1995). "Triaxial A-Value Versus Swell or Collapse for Compacted Soil." *Journal of Geotechnical Engineering,* ASCE, vol. 121, no. 7, pp. 566–570.

Day, R. W., and Poland, D. M. (1996). "Damage Due to Northridge Earthquake Induced Movement of Landslide Debris." *Journal of Performance of Constructed Facilities,* ASCE, vol. 10, no. 3, pp. 96–108.

DeGaetano, A. T., Wilks, D. S., and McKay, M. (1997). "Extreme-Value Statistics for Frost Penetration Depths in Northeastern United States." *Journal of Geotechnical and Geoenvironmental Engineering,* ASCE, vol. 123, no. 9, pp. 828–835.

DeMello, V. (1971). "The Standard Penetration Test: A State-of-the-Art Report." *Fourth PanAmerican Conference on Soil Mechanics and Foundation Engineering,* vol. 1, pp. 1–86.

Department of the Army (1970). *Engineering and Design, Laboratory Soils Testing* (Engineer Manual EM 1110-2-1906). Prepared at the U.S. Army Engineer Waterways Experiment Station, published by the Department of the Army, Washington, D.C., 282 pp.

Design and Control of Concrete Mixtures (1994), 4th printing of 13th ed., Portland Cement Association, Skokie, Ill., 205 pp.

Diaz, C. F., Hadipriono, F. C., and Pasternack, S. (1994). "Failures of Residential Building Basements in Ohio." *Journal of Performance of Constructed Facilities,* ASCE, vol. 8, no. 1, pp. 65–80.

Dounias, G. T., and Potts, D. M. (1993). "Numerical Analysis of Drained Direct and Simple Shear Tests." *Journal of Geotechnical Engineering,* ASCE, vol. 119, no. 12, pp. 1870–1891.

Driscoll, R. (1983). "The Influence of Vegetation on the Swelling and Shrinkage Caused by Large Trees." *Geotechnique,* London, vol. 33, no. 2, pp. 1–67.

Dudley, J. H. (1970). "Review of Collapsing Soils." *Journal of Soil Mechanics and Foundation Engineering Division,* ASCE, vol. 96, no. SM3, pp. 925–947.

Duke, C. M. (1960). "Foundations and Earth Structures in Earthquakes." *Proceedings of the Second World Conference on Earthquake Engineering,* vol. 1, Tokyo and Kyoto, pp. 435–455.

Duncan, J. M. (1993). "Limitations of Conventional Analysis of Consolidation Settlement." *Journal of Geotechnical Engineering,* ASCE, vol. 119, no. 9, 1331–1359.

Duncan, J. M. (1996). "State of the Art: Limit Equilibrium and Finite-Element Analysis of Slopes." *Journal of Geotechnical and Geoenvironmental Engineering,* ASCE, vol. 122, no. 7, pp. 577–596.

Duncan, J. M. and Buchignani, A. L. (1976). "An Engineering Manual for Settlement Studies." *Geotechnical Engineering Report,* University of California at Berkeley, 94 pp.

Duncan, J. M., Williams, G. W., Sehn, A. L., and Seed, R. B. (1991). "Estimation Earth Pressures due to Compaction." *Journal of Geotechnical Engineering,* ASCE, vol. 117, no. 12, pp. 1833–1847.

Dyni, R. C. and Burnett, M. (1993). "Speedy Backfilling for Old Mines." *Civil Engineering Magazine,* ASCE, vol. 63, no. 9, pp. 56–58.

Earth Manual (1985). A water resources technical publication, 2nd ed. U.S. Department of the Interior, Bureau of Reclamation, Denver, 810 pp.

Ehlig, P. L. (1986). "The Portuguese Bend Landslide: Its Mechanics and a Plan for Its Stabilization." *Landslides and Landslide Mitigation in Southern California.* 82d Annual Meeting Of the Cordilleran Section of the Geological Society of America, Los Angeles, pp. 181–190.

Ehlig, P. L. (1992). "Evolution, Mechanics, and Migration of the Portuguese Bend Landslide, Palos Verdes Peninsula, California." *Engineering Geology Practice in Southern California,* B. W. Pipkin and R. J. Proctor (eds.). Star Publishing Company, Association of Engineering Geologists, Southern California Section, Special Publication no. 4, pp. 531–553.

Ellen, S. D., and Fleming, R. W. (1987). "Mobilization of Debris Flows from Soil Slips, San Francisco Bay Region, California." *Debris Flows/Avalanches: Process, Recognition, and Mitigation,* The Geological Society of America, Boulder, Colo., pp. 31–40.

Evans, D. A. (1972). *Slope Stability Report,* Slope Stability Committee, Department of Building and Safety, Los Angeles.

"Excavations, Final Rule." (1989). *29CFR Part 1926, Federal Register,* vol. 54, no. 209, pp. 45894–45991.

Fairweather, V. (1992). "L'Ambiance Plaza: What Have We Learned?" *Civil Engineering Magazine,* ASCE, vol. 62, no. 2, pp. 38–41.

Feld, J. (1965). "Tolerance of Structures to Settlement." *Journal of Soil Mechanics,* ASCE, vol. 91, no. SM3, pp. 63–77.

Feld, J., and Carper, K. L. (1997). *Construction Failure,* 2d ed. John Wiley & Sons, New York, 512 pp.

Fellenius, W. (1936). "Calculation of the Stability of Earth Dams." *Proceedings of the Second Congress on Large Dams,* vol. 4, Washington, D.C., pp. 445–463.

Fields of Expertise (undated). Joint Publication Prepared by the Joint Civil Engineers/Engineering Geology Committee, Appointed by the Professional Engineers and Geologist and Geophysicist Boards of Registration, Calif., 5 pp.

Florensov, N. A., and Solonenko, V. P. (eds.) (1963). "Gobi-Altayskoye Zemletryasenie." *Iz. Akad. Nauk SSSR.*; also 1965, *The Gobi-Altai Earthquake,* U.S. Department of Commerce (English translation), Washington, D.C.

Foott, R., and Ladd, C. C. (1981). "Undrained Settlement of Plastic and Organic Clays." *Journal of the Geotechnical Engineering Division,* ASCE, vol. 107, GT8, pp. 1079–1094.

Foshee, J., and Bixler, B. (1994). "Cover-Subsidence Sinkhole Evaluation of State Road 434, Longwood, Florida." *Journal of Geotechnical Engineering,* ASCE, vol. 120, no. 11, pp. 2026–2040.

Foster, C. R., and Ahlvin, R. G. (1954). "Stresses and Deflections Induced by a Uniform Circular Load." *Proceedings of the Highway Research Board,* vol. 33, Washington, D.C., pp. 467–470.

Fourie, A. B. (1989). "Laboratory Evaluation of Lateral Swelling Pressures." *Journal of Geotechnical Engineering,* ASCE, vol. 115, no. 10, pp. 1481–1486.

Franklin, A. G., Orozco, L. F., and Semrau, R. (1973). "Compaction and Strength of Slightly Organic Soils." *Journal of Soil Mechanics and Foundations Division,* ASCE, vol. 99, no. 7, pp. 541–557.

Fredlund, D. G. (1983). "Prediction of Ground Movements in Swelling Clays." *Thirty-first Annual Soil Mechanics and Foundation Engineering Conference,* ASCE, invited lecture, University of Minnesota.

Fredlund, D. G., and Rahardjo, H. (1993). *Soil Mechanics for Unsaturated Soil.* John Wiley & Sons, New York, 517 pp.

Geologist and Geophysicist Act (1986). Prepared by the Board of Registration for Geologists and Geophysicists, Department of Consumer Affairs, State of California, Sacramento, 55 pp.

Geo-Slope (1991). *User's Guide, SLOPE/W for Slope Stability Analysis,* version 2. Geo-Slope International, Calgary, Canada, 444 pp.

Geo-Slope (1992). *User's Guide, SEEP/W for Finite Element Seepage Analysis,* version 2. Geo-Slope International, Calgary, Canada, 349 pp.

Geo-Slope (1993). *User's Guide, SIGMA/W for Finite Element Stress/Deformation Analysis,* version 2. Geo-Slope International, Calgary, Canada, 407 pp.

Gill, L. D. (1967). "Landslides and Attendant Problems." *Mayor's Ad Hoc Landslide Committee Report,* Los Angeles.

gINT (1991). *Geotechnical Integrator Software Manual.* Computer software and manual developed by Geotechnical Computer Applications (GCA), Santa Rosa, Calif., 800 pp.

Goh, A. T. C. (1993). "Behavior of Cantilever Retaining Walls." *Journal of Geotechnical Engineering,* ASCE, vol. 119, no. 11, pp. 1751–1770.

Goldman, S. J., Jackson, K., and Bursztynsky, T. A. (1986). *Erosion and Sediment Control Handbook.* McGraw-Hill, New York, 454 pp.

Gould, J. P. (1995). "Geotechnology in Dispute Resolution," The Twenty-sixth Karl Terzaghi Lecture. *Journal of Geotechnical Engineering,* ASCE, vol. 121, no. 7, pp. 521–534.

Graf, E. D. (1969). "Compaction Grouting Techniques," *Journal of the Soil Mechanics and Foundations Division,* ASCE, vol. 95, no. SM5, pp. 1151–1158.

Grant, R., Christian, J. T., and Vanmarcke, E. H. (1974). "Differential Settlement of Buildings." *Journal of Geotechnical Engineering,* ASCE, vol. 100, no. 9, pp. 973–991.

Grantz, A., Plafker, G., and Kachadoorian, R. (1964). *Alaska's Good Friday Earthquake,* March 27, 1964. Department of the Interior, Geological Survey Circular 491, Washington, D.C.

Gray, R. E. (1988). "Coal Mine Subsidence and Structures." *Mine Induced Subsidence: Effects on Engineered Structures,* Geotechnical Special Publication 19, ASCE, New York, pp. 69–86.

Greenfield, S. J., and Shen, C. K. (1992). *Foundations in Problem Soils.* Prentice-Hall, Englewood Cliffs, N.J., 240 pp.

Griffin, D. C. (1974). "Kentucky's Experience with Dams and Dam Safety." *Safety of Small Dams.* Proceedings of the Engineering Foundation Conference, Henniker, N.H., ASCE, New York, pp. 194–207.

Gromko, G. J. (1974). "Review of Expansive Soils." *Journal of Geotechnical Engineering Division,* ASCE, vol. 100, no. 6, pp. 667–687.

"Guajome Ranch House, Vista, California." (1986). *National Historic Landmark Condition Assessment Report,* Preservation Assistance Division, National Park Service, Washington, D.C.

Hammer, M. J., and Thompson, O. B. (1966). "Foundation Clay Shrinkage Caused by Large Trees." *Journal of Soil Mechanics and Foundation Division,* ASCE, vol. 92, no. 6, pp. 1–17.

Hansbo, S. (1975). *Jordmateriallära,* Almqvist & Wiksell Förlag AB, Stockholm, 218 pp.

Hansen, M. J. (1984). "Strategies for Classification of Landslides," in *Slope Instability.* John Wiley & Sons, New York, pp. 1–25.

Hansen, W. R. (1965). *Effects of the Earthquake of March 27, 1964, at Anchorage, Alaska.* Geological Survey Professional Paper 542-A, U.S. Department of the Interior, Washington, D.C.

Harr, E. (1962). *Groundwater and Seepage.* McGraw-Hill, New York, 315 pp.

Hawkins, A. B., and Pinches, G. (1987). "Expansion due to Gypsum Crystal Growth." *Proceedings of the Sixth International Conference on Expansive Soils,* vol. 1, New Delhi, pp. 183–188.

Hawkins, A. B., and Privett, K. D. (1985). "Measurement and Use of Residual Shear Strength of Cohesive Soils." *Ground Engineering,* vol. 18, no. 8, pp. 22–29.

Hollingsworth, R. A., and Grover, D. J. (1992). "Causes and Evaluation of Residential Damage in Southern California." *Engineering Geology Practice in Southern California,* Pipkin, B. W., and Proctor, R. J. (eds.). Star Publishing, Belmont, Calif., pp. 427–441.

Holtz, R. D., and Kovacs, W. D. (1981). *An Introduction to Geotechnical Engineering.* Prentice-Hall, Englewood Cliffs, N.J., 733 pp.

Holtz, W. G. (1959). "Expansive Clays—Properties and Problems." *Quarterly of the Colorado School of Mines,* vol. 54, no. 4, pp. 89–125.

Holtz, W. G. (1984). "The Influence of Vegetation on the Swelling and Shrinking of Clays in the United States of America." *The Influence of Vegetation on Clays,* Thomas Telford, London, pp. 69–73.

Holtz, W. G., and Gibbs, H. J. (1956a). "Engineering Properties of Expansive Clays." *Transactions,* ASCE, vol. 121, pp. 641–677.

Holtz, W. G., and Gibbs, H. J. (1956b). "Shear Strength of Pervious Gravelly Soils." *Proceedings,* ASCE, paper no. 867.

Horn, H. M., and Deere, D. U. (1962). "Frictional Characteristics of Minerals." *Geotechnique,* vol. 12, London, pp. 319–335.

Hough, B. K. (1969). *Basic Soils Engineering.* Ronald Press, New York, 634 pp.

Housner, G. W. (1970). "Strong Ground Motion." Chapter 4 of *Earthquake Engineering,* R. L. Wiegel, coordinating editor. Prentice-Hall, Englewood Cliffs, N.J., pp. 75–92.

Houston, S. L., and Walsh, K. D. (1993). "Comparison of Rock Correction Methods for Compaction of Clayey Soils." *Journal of Geotechnical Engineering,* ASCE, vol. 119, no. 4, pp. 763–778.

Howard, A. K. (1977). *Laboratory Classification of Soils—Unified Soil Classification System.* Earth Sciences Training Manual no. 4, U.S. Bureau of Reclamation, Denver, Colo., 56 pp.

Hurst, W. D. (1968). "Experience in the Winnipeg Area with Sulphate-Resisting Cement Concrete." *Performance of Concrete, Resistance of Concrete to Sulphate and Other Environmental Conditions,* University of Toronto Press, Toronto, Canada, pp. 125–134.

Hvorslev, M. J. (1949). *Subsurface Exploration and Sampling of Soils for Civil Engineering Purposes.* Waterways Experiment Station, Vicksburg, Miss., 465 pp.

Hvorslev, M. J. (1951). "Time Lag and Soil Permeability in Ground Water Measurements," Corps of Engineers Waterways Experiment Station, Vicksburg, Miss., *Bulletin 36,* 50 pp.

Ishihara, K. (1985). "Stability of Natural Deposits During Earthquakes." *Proceedings of the Eleventh International Conference on Soil Mechanics and Foundation Engineering,* San Francisco, vol.1 pp. 321–376.

Ishihara, K. (1993). "Liquefaction and Flow Failure During Earthquakes." *Geotechnique,* vol. 43, no. 3, London, pp. 351–415.

Jaky, J. (1944). "The Coefficient of Earth Pressure at Rest" (in Hungarian). *Journal of the Society of Hungarian Architects and Engineers,* vol. 78, no. 22, pp. 355–358.

Jaky, J. (1948). "Earth Pressure in Silos." *Proceedings of the Second International Conference on Soil Mechanics and Foundation Engineering,* Rotterdam, Netherlands, vol. 1, pp. 103–107.

Jamiolkowski, M., Ladd, C. C., Germaine, J. T., and Lancellotta, R. (1985). "New Developments in Field and Laboratory Testing of Soils." *Eleventh International Conference on Soil Mechanics and Foundation Engineering,* San Francisco, vol. 1, pp. 57–153.

Janbu, N. (1957). "Earth Pressure and Bearing Capacity Calculation by Generalized Procedure of Slices." *Proceedings of the Fourth International Conference on Soil Mechanics and Foundation Engineering,* London, vol. 2, pp. 207–212.

Janbu, N. (1968). "Slope Stability Computations." *Soil Mechanics and Foundation Engineering Report.* The Technical University of Norway, Trondheim.

Janney, J. R., Vince, C. R., and Madsen, J. D. (1996). "Claims Analysis from Risk-Retention Professional Liability Group." *Journal of Performance of Constructed Facilities,* ASCE, vol. 10, no. 3, pp. 115–122.

Jennings, J. E. (1953). "The Heaving of Buildings on Desiccated Clay." *Proceedings of the Third International Conference on Soil Mechanics and Foundation Engineering,* vol. 1, Zurich, pp. 390–396.

Jennings, J. E., and Knight, K. (1957). "The Additional Settlement of Foundations due to a Collapse of Structure of Sandy Subsoils on Wetting." *Proceedings of the Fourth International Conference on Soil Mechanics and Foundation Engineering,* vol. 1, London, pp. 316–319.

Johnpeer, G. D. (1986). "Land Subsidence Caused by Collapsible Soils in Northern New Mexico." *Ground Failure,* National Research Council, Committee on Ground Failure Hazards, vol. 3, Washington, D.C., 24 pp.

Johnson, A. M., and Hampton, M. A. (1969). "Subaerial and Subaqueous Flow of Slurries." Final Report. U.S. Geological Survey (USGS) Contract no. 14-08-0001-10884, USGS, Boulder, Colo.

Johnson, A. M., and Rodine, J. R. (1984). "Debris Flow." *Slope Instability,* John Wiley & Sons, New York, pp. 257–361.

Johnson, L. D. (1980). "Field Test Sections on Expansive Soil." *Fourth International Conference on Expansive Soils,* ASCE, Denver, vol. 1, pp. 262–283.

Jones, D. E., and Holtz, W. G. (1973). "Expansive Soils—The Hidden Disaster." *Civil Engineering Magazine,* ASCE, vol. 43, No. 11, November.

Jones, D. E., and Jones, K. A. (1987). "Treating Expansive Soils." *Civil Engineering Magazine,* ASCE, vol. 57, no. 8, August.

Jubenville, D. M., and Hepworth, R. (1981). "Drilled Pier Foundation in Shale, Denver, Colorado Area." *Proceedings of the Session on Drilled Piers and Caissons,* ASCE, St. Louis.

Kaplar, C. W. (1970). "Phenomenon and Mechanism of Frost Heaving." *Highway Research Record 304,* pp. 1–13.

Kassiff, G., and Baker, R. (1971). "Aging Effects on Swell Potential of Compacted Clay." *Journal of the Soil Mechanics and Foundation Division,* ASCE, vol. 97, no. SM3, pp. 529–540.

Kay, B. L. (1983). "Straw as an Erosion Control Mulch." Agronomy Progress Report no. 140, University of California, Davis, Agricultural Experiment Station Cooperative Extension, Davis.

Kayan, R. E., Mitchell, J. K., Seed, R. B., Lodge, A., Nishio, S., and Coutinho, R. (1992). "Evaluation of SPT-, CPT-, and Shear Wave-Based Methods for Liquefaction Potential Assessments Using Loma Prieta Data." *Proceedings, Fourth Japan-U.S. Workshop on Earthquake Resistant Design of Lifeline Facilities and Countermeasures for Soil Liquefaction; NCEER-92-0019,* National Center for Earthquake Engineering, Buffalo, N.Y., pp. 177–192.

Kennedy, M. P. (1975). "Geology of Western San Diego Metropolitan Area, California." *Bulletin 200,* California Division of Mines and Geology, Sacramento, 39 pp.

Kennedy, M. P., and Tan, S. S. (1977). "Geology of National City, Imperial Beach and Otay Mesa Quadrangles, Southern San Diego Metropolitan Area, California." Map Sheet 29, California Division of Mines and Geology, Sacramento, 1 sheet.

Kenney, T. C. (1964). "Sea-Level Movements and the Geologic Histories of the Post-Glacial Marine Soils at Boston, Nicolet, Ottawa, and Oslo." *Geotechnique,* vol. 14, no. 3, London, pp. 203–230.

Kerwin, S. T., and Stone, J. J. (1997). "Liquefaction Failure and Remediation: King Harbor, Redondo Beach, California." *Journal of Geotechnical and Geoenvironmental Engineering,* ASCE, vol. 123, no. 8, pp. 760–769.

Koerner, R. M. (1990). *Design with Geosynthetics,* 2d ed., Prentice-Hall, Englewood Cliffs, N.J.

Kononova, M. M. (1966). *Soil Organic Matter,* 2d ed. Pergamon Press, Oxford, England.

Kratzsch, H. (1983). *Mining Subsidence Engineering.* Springer-Verlag, Berlin, 543 pp.

Ladd, C. C. (1971). "Strength Parameters and Stress-Strain Behavior of Saturated Clays." Massachusetts Institute of Technology, Research Report R71-23.

Ladd, C. C. (1973). "Settlement Analysis for Cohesive Soil." Research Report R71-2, no. 272, Department of Civil Engineering, Massachusetts Institute of Technology, Cambridge, Mass., 115 pp.

Ladd, C. C., Azzouz, A. S., Martin, R. T., Day, R. W., and Malek, A. M. (1980). *Evaluation of Compositional and Engineering Properties of Offshore Venezuelan Soils.* vol. 1, MIT Research Report R80-14, no. 665, 286 pp.

Ladd, C. C., Foott, R., Ishihara, K., Schlosser, F., and Poulos, H. G. (1977). "Stress-Deformation and Strength Characteristics." *State-of-the-Art Report, Proceedings, Ninth International Conference on Soil Mechanics and Foundation Engineering,* International Society of Soil Mechanics and Foundation Engineering, Tokyo, vol. 2, pp. 421–494.

Ladd, C. C., and Lambe, T. W. (1961). "The Identification and Behavior of Expansive Clays." *Proceedings, Fifth International Conference on Soil Mechanics and Foundation Engineering,* Paris, vol. 1.

Ladd, C. C., and Lambe T. W. (1963). "The Strength of 'Undisturbed' Clay Determined from Undrained Tests," Special Technical Publication no. 361, ASTM, Philadelphia, pp. 342–371.

Lambe, T. W. (1951). *Soil Testing for Engineers.* John Wiley & Sons, New York, 165 pp.

Lambe, T. W. (1958a). "The Structure of Compacted Clay." *Journal of the Soil Mechanics and Foundations Division,* ASCE, vol. 84, no. SM2, pp. 1654-1 to 1654-34.

Lambe, T. W. (1958b). "The Engineering Behavior of Compacted Clay." *Journal of the Soil Mechanics and Foundations Division,* ASCE, vol. 84, no. SM2, pp. 1655-1 to 1655-35.

Lambe, T. W. (1967). "Stress Path Method." *Journal of the Soil Mechanics and Foundations Division,* ASCE, vol. 93, no. SM6, pp. 309–331.

Lambe, T. W., and Whitman, R. V. (1969). *Soil Mechanics.* John Wiley & Sons, New York, 553 pp.

Lane et al. (1947). "Modified Wentworth Scale." *Transactions of the American Geophysical Union,* vol. 28, pp. 936–938.

Lawson, A. C., et al. (1908). *The California Earthquake of April 18, 1906—Report of the State Earthquake Investigation Commission,* vol. 1, part 1, pp. 1–254; part 2, pp. 255–451. Carnegie Institution of Washington, Publication 87.

Lawton, E. C. (1996). "Nongrouting Techniques." *Practical Foundation Engineering Handbook.* Robert W. Brown (ed.). McGraw-Hill, New York, sec. 5, pp. 5.3–5.276.

Lawton, E. C., Fragaszy, R. J., and Hardcastle, J. H. (1989). "Collapse of Compacted Clayey Sand." *Journal of Geotechnical Engineering,* ASCE, vol. 115, no. 9, pp. 1252–1267.

Lawton, E. C., Fragaszy, R. J., and Hardcastle, J. H. (1991). "Stress Ratio Effects on Collapse of Compacted Clayey Sand." *Journal of Geotechnical Engineering,* ASCE, vol. 117, no. 5, pp. 714–730.

Lawton, E. C., Fragaszy, R. J., and Hetherington, M. D. (1992). "Review of Wetting-Induced Collapse in Compacted Soil." *Journal of Geotechnical Engineering,* ASCE, vol. 118, no. 9, pp. 1376–1394.

Lea, F. M. (1971). *The Chemistry of Cement and Concrete,* 1st American ed. Chemical Publishing Company, New York.

Leonards, G. A. (1962). *Foundation Engineering.* McGraw-Hill, New York, 1136 pp.

Leonards, G. A., and Altschaeffl, A. G. (1964). "Compressibility of Clay." *Journal of the Soil Mechanics and Foundations Division,* ASCE, vol. 90, no. SM5, pp. 133–156.

Leonards, G. A., and Ramiah, B. K. (1959). "Time Effects in the Consolidation of Clay." Special Technical Publication no. 254, ASTM, Philadelphia, pp. 116–130.

Lin, G., Bennett, R. M., Drumm, E. C., and Triplett, T. L. (1995). "Response of Residential Test Foundations to Large Ground Movements." *Journal of Performance of Constructed Facilities,* ASCE, vol. 9, no. 4, pp. 319–329.

Lowe, J., and Zaccheo, P. F. (1975). "Subsurface Explorations and Sampling." Chapter 1 of *Foundation Engineering Handbook,* edited by Hans F. Winterkorn and Hsai-Yang Fang, Van Nostrand Reinhold, New York, pp. 1–66.

Lytton, R. L., and Dyke, L. D. (1980). "Creep Damage to Structures on Expansive Clay Slopes." *Fourth International Conference on Expansive Soils,* Denver, ASCE, vol. 1, pp. 284–301.

Mabsout, M. E., Reese, L. C., and Tassoulas, J. L. (1995). "Study of Pile Driving by Finite-Element Method." *Journal of Geotechnical Engineering,* ASCE, vol. 121, no. 7, pp. 535–543.

Mahtab, M. A., and Grasso, P. (1992). *Geomechanics Principles in the Design of Tunnels and Caverns in Rock.* Developments in Geotechnical Engineering, 72, Elsevier Science, New York, 250 pp.

Maksimovic, M. (1989a). "On the Residual Shearing Strength of Clays." *Geotechnique,* London, vol. 39, no. 2, pp. 347–351.

Maksimovic, M. (1989b). "Nonlinear Failure Envelope for Soils." *Journal of Geotechnical Engineering,* ASCE, vol. 115, no. 4, pp. 581–586.

Marino, G. G., Mahar, J. W., and Murphy, E. W. (1988). "Advanced Reconstruction for Subsidence-Damaged Homes." *Mine Induced Subsidence: Effects on Engineered Structures.* H. J. Siriwardane (ed.). ASCE, New York, pp. 87–106.

Marsh, E. T., and Walsh, R. K. (1996). "Common Causes of Retaining-Wall Distress: Case Study." *Journal of Performance of Constructed Facilities,* ASCE, vol. 10, no. 1, 35–38.

Massarsch, K. R. (1979). "Lateral Earth Pressure in Normally Consolidated Clay." *Proceedings of the Seventh European Conference on Soil Mechanics and Foundation Engineering,* Brighton, England, vol. 2, pp. 245–250.

Massarsch, K. R., Holtz, R. D., Holm, B. G., and Fredricksson, A. (1975). "Measurement of Horizontal In Situ Stresses." *Proceedings of the ASCE Specialty Conference on In Situ Measurement of Soil Properties,* Raleigh, N.C. vol. 1, pp. 266–286.

Mather, B. (1968). "Field and Laboratory Studies of the Sulphate Resistance of Concrete." *Performance of Concrete, Resistance of Concrete to Sulphate and Other Environmental Conditions,* University of Toronto Press, Toronto, pp. 66–76.

McCarthy, D. F. (1977). *Essentials of Soil Mechanics and Foundations.* Reston Publishing, Reston Va., 505 pp.

McElroy, C. H. (1987). "The Use of Chemical Additives to Control the Erosive Behavior of Dispersed Clays." *Engineering Aspects of Soil Erosion, Dispersive Clays and Loess,* Geotechnical Special Publication No. 10, C. W. Lovell and R. L. Wiltshire (eds.). ASCE, New York, pp. 1–16.

Meehan, R. L., Chun, B., Sang-wuk, J., King, S., Ronold, K., and Yang, F. (1993). "Contemporary Model of Civil Engineering Failures." *Journal of Professional Issues in Engineering Education and Practice,* ASCE, vol. 119, no. 2, pp. 138–146.

Meehan, R. L., and Karp, L. B. (1994). "California Housing Damage Related to Expansive Soils." *Journal of Performance of Constructed Facilities,* ASCE, vol. 8, no. 2, pp. 139–157.

Mehta, P. K. (1976). Discussion of "Combating Sulfate Attack in Corps of Engineers Concrete Construction" by Thomas J. Reading, *ACI Journal Proceedings,* vol. 73, no. 4, pp. 237–238.

Merfield, P. M. (1992). "Surficial Slope Failures: The Role of Vegetation and Other Lessons from Rainstorms." *Engineering Geology Practice in Southern California,* B. W. Pipkin and R. J. Proctor (eds.). Star Publishing, Association of Engineering Geologists, Southern California section, Special Publication no. 4, pp. 613–627.

Meyerhof, G. G. (1951). "The Ultimate Bearing Capacity of Foundations." *Geotechnique,* vol. 2, no. 4, pp. 301–332.

Meyerhof, G. G. (1953). "Bearing Capacity of Foundations Under Eccentric and Inclined Loads." *Proceedings of the Third International Conference on Soil Mechanics and Foundation Engineering,* vol. 1, Zurich, pp. 440–445.

Meyerhof, G. G. (1961). Discussion of "Foundations Other Than Piled Foundations." *Proceedings of Fifth International Conference on Soil Mechanics and Foundation Engineering,* vol. 3, Paris, p. 193.

Meyerhof, G. G. (1965). "Shallow Foundations." *Proceedings,* ASCE, vol. 91, no. SM2, pp. 21–31.

Middlebrooks, T. A. (1953). "Earth Dam Practice in the United States." *Transactions, American Society of Civil Engineers,* centennial volume, pp. 697.

Mitchell, J. K. (1970). "In-Place Treatment of Foundation Soils," *Journal of the Soil Mechanics and Foundation Division,* ASCE, vol. 96, no. SM1, pp. 73–110.

Mitchell, J. K. (1976). *Fundamentals of Soil Behavior.* John Wiley & Sons, New York, 422 pp.

Mitchell, J. K. (1978). "In Situ Techniques for Site Characterization." *Site Characterization and Exploration. Proceedings of Specialty Workshop,* Northwestern University, edited by C. H. Dowding, ASCE, New York, pp. 107–130.

Monahan, E. J. (1986). *Construction of and on Compacted Fills.* John Wiley & Sons, New York, 200 pp.

Morgenstern, N. R., and Price, V. E. (1965). "The Analysis of the Stability of General Slip Surfaces." *Geotechnique,* vol. 15, London, pp. 79–93.

Mottana, A., Crespi, R., and Liborio, G. (1978). *Rocks and Minerals.* Simon & Schuster, New York, 607 pp.

Munsell Soil Color Charts (1975). Prepared by Munsell Color, a Division of Kollmorgen Corporation, Baltimore, Md.

Myslivec, A., and Kysela, Z. (1978). *The Bearing Capacity of Building Foundations,* Developments in Geotechnical Engineering 21, Elsevier Scientific, New York, 237 pp.

Nadjer, J., and Werno, M. (1973). "Protection of Buildings on Expansive Clays." *Proceeding of the Third International Conference on Expansive Soils,* Haifa, Israel, vol. 1, pp. 325–334.

Narver (1993). *A/E Risk Review.* Professional Liability Agents Network, vol. 3, no. 7.

National Coal Board (1975). *Subsidence Engineers Handbook.* National Coal Board Mining Department, National Coal Board, Hobart House, Grosvenor Square, London, 111 pp.

National Research Council (1985). *Reducing Losses from Landsliding in the United States.* Committee on Ground Failure Hazards, Commission on Engineering and Technical Systems, National Academy Press, Washington, D.C., 41 pp.

National Science Foundation (NSF). (1992). "Quantitative Nondestructive Evaluation for Constructed Facilities." *Announcement Fiscal Year 1992.* Directorate for Engineering, Division of Mechanical and Structural Systems, Washington, D.C.

NAVFAC DM-7 (1971). *Soil Mechanics, Foundations, and Earth Structures,* Design Manual DM-7. Department of the Navy, Naval Facilities Engineering Command, Alexandria, Va., 280 pp.

NAVFAC DM-7.1 (1982). *Soil Mechanics,* Design Manual 7.1, Department of the Navy, Naval Facilities Engineering Command, Alexandria, Va., 364 pp.

NAVFAC DM-7.2 (1982). *Foundations and Earth Structures,* Design Manual 7.2, Department of the Navy, Naval Facilities Engineering Command, Alexandria, Va., 253 pp.

NAVFAC DM-7.3 (1983). *Soil Dynamics, Deep Stabilization, and Special Geotechnical Construction,* Design Manual 7.3, Department of the Navy, Naval Facilities Engineering Command, Alexandria, Va., 106 pp.

NAVFAC DM-21.3 (1978). *Flexible Pavement Design for Airfields,* Design Manual 21.3, Department of the Navy, Naval Facilities Engineering Command, Alexandria, Va., 98 pp.

Neary, D. G., and Swift, L. W. (1987). "Rainfall Thresholds for Triggering a Debris Avalanching Event in the Southern Appalachian Mountains." *Debris Flows/Avalanches: Process, Recognition, and Mitigation,* The Geological Society of America, Boulder, Colo., pp. 81–92.

Nelson, J. D., and Miller, D. J. (1992). *Expansive Soils, Problems and Practice in Foundation and Pavement Engineering.* John Wiley & Sons, New York, 259 pp.

Newmark, N. M. (1935). "Simplified Computation of Vertical Pressures in Elastic Foundations." University of Illinois Engineering Experiment Station Circular 24, Urbana, Ill., 19 pp.

Newmark, N. M. (1942). "Influence Charts for Computation of Stresses in Elastic Foundations." University of Illinois Engineering Experiment Station Bulletin, Series no. 338, vol. 61, no. 92, Urbana, Ill., reprinted 1964, 268 pp.

Nichols, H. L., and Day, D. A. (1999). *Moving the Earth: The Workbook of Excavation,* 4th ed. McGraw-Hill, New York, 1400 pp.

Norris, R. M., and Webb, R. W. (1990). *Geology of California,* 2d ed. John Wiley & Sons, New York, 541 pp.

Oldham, R. D. (1899). "Report on the Great Earthquake of 12th June, 1897." India Geologic Survey Memorial, Publication 29, 379 pp.

Oliver, A. C. (1988). *Dampness in Buildings.* Internal and Surface Waterproofers, Nichols Publishing, New York, 221 pp.

Orange County Grading Manual (1993). *Orange County Grading Manual,* part of the *Orange County Grading and Excavation Code,* prepared by Orange County, Calif.

Ortigao, J. A. R., Loures, T. R. R., Nogueiro, C., and Alves, L. S. (1997). "Slope Failures in Tertiary Expansive OC Clays." *Journal of Geotechnical and Geoenvironmental Engineering,* ASCE, vol. 123, no. 9, pp. 812–817.

Osterberg, J. O. (1957). "Influence Values for Vertical Stresses in a Semi-infinite Mass due to an Embankment Loading." *Proceedings of the Fourth International Conference on Soil Mechanics and Foundation Engineering,* vol. 1, London, pp. 393–394.

Owens, D. T. (1993). "Red Line Cost Overruns." *Los Angeles Times,* Editorial Section, Nov. 20, 1993, Los Angeles.

Patton, J. H. (1992). "The Nuts and Bolts of Litigation." *Engineering Geology Practice in Southern California,* Pipkin and Proctor (eds.). Star Publishing Company, Belmont, Calif. pp. 339–359.

Peck, R. B., Hanson, W. E., and Thornburn, T. H. (1974). *Foundation Engineering.* John Wiley & Sons, New York, 514 pp.

Peckover, F. L. (1975). "Treatment of Rock Falls on Railway Lines." *American Railway Engineering Association,* Bulletin 653, Chicago, Ill. pp. 471–503.

Peng, S. S. (1986). *Coal Mine Ground Control.* 2d ed. John Wiley & Sons, New York, 491 pp.

Peng, S. S. (1992). *Surface Subsidence Engineering.* Society for Mining, Metallurgy and Exploration, Inc., Littleton, Colo.

Perloff, W. H., and Baron, W. (1976). *Soil Mechanics, Principles and Applications.* John Wiley & Sons, New York, 745 pp.

Perry, D., and Merschel, S. (1987). "The Greening of Urban Civilization." *Smithsonian,* vol. 17, no. 10, pp. 72–79.

Perry, E. B. (1987). "Dispersive Clay Erosion at Grenada Dam, Mississippi." *Engineering Aspects of Soil Erosion, Dispersive Clays and Loess,* Geotechnical Special Publication no. 10, C. W. Lovell and R. L. Wiltshire (eds.). ASCE, New York, pp. 30–45.

Petersen, E. V. (1963). "Cave-in!" *Roads and Engineering Construction,* November, pp. 25–33.

Piteau, D. R., and Peckover, F. L. (1978). "Engineering of Rock Slopes." *Landslides, Analysis and Control,* Special Report 176, Transportation Research Board, National Academy of Sciences, Chap. 9, pp. 192–228.

Poh, T. Y., Wong, I. H., and Chandrasekaran, B. (1997). "Performance of Two Propped Diaphragm Walls in Stiff Residual Soils." *Journal of Performance of Constructed Facilities,* ASCE, vol. 11, no. 4, pp. 190–199.

Post-Tensioning Institute (1996). "Design and Construction of Post-tensioned Slabs-on-Ground," 2d ed. Report, Phoenix, Ariz., 101 pp.

Poulos, H. G., and Davis, E. H. (1974). *Elastic Solutions for Soil and Rock Mechanics.* John Wiley & Sons, New York, 411 pp.

Pradel, D., and Raad, G. (1993). "Effect of Permeability of Surficial Stability of Homogeneous Slopes." *Journal of Geotechnical Engineering,* ASCE, vol. 119, no. 2, pp. 315–332.

Prentis, E. A., and White, L. (1950). *Underpinning,* 2d ed. Columbia University Press, New York.

Price, N. J. (1966). *Fault and Joint Development in Brittle and Semi-Brittle Rock.* Pergamon Press, Oxford, England.

Proctor, R. R. (1933). "Fundamental Principles of Soil Compaction." *Engineering News-Record,* vol. 111, nos. 9, 10, 12, and 13.

Purkey, B. W., Duebendorfer, E. M., Smith, E. I., Price, J. G., and Castor, S. B. (1994). *Geologic Tours in the Las Vegas Area,* Nevada Bureau of Mines and Geology, Special Publication 16, Las Vegas, Nev., 156 pp.

Raschke, S. A. and Hryciw, R. D. (1997). "Vision Cone Penetrometer for Direct Subsurface Soil Observation." *Journal of Geotechnical and Geoenvironmental Engineering,* ASCE, vol. 123, no. 11, 1074–1076.

Rathje, W. L., and Psihoyos, L. (1991). "Once and Future Landfills." *National Geographic,* vol. 179, no. 5, pp. 116–134.

Ravina, I. (1984). "The Influence of Vegetation on Moisture and Volume Changes." *The Influence of Vegetation on Clays,* Thomas Telford, London, pp. 62–68.

Reading, T. J. (1975). "Combating Sulfate Attack in Corps of Engineering Concrete Construction." *Durability of Concrete, SP47,* American Concrete Institute, Detroit, pp. 343–366.

Reed, M. A., Lovell, C. W., Altschaeffl, A. G., and Wood, L. E. (1979). "Frost Heaving Rate Predicted from Pore Size Distribution." *Canadian Geotechnical Journal*, vol. 16, no. 3, pp. 463–472.

Reese, L. C., Owens, M., and Hoy, H. (1981). "Effects of Construction Methods on Drilled Shafts." *Drilled Piers and Caissons*, M. W. O'Neill (ed.). ASCE, New York, pp. 1–18.

Reese, L. C., and Tucker, K. L. (1985). "Bentonite Slurry in Concrete Piers." *Drilled Piers and Caisson II*, C. N. Baker (ed.). ASCE, New York, pp. 1–15.

Rice, R. J. (1988). *Fundamentals of Geomorphology*, 2d ed. John Wiley & Sons, New York.

Ritchie, A. M. (1963). "Evaluation of Rockfall and Its Control." *Highway Research Record 17*, Highway Research Board, Washington, D.C., pp. 13–28.

Robertson, P. K., and Campanella, R. G. (1983). "Interpretation of Cone Penetration Tests: Parts 1 and 2." *Canadian Geotechnical Journal*, vol. 20, pp. 718–745.

Rogers, J. D. (1992). "Recent Developments in Landslide Mitigation Techniques." Chapter 10 of *Landslides/Landslide Mitigation*, J. E. Slosson, G. G. Keene, and J. A. Johnson (eds.). The Geological Society of America, Boulder, Colo., pp. 95–118.

Rollins, K. M., Rollins, R. C., Smith, T. D., and Beckwith, G. H. (1994). "Identification and Characterization of Collapsible Gravels." *Journal of Geotechnical Engineering*, ASCE, vol. 120, no. 3, pp. 528–542.

Ross, C. S., and Smith, R. L. (1961). *Ash-Flow Tuffs, Their Origin, Geologic Relations and Identification*, U.S. Geological Survey Professional Paper 366: U. S. Geological Survey, Denver.

Rutledge, P. C. (1944). "Relation of Undisturbed Sampling to Laboratory Testing." *Transactions*, ASCE, vol. 109, pp. 1162–1163.

Sandström, G. E. (1963a). *History of Tunneling*. Barrie and Rockcliff, London, 427 pp.

Sandström, G. E. (1963b). *Tunnels*. Holt, Rinehart, and Winston, New York, 427 pp.

Sanglerat, G. (1972). *The Penetrometer and Soil Exploration*. Elsevier Scientific, New York, 464 pp.

Savage, J. C., and Hastie, L. M. (1966). "Surface Deformation Associated with Dip-Slip Faulting." *Journal of Geophysical Research*, vol. 71, no. 20, pp. 4897–4904.

Saxena, S. K., Lourie, D. E., and Rao, J. S. (1984). "Compaction Criteria for Eastern Coal Waste Embankments." *Journal of Geotechnical Engineering*, ASCE, vol. 110, no. 2, pp. 262–284.

Schlager, N. (1994). *When Technology Fails*. "St. Francis Dam Failure." Gale Research Inc., Detroit, pp. 426–430.

Schmertmann, J. H. (1970). "Static Cone to Compute Static Settlement Over Sand." *Journal of the Soil Mechanics and Foundations Division*, ASCE, vol. 96, no. SM3, pp. 1011–1043.

Schmertmann, J. H. (1975). "Measurement of In Situ Shear Strength." State-of-the-Art Report, *Proceedings of the ASCE Specialty Conference on In Situ Measurement of Soil Properties*, Raleigh, N.C. vol. 2, pp. 57–138.

Schmertmann, J. H. (1977). *Guidelines for Cone Penetration Test, Performance and Design*. U.S. Department of Transportation, Federal Highway Administration, Washington, D.C., 145 pp.

Schmertmann, J. H., Hartman, J. P., and Brown, P. R. (1978). "Improved Strain Influence Factor Diagrams." *Journal of the Geotechnical Engineering Division*, vol. 104, no. GT8, pp. 1131–1135.

Schnitzer, M., and Khan, S. U. (1972). *Humic Substances in the Environment*. Marcel Dekker, New York.

Schuster, R. L. (1986). *Landslide Dams: Processes, Risk, and Mitigation*, Geotechnical Special Publication no. 3. Proceedings of Geotechnical Session, Seattle. ASCE, New York, 164 pp.

Schuster, R. L., and Costa, J. E. (1986). "A Perspective on Landslide Dams." *Landslide Dams: Processes, Risk, and Mitigation*, Geotechnical Special Publication no. 3. Proceedings of Geotechnical Session, Seattle. ASCE, New York, pp. 1–20.

Schutz, R. J. (1984). "Properties and Specifications for Epoxies Used in Concrete Repair." *Concrete Construction Magazine*, Concrete Construction Publications, Addison, Ill., pp. 873–878.

Seed, H. B. (1970). "Soil Problems and Soil Behavior." Chapter 10 of *Earthquake Engineering*, R. L. Wiegel, coordinating editor. Prentice-Hall, Englewood Cliffs, N.J., pp. 227–252.

Seed, H. B., and DeAlba (1986). "Use of SPT and CPT Tests for Evaluating the Liquefaction Resistance of Sands." *Proceedings, In Situ 1986, ASCE Specialty Conference on Use of In Situ Testing in Geotechnical Engineering*, Special Publication no. 6, ASCE, New York.

Seed, H. B., and Idriss, I. M. (1971). "Simplified Procedure for Evaluating Soil Liquefaction Potential." *Journal of Geotechnical Engineering Division,* ASCE, vol. 97, no. 9, pp. 1249–1273.

Seed, H. B., Idriss, I. M., and Arango, I. (1983). "Evaluations of Liquefaction Potential Using Field Performance Data." *Journal of Geotechnical Engineering,* ASCE, vol. 109, no. 3, pp. 458–482.

Seed, H. B., Tokimatsu, K., Harder, L. F., and Chung, R. (1985). "Influence of SPT Procedures in Soil Liquefaction Resistance Evaluations." *Journal of Geotechnical Engineering,* ASCE, vol. 111, no. 12, pp. 861–878.

Seed, H. B., and Whitman, R. V. (1970). *Design of Earth Structures for Dynamic Loads.* Lateral Stresses in the Ground and Design of Earth Retaining Structures, ASCE, Cornell University.

Seed, H. B., Woodward, R. J., and Lundgren, R. (1962). "Prediction of Swelling Potential for Compacted Clays." *Journal of Soil Mechanics and Foundations Division,* ASCE, vol. 88, no. SM3, pp. 53–87.

Seismic Safety Study (1995). City of San Diego, Development Services Department, San Diego, Calif.

Shannon and Wilson, Inc. (1964). *Report on Anchorage Area Soil Studies, Alaska, to U.S. Army Engineer District, Anchorage, Alaska.* Seattle, Wash.

Sherard, J. L. (1972). "Study of Piping Failures and Eroding Damage from Rain in Clay Dams in Oklahoma and Mississippi." U.S. Department of Agriculture, Soil Conservation Service, Washington, D.C.

Sherard, J. L., Decker, R. S., and Ryker, N. L. (1972). "Piping in Earth Dams of Dispersive Clay." *Proceedings of the Specialty Conference on Performance of Earth and Earth-Supported Structures.* vol. 1, Part 1, cosponsored by ASCE and Purdue University, Lafayette, Ind., pp. 589–626.

Sherard, J. L., Woodward, R. J., Gizienski, S. F., and Clevenger, W. A. (1963). *Earth and Earth-Rock Dams.* John Wiley and Sons, New York, 725 pp.

Sinha, R. S. (1989). *Underground Structures, Design and Instrumentation.* Developments in Geotechnical Engineering, 59A, Elsevier Science Publishers, New York, 480 pp.

Sinha, R. S. (1991). *Underground Structures, Design and Construction.* Developments in Geotechnical Engineering, 59B, Elsevier Science Publishers, New York, 529 pp.

Skempton, A. W. (1953). "The Colloidal Activity of Clays." *Proceeding of the Third International Conference on Soil Mechanics and Foundation Engineering,* Zurich, Switzerland, vol. 1, pp. 57–61.

Skempton, A. W. (1954). "The Pore-Pressure Coefficients A and B." *Geotechnique,* vol. 4, pp. 143–147.

Skempton, A. W. (1961). "Effective Stress in Soils, Concrete and Rock." *Pore Pressure and Suction in Soils.* Butterworths, London, p. 4.

Skempton, A. W. (1964). "Long-term Stability of Clay Slopes." *Geotechnique,* London, vol. 14, no. 2, pp. 75–101.

Skempton, A. W. (1985). "Residual Strength of Clays in Landslides, Folded Strata and the Laboratory." *Geotechnique,* London, vol. 35, no. 1, pp. 3–18.

Skempton, A. W. (1986). "Standard Penetration Test Procedures and the Effects in Sands of Overburden Pressure, Relative Density, Particle Size, Aging and Overconsolidation." *Geotechnique,* vol. 36, no. 3, pp. 425–447.

Skempton, A. W., and Bjerrum, L. (1957). "A Contribution to the Settlement Analysis of Foundations on Clay." *Geotechnique,* vol. 7, p. 168.

Skempton, A. W., and Henkel, D. J. (1953). "The Post-Glacial Clays of the Thames Estuary at Tilbury and Shellhaven." *Proceeding of the Third International Conference on Soil Mechanics and Foundation Engineering,* Switzerland, vol. 1, p. 302.

Skempton, A. W., and Hutchinson, J. (1969). "State of-the-Art Report: Stability of Natural Slopes and Embankment Foundations." *Seventh International Conference on Soil Mechanics and Foundation Engineering,* Mexico, pp. 291–340.

Skempton, A. W., and MacDonald, D. H. (1956). "The Allowable Settlement of Buildings." *Proceedings of the Institution of Civil Engineers,* Part III. The Institution of Civil Engineers, London, no. 5, pp. 727–768.

Slope Indicator (1998). *Geotechnical and Structural Instrumentation,* prepared by Slope Indicator Company, Bothell, Wash., 92 pp.

Smith, D. D., and Wischmeier, W. H. (1957). "Factors Affecting Sheet and Rill Erosion." *Transactions of the American Geophysical Union,* vol. 38, no. 6, pp. 889–896.

Smith, R. L. (1960). *Zones and Zonal Variations in Welded Ash Flows,* U.S. Geological Survey Professional Paper 354-F: U.S. Geological Survey, Denver.

Snethen, D. R. (1979). *Technical Guidelines for Expansive Soils in Highway Subgrades.* U.S. Army Engineering Waterway Experiment Station, Vicksburg, Miss. Report No. FHWA-RD–79-51.

Soils and Geology, Procedures for Foundation Design of Buildings and Other Structures (Except Hydraulic Structures) (1979). Departments of the Army and Air Force, TM 5-818-1/AFM 88-3, chap. 7, Washington, D.C.

Sowers, G. B., and Sowers, G. F. (1970). *Introductory Soil Mechanics and Foundations,* 3d ed. Macmillan, New York, 556 pp.

Sowers, G. F. (1962). "Shallow Foundations," Chap. 6 from *Foundation Engineering,* G. A. Leonards, ed. McGraw-Hill, New York

Sowers. G. F. (1974). "Dam Safety Legislation: A Solution or a Problem." *Safety of Small Dams.* Proceedings of the Engineering Foundation Conference, Henniker, N.H., ASCE, New York, pp. 65–100.

Sowers, G. F. (1979). *Soil Mechanics and Foundations: Geotechnical Engineering,* 4th ed. Macmillan, New York.

Sowers, G. F. (1997). *Building on Sinkholes: Design and Construction of Foundations in Karst Terrain,* ASCE Press, New York.

Sowers, G. F., and Royster, D. L. (1978). "Field Investigation." Chapter 4 of *Landslides, Analysis and Control,* Special Report 176, Transportation Research Board, National Academy of Sciences, R. L. Schuster and R. J. Krizek (eds.). Washington, D.C., pp. 81–111.

Spencer, E. (1967). "A Method of Analysis of the Stability of Embankments Assuming Parallel Interslice Forces." *Geotechnique,* vol. 17, London, pp. 11–26.

Spencer, E. (1968). "Effect of Tension on Stability of Embankments." *Journal of the Soil Mechanics and Foundations Division,* ASCE, vol. 94, no. SM5, pp. 1159–1173.

Spencer, E. W. (1972). *The Dynamics of the Earth: An Introduction to Physical Geology.* Thomas Y. Crowell Company, New York, 649 pp.

Standard Specifications for Public Works Construction (1997), 11th ed. Published by BNI Building News, Anaheim, Calif., commonly known as the "Green Book," 761 pp.

Standards Presented to California Occupational and Safety and Health Standard Board, Sections 1504 and 1539-1547 (1991). California Occupational Safety and Health Standard Board, San Francisco, July.

Stapledon, D. H., and Casinader, R. J. (1977). "Dispersive Soils at Sugarloaf Dam Site Near Melbourne, Australia." *Dispersive Clays, Related Piping, and Erosion in Geotechnical Projects.* STP 623, American Society for Testing and Materials, Philadelphia, pp. 432–466.

Stark, T. D., and Eid, H. T. (1994). "Drained Residual Strength of Cohesive Soils." *Journal of Geotechnical Engineering,* ASCE, vol. 120, no. 5, pp. 856–871.

Stark, T. D., and Olson, S. M. (1995). "Liquefaction Resistance Using CPT and Field Case Histories." *Journal of Geotechnical Engineering,* ASCE, vol. 121, no. 12, pp. 856–869.

State of California Special Studies Zones Maps (1982). Prepared by the State of California on the basis of the Alquist-Priolo Special Studies Zones Act.

Steinbrugge, K. V. (1970). "Earthquake Damage and Structural Performance in the United States." Chapter 9 of *Earthquake Engineering,* R. L. Wiegel, coordinating editor. Prentice-Hall, Englewood Cliffs, N.J., pp. 167–226.

Stokes, W. C., and Varnes, D. J. (1955). *Glossary of Selected Geologic Terms.* Colorado Scientific Society Proceedings, vol. 116, Denver, 165 pp.

Sweet, J. (1970). *Legal Aspects of Architecture, Engineering and Construction Process.* West Publishing Co., St. Paul, Minn., 953 pp.

Szechy, K. (1973). *The Art of Tunneling.* Akkdemiaikido, Budapest, 1097 pp.

Tadepalli, R., and Fredlund, D. G. (1991). "The Collapse Behavior of Compacted Soil During Inundation." *Canadian Geotechnical Journal,* vol. 28, no. 4, pp. 477–488.

Taylor, D. W. (1948). *Fundamentals of Soil Mechanics.* John Wiley & Sons, New York, 700 pp.

Terzaghi, K. (1925). *Erdbaumechanik.* Franz Deuticke, Vienna.

Terzaghi, K. (1938). "Settlement of Structures in Europe and Methods of Observation," *Transactions,* ASCE, vol. 103, p. 1432.

Terzaghi, K. (1943). *Theoretical Soil Mechanics.* John Wiley & Sons, New York, 510 pp.

Terzaghi, K., and Peck, R. B. (1967). *Soil Mechanics in Engineering Practice,* 2d ed. John Wiley and Sons, New York, 729 pp.

Thompson, L. J., and Tanenbaum, R. J. (1977). "Survey of Construction Related Trench Cave-Ins." *Journal of the Construction Division,* ASCE, vol. 103, no. CO3, September.

Thorburn, S., and Hutchison, J. F., eds. (1985). *Underpinning.* Surrey University Press, London, 296 pp.

Tokimatsu, K., and Seed, H. B. (1984). *Simplified Procedures for the Evaluation of Settlements in Sands due to Earthquake Shaking.* Report No. UCB/EERC-84/16. Report sponsored by the National Science Foundation, Earthquake Engineering Research Center, College of Engineering, University of California, Berkeley, Calif., 41 pp.

Tomlinson, M. J. (1986). *Foundation Design and Construction,* 5th ed., Longman Scientific & Technical, Essex, England, 842 pp.

Transportation Research Board (1977). *Rapid-Setting Materials for Patching Concrete.* National Cooperative Highway Research Program Synthesis of Highway Practice 45, National Academy of Sciences, Washington, D.C.

"Trench and Excavation Safety Guide." (1984). *Publication S-358,* California Department of Industrial Relations/California Occupational Safety and Health Administration, San Francisco.

Tucker, R. L., and Poor, A. R. (1978). "Field Study of Moisture Effects on Slab Movements." *Journal of Geotechnical Engineering Division,* ASCE, vol. 104, no. 4, pp. 403–414.

Turnbull, W. J., and Foster, C. R. (1956). "Stabilization of Materials by Compaction." *Journal of Soil Mechanics and Foundation Division,* ASCE, vol. 82, no. 2, pp. 934.1–934.23.

Tuthill, L. H. (1966). "Resistance to Chemical Attack-Hardened Concrete." *Significance of Tests and Properties of Concrete and Concrete-Making Materials,* STP-169A, ASTM, Philadelphia, pp. 275–289.

Uniform Building Code (1997). International Conference of Building Officials, 3 volumes, Whittier, Calif.

U.S. Army Engineer Waterways Experiment Station (1960). *The Unified Soil Classification System.* Technical Memorandum no. 3-357. U.S. Army Engineer Waterways Experiment Station, Vicksburg, Miss.

U.S. Department of Agriculture (1975). *Agriculture Handbook No. 436.* Published by the U.S. Department of Agriculture, Washington, D.C.

U.S. Department of Agriculture (1977). *Guides for Erosion and Sediment Control in California.* U.S. Department of Agriculture, SCS, Davis, Calif.

U.S. Department of the Interior (1987). *Engineering Geology Field Manual,* Published by the U.S. Department of the Interior, Bureau of Reclamation, Washington, D.C., 598 pp.

U.S. Geological Survey (1975). *Encinitas Quadrangle, San Diego, California.* Topographic map mapped, edited, and published by the U.S. Geological Survey, Denver.

U.S. Geological Survey (1997). *Index of Publications of the Geological Survey.* U.S. Geological Survey, Department of the Interior, Washington, D.C.

USS Steel Sheet Piling Design Manual (1984). U.S. Department of Transportation, Washington, D.C., 132 pp.

Van der Merwe, C. P., and Ahronovitz, M. (1973). "The Behavior of Flexible Pavements on Expansive Soils." *Third International Conference on Expansive Soil,* Haifa, Israel.

Varnes, D. J. (1978). "Slope Movement and Types and Processes." *Landslides: Analysis and Control,* Transportation Research Board, National Academy of Sciences, Washington, D.C., Special Report 176, chap. 2, pp. 11–33.

Vesic, A. S. (1963). "Bearing Capacity of Deep Foundations in Sand." *Highway Research Record,* 39, National Academy of Sciences, National Research Council, Washington, D.C., pp. 112–153.

Vesic, A. S. (1967). "Ultimate Loads and Settlements of Deep Foundations in Sand." *Proceedings of the Symposium on Bearing Capacity and Settlement of Foundations,* Duke University, Durham, N.C., p. 53.

Vesic, A. S. (1975). "Bearing Capacity of Shallow Foundations." Chapter 3 of *Foundation Engineering Handbook,* Hans F. Winterkorn and Hsai-Yang Fang (eds.). Van Nostrand Reinhold, New York, pp. 121–147.

Virginia Soil and Water Conservation Commission (1980). *Virginia Erosion and Sediment Control Handbook,* 2d ed. Richmond, Va.

Wahls, H. E. (1994). "Tolerable Deformations." *Vertical and Horizontal Deformations of Foundations and Embankments,* Geotechnical Special Publication no. 40, ASCE, New York, pp. 1611–1628.

Wahlstrom, E. E. (1973). *Tunneling in Rock.* Developments in Geotechnical Engineering 3, Elsevier Scientific, New York, 250 pp.

Waldron, L. J. (1977). "The Shear Resistance of Root-Permeated Homogeneous and Stratified Soil." *Soil Science Society of America,* vol. 41, no. 5, pp. 843–849.

Wallace, T. (1981). "Preparation of Reports, Field Notes, and Documentation." *Proceedings Geotechnical Construction Loss Prevention Seminar,* Santa Clara, Calif.

Warner, J. (1982). "Compaction Grouting—The First Thirty Years." *Proceedings of the Conference on Grouting in Geotechnical Engineering,* W. H. Baker (ed.). ASCE, New York, pp. 694–707.

Warriner, J. E. (1957). *English Grammar and Composition.* Harcourt, Brace and Company, New York, 692 pp.

Watry, S. M., and Ehlig, P. L. (1995). "Effect of Test Method and Procedure on Measurements of Residual Shear Strength of Bentonite from the Portuguese Bend Landslide." *Clay and Shale Slope Instability,* edited by W. C. Haneberg and S. A. Anderson, Geological Society of America, *Reviews in Engineering Geology,* vol. 10, Boulder, Colo., pp. 13–38.

Wellington, A. M. (1888). "Formulae for Safe Loads of Bearing Piles." *Engineering News,* no. 20, pp. 509–512.

Westergaard, H. M. (1938). "A Problem of Elasticity Suggested by a Problem in Soil Mechanics: A Soft Material Reinforced by Numerous Strong Horizontal Sheets." In *Contributions to the Mechanics of Solids, Stephen Timoshenko Sixtieth Anniversary Volume,* Macmillan, New York, pp. 268–277.

WFCA (1984). "Moisture Guidelines for the Floor Covering Industry." WFCA Management Guidelines, Western Floor Covering Association, Los Angeles.

Whitaker, T. (1957). "Experiments With Model Piles in Groups." *Geotechnique,* London, England.

Whitlock, A. R., and Moosa, S. S. (1996). "Foundation Design Considerations for Construction on Marshlands." *Journal of Performance of Constructed Facilities,* ASCE, vol. 10, no. 1, pp. 15–22.

Whitman, R. V., and Bailey, W. A. (1967). "Use of Computers for Slope Stability Analysis." *Journal of the Soil Mechanics and Foundations Division,* ASCE, vol. 93, no. SM4, pp. 475–498.

Williams, A. A. B. (1965). "The Deformation of Roads Resulting From Moisture Changes in Expansive Soils in South Africa." *Moisture Equilibria and Moisture Changes in Soils Beneath Covered Areas,* G. D. Aitchison (ed.). Symposium Proceedings, Butterworths, Australia, pp. 143–155.

Winterkorn, H. F., and Fang, H. (1975). *Foundation Engineering Handbook.* Van Nostrand Reinhold, New York, 751 pp.

Wischmeier, W. H., and Smith, D. D. (1965). *Predicting Rainfall Erosion Losses from Cropland East of the Rocky Mountains.* Agriculture Handbook no. 282, U.S. Department of Agriculture, Washington D.C.

Wischmeier, W. H., and Smith, D. D. (1978). *Predicting Rainfall Erosion Losses—A Guide to Conservation Planning.* Agriculture Handbook no. 537, U.S. Department of Agriculture, Science and Education Administration, Washington D.C.

Wolman, M. G., and Schick, A. P. (1967). "Effects of Construction on Fluvial Sediment, Urban and Suburban Areas of Maryland." *Water Resources Research,* vol. 3, pp. 451–464.

Woodward, R. J., Gardner, W. S., and Greer, D. M. (1972). "Design Considerations," *Drilled Pier Foundations,* D. M. Greer (ed.). McGraw-Hill, New York, pp. 50–52.

Wu, T. H., Randolph, B. W., and Huang, C. (1993). "Stability of Shale Embankments." *Journal of Geotechnical Engineering,* ASCE, vol. 119, no. 1, pp. 127–146.

Yegian, M. K., Ghahraman, V. G., and Gazetas, G. (1994). "1988 Armenia Earthquake. II: Damage Statistics Versus Geologic and Soil Profiles." *Journal of Geotechnical Engineering,* ASCE, vol. 120, no. 1, pp. 21–45.

Young, R. N., and Warkentin, B. P. (1975). *Soil Properties and Behavior,* Elsevier Scientific, New York, 449 pp.

Zaruba, Q., and Mencl, V. (1969). *Landslides and Their Control.* Elsevier, New York, 205 pp.

Zornberg, J. G., Sitar, N., and Mitchell, J. K. (1998). "Limit Equilibrium as Basis for Design of Geosynthetic Reinforced Slopes." *Journal of Geotechnical and Geoenvironmental Engineering,* ASCE, vol. 124, no. 8, pp. 684–698.

INDEX

A-line, **5.32–5.34**, **14.**7
A-value, **6.38–6.41**
Absorbed water layer, **5.**30
Absorption, **A.**8
Acceleration of gravity, **6.**7, **11.**9
Achievable Density Criteria (ADC), **14.**7, **14.**22
Acid mine water, **13.**6
Acidic environment, **1.**2, **9.**33
Active earth pressure (*see* Retaining walls)
Activity of clay, **5.33–5.34**, **9.2–9.**3, **A.**8
Adhesion, **8.20–8.**21, **9.21–9.**22, **A.**12
Adobe, **13.11–13.**17, **A.**3
Aeolian, **1.**1, **A.**3
Aerial photographs, **4.**5
Aerobic conditions, **9.**33
Aggregate, **8.23–8.**27, **13.**7, **A.**17
Aging, **7.**20, **7.**33, **11.**8
Agreement (*see* Contract and proposal)
Agriculture, **12.**3
Agronomy, **1.**1
Air track, **4.**16
Airport, **8.24–8.**25
Allowable bearing pressure (*see* Bearing capacity)
Alluvial fans, **10.**50, **14.2–14.**4
Alluvium (*see* Soil deposits)
American Concrete Institute (ACI), **13.**5, **13.**8
American Society for Testing and Materials (ASTM), **4.**17, **5.**2, **13.**7
Anaerobic microorganisms, **7.**42
Anchoring systems, **10.**7
Andesite, **5.**44, **10.**5
Angular distortion, **7.4–7.**8, **10.**2, **10.**4
Animal wastes, **13.**6
Anisotropic soil, **6.**44, **6.**56, **A.**12
Anthracite coal, **5.**44
Appalachians, **10.**50
Apparent opening size, **A.**17
Appendixes:
 Appendix A (glossary), **A.1–A.**22
 Appendix B (technical guidelines), **B.1–B.**11
 Appendix C (grading specifications), **C.1–C.**19
 Appendix D (percolation test), **D.1–D.**6

Appendixes (*Cont.*):
 Appendix E (example of a geotechnical report), **E.1–E.**13
 Appendix F (solution to problems), **F.1–F.**52
 Appendix G (references), **G.1–G.**22
Approval, **A.**17
Approved plans, **A.**17
Approved testing agency, **A.**17
Aquiclude, **A.**3
Aquifer, **A.**3
Arching, **15.**50, **15.**57, **A.**12
Architects, **1.**9, **16.**9
Area ratio, **4.**23
Artesian condition, **6.**49, **16.**13, **A.**3
As-graded (definition), **A.**17
Ash, **14.1–14.**2, **A.**3
Asphalt (*see* Pavements)
Asphalt concrete (*see* Pavements)
Assignment, **2.3–2.**14
Atterberg limits, **5.3–5.**4, **5.**6, **5.30–5.**31, **A.**8
Auger borings, **4.**14, **4.17–4.**21
 bucket auger, **4.**17
 flight auger, **4.**17
 hollow-stem flight auger, **4.**14, **4.**17
Avongard, **18.**6, **18.**9

B-value, **6.38–6.**39
Bacillariophyceae, **14.**3
Backdrain (definition), **A.**17
Backfill, **9.12–9.**13, **A.**17
Backhoe, **4.39–4.**40
Badlands, **12.17–12.**18, **A.**3
Barium chloride, **13.**1
Barium sulfate, **13.**1
Basalt, **5.**44
Base (*see* Pavements)
Basement walls (*see* Walls)
Bearing capacity, **1.**6, **8.1–8.**31
 adjustment for groundwater table, **8.9–8.**10
 allowable bearing pressure, **8.**4, **8.6–8.**12, **A.**12
 bearing capacity factors, **8.8–8.**9, **8.16–8.**17
 for cohesionless soil, **8.9–8.**10, **8.16–8.**19

Bearing capacity (*Cont.*):
  for cohesive soil, **8.**10–**8.**12, **8.**20–**8.**23
  for deep foundations, **8.**14–**8.**22
  depth of bearing capacity failure, **8.**5
  earthquake loading, **8.**13, **11.**14
  eccentric loads, **8.**12–**8.**13
  end bearing, **8.**14–**8.**18
  failure, **8.**2–**8.**6, **A.**12, **A.**15
  for footings at the top of slopes, **8.**13
  frictional resistance, **8.**18–**8.**19
  general shear failure, **8.**2–**8.**3
  inclined base of footing, **8.**3
  lateral loads, **8.**12
  load settlement curve, **8.**2–**8.**5
  local shear failure, **8.**2, **8.**5
  moments, **8.**12
  other considerations, **8.**12–**8.**13
  for pile groups, **8.**19, **8.**22–**8.**23
  plane strain condition, **8.**9
  punching shear failure, **8.**2, **8.**4
  for retaining wall footings, **15.**7
  for roads (*see* Pavements)
  for shallow foundations, **8.**7–**8.**14
  soil rupture, **8.**2–**8.**3
  for square footings, **8.**9
  for strip footings, **8.**7
  Terzaghi bearing capacity equation, **8.**7
  Transcona grain-elevator, **8.**5–**8.**6
  ultimate bearing capacity, **8.**7–**8.**11, **8.**16–**8.**23, **A.**11
Bedding, **10.**27, **10.**36, **A.**3
Bedrock, **A.**3
Bell, **15.**49, **A.**12
Bench excavation, **17.**4, **A.**17
Bentonite, **8.**15, **9.**33, **14.**3, **A.**3
Berm, **A.**17
Bernoulli's energy equation, **6.**49
Biologic activity, **9.**33
Biotite, **5.**6
Bishop method of slices, **10.**31
Bit, **4.**15, **A.**3
Black Mountains, **11.**4–**11.**5
Block sample, **4.**39
Blowout condition, **6.**56
Bog, **14.**1, **A.**3
Boiling condition, **6.**56
Boring, **4.**14–**4.**39
  auger, **4.**14, **4.**17–**4.**21
  boring, definition of, **A.**3
  camera, **4.**16
  depth, **4.**38–**4.**39
  down-hole logging, **4.**17, **4.**19–**4.**21

Boring (*Cont.*):
  dynamic soundings, **4.**16
  hollow stem auger, **4.**14, **4.**17
  large diameter, **4.**17–**4.**21
  layout, **4.**35–**4.**37
  log (*see* Logs)
  percussion drilling, **4.**16–**4.**17
  rotary coring, **4.**14–**4.**17
  wash type, **4.**17
Borrow, **17.**15–**17.**18, **A.**17
Boston, Massachusetts, **18.**4
Boulders, **5.**25, **A.**4
Braced excavation, **15.**43–**15.**45
Breccia, **5.**44
Breed's Hill, **13.**10
Bridge abutments, **15.**19
Bridgeport, Connecticut, **19.**10
Bridges, **2.**11, **8.**14, **11.**3
Britain, **9.**14
Bromhead ring shear apparatus, **5.**14
Brooming, **A.**17
Budget, **3.**5–**3.**6
Building code, **4.**4–**4.**5, **8.**4, **8.**6
Building department, **2.**14, **18.**2, **18.**4
Building official, **A.**18
Buildings:
  allowable movement, **7.**3–**7.**8
  conventional brick, **7.**5
  definition, **A.**21
  reinforced concrete-frame, **2.**4, **2.**8, **7.**7
  steel frame, **2.**4, **2.**7–**2.**8, **7.**7
  tilt-up, **2.**2, **2.**4, **2.**9–**2.**11, **10.**4
  wood frame, **2.**4–**2.**5
Bulking, **A.**18
Bull's liver, **14.**1
Bunker Hill Monument, **13.**10
Bureau of Workers' Compensation, **15.**46
Buttress, **10.**48, **A.**18

Caissons, **1.**4, **7.**47, **A.**18
CAL/OSHA, **15.**49
Calcite, **5.**6
Calcium aluminate, **13.**2
Calcium carbonate, **7.**14, **14.**2
Calcium hydroxide, **13.**2
Calcium sulfate, **13.**2
Calcium sulfoaluminate, **13.**2
Caliche, **14.**2

California, **1.4–1.7**, **4.43**, **5.20**, **5.39**, **7.38**, **7.44**, **8.24**, **9.2**, **9.16**, **10.10–10.11**, **10.14**, **10.18**, **10.37**, **10.50**, **10.59**, **11.1–11.7**, **11.13–11.17**, **12.11**, **12.16–12.17**, **13.12**, **14.5**, **15.49–15.50**, **16.8**, **16.14–16.15**, **17.8**, **17.13**, **17.15**, **18.11**, **19.3**, **19.12**
California bearing ratio (*see* Pavements)
Canada, **13.8**
Canal liners, **9.2**
Canyon, **7.8**, **7.10**
Canyon Estates, **14.5**
Capillarity, **6.11–6.12**, **13.8**, **16.9**, **16.15**, **A.8**
Capillary:
   action, **6.11**, **16.9**, **16.15**
   rise, **6.11**
   tension, **6.11**
Carmel Valley fault, **11.3–11.4**
Cartoon, **19.6**
Case studies:
   creep, **10.54–10.58**
   debris flow, **10.50–10.53**
   fill settlement, **18.10–18.14**
   historic structures, **13.10–13.17**
   landslides, **4.43**, **4.46–4.51**, **4.54–4.61**, **10.37–10.43**, **15.50**, **15.53–15.54**
   repair of landslide, **15.50**, **15.53–15.54**
   repair of slope failure, **15.50–15.52**
   rockfall, **10.5–10.12**
   slope softening, **10.54–10.58**
   surficial slope failure, **10.14**, **10.16–10.22**
   unusual soil, **14.5–14.22**
Casing, **A.4**
Cation exchange capacity, **5.30**
Cement, **8.23**, **9.25**, **13.5**
Cement slurry, **16.3**
Cement treatment (expansive soil), **9.25**
Center lift (*see* Expansive soil)
Certificate of occupancy, **18.4**
Chemical reactions (sulfate attack of concrete), **13.2**, **13.5**
Chert, **5.44**
Chicago, Illinois, **13.8**
Chloride, **5.20**
Chlorite, **5.32**, **5.34**
Civil engineer, **1.1**, **A.1**
Civil engineering, **1.1**, **A.1**
Classification (*see* Soil classification; Rock classification)
Clay:
   activity, **5.33–5.34**, **9.2–9.3**, **A.8**
   attractive forces, **5.30**
   backfill, **9.12–9.13**, **15.18**

Clay (*Cont.*):
   bearing capacity failure, **8.5–8.6**
   bonds, **5.30**
   classification of, **5.34–5.42**
   clods, **9.12**
   consistency, **5.10**, **5.35**, **5.38**, **A.9**
   content, **9.2**
   desiccated, **9.7–9.12**, **10.11**
   dispersive, **10.60**, **12.1**, **12.8**
   expansion of (*see* Expansive soil)
   fraction, **5.25**
   hydration of clay minerals, **9.33**
   Leda clay, **14.3**
   London clay, **10.13**, **10.16**
   Mexico City, **7.38**
   mineralogy, **5.6**, **5.33–5.34**
   quick (*see* Quick clay)
   repulsive forces, **10.60**
   sensitive, **5.9**, **11.12**, **11.14**
   shrinkage (*see* Expansive soil)
   size of the particles, **5.25**, **A.8**
   stiff-fissured, **5.14**, **10.35**
Clayey sand, **5.27**, **5.36**, **5.38**, **5.42**, **6.10**, **13.16**
Clayey silt, **14.6**
Claystone, **5.38**, **5.44**, **9.33**, **10.16**, **10.18**, **10.21**, **14.6**
Clearing, brushing, and grubbing, **17.4**, **A.18**
Clients, **1.9–1.10**
   agreement (*see* Contract and proposal)
   contractors, **1.9**
   design professionals, **1.9**
   governmental agencies, **1.10**
   high-budget speculator, **19.11**
   low-budget speculator, **19.11**
   mass builders, **1.9**, **19.11**
   owner, **A.20**
   professional developers, **1.9**, **19.11**
   property owners, **1.9**
   remodelers, **19.11**
   responsibilities, **10.27**
   types, **1.9–1.10**
Clogging, **A.18**
Close sheeting, **15.47–15.48**
Coal, **5.44**
Coal mines, **7.37–7.38**
Coarse-grained soil, **5.34**, **5.36**, **A.8**
Cobbles, **5.25**, **A.4**
Codes (*see* Uniform Building Code)
Coefficient of consolidation (*see* Consolidation)
Coefficient of curvature, **5.28**
Coefficient of earth pressure at rest, **5.43**, **6.12**, **15.19**

Coefficient of permeability, **5.**15–**5.**18, **6.**47–**6.**55, **9.**10–**9.**12, **16.**1, **16.**23, **A.**9
Coefficient of uniformity, **5.**28
Cofferdams, **15.**42–**15.**43
Cohesion (*see* Shear strength)
Cohesionless soil, **6.**23–**6.**27, **11.**8, **A.**4
Cohesive soil, **6.**27–**6.**46, **A.**4
Collapse:
  of buildings, **11.**1–**11.**2
  of mines, **7.**37–**7.**38
  of tunnels, **7.**37
Collapse potential, **7.**12–**7.**13
Collapsible formations, **A.**12
Collapsible soil, **7.**8–**7.**16
  alluvium, **7.**11
  colluvium, **7.**11
  debris fill, **7.**11
  deep fill, **7.**11
  deep foundation system, **7.**14
  definition, **7.**10, **A.**12
  design and construction on, **7.**13–**7.**14
  due to soluble soil particles, **7.**14–**7.**16
  dumped fill, **7.**11
  foundation options, **7.**14
  hydraulically placed fill, **7.**11
  laboratory testing, **7.**11–**7.**12
  pipe break, **7.**14
  removal and replacement, **7.**13
  settlement analyses, **7.**12–**7.**13
  stabilization, **7.**13–**7.**14
  uncontrolled fill, **7.**11
  variables that govern collapse potential, **7.**10–**7.**11
  wetting front, **7.**13, **16.**2
Colloidal soil particles, **A.**8
Colluvium (*see* Soil deposits)
Colorado, **9.**2
Compacted soil:
  base, **8.**23–**8.**25
  decomposed granite, **6.**8, **17.**7
  London clay, **10.**13, **10.**16
  lowest void ratio, **6.**8
Compaction:
  backfill compaction, **17.**18–**17.**22
  checking field compaction, **17.**18
  compaction energy, **17.**8
  for diatomaceous soil, **14.**7
  definition, **17.**7, **A.**18
  for expansive soil, **9.**23–**9.**24
  factors that affect compaction, **17.**7–**17.**8
  field compaction, **17.**9–**17.**13
  field density tests, **17.**13–**17.**14

Compaction (*Cont.*):
  fundamentals, **17.**7–**17.**14
  impact or sharp blow, **17.**8
  kneading action, **17.**8
  for mechanically stabilized earth retaining walls, **15.**27, **15.**30–**15.**31
  one-point Proctor test, **17.**17, **17.**19
  pumping, **17.**22–**17.**26
  relative compaction, **17.**13–**17.**14, **A.**20
  for retaining wall backfill, **15.**18
  specifications, **17.**4–**17.**7, **C.**1–**C.**19
  static weight or pressure, **17.**8
  utility trench compaction, **17.**18–**17.**22
  vibration or shaking, **17.**8
  (*See also* Grading; Fill)
Compaction curve, **17.**8, **17.**10, **A.**8
Compaction equipment, **4.**45, **15.**18, **15.**30–**15.**31, **17.**9–**17.**13, **17.**20, **A.**18
  bulldozer, **17.**9
  scraper, **17.**9
  water trucks, **17.**9
Compaction grouting, **18.**18
Compaction production, **A.**18
Compaction test (in the laboratory):
  compaction energy, **17.**8, **17.**10
  definition of, **A.**8–**A.**9
  maximum dry density, **17.**8, **A.**9
  modified Proctor, **9.**27, **17.**8
  optimum moisture content, **9.**23, **14.**7, **17.**8, **A.**9
  standard Proctor, **17.**8
  zero air voids curve, **17.**8
Compensation (*see* Payment)
Compressibility:
  definition, **A.**12
  of subgrade, **4.**44
Compression index (*see* Consolidation)
Computer programs, **5.**28, **6.**56, **7.**36, **10.**43–**10.**45, **11.**14
Concrete:
  admixtures, **13.**8, **14.**2
  alkali reaction, **14.**2
  consolidation, **13.**6
  construction, **13.**6–**13.**7
  corrosion of reinforcement, **13.**6
  cracking, **13.**6
  curing, **13.**6
  definition, **A.**18
  design, **13.**6–**13.**7
  deterioration, **13.**1–**13.**8
  discoloration, **13.**1, **13.**4
  durability, **13.**8

Concrete (*Cont.*):
  moisture migration, **16.9–16.18**
  pavements, **8.23**
  quality, **13.6**
  sulfate attack, **13.1–13.7**
  surface preparation, **13.6**
Condominiums, **2.3**, **2.5**, **19.**12
Cone penetration test, **4.17**, **4.28–4.33**, **A.4**
  electric cone, **4.28**, **4.31**, **A.4**
  mechanical cone, **4.28–4.29**, **A.4**
  mechanical-friction cone, **4.28**, **A.4**
  piezocone, **4.28**, **A.4**
  used to predict drained modulus, **7.36**
  used to predict liquefaction potential, **11.8**
Confining pressure (for liquefaction analysis), **11.8**
Conflict of interest, **2.3**
Conglomerate, **5.44**
Consistency of clay, **5.10**, **5.35**, **5.38**, **A.9**
Consolidation, **7.9–7.33**, **16.2**
  average degree of consolidation settlement, **7.29–7.32**
  Casagrande construction technique, **7.21**, **7.23**
  coefficient of compressibility, **7.27**
  coefficient of consolidation, **7.22**, **7.27–7.28**, **7.30–7.32**, **A.8**
  compression index, **7.24–7.26**, **A.9**
  consolidation curve, **7.21–7.23**
  consolidation ratio, **7.28**
  definition of primary consolidation, **7.19**, **A.13**
  double drainage, **7.29**
  effects of sample disturbance, **7.21–7.22**, **7.31**
  excess pore water pressure, **7.19**, **7.27–7.28**, **16.2**
  geologic definition of consolidation, **7.19**
  laboratory testing, **7.21–7.26**, **7.30–7.32**
  limitations of consolidation theory, **7.32**
  load increment ratio, **7.21**
  maximum past pressure, **6.3**, **7.21–7.26**
  mechanisms causing an overconsolidated soil, **7.20**
  modified compression index, **7.26**
  modified recompression index, **7.26**
  normally consolidated, **6.30**, **7.20**, **A.14**
  overconsolidated, **7.20**, **A.14**
  overconsolidation ratio, **6.12**, **6.30**, **7.19–7.20**, **A.9**
  preconsolidation pressure, **6.12**, **6.30**, **7.21–7.26**, **A.14**
  primary consolidation, **7.19–7.32**
  rate of one-dimensional consolidation, **7.27–7.32**

Consolidation (*Cont.*):
  recompression curve, **7.23**
  recompression index, **7.24**
  settlement analyses, **7.23–7.26**
  single drainage, **7.29**
  stress history, **7.23**
  Terzaghi one-dimensional consolidation equation, **7.27–7.32**
  test (definition), **A.9**
  time factor, **7.28–7.32**, **A.16**
  underconsolidated, **7.19**, **A.16**
  virgin consolidation curve, **7.23–7.24**
Consolidometer (*see* Oedometer)
Construction, **17.3**
Construction services, **18.1–18.18**
  field observations, **18.1–18.4**
  footing observations, **18.2–18.4**
  load tests, **18.1**
  monitoring, **18.4–18.9**
  observational method, **18.7**, **18.9–18.10**
  percolation tests, **18.1**
  performance tests, **18.1**
  underpinning, **18.10–18.18**
Contract:
  disclaimer of warranties, **3.5**
  jurisdiction, **3.5**
  limitation of liability, **3.5**
  modification of, **3.5**
  ownership of documents, **3.4–3.5**
  payment, **3.5**
  retainers, **3.5**
  safety, **3.4**
  signature page, **3.4**
  termination of, **3.5**
  time limit, **3.5**
  (*See also* Proposal)
Contraction, **6.23**, **6.39**, **6.47**, **11.8**, **A.9**
Contractors, **1.9**, **16.3**, **A.18**
Core drilling, **4.14–4.15**, **A.4**
Core recovery, **4.14–4.15**, **5.45**, **A.4**
Corrosive environmental, **13.6**
Couts family, **13.13**
Cracks (asphalt):
  alligator, **9.26**, **13.7**
  block cracking, **13.7**
Cracks (concrete):
  due to sulfate attack, **13.6**
  flaking, **13.5**
  flatwork, **9.29**, **9.31**
  repair, **18.15–18.18**
  severity, **7.9**
  shrinkage, **16.11**

Cracks (concrete) (*Cont.*):
  spalling, **13.5**
  spider crack pattern, **9.29, 9.31**
  width, **7.9**
  x-type crack pattern, **9.29, 9.31**
Cracks (in outlet pipes), **10.**60
Cracks (soil):
  dam, **10.**60
  desiccation, **9.7–9.**12, **16.**23
  due to earthquakes, **11.**3
  due to landslide movement, **10.**38
  ground, **9.7–9.**12, **10.**20, **10.**22
  in earth embankments, **10.**60
  micro, **9.**12
  tension, **10.**34
Creep, **8.**13, **10.**2, **10.53–10.**59
  at faults, **11.**5
  definition, **A.**13
  depth of creep, **10.**59
  due to loss of peak shear strength, **10.57–10.**58
  due to seasonal moisture changes, **10.58–10.**59
  example of creep, **10.54–10.**58
  method of analysis, **10.58–10.**59
  primary or transient, **10.54–10.**55
  secondary or steady-state, **10.54–10.**55
  slope creep, **10.53–10.**59, **15.**13, **15.**17
  tertiary, **10.54–10.**55
  undrained creep, **14.**15, **14.**22
Creosote, **5.**20
Critical height or critical slope, **A.**13
Crown, **A.**13
Crystallization (of salt in concrete pores), **13.**5
Curbs (*see* Pavements)
Curvature, coefficient of, **5.**28
Cut, **7.**8, **7.**10
Cut-fill transition, **7.**8, **7.**10, **A.**18

Dacite, **10.**5
Daily field reports, **19.5–19.**6
Damage:
  caused by frost action, **13.**8
  caused by the Northridge earthquake, **11.16–11.**17
  classification, **7.**9
  due to lateral movement, **10.**4
  due to settlement, **7.**9
  to concrete, **13.1–13.**5
Dams:
  causes of failure, **10.59–10.**63
  definition, **A.**18
  dispersive clays, **10.**60
  flow net, **6.56–6.**57, **10.**62

Dams (*Cont.*):
  homogeneous, **A.**18
  hydrological design, **10.**63
  landslide dams, **10.**63
  large, **10.59–10.**62
  maintenance, **10.**63
  overtopping, **10.**59, **10.**63
  pinhole test, **10.**60
  piping, **10.59–10.**60
  sand boils, **10.**60, **10.**62
  slope instability, **10.**60, **10.**62
  small, **10.62–10.**63
  stability analysis, **10.**60, **16.**2
  St. Francis dam failure, **10.**59
  surficial failures, **10.**20
  zoned, **A.**18
Darcy's law, **5.**16, **6.**47, **6.**49, **7.**27, **16.**1
Dead load, **7.2–7.**3, **8.**4, **9.**22, **A.**13
Death Valley, California, **5.**4, **9.**8, **12.**17
Death Valley Junction, **16.**8
Debris, **5.**38, **A.**18
Debris flow, **10.**2, **10.**48, **10.50–10.**53, **14.2–14.**4
  definition, **10.**2, **10.**50, **A.**13
  depositional area, **10.**50, **10.**53
  erosion of, **12.**18, **12.**21
  example, **10.**50, **10.52–10.**53
  historical method, **10.**50
  main tract, **10.**50, **10.**52
  measures to protect structures, **10.**50, **10.**53
  mobilization from a surficial failure, **10.**19, **10.**21, **16.**2
  prediction of a debris flow, **10.**50
  source area, **10.**50, **10.**52
Decomposed granite, **6.**8, **17.**7
Decomposition, **7.39–7.**42
Deep foundation (*see* Foundation)
Deflocculation, **10.**60
Degree of saturation, **6.**9
Densification of soil, **7.**46
Density, **5.**4, **5.**38, **6.6–6.**7
  definition, **6.6–6.**7, **A.**9
  low dry density, **14.2–14.**3
  of subgrade soil, **4.44–4.**45
Deposition of soil, **4.**43, **4.**52, **A.**4
Desiccated clay, **9.7–9.**12, **10.**11, **A.**13
Design (*see* Engineering analyses)
Deterioration, **13.1–13.**17
  causes:
    corrosion, **13.**6
    frost, **5.**20, **13.8–13.**9
    growing tree roots, **13.9–13.**12
    sulfate attack, **5.**20, **13.1–13.**7

Deterioration (causes) (*Cont.*):
    wood rot, **13.**16–**13.**17
    laboratory testing, **5.**20
    of adobe, **13.**11–**13.**17
    of basement walls, **16.**15–**16.**18
    of concrete, **13.**1–**13.**8
    of historic structures, **13.**10–**13.**17
    of pavements, **13.**7, **16.**3
    of rock, **13.**8
    of timber piles, **5.**20
Detritus, **12.**17, **A.**4
Developers, **1.**9, **14.**7
Deviator stress, **6.**30, **6.**36, **A.**9
Dewatering, **16.**1, **A.**18
Diatomaceous earth, **14.**3–**14.**22, **A.**4
Diatomaceous rock, **4.**43, **4.**50
Diatomite, **14.**5
Diatoms, **7.**38, **14.**3, **14.**5–**14.**7
Differential settlement, **7.**1, **7.**2, **11.**13
Dilation, **6.**39, **6.**47, **11.**8, **A.**9
Diorite, **5.**44
Direct shear apparatus, **5.**10–**5.**13
Dispersing agent, **5.**27
Dissolution, **7.**14–**7.**16
Document review, **4.**4–**4.**13
    aerial photographs, **4.**5
    building codes, **4.**5
    correspondence, **4.**4
    field change orders, **4.**4
    geologic data, **4.**5–**4.**9
    history of the site, data on, **4.**4–**4.**5
    information bulletins, **4.**4
    plans, **4.**4
    preliminary design information, **4.**4
    reference materials, **4.**4, **4.**9
    reports, **4.**9
    seismic data, **4.**4
    special study data, **2.**12–**2.**14, **4.**4
    specifications, **4.**4, **4.**9
    standard drawings, **4.**4, **4.**9, **4.**12–**4.**13
    topographic maps, **4.**5, **4.**10–**4.**11
Dolomite, **5.**6
Double layer, **5.**30, **A.**9
Dowels, **18.**14–**18.**16
Down-hole logging, **4.**17, **4.**19–**4.**21
Downdrag, **8.**3b, **A.**13
Dozer, **4.**40, **A.**18
Drag effect, **7.**8, **7.**10
Drainage:
    canyon subdrain, **16.**3–**16.**4, **17.**4
    definition, **A.**18
    ditches, **4.**13

Drainage(*Cont.*):
    drainage buttress, **16.**7
    galleries, **16.**7
    horizontal drains, **16.**7
    of clay backfill, **9.**12
    of pavements, **16.**3
    relief wells, **16.**7
    repairs, **13.**16
    retaining wall backfill, **15.**3, **15.**18, **16.**16–**16.**18
    of slopes, **16.**4, **16.**6–**16.**8
    sump, **16.**13–**16.**14
    surface, **16.**19
    swale, **10.**35
    system, **10.**48, **13.**16, **15.**1, **15.**3, **15.**4, **16.**3, **16.**7
    trenches, **16.**7
Drainage properties of soil, **5.**17–**5.**19, **6.**47, **11.**8
Drained modulus, **7.**35–**7.**36
Drains, **6.**56, **7.**46
Drawdown, **A.**18
Drill rig, **4.**14–**4.**35
Drilling accidents, **4.**21
Drought, **9.**14, **9.**18
Dry density (*see* Density)
Dynamic sounding, **4.**16
Dynamite, **14.**5

Earth flow, **10.**2, **10.**50
Earth material, **A.**18
Earth pressure, **A.**13
    at rest, coefficient of, **5.**43, **6.**12, **15.**19
    (*See also* Retaining walls)
Earth pressure theory (for pier walls), **15.**56
Earthquakes:
    Alaskan (1964), **11.**4, **11.**7, **11.**12
    Alquist-Priolo special studies zone map, **11.**5
    Assam (1897), **11.**3
    California Northridge (1994), **11.**1–**11.**3, **11.**7, **11.**12–**11.**13, **11.**16–**11.**17, **15.**56
    Chile (1960), **11.**7
    duration, **11.**7–**11.**8
    effect on earth dams, **11.**1, **11.**3
    effect on landslides, **10.**35–**10.**36, **11.**12
    effect on retaining walls, **15.**13, **15.**17
    effect on sensitive soil, **11.**12, **11.**14
    energy waves, **11.**8
    epicenter, **11.**8
    fault creep, **11.**5
    fault displacement, **11.**3–**11.**4
    fault zones, **11.**3–**11.**4
    flow slides, **11.**6–**11.**7

Earthquakes (*Cont.*):
  foundation behavior, **11.15–11.17**
  Gobi-Altai (1957), **11.4**
  ground movement, **11.3–11.17**
  ground rupture, **11.3–11.5**
  intensity, **11.7–11.8**
  liquefaction, **11.5, 11.15**
  Niigata (1964), **11.7**
  peak acceleration, **11.9**
  psuedo-static analysis, **10.35, 11.13–11.14**
  rotation of objects, **11.12–11.13**
  San Francisco (1906), **10.59**
  seismic shear stress ratio, **11.8–11.15**
  seismic study maps, **11.5**
  settlement, **11.1, 11.12–11.15**
  slope movement, **11.1, 11.12**
  soil strength loss, **11.2, 11.14**
  surface faulting, **11.3–11.5**
  translation, **11.12–11.13**
Easements, **17.4**
Economic losses:
  collapse of mines and tunnels, **7.37**
  due to saturation and seepage, **16.2**
  earthquakes, **11.16**
  expansive soil, **9.1**
  frost action, **13.8**
  landslides, **10.36**
  pavements, **13.8, 16.3**
Edge lift (*see* Expansive soil)
Effective stress:
  calculation of, **6.10–6.12**
  definition, **6.10, A.13**
  of the Orinoco Clay, **6.29**
Effective stress analysis, **5.8, 6.23, 6.45–6.46**
  for bearing capacity, **8.12**
  for dam design, **10.62**
  for gross stability, **10.28–10.30**
  for landslides, **10.43–10.48**
  for pier walls to stabilize slopes, **15.55**
  for piles in cohesive soil, **8.21–8.22**
  for retaining walls, **15.6**
  for surficial stability, **10.12–10.17**
Efflorescence, **16.11, 16.15**
Elastic method, **7.17–7.18, 7.35–7.36**
Electroosmosis, **A.19**
Elimination method, **17.17**
El Niño, **4.43, 5.39, 10.37, 12.2**
Engineer (*see* Geotechnical engineer)
Engineering analyses:
  basic principles, **6.2–6.57**
  bearing capacity (*see* Bearing capacity)
  design load, **A.13**

Engineering analyses (*Cont.*):
  deterioration (*see* Deterioration)
  earthquakes (*see* Earthquakes)
  effective stress analysis (*see* Effective stress analysis)
  erosion (*see* Erosion)
  expansive soil (*see* Expansive soil)
  groundwater (*see* Groundwater)
  moisture migration (*see* Moisture problems)
  retaining walls (*see* Retaining walls)
  settlement (*see* Settlement)
  short-term analysis (*see* Total stress analysis)
  slope stability (*see* Slope movement)
  total stress analysis (*see* Total stress analysis)
  unusual soil (*see* Unusual soil)
Engineering geologist, **1.6–1.9, A.2**
  analysis for debris flow, **10.50**
  analysis for landslides, **4.43, 4.46–4.47, 4.53–4.61, 10.45**
  analysis for potential rockfall, **10.6**
  areas of responsibility, **1.7–1.9**
  education, training, and practice, **1.7**
  fields of expertise, **1.8**
  missed geologic features, **4.43**
  preparation of logs, **4.43, 4.46, 4.48–4.51**
  role of, **4.43**
Engineering geology, **1.6, A.2**
Engineering-in-training, **19.8**
Engineering jargon, **19.5, 19.7–19.8**
  certification, **19.7–19.8**
  control, **19.7–19.8**
  ensure, **19.7–19.8**
  guarantee, **19.7–19.8**
  inspect, **19.7–19.8**
  supervise, **19.7–19.8**
  warrant, **19.7–19.8**
Engineering News formula, **8.15**
Eocene, **10.16, 16.8**
Epoxy, **18.16–18.18**
Equipotential line, **6.53–6.56, A.13**
Equivalent fluid pressure, **15.4, 15.56, A.13**
Equivalent wheels loads (*see* Pavements)
Erosion:
  of adobe, **13.16–13.17**
  of badlands, **12.17–12.21**
  causing landslides, **10.36**
  control devices (temporary), **12.13–12.15, A.19**
  control fabric, **10.25–10.26**
  control system, **12.13–12.15, 17.5, A.19**
  of dams, **10.59–10.60**
  definition, **A.4**
  design to reduce soil erosion, **12.12–12.15**

Erosion (*Cont.*):
  of dispersive clays, **10.**60, **12.**1–**12.**2, **12.**8
  due to overtopping of earth dams, **10.**59
  effect of vegetation, **12.**12
  factors causing erosion, **12.**1
  gullies, **12.**1–**12.**3, **12.**11
  index test, **13.**16
  jugs, **12.**2
  levels of erosion, **12.**2
  principles of erosion control, **12.**12–**12.**13
  rills, **12.**1–**12.**2, **12.**16
  of sand, **10.**60
  of sea cliffs, **12.**16–**12.**18
  sediment basin, **12.**13, **12.**15
  sheet erosion, **12.**1
  of slopes, **12.**1–**12.**17
  slopewash, **12.**1
  stream flow, **12.**1
  undermining of structures, **12.**5
  universal soil loss equation, **12.**5–**12.**13
  unusual landforms, **12.**18, **12.**21
Ettringite, **13.**2
Eucalyptus tree, **13.**9
Evaporites, **16.**8
  anhydrite, **16.**8
  gypsum, **16.**8
  sodium chloride, **16.**8
Examples:
  bearing capacity of a strip footing on cohesionless soil, **8.**10
  bearing capacity of a strip footing on cohesive soil, **8.**11
  combined slope length and steepness factor, **12.**10
  consolidation of a clay deposit, **7.**26
  end-bearing capacity of a pile, **8.**16–**8.**18
  frictional capacity of a pile, **8.**18–**8.**19
  liquefaction analysis, **11.**10–**11.**12, **11.**14
  pavement design, **8.**26–**8.**27
  pore water pressure using a flow net, **6.**55
  seepage into a foundation pit, **6.**54
  sheet pile wall, **15.**40–**15.**41
  surficial slope stability analysis, **10.**17
  time for end of primary consolidation, **7.**31–**7.**32
  universal soil loss equation, **12.**11–**12.**12
  wedge slope stability analysis, **10.**29–**10.**30
Excavation, **A.**19
Exit gradient, **6.**55–**6.**56, **A.**13
Expansion index, **9.**3–**9.**5, **9.**16–**9.**17, **9.**34
Expansion index test, **9.**4–**9.**5
Expansion potential, **5.**43, **9.**2–**9.**3

Expansive rock, **9.**32–**9.**34
  expansion due to physical factors, **9.**33
  expansion due to weathering of rock, **9.**33
  rebound, **9.**32–**9.**33
Expansive soil, **9.**1–**9.**35
  active zone, **9.**7, **9.**22, **A.**13
  bearing capacity of, **8.**10
  calculating foundation heave, **9.**19–**9.**21
  capillary action, **9.**13–**9.**14, **16.**2
  cement treatment, **9.**25
  center-lift, **9.**13–**9.**15, **9.**18–**9.**19, **16.**2
  chemical injection, **18.**18
  classification, **9.**2–**9.**3
  clay size particles, **9.**2–**9.**3
  coefficient of swell, **9.**6
  compaction control, **9.**23
  correlations with index tests, **9.**6
  cost of damage, **9.**1
  crack patterns, **9.**29, **9.**31
  cyclic heave and shrinkage, **9.**13, **9.**16, **9.**18–**9.**19
  depth of seasonal moisture changes, **9.**7, **9.**21, **A.**13
  depth of the active zone, **9.**7
  desiccated clay, **9.**7–**9.**12, **10.**11, **16.**2
  desiccation cracks, **9.**7–**9.**12
  downward displacement of foundation, **7.**3
  edge lift, **9.**18–**9.**19
  effect of drainage, **16.**19
  effect of roots, **9.**14–**9.**16
  effect of vegetation, **9.**7, **9.**14–**9.**16
  effect on concrete pavements, **9.**27
  effect on retaining walls, **9.**12–**9.**13, **15.**18
  factors causing expansion, **9.**2–**9.**3
  flatwork, **9.**23, **9.**26–**9.**32
  fly ash treatment, **9.**25
  foundation design, **9.**16–**9.**26
  foundation repair, **18.**10–**18.**18
  heave caused by water from pipe breaks, **16.**19
  horizontal barriers, **9.**26
  hydraulic conductivity, **9.**10–**9.**12
  identification, **9.**4
  laboratory testing, **5.**6, **9.**4–**9.**7
  lateral movement, **9.**12–**9.**13
  lime treatment, **9.**25
  mitigation options, **9.**23–**9.**26
  moisture variation, **9.**4, **9.**18–**9.**19
  pavements, **9.**23, **9.**26–**9.**32
  prewetting, **9.**16, **9.**24
  progressive swelling, **9.**13–**9.**14, **9.**16, **9.**18
  rate of swell, **9.**5–**9.**6

Expansive soil (*Cont.*):
  salt treatment, **9.**25
  shrinkage, **9.**7–**9.**9, **9.**14
  slaking, **9.**12
  suction pressure, **9.**12
  surcharge pressure, **9.**2–**9.**3
  swelling pressure, **9.**6, **9.**20
  swell test, **9.**6–**9.**7
  thermo osmosis, **9.**14, **16.**2
  total heave, **9.**18–**9.**20
  treatment alternatives, **9.**23–**9.**26
  vertical movement, **9.**13–**9.**23
  walking of flatwork, **9.**29, **9.**32
  wetting and drying, **14.**13
  (*See also* Swell)
Expansive soil analysis, **9.**19–**9.**21

Fabric (of soil), **A.**9
Factor of safety:
  for bearing capacity failure, **8.**3, **8.**8, **8.**12, **8.**17, **8.**19
  for gross stability, **10.**27–**10.**35
  for landslides, **10.**43–**10.**48
  for pier walls, **15.**55–**15.**56
  for retaining walls, **15.**7, **15.**10–**15.**56
  for surficial stability, **10.**12–**10.**23
  for toe kick-out of sheet pile walls, **15.**40
Faults, **7.**38, **10.**6, **10.**36, **11.**1, **11.**3–**11.**6, **A.**4
Fee schedule, **3.**2
Feldspar, **5.**6, **5.**33
Fellenius method, **10.**30
Fertilizers, **13.**6
Field exploration, **4.**1–**4.**73
  borings, **4.**14–**4.**39
  document review, **4.**4–**4.**13
  during construction, **18.**1–**18.**4
  field meeting, **4.**4
  geophysical techniques, **4.**47, **4.**52, **4.**62–**4.**63
  preparation of logs, **4.**43, **4.**46–**4.**52
  role of engineering geologist, **4.**43, **4.**46–**4.**53
  standard penetration test, **4.**25–**4.**28
  subsoil profile, **4.**53–**4.**54, **4.**66–**4.**73
  test pits, **4.**39–**4.**42
  trenches, **4.**39–**4.**42
  (*See also* Subsurface exploration)
Field services (*see* Construction services)
Files:
  calculations, **19.**6
  confidential material, **19.**6
  correspondence, **19.**6
  daily field reports, **19.**5–**19.**6
  management, **19.**6–**19.**7

Files (*Cont.*):
  maps, **19.**6
  notes, **19.**6
  photographs, **19.**6
Fill:
  borrow with oversize particles, **17.**17–**17.**18
  definition, **A.**19
  dumped fill, **7.**11, **17.**15
  engineered fill, **10.**43, **17.**14
  field density tests, **17.**13–**17.**14
  fill lifts, **17.**10
  hydraulic fill, **7.**11, **11.**8, **17.**14, **A.**19
  mixed borrow, **17.**17
  select import, **17.**15
  structural fill, **7.**41, **17.**2, **17.**14
  uncompacted fill, **17.**2, **17.**4
  uniform borrow, **17.**15–**7.**17
  (*See also* Compaction and grading)
Filters, **6.**56–**6.**57
Fine-grained soil, **5.**34–**5.**35, **5.**37–**5.**38, **A.**9
Fines, **5.**27, **5.**37, **6.**27, **A.**4
Finite element analyses, **6.**5, **A.**13
Fire, **10.**20, **10.**50
Fissures, **5.**43, **7.**38, **7.**40
Flatwork, **9.**23, **9.**29–**9.**32
Flocculation, **5.**27, **A.**9
Floor level survey, **18.**11
Flow (laminar), **5.**16
Flow line, **6.**53–**6.**56, **A.**14
Flow net, **6.**53–**6.**57, **10.**14, **10.**62, **16.**1, **A.**14
Flow slides, **11.**6–**11.**7
Flow (turbulent), **5.**16
Flows, **10.**48, **10.**50–**10.**53
Fly ash, **9.**25
Fold (of rock layers), **A.**4
Footing (*see* Foundation)
Foraminifera, **14.**5
Force equilibrium equations, **10.**30
Forms, **1.**12, **1.**16, **A.**19
Foundation, **1.**2–**1.**4
  ability to resist lateral movement, **10.**2, **10.**4
  allowable settlement of, **7.**3–**7.**8
  basement types, **1.**4
  bearing capacity failure (*see* Bearing capacity)
  behavior (*see* Foundation design)
  caissons, **1.**4
  California slab, **9.**18
  center lift due to expansive soil, **9.**13–**9.**15, **9.**18–**9.**19
  combined footing, **1.**3, **1.**12, **1.**15, **7.**43
  construction example, **1.**10–**1.**16
  cracks, **7.**9

INDEX
I.11

Foundation (*Cont.*):
  damage due to frost, **13.**8–**13.**9
  deep, **1.**4, **7.**46–**7.**49, **8.**14–**8.**15, **11.**15, **A.**19
  definition, **1.**1, **A.**19
  design (*see* Foundation design)
  engineering, **1.**2
  excavation of, **1.**10, **17.**5
  on expansive soil, **9.**13–**9.**23
  flexible, **7.**4
  floating, **1.**4, **7.**46
  floor slabs, **16.**9–**16.**15
  footings, **1.**3, **1.**10–**1.**16, **9.**16–**9.**17, **A.**19
  footing trenches, **18.**2–**18.**4
  mat, **1.**3–**1.**4, **7.**14, **7.**44, **10.**4, **11.**16, **18.**10–**18.**14
  minimum footing sizes, **2.**1, **8.**3
  pier and grade beam, **1.**4, **9.**21–**9.**23
  piers, **1.**4, **7.**47, **8.**14–**8.**23
  piles, **1.**4, **7.**46–**7.**47, **8.**14–**8.**23
  pit, **6.**54
  post-tensioned, **1.**3, **7.**44–**7.**46, **9.**16, **9.**18–**9.**19, **10.**4, **11.**16, **14.**11
  raft, **1.**4
  raised wood floor, **1.**3, **9.**23, **9.**27, **11.**16–**11.**17
  ribbed, **9.**18
  rigid, **7.**7
  selection of foundation type, **7.**42–**7.**49
  shallow, **1.**2–**1.**3, **7.**43–**7.**46, **A.**19
  slab-on-grade, **1.**3, **7.**5–**7.**6, **7.**44–**7.**46, **9.**16–**9.**17, **11.**17
  spread footings, **1.**3, **1.**10, **1.**14, **7.**43–**7.**44
  strip footing, **1.**3, **1.**10, **1.**12, **1.**15–**1.**16, **7.**43
  sulfate attack, **13.**1–**13.**7
  types of deep foundations, **1.**4, **8.**14–**8.**15
  types of foundations, **1.**2–**1.**4
Foundation design:
  adequate depth, **7.**42
  adequate strength, **7.**42
  adjacent top of slopes, **10.**35, **15.**13
  adverse soil changes, **7.**43, **9.**1–**9.**32
  bearing capacity failure, **7.**43, **8.**1–**8.**23
  calculating foundation heave, **9.**18–**9.**21
  footing dimensions, **7.**3
  for foundations on expansive soil, **9.**16–**9.**23
  pier and grade beam support, **9.**21–**9.**23
  quality of concrete, **7.**42, **13.**1–**13.**8
  seismic forces, **7.**43, **11.**14
  selection of foundation type, **7.**42–**7.**49, **11.**15–**11.**17
  settlement, **7.**1–**7.**56
  undermining due to erosion, **12.**5
Fracture (of rock), **5.**43, **A.**5

Fredonia, New York, **13.**8
Freeze, **8.**20, **A.**19
Fresh Kills landfill, **7.**42
Friars formation, **5.**14, **10.**16, **10.**18
Friction angle (*see* Shear strength)
Frost:
  action, **4.**44
  depth of frost action, **13.**8–**13.**9
  effect on buildings, **13.**8–**13.**9
  effect on concrete, **13.**8
  effect on pavements, **13.**7
  effect on retaining walls, **15.**3
  effect on rock, **9.**33, **13.**8
  effect on rock slopes, **10.**5, **13.**8
  expansive forces, **13.**8
  freezing of water in cracks, **13.**8
  ice lenses, **13.**8–**13.**9, **16.**2
  permafrost, **13.**9
  spring thaw, **16.**2
Frustules, **14.**3

Gabbro, **5.**44
Gas, **5.**21, **6.**6
Geofabric, **9.**29, **15.**22, **16.**3, **16.**7
Geogrid, **10.**23–**10.**25, **15.**22–**15.**24
Geologic:
  analyses, **6.**5, **10.**59
  faults (*see* Faults)
  features, **4.**43
  hazards, **2.**12–**2.**13, **4.**5
  maps, **4.**5–**4.**7
  symbols, **4.**9
Geologist (*see* Engineering geologist)
Geology, **10.**3
Geophysical techniques, **4.**47, **4.**52, **4.**62–**4.**63, **A.**5
  drop in potential, **4.**63
  electrical methods, **4.**63
  E-logs, **4.**63
  gravity measurements, **4.**63
  high resolution refraction, **4.**62
  magnetic methods, **4.**63
  resistivity, **4.**63
  seismic refraction, **4.**47, **4.**62
  seismic wave velocity, **4.**47, **4.**62, **4.**64–**4.**65
  uphole, downhole, and cross-hole surveys, **4.**62
  vibration, **4.**62
Geostatic condition, **6.**10
Geosynthetic, **15.**22, **A.**19
Geosynthetic clay liners, **14.**3

Geotechnical engineer, **1.1–1.6**, **A.2**
  analysis (*see* Engineering analyses)
  areas of responsibility, **1.9**
  education, training, and practice, **1.7**
  fields of expertise, **1.8**
  qualifying experience, **1.5**
Geotechnical engineering, **1.1**, **A.2**
Geotextile, **8.24**, **17.23**, **A.19**
Glacial material, **1.1**
Glacier, **6.8**, **14.1**
Glass, **5.38**
Glossary:
  basic terms, **A.1–A.2**
  construction terminology, **A.17–A.22**
  engineering analyses terminology, **A.12–A.16**
  engineering geology terminology, **A.3–A.7**
  grading terminology, **A.17–A.22**
  laboratory testing terminology, **A.8–A.11**
  references for the glossary, **A.1**
  subsurface exploration terminology, **A.3–A.7**
  terminology for engineering computations, **A.12–A.16**
Gneiss, **1.2**, **5.44**
Golden Canyon, **12.18**
Graben, **4.47**, **4.61**
Grade:
  existing grade, **A.19**
  finished grade, **A.19**
  lowest adjacent grade, **A.19**
  natural grade, **A.19**
  rough grade, **A.19**
Grading, **17.2–17.30**
  adjacent property damage, **17.23**, **17.27–17.29**
  back-rolling technique, **17.5**
  benching, **17.4**
  blasting of rock, **17.4**
  buttress fill, **17.4**
  canyon subdrain, **17.4**
  cleanout, **17.4**
  clearing, brushing, and grubbing, **17.4**
  cut, **7.8**, **7.10**, **17.4**
  cut/fill transition, **7.8**, **7.10**, **17.4**
  definition, **17.3**, **A.19**
  earthwork operations, **17.4–17.5**, **17.9–17.14**
  equipment, **4.45**, **15.18**, **15.30–15.31**, **17.9–17.13**, **17.20**
  for fill slopes, **17.4–17.6**
  final grade, **17.7**
  fine grading, **17.5**
  inspection during grading, **17.5–17.7**
  overbuilding and cutting back slopes faces, **17.5–17.6**

Grading (*Cont.*):
  pre-grading meeting, **17.6**
  process of grading, **17.4–17.5**
  recompaction, **17.4**
  revision of grading operations, **17.5**
  ripping of rock, **4.52**, **4.64–4.65**, **17.4**
  rough grading operations, **17.4–17.5**
  scarifying, **17.4**
  shear key, **10.48–10.49**, **17.4**
  specifications for grading, **17.4–17.17**, **17.13–17.14**, **C.1–C.19**
  stabilization fill, **17.4**
  terms and definitions, **A.19**
  windrow, **17.4**
  (*See also* Compaction; Fill)
Grading contractor, **A.19**
Grading permit, **A.19**
Grading plan, **17.4**
Grain size analysis (*see* Soil classification)
Gram, **6.6**
Granite, **5.44**, **13.11**
Gravel, **5.25**, **5.34–5.36**, **5.42**, **9.29**, **12.2**, **16.7**, **16.12–16.13**, **16.18**
Gravel equivalent factor (*see* Pavements)
Gravel size particles, **5.25**, **A.9**
Gross slope failures, **10.2**, **10.27–10.35**
Groundwater, **4.13**, **16.1–16.24**
  effect of surface drainage, **16.19**
  effect on bearing capacity, **8.9–8.10**, **16.2**
  effect on foundations, **16.2**
  effect on landslides, **10.36–10.38**
  effect on liquefaction, **11.8**
  effect on pavements, **13.7**, **16.3**
  effect on retaining walls, **9.12**, **16.15–16.18**
  effect on sewage disposal system, **16.22–16.24**
  effect on shear strength, **6.47**, **16.1**
  effect on slopes, **10.3**, **16.2**, **16.4**, **16.6–16.10**
  effect on vegetation, **16.8–16.10**
  evaporation at ground surface, **16.8–16.10**
  laboratory testing, **5.15–5.19**, **16.1**
  lowering of groundwater table, **10.48**, **16.2**
  methods to control groundwater, **16.13–16.15**
  perched condition, **9.12**, **16.1**
  permeability (*see* Permeability)
  phreatic surface, **16.1**, **A.5**
  piezometers, **6.11**, **16.1**
  pore water pressure, **6.11**, **16.1**
  pumping of groundwater, **7.38**, **16.2**
  seepage, **6.47**, **6.49**, **6.53**, **16.1**, **A.15**
  seepage forces, **6.49**, **6.53**, **16.6**, **A.15**
  seepage pressures, **16.3**
  seepage velocity, **6.49**, **A.15**

Groundwater (*Cont.*):
  settlement related to groundwater extraction, **7.38–7.39**
  for slope stability analyses, **10.33–10.34**
  steady-state flow conditions, **10.12**
  subsurface exploration, **4.12–4.43, 4.46, 16.1**
  sulfate concentration, **13.1, 13.6**
  superficial velocity, **6.49**
  table, **6.46, 16.1–16.24, A.5**
  transportation of contaminants, **16.3**
  uplift forces, **16.3**
Grout, **12.17–12.18, 16.3, 18.18**
Grouting, **18.18, A.20**
Guajome Ranch House, **13.12–13.17**
Gunite, **10.23–10.24, 12.17**
Gypsiferous soil, **7.14**
Gypsum, **5.6, 5.44, 7.14, 9.33, 13.2, 13.6**

Halite, **7.14**
Halloysite, **5.32, 5.34**
Hanover Park, Illinois, **7.41**
Head (fluid flow), **5.16, 6.49, A.14**
Heave, **A.14**
Hematite, **5.6, 5.44**
Highway, **8.25–8.26**
Hillside site, **17.4**
Historical method, **10.50**
Historic structures:
  Bunker Hill Monument, **13.10–13.11**
  Guajome Ranch House, **13.12–13.17**
  monitoring of, **18.4–18.6**
  repair of, **13.6**
  Trinity Church, **18.4**
History of slope changes, **10.3**
Homogeneous soil, **6.53, 6.56, 10.30, A.14**
Horizon (of soil), **A.5**
Hornblende granite, **13.11**
Hornfels, **5.44**
Humic substances, **7.40–7.41**
Humidity, **9.7, 10.11**
Hydraulic conductivity (*see* Coefficient of permeability)
Hydraulic gradient, **5.16, 6.49, 6.56**
Hydrogen bonding, **5.30**
Hydrometer analysis (*see* Soil classification)

Ibrid, Jordan, **9.7**
Igneous rock, **5.43–5.44, 12.17**
Ignition test, **7.41**
Illite, **5.6, 5.34**
Immediate settlement, **7.16–7.19**
  caused by earthquakes, **11.14**

Immediate settlement (*Cont.*):
  definition, **A.13**
  plastic flow, **7.16–7.18**
  plate load tests, **7.18**
  Poisson's ratio, **7.17–7.18**
  stress path method, **7.18–7.19**
  theory of elasticity, **7.17–7.18**
  types of loading, **7.16**
  undrained modulus, **5.43, 7.17–7.18**
Inch-pound units, **1.18**
Inclinometer, **10.6, 10.43, 10.54, 14.12–14.13, 14.17–14.20, 18.2, A.5**
Index tests:
  Atterberg limits, **5.6, 5.30–5.31**
  for collapse potential, **7.12–7.13**
  expansion index test, **5.3–5.5**
  listed on boring logs, **4.43**
  moisture content, **5.3–5.4**
  particle size distribution, **5.6, 5.25–5.28**
  specific gravity tests, **5.5–5.6**
  water content, **5.3–5.4**
  wet density determinations, **5.4–5.5**
  (*See also* Laboratory testing; Phase relationships)
Initial settlement (*see* Immediate settlement)
Initial site visit, **3.1**
Inorganic Soil Classification System Based on Plasticity (ISBP), **5.38–5.43**
Inside clearance ratio, **4.23**
In situ (definition), **A.5**
Inspector, **19.7**
Insurance:
  claim, **19.12**
  companies, **11.16, 19.12**
  errors and omissions, **19.12**
  policy, **19.12**
International System of Units (SI), **1.18, 5.5, 6.7, 6.10**
Investigation:
  conclusions, **19.5**
  document review, **4.4–4.13**
  initial site visit, **4.4**
  laboratory testing, **5.1–5.47**
  monitoring, **18.4–18.9**
  planning, **3.5–3.6**
  report preparation, **19.1–19.10**
  steps, **1.17**
  subsurface exploration, **4.12–4.73**
  (*See also* Project)
Ion exchange, **5.30**
Iowa borehole shear test, **4.31, 4.34, A.5**
Irrigation, **10.4, 10.25, 10.27, 10.45**

Isomorphous substitution, **5.**34
Isotropic soil, **6.**53, **6.**56, **A.**14

Janbu method of slices, **10.**31, **10.**57, **14.**15
Jetting, **A.**20
John Hancock tower, **18.**4
Joints in floor slabs, **10.**4–**10.**5
Joints in rock slopes, **10.**36
Julian schist, **10.**53
Jute netting, **12.**10

Kaolinite, **5.**6, **5.**33, **9.**2, **10.**18
Karst topography, **7.**37, **A.**5
Kelly, **A.**5
Key, **10.**48–**10.**49, **16.**8, **A.**20
Keyway, **10.**48–**10.**49, **A.**20

Laboratory testing, **5.**1–**5.**47
  Atterberg limits, **5.**3–**5.**4, **5.**6, **5.**30–**5.**31
  common laboratory tests, **5.**3
  compressibility, **5.**6–**5.**8
  consolidated undrained triaxial compression test, **5.**10
  constant head permeameter, **5.**16–**5.**17
  direct shear test, **5.**10–**5.**13
  drained residual shear strength, **5.**13–**5.**15
  expansion index test, **5.**3–**5.**4, **9.**4–**9.**5
  falling head permeameter, **5.**16, **5.**18
  hydraulic conductivity, **5.**15–**5.**18
  hydrometer analysis, **5.**27–**5.**29
  index tests, **5.**3–**5.**6, **9.**4–**9.**6
  laboratory tests for deterioration, **5.**18–**5.**20
  laboratory tests for expansive rock, **9.**33–**9.**34
  laboratory tests for pavements, **5.**18–**5.**20
  maximum dry density, **17.**8, **17.**10
  oedometer, **5.**6–**5.**8
  optimum moisture content, **17.**8, **17.**10
  particle size analysis, **5.**25–**5.**29
  particle size distribution, **5.**28
  plasticity, **5.**30–**5.**34
  rock classification (*see* Rock classification)
  sample disturbance, **5.**20–**5.**24
  shear strength tests, **5.**8–**5.**15, **A.**10
  sieve analysis, **5.**25–**5.**27
  soil classification (*see* Soil classification)
  specific gravity, **5.**5–**5.**6
  swell tests, **9.**6–**9.**7
  testing program, **5.**2
  total density, **5.**4–**5.**5
  total unit weight, **5.**4–**5.**5
  unconfined compression test, **5.**8–**5.**10

Laboratory testing (*Cont.*):
  unconsolidated undrained triaxial compression test, **5.**10
  vane shear test, **5.**8
  water content tests, **5.**3–**5.**4
Laguna Niguel, California, **4.**43, **10.**43
Laguna Niguel landslide, **4.**43, **4.**46–**4.**47, **4.**53–**4.**61, **10.**43–**10.**47, **15.**53–**15.**54
L'Ambiance Plaza, **19.**10
Laminar flow, **5.**16, **6.**49, **A.**14
Landfill, **7.**41–**7.**42
Landfill liners, **9.**7
Landscape architect, **10.**25, **13.**9
Landslide, **10.**2, **10.**35–**10.**49
  active, **10.**36
  ancient, **10.**36
  computer analysis, **10.**43–**10.**45
  cross section of landslide, **10.**44–**10.**45, **10.**47
  crown of the landslide, **10.**37–**10.**38
  debris, **A.**5
  definition, **10.**36, **A.**5
  described on boring log, **4.**48–**4.**51
  destabilizing effects, **10.**36
  displaced landslide material, **10.**37–**10.**38
  example, **10.**37–**10.**43
  flank of the landslide, **10.**37–**10.**38
  foot of the landslide, **10.**37–**10.**38
  fossil landslides, **10.**36
  geologic symbol, **4.**9
  groundwater, **10.**33–**10.**34, **10.**45
  head of the landslide, **10.**37–**10.**38
  Laguna Niguel landslide, **4.**43, **4.**46–**4.**47, **4.**53–**4.**61, **10.**43–**10.**47, **15.**53–**15.**54
  main body, **10.**37–**10.**38
  main scarp, **10.**37–**10.**38
  method of analysis, **10.**43–**10.**48
  minor scarp, **10.**37–**10.**38
  nomenclature, **10.**37–**10.**38
  Portuguese Bend, **15.**50, **18.**17
  residual shear strength, **10.**44, **10.**46–**10.**47
  rotational, **10.**36
  rupture surface, **10.**36
  stabilization, **10.**48
  surface of landslide separation, **10.**37–**10.**38
  tip of the landslide, **10.**37–**10.**38
  toe of the landslide, **10.**37–**10.**38
  top of the landslide, **10.**37–**10.**38
  translational, **10.**36
  transverse ridges, **10.**36
  triggering conditions, **10.**36
  zone of landslide accumulation, **10.**37–**10.**38
  zone of landslide depletion, **10.**37–**10.**38

Lapilli, **14.**1
Las Vegas valley, **7.**38–**7.**39
Lateral fill extension, **10.**54
Lateral movement:
　allowable, **10.**2, **10.**4
　causes, **10.**1–**10.**65
　dams, **10.**59–**10.**63
　due to earthquakes, **11.**12–**11.**14
　due to expansive soil, **9.**12
　slopes, **10.**1–**10.**65
Lawsuit, **17.**23, **18.**4, **19.**6, **19.**13
Leaching, **A.**5
Leakage, **10.**60, **16.**9
Letter of authorization, **3.**4
Leyte, Philippines, **10.**48
Liability:
　assessing risk, **19.**11–**19.**12
　avoiding, **19.**10–**19.**13
　insurance, **19.**12
　limitation clauses, **3.**5, **19.**12–**19.**13
　reducing, **19.**10–**19.**13
　self-incriminating memos, **19.**7
　strict, **14.**7, **19.**12
Lift (of fill), **A.**20
Lime, **8.**24, **9.**25
Limestone, **5.**44, **7.**37
Liquefaction, **11.**5–**11.**15, **16.**3
　analysis, **11.**5–**11.**15
　definition, **A.**14
　factors causing liquefaction, **11.**7–**11.**8
Liquidity index, **5.**31, **14.**3
Liquid limit, **5.**30, **A.**8
Liquids, **6.**5–**6.**6
Live load, **7.**2–**7.**3, **8.**4, **A.**14
Load test, **8.**15, **18.**1
Loess, **12.**17, **14.**2, **A.**5
Logging, **10.**50
Logs:
　definition, **A.**4
　description of geologic soil deposits, **4.**52
　description of man-made soil deposits, **4.**52
　example of, **4.**48–**4.**51
　preparation of, **4.**43
　type of information recorded on, **4.**46
Long Beach naval shipyard, **7.**38
Longwall mining, **7.**37
Los Altos Hills, California, **10.**50
Los Angeles, California, **16.**17
Louisiana, **9.**16

Maintenance, **10.**27, **10.**63, **12.**13
Mallard North landfill, **7.**41

Marble, **5.**44
Mass, **6.**6–**6.**7
Maximum dry density (*see* Compaction test)
Meadowlands, **7.**47
Mechanically stabilized earth retaining walls,
　　15.22–**15.**24, **15.**27–**15.**35
　compaction of fill, **15.**27, **15.**30–**15.**31
　construction of, **15.**22–**15.**24, **15.**27–**15.**33
　design analysis, **15.**31–**15.**35
　drainage system, **15.**22, **15.**27
　external stability, **15.**31, **15.**33–**15.**34
　geogrid, **15.**23–**15.**24, **15.**29
　geosynthetic, **15.**22
　internal stability, **15.**32–**15.**34
　wall facing element, **15.**22–**15.**23, **15.**28
Memorandum, **19.**7
Metamorphic rock, **5.**44
Method of slices, **10.**30–**10.**35, **10.**43
Mexican architecture, **13.**13
Mexico City, **7.**38
Mica, **6.**26
Microfossils, **7.**38
Microorganisms, **7.**39–**7.**42
Mildew, **16.**10, **16.**15
Mineral, **5.**5–**5.**6, **A.**5
Mines:
　collapse of, **7.**37–**7.**38
　spoil, **7.**37
　strip mining, **7.**37
Minutemen, **13.**10
Miocene, **15.**50
Mission Viejo, California, **14.**5
Mississippi, **10.**60
Modified Mercalli intensity, **11.**16
Modified Proctor (*see* Proctor)
Modified Wentworth Scale, **5.**28, **5.**44
Modulus of elasticity of soil, **7.**17
Modulus of subgrade reaction, **4.**42
Mohr circle, **6.**30–**6.**43, **A.**14
Moisture content (*see* Water content)
Moisture migration, **16.**1–**16.**18
　below slab barriers, **16.**12–**16.**13
　capillary break, **16.**12–**16.**13
　construction details, **16.**12–**16.**15, **16.**18
　damage caused by, **16.**9–**16.**12, **16.**15–**16.**18
　due to capillary action, **16.**9, **16.**15
　due to hydrostatic pressure, **16.**9, **16.**15
　due to leakage, **16.**9
　due to water vapor, **16.**9, **16.**15
　effect on pavements, **16.**3–**16.**5
　effect on slopes, **16.**4, **16.**6–**16.**10
　through basement walls, **16.**15–**16.**18

Moisture migration (*Cont.*):
   through floor slabs, **16.**9–**16.**15
   waterproofing, **16.**8
Mold, **16.**11
Monitoring devices:
   Avongard crack monitor, **18.**6, **18.**9
   beam sensors, **18.**6
   borehole and tape extensometers, **18.**6
   crack pins, **18.**6
   inclinometers, **10.**43, **10.**54, **10.**56, **14.**3, **18.**4–**18.**5
   piezometers, **6.**11, **16.**1, **18.**5–**18.**7, **A.**6
   pressure and load cells, **6.**10, **6.**12, **18.**5–**18.**6, **18.**8
   settlement monuments or cells, **18.**5, **18.**8
   soil strainmeters, **18.**6
   strain gauges, **18.**6
   tiltmeters, **18.**6
Monterey Formation, **14.**6, **15.**8
Montmorillonite, **5.**6, **5.**33, **6.**8, **9.**2, **9.**10, **9.**33, **10.**18, **14.**3, **14.**7, **15.**3
Morgenstern-Price method of slices, **10.**31
Mud:
   flow, **10.**2, **10.**50
   particle size, **5.**28
   slide, **10.**2, **10.**50
   used for adobe bricks, **13.**11
Mudjacking, **18.**18
Mudstone, **5.**38, **9.**32
Mulch, **12.**10
Muscovite, **5.**6

National Historic Landmark, **13.**13
National Park Service, **12.**18
National Research Council, **10.**35
Necking, **A.**20
Negative skin friction (*see* Downdrag)
New England, **13.**11
New Jersey, **7.**47
New York City, **7.**42, **7.**47
Nitroglycerin, **14.**5
Nonhumic substances, **7.**40–**7.**42
Nonplastic soil, **5.**39, **5.**42
Normally consolidated (*see* Consolidation)

Observation method, **18.**7, **18.**9–**18.**10
Obsidian, **5.**44
Oedometer, **5.**6–**5.**8, **7.**11–**7.**12, **9.**4, **9.**6, **9.**19, **9.**33, **14.**5–**14.**6
Ohio, **15.**46
Oil platforms, **6.**28
O'Kon and Company, **19.**10

Olivenhain, California, **10.**37
Optimum moisture content (*see* Compaction test)
Ordinary method of slices, **10.**31
Organic:
   compounds, **7.**40, **9.**26
   matter, **7.**39–**7.**42, **12.**2, **14.**1
   settlement of, **7.**39–**7.**42
   soil, **5.**37
Orinoco clay:
   consolidation properties of, **7.**21–**7.**22, **7.**24
   factors that effect the undrained shear strength, **6.**44–**6.**45
   radiographs of, **5.**21–**5.**24
   undrained shear strength of, **6.**28–**6.**30
   water content of, **6.**28–**6.**29
Orinoco river, **6.**28
Otay Mesa, **9.**10
Ottawa, Ontario, **14.**3
Overburden (definition), **A.**5
Overconsolidated soil (*see* Consolidation)
Overconsolidation ratio (*see* Consolidation)
Oversize particles, **14.**2, **17.**17–**17.**18
Overtopping, **10.**59, **10.**63
Owner (*see* Client)
Oxidation, **9.**33

Palos Verdes, California, **15.**50
Particle size, **5.**25–**5.**29
Particle size distribution (*see* Soil classification)
Passive earth pressure (*see* Retaining walls)
Pauma Indian Reservation, **10.**50, **10.**52
Pavements:
   aggregates, **8.**23–**8.**27
   alligator cracking, **9.**5, **13.**7
   asphalt, **8.**23, **13.**7, **A.**17
   asphalt concrete, **8.**23, **A.**17
   base course, **8.**23–**8.**24, **13.**7, **A.**17
   bearing capacity, **8.**23
   black top, **8.**23
   block cracking, **13.**7
   California bearing ratio, **4.**40, **4.**42, **5.**18, **5.**20, **8.**27, **13.**7, **17.**22, **17.**24, **A.**4
   California method of design, **8.**24–**8.**27
   cement treated base, **8.**23
   concrete, **9.**27–**9.**28
   corrugation, **13.**7
   crushed stone, **8.**23
   curbs, **8.**24, **9.**23
   depressions, **13.**7
   design, **8.**24–**8.**27, **13.**7
   design life of pavements, **8.**24

Pavements (*Cont.*):
  deterioration, **13.**7
  drainage properties of the subgrade, **4.**45
  effect of expansive soil, **9.**23, **9.**26–**9.**29
  effect of groundwater, **16.**3, **16.**5
  equivalent wheel loads, **8.**25
  erosion, **12.**17, **12.**19–**12.**20
  on expansive soil, **9.**13, **9.**23, **9.**26–**9.**29
  flexible, **8.**23
  frost, **4.**44, **13.**7–**13.**8
  gravel, **8.**23
  gravel equivalent factor, **8.**25–**8.**27
  hot mix, **8.**23
  laboratory tests, **5.**18, **5.**20
  macadam, **8.**23
  mineral filler, **8.**23
  potholes, **13.**7
  raveling, **13.**7
  rigid, **8.**23
  root damage, **13.**9–**13.**12
  rutting, **8.**23, **9.**26, **13.**7
  R-value, **5.**18, **5.**20, **8.**25–**8.**27, **9.**27, **13.**7
  subbase, **8.**24, **13.**7
  subgrade, **8.**24, **9.**5, **13.**7, **A.**21
  subgrade modulus, **4.**45
  surface, **8.**23
  traffic index, **8.**24–**8.**27
  traffic loads, **8.**24–**8.**27, **13.**7
Payment, **3.**5
Peat:
  bogs, **14.**1
  classification, **5.**35, **5.**37
  composition, **14.**1
  definition, **A.**5
  moors, **14.**1
  water content of, **5.**4, **6.**6
Pebbles, **5.**28
Penetration resistance, **4.**14, **A.**5
Percolation tests, **16.**22–**16.**24, **18.**1, **D.**1–**D.**6
Percussion drilling, **4.**16–**4.**17, **A.**5
Permafrost, **13.**9, **A.**6
Permanent erosion control devices, **A.**20
Permeability:
  coefficient of permeability, **5.**15–**5.**19, **6.**46–**6.**55
  of concrete, **13.**6
  Darcy's law, **5.**16, **6.**47, **6.**49, **7.**27, **16.**1
  definition, **5.**15, **A.**10
  determined from field measurements, **6.**47–**6.**51
  of drainage buttress, **16.**7
  effect on sewage disposal system, **16.**22–**16.**24

Permeability (*Cont.*):
  effect on surficial stability, **10.**12
  factors that affect permeability, **6.**47, **6.**49
  hydraulic conductivity, **5.**15–**5.**19, **6.**47
  of loess, **14.**2
  of soil, **5.**15–**5.**19, **6.**46–**6.**57
  of swelling clay, **9.**10–**9.**11
  of trench backfill, **16.**3, **16.**5
Permeameter test:
  constant head permeameter, **5.**16–**5.**17, **6.**53, **16.**1
  falling head permeameter, **5.**16, **5.**18, **6.**53, **9.**10–**9.**11, **16.**1
Permit, **2.**14, **A.**20
pH, **5.**20
Phase relationships, **6.**5–**6.**10
  buoyant unit weight, **6.**9
  degree of saturation, **6.**9
  density, **6.**6–**6.**7
  direct relationships, **6.**6–**6.**7
  dry unit weight, **6.**8
  elements of soil, **6.**5–**6.**6
  indirect relationships, **6.**7–**6.**9
  porosity, **6.**7–**6.**8
  relative density, **6.**8
  specific gravity, **6.**7
  unit weight, **6.**6–**6.**7
  useful relationships, **6.**9
  void ratio, **6.**7–**6.**8
  water content, **6.**6
  (*See also* Index tests; Laboratory testing)
Phreatic surface, **16.**1, **A.**5
Phyllite, **5.**44
Piers:
  belled, **9.**21
  for bridges, **12.**5
  definition, **7.**47, **8.**15, **A.**20
  design of, **8.**15–**8.**23
  end bearing, **8.**16–**8.**18, **8.**20–**8.**23
  for expansive soil, **9.**21–**9.**23
  skin friction, **8.**18–**8.**19, **8.**20–**8.**23
  for underpinning, **18.**12–**8.**14
  (*See also* Foundation; Pier walls)
Pier walls, **15.**49–**15.**59
  construction of pier walls, **15.**49–**15.**59
  design of pier walls, **15.**54–**15.**59
  for landslide stabilization, **10.**48, **15.**49–**15.**59
  for slope stabilization, **15.**50–**15.**59
  Laguna Niguel landslide, **15.**53–**15.**54
  Portuguese Bend landslide, **15.**50
  soil arching, **15.**50, **15.**57
Piezometer, **6.**11, **16.**1, **18.**5–**18.**7, **A.**6

Pile groups, **8.19**, **8.22**–**8.**23
Piles:
  batter pile, **7.**47, **8.**14, **A.**20
  combination end-bearing and friction pile, **7.**47, **8.**14, **A.**20
  definition, **A.**20
  design of, **8.15**–**8.**23
  driving resistance, **8.**15
  end-bearing pile, **7.**47, **8.16**–**8.**19, **8.20**–**8.**23, **A.**20
  engineering analysis, **8.14**–**8.**23
  experience, **8.**15
  field load tests, **8.**15
  for landslide stabilization, **10.**48
  friction pile, **7.**47, **8.18**–**8.**19, **8.20**–**8.**23, **A.**20
  group efficiency, **8.22**–**8.**23
  high displacement, **7.46**–**7.**47
  installation procedures, **7.46**–**7.**47
  load tests, **18.**1
  low displacement, **7.46**–**7.**47
  material types, **7.**46
  mixed in place, **8.**15
  pile cap, **7.**47
  pile driving equations, **8.**15
  specifications, **8.**15
  timber, **5.**20
  uplift capacity, **8.**20
  (*See also* Foundation)
Pinhole porosity, **5.**35
Pinhole test, **10.**60
Pinnacles, **14.**2
Pipe:
  break, **16.19**–**16.**21
  clogs, **16.19**–**16.**21
  leaks, **10.**45, **16.19**–**16.**21
  non-pressurized, **16.**19
  pressurized, **16.**19
  sewer, **16.**20
Piping, **6.55**–**6.**56, **10.59**–**10.**62, **16.**2, **A.**14
Pit (*see* Test pit)
Plastic equilibrium, **A.**14
Plasticity, **5.30**–**5.**43, **A.**10
Plasticity chart, **5.31**–**5.**32
Plasticity index, **5.31**–**5.**34, **5.**37, **5.**38, **5.42**–**5.**43, **A.**10
Plastic limit, **5.**30, **A.**8
Plastic soil, **5.30**–**5.**43
Plate load test, **4.**40, **4.**42, **7.**18, **7.33**–**7.**34
Poisson's ratio, **7.17**–**7.**18
Poplars, **9.**14
Pore water pressure:
  below groundwater table, **6.**11, **16.**1

Pore water pressure (*Cont.*):
  causing liquefaction, **11.5**–**11.**6
  definition, **6.10**–**6.**11, **A.**14
  determined from a flow net, **6.**55
  dissipation of excess pore water pressure, **7.**19, **7.**27
  effect on landslides, **10.**36
  excess, **6.**3, **7.**19, **7.**27, **16.**2, **A.**14
  excess caused by earthquakes, **11.5**–**11.**6
  excess caused by pile driving, **8.**20
  for slope stability analysis, **10.33**–**10.**34
  hydrostatic, **6.**11, **6.**29, **15.**3, **16.**9, **16.**15, **A.**14
  negative, **6.**11, **6.**27, **7.**11, **9.**26
  pore water pressure parameters A and B, **6.38**–**6.**41
  positive, **6.**27
  reduction of pore water pressure due to pumping of groundwater, **7.**38
  shear induced, **5.**13, **6.**27
  uplift pressure, **7.**44
Pore water pressure ratio, **10.**33
Porosity, **5.**35, **6.7**–**6.**8, **A.**14
Portland cement, **8.**23, **13.**5
Portland Cement Association, **13.**2
Portuguese Bend Landslide, **15.**50, **18.**17
Post-Tensioning Institute, **7.**45, **9.18**–**9.**19
Poway, California, **10.**14
Pozzolan, **A.**20
Pozzolanic cement, **14.**2
Precise grading permit, **A.**20
Preconsolidation pressure (*see* Consolidation)
Preliminary information, **2.**3
Presoaking (for expansive soil), **9.16**–**9.**17
Presoaking (for percolation tests), **16.**23
Pressure, **6.**10, **A.**14
Pressuremeter test, **4.**31, **4.**34, **A.**6
Primary consolidation (*see* Consolidation)
Principal planes, **6.**30, **A.14**–**A.**15
Principal stresses, **6.29**–**6.**43, **A.**15
Problems:
  basic geotechnical and foundation principles, **6.58**–**6.**62
  bearing capacity, **8.28**–**8.**31
  compaction, **17.**27, **17.29**–**17.**30
  earthquake, **11.17**–**11.**18
  erosion, **12.18**–**12.**19
  expansive soil, **9.34**–**9.**35
  field exploration, **4.54**–**4.**56
  groundwater, **16.**24
  laboratory testing, **5.45**–**5.**47
  pavements, **8.**31

Problems (*Cont.*):
  retaining walls, **15**.59, **15**.61–**15**.66
  settlement, **7**.49–**7**.56
  sewage disposal system, **16**.24
  slope stability, **10**.63–**10**.65
  solution to problems, **F**.1–**F**.52
Proctor, **6**.8, **9**.27, **17**.8, **17**.10
Progressive failure, **10**.35, **A**.15
Project:
  client, **2**.3
  conflict of interest, **2**.3
  information, **2**.3
  investigation (*see* Investigation)
  location, **2**.3
  planning, **3**.5–**3**.6
  recommendations, **7**.47, **7**.49
  requirement, **2**.11–**2**.14
  scope of work, **2**.3
Project types, **1**.6, **2**.3–**2**.11
  commercial, **2**.4–**2**.10
  condominiums, **2**.3–**2**.5
  essential facilities, **2**.10–**2**.11, **11**.5, **18**.2
  flatland, **2**.4
  hillside, **2**.4
  industrial, **2**.4
  private sector, **2**.6
  public works, **2**.6
  single family dwellings, **2**.3–**2**.4
  special considerations, **1**.2
Proposal:
  cost estimate, **3**.1–**3**.4
  cost estimating sheet, **3**.3
  extra work, **3**.2
  schedule of fees, **3**.2
  scope of services, **3**.1–**3**.2, **3**.4
  (*See also* Contract)
Pumice, **5**.44, **14**.1–**14**.2
Pumping:
  of groundwater, **7**.38–**7**.39
  of oil, **7**.38
  of soil, **17**.22–**17**.26
Pyritic shale, **1**.2, **9**.33
Pyroclastic rock, **14**.1

Quartz, **5**.5–**5**.6, **14**.1
Quartzite, **5**.44
Questions (*see* Problems)
Quick clay, **5**.9, **5**.31, **A**.15
Quick condition (quicksand), **6**.56, **A**.15

Radiograph, **5**.21–**5**.24
Radiolarians, **14**.5

Rainfall, **10**.11–**10**.12, **10**.20, **10**.48–**10**.49, **12**.1, **12**.5
Rankine earth pressure states, **15**.4, **15**.6
Raveling, **7**.37
Ravines, **12**.17
Redondo Beach King Harbor, **11**.7
Refusal (during drilling), **4**.46, **A**.6
Reimer, Peter, **15**.46
Relative compaction (*see* Compaction)
Relative density, **4**.27, **6**.8, **11**.8, **A**.15
Reno/Tahoe International Airport, **8**.24–**8**.25
Repair:
  for concrete cracks, **18**.15–**18**.18
  foundation strengthening, **18**.10–**18**.18
  for gross slope failures, **15**.50–**15**.59
  historic structures, **13**.16
  for landslides, **15**.50–**15**.59
  strip replacement, **18**.15–**18**.16
  for surficial slope failures, **10**.23–**10**.26
Report:
  as-built reports, **19**.9
  contents, **19**.4–**19**.5
  engineering evaluation of reports, **19**.9–**19**.10
  feasibility report, **17**.3, **19**.3
  final inspection report, **18**.3–**18**.4
  jargon, **19**.5, **19**.7–**19**.8
  preliminary report, **17**.3, **19**.3
  preparation, **19**.1–**19**.10
  purpose of report, **19**.4
  recommendations, **7**.47, **7**.49, **8**.14, **10**.25, **10**.27, **10**.35, **13**.7, **15**.20–**15**.21
  standardized formats, **19**.5
  superlatives, **19**.8–**19**.9
  technical words, **19**.9
Reservoir, **10**.59
Residual shear strength (*see* Shear strength)
Residual soil (*see* Soil deposits)
Retaining walls, **15**.1–**15**.66
  active earth pressure, **15**.3–**15**.8, **A**.13
  active earth pressure coefficient, **15**.2, **15**.7–**15**.8
  active wedge, **15**.5–**15**.6
  adhesion, **15**.7, **15**.13
  analysis, **15**.3–**15**.66
  at-rest earth pressure, **15**.19, **A**.13
  back-cut, **15**.18–**15**.19
  backfill, **15**.1–**15**.2, **15**.19
  basement, **13**.2, **15**.19
  bearing capacity failure, **15**.7
  bearing pressure of footing, **15**.7, **15**.10
  cantilevered, **15**.2–**15**.6, **15**.13, **15**.15
  common causes of failure, **15**.17–**15**.18

Retaining walls (*Cont.*):
  compaction of backfill, **15.**18
  construction, **15.**18–**15.**33
  construction at the top of slopes, **15.**13
  Coulomb's equation, **15.**7–**15.**8
  counterfort, **15.**2, **15.**13, **15.**15
  crib walls, **15.**2
  definition, **15.**2, **A.**22
  design, **15.**2–**15.**66
  drainage system, **15.**3, **15.**23, **15.**27
  effect of earthquakes, **15.**13, **15.**17
  effect of expansive soil, **15.**18
  effect of frost, **15.**3
  effect of groundwater, **15.**3, **16.**15–**16.**18
  effective stress analysis, **15.**6
  equivalent fluid pressure, **15.**4, **15.**56
  factor of safety, **15.**7–**15.**66
  for rock slopes, **10.**7
  gravity, **15.**2, **15.**13–**15.**14, **15.**31
  gunite, **12.**17
  inclined slope, **15.**7–**15.**8, **15.**24
  mechanically stabilized earth retaining walls (*see* Mechanically stabilized)
  overturning analysis, **15.**7, **15.**11–**15.**12
  passive earth pressure, **15.**5–**15.**7, **15.**10–**15.**13, **15.**39, **15.**56, **A.**13
  passive earth pressure coefficient, **15.**6, **15.**13
  passive wedge, **15.**5–**15.**6
  pier walls (*see* Pier walls)
  plane strain condition, **15.**3
  reduction factor for passive pressure, **15.**6
  restrained retaining walls, **15.**19
  semigravity, **15.**14
  shear strength of backfill, **15.**3
  sheet pile walls (*see* Sheet pile)
  sliding analysis, **15.**7, **15.**10–**15.**11
  slope creep, **15.**13
  surcharge, **15.**5
  temporary retaining walls (*see* Temporary retaining walls)
  translation, **15.**6
  with expansive soil backfill, **9.**12–**9.**13
  without wall friction, **15.**3–**15.**7
  with wall friction, **15.**7–**15.**12
Rhyolite, **5.**44
Rillwash, **12.**1
Rippability, **4.**52, **4.**64–**4.**65, **17.**9, **A.**20
Riprap, **12.**16, **A.**20
Rock classification, **5.**43–**5.**45, **A.**6
Rock (expansive), **9.**32–**9.**34
Rock flour, **5.**39, **11.**8, **14.**1
Rock mechanics, **1.**1, **A.**2

Rock quality, **5.**43, **5.**45
Rock quality designation (RQD), **5.**45
Rock salt, **5.**44
Rock strata, **4.**13
Rock topples, **10.**2
Rock types, **1.**1, **5.**43–**5.**44
  igneous, **1.**1, **5.**44
  metamorphic, **1.**1, **5.**44
  sedimentary, **1.**1, **5.**44
Rockfall, **10.**1–**10.**2, **10.**4–**10.**12
  damage caused by rockfall, **10.**5–**10.**12
  factors that govern a rockfall, **10.**4–**10.**5
  remedial measures, **10.**6–**10.**7, **10.**10
  rockfall in tunnel excavations, **10.**10
Rockslide, **10.**4
Roots, **9.**14–**9.**16, **10.**11, **10.**21, **13.**9–**13.**12
Rose Canyon fault zone, **4.**5, **4.**8, **10.**6
Rotary drilling, **4.**14–**4.**17, **A.**6
Rubble, **A.**6
Running ground, **A.**21
Running soil, **15.**49
Rupture surface (*see* Shear plane)

Sabkhas, **7.**14
Salinas, **7.**14
Salt, **7.**14–**7.**15, **12.**16, **13.**5, **16.**8
Salt marshes, **7.**14
Salt playas, **7.**14
Sample disturbance, **5.**20–**5.**24
  altered soil, **4.**21–**4.**22, **5.**20
  disturbed soil, **4.**22–**4.**23, **5.**20–**5.**21
  due to gas coming out of solution, **5.**21, **5.**23
  due to tube friction, **5.**21, **5.**23
  effect on coefficient of consolidation, **7.**22, **7.**31
  effect on consolidation test, **7.**21–**7.**23
  effect on shear strength, **6.**44–**6.**45
  effect on swelling of clay, **9.**7
  of the Orinoco clay, **5.**21–**5.**24, **6.**44–**6.**45
  of quick clay, **6.**44
  soil cracks, **5.**21, **5.**23–**5.**24
  turning of edges, **5.**21, **5.**23
  undisturbed soil, **4.**23–**4.**25, **5.**21, **5.**24
  voids, **5.**21–**5.**22
Samplers:
  Acker, **4.**14
  definition, **A.**6
  Denison, **4.**14
  Pitcher, **4.**14
  rotary coring, **4.**14–**4.**17
  Shelby tube, **4.**21, **4.**43
  Swedish foil, **4.**16

Samplers (*Cont.*):
   thin wall tube, **4.**15
   (*See also* Soil sampling)
Sampling (*see* Soil sampling)
Sand:
   boils, **10.**60–**10.**61, **11.**6, **A.**21
   classification of, **5.**35, **5.**42
   liquefaction of, **11.**5–**11.**15
   loss of shear strength, **11.**5–**11.**15
   size particles, **5.**25, **A.**10
Sand cone tests, **17.**13–**7.**14
Sand equivalent, **5.**3, **A.**10
San Diego, California, **5.**29, **16.**16
San Diego County, California, **10.**50, **13.**13, **16.**11, **16.**22
Sand-size particles, **5.**25, **A.**10
Sandstone, **5.**38, **5.**44, **8.**14, **10.**16, **10.**18
Santiago Formation, **16.**8
Santiago Peak Volcanics, **10.**5
Saturation (degree of), **6.**9, **A.**15
Scanning electron microscope, **7.**41
Scheduling, **3.**6
Schist, **1.**2, **5.**44
Scope of work, **3.**1–**3.**2, **3.**4
Scoria, **5.**44
Screw plate compressometer, **4.**31, **4.**34, **A.**6
Scripps Ranch, California, **18.**11
Sea cliffs, **12.**16–**12.**18
Seasonal moisture change, **9.**7, **9.**21
Sea walls, **12.**16–**12.**17
Seawater, **13.**7
Secondary compression, **7.**32–**7.**33, **A.**13
Secondary compression ratio, **7.**33
Sedimentary rock, **5.**44, **9.**33, **10.**18, **14.**5
Sedimentation, **5.**27, **6.**12
Seep, **A.**6
Seepage (*see* Groundwater)
Seismic activity (*see* Earthquakes)
Sensitivity, **5.**9, **11.**12, **11.**14, **14.**3, **A.**15
Serpentine (mineral), **5.**6
Serpentine (rock), **5.**44
Settlement, **1.**6, **7.**1–**7.**56
   allowable, **7.**3–**7.**9
   angular distortion, **7.**4–**7.**9
   caused by groundwater extraction, **7.**38–**7.**39
   caused by oil extraction, **7.**38–**7.**39
   caused by organic decomposition, **7.**39–**7.**42
   caused by undermining, **7.**3
   collapse of underground mines and tunnels, **7.**37–**7.**38
   collapsible soil (*see* Collapsible soil)
   component of lateral movement, **7.**8

Settlement (*Cont.*):
   compression features, **7.**8, **7.**10, **7.**37–**7.**38
   consolidation (*see* Consolidation)
   cut/fill transition, **7.**8, **7.**10
   definition, **7.**2, **A.**15
   differential, **7.**3–**7.**9, **A.**15
   drag effect, **7.**8, **7.**10
   due to earthquakes, **7.**3, **11.**5–**11.**15
   elastic method, **7.**17–**7.**18, **7.**35–**7.**36
   empirical correlations, **7.**4–**7.**9, **7.**34–**7.**35
   immediate (*see* Immediate settlement)
   landfills, **7.**39–**7.**42
   limestone cavities, **7.**37
   of cohesionless soil, **7.**33–**7.**36
   of cohesive soil, **7.**16–**7.**33
   of diatoms, **14.**5–**14.**6
   of organic soil, **7.**16–**7.**33
   of sands, **4.**42
   other considerations, **7.**36
   rate, **7.**3, **7.**5, **7.**27–**7.**33
   Schmertmann's method, **7.**36
   secondary compression, **7.**32–**7.**33
   selection of foundation type, **7.**42–**7.**49
   severity of cracking damage, **7.**9
   sinkholes, **7.**37
   tensional features, **7.**8, **7.**10, **7.**37–**7.**38
   tilting, **7.**7
   total, **7.**3, **7.**13. **7.**33, **A.**15
   two- or three-dimensional, **7.**16
   two types of settlement, **7.**2–**7.**3
Setup (*see* Freeze)
Sewage disposal system, **16.**22–**16.**24, **D.**1–**D.**6
   design, **16.**22–**16.**24, **D.**1–**D.**6
   disposal field, **16.**22–**16.**24
   effluent, **16.**22–**16.**24
   leach field, **16.**22–**16.**24
   lot size, **16.**23, **D.**6
   nitrate impacts, **16.**24
   percolation rate, **16.**22–**16.**24
   percolation test procedures, **16.**23–**16.**24, **D.**1–**D.**6
   percolation tests, **16.**22–**16.**24, **D.**1–**D.**6
   primary disposal trench length, **16.**23, **D.**5
   septic tank, **16.**22–**16.**23, **D.**5
Sewer, **16.**19–**16.**21
Shafts, **7.**47, **8.**15
Shale, **1.**2, **5.**44, **9.**32, **12.**17
Shallow foundation (*see* Foundation)
Shear failure, **A.**15
Shear key, **10.**48–**10.**49, **17.**4, **A.**21
Shear plane (*also known as* rupture surface or slip surface), **10.**2, **10.**11, **10.**36, **A.**15

Shear strain (due to earthquakes), **11**.7–**11**.8
Shear strength:
  ɸ = 0 concept, **6**.32, **8**.20
  at soil-pile interface, **8**.18
  cohesion, **5**.8, **6**.23–**6**.46, **A**.8
  consolidated undrained triaxial compression test, **5**.10, **6**.30–**6**.34
  contraction during shear, **6**.23
  curved (nonlinear) failure envelope, **6**.26, **6**.38, **10**.13–**10**.14, **10**.16, **10**.44, **10**.46
  definition, **5**.8, **6**.23, **A**.10
  dilation during shear, **6**.23
  direct shear, **5**.10–**5**.13, **6**.23–**6**.25, **6**.46
  displacement rate, **5**.13, **6**.44
  drained, **5**.8, **6**.23, **6**.46
  effective cohesion, **6**.23, **6**.37–**6**.46
  effective friction angle, **6**.23–**6**.27, **6**.35–**6**.46
  effective stress analysis, **5**.8, **6**.23, **6**.46, **A**.10
  effect of anisotropy, **6**.44
  effect of sample disturbance (*see* Sample disturbance)
  effect of strain rate, **6**.44
  effect of water on friction angle, **6**.24
  estimated from the cone penetration test, **4**.30, **6**.27
  estimated from the standard penetration test, **4**.28, **6**.27
  factors that cause a reduction in shear strength, **6**.26, **6**.44, **10**.36
  factors that effect shear strength, **6**.23–**6**.46
  failure envelope, **6**.23–**6**.26, **6**.32–**6**.44
  failure surfaces, **5**.11
  for earthquake analysis, **11**.13–**11**.14
  friction angle, **4**.28, **4**.30, **5**.8, **5**.43, **6**.23–**6**.27, **6**.30–**6**.46, **A**.9
  laboratory testing, **5**.8–**5**.15
  maximum obliquity, **6**.36
  Mohr-Coulomb failure law, **6**.26, **6**.35
  normalized undrained shear strength **6**.30
  of compacted London clay, **10**.16
  of diatomaceous soil, **14**.7–**14**.10
  of nonplastic (cohesionless) soil, **6**.23–**6**.27
  of plastic (cohesive) soil, **6**.23, **6**.27–**6**.46
  of rock joints and discontinuities, **10**.5
  peak, **5**.13, **6**.24–**6**.26, **A**.10
  pore water pressure parameters A and B, **6**.38–**6**.41
  remolded, **5**.14
  residual, **5**.13–**5**.15, **6**.44, **10**.44, **10**.46–**10**.47
  ring shear, **5**.14–**5**.15, **10**.44
  root-permeated, **10**.21
  sample disturbance (*see* Sample disturbance)

Shear strength (*Cont.*):
  secant friction angle, **10**.44, **10**.46
  stress paths, **6**.40–**6**.43
  tests, **5**.8–**5**.15
  total stress analysis, **5**.8, **6**.23, **6**.45–**6**.46, **A**.10
  ultimate, **6**.24–**6**.26
  unconfined compression test, **5**.8–**5**.10, **6**.27–**6**.29, **6**.46
  unconsolidated undrained triaxial compression test, **5**.10, **6**.30, **6**.32–**6**.34, **14**.7–**14**.10
  undisturbed, **4**.23–**4**.25, **5**.21, **5**.24
  undrained, **4**.31, **5**.8, **5**.43, **6**.23, **6**.45–**6**.46, **9**.26
  vane tests, **4**.31, **4**.34, **5**.8, **6**.27–**6**.28, **6**.44
  (*See also* Laboratory testing)
Shear stress, **6**.23, **10**.36, **A**.16
Shear stress ratio, **11**.8–**11**.15
Shear walls, **11**.17
Sheet pile walls, **6**.53–**6**.54, **15**.35–**15**.43
  anchored bulkhead, **15**.35, **15**.39–**15**.40
  anchor pull, **15**.35, **15**.39–**15**.40
  cantilevered, **15**.35, **15**.40–**15**.41
  cofferdams, **15**.42–**15**.43
  cohesive soil, **15**.42
  design analysis, **15**.35–**15**.43
  factor of safety for toe kick–out, **15**.40
  factors increasing the stability, **15**.42
  loading conditions, **15**.42
  penetration depth, **15**.41
  plane strain condition, **15**.42
  soil layers, **15**.41
  surcharge loads, **15**.41
  tieback anchors, **15**.35
  unbalanced hydrostatic and seepage forces, **15**.41–**15**.42
Sheffield Dam, **11**.3
Shelby tube (*see* Samplers)
Shells, **5**.21, **5**.38
Shoring, **15**.45–**15**.49
Shotcrete, **10**.7, **A**.21
Shrinkage factor, **A**.21
Shrinkage limit, **5**.30, **9**.7, **9**.9, **9**.12, **A**.8
Shrinkage of soil (*see* Expansive soil)
Sieve, **5**.25–**5**.27, **A**.10
Sieve analysis (*see* Soil classification)
Silicates, **5**.6
Sill plate, **11**.16–**11**.17
Silt, **5**.37, **5**.42, **13**.8
Silt-size particles, **5**.25, **A**.10
Siltstone, **5**.38, **5**.44, **8**.14, **9**.32, **14**.6

Silty clay, **14**.6
Silty sand, **7**.11–**7**.12, **13**.16
Sinkhole activity, **7**.37
Site, **A**.21
Site investigation (*see* Investigation)
Slab-on-grade (*see* Foundation)
Slag, **5**.38
Slaking, **9**.12, **A**.6
Slate, **5**.44, **9**.32
Slickensides, **10**.43, **A**.6
Slides, **10**.27
Slip surface (*see* Shear plane)
Slope:
 definition, **A**.21
 fill slopes, **17**.4–**17**.5
 permanent, **10**.27
 temporary, **10**.35
Slope Indicator Company, **18**.4–**18**.9
Slope movement:
 allowable, **10**.2, **10**.4
 causes, **10**.1–**10**.2
 creep, **10**.2, **10**.53–**10**.59
 debris flow, **10**.2, **10**.48–**10**.53
 due to flow slides, **11**.6–**11**.7
 due to freezing of water, **13**.8
 due to gross instability, **10**.2, **10**.27–**10**.35
 due to liquefaction of seams, **11**.12–**11**.14
 due to reservoir drawdown, **10**.62
 due to slope softening, **10**.53–**10**.59
 due to surficial instability, **10**.2, **10**.10–**10**.27
 due to unusual soil, **14**.5–**14**.22
 during construction, **17**.23, **17**.27–**17**.29, **18**.1–**18**.2
 effect of groundwater, **16**.4, **16**.6–**16**.10
 landslides, **10**.2, **10**.35–**10**.49
 of dams, **10**.2, **10**.59–**10**.63
 of diatomaceous soil, **14**.5–**14**.22
 rock topples, **10**.2
 rockfall, **10**.1–**10**.2, **10**.4–**10**.11
 rotational slope failure, **10**.30, **10**.36
 stabilization measures, **16**.6–**16**.7, **17**.4
 translational slope failure, **10**.30, **10**.36
 types of slope movement, **10**.1–**10**.2
Slope softening, **10**.2, **10**.53–**10**.59, **14**.15, **16**.2
Slope stability analysis, **10**.1–**10**.65, **14**.15, **14**.21–**14**.22
 different types of analyses, **10**.27, **A**.16
 for plane strain condition, **10**.34
 for retaining walls, **10**.35
 including pore water pressures, **10**.33
 progressive failure, **10**.35
 pseudo-static analysis, **10**.35, **11**.13–**11**.14

Slope stability analysis (*Cont.*):
 slip surfaces, **10**.34
 using computer programs, **10**.30–**10**.31
 using different soil layers, **10**.34
 using nonlinear shear strength envelope, **10**.34
 with surcharge loads, **10**.34
 with tension cracks, **10**.34
Slope stabilization (to resist earthquakes), **11**.15
Slope wash, **A**.6
Slough, **A**.21
Slumps, **12**.1, **17**.22, **17**.27, **A**.21
Slurry seal, **A**.21
Soapstone, **5**.44
Sodium hexametaphosphate, **5**.27
Soil arching, **15**.50, **15**.57
Soil behavior (*see* Engineering analyses)
Soil cementing agents, **12**.8
Soil classification, **5**.21, **5**.25–**5**.43
 color, **5**.35
 consistency, **5**.35, **5**.38
 descriptive terms, **5**.35, **5**.38
 grain size analysis, **5**.25–**5**.29
 group symbols, **5**.34–**5**.37, **5**.42
 hydrometer tests, **5**.27–**5**.29
 Inorganic Soil Classification System Based on Plasticity, **5**.38–**5**.43
 Modified Wentworth Scale, **5**.28, **5**.44
 particle size distribution, **5**.25–**5**.29
 plasticity, **5**.30–**5**.33
 porosity, **5**.35
 sieve analysis, **5**.25–**5**.29
 structure, **5**.35
 texture, **5**.35
 Unified Soil Classification System, **5**.21, **5**.34–**5**.38
 (*See also* Collapsible soil; Compacted soil)
Soil containing gypsum, **7**.14
Soil containing soluble minerals, **7**.14–**7**.16
Soil containing sulfate (*see* Sulfate)
Soil (definition), **1**.1, **A**.6
Soil deposits:
 aeolian, **4**.52, **A**.3
 alluvium, **1**.1, **4**.9, **4**.52, **10**.43, **17**.4, **A**.3
 artificial fill, **4**.9
 beach sand, **4**.9
 colluvium, **4**.52, **17**.4, **A**.4
 debris fill, **4**.52
 glacial, **4**.52
 gypsiferous, **7**.14
 lacustrine, **4**.52
 marine, **4**.52
 municipal dump, **4**.52

Soil deposits (*Cont.*):
  organic, **4.**52
  pyroclastic, **4.**52
  residual, **1.**1, **4.**52, **A.**6
  slope wash, **4.**9
  structural fill, **4.**52
  uncompacted fill, **4.**52
Soil (hard compact), **15.**49
Soil imperfections, **6.**47, **6.**49
Soil matrix, **14.**3, **17.**17–**17.**18
Soil mechanics, **1.**1, **A.**2
Soil nailing, **10.**48
Soil profile (*see* Subsoil profile)
Soil report (*see* Report)
Soil (running), **15.**49
Soil samples:
  altered soil, **4.**21–**4.**22
  disturbed samples, **4.**22–**4.**23
  overdriven samples, **4.**22–**4.**23
  undisturbed soil samples, **4.**23–**4.**25
Soil sampling:
  block samples, **4.**39
  California sampler, **4.**21–**4.**22, **4.**25
  drive cylinders, **4.**39
  Shelby tube, **4.**21, **4.**43
  split-spoon sampler, **4.**21
  standard penetration test, **4.**21–**4.**22, **4.**25–**4.**28
  (*See also* Samplers)
Soil scientists, **5.**29
Soil stabilization, **7.**13–**7.**14, **7.**45–**7.**46, **A.**21
Soil strata, **4.**12
Soil structure, **5.**35, **A.**10
  different types of soil structure, **A.**10–**A.**11
  effect on permeability, **6.**47
Soils engineer (*see* Geotechnical engineer)
Soils engineering (*see* Geotechnical engineering)
Solana Beach, California, **12.**16–**12.**17
Solids, **6.**5–**6.**6
Solution to problems, **F.**1–**F.**52
South Carolina, **12.**11
Specification, **4.**5, **A.**21
Specific gravity, **5.**5–**5.**6, **6.**7, **A.**11
Spencer method of slices, **10.**31, **10.**44–**10.**45
Spicules of sponges, **14.**5
Spillway, **10.**59
Spring thaw, **13.**8, **16.**2
Stabilization fill, **16.**6, **A.**21
Staking, **A.**21
Standard penetration test, **4.**21–**4.**22, **4.**25–**4.**28
  correlation factors for field testing procedures, **4.**27
Standard penetration test (*Cont.*):
  definition, **A.**6
  N-value, **4.**26–**4.**28
  used to predict drained modulus, **7.**36
  used to predict liquefaction potential, **11.**9–**11.**12
  used to predict settlement, **7.**36, **11.**14–**11.**15
Standpipe piezometer, **16.**1, **18.**5, **18.**7
Staten Island, **7.**42
Stereoscope, **4.**5
St. Francis Dam, **10.**59
Stiffening beams, **9.**18
Stokes law, **5.**28
Stone columns, **8.**15
Storm drain, **15.**45, **15.**47
Strain, **10.**2, **A.**16
Stress, **6.**10, **A.**14
Stress concentrations, **15.**31
Stress difference, **6.**30, **6.**36
Stress-displacement plots, **6.**25, **10.**46, **10.**55
Stress distribution, **6.**12–**6.**22
  2:1 approximation, **6.**13–**6.**14, **8.**19, **8.**23
  based on the theory of elasticity, **6.**14–**6.**22
  Boussinesq equations, **6.**16
  for layered soil, **6.**18
  for one-dimensional loading, **6.**13
  for three-dimensional loading, **6.**13
  for two-dimensional loading, **6.**13
  Newmark charts, **6.**16–**6.**17, **6.**19, **6.**22
  plane strain condition, **6.**13
  pressure bulbs, **6.**13, **6.**15
Stress path method, **7.**18–**7.**19, **10.**59
Stress paths, **6.**40–**6.**43
Stress strain plot, **6.**33
Strict liability, **14.**7
Strike and dip, **A.**6
Structural engineer, **4.**4, **7.**4, **8.**4, **9.**18
Structure (*see* Buildings)
Stucco, **9.**29, **9.**32
Subdrain, **16.**3–**16.**4, **A.**21
Subgrade:
  clay, **9.**23, **9.**26–**9.**27
  definition, **A.**21
  modulus, **4.**42, **4.**44–**4.**45, **A.**6
  properties, **4.**44–**4.**45
  (*See also* Pavements)
Subsidence, **7.**3, **7.**38–**7.**39, **16.**2, **A.**16
Subsoil profile, **4.**8, **A.**6
Substructure, **A.**21
Subsurface exploration:
  boring layout, **4.**35–**4.**36

cone penetration tests, **4.28–4.31**
Subsurface exploration (*Cont.*):
   depth of subsurface exploration, **4.38–4.39**
   dozer cuts, **4.40**
   general purpose, **4.12–4.13**
   standard penetration test, **4.21–4.22**, **4.25–4.28**
   test pits, **4.13**, **4.39–4.40**
   trenches, **4.39–4.41**
   (*See also* Field exploration)
Suction, **9.12**
Sulfate attack, **13.1–13.7**
Sulfates, **5.20**, **13.1–13.7**, **16.15**, **A.21**
Sump, **16.13–16.14**, **A.21**
Sump pump, **16.13–16.14**
Superficial velocity (*see* Groundwater)
Superior Court, **14.7**
Superlatives, **19.8–19.9**
Superstructure, **A.21**
Surcharge pressure, **7.46**, **10.6**, **15.5**, **15.31**, **15.49**
Surficial slope stability, **10.2**, **10.10–10.27**, **16.2**
   cause of, **10.10–10.12**
   for cut slope, **10.4**, **10.16–10.18**
   design for, **10.23–10.27**
   effect of seepage, **10.2**, **10.11–10.12**, **16.2**
   effect of vegetation, **10.20–10.21**
   for fill slope, **10.18–10.21**
   geogrid, **10.23–10.24**
   repair of, **10.23–10.27**
   shallow surface slips, **10.16**
   stability analysis, **10.12–10.17**
   surficial failures, **10.13**, **10.17–10.22**
Surveyor, **1.11**
Swedish circle method, **10.30**
Swell:
   definition, **9.6**, **A.16**
   effect on surficial stability, **10.11**
   pressure, **9.6**, **9.20**, **15.18**
   primary swell, **9.4–9.5**, **9.10–9.11**
   progressive swell, **9.13–9.14**
   secondary swell, **9.10–9.11**
   steady-state swell, **9.11–9.12**
   tests, **9.6–9.7**
   total swell, **9.18**
   (*See also* Expansive soil)
Swelling index, **9.19–9.20**

Tack coat, **A.22**
Tailings, **11.8**, **A.22**
Technical words, **19.9**
Technician, **5.2**, **17.14–17.15**
Temperature, **9.7**, **9.33**, **10.11**
Temporary retaining walls, **15.43–15.49**

braced, **15.43–15.45**
Temporary retaining walls (*Cont.*):
   close sheeting, **15.47–15.48**
   design analysis, **15.43–15.49**
   earth pressure distribution, **15.44**
   excavation cave-ins, **15.46**
   steel I-beam and wood lagging, **15.45**
   struts, **15.48**
   surcharge pressures, **15.43**, **15.49**
   tieback anchors, **15.45–15.46**
   utility trench shoring, **15.45–15.49**
Tendons, **7.45**
Tensile test, **A.11**
Terrace, **A.22**
Testing:
   borings (*see* Boring)
   cone penetration tests (*see* Cone penetration test)
   core drilling (*see* Core drilling)
   field load tests (*see* Field exploration Subsurface exploration)
   in-place testing (*see* Field exploration; Subsurface exploration)
   laboratory (*see* Laboratory testing)
   subsurface exploration (*see* Field exploration; Subsurface exploration)
   test pits (*see* Test pits)
   trenches (*see* Trenches)
Testing requirements (for laboratory tests), **5.2**
Test pits, **4.13**, **4.39–4.40**, **A.6**
Texas, **9.2**, **9.16**
Texture (of soil), **5.35**, **A.11**
Thermal osmosis, **9.14**, **16.2**
Thixotropy, **5.10**, **A.11**
Tieback anchors, **10.34**, **12.17**, **15.45–15.46**, **15.59–15.60**
Till, **6.8**, **A.7**
Timberlane, **18.11–18.12**
Time factor (*see* Consolidation)
Topographic map, **4.5**, **4.10–4.11**, **10.49**
Topography, **10.3**, **13.13**
Topple (of rocks), **10.2**
Topsoil, **A.7**
Torry Pines, California, **11.3–11.4**
Torvane, **5.8**, **6.27**
Total stress (definition), **6.10**, **A.16**
Total stress analysis, **5.8**, **6.23**, **6.45–6.46**
   for bearing capacity, **8.11**
   for dam stability, **10.62**
   for gross slope stability, **10.28–10.33**
   for liquefaction, **11.5–11.15**
   for piles in cohesive soil, **8.20–8.23**

for temporary retaining walls, **15.**45
Toxic waste, **19.**12
Traffic index (*see* Pavements)
Transcona, Canada, **8.**5
Transpiration, **9.**14
Transportation engineer, **8.**23, **8.**25
Trench cave-ins, **15.**46
Trenches, **4.**39–**4.**41, **11.**5, **15.**45–**15.**49, **17.**5, **17.**18–**7.**20, **19.**12
Triaxial tests, **5.**10, **6.**30–**6.**43, **A.**11
Tricalcium aluminate, **13.**5
Trinity Church, **18.**4
Tropical Storm Thelma, **10.**48
Tuff, **5.**44, **14.**1–**14.**2
Tunnels:
 collapse of, **7.**37
 coring and bring machines, **10.**10
 damage due to ground rupture, **11.**3
 rockfall, **10.**10
Turbidity, **13.**1
Type of project, **2.**3–**2.**11

U-line, **5.**32–**5.**33, **14.**7
Unconfined compressive strength:
 definition, **A.**11
 soil, **5.**8–**5.**9, **6.**28–**6.**29
 rock, **5.**43, **5.**46
Underconsolidated (*see* Consolidation)
Underpinning, **18.**10–**18.**18, **A.**22
Undrained loading (due to earthquakes), **11.**6
Undrained modulus, **5.**43, **7.**17–**7.**18
Undrained shear strength (*see* Shear strength)
Unified Soil Classification System (USCS), **5.**21, **5.**34–**5.**38, **14.**6
 group symbols, **5.**34–**5.**37
Uniform Building Code, **2.**6, **4.**5, **8.**4, **8.**6, **8.**8, **9.**4, **9.**18, **11.**12–**11.**13, **13.**5, **15.**6, **17.**13
Uniformity, coefficient of, **5.**28
United States Customary System units (USCS), **1.**18, **5.**5, **6.**7, **6.**10
United States Geological Survey (USGS), **4.**5, **4.**10–**4.**11
Unit weight, **5.**4–**5.**5, **6.**6–**6.**9, **10.**13, **A.**11
Universal soil loss equation, **12.**5–**12.**13
 combined slope length and steepness factor, **12.**5, **12.**8–**12.**10
 erosion control practice factor, **12.**5, **12.**11
 limitations, **12.**11
 rainfall erosion index, **12.**5–**12.**7
 soil erodibility factor, **12.**5–**12.**8
 soil loss, **12.**5–**12.**13

 vegetative cover factor, **12.**5, **12.**10
Unusual soil, **14.**1–**14.**22
 bentonite, **14.**3
 bull's liver, **14.**1
 caliche, **14.**2
 diatomaceous earth, **14.**3, **14.**5–**14.**22
 loess, **14.**2
 non-welded tuff, **14.**1–**14.**2
 peat, **14.**1
 quick clay, **14.**3
 rock flour, **14.**1
 sensitive clay, **14.**3
 varved clay, **14.**3
 volcanic ash, **14.**1–**14.**2
Uplift loads, **11.**14
Utilities, **4.**21, **9.**29, **9.**32, **11.**3, **17.**4
Utility trenches, **15.**45–**15.**49, **16.**3, **16.**5, **17.**18–**17.**20, **19.**2
UUC tests (unconfined compression tests), **6.**28–**6.**29

Vane shear test, **4.**31, **4.**34, **A.**7
 miniature vane, **4.**31
 tapered vane, **4.**31
 torvane, **4.**31
 undrained shear strength, **4.**31
Varved clay, **14.**3, **A.**7
Varved silt, **A.**7
Vegetation, **9.**7, **9.**14–**9.**16, **10.**20–**10.**21, **12.**10, **12.**12, **17.**2
Venezuela, **6.**28
Vernal pool, **5.**38–**5.**39, **5.**41
Vertical pressure, **6.**10
Vibrations, **10.**3
Vibro columns, **8.**15
Vibrodensification, **A.**22
Vibroflotation, **8.**15
Vista, California, **13.**12
Void ratio, **6.**6–**6.**7, **7.**23, **A.**16
Volcanic:
 ash, **1.**1, **14.**1–**14.**2
 dust, **14.**1
 eruptions, **14.**1
 rock, **10.**5
Volume, **6.**6–**6.**9
Volumetric strain (caused by an earthquake), **11.**14–**11.**15

Walls:
 basement, **9.**12, **13.**5, **15.**19, **16.**15–**16.**18
 bearing wall, **A.**22
 crib wall, **12.**16

cutoff wall, **16.**1, **A.**22
Walls (*Cont.*):
   definition, **A.**22
   mechanically stabilized earth retaining walls
      (*see* Mechanically stabilized)
   retaining wall (*see* Retaining walls)
   sheet pile walls (*see* Sheet pile)
   temporary retaining walls (*see* Temporary
      retaining walls)
   utility trench shoring (*see* Temporary retain-
      ing walls)
Water cement ratio, **13.**5–**13.**6, **A.**22
Water content, **5.**3–**5.**4, **6.**6, **17.**7,
   A.9
Water content profile, **9.**9
Waterproofing, **16.**13, **16.**15, **16.**18
   membrane, **16.**18
   mastic, **16.**18
   primer, **16.**18
   protection board, **16.**18
   self-adhering, **16.**18
Water table (*see* Groundwater)
Water vapor, **16.**9, **16.**15

Weather, **10.**3
Weathering, **9.**33
   causing surficial failures, **10.**18
   chemical methods, **9.**33
   physical methods, **9.**33
Wedge method, **10.**28–**10.**30, **15.**55–**15.**56
Welding, **14.**1–**14.**2
Well point, **A.**22
Wells, **16.**7
Wetland, **A.**7
Wetting (*see* Expansive soil)
Windrow, **17.**4, **A.**22
Work product, **3.**4
Workability of concrete, **A.**22
Workers' Compensation, Bureau of, **15.**46
Wyoming, **9.**2

X-ray diffraction tests, **5.**33, **14.**7
X-ray radiograph, **5.**21–**5.**24
X-rays, **5.**21

Yonkers, New York, **13.**8

Zero air voids curve, **17.**8, **A.**11

ABOUT THE AUTHOR

Robert W. Day is a leading geotechnical engineer and the chief engineer at American Geotechnical in San Diego, California. The author of over 200 published technical papers and the textbook *Forensic Geotechnical and Foundation Engineering,* he serves on advisory committees for several professional associations, including ASCE, ASTM, and NCEES. He holds four college degrees: two from Villanova University, bachelor's and master's degrees majoring in structural engineering, and two from the Massachusetts Institute of Technology, master's and the *Civil Engineer* degree (highest degree) majoring in geotechnical engineering. He is also a registered civil engineer in several states and a registered geotechnical engineer in California.